Empirical Issues in The Decision Process For Multivariate Analysis

Stage 1: Define The Research Problem, Objectives and Multivariate to be Used
· Is the problem suitable for multivariate analysis?
· Can specific measures be identified for the concepts of interest?
· Which multivariate technique is best suited to the research problem?

Stage 2: Develop an Analysis Plan
· How does sample size affect your results?
· Are the variables of the correct measurement type? If not, can they be transformed?
· Can nonlinear relationships be identified and represented in the variate?

Stage 3: Evaluate the Assumptions of the Multivariate Technique
· Have the missing data characteristics of the data been assessed?
· Are there any outliers that might affect the results?
· Have the underlying assumptions been tested empirically?

Stage 4: Estimate the Multivariate Model and Assess Overall Model Fit
· What is the statistical power of the multivariate technique?
· What are the measures of overall model fit and how are they interpreted?
· How do you interpret the errors of prediction or explanation?
· What empirical bases are available for possibly respecifying the multivariate model?

Stage 5: Interpret the Variate
· Are the results evaluated with some measure of statistical significance?
· What results come from evaluating the variate versus evaluating individual variable(s)?
· How do you compare the impact of different variables on the results?

Stage 6: Validate the Model
· Can you form estimation and holdout samples from the original sample?
· Can you use techniques which validate through the omission of single cases, such as bootstrapping?
· How do you compare and evaluate the differing results obtained in your validation efforts?

MULTIVARIATE DATA ANALYSIS

MULTIVARIATE DATA ANALYSIS

with **Readings**

Fourth Edition

JOSEPH F. HAIR, JR.
Louisiana State University

ROLPH E. ANDERSON
Drexel University

RONALD L. TATHAM
Burke Marketing Research

WILLIAM C. BLACK
Louisiana State University

PRENTICE HALL, Englewood Cliffs, New Jersey 07632

Library of Congress Cataloging-in-Publication Data
Multivariate data analysis: with readings/Joseph F. Hair, Jr. . . .
[et al.].—4th ed.
p. cm.
Includes bibliographical references and index.
ISBN 0-02-349020-9
1. Multivariate analysis. I. Hair, Joseph F.
QA278.M85 1995
519.5'35—dc20   94-4217
                                                       CIP

Editor: David Borkowsky
Production Supervisor: Dora Rizzuto
Production Manager: Francesca Drago
Text and Cover Designer: Brian Deep
Cover Photo by Richard Megna of *Fundamental Photos*

© 1995, 1992, 1987, 1984 by Prentice-Hall, Inc.
A Simon & Schuster Company
Englewood Cliffs, New Jersey 07632

The author and publisher of this book have used their best efforts in
preparing this book. These efforts include the development, research,
and testing of the theories and programs to determine their effectiveness.
The author and publisher shall not be liable in any event for incidental
or consequential damages in connection with, or arising out of, the
furnishing, performance, or use of these programs.

Printed in the United States of America

10 9 8 7 6 5 4 3 2 1

ISBN:0-02-349020-9

PRENTICE-HALL INTERNATIONAL (UK) LIMITED, London
PRENTICE-HALL OF AUSTRALIA PTY. LIMITED, Sydney
PRENTICE-HALL CANADA INC., Toronto
PRENTICE-HALL HISPANOAMERICANA, S.A., Mexico
PRENTICE-HALL OF INDIA PRIVATE LIMITED, New Delhi
PRENTICE-HALL OF JAPAN, INC., Tokyo
SIMON & SCHUSTER ASIA PTE. LTD., Singapore
EDITORA PRENTICE-HALL DO BRASIL, LTDA., Rio de Janeiro

Contents

Preface   xv

Chapter 1   Introduction   xx

Learning Objectives • Chapter Preview • Key Terms   1
What Is Multivariate Analysis?   3
Impact of the Computer Revolution   4
Multivariate Analysis Defined   5
Some Basic Concepts of Multivariate Analysis   6
    The Variate   6
    Measurement Scales   6
    Measurement Error and Multivariate Measurement   8
    Statistical Significance versus Statistical Power   10
Types of Multivariate Techniques   13
    Multiple Regression   13
    Multiple Discriminant Analysis   13
    Multivariate Analysis of Variance   14
    Canonical Correlation   14
    Linear Probability Models   14
    Conjoint Analysis   15
    Structural Equation Modeling   15
    Factor Analysis   16
    Cluster Analysis   16
    Multidimensional Scaling   16
    Correspondence Analysis   16
A Classification of Multivariate Techniques   17
Guidelines for Multivariate Analyses and Interpretation   21
    Establish Practical Significance as Well as Statistical Significance   22
    Sample Size Affects All Results   22
    Know Your Data   23
    Strive for Model Parsimony   23
    Look at Your Errors   23
    Validate Your Results   24
A Structured Approach to Multivariate Model Building   24
    Stage One: Define the Research Problem, Objectives, and Multivariate
    Technique to be Used   25
    Stage Two: Develop the Analysis Plan   25
    Stage Three: Evaluate the Assumptions Underlying the Multivariate
    Technique   25
    Stage Four: Estimate the Multivariate Model and Assess Overall Model Fit . .   26
    Stage Five: Interpret the Variate(s)   26
    Stage Six: Validate the Multivariate Model   26
Databases   27
    Perceptions of HATCO   28
    Purchaser Characteristics   28
Summary • Questions • References   29

Chapter 2 Examining Your Data 32

Learning Objectives • Chapter Preview • Key Terms 32
Introduction 35
Graphical Examination of the Data 36
    The Nature of the Variable: Examining the Shape of the Distribution 37
    Examining the Relationship Between Variables 38
    Examining Group Differences 40
    Multivariate Profiles 40
    Summary 42
Missing Data 43
    Understanding the Reasons Leading to Missing Data 43
    Examining the Patterns of Missing Data 45
    Diagnosing the Randomness of the Missing Data Process 45
    Approaches for Dealing with Missing Data 46
    An Illustration of Missing Data Diagnosis 50
    Summary 57
Outliers 57
    Detecting Outliers 58
    Outlier Description and Profiling 59
    Retention or Deletion of the Outlier 60
    An Illustrative Example of Analyzing Outliers 60
Testing the Assumptions of Multivariate Analysis 62
    Assessing Individual Variables Versus the Multivariate Model 64
    Normality 64
    Homescedasticity 66
    Linearity 68
    Absence of Correlated Errors 69
    Data Transformations 69
    An Illustration of Testing the Assumptions Underlying Multivariate Analysis . . . 71
Summary • Questions • References 75

Chapter 3 Multiple Regression Analysis 78

Learning Objectives • Chapter Preview • Key Terms 78
What is Multiple Regression Analysis? 85
Relating Independent to Dependent Variables with Regression 86
    Prediction Without an Independent Variable 86
    Prediction Using a Single Independent Variable—Simple Regression 88
    Prediction Using Several Independent Variables: Multiple Regression
    Analysis 92
A Decision Process for Multiple Regression Analysis 97
Stage One: Objectives of Multiple Regression 97
    Research Problems Appropriate for Multiple Regression 97
    Specifying a Statistical Relationship 101
    Selection of Dependent and Independent Variables 101
Stage Two: Research Design of a Multiple Regression Analysis 103
    Sample Size 103
    Fixed Versus Random Effects Predictors 105
    Creating Additional Variables 106

Stage Three: Assumptions in Multiple Regression Analysis 110
   Assessing Individual Variables Versus the Variate 110
   Linearity of the Phenomenon 111
   Constant Variance of the Error Term...................................... 113
   Independent of the Error Terms 113
   Normality of the Error Term Distribution 114
   Summary.. 114
Stage Four: Estimating the Regression Model and Assessing Overall Fit 114
   General Approaches to Variable Selection.................................. 115
   Testing the Regression Variate for Meeting the Regression Assumptions 118
   Examining the Statistical Significance of Our Model 118
   Identifying Influential Observations..................................... 120
Stage Five: Interpreting the Regression Variate 124
   Using the Regression Coefficients.. 125
   Standardizing the Regression Coefficients: Beta Coefficients 125
   Assessing Multicollinearity 126
Stage Six: Validation of the Results 128
   Additional or Split Samples 128
   Calculating the PRESS Statistic.. 128
   Comparing Regression Models.. 129
   Predicting with the Model.. 129
Regression with a Binary Dependent Variable............................... 129
   Unique Characteristics of Logit Analysis................................. 130
   Other Similarities 133
Illustration of a Regression Analysis 133
   Stage One: Objectives of the Multiple Regression 133
   Stage Two: Research Design of the Multiple Regression Analysis 134
   Stage Three: Assumptions of the Multiple Regression Analysis 134
   Stage Four: Estimating the Regression Model and Assessing Overall Fit 135
   Stage Five: Interpreting the Variate.................................... 145
   Stage Six: Validating the Results....................................... 147
Summary • Questions • References 148
Appendix 3A Advanced Diagnostics for Multiple Regression Analysis 150
Learning Objectives • Preview • Key Terms 150
Assessing Multicollinearity.. 152
   A Two-Part Process 153
   An Illustration of Assessing Multicollinearity........................... 153
Identifying Influential Observations..................................... 154
   Step 1: Examining Residuals 154
   Step 2: Identifying Leverage Points from the Predictors.................. 155
   Step 3: Single-Case Diagnostics Identifying Influential Observations 156
   Step 4: Selecting and Accommodating Influential Observations............. 157
   Example from the HATCO Database.. 157
Summary • Questions • References 165

Chapter 4 Multiple Discriminant Analysis 178
   Learning Objectives • Chapter Preview • Key Terms 178
   What Is Discriminant Analysis? 181
   Analogy with Regression and ANOVA 183

Hypothetical Example of Discriminant Analysis for Two Groups 184
A Geometric Representation of the Discriminant Function 187
A Hypothetical Example of a Three-group Discriminant Analysis 188
The Decision Process for Discriminant Analysis 191
Stage One: Objectives of Discriminant Analysis 191
Stage Two: Research Design for Discriminant Analysis 194
    Selection of Dependent and Independent Variables 194
    Sample Size 195
    Division of the Sample 195
Stage Three: Assumptions of Discriminant Analysis 196
Stage Four: Estimation of the Discriminant Model and Assessing Overall Fit . . 197
    Computational Method 197
    Statistical Significance 198
    Assessing Overall Fit 199
Stage Five: Interpretation of the Results 205
    Discriminant Weights 206
    Discriminant Loadings 206
    Partial F values 206
    Interpretation of Two or More Functions 207
    Which Interpretive Method to Use? 209
Stage Six: Validation of the Results 209
    Split-Sample or Cross-Validation Procedures 209
    Profiling Group Differences 210
A Two-group Illustrative Example 211
    Stage One: Objectives of the Discriminant Analysis 211
    Stage Two: Research Design of the Discriminant Analysis 211
    Stage Three: Assumptions of Discriminant Analysis 212
    Stage Four: Estimation of the Discriminant Function and Assessing Overall
    Fit 212
    Stage Five: Interpretation of the Discriminant Function 220
    Stage Six: Validation of the Discriminant Results 221
A Three-group Illustrative Example 221
    Stage One: Objectives of the Discriminant Analysis 222
    Stage Two: Research Design of the Discriminant Analysis 222
    Stage Three: Assumptions of Discriminant Analysis 222
    Stage Four: Estimation of the Discriminant Function and Assessing Overall
    Fit 223
    Stage Five: Interpretation of Three-group Discriminant Analysis Results 229
    Stage Six: Validation of the Discriminant Results 237
Summary • Questions • References 237

Chapter 5 Multivariate Analysis of Variance 256
Learning Objectives • Chapter Preview • Key Terms 256
What is Multivariate Analysis of Variance? 261
    The *t* test 261
    Analysis of Variance (ANOVA) 262
    Multivariate Analysis of Variance (MANOVA) 263

When Should We Use MANOVA? 265
    Control of Experimentwide Error Rate 266
    Differences Among a Combination of Dependent Variables 266
A Decision Process for MANOVA 266
Stage One: Objectives of MANOVA 268
    Types of Multivariate Questions Suitable for MANOVA 268
    Selecting the Dependent Measures 269
Stage Two: Issues in the Research Design of MANOVA 269
    Sample Size Requirements—Overall and by Group 269
    Factorial Designs—Two or More Treatments 270
    Using Covariates—ANCOVA and MANCOVA 273
    A Special Case of MANOVA: Repeated Measures 274
Stage Three: Assumptions of ANOVA and MANOVA 274
    Independence 275
    Equality of Variance-Covariance Matrices 275
    Normality 276
    Linearity and Multicollinearity Among the Dependent Variables 276
    Sensitivity to Outliers 276
Stage Four: Estimation of the MANOVA Model and Assessing Overall Fit 277
    Criteria for Significance Testing 277
    Statistical Power of the Multivariate Tests 278
Stage Five: Interpretation of the MANOVA Results 280
    Evaluating Covariates 280
    Assessing the Dependent Variate 281
    Identifying Differences Between Individual Groups 282
Stage Six: Validation of the Results 283
Summary 284
Example 1: Difference Between Two Independent Groups 284
    A Univariate Approach: The *t* test 285
    A Multivariate Approach: Hotelling's T^2 288
Example 2: Difference Between k Independent Groups 293
    A Univariate Approach: *k*-Groups ANOVA 294
    A Multivariate Approach: *k*-Groups MANOVA 297
Example 3: A Factorial Design for MANOVA with Two Independent Variables . . 302
    Stage One: Objectives of the MANOVA 302
    Stage Two: Research Design of the MANOVA 303
    Stage Three: Assumptions in MANOVA 305
    Stage Four: Estimation of the MANOVA Model and Assessing Overall Fit . . . 306
    Stage Five: Interpretation of the Results 308
Summary • Questions • References 311

Chapter 6   Canonical Correlation Analysis 326
Learning Objectives • Chapter Preview • Key Terms 326
What Is Canonical Correlation? 328
Hypothetical Example of Canonical Correlation 328
Analyzing Relationships with Canonical Correlation 329
Stage One: Objectives of Canonical Correlation Analysis 330
Stage Two: Designing a Canonical Correlation Anaylsis 330

Stage Three: Assumptions in Canonical Correlation............................... 332
Stage Four: Deriving the Canonical Functions and Assessing Overall Fit 333
    Which Canonical Functions Should Be Interpreted? 333
Stage Five: Interpreting the Canonical Variate.............................. 336
    Canonical Weights... 336
    Canonical Loadings.. 337
    Canonical Cross-Loadings 337
    Which Interpretation Approach to Use 338
Stage Six: Validation and Diagnosis.................................... 338
An Illustrative Example 339
    Stage 1: Objectives of Canonical Correlation Analysis 339
    Stages 2 and 3: Designing a Canonical Correlation Analysis and Testing the
    Assumptions 339
    Stage 4: Deriving the Canonical Functions and Assessing Overall Fit 339
    Stage 5: Interpreting the Canonical Variates 342
    Stage 6: Validation... 344
Summary • Questions • References 344

Chapter 7   Factor Analysis.. 364
Learning Objectives • Chapter Preview • Key Terms 364
What Is Factor Analysis? 366
A Hypothetical Example of Factor Analysis 368
Factor Analysis Decision Diagram....................................... 368
    Stage One: Objectives of Factor Analysis.............................. 368
    Stage Two: Designing a Factor Analysis 372
    Stage Three: Assumptions in Factor Analysis 374
    Stage Four: Deriving Factors and Assessing Overall Fit 375
    Stage Five: Interpreting the Factors 379
    Stage Six: Validation of Factor Analysis.............................. 388
    Stage Seven: Additional Uses of the Factor Analysis Results 389
An Illustrative Example 391
    Stage One: Objectives of Factor Analysis.............................. 391
    Stage Two: Designing a Factor Analysis 391
    Stage Three: Assumptions in Factor Analysis 392
    Stage Four: Deriving Factors and Assessing Overall Fit 394
    Stage Five: Interpreting the Factors 394
    Stage Six: Validation of Factor Analysis.............................. 398
    Stage Seven: Additional Uses of the Factor Analysis Results 400
Common Factor Analysis: Stages 4 and 5 402
    Stage Four: Deriving Factors and Assessing Overall Fit 402
    Stage Five: Interpreting the Factors 404
Summary • Questions • References 404

Chapter 8   Cluster Analysis... 420
Learning Objectives • Chapter Preview • Key Terms 420
What Is Cluster Analysis? 423
How Does Cluster Analysis Work? 424
Cluster Analysis Decision Process 425

Stage One: Objectives of Cluster Analysis 426
Stage Two: Research Design in Cluster Analysis 428
    Detecting Outliers 429
    Similarity Measures 429
    Standardizing the Data 434
Stage Three: Assumptions in Cluster Analysis 435
Stage Four: Deriving Clusters and Assessing Overall Fit 436
    Clustering Algorithms 437
    How Many Clusters Should Be Formed? 442
    Should the Cluster Analysis Be Respecified? 443
Stage Five: Interpretation of the Clusters 443
Stage Six: Validation and Profiling of the Clusters 444
An Illustrative Example 445
    Stage One: Objectives of the Cluster Analysis 445
    Stage Two: Research Design of the Cluster Analysis 445
    Stage Three: Assumptions in Cluster Analysis 445
    Stage Four: Deriving Clusters and Assessing Overall Fit 445
    Stage Five: Interpretation of the Clusters 453
    Stage Six: Validation and Profiling of the Clusters 454
Summary • Questions • References 455

Chapter 9   Multidimensional Scaling 484

Learning Objectives • Chapter Preview • Key Terms 484
What Is Multidimensional Scaling? 488
A Simplified Look at How Multidimensional Scaling Works 489
Comparing MDS to Other Interdependence Techniques 491
A Decision Framework for Perceptual Mapping 492
Stage One: Objectives of Multidimensional Scaling 492
    Identification of All Relevant Objects to Be Evaluated 494
    Similarities Versus Preference Data 495
    Aggregate Versus Disaggregate Analysis 495
Stage Two: Research Design of Multidimensional Scaling 496
    Selection of Either a Decompositional (Attribute-free) or Compositional
    (Attribute-based) Approach 496
    Objects: Their Number and Selection 499
    Nonmetric Versus Metric Methods 499
    Collection of Similarity or Preference Data 500
Stage Three: Assumptions of Multidimensional Scaling Analysis 502
Stage Four: Deriving the MDS Solution and Assessing Overall Fit ... 503
    Determining an Object's Position in the Perceptual Map 503
    Selecting the Dimensionality of the Perceptual Map 504
    Incorporating Preferences into Multidimensional Scaling 506
Stage Five: Interpreting the MDS Results 510
    Identifying the Dimensions 511
Stage Six: Validating the MDS Results 512
Correspondence Analysis 513
    Stage One: Objectives of Correspondence Analysis 513
    Stage Two: Research Design of Correspondence Analysis 514
    Stage Three: Assumptions in Correspondence Analysis 514

Stage Four: Deriving the Correspondence Analysis Results and Assessing
Overall Fit **514**
Stage Five: Interpretation of Results. **515**
Stage Six: Validation of the Results **515**
Overview of Correspondence Analysis **515**
Illustration of Multidimensional Scaling and Correspondence Analysis **516**
Stage One: Objectives of Perceptual Mapping. **516**
Stage Two: Research Design of the Perceptual Mapping Study **517**
Stage Three: Assumptions in Perceptual Mapping **518**
Multidimensional Scaling: Stages Four and Five **518**
Correspondence Analysis: Stages Four and Five. **523**
Stage Five: Interpreting the Correspondence Analysis Results **525**
Stage Six: Validation of the Results **526**
Summary • Questions • References **527**

Chapter 10   Conjoint Analysis 556

Learning Objectives • Chapter Preview • Key Terms **556**
What Is Conjoint Analysis? **560**
A Hypothetical Example of Conjoint Analysis **561**
The Managerial Uses of Conjoint Analysis **562**
Comparing Conjoint Analysis with Other Multivariate Methods **562**
Compositional Versus Decompositional Techniques. **562**
Specifying the Conjoint Variate. **563**
Separate Models for Each Individual. **563**
Types of Relationships **564**
Designing a Conjoint Analysis Experiment **564**
Stage One: The Objectives of Conjoint Analysis **564**
Defining the Total Worth of the Object **567**
Specifying the Determinant Factors. **567**
Stage Two: The Design of a Conjoint Analysis **568**
Designing Stimuli. **568**
Data Collection. **572**
Stage Three: Assumptions of Conjoint Analysis **577**
Stage Four: Estimating the Conjoint Model and Assessing Overall Fit **577**
Selecting an Estimation Technique **578**
Evaluating the Results **578**
Stage Five: Interpreting the Results **579**
Aggregate Versus Disaggregate Analysis. **579**
Assessing the Relative Importance of Attributes. **579**
Stage Six: Validation of the Conjoint Results **579**
Stage Seven: Applying Conjoint Analysis Results **580**
Segmentation **580**
Profitability Analysis. **580**
Conjoint Simulators **580**
Special Topics in Conjoint Analysis **581**
Conjoint Analysis with a Large Number of Factors **581**
Choice-based Conjoint **582**

An Illustration of Conjoint Analysis 585
 Stage One: Objectives of the Conjoint Analysis 585
 Stage Two: Design of the Conjoint Analysis 586
 Stage Three: Assumptions in Conjoint Analysis 587
 Stage Four: Estimating the Conjoint Model and Assessing Overall Model Fit 587
 Stage Five: Interpreting the Results 589
 Stage Six: Validation of the Results 590
 Stage Seven: Application of a Choice Simulator 591
Summary 592
Appendix 10A: An Illustration of Conjoint Analysis 594
Preview 594
Assuming an Additive Model 595
When Is an Interactive Composition Rule Appropriate? 598
Checking for Interactions 598
Summary • Questions • References 599

Chapter 11 Structural Equation Modeling 616
Learning Objectives • Chapter Preview • Key Terms 616
What Is Structural Equation Modeling? 622
 Accommodating Multiple Interrelated Dependence Relationships 622
 Incorporating Variables That We Do Not Measure Directly 623
The Role of Theory in Structural Equation Modeling 624
Developing a Modeling Strategy 625
Steps in Structural Equation Modeling 626
 Step One: Developing a Theoretically Based Model 626
 Step Two: Constructing a Path Diagram of Causal Relationships 629
 Step Three: Converting the Path Diagram Into a Set of Structural Equations
 and Specifying the Measurement Model 631
 Step Four: Choosing the Input Matrix Type and Estimating the Proposed
 Model 635
 Step Five: Assessing the Identification of the Structural Model 638
 Step Six: Evaluating Goodness-of-Fit Criteria 639
 Step Seven: Interpreting and Modifying the Model 644
 A Recap of the Seven-Step Process 645
Two Illustrations of Structural Equation Modeling 645
A Confirmatory Factor Analysis 645
 Step One: Developing a Theoretically Based Model 646
 Step Two: Constructing a Path Diagram of Causal Relationships 646
 Step Three: Converting the Path Diagram Into A Set of Structural Equations
 and Specifying the Measurement Model 647
 Step Four: Choosing the Input Matrix Type and Estimating the Proposed
 Model 648
 Step Five: Assessing the Identification of the Structural Model 648
 Step Six: Evaluating Goodness-of-Fit Criteria 649
 Step Seven: Interpreting and Modifying the Model 654
 Summary 654

Estimating a Path Model With Structural Equation Modeling 655

Step One: Developing a Theoretically Based Model 656

Step Two: Constructing a Path Diagram of Causal Relationships 657

Step Three: Converting the Path Diagram Into A Set of Structural Equations and Specifying the Measurement Model 658

Step Four: Choosing the Input Matrix Type and Estimating the Proposed Model 658

Step Five: Assessing the Identification of the Structural Model 659

Step Six: Evaluating Goodness-of-Fit Criteria 659

Step Seven: Interpreting and Modifying the Model 668

Review of the Structural Equation Modeling Process 670

Summary • Questions • References 670

Appendix 11A   A Mathematical Representation in LISREL Notation 672

LISREL Notation 672

From a Path Diagram to LISREL Notation 674

Constructing Structural Equations from the Path Diagram 675

Summary 679

Appendix 11B   Path Analysis: A Method of Computing Structural Coefficients 680

Appendix 11C   Overall Goodness-of-fit Measures for Structural Equation Modeling 682

Measures of Absolute Fit 683

Incremental Fit Measures 685

Parsimonious Fit Measures 686

Summary • References 688

Appendix A: Applications of Multivariate Data Analysis 708

Index **I-1**

Preface

The fourth edition of *Multivariate Data Analysis* contains a number of substantive additions and notable modifications that both extend and expand its coverage of multivariate analytical techniques. These changes are directed toward our primary objective: an applications-oriented introduction to multivariate analysis for the nonstatistician. We have continually striven to reduce our reliance on statistical notation and terminology and instead identify the fundamental concepts that affect our use of these techniques and express them in simple terms.

One change throughout this edition is the use of a model-building paradigm as the organizational framework for each chapter. Introduced in Chapter 1, the six-stage process provides the researcher with a structured approach for applying each multivariate technique. The model-building process starts with defining the research problems appropriately addressed by the technique, then proceeds through the design issues and assumptions underlying the techniques. When assured that the study is designed properly and meets the necessary assumptions, the process then proceeds to estimation issues of the multivariate model and then to the interpretation of the results. The concept of a variate as the "building block" of multivariate analysis is introduced as a common element among all the techniques that aid in the interpretation of the results. The final step is to validate the results to ensure generalizability to the general population. The model-building perspective intends to ensure that the conceptual issues necessary for an appropriate application of each technique are addressed aong with the statistical considerations. To assist in utilizing the model-building approach, each chapter now has a decision process diagram depicting the model-building stages and the decisions at each stage for that specific multivariate technique. In each chapter the issues in each of the model-building stages are first discussed. Then the results for the empirical example are discussed by stage as well.

To complement the model-building orientation, a new chapter, Chapter 2—Examining Your Data—has been added to address the need to understand the properties of your data *before* you undertake any multivariate analysis. The chapter focuses on three primary issues: assessing the extent, potential impact, and remedies for missing data; identification and handling of outliers; and the evaluation of the basic statistical assumptions underlying multivariate techniques. The chapter brings together material covered previously in several chapters and adds substantial information on handling missing data and outlier identification. Our hope is to encourage analysts to integrate these analyses into their standard routine of multivariate analysis.

Five chapters—1, 3, 5, 6, and 7—have been substantially revised both to incorporate new material and to reorganize their presentation. Chapter 1, Introduction, has several new features in addition to the model-building paradigm discussed above. Principal among these are discussions of such fundamental issues as practical versus statistical significance, the role of model validation, and parsimony in multivariate models. Of particular note is the introduction of statistical power, the complementary statistical probability to the traditional significance level. The reader is provided a perspective on the role of power and the factors that influence its level, particularly sample size. Power is then discussed in subsequent chapters for several of the multivariate techniques where it plays a key role.

Chapter 3, Multiple Regression Analysis, has been extensively reorganized and expanded to include diagnostic techniques concerning influential observations and alternative regression models to represent effects such as interaction and nonlinearity among the variables. Also, the section on logistic regression has been expanded. Chapter 5, Multivariate Analysis of Variance, also has been extensively reorganized and several notable additions made. Among these are the extension of the empirical examples to include two independent variables and the associated discussion on interaction terms and their interpretation. Also, statistical power in multivariate analysis of variance is introduced and an extensive discussion provided concerning its impact on the ''success or failure'' of the analysis.

Chapter 6, Canonical Correlation, and Chapter 7, Factor Analysis, have incorporated not only the model-building perspective but also a more structured approach to model estimation and interpretation. Guidelines for model estimation and interpretation have been added, along with a discussion of statistical assumptions and model validation not found in previous editions.

The remaining chapters, covering Discriminant Analysis (Chapter 4), Cluster Analysis (Chapter 8), Multidimensional Scaling (Chapter 9), Conjoint Analysis (Chapter 10), and Structural Equation Modeling (Chapter 11), have all been reorganized to reflect the model-building paradigm. This provides the researcher a consistent approach applicable to all the multivariate techniques in the fourth edition.

Along with the reorganization of the chapters, the new chapter, and additional material, Appendix A—Applications of Multivariate Data Analysis—now provides specific examples of the control language necessary to perform any of the analyses in the text using SAS, SPSS, and now BMDP, plus the special programs to perform multidimensional scaling, conjoint analysis, and structural equation modeling. Annotated examples explain the basic structure of the control commands for each technique to allow the researcher to execute the techniques in conjunction with the text and extend them easily to specific situations.

Although the book is intended as a basic text for an introductory course in multivariate data analysis, some schools may want to use it as a supplementary text for a research course, a quantitative models course, or a second course in statistics. Feedback from earlier editions indicates that even readers with advanced statistical backgrounds find the book useful for review and convenient reference.

The format for each chapter in this edition consists of a text portion followed by an article carefully selected from the business literature. The text portion introduces the fundamentals of the technique through the model-building paradigm and explains the various issues involved in applying it. The article then sets forth the theory and illustrates its practical application. We believe it is both instructive and reassuring for the reader to see how others have applied the multivariate tools. Finally, Chapters 10 and 11 have appendixes that address several of the more complex statistical issues that may not be of interest to all readers.

The widespread use of personal computers and the availability of statistical programs for the personal computer have made these techniques available to everyone. With this continually improving availability, every data analyst may now encounter the situation where these techniques are applicable or at least needed to evaluate their use by another researcher. We wrote this book for people who want an overall understanding of what multivariate techniques can do for them, when and how they can apply them, and how results are interpreted so that they can use the tools successfully and read the technical literature with confidence. With each edition we have further refined the initial efforts to achieve these objectives, and we believe the present edition continues this effort.

A number of individuals assisted us in completing the fourth edition. Laura Williams, doctoral candidate at Louisiana State University, provided invaluable assistance in all phases of the revision. Maresha Leeds of BMDP Statistical Software was instrumental in the introduction of the BMDP software into this edition, and Ann Glynn of SPSS Inc. provided invaluable assistance in making the transfer to SPSS for Windows.

We are indebted to the following reviewers for their invaluable assistance in the additions to the fourth edition:

Robert Bush, Fogelman College of Business, Memphis State University
Margaret Liebman, La Salle University
Ronald D. Taylor, Mississippi State University

We would also like to acknowledge the assistance of the following individuals on earlier editions: Bruce Alford, University of Evansville; David Andrus, Kansas State University; Alvin C. Burns, Louisiana State University; Alan J. Bush, Memphis State University; Chaim Ehrman, University of Illinois at Chicago; Joel Evans, Hofstra University; Thomas L. Gillpatrick, Portland State University; Dipak Jain, Northwestern University; John Lastovicka, University of Kansas; Richard Netemeyer, Louisiana State University; Scott Roach, Northwestern Louisiana University; Walter A. Smith, Tulsa University; Ronald D. Taylor, Mississippi State University; and Jerry L. Wall, Northeast Louisiana University.

J.F.H.
R.E.A.
R.L.T.
W.C.B.

MULTIVARIATE DATA ANALYSIS

CHAPTER 1

Introduction

LEARNING OBJECTIVES

Upon completing this chapter, you should be able to do the following:

- Explain what multivariate analysis is and when its application is appropriate.

- Define and discuss the specific techniques included in multivariate analysis.

- Determine which multivariate technique is appropriate for a specific research problem.

- Discuss the nature of measurement scales and their relationship to multivariate techniques.

- Describe the conceptual and statistical issues inherent in multivariate analyses.

Chapter 1 presents a simplified overview of multivariate data analysis. It stresses that multivariate analysis methods will increasingly influence not only the analytical aspects of research but also the design and approach to data collection for decision making and problem solving. While multivariate techniques share many characteristics with their univariate and bivariate counterparts, several key differences arise in the transition to a multivariate analysis. To illustrate this transition, a classification of multivariate techniques is presented. Then general guidelines for their application are provided and a structured approach to the formulation, estimation, and interpretation of multivariate results is presented. The chapter concludes with a discussion of the database utilized throughout most of the text to illustrate application of the techniques.

KEY TERMS

Before starting the chapter, review the key terms to develop an understanding of the concepts and terminology used. Throughout the chapter the key terms appear in **boldface.** Other points of emphasis in the chapter are *italicized*. Also, cross-references within the key terms are in *italics*.

Alpha (α)   See *Type I error*.

Beta (β)   See *Type II error*.

Bivariate partial correlation   Simple (two-variable) correlations between two sets of residuals (unexplained variances) that remain after the association of other independent variables is removed.

Dependence technique   Classification of statistical techniques distinguished by having a variable or set of variables identified as the dependent variable(s) and the remaining variables as independent. The objective is prediction of the dependent variable(s) by the independent variable(s). An example is regression analysis.

Dependent variable   Presumed effect of, or response to, a change in the independent variable(s).

Dummy variable   Nonmetrically measured variable transformed into a metric variable by assigning a 1 or a 0 to a subject, depending on whether it possesses a particular characteristic.

Effect size   Estimate of the degree to which the phenomenon being studied (e.g., correlation or difference in means) exists in the population.

Independent variable   Presumed cause of any change in a response or dependent variable.

Indicator   Single variable used in conjunction with one or more other variables to form a composite measurement.

Interdependence technique   Classification of statistical techniques for which the variables are not divided into dependent and independent groups (e.g., factor analysis); rather, all variables are analyzed as a single set.

Measurement error   Inaccuracies of measuring the "true" variable values due to the fallibility of the measurement instrument (i.e., inappropriate response scales), data entry errors, or respondent errors.

Metric data   Also called *quantitative data*, and *interval* and *ratio data*, these measurements identify or describe subjects (or objects) not only on the possession of an attribute but also by the amount or degree to which the subject may be characterized by the attribute. For example, a person's age and weight are metric data.

Multicollinearity Extent to which a variable can be explained by the other variables in the analysis. As multicollinearity increases, it complicates the interpretation of the *variate* as it is more difficult to ascertain the effect of any single variable, owing to their interrelationships.

Multivariate analysis Analysis of multiple variables in a single relationship or set of relationships.

Multivariate measurement Use of two or more variables as *indicators* of a single composite measure. For example, a personality test may provide the answers to a series of individual questions (indicators), which are then combined to form a single score representing the personality trait.

Nonmetric data Also called *qualitative data*, these are attributes, characteristics, or categorical properties that identify or describe a subject or object. Examples are occupation (physician, attorney, professor) or consumer confidence (high, medium, low). Also called *nominal* and *ordinal data*.

Power Probability of correctly rejecting the null hypothesis when it is false, that is, correctly finding a hypothesized relationship when it exists. Determined as a function of (1) the statistical significance level (α) set by the researcher for a *Type I error*, (2) the sample size used in the analysis, and (3) the *effect size* being examined.

Practical significance Means of assessing multivariate analysis results based on their substantive findings rather than their statistical significance. While statistical significance determines whether the result is attributable to chance, practical significance assesses whether the result is useful (i.e., substantial enough to warrant action).

Reliability Extent to which a variable or set of variables is consistent in what it is intended to measure. If multiple measurements are taken, the reliable measures will all be very consistent in their values. It differs from *validity* in that it does not relate to what should be measured, but instead how it is measured.

Specification error Omitting a key variable from the analysis, thus impacting the estimated effects of included variables.

Summated scales Method of combining several variables that measure the same concept into a single variable in an attempt to increase the *reliability* of the measurement through *multivariate measurement*. In most instances, the variables are summed and the total or its average is used in the analysis.

Treatment Independent variable the researcher manipulates to see the effect (if any) on the dependent variable(s).

Type I error Probability of incorrectly rejecting the null hypothesis—in most cases, this means saying a difference or correlation exists when it actually does not. Also termed *alpha* (α). Typical levels are 5 or 1 percent, termed the .05 or .01 level, respectively.

Type II error Probability of incorrectly failing to reject the null hypothesis—in simple terms, the chance of not finding a correlation or mean difference when it does exist. Also termed *beta* (β), it is inversely related to *Type I error*. Also, 1 minus the Type II error is defined as *power*.

Univariate analysis of variance (ANOVA) Statistical technique to determine, on the basis of one dependent measure, whether samples are from populations with equal means.

Validity   Extent to which a measure or set of measures correctly represents the concept of study—the degree to which it is free from any systematic or nonrandom error. Validity is concerned with how well the concept is defined by the measure(s), while *reliability* relates to the consistency of the measure(s).

Variate   Linear combination of variables formed in the multivariate technique by deriving empirical weights applied to a set of variables specified by the researcher.

What Is Multivariate Analysis?

The computer technology available today, almost unimaginable just two short decades ago, has made possible extraordinary advances in the analysis of psychological, sociological, and other types of behavioral data. This impact is most evident in the relative ease with which computers can analyze large quantities of complex data. Almost any problem today is easily analyzed by any number of statistical programs, even on microcomputers. In addition, the effects of technological progress have extended beyond the ability to manipulate data, releasing researchers from past constraints on data analysis and affording them the ability to engage in more substantive development and testing of their theoretical models. No longer are methodological limitations a critical concern to the theorist striving for empirical support. Much of this increased understanding and mastery of data analysis has come about through the study of statistics and statistical inference. Equally important, however, has been the expanded understanding and application of a group of statistical techniques known as **multivariate analysis.**

Multivariate analytical techniques are being widely applied in industry, government, and university-related research centers. Moreover, few fields of study or research have failed to integrate multivariate techniques into their analytical "toolbox." To serve this increased interest, numerous books and articles have been published on the theoretical and mathematical aspects of these tools, and introductory texts have appeared in almost every field as well. Few books, however, have been written for the researcher who is not a specialist in math or statistics. Still fewer books discuss *applications* of multivariate statistics while also providing a conceptual discussion of the statistical methods. To fill this gap, this book was written.

Applications-oriented books are of crucial interest to behavioral scientists and business or government managers of all backgrounds who have to expand their knowledge of multivariate analysis to gain a better understanding of the complex phenomena in their work environment. Any researcher who examines only two-variable relationships and avoids multivariate analysis is ignoring powerful tools that can provide potentially useful information. As one researcher states, "For the purposes of . . . any . . . applied field, most of our tools are, or should be, multivariate. One is pushed to a conclusion that unless a . . . problem is treated as a multivariate problem, it is treated superficially" [11, p. 158]. According to statisticians Hardyck and Petrinovich [12, p. 7]:

> Multivariate analysis methods will predominate in the future and will result in drastic changes in the manner in which research workers think about problems and how

they design their research. These methods make it possible to ask specific and precise questions of considerable complexity in natural settings. This makes it possible to conduct theoretically significant research and to evaluate the effects of naturally occurring parametric variations in the context in which they normally occur. In this way, the natural correlations among the manifold influences on behavior can be preserved and separate effects of these influences can be studied statistically without causing a typical isolation of either individuals or variables.

As just one example, businesspeople in most markets today are not able to follow past approaches whereby consumers were considered homogeneous and characterized by a small number of demographic variables. Instead, they must develop strategies to appeal to numerous segments of customers with varied demographic and psychographic characteristics in a marketplace with multiple constraints (legal, economic, competitive, technological, etc.). It is only through multivariate techniques that multiple relationships of this type can be adequately examined to obtain a more complete, realistic understanding for decision making.

Impact of the Computer Revolution

It is almost impossible to discuss the application of multivariate techniques without a discussion of the impact of the computer. As mentioned earlier, the widespread application of computers (first mainframe and, more recently, personal or microcomputers) to process large, complex databases has dramatically spurred the use of multivariate statistical methods. All the statistical theory for today's multivariate techniques were developed well *before* the appearance of computers, but only when the computational power became available to perform the increasingly complex calculations did the techniques' existence become known outside the field of theoretical statistics. The continued technological advances in computing, particularly personal computers, have provided any interested researcher ready access to all the resources needed to address almost any size multivariate problem. In fact, many researchers call themselves *data analysts* instead of statisticians or (in the vernacular) "quantitative types." These data analysts have contributed substantially to the increase in usage and acceptance of multivariate statistics in the business and government sectors. Within the academic community, disciplines in all fields have embraced multivariate techniques, and academicians increasingly must be versed in the appropriate multivariate techniques for their empirical research. Even for people with strong quantitative training, the availability of prepackaged programs for multivariate analysis has facilitated the complex manipulation of data matrices that has long hampered the growth of multivariate techniques.

With many major universities already requiring entering students to purchase their own microcomputers before matriculating, students and professors routinely analyze multivariate data for answers to questions in fields of study from anthropology to zoology. Today a number of prepackaged computer programs are available for multivariate data analysis [4, 5, 14, 16, 17, 22, 23, 24, 25, 26, 27, 29, 30]. All the comprehensive statistical packages designed for mainframe computers (e.g., SPSS, SAS, and BMDP) are also now available for personal computers [4, 5, 16, 17, 23, 24, 25, 26]. Specialized programs for all types of multivariate analysis, including multidimensional scaling (e.g., ALSCAL, INDSCAL, KYST, PREFMAP) [21, 26, 29], simultaneous/structural equation modeling (LISREL, EQS) [2, 3, 13, 14], and conjoint analysis (ACA, CVA, and CBC) [18, 19, 20] once were available only, if at all, on mainframe computers but today are personal computer–compatible. Expert systems are being developed to address even such issues as selecting a statistical technique [8] or designing a sampling plan to ensure desired statistical and practical objectives [7].

No longer are the statistical programs first developed on mainframe systems and then migrated to micro- and personal computers, but instead programs are now initially developed for the personal computer. Perhaps the fastest-growing category of statistical programs are the statistical packages designed specifically to take advantage of the flexibility of the personal computer [3, 5, 14, 26, 27, 30]. Multivariate techniques are so widespread that *all* the techniques illustrated in this text can be estimated with statistical packages readily available for either a mainframe, microcomputer, or personal computer.

Multivariate Analysis Defined

Multivariate analysis is not easy to define. Broadly speaking, it refers to all statistical methods that simultaneously analyze multiple measurements on each individual or object under investigation. Any simultaneous analysis of more than two variables can be loosely considered multivariate analysis. As such, many multivariate techniques are extensions of univariate analysis (analysis of single-variable distributions) and bivariate analysis (cross-classification, correlation, analysis of variance, and simple regression used to analyze two variables). For example, simple regression (with one predictor variable) is extended in the multivariate case to include several predictor variables. Likewise, the single dependent variable found in analysis of variance is extended to include multiple dependent variables in multivariate analysis of variance. As you will come to see, in many instances multivariate techniques are a means of performing in a single analysis what once took multiple analyses using univariate techniques. Other multivariate techniques, however, are uniquely designed to deal with multivariate issues, such as factor analysis to identify the structure underlying a set of variables or discriminant analysis to differentiate among groups based on a set of variables.

One reason for the difficulty of defining multivariate analysis is that the term *multivariate* is not used consistently in the literature. To some researchers, *multivariate* simply means examining relationships between or among more than two variables. Others use the term only for problems where all the multiple variables are assumed to have a multivariate normal distribution. To be considered truly multivariate, however, all the variables must be random and interrelated in such ways that their different effects cannot meaningfully be interpreted separately. Some authors state that the purpose of multivariate analysis is to measure, explain, and predict the degree of relationship among variates (weighted combinations of variables). Thus the multivariate character lies in the multiple variates (multiple combinations of variables), not only in the number of variables or observations. For the purposes of this book, we do not insist on a rigid definition of multivariate analysis. Instead, **multivariate analysis** will include both multivariable techniques and truly multivariate techniques, because the authors believe that knowledge of multivariable techniques is an essential first step in understanding multivariate analysis.

Some Basic Concepts of Multivariate Analysis

While multivariate analysis has its roots in univariate and bivariate statistics, the extension to the multivariate domain introduces additional concepts and issues that have particular relevance. These concepts range from the need for a conceptual understanding of the basic building block of multivariate analysis—the variate—to specific issues dealing with the types of measurement scales used and the statistical issues of significance testing and confidence levels. Each concept plays a significant role in the successful application of any multivariate technique.

The Variate

As mentioned above, the building block of multivariate analysis is the **variate,** a linear combination of variables with empirically determined weights. The variables are specified by the researcher, while the weights are determined by the specific objective of the multivariate technique. A variate of n weighted variables (X_1 to X_n) can be stated mathematically as:

$$\text{Variate value} = w_1X_1 + w_2X_2 + w_3X_3 + \ldots + w_nX_n$$

where X_n is the observed variable and w_n is the weight determined by the multivariate technique.

The result is a single value representing a combination of the *entire set* of variables that best achieves the objective of the specific multivariate analysis. In multiple regression, the variate is determined so as to best correlate with the variable being predicted. In discriminant analysis, the variate is formed so as to create scores for each observation that maximally differentiates between groups of observations. And in factor analysis, variates are formed to best represent the underlying structure or dimensionality of the variables as represented by their intercorrelations.

In each instance, the variate captures the multivariate character of the analysis. Thus, in our discussions of each technique, the variate is a focal point of the analysis in many respects. We must understand not only its collective impact in meeting the technique's objective but also the contribution of each separate variable to the overall variate effect.

Measurement Scales

Data analysis involves the partitioning, identification, and measurement of variation in a set of variables, either among themselves or between a dependent variable and one or more independent variables. The key word here is *measurement* because the researcher cannot partition or identify variation unless it can be measured. Measurement is important in accurately representing the concept of interest and is instrumental in the selection of the appropriate multivariate method of analysis. In the next few paragraphs we discuss the concept of measurement as it relates to data analysis and particularly to the various multivariate techniques.

There are two basic kinds of data: **nonmetric** (qualitative) and **metric** (quantitative). Nonmetric data are attributes, characteristics, or categorical properties that identify or describe a subject. Nonmetric data describe differences in type or kind by indicating the presence or absence of a characteristic or property. While often portrayed as less precise or rigorous, many properties are discrete in that by having a particular feature, all other features are excluded. For example, if one is male, one cannot be female. There is no "amount" of gender, just the state of being male or female. In contrast, metric data measurements are made so that subjects may be identified as differing in amount or degree. Metrically measured variables reflect relative quantity or distance. Where one can make statements as to the amount or magnitude, such as the level of satisfaction or commitment to a job, metric measurements are appropriate.

Nonmetric Measurement Scales

Nonmetric measurements can be made with either a nominal or an ordinal scale. Measurement with a nominal scale assigns numbers used to label or identify subjects or objects. Nominal scales, also known as categorical scales, provide the number of occurrences in each class or category of the variable being studied. Therefore, the numbers or symbols assigned to the objects have no quantitative meaning beyond indicating the presence or absence of the attribute or characteristic under investigation. Examples of nominally scaled data include an individual's sex, religion, or political party. In working with these data, the analyst might assign numbers to each category, for example, 2 for females and 1 for males. These numbers only represent categories or classes and do not imply amounts of an attribute or characteristic.

Ordinal scales are the next higher level of measurement precision. Variables can be ordered or ranked with ordinal scales in relation to the amount of the attribute possessed. Every subclass can be compared with another in terms of a "greater than" or "less than" relationship. For example, different levels of an individual consumer's satisfaction with several new products can be illustrated on an ordinal scale. The following scale shows a respondent's view of three products. The respondent is more satisfied with A than B and more satisfied with B than C.

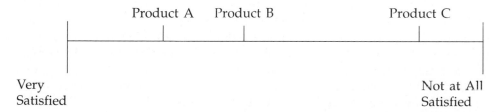

Numbers utilized in ordinal scales like these are nonquantitative, because they indicate only relative positions in an ordered series. There is no measure of how much satisfaction the consumer receives in absolute terms, nor does the researcher know the exact difference between points on the scale of satisfaction. Many scales in the behavioral sciences fall into this ordinal category.

Metric Measurement Scales

Interval scales and ratio scales (both metric) provide the highest level of measurement precision, permitting nearly all mathematical operations to be performed. These two scales have constant units of measurement, so differences between any two adjacent points on any part of the scale are equal. The only real difference between interval and ratio scales is that interval scales have an arbitrary zero point, while ratio scales have an absolute zero point. The most familiar interval scales are the Fahrenheit and Celsius temperature scales. Both have a different arbitrary zero point, and neither indicates a zero amount or lack of temperature, because we can register temperatures below the zero point on each scale. Therefore, it is not possible to say that any value on an interval scale is a multiple of some other point on the scale. For example, an 80°F day cannot correctly be said to be twice as hot as a 40°F day because we know that 80°F, on a different scale, such as Celsius, is 26.7°C. Similarly, 40°F, on Celsius, is 4.4°C. Although 80°F is indeed twice 40°F, one cannot state that the heat of 80°F is twice the heat of 40°F because, using different scales, the heat is not twice as great; that is, $4.4°C \times 2 \neq 26.7°C$.

Ratio scales represent the highest form of measurement precision, because they possess the advantages of all lower scales plus an absolute zero point. All mathematical operations are permissible with ratio scale measurements. The bathroom scale or other common weighing machines are examples of these scales, for they have an absolute zero point and can be spoken of in terms of multiples when relating one point on the scale to another; for example, 100 pounds is twice as heavy as 50 pounds.

Understanding the different types of measurement scales is important for two reasons. First, the researcher must identify the measurement scale of each variable used, so that nonmetric data are not incorrectly used as metric data and vice versa. Second, the measurement scale is critical in determining which multivariate techniques are the most applicable to the data, with considerations made for both independent and dependent variables. In the discussion of techniques and their classification in later sections of this chapter, the metric or nonmetric properties of independent and dependent variables are the determining factors in selecting the appropriate technique.

Measurement Error and Multivariate Measurement

The use of multiple variables and the reliance on their combination (the variate) in multivariate techniques also focuses attention on a complementary issue—measurement error. **Measurement error** is the degree to which the observed values are not representative of the "true" values. Measurement error have many sources, ranging from data entry errors to the imprecision of the measurement (e.g., imposing seven-point rating scales for attitude measurement when the researcher knows the respondents can accurately respond only to a three-point rating) to the inability of respondents to accurately provide information (e.g., responses as to household income may be reasonably accurate but rarely totally precise). Thus, all variables used in multivariate techniques must be assumed to have some degree of measurement error. The impact of measurement error is to add "noise" to the observed or measured variables. Thus, the observed value obtained represents both the "true" level and the "noise." When used to compute correlations or means, the "true" effect is partially masked by the measurement error, causing the correlations to weaken and the means to be less precise. The specific impact of measurement error in dependence relationships is covered in more detail in Chapter 11.

The researcher's goal of reducing measurement error can follow several paths. In assessing the degree of measurement error present in any measure, the analyst must address both the **validity** and **reliability** of the measure. Validity is the degree to which a measure accurately represents what it is supposed to. For example, if we want to measure discretionary income, we should not ask about total household income. Ensuring validity starts with a thorough understanding of what is to be measured and then making the measurement as "correct" and accurate as possible. However, accuracy does not ensure validity. In our income example, the researcher could very precisely define total household income but still have an invalid measure of discretionary income because the "correct" question was not being asked.

If validity is assured, the researcher must still consider the reliability of the measurements. Reliability is the degree to which the observed variable measures the "true" value and is "error free"; thus it is the opposite of measurement error. If the same measure is asked repeatedly, for example, more reliable measures will show greater consistency than less reliable measures. The researcher should always assess the variables being used and if valid alternative measures are available, choose the variable with the higher reliability.

The researcher may also choose to develop **multivariate measurements,** also known as **summated scales,** where several variables are joined to represent a composite variable (e.g., multiple-item personality scales or summed ratings of a product). The objective is to avoid using only a single variable to represent a concept, and instead use several variables as **indicators** (see Key Terms), all representing differing facets of the concept to obtain a more "well-rounded" perspective. The use of multiple indicators allows the researcher to more precisely specify the responses desired and does not place total reliance on a single response but instead on the "average" or "typical" response to a set of related responses. For example, in measuring satisfaction, one could ask a single question, "How satisfied are you?," and base the analysis on the single response. Or a summated scale could be developed that combined several responses of satisfaction, perhaps in different response formats and in differing areas of interest thought to comprise overall satisfaction. The guiding premise is that multiple responses more accurately reflect the "true" response than does a single response. Assessing reliability and incorporating scales in the analysis are methods any researcher should employ. For a more detailed introduction to multiple measurement models and scale construction, see [28]. In addition, compilations of scales that can provide the researcher a "ready-to-go" scale with demonstrated reliability have been published in recent years [1, 9].

The impact of measurement error and poor reliability cannot be directly seen, because they are embedded in the observed variables. The researcher must therefore always work to increase reliability and validity, which in turn will result in a "truer" portrayal of the variables of interest. Poor results are not always due to measurement error, but the presence of measurement error is guaranteed to distort the observed relationships and make multivariate techniques less powerful. Reducing measurement error, while taking effort, time, and additional resources, may improve weak or marginal results and strengthen proven results as well.

Statistical Significance Versus Statistical Power

All the multivariate techniques, except for cluster analysis and multidimensional scaling, are based on the statistical inference of a population's values or relationships among variables from a randomly drawn sample of that population. If we have conducted a census of the entire population, then statistical inference is unnecessary, because any difference or relationship, however small, is "true" and does exist. But rarely, if ever, is a census conducted; therefore, the researcher is forced to draw inferences from a sample.

Interpreting statistical inferences requires that the researcher specify the acceptable levels of statistical error. The most common approach is to specify the level of **Type I error,** also known as **alpha (α).** The **Type I error** is the probability of rejecting the null hypothesis when actually true, or in simple terms, the chance of the test showing statistical significance when it actually is not present, the case of a "false positive." By specifying an alpha level, the researcher sets the allowable limits for error by specifying the probability of concluding that significance exists when it really does not.

But in specifying the level of Type I error, the researcher also determines an associated error, termed the **Type II error** or **beta (β). Type II error** is the probability of failing to reject the null hypothesis when it is actually false. An even more interesting probability is $1 - \beta$, termed the power of the statistical inference test. **Power** is the probability of correctly rejecting the null hypothesis when it should be rejected. Thus, power is the probability that statistical significance will be indicated if it is present. The relationship of the different error probabilities in the hypothetical setting of testing for the difference in two means is shown below:

		Reality	
		H_0: No Difference	H_a: Difference
	H_0: No Difference	$1 - \alpha$	β Type II error
Statistical Decision	H_a: Difference	α Type I error	$1 - \beta$ Power

One can see that while specifying alpha establishes the level of acceptable statistical significance, it is the level of power that dictates the probability of "success" in finding the differences if they actually exist. Then why not set both alpha and beta at acceptable levels? Because Type I and Type II errors are negatively related, and as the Type I error becomes more restrictive (moves closer to zero), the Type II error increases. Reducing the Type I errors also reduces the power of the statistical test. Thus, the analyst must strike a balance between the level of alpha and the resulting power.

Power is not solely a function of α. It is actually determined by three factors:

1. Effect size—The probability of achieving statistical significance is based not only on statistical considerations but also on the actual magnitude of the effect of interest (e.g., a difference of means between two groups or the correlation between variables) in the population, termed the **effect size** (see Key Terms). As one would expect, a larger effect is more likely to be found than a smaller effect and thus impact the power of the statistical test. To assess the power of

any statistical test, the researcher must first understand the effect being examined. Effect sizes are defined in standardized terms for ease of comparison. Mean differences are stated in terms of standard deviations, so that an effect size of .5 indicates that the mean difference is one-half of a standard deviation. For correlations, the effect size is based on the actual correlation between the variables.

2. Alpha—As already discussed, as alpha becomes more restrictive, power decreases. This means that as the analyst reduces the chance of finding an incorrect significant effect, the probability of correctly finding an effect also decreases. Conventional guidelines suggest alpha levels of .05 or .01. But the analyst must consider the impact of this decision on the power before selecting the alpha level. In later discussions the relationship of these two probabilities will be illustrated.

3. Sample size—At any given alpha level, increased sample sizes always produce greater power of the statistical test. But increasing sample sizes can also produce "too much" power. By this we mean that by increasing sample sizes, smaller and smaller effects will be found to be statistically significant, until at very large sample sizes almost any effect is significant. The researcher must always be aware that sample size can impact the statistical test by either making it insensitive (at small sample sizes) or overly sensitive (at very large sample sizes).

The relationships among alpha, sample size, effect size, and power are quite complicated, but a number of sources of guidance are available. Cohen [10] has examined power for most statistical inference tests and provided guidelines for acceptable levels of power, suggesting that studies be designed to achieve alpha levels of at least .05 with power levels of 80 percent. To achieve such levels, all three factors must be considered simultaneously. These interrelationships can be illustrated by two simple examples. The first involves testing for the difference between the mean scores of two groups. Assume that the effect size is thought to range between small (.2) and moderate (.5). The researcher must now determine the necessary alpha level and sample size of each group. Table 1.1 illustrates the impact of both sample size and alpha levels on power. As can be seen, power

TABLE 1.1 Power Levels for the Comparison of Two Means: Variations by Sample Size, Significance Level, and Effect Size

	alpha (α) = .05 Effect Size (ES)		alpha (α) = .01 Effect Size (ES)	
Sample Size	Small (.2)	Moderate (.5)	Small (.2)	Moderate (.5)
20	.095	.338	.025	.144
40	.143	.598	.045	.349
60	.192	.775	.067	.549
80	.242	.882	.092	.709
100	.290	.940	.120	.823
150	.411	.990	.201	.959
200	.516	.998	.284	.992

Source: *Solo Power Analysis,* BMDP Statistical Software, Inc.

becomes acceptable at sample sizes of 100 or more in situations with a moderate effect size at both alpha levels. But when small effect sizes occur, the statistical tests have little power, even with expanded alpha levels or samples of 200 or more. For example, a sample of 200 in each group with an alpha of .05 still only has a 50 percent chance of significant differences being found if the effect size is small. This suggests that the analyst, if anticipating the effects to be small, must design the study with much larger sample sizes and/or less restrictive alpha levels (.05 or .10). In the second example, Figure 1.1 graphically presents the power for significance levels of .01, .05, and .10 for sample sizes of 20 to 300 per group when the effect size (.35) falls between small and moderate. Faced with such prospects, specification of a .01 significance level requires a sample of 200 per group to achieve the desired level of 80 percent power. But if the alpha level is relaxed, 80 percent power is reached at samples of 130 for a .05 alpha level and samples of 100 for a .10 significance level.

Such analyses allow the researcher to make more informed decisions in study design and interpretation of the results. In planning the research, the researcher must estimate the expected effect size and then select the sample size and alpha to achieve the desired power level. In addition to its uses for planning, power analysis is also utilized after the analysis is completed to determine the actual power achieved, so that the results can be properly interpreted. Are the results due to effect sizes, sample sizes, or significance levels? The analyst can assess each of these factors for their impact on the significance or nonsignificance of the results. The researcher today can refer to published studies detailing the specifics of power determination [10] or turn to several personal computer–based programs that assist in planning studies to achieve the desired power or calculate the power of actual results [6, 7]. Specific guidelines for multiple regression and multivariate analysis of variance, the most common applications of power analysis, are discussed in more detail in Chapters 3 and 5.

Having addressed the issues in extending multivariate techniques from their univariate and bivariate origins, we now briefly introduce each multivariate

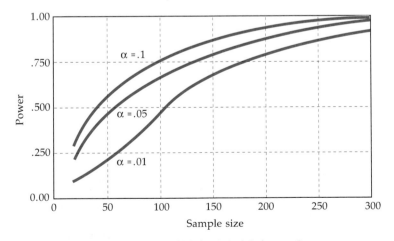

Source: *SOLO Power Analysis*, BMDP Statistical Software, Inc.

FIGURE 1.1 Impact of sample size on power for various alpha levels (.01, .05, .10) with effect size of .35.

method. Then following the introductions of the techniques, we present a classification schema to assist in the selection of the appropriate technique by identifying the research objectives (independence or dependence relationship) and the data type (metric or nonmetric).

Types of Multivariate Techniques

Multivariate analysis is an ever-expanding set of techniques for data analysis. Among the more established techniques discussed in this text are (1) multiple regression and multiple correlation; (2) multiple discriminant analysis; (3) principal components and common factor analysis; (4) multivariate analysis of variance and covariance; (5) canonical correlation; (6) cluster analysis; (7) multidimensional scaling; and (8) conjoint analysis. Among the emerging techniques also included are (9) correspondence analysis; (10) linear probability models such as logit and probit; and (11) simultaneous/structural equation modeling. At this point we introduce each of the multivariate techniques, briefly defining the technique and the objective for its application.

Multiple Regression

Multiple regression is the appropriate method of analysis when the research problem involves a single metric dependent variable presumed to be related to one or more metric independent variables. The objective of multiple regression analysis is to predict the changes in the dependent variable in response to changes in the several independent variables. This objective is most often achieved through the statistical rule of least squares.

Whenever the researcher is interested in predicting the amount or magnitude of the dependent variable, multiple regression is useful. For example, monthly expenditures on dining out (dependent variable) might be predicted from information regarding a family's income, its size, and the age of the head of household (independent variables). Similarly, the researcher might attempt to predict a company's sales from information on its expenditures for advertising, the number of salespeople, and the number of stores carrying its products.

Multiple Discriminant Analysis

If the single dependent variable is dichotomous (e.g., male-female) or multichotomous (e.g., high-medium-low) and therefore nonmetric, the multivariate technique of multiple discriminant analysis (MDA) is appropriate. As with multiple regression, the independent variables are assumed to be metric. Discriminant analysis is useful in situations where the total sample can be divided into groups based on a dependent variable characterizing several known classes. The primary objectives of multiple discriminant analysis are to understand group differences and to predict the likelihood that an entity (individual or object) will belong to a particular class or group based on several metric independent variables. For example, discriminant analysis might be used to distinguish innovators from noninnovators according to their demographic and psychographic profiles. Other applications include distinguishing heavy product users from light users, males from females, national-brand buyers from private-label buyers, and good credit risks from poor credit risks. Even the Internal Revenue Service uses discriminant analysis to compare selected federal tax returns with a composite, hypothetical, normal taxpayer's return (at different income levels) to identify the most promising returns and areas for audit.

Multivariate Analysis of Variance

Multivariate analysis of variance (MANOVA) is a statistical technique that can be used to simultaneously explore the relationship between several categorical independent variables (usually referred to as **treatments**) and two or more metric dependent variables. As such, it represents an extension of **univariate analysis of variance (ANOVA).** Multivariate analysis of covariance (MANCOVA) can be used in conjunction with MANOVA to remove (after the experiment) the effect of any uncontrolled independent variables on the dependent variables. The procedure is similar to that involved in **bivariate partial correlation.** MANOVA is useful when the researcher designs an experimental situation (manipulation of several nonmetric treatment variables) to test hypotheses concerning the variance in group responses on two or more metric dependent variables.

Canonical Correlation

Canonical correlation analysis can be viewed as a logical extension of multiple regression analysis. Recall that multiple regression analysis involves a single metric dependent variable and several metric independent variables. With canonical analysis the objective is to correlate simultaneously several metric dependent variables and several metric independent variables. Whereas multiple regression involves a single dependent variable, canonical correlation involves multiple dependent variables. The underlying principle is to develop a linear combination of each set of variables (both independent and dependent) to maximize the correlation between the two sets. Stated differently, the procedure involves obtaining a set of weights for the dependent and independent variables that provide the maximum simple correlation between the set of dependent variables and the set of independent variables.

Linear Probability Models

Linear probability models, often referred to as *logit analysis,* are a combination of multiple regression and multiple discriminant analysis. This technique is similar to multiple regression analysis in that one or more independent variables are used to predict a single dependent variable. What distinguishes a linear probability model from multiple regression is that the dependent variable is nonmetric, as in discriminant analysis. The nonmetric scale of the dependent variable requires differences in the estimation method and assumptions about the type of underlying distribution, yet in most other facets it is quite similar to multiple regression. Thus, once the dependent variable is correctly specified and the appropriate estimation technique employed, the basic factors considered in multiple regression are used here as well. Linear probability models are distinguished from discriminant analysis primarily in that they accommodate all types of independent variables (metric and nonmetric) and do not require the assumption of multivariate normality. However, in many instances, particularly with more than two levels of the dependent variable, discriminant analysis is the more appropriate technique.

Conjoint Analysis

Conjoint analysis is an emerging dependence technique that has brought new sophistication to the evaluation of objects, whether they be new products, services, or ideas. The most direct application is in new product or service development, allowing for the evaluation of complex products while maintaining a realistic decision context for the respondent. The market researcher is able to assess the importance of attributes as well as the levels of each attribute while consumers evaluate only a few product profiles, which are combinations of product levels. For example, assume a product concept has three attributes (price, quality, and color), each at three possible levels (e.g., red, yellow, and blue). Instead of having to evaluate all 27 ($3 \times 3 \times 3$) possible combinations, a subset (9 or more) can be evaluated for their attractiveness to consumers, and the researcher knows not only how important each attribute is but also the importance of each level (the attractiveness of red versus yellow versus blue). Moreover, when the consumer evaluations are completed, the results of conjoint analysis can also be used in product design simulators, which show customer acceptance for any number of product formulations and aid in the design of the optimal product.

Structural Equation Modeling

Structural equation modeling, often referred to simply as LISREL (the name of one of the more popular software packages), is a technique that allows separate relationships for each of a set of dependent variables. In its simplest sense, structural equation modeling provides the appropriate and most efficient estimation technique for a series of separate multiple regression equations estimated simultaneously. It is characterized by two basic components: (1) the structural model and (2) the measurement model. The *structural model* is the "path" model, which relates independent to dependent variables. In such situations, theory, prior experience, or other guidelines allow the researcher to distinguish which independent variables predict each dependent variable. Models discussed previously that accommodate multiple dependent variables—multivariate analysis of variance and canonical correlation—are not appropriate in this situation because they allow only a *single* relationship between dependent and independent variables.

The *measurement model* allows the researcher to use several variables (**indicators,** see Key Terms) for a single independent or dependent variable. For example, the dependent variable might be a concept represented by a summated scale, such as self-esteem. In the measurement model the researcher can assess the contribution of each scale item as well as incorporate how well the scale measures the concept (reliability) into the estimation of the relationships between dependent and independent variables. This procedure is similar to performing a factor analysis (discussed in a later section) of the scale items and using the factor scores in the regression.

The techniques discussed thus far have focused on multivariate methods applied to data that contain both dependent and independent variables. However, if the researcher is investigating the interrelations, and therefore the interdependence among all the variables, without regard to whether they are dependent or independent variables, several other multivariate methods are appropriate. These methods include factor analysis, cluster analysis, multidimensional scaling, and correspondence analysis.

Factor Analysis

Factor analysis, including variations such as component analysis and common factor analysis, is a statistical approach that can be used to analyze interrelationships among a large number of variables and to explain these variables in terms of their common underlying dimensions (factors). The objective is to find a way of condensing the information contained in a number of original variables into a smaller set of variates (factors) with a minimum loss of information.

Cluster Analysis

Cluster analysis is an analytical technique for developing meaningful subgroups of individuals or objects. Specifically, the objective is to classify a sample of entities (individuals or objects) into a small number of mutually exclusive groups based on the similarities among the entities. In cluster analysis, unlike discriminant analysis, the groups are not predefined. Instead, the technique is used to identify the groups.

Cluster analysis usually involves at least two steps. The first is the measurement of some form of similarity or association between the entities to determine how many groups really exist in the sample. The second step is to profile the persons or variables to determine their composition. This step may be accomplished by applying discriminant analysis to the groups identified by the cluster technique.

Multidimensional Scaling

In multidimensional scaling, the objective is to transform consumer judgments of similarity or preference (e.g., preference for stores or brands) into distances represented in multidimensional space. If objects A and B are judged by respondents as being the most similar compared with all other possible pairs of objects, multidimensional scaling techniques will position objects A and B in such a way that the distance between them in multidimensional space is smaller than the distance between any other two pairs of objects. The resulting perceptual maps show the relative positioning of all objects, but additional analysis is needed to assess which attributes predict the position of each object.

Correspondence Analysis

Finally, correspondence analysis is a recently developed interdependence technique that facilitates both dimensional reduction of object ratings (e.g., products, persons, etc.) on a set of attributes and the perceptual mapping of objects relative to these attributes. Researchers are constantly faced with the need to "quantify the qualitative data" found in nominal variables. Correspondence analysis differs from the other interdependence techniques discussed earlier in its ability to accommodate both nonmetric data and nonlinear relationships.

In its most basic form, correspondence analysis employs a contingency table, which is the cross-tabulation of two categorical variables. It then transforms the nonmetric data to a metric level and performs dimensional reduction (similar to factor analysis) and perceptual mapping (similar to multidimensional analysis). As an example, respondents' brand preferences can be cross-tabulated on demographic variables (e.g., gender, income categories, occupation) by indicating how many people preferring each brand fall into each category of the demographic variables. Through correspondence analysis, the association, or "correspondence," of brands and the distinguishing characteristics of those preferring each brand are then shown in a two- or three-dimensional map of both brands and respondent characteristics. Brands perceived as similar are located in close proximity to one another. Likewise, the most distinguishing characteristics of respondents preferring each brand are also determined by the proximity of the demographic variable categories to the brand's position. Correspondence analysis provides a multivariate representation of interdependence for nonmetric data not possible with other methods.

A Classification of Multivariate Techniques

To assist you in becoming familiar with specific multivariate techniques, we present a classification of multivariate methods in Figure 1.2. This classification is based on three judgments the analyst must make about the research objective and nature of the data: (1) Can the variables be divided into independent and dependent classifications based on some theory? (2) If they can, how many variables are treated as dependent in a single analysis? (3) How are the variables measured? Selection of the appropriate multivariate technique to be utilized depends on the answers to these three questions.

When considering the application of multivariate statistical techniques, the first question to be asked is, Can the data variables be divided into independent and dependent classifications? The answer to this question indicates whether a dependence or interdependence technique should be utilized. Note that in Figure 1.2, the dependence techniques are on the left side and the interdependence techniques are on the right. A **dependence technique** may be defined as one in which a variable or set of variables is identified as the **dependent variable** to be predicted or explained by other variables known as **independent variables.** An example of a dependence technique is multiple regression analysis. In contrast, an **interdependence technique** is one in which no single variable or group of variables is defined as being independent or dependent. Rather, the procedure involves the analysis of all variables in the set simultaneously. Factor analysis is an example of an interdependence technique. Let us focus on dependence techniques first and use the classification in Figure 1.2 to select the appropriate multivariate method.

The different methods that constitute the analysis of dependence can be categorized by two things: (1) the number of dependent variables and (2) the type of

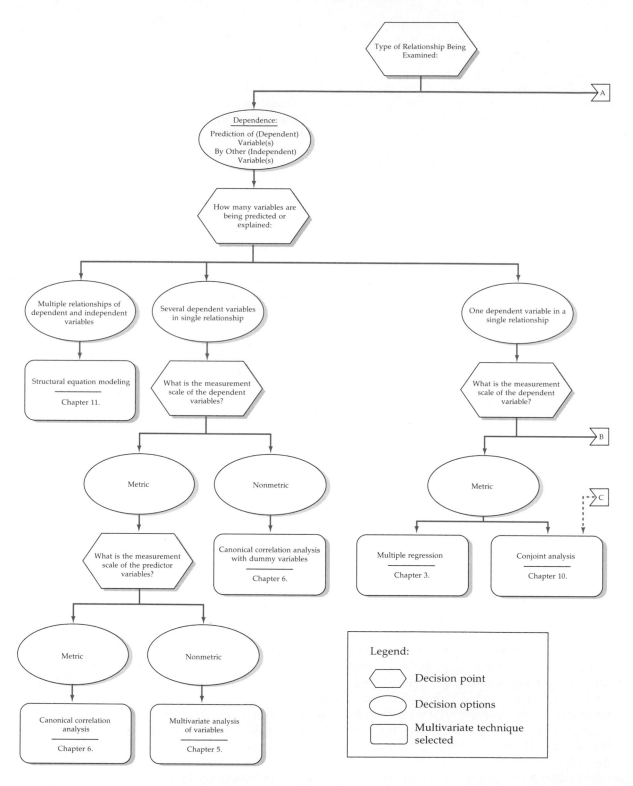

FIGURE 1.2 Selecting a multivariate technique.

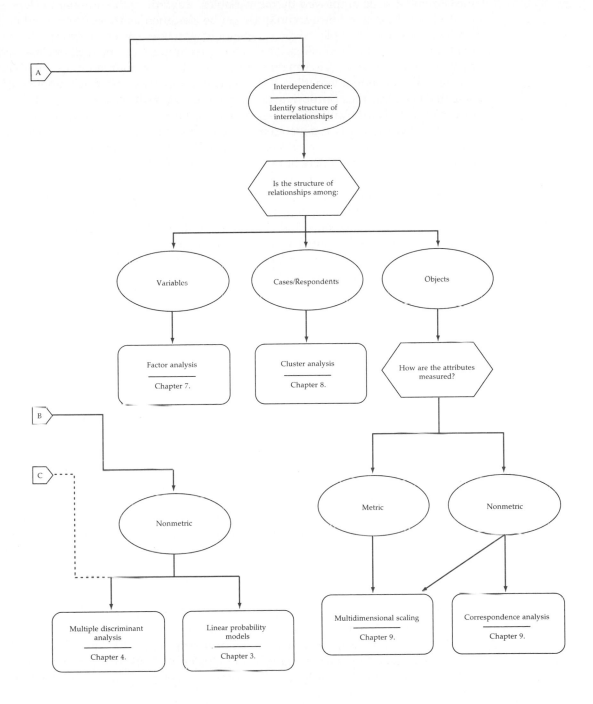

measurement scale employed by the variables. Regarding the number of dependent variables, dependence techniques can be classified as those having either a single dependent variable, several dependent variables, or even several dependent/independent relationships. Dependence techniques can be further classified as those with either metric (quantitative/numerical) or nonmetric (qualitative/categorical) dependent variables. If the analysis involves a single dependent variable that is metric, the appropriate technique is either multiple regression analysis or conjoint analysis. Conjoint analysis is a special case, as denoted by the dashed lines in Figure 1.2. It is a dependence procedure that may treat the dependent variable as either nonmetric or metric, depending on the circumstances. On the other hand, if the single dependent variable is nonmetric (categorical), then the appropriate techniques are multiple discriminant analysis and linear probability models. In contrast, when the research problem involves several dependent variables, four other techniques of analysis are appropriate. If the several dependent variables are metric, we must then look to the independent variables. If the independent variables are nonmetric, the technique of multivariate analysis of variance should be selected. If the independent variables are metric, canonical correlation is appropriate. If the several dependent variables are nonmetric, then they can be transformed through dummy variable coding (0–1) and canonical analysis can again be used.* Finally, if a set of dependent/independent variable relationships is postulated, then structural equation modeling is appropriate.

There is a close relationship between the various dependence procedures, and they can be viewed as a family of techniques. Table 1.2 defines the various multivariate dependence techniques in terms of the nature and number of dependent and independent variables. As we can see, canonical correlation can be considered to be the general model upon which many other multivariate techniques are based, because it places the least restrictions on the type and number of variables in both the dependent and independent variates. As restrictions are placed on the variates, more precise conclusions can be reached based on the specific scale of data measurement employed. Thus multivariate techniques range from the quite general method of canonical analysis to the quite specialized technique of structural equation modeling.

Interdependence techniques are shown on the right side of Figure 1.2. Readers will recall that with interdependence techniques the variables cannot be classified as either dependent or independent. Instead, all the variables are analyzed simultaneously in an effort to find an underlying structure to the entire set of variables or subjects. If the structure of variables is to be analyzed, then factor analysis is the appropriate technique. If cases or respondents are to be grouped to represent structure, then cluster analysis is selected. Finally, if the interest is in the structure of objects, the techniques of multidimensional scaling should be applied. As with dependence techniques, the measurement properties of the techniques should be considered. Generally, factor analysis and cluster analysis are considered to be metric interdependence techniques. However, nonmetric data may be trans-

---

*Dummy variables (see Key Terms) are discussed in greater detail later. Briefly, dummy variable coding is a means of transforming nonmetric data into metric data. It involves the creation of so-called dummy variables, in which 1s and 0s are assigned to subjects depending on whether they do or do not possess a characteristic in question. For example, if a subject is a male, assign him a 1; if the subject is a female, assign her a 0; or the reverse.

TABLE 1.2 The Relationship Between Multivariate Dependence Methods

Canonical Correlation
$$Y_1 + Y_2 + Y_3 + \ldots + Y_n = X_1 + X_2 + X_3 + \ldots + X_n$$
(metric, nonmetric)   *(metric, nonmetric)*

Multivariate Analysis of Variance
$$Y_1 + Y_2 + Y_3 + \ldots + Y_n = X_1 + X_2 + X_3 + \ldots + X_n$$
(metric)   *(nonmetric)*

Analysis of Variance
$$Y_1 - X_1 + X_2 + X_3 + \ldots + X_n$$
(metric)   *(nonmetric)*

Multiple Discriminant Analysis
$$Y_1 = X_1 + X_2 + X_3 + \ldots + X_n$$
(nonmetric)   *(metric)*

Multiple Regression Analysis
$$Y_1 = X_1 + X_2 + X_3 + \ldots + X_n$$
(metric)   *(metric, nonmetric)*

Conjoint Analysis
$$Y_1 = X_1 + X_2 + X_3 + \ldots + X_n$$
(nonmetric, metric)   *(nonmetric)*

Structural Equation Modeling
$$Y_1 = X_{11} + X_{12} + X_{13} + \ldots + X_{1n}$$
$$Y_2 = X_{21} + X_{22} + X_{23} + \ldots + X_{2n}$$
$$Y_m - X_{m1} + X_{m_2} + X_{m3} + \ldots + X_{mn}$$
(metric)   *(metric, nonmetric)*

formed through dummy variable coding for use with factor analysis and cluster analysis. Both metric and nonmetric approaches to multidimensional scaling have been developed. If the interdependencies of objects measured by nonmetric data are to be analyzed, correspondence analysis is also an appropriate technique.

Guidelines for Multivariate Analyses and Interpretation

As you have just seen, multivariate analyses have a very diverse character and can be quite powerful. This power is especially tempting when the researcher is unsure of the most appropriate analysis design and relies instead on the multivariate technique as a substitute for the necessary conceptual development. Even applied correctly, the strengths of accommodating multiple variables and relationships create complexity in the results and their interpretation. We have also already discussed several issues particularly applicable to multivariate analyses. Therefore, while no single "answer" exists, we have found that analysis and interpretation of any multivariate problem can be helped by following a set of general guidelines. By no means an exhaustive list of considerations, the list instead represents more of a "philosophy of multivariate analysis" that has served us well. The following sections discuss these points in no particular order, but with equal emphasis on all.

Establish Practical Significance as Well as Statistical Significance

The strength of multivariate analysis is its seemingly magic means of sorting through a myriad number of possible alternatives and finding those that have statistical significance. But with this power must come caution. Many researchers become myopic in focusing solely on the achieved significance of the results without understanding their interpretations, good or bad. A researcher must instead look not only at the statistical significance of the results but at their practical significance as well. **Practical significance** asks the question, "So what?" For any managerial application, the results must have a demonstrable effect that justifies action. In academic settings, research is becoming more focused on not only the statistically significant results but their substantive and theoretical implications as well, which are many times drawn from their practical significance.

An example illustrates this well. A regression analysis is undertaken to predict repurchase intentions, measured as the probability between 0 and 100 that the customer will shop with the firm next time. The study is conducted and the results come back significant at the .01 significance level. Executives rush to embrace the results and modify firm strategy accordingly. But what goes unnoticed is that while the relationship was significant, the predictive ability was poor, so poor that the estimate of repurchase probability could vary by as much as ±20 percent at the .05 significance level. The "statistically significant" relationship could then have a range of error of 40 percentage points! A customer predicted to have a 50/50 chance of return could really have probabilities from 10 percent to 90 percent, representing unacceptable levels upon which to take action. Researchers and managers did not probe the practical or managerial significance of the results, wherein they would have seen that the relationship still needed refinement if it was to be relied upon to guide strategy in any substantive sense.

Sample Size Affects All Results

The discussion of statistical power demonstrated the substantial impact sample size plays in achieving statistical significance, both in small and large sample sizes. For smaller samples, the sophistication and complexity of the multivariate technique may easily result in either (1) too little statistical power for the test to realistically identify significant results or (2) too easily an "overfitting" of the data such that the results are artificially good because they fit the sample very well, yet have no generalizability. A similar impact also occurs for large sample sizes, which as discussed earlier can make the statistical tests overly sensitive. Any time sample sizes exceed 200 to 300 respondents, the researcher should examine all significant results to ensure that they have practical significance due to the increased statistical power from the sample size. Sample sizes also affect the results when the analyses involve groups of respondents, such as discriminant analysis or MANOVA. Unequal sample sizes between groups influence the results and require additional interpretation and/or analysis. Thus, a researcher or user of multivariate techniques should always assess the results in light of the sample used in the analysis.

Know Your Data

Multivariate techniques, by their very nature, identify complex relationships that are very difficult to represent simply. As a result, the tendency is to accept the results without the typical examination one undertakes in univariate and bivariate analyses (e.g., scatterplots of correlations and boxplots of mean comparisons). But such "shortcuts" can be a prelude to disaster. Multivariate analyses require an even *more rigorous* examination of the data because the influence of outliers, violations of assumptions, and missing data can be compounded across several variables to have quite substantial effects. To utilize the full benefits of multivariate techniques, the analyst also must "know where to look" with alternative formulations of the original model, such as nonlinear and interactive relationships. The analyst has, however, an ever-expanding set of diagnostic techniques that enable these multivariate relationships to be discovered in means quite similar to the univariate and bivariate methods. The multivariate researcher must take the time to utilize these diagnostic measures for a greater understanding of the data and the basic relationships that exist.

Strive for Model Parsimony

Multivariate techniques are designed to accommodate multiple variables in the analysis. This feature, however, should not substitute for conceptual model development *before* the multivariate techniques are applied. While it is always more important to avoid omitting a critical predictor variable, termed **specification error,** for several reasons the analyst must also avoid inserting variables indiscriminately and letting the multivariate technique "sort out" the relevant variables. First, irrelevant variables usually increase a technique's ability to fit the sample data, but at the expense of overfitting the data and making them less generalizable to the population. Second, irrelevant variables do not typically bias the estimates of the relevant variables, but they can mask the true effects owing to multicollinearity. **Multicollinearity** represents the degree to which any variable's effect can be predicted or accounted for by the other variables in the analysis. As multicollinearity rises, the ability to define any variable's effect is diminished. Thus, including variables that are conceptually not relevant can have several potentially harmful effects, even if the additional variables do not directly bias the model results.

Look at Your Errors

Even with the statistical prowess of multivariate techniques, rarely do we achieve the best prediction in the first analysis. The analyst is then faced with the question, "Where does one go from here?" The best answer is to look at the errors in prediction, whether they be the residuals from regression analysis, the misclassification of observations in discriminant analysis, or the outliers in cluster analysis.

In each case, the analyst should use the errors in prediction not as a measure of failure or merely something to eliminate but as a starting point for diagnosing the validity of the obtained results and an indication of the remaining unexplained relationships.

Validate Your Results

The ability of multivariate analyses to identify complex interrelationships also means that results can be found that are specific only to the sample and not generalizable to the population. The researcher must always ensure that there are sufficient observations per estimated parameter to avoid "overfitting" the sample, as discussed earlier. But just as important are the efforts to validate the results by one of several methods, including (1) splitting the sample and using one subsample to estimate the model and using the second subsample to estimate the predictive accuracy, (2) employing a bootstrapping technique [15], or (3) even gathering a separate sample to ensure that the results are appropriate for other samples. Whatever multivariate technique is employed, the researcher must strive not only to estimate a significant model but to ensure that it is representative of the population as a whole. Remember, the objective is not to find the best "fit" just to the sample data but instead to develop a model that best describes the population as a whole.

A Structured Approach to Multivariate Model Building

As we discuss the numerous multivariate techniques available to the researcher and the myriad set of issues involved in their application, it becomes apparent that the successful completion of a multivariate analysis involves more than just the selection of the correct method. Issues ranging from problem definition to a critical diagnosis of the results must be addressed. To aid the researcher or user in applying multivariate methods, a six-step approach to multivariate analysis is presented. The intent is not to provide a rigid set of procedures to follow but, instead, to provide a series of guidelines that emphasize a model-building approach. A model-building approach focuses the analysis on a well-defined research plan, starting with a conceptual model detailing the relationships to be examined. Once defined in conceptual terms, the empirical issues can be addressed, including the selection of the specific multivariate technique and the implementation issues. After significant results have been obtained, their interpretation becomes the focus, with special attention directed toward the variate. Finally, the diagnostic measures ensure that the model is not only valid for the sample data but that it is as generalizable as possible. The following discussion briefly describes each step in this approach.

This six-step model-building process provides a framework for developing, interpreting and validating any multivariate analysis. Each researcher must develop his or her own criteria for "success or failure" at each stage, but the discussions of each technique provide guidelines whenever available. It is our hope that emphasis on a model-building approach rather than just the specifics of each technique will provide a broader base of model development, estimation, and interpretation that will improve the multivariate analyses of practitioner and academician alike.

Stage One: Define the Research Problem, Objectives, and Multivariate Technique to Be Used

The starting point for any multivariate analysis is to define the research problem and analysis objectives in conceptual terms before specifying any variables or measures. The role of conceptual model development, or theory, cannot be overstated. No matter whether in academic or applied research, the researcher must first view the problem in conceptual terms by defining the concepts and identifying the fundamental relationships to be investigated. And developing a conceptual model is not the exclusive domain of academicians; it is just as suited to the application of real-world experience.

A conceptual model need not be complex and detailed but a simple representation of the relationships to be studied. If a dependence relationship is proposed as the research objective, the researcher needs to specify the dependent and independent concepts. Note that a concept, rather than a variable, is defined. For an application of an interdependence technique, the dimensions of structure or similarity should be specified. In both dependence and interdependence situations, the researcher first identifies the ideas or topics of interest rather than focusing on the specific measures to be used. This minimizes the chance that relevant concepts will be omitted in the effort to develop measures and to define the specifics of the research design. Readers interested in conceptual model development should see Chapter 11.

Stage Two: Develop the Analysis Plan

With the conceptual model established, attention turns to the implementation issues for the selected technique. For each technique the researcher must now develop a specific analysis plan that addresses the set of issues particular to its purpose and design. The issues range from the general considerations of minimum or desired sample sizes, to allowable or required types of variables (metric versus nonmetric) and estimation methods, to such specific issues as the type of association measures used in multidimensional scaling, the estimation of aggregate or disaggregate results in conjoint, or the use of special variable formulations to represent nonlinear or interactive effects in regression. In each instance, these issues resolve specific details and finalize the model formulation and the requirements for the data collection effort.

Stage Three: Evaluate the Assumptions Underlying the Multivariate Technique

With data collected, the first analysis is not to estimate the multivariate model but to evaluate the underlying assumptions. All multivariate techniques have underlying assumptions, both statistical and conceptual, that substantially impact their ability to represent multivariate relationships. For the techniques based on statistical inference, the assumptions of multivariate normality, linearity, independence of the error terms, and equality of variances in a dependence relationship must all be met. Chapter 2 discusses assessing these assumptions in more detail. Each technique also has a series of conceptual assumptions dealing with such issues as model formulation and the types of relationships represented. Before any model estimation is attempted, the researcher must ensure that both statistical and conceptual assumptions are met.

Stage Four: Estimate the Multivariate Model and Assess Overall Model Fit

With the assumptions satisfied, the analysis proceeds to the actual estimation of the multivariate model and an assessment of overall model fit. In the estimation process, the analyst may choose among options to meet specific characteristics of the data (e.g., use of covariates in MANOVA) or to maximize the fit to the data (e.g., rotation of factors or discriminant functions). After the model is estimated, the overall model fit is evaluated to ascertain whether it achieves acceptable levels on the statistical criteria (e.g., level of significance), identifies the proposed relationships, and achieves practical significance. Many times, the model will be respecified in an attempt to achieve better levels of overall fit and/or explanation. An acceptable model must be obtained before proceeding.

No matter what level of overall model fit is found, the analyst must also determine if the results are unduly affected by any single or small set of observations that indicate the results may be unstable. These efforts ensure that the results are "robust" and stable by applying reasonably well to all observations in the sample. Ill-fitting observations may be identified as outliers, influential observations, or other disparate results (i.e., single-member clusters or seriously misclassified cases in discriminant analysis).

Stage Five: Interpret the Variate(s)

With an acceptable level of model fit, interpreting the variate(s) reveals the nature of the multivariate relationship. The interpretation of effects for individual variables is made by examining the estimated coefficients (weights) for each variable in the variate (e.g., regression weights, factor loadings, or conjoint utilities). Moreover, some techniques also estimate multiple variates that represent underlying dimensions of comparison or association (i.e., discriminant functions or principal components). The interpretation may lead to additional respecifications of the variables and/or model formulation, wherein the model is reestimated and then interpreted again. The objective is to identify empirical evidence of multivariate relationships in the sample data that can be generalized to the total population.

Stage Six: Validate the Multivariate Model

Before accepting the results, the researcher must subject them to one final set of diagnostic analyses that assess the degree of generalizability of the results by the available validation methods. The attempts to validate the model are directed toward demonstrating the generalizability of the results to the total population (see the earlier discussion of validation techniques). Both diagnostic analyses add little to the interpretation of the results but can be viewed as "insurance" that the results are the most descriptive of the data and generalizable to the population.

Databases

To explain and illustrate each of the multivariate techniques more fully, we use hypothetical data sets throughout the book. These data sets were all obtained from the Hair, Anderson, and Tatham Company (HATCO), a large (though non-existent) industrial supplier. The primary database, consisting of 100 observations on 14 separate variables, is an example of a segmentation study for a business-to-business situation, specifically a survey of existing customers of HATCO. Two classes of information were collected. The first class is the perception of HATCO on seven attributes identified in past studies as the most influential in the choice of suppliers. The respondents, purchasing managers of firms buying from HATCO, rated HATCO on each attribute. The second class of information contains evaluations of each respondent's satisfaction with HATCO, the percentage of his or her product purchases from HATCO, and general characteristics of the purchasing companies (e.g., firm size, industry type). The data provided should give HATCO a better understanding of both the characteristics of its customers and the relationships between their perceptions of HATCO and their actions toward HATCO (purchases and satisfaction). A brief description of the database variables is provided in Table 1.3, in which the variables are classified as either independent or dependent and either metric or nonmetric. A listing of the database is provided in Appendix A for those who wish to reproduce the solutions reported in this book. A definition of each variable and an explanation of its coding is given in the following sections.

TABLE 1.3 Description of Database Variables

	Variable Description	*Variable Type*
	Perceptions of HATCO	
X_1	Delivery speed	Independent/Metric
X_2	Price level	Independent/Metric
X_3	Price flexibility	Independent/Metric
X_4	Manufacturer's image	Independent/Metric
X_5	Overall service	Independent/Metric
X_6	Salesforce's image	Independent/Metric
X_7	Product quality	Independent/Metric
	Purchaser Characteristics	
X_8	Size of firm	Independent or Dependent/Nonmetric
X_9	Usage level	Dependent/Metric
X_{10}	Satisfaction level	Dependent/Metric
X_{11}	Specification buying	Independent or Dependent/Nonmetric
X_{12}	Structure of procurement	Independent or Dependent/Nonmetric
X_{13}	Type of industry	Independent or Dependent/Nonmetric
X_{14}	Type of buying situation	Independent or Dependent/Nonmetric

Perceptions of HATCO

Each of these variables was measured on a graphic rating scale, where a ten-centimeter line was drawn between the endpoints, labeled "Poor" and "Excellent":

Poor   Excellent

Respondents indicated their perceptions by making a mark anywhere on the line. The mark was then measured and the distance from zero (in centimeters) was recorded. The result was a scale ranging from zero to ten, rounded to a single decimal place. The seven attributes of HATCO rated are as follows:

X_1 Delivery speed—amount of time it takes to deliver the product once an order has been confirmed

X_2 Price level—perceived level of price charged by product suppliers

X_3 Price flexibility—perceived willingness of HATCO representatives to negotiate price of all types of purchases

X_4 Manufacturer's image—overall image of the manufacturer/supplier

X_5 Service—overall level of service necessary for maintaining a satisfactory relationship between supplier and purchaser

X_6 Salesforce's image—overall image of the manufacturer's sales force

X_7 Product quality—perceived level of quality of a particular product (e.g., performance or yield)

Purchaser Characteristics

The seven characteristics used in the study, some metric and some nonmetric, are as follows:

X_8 Size of firm—size of the firm relative to others in this market. This variable has two categories: 1 = large, and 0 = small

X_9 Usage level—how much of the firm's total product is purchased from HATCO, measured on a 100-point percentage scale, ranging from 0 to 100 percent

X_{10} Satisfaction level—how satisfied the purchaser is with past purchases from HATCO, measured on the same graphic rating scale as the perceptions X_1 to X_7

X_{11} Specification buying—extent to which a particular purchaser evaluates each purchase separately (total value analysis) versus the use of specification buying, which details precisely the product characteristics desired. This variable has two categories: 1 = employs total value analysis approach, evaluating each purchase separately, and 0 = use of specification buying

X_{12} Structure of procurement—method of procuring/purchasing products within a particular company. This variable has two categories: 1 = centralized procurement, and 0 = decentralized procurement

X_{13} Type of industry—industry classification in which a product purchaser belongs. This variable has two categories: 1 = industry A classification, and 0 = other industries

X_{14} Type of buying situation—type of situation facing the purchaser. This
    variable has three categories: 1 = new task, 2 = modified rebuy, and
    3 = straight rebuy

Three other specialized databases are used in the text. Chapter 2 employs a
smaller database of many of these variables obtained in some pretest surveys. The
purpose is to illustrate the identification of outliers, handling of missing data, and
testing of statistical assumptions. Chapters 9 and 10 examine databases with the
unique data needed for these techniques. In each instance, the database is described more fully in those chapters. A full listing of these databases is given in
Appendix A as well.

Summary

This chapter has introduced the exciting, challenging topic of multivariate data
analysis. The following chapters discuss each of the techniques in sufficient detail
to enable the novice data analyst to understand what a particular technique can
achieve, when and how it should be applied, and how the results of its application are to be interpreted. End-of-chapter readings from academic literature further demonstrate the application and interpretation of the techniques.

Questions

1. In your own words, define multivariate analysis.
2. Name several factors that have contributed to the increased application of
   techniques for multivariate data analysis in recent years.
3. List and describe the 11 multivariate data analysis techniques described in this
   chapter. Cite examples where each technique is appropriate.
4. Explain why and how the various multivariate methods can be viewed as a
   family of techniques.
5. Why is knowledge of measurement scales important to an understanding of
   multivariate data analysis?
6. What are the differences between statistical and practical significance? Is one a
   prerequisite for the other?
7. What are the implications of low statistical power? How can the power be
   improved if it is deemed too low?
8. Detail the model-building approach to multivariate analysis, focusing on the
   major issues at each step.

References

1. Bearden, William O., Richard G. Netemeyer, and Mary F. Mobley. *Handbook of
   Marketing Scales, Multi-Item Measures for Marketing and Consumer Behavior*.
   Newbury Park, Calif.: Sage, 1993.

2. Bentler, Peter M. *EQS Structural Equations Program Manual*. Los Angeles: BMDP Statistical Software, 1992.

3. Bentler, Peter M., and Eric J. C. Wu. *EQS/Windows User's Guide*. Los Angeles: BMDP Statistical Software, 1993.

4. BMDP Statistical Software, Inc. *BMDP Statistical Software Manual, Release 7*, vols. 1 and 2. Los Angeles: 1992.

5. BMDP Statistical Software, Inc. *BMDP/PC User's Guide, Release 7*, vols. 1 and 2. Los Angeles, 1993.

6. BMDP Statistical Software, Inc. *SOLO Power Analysis*. Los Angeles, 1991.

7. Brent, Edward E., Edward J. Mirielli, and Alan Thompson. *Ex-Sample™: An Expert System to Assist in Determining Sample Size, Version 3.0*. Columbia, Mo.: Idea Works, 1993.

8. Brent, Edward E. et al. *Statistical Navigator Professional™: An Expert System to Assist in Selecting Appropriate Statistical Analyses, Version 1.0*. Columbia, Mo.: Idea Works, 1991.

9. Brunner, Gordon C., and Paul J. Hensel. *Marketing Scales Handbook, A Compilation of Multi-Item Measures*. Chicago: American Marketing Association, 1993.

10. Cohen, J. *Statistical Power Analysis for the Behavioral Sciences*. New York: Academic Press, 1977.

11. Gatty, R. "Multivariate Analysis for Marketing Research: An Evaluation." *Applied Statistics* 15 (November 1966): 157–72.

12. Hardyck, C. D., and L. F. Petrinovich. *Introduction to Statistics for the Behavioral Sciences*, 2d ed. Philadelphia: Saunders, 1976.

13. Joreskog, K. G., and D. Sorbom. *LISREL VII: A Guide to the Program and Applications*. Mooresville, Ind.: Scientific Software International, 1989.

14. Joreskog, K. G., and D. Sorbom. *LISREL 8: Structural Equation Modeling with the SIMPLIS Command Language*. Mooresville, Ind.: Scientific Software International, 1993.

15. Mooney, Christopher Z., and Robert D. Duval. *BOOTSTRAPPING: A Nonparametric Approach to Statistical Inference*. Beverly Hills, Calif.: Sage, 1993.

16. SAS Institute, Inc. *SAS User's Guide: Basics, Version 6*. Cary, N.C.: 1990.

17. SAS Institute, Inc. *SAS User's Guide: Statistics, Version 6*. Cary, N.C.: 1990.

18. Sawtooth Software. *Adaptive Conjoint Analysis*. Evanston, Ill.: Sawtooth Software, 1993.

19. Sawtooth Software. *Choice-Based Conjoint*. Evanston, Ill.: Sawtooth Software, 1993.

20. Sawtooth Software. *Conjoint Value Analysis*. Evanston, Ill.: Sawtooth Software, 1993.

21. Smith, Scott M. *PC-MDS: A Multidimensional Statistics Package*. Provo, Utah: Brigham Young University Press, 1989.

22. Smith, Scott M. *VISA—General Purpose Statistics for the IBM PC*. Provo, Utah: Brigham Young University Press, 1989.

23. SPSS, Inc. *SPSS User's Guide*, 4th ed. Chicago, 1990.

24. SPSS, Inc. *SPSS Advanced Statistics Guide*, 4th ed. Chicago, 1990.

25. SPSS, Inc. *SPSS/PC+, Version 5.0*. Chicago, 1992.

26. SPSS, Inc. *SPSS for Windows, Version 6.0*. Chicago, 1993.

27. StatSoft, Inc. *STATISTICA/w*. Tulsa, Okla.: 1993.

28. Sullivan, John L., and Stanley Feldman. *Multiple Indicators: An Introduction.* Beverly Hills, Calif.: Sage, 1979.

29. SYSTAT, Inc. *SYSTAT for DOS*. Evanston, Ill.: 1992.

30. SYSTAT, Inc. *SYSTAT for Windows, version 5*. Evanston, Ill.: 1992.

CHAPTER 2

Examining Your Data

LEARNING OBJECTIVES

Upon completing this chapter, you should be able to:

- Select the appropriate graphical method to examine the characteristics of the data or relationships of interest.
- Understand the different types of missing data processes.
- Assess the type and potential impact of missing data.
- Explain the advantages and disadvantages of the approaches available for dealing with missing data.
- Identify univariate, bivariate and multivariate outliers.
- Test your data for the assumptions of most multivariate techniques.
- Determine the best method of data transformation given a specific problem.

In this chapter, we review and describe the methods currently available to examine your data. Examination of data is a time-consuming but necessary step that is sometimes overlooked by data analysts. Careful analysis of data leads to better prediction and more accurate assessment of dimensionality. The introductory section of this chapter offers a summary of various graphical techniques available to the researcher as a means of representing data. These techniques provide the analyst with a set of simple yet comprehensive ways to examine both the individual variables and the relationships among them. Other important concerns to the analyst when examining data are how to assess and overcome pitfalls in research design and data collection. Specifically, this chapter addresses the evaluation of missing data, identification of outliers, and testing of the assumptions underlying most multivariate techniques. Missing data are a nuisance to researchers, which may result from data entry errors or from the omission of answers by respondents. Classification of missing data and the processes, or reasons, underlying their presence will be discussed in this chapter. Outliers, or extreme responses, may unduly influence the outcome of any multivariate analysis. For this reason, methods to assess their impact are discussed below. Finally, the assumptions underlying most multivariate analyses are reviewed. To apply any multivariate technique, the researcher must assess the fit of the sample data with the assumptions underlying that multivariate technique. For example, researchers wishing to apply regression analysis (Chapter 3) would be particularly interested in assessing the assumptions of normality, homoscedasticity, independence of error, and linearity. The applicability of these issues varies with each data analysis project. However, they represent a process that should be performed to some extent for each application of a multivariate technique.

KEY TERMS

Before starting the chapter, review the key terms to develop an understanding of the concepts and terminology to be used. Throughout the chapter the key terms appear in **boldface**. Other points of emphasis in the chapter are *italicized*. Also, cross-references within the key terms are in *italics*.

All-available approach Imputation technique for missing data that computes values based on all available valid observations.

Box plot Method of representing the distribution of a variable. A box represents the major portion of the distribution and the extensions, called whiskers, reach to the extreme of the distribution. Very useful in making comparisons of a single variable across groups or between several variables.

Censored data Data or samples that are incomplete in some way, as when certain values are unknown or ignored.

Complete case approach Imputation technique for missing data that computes values based on only data from complete cases, or cases with no *missing data*.

Data transformations A variable may have an undesirable characteristic, such as nonnormality, that detracts from the ability of the correlation coefficient to represent the relationship between it and another variable. A transformation, such as taking the logarithm or square root of the variable, creates a better measure of the relationship. Transformations may be applied to either the dependent or independent variables, or both. The need and specific type of transformation may be based on theoretical reasons (such as transforming a known nonlinear relationship) or empirical reasons (identified through graphical or statistical means).

Heteroscedasticity   See *homoscedasticity*.

Histogram   Graphical display of the distribution of a single variable. By forming frequency counts in categories, the shape of the variable's distribution can be shown. Used to make a visual comparison to the *normal distribution*.

Homoscedasticity   When the variance of the error terms (ϵ) appears constant over a range of predictor variables, the data are said to be homoscedastic. The assumption of equal variance of the population error $\boldsymbol{\epsilon}$ (where $\boldsymbol{\epsilon}$ is estimated from ϵ) is critical to the proper application of linear regression. When the error terms have increasing or modulating variance, the data are said to be heteroscedastic. Analysis of *residuals* best illustrates this point.

Ignorable missing data   *Missing data processes* that are explicitly identifiable and/or are under the control of the researcher. Ignorable missing data do not require remedy because the missing data are inherent in the technique used.

Imputation methods   Process of estimating the *missing data* based on valid values of other variables and/or cases in the sample. The objective is to employ known relationships that can be identified in the valid values of the sample to assist in estimating the missing values.

Kurtosis   Measure of the peakedness or flatness of a distribution when compared with a normal distribution. A positive value indicates a relatively peaked distribution, and a negative value indicates a relatively flat distribution.

Linearity   Used to express the concept that the model possesses the properties of additivity and homogeneity. In a simple sense, linear models predict values that fall in a straight line by having a constant unit change of the dependent variable (slope) for a constant unit change of the independent variable. In the population model $Y = b_0 + b_1X_1 + E$, the effect of a change of 1 in X_1 is to add b_1 (a constant) units of Y. The model $Y = b_0 + b_2(X_2)$ is not additive because a unit change in X_2 does not increase Y by b_2 units; rather, it increases Y by $(X_2)b_2$ units, an amount that varies for different levels of X.

Missing at random (MAR)   Classification of *missing data* applicable when missing values of Y depend on X, but not on Y. When missing data are MAR, observed data for Y do not necessarily represent a truly random sample of all Y values.

Missing completely at random (MCAR)   Classification of missing data applicable when missing values of Y are not dependent on X. When missing data are MCAR, observed values of Y are truly a random sample of all Y values, with no underlying process that lends bias to the observed data.

Missing data   Information not available for a subject (or case) about whom other information is available. Missing data often occur when a respondent fails to answer one or more questions in a survey.

Missing data process   Any systematic event external to the respondent (such as data entry errors or data collection problems) or any action on the part of the respondent (such as refusal to answer a question) that leads to *missing data*.

Multivariate graphical display   Method of presenting a multivariate profile of an observation on three or more variables. The methods include such approach as glyphs, mathematical transformations, and even iconic representations such as faces.

Normal distribution Purely theoretical continuous probability distribution in which the horizontal axis represents all possible values of a variable and the vertical axis represents that probability of those values occurring. The scores on the variable are clustered around the mean in a symmetrical, unimodal pattern known as the bell-shaped, or normal, curve.

Normality Degree to which the distribution of the sample data corresponds to a *normal distribution*.

Normal probability plot Graphical comparison of the form of the distribution to the *normal distribution*. In the graph, the normal distribution is represented by a straight line angled at 45 degrees. The actual distribution is plotted against this line, so that any differences are shown as deviations from the straight line, making identification of differences quite simple.

Residual Portion of a dependent variable not explained by a multivariate technique. Associated with dependent methods that attempt to predict the dependent variable, the residual represents the unexplained portion of the dependent variable. Residuals can be used in diagnostic procedures that examine the unexplained portions to identify problems in the estimation technique or unidentified relationships.

Scatterplot Representation of the relationship between two metric variables portraying the joint values of each observation in a two-dimensional graph.

Skewness Measure of the symmetry of a distribution; in most instances the comparison is made to a *normal distribution*. A distribution is positively skewed when it contains relatively few large values and tails off to the right. A negatively skewed distribution has relatively few small values. Skewness values falling outside the range of −1 to +1 indicate a substantially skewed distribution.

Stem and leaf diagram A variant of the *histogram*, which provides a visual depiction of the variable's distribution as well as an enumeration of the actual data values.

Variate Linear combination of variables formed in the multivariate technique by deriving empirical weights applied to a set of variables specified by the researcher.

Introduction

The tasks involved in examining your data may seem mundane and inconsequential but are an essential part of any multivariate analysis. While multivariate techniques place tremendous analytical power in the analyst's hands, they also place a greater burden on the analyst to ensure that the statistical and theoretical underpinning on which they are based are also supported. By examining the data before the application of a multivariate technique, the analyst gains a basic understanding of the data and relationships between variables. This knowledge can aid immeasurably in the specification and refinement of the multivariate model as well as providing a reasoned perspective for interpretation of the results. Multivariate techniques place great demands on the analyst to understand, interpret, and articulate results based on relationships that are ever increasing in complexity. In addition to the demands on the analysts, multivariate techniques demand much more of the data they are to analyze. The statistical power of the multivariate techniques requires larger data sets and more complex assumptions than encountered with univariate analyses. The analytical sophistication needed to ensure that the statistical requirements are met when applying the chosen multivariate technique has forced the analyst to use a series of data examination techniques that in many instances match the complexity of the multivariate techniques. The purpose of this chapter is to provide an overview of the data examination techniques available, ranging from the simple process of visual inspection of graphical displays to the multivariate statistical process involved in missing data handling and testing of the assumptions underlying all multivariate methods.

Both novice and experienced analysts may be tempted to skim or even skip this chapter to gain knowledge concerning the application of a multivariate technique(s). Although the time, effort, and resources devoted to the data examination process may seem almost wasted because many times no features requiring corrective action are found, the analyst should view these techniques as an "investment in multivariate insurance." Even though results may indicate that everything is fine in many instances, the potentially catastrophic problems that can be avoided by following these analyses each and every time a multivariate technique is applied will more than pay for themselves in the long run. The occurrence of one serious and possibly fatal problem in a multivariate analysis will make a convert of any analyst. We encourage you to embrace these techniques before you are forced to because of problems raised during analysis.

The chapter addresses four separate phases of examining your data. These include (1) a graphical examination of the nature of the variables in the analysis and the relationships that form the basis of the multivariate analysis, (2) an evaluation process for understanding the impacts missing data can have on the analysis, plus alternatives for retaining cases with missing data in the analysis, (3) the techniques best suited for identifying outliers, those cases that may distort the relationships by their uniqueness on one or more of the variables under study, and (4) the analytical methods necessary to adequately assess the ability of the data to meet the statistical assumptions specific to the selected multivariate technique.

Graphical Examination of the Data

As discussed earlier, the use of multivariate techniques places an increased burden on the researcher to understand, evaluate, and interpret the findings from multivariate analysis. One aid in these tasks is a thorough understanding of the basic characteristics of the underlying data and relationships. When univariate analyses are considered, the level of understanding is fairly simple. But as the analyst moves to more complex multivariate analyses, the need and level of understanding increases dramatically. This section reviews some of the basic graphical methods available to assist in gaining a basic understanding of the characteristics of the data, particularly in a multivariate sense.

The advent and widespread use of statistical programs designed for the personal computer has led to increased access to such methods. Programs such as SYSTAT and STATISTICA [15, 16] have particularly comprehensive modules of graphical techniques available for data examination. The larger statistical programs, such as SPSS, SAS, and BMDP [2, 11, 12, 14] also have the basic graphical techniques that are augmented many times with more detailed statistical measures of data description. The following sections will detail some of the more widely used techniques for examining the characteristics of the distribution, bivariate relationships, group differences, and even multivariate profiles.

The Nature of the Variable: Examining the Shape of the Distribution

The starting point for understanding the nature of any variable is to characterize the shape of its distribution. While a number of statistical measures are discussed in a later section on normality, many times the analyst can gain an adequate perspective on the variable through a **histogram.** A histogram is a graphical representation of data that represents the frequency of occurrences (data values) within data categories. For example, if a set of values ranged from one to ten, the analyst could construct a histogram by counting the number of responses that were a one, a two, and so on. These frequencies are then plotted to examining the shape of the distribution of responses (see Figure 2.1). Here the responses for X_1 from the database introduced in Chapter 1 are represented. Categories with midpoints of 0.0, .5, 1.0, 1.5, . . . , up to 6.0 are used. The height of the bars represents the frequencies of data values within each category. If examination of the distribution is to assess its normality (see section below on testing assumptions for details on this issue), the normal curve can be superimposed on the distribution as well, as was done in Figure 2.1. The histogram can be used to examine any type of variable, from original values to residuals from a multivariate technique.

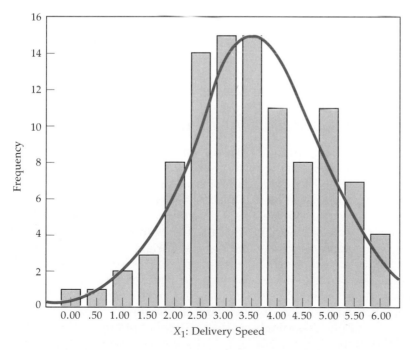

FIGURE 2.1 A graphical representation of a univariate
distribution: The histogram.

Frequency			Stem and Leaf
1.00	0	*	0
1.00	0	.	6
3.00	1	*	013
7.00	1	.	6688999
12.00	2	*	001333444444
10.00	2	.	5566788899
18.00	3	*	000001111233444444
10.00	3	.	5666777889
10.00	4	*	001122233
10.00	4	.	556778999
11.00	5	*	00112223344
5.00	5	.	55689
2.00	6	*	01

Stem width: 1.0
Each leaf: 1 case (s)

Valid cases: 100.0   Missing cases: .0   Percent missing: .0

FIGURE 2.2 Stem and leaf plot of X_1 (delivery speed).

A variant of the histogram is the **stem and leaf diagram,** which presents the same graphical picture as the histogram but also provides an enumeration of the actual data values. The stem and leaf diagram in Figure 2.2 is composed of *stems* and *leaves*. The stem is the root value, to which the leaves are added. For example, in Figure 2.2, the first stem is 0.0. To this is added the leaf of 0, resulting in a value of zero. In the next stem, a value of .6 is added to the stem of zero, resulting in a value of .6. If the frequencies of X_1 are compiled, 0.0 and .6 are the first two values. At the other end of the figure, the stem is 6.0. It is associated with two leaves (0 and 1), representing the values 6.0 and 6.1. These are the two highest values for X_1. The stem and leaf diagram provides a general shape of the distribution as found with the histogram, as well as providing the actual data values.

Examining the Relationship Between Variables

While examining the distribution of a variable is essential, many times the analyst is also interested in examining relationships between two or more variables. The most popular method for examining bivariate relationships is the **scatterplot,** a graph of data points based on two variables. One variable defines the horizontal axis and the other variable defines the vertical axis. Variables may be observations, expected values, or even **residuals.** The points found on the graph represent the corresponding joint values of the variables for any given case. The patterns of points represent the relationship between variables. A strong organization of points along a straight line characterizes a linear relationship of correlation. A curved set of points may denote a nonlinear relationship, which can be accommodated in many ways (see later discussion on linearity). Or there may be no patterns, only a seemingly random pattern of points. In this instance, there is no relationship.

There are many types of scatterplots, but one format particularly suited to multivariate techniques is the scatterplot matrix (see Figure 2.3). Here the scat-

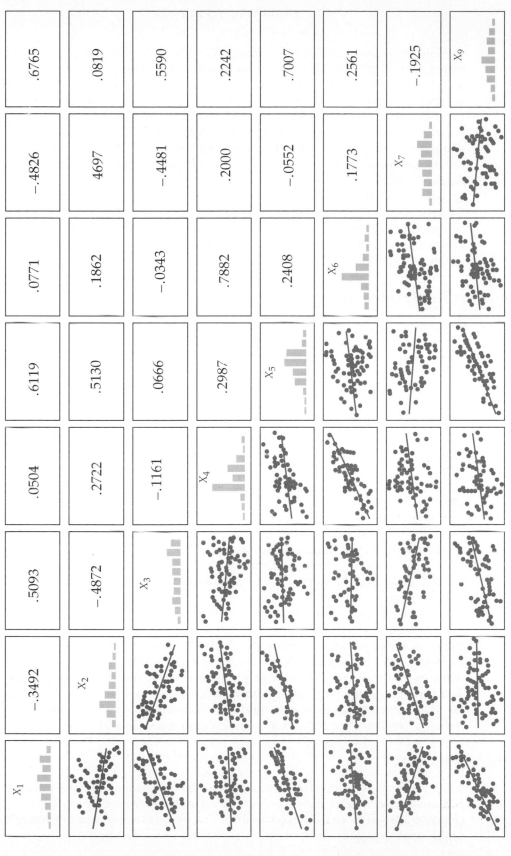

FIGURE 2.3 Scatterplot matrix of X_1, X_2, X_3, X_4, X_5, X_6, X_7, and X_9.

Note: Values above the diagonal are bivariate correlations, with corresponding scatterplots below the diagonal. Diagonal partrays distribution of each variable.

39

terplots are represented for all combinations of variables. The diagonals contain histograms of the variables. Figure 2.3 presents the scatterplots for a set of variables from the HATCO database (X_1, X_2, X_3, X_4, X_5, X_6, X_7, and X_9). Also included are the corresponding correlations so that the reader can assess the correlation represented in each scatterplot. For example, the scatterplot in the bottom left corner between X_1 and X_9 represents a correlation of .6765. The points are closely aligned around a straight line, indicative of a high correlation. The scatterplot in the leftmost column, third from the top (X_1 versus X_4) demonstrates the opposite, an almost total lack of relationship as evidenced by the widely dispersed pattern of points and the correlation .0504. Scatterplot matrices and individual scatterplots are now available in all popular statistical programs. A variant of the scatterplot is discussed in the following section on outlier detection, where an ellipse representing a specified confidence interval for the bivariate normal distribution is superimposed to allow for outlier identification.

Examining Group Differences

The analyst is also faced with understanding the extent and character of differences between two or more groups for one or more metric variables, such as found in discriminant analysis, analysis of variance, and multivariate analysis of variance. In these cases, the analyst needs to understand how the values are distributed for each group and if there are sufficient differences between the groups to support statistical significance. Another important aspect is to identify outliers that may become apparent only when the data values are separated into groups. The method used for this task is the **box plot,** a pictorial representation of the data distribution. The upper and lower boundaries of the box mark the upper and lower quartiles of the data distribution. Thus, the box length is the distance between the 25 percent percentile and the 75 percent percentile, so that the box contains the middle 50 percent of the data values. The asterisk (*) inside the box identifies the median. The larger the box, the greater the spread of the observations. The lines extending from each box (called *whiskers*) represent the distance to the smallest and the largest observations that are less than one quartile range from the box. These values are marked with an X. Outliers (marked O) are observations that range between 1.0 and 1.5 quartiles away from the box. Extreme values are marked E and represent those observations greater than 1.5 quartiles away from the end of the box.

Figure 2.4 shows the box plots for X_1 (delivery speed) for each group of X_{14} (types of buying situations) from the HATCO database. The three groups have markedly different sets of values, which indicates that there are true differences between the groups in terms of perceptions of delivery speed. The box plot for the first type of buying situation also indicates that an outlier exists. The analyst should examine this observation and then consider the possible remedies. The remedies available for outliers are discussed in more detail below.

Multivariate Profiles

To this point the graphical methods have been restricted to univariate or bivariate portrayals. But in many instances, the analyst may desire to compare observations characterized on more than two variables. The need is for a means of presenting a multivariate profile of an observation, whether it be for descriptive purposes or as

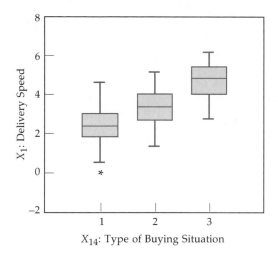

FIGURE 2.4 Box and whiskers plot.

a complement to analytical procedures. To overcome these limitations, a number of multivariate graphical methods have been devised that center around one of three approaches [8]. The first are glyphs or metroglyphs, which are some form of circle with radii that correspond to a data value or a multivariate profile, which portrays a barlike profile for each observation. A second form of multivariate display involves a mathematical transformation of the original data into a mathematical relationship, which can then be portrayed graphically. The final approach is the use of graphical displays with iconic representativeness, the most popular being a face [4]. The value of this type of display is the inherent processing capacity humans have for their interpretation. As noted by Chernoff [4, p. 9]:

> I believe that we learn very early to study and react to real faces. Our library of responses to faces exhausts a large part of our dictionary of emotions and ideas. We perceive the faces as a gestalt and our built-in computer is quick to pick out the relevant information and to filter out the noise when looking at a limited number of faces.

Facial representations provide a potent graphical format but also give rise to a number of considerations that impact the assignment of variables to facial features, unintended perceptions, and the quantity of information that can actually be accommodated. Discussion of these issues is beyond the scope of this text and interested readers are encouraged to review them before attempting to use these methods [17, 19].

Figure 2.5 contains an illustration of three types of **multivariate graphical displays** that were produced from SYSTAT but are available in many personal computer-based statistical programs. The upper portion of Figure 2.5 contains the data values for six observations on seven variables. In this instance, the data are profiles of six customer groups for the seven performance factors from the HATCO database. If the analyst were to examine the actual data values, similarities and differences are hard to distinguish, even to the extent that there may not be any differences. The objective of the multivariate profiles is to portray the data in a manner that enables differences and similarities to be easily identified. The first display are multivariate profiles, while the second are blobs, a form of Fourier transformation. Finally, faces were constructed with the seven variables as-

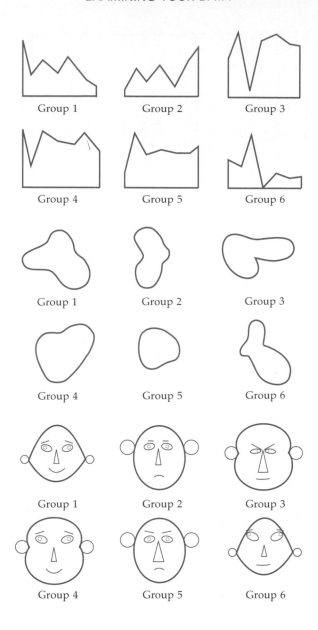

FIGURE 2.5 Examples of multivariate graphical displays

			Actual Data				
	X_1	X_2	X_3	X_4	X_5	X_6	X_7
Group 1	4.794	1.622	8.267	4.717	3.222	2.067	5.044
Group 2	2.011	2.133	6.544	5.267	2.039	2.672	8.483
Group 3	3.700	4.158	6.008	6.242	3.900	3.233	8.258
Group 4	4.810	1.510	9.319	5.691	3.148	3.195	6.981
Group 5	2.395	3.465	7.725	5.440	2.925	2.705	7.505
Group 6	3.246	1.627	9.136	3.809	2.318	1.927	5.355

signed to various facial features. The analyst can employ any of these methods when examining multivariate data to provide a format that is many times more insightful than just a review of the actual data values.

Summary

The graphical displays in this section are not intended as a replacement for the statistical diagnostic measures discussed in later sections of this chapter and in other chapters. But they do provide an alternative means of developing a perspective on the character of the data and the interrelationships that exist, even if multivariate in nature.

Missing Data

Missing data are a fact of life in multivariate analysis; in fact, it is rare for the analyst to avoid facing some form of missing data problem in almost every situation. For this reason, the researcher's challenge is to address the issues raised by missing data in the estimation process that affect the generalizability of the results. To do so, the analyst's primary concern is to determine the reasons underlying the missing data, with the extent of missing data being a secondary issue in most instances. This need to focus on the reasons for missing data comes from the fact that the analyst must understand the processes leading to the missing data in order to select the appropriate course of action. A **missing data process** is any (a) systematic event external to the respondent (such as data entry errors or data collection problems) or (b) action on the part of the respondent (such as refusal to answer) that leads to missing values. The effects of some missing data processes are known and directly accommodated in the research plan. But others, particularly those based on actions by the respondent, are rarely known. When the missing data processes are unknown, the analyst attempts to identify any patterns in the missing data that would characterize the missing data process. In doing so, the analyst asks such questions as, are the missing data scattered randomly throughout the observations or are distinct patterns identifiable? How prevalent are the missing data? If patterns are found and the extent of missing data is sufficient to warrant action, then it is assumed that some missing data process is in action and that any statistical results based on these data would be biased to the extent that the variables included in the analysis are influenced by the missing data process. The concern for understanding the missing data processes is similar to the need to understand the causes of nonresponse in the data collection process. For example, are those individuals who did not respond different from those who did? If so, do these differences have any impact on the analysis, the results, or their interpretation? Concerns similar to these also arise from missing responses to individual variables.

The impact of missing data is detrimental not only through its potential "hidden" biases of the results but also in its practical impact on the sample size available for analysis. For example, if remedies for missing data are not applied, any observation with missing data on any of the variables will be excluded from the analysis. In many multivariate analyses, particularly survey research applications, missing data eliminate so many observations that what was an adequate sample is reduced to an inadequate sample. In such situations, the researcher must either gather additional observations or find a remedy for the missing data in the original sample. Although finding a remedy for missing data is the most practical solution, few guidelines exist pertaining to the diagnosis and remedy of missing data. For this reason, the following sections discuss the different types of missing data processes, methods to identify the nature of the missing data processes, and available remedies for accommodating missing data into multivariate analyses.

Understanding the Reasons Leading to Missing Data

Before any missing data remedy can be implemented, the analyst must first diagnose and understand the missing data processes underlying the missing data. Sometimes these processes are under the control of the analyst and can be explicitly identified. In such instances, the missing data are termed **ignorable,** which means that specific remedies for missing data are not needed because the allowances for missing data are inherent in the technique used [10].

Ignorable Missing Data

One example of an ignorable missing data process is the "missing data" of those observations in a population that are not included in the sample. While the analyst has a priori knowledge of certain characteristics about the entire population (i.e., the variables that define the sampling frame), the purpose of the multivariate techniques is to generalize from the sample observations to the entire population. The process of generalizing to the population is really an attempt to overcome the missing data of observations not in the sample. The researcher makes this missing data ignorable by using probability sampling to select respondents. Probability sampling allows the analyst to specify that the missing data process leading to the omitted observations is random and that the missing data can be accounted for as sampling error in the statistical procedures. Thus, the "missing data" of the nonselected observations is ignorable.

Another instance of ignorable missing data occurs when the data are censored. **Censored data** are observations not complete because of their stage in the missing data process. A typical example is an analysis of the causes of death. Respondents who are still living cannot provide complete information (i.e., cause of death) and are thus censored. Another interesting example of censored data is found in the attempt to estimate the heights of the American population based on the heights of the armed services recruits [as cited in 10]. The data are censored because the armed services had height restrictions that varied in level and enforcement. Thus, the researchers are faced with the task of estimating the heights of the entire population when it is known that certain individuals (i.e., all those below the height restrictions) are not included in the sample. In both instances the researcher's knowledge of the missing data process allows for the use of specialized methods, such as event history analysis, to accommodate censored data [10].

The justification for designating missing data as ignorable is that the missing data process is operating at random (i.e., the observed values are a random sample of the total set of values, observed and missing) and that these random effects are identifiable and explicitly accommodated in the technique used. However, in most instances, the missing data process is not explicitly addressed by the techniques used. Thus, the analyst must assess the extent and impact of the missing data to determine whether it is a random process or, if not, is it amenable to one of the available remedies.

Other Types of Missing Data Processes

Missing data can occur for many reasons and in many situations. One type of missing data process that may occur in any situation is due to procedural factors, such as errors in data entry that create invalid codes, disclosure restrictions (e.g., small counts in U.S. Census data), failure to complete the entire questionnaire, or even the morbidity of the respondent. In these situations, the researcher has little control over the missing data processes, but some remedies may be applicable if the missing data are found to be random. Another type of missing data process occurs when the response is inapplicable, such as questions regarding the years of marriage for adults who have never been married. Again, the analyses can be specifically formulated to accommodate these respondents.

Other types of missing data processes may be less easily identified and accommodated. Most often these types are related directly to the respondent. One example is the refusal to respond to certain questions. This is common in questions of a sensitive nature, such as those concerning income or particularly controversial issues. Another example is when the respondent has no opinion or insufficient knowledge to answer the question. The researcher should anticipate these problems and attempt to minimize them in the research design and data collection stages of the research. However, they still may occur and the analyst must now deal with the resulting missing data. But all is not lost. When the missing data do occur in a random pattern, remedies may be available to mitigate their effect.

Examining the Patterns of Missing Data

To decide whether a remedy for missing data can be applied, the researcher must first ascertain the degree of randomness present in the missing data. Assume for the purposes of illustration that two variables (X and Y) are collected. Whereas X has no missing data, Y has some missing data. If a distinct relationship is found between X and Y, such as a strong correlation between certain values of X and the occurrence of missing data for Y, then the missing data are not at random and any remedy must explicitly accommodate the missing data process between X and Y.

Missing data are termed **missing at random (MAR)** if the missing values of Y depend on X, but not on Y. By this we mean that the observed Y values represent a random sample of the actual Y values for each value of X, but the observed data for Y do not necessarily represent a truly random sample of all Y values. For example, assume that we know the gender of respondents (the X variable) and are asking about household income (the Y variable). We find that the missing data are random for both males and females but occur at a much higher frequency for males than females. While the missing data process is operating in a random manner, any remedy applied to the missing data must still take into account the gender of the respondent because gender affects the ultimate distribution of the household income values.

A higher level of randomness is termed **missing completely at random (MCAR).** In these instances the observed values of Y are truly a random sample of all Y values, with no underlying process that lends bias to the observed data. In our earlier example, this would be shown by the fact that the missing data for household income were randomly missing in equal proportions for both males and females. If this is the form of the missing data process, any of the remedies can be applied without making allowances for the impact of any other variable.

Diagnosing the Randomness of the Missing Data Process

As noted above, the researcher must ascertain whether the missing data process occurs in a random manner. Three methods are available for assisting in this diagnosis. The first assesses the missing data for a single variable Y by forming two groups—observations with missing data for Y and those with valid values of Y. Tests are then performed to determine whether significant differences between the two groups exist on other variables of interest. If patterns of significant differences are found, it would indicate a nonrandom missing data process was operating. Let's use our earlier example of household income and gender again. We would first form two groups of respondents, those with missing data on the household income question and those who answered the question. We would then compare the percentages of gender for each group. If one gender (e.g., males) was found in greater proportion in the missing data group, we would suspect the process was not operating at random. If the variable we were comparing was metric (e.g., an attitude or perception) instead of categorical (gender), then t tests are the appropriate test. The analyst should examine a number of variables to see whether any consistent pattern emerges. Remember that some differences will occur by chance, but any series of differences may indicate an underlying pattern.

A second approach utilizes dichotomized correlations to assess the correlation of missing data for any pair of variables. For each variable, valid values are represented by the value of one, while missing data are replaced by values of zero. These missing value indicators for each variable are then correlated. The correlations indicate the degree of association between the missing data on each variable pair. Low correlations denote randomness in the pair of variables. Although no guidelines exist for identifying the level of correlation needed to indicate that the missing data are not random, statistical significance tests of the correlations provide a conservative estimate of the degree of randomness. If randomness is indicated for all variable pairs, then the analyst can assume that the missing data can be classified as MCAR. If significant correlations exist between some pairs of variables, then the analyst may have to assume that the data are only MAR and these relationships must be accommodated in any remedies that are applied.

Finally, an overall test of randomness can be performed that determines whether the missing data can be classified as MCAR. This test analyzes the pattern of missing data on all variables and compares it with the pattern expected for a random missing data process. If no significant differences are found, the missing data can be classified as MCAR. If significant differences are found, however, the analyst must use the approaches described above to identify the specific missing data processes that are nonrandom.

Approaches for Dealing with Missing Data

The approaches or remedies for dealing with missing data can be classified into one of four categories based on the randomness of the missing data process and the method used to estimate the missing data [10]. If nonrandom or MAR missing data processes are found, the analyst should apply only one—the specifically designed modeling approach. Application of any other methods introduces bias into the results. Instead, the analyst may wish to delete cases and/or variables if possible. Only if the analyst determines that the missing data process can be classified as MCAR can all of the approaches discussed in the following sections be used.

Yet how often do analysts make the assessment for randomness of the missing data process before applying one of the missing data remedies? And even if the remedy is appropriate, the analyst must note the specific impacts on the results associated with that remedy. Too often a remedy is applied without an assessment of the missing data processes, the appropriateness of the selected remedy, or the consequences it will have. Thus, the analyst never realizes the effects because they are hidden in the overall results.

Use of Only Observations with Complete Data

The simplest and most direct approach for dealing with missing data is to include only those observations with complete data. This method is available in all statistical programs and is the default method in many programs. Yet this approach should be used only if the missing data are MCAR, because missing data that are not MCAR have nonrandom elements that bias the results. Thus, even though only valid observations are used, the results are not generalizable to the population. Moreover, in many situations, the resulting sample size is reduced to a size inappropriate for the proposed analysis. The use of only complete data is best suited for instances in which the extent of missing data is small and the intercorrelations in the data are strong.

Delete Case(s) and/or Variable(s)

Another simple remedy for missing data is to delete the offending case(s) and/or variable(s). In this approach, the analyst determines the extent of missing data on each case and variable and then deletes the case(s) or variable(s) exceeding a specified level. In many cases where a nonrandom pattern of missing data is present, this may be the most efficient solution. The analyst may find that the missing data are concentrated in a small subset of cases and/or variables, with their exclusion substantially reducing the extent of the missing data. No firm guidelines exist on the necessary level for exclusion, but any decision should be based on both empirical and theoretical considerations. If missing values are found for what will be a dependent variable in the proposed analysis, the case is usually excluded. This avoids any artificial increases in the explanatory power of the analysis that can occur when the analyst first estimates the missing data for the dependent variable by one of the imputation processes described below and then uses the estimated values in the analysis of the dependence relationship. If a variable other than a dependent variable has missing values and is a candidate for deletion, the analyst should be sure that alternative variables, one hopes highly correlated, are available to represent the intent of the original variable. The analyst must always consider the gain of eliminating a source of missing data versus the deletion of a variable in the multivariate analysis.

Imputation Methods

A third approach to handling missing data is through one of the many **imputation methods.** Imputation is the process of estimating the missing value based on valid values of other variables and/or cases in the sample. The objective is to employ known relationships that can be identified in the valid values of the sample to assist in estimating the missing values. However, the analyst should carefully consider the use of imputation in each instance because of its potential impacts on the analysis [7]:

> The idea of imputation is both seductive and dangerous. It is seductive because it can lull the user into the pleasurable state of believing that the data are complete after all, and it is dangerous because it lumps together situations where the problem is sufficiently minor that it can be legitimately handled in this way and situations where standard estimators applied to the real and imputed data have substantial biases.

Imputation methods require that two issues be addressed: (1) the values used in the imputation process, and (2) the methods of estimating the replacement values for the missing data, which are then analyzed by standard multivariate techniques.

Information Available for the Imputation Technique   The analyst must first determine which values and observations are to be used in the imputation process. The most common alternatives are the complete case and all-available approaches. The **complete case approach** uses only data from observations that have no missing data. The imputation process occurs for missing data on the remaining cases. Just as seen in using only complete cases for the entire process, this approach can introduce bias if the missing data process is not MCAR.

The **all-available approach** relies on imputating missing data based on all available valid observations. This approach maximizes the pairwise information available in the sample and is known as the PAIRWISE option in SPSS and the CORPAIR, COVPAIR, or ALLVALUE options in BMDP. The distinguishing characteristic of this approach is that each correlation is based on a potentially unique set of observations and the number of observations used in the calculations can vary for each correlation. While the all-available method maximizes the data utilized and overcomes the problem of missing data on a single variable eliminating a case from the entire analysis, several problems can also arise from this approach. First, correlations may be calculated that are "out of range" and inconsistent with the other correlations in the correlation matrix. Any correlation between X and Y is constrained by their correlation to a third variable Z. The correlation between X and Y can range only from $+1$ to -1 if both X and Y have zero correlation with all other variables in the correlation matrix. Yet rarely are the correlations with other variables zero. As the correlations with other variables increase, the range of possible correlations between X and Y decreases, thus increasing the potential for the correlation in a unique set of cases to be inconsistent with correlations derived from other sets of cases. An associated problem is that the eigenvalues in the correlation matrix can become negative, thus altering the variance properties of the correlation matrix. Although the correlation matrix can be adjusted to eliminate this problem (e.g., the ALLVALUE option in BMDP), many procedures do not include this adjustment process. In extreme cases, the estimated variance/covariance matrix is not positive definite. All these problems must be considered in selecting this approach versus excluding cases with missing data.

The Replacement of Missing Data   The issue in imputation is the actual method for replacement of the missing values with some estimated value based on other information available in the sample. This can take many forms, varying from a direct substitution of values to estimation processes based on relationships among the variables. The following discussion focuses on the more widely used methods, although many other forms of imputation are available [10].

- *Case substitution*. In this method, observations with missing data are replaced by choosing another nonsampled observation. A common example is to replace a sampled household that cannot be contacted or that has extensive missing data with another household not in the sample, preferably very similar to the original observation. This method is most widely used to replace observations with complete missing data, although it can be used to replace observations with lesser amounts of missing data as well.
- *Mean substitution*. One of the more widely used methods, this replaces the missing values for a variable with the mean value of that variable based on all valid responses. In this manner, the valid sample responses are used to calculate the replacement value. The rationale of this approach is that the mean is the best single replacement value. This approach, while used extensively, has three disadvantages. First, it makes the variance estimates derived from the standard variance formulas invalid by understating the true variance in the data. Second, the actual distribution of values is distorted by substituting the mean for the missing values. Third, this method depresses the observed correlation because all missing data will have a single constant value.
- *Cold deck imputation*. In this method, the analyst substitutes a constant value derived from external sources or previous research for the missing values. This is similar in nature to the mean substitution method, differing only in the source of the substitution value. Cold deck imputation has the same disadvantages as the mean substitution method, and the analyst must be sure that the replacement value from an external source is more valid than the internally generated value, the mean.
- *Regression imputation*. In this method, regression analysis is used to predict the missing values of a variable based on its relationship to other variables in the data set. The regression analysis procedure, described more fully in Chapter 3, can be used to make simple estimates, or stochastic (random) components can be added to better reflect the accuracy of the prediction process. While having the appeal of using relationships already existing in the sample as the basis of prediction, several disadvantages are also associated with this method. First, it reinforces the relationships already in the data. As the use of this method increases, the resulting data become more characteristic of the sample and less generalizable. Second, unless stochastic terms are added to the estimated values, the variance of the distribution is understated. Third, this method assumes that the variable with missing data has substantial correlations with the other variables. If these correlations are not sufficient to produce a meaningful estimate, then other methods, such as mean substitution, are preferable. Finally, the regression procedure is not constrained in the estimates it makes. Thus, the predicted values may not fall in the valid ranges for variables (e.g., predicting a value of 11 for a 10-point scale), thereby requiring some form of additional adjustment. Even with all of these potential problems, the regression method of imputation holds promise in those instances where moderate levels of widely scattered missing data are present and where the relationships between vari-

ables are sufficiently established so that the analyst is confident that the use of the method will not impact the generalizability of the results.

• *Multiple imputation*. The final imputation method is actually a combination of several methods. In this approach, two or more methods of imputation are used to derive a composite estimate, usually the mean of the various estimates, for the missing value. The rationale of this approach is that the use of multiple approaches minimizes the specific concerns with any single method and the composite will be the best possible estimate. The choice of this approach is primarily based on the trade-off between the analyst's perception of the potential benefits weighed against the substantially higher effort required in making and combining the multiple estimates.

Model-based Procedures

The final set of procedures explicitly incorporates the missing data into the analysis, either through a process specifically designed for missing data estimation or as an integral portion of the standard multivariate analysis. The first approach involves maximum likelihood estimation models that attempt to model the processes underlying the missing data and to make the most accurate and reasonable estimates possible [10]. This modeling approach, however, involves a level of complexity beyond the scope of this text. The second approach involves the inclusion of missing data directly into the analysis, defining observations with missing data as a select subset of the sample. This approach is most applicable for dealing with missing values on the independent variables of a dependent relationship. Its premise is best characterized by this passage from Cohen and Cohen:

> We thus view missing data as a pragmatic fact that must be investigated, rather than a disaster to be mitigated. Indeed, implicit in this philosophy is the idea that like all other aspects of sample data, missing data are a property of the population to which we seek to generalize. [5, p. 299]

When the missing values occur on a nonmetric variable, the analyst can easily define those observations as a separate group and then include them in any analysis, such as ANOVA or MANOVA or even discriminant analysis. When the missing data are present on a metric independent variable in a dependence relationship, a procedure has been developed to incorporate the observations into the analysis while maintaining the relationships among the valid values [5]. This procedure is best illustrated in the context of regression analysis, although it can be used in other dependence relationships as well. The first step is to code all observations with missing data as dummy variables (where the cases with missing data receive a value of one and the other cases have a value of zero). The missing values are then imputed by the mean substitution method. Finally, the relationship is estimated by normal means. The dummy variable represents the difference on the dependent variable between those observations with missing data and those observations with valid data. The test of the dummy variable coefficient assesses the statistical significance of this difference. The coefficient of the original variable represents the relationship for all cases with nonmissing data. This method allows the analyst to retain all the observations in the analysis for purposes of maintaining the sample size, while also providing a direct test for the differences between the two groups along with the estimated relationship between the dependent and independent variables.

TABLE 2.1 Summary Statistics of Pretest Data

	Variable	Number of Cases with Valid Data	Mean	Standard Deviation	Coefficient of Variation	Smallest Value	Largest Value	Percentage of Missing Data
X_1	Delivery Speed	45	4.01	0.97	0.24	2.8	6.5	29.7
X_2	Price Level	54	1.90	0.86	0.45	0.4	3.8	15.6
X_3	Price Flexibility	50	8.13	1.32	0.16	5.2	9.9	21.9
X_4	Manufacturer's Image	60	5.15	1.19	0.23	2.5	7.8	6.2
X_5	Overall Service	59	2.84	0.75	0.27	1.1	4.6	7.8
X_6	Sales Force's Image	63	2.60	0.72	0.28	1.1	4.0	1.6
X_7	Product Quality	60	6.79	1.68	0.25	1.7	9.9	6.2
X_9	Usage Level	60	45.97	9.42	0.20	25.0	65.0	6.2
X_{10}	Satisfaction Level	60	4.80	0.82	0.17	3.3	6.2	6.2

Note: Six of the original 70 cases had more than 90% missing data and were excluded from the analysis. All analyses are based on the remaining 64 cases. Twenty-six cases had no missing data.

An Illustration of Missing Data Diagnosis

To illustrate the process of diagnosing the patterns of missing data and the application of possible remedies, a new data set will be introduced (see Appendix A for a complete listing of the observations). This data set was collected during the pretest of the questionnaire used to collect the data described in Chapter 1. The pretest involved 70 individuals and collected responses on all 14 variables. In the course of pretesting, however, missing data occurred. The following sections detail the diagnosis of the extent of missing data in the data set and the analyses available for selecting and applying the various missing data remedies available in most statistical programs.

Examining the Patterns of Missing Data

The AM procedure of BMDP [2] was used in the analysis of the missing data patterns in the pretest data. Table 2.1 contains the descriptive statistics for the observations with valid values, including the percentage of cases with missing data on each variable. Six cases were eliminated from the analysis owing to what was essentially missing data on all variables of interest. The extent of missing data for the remaining 64 observations ranges from a high of 30 percent of the cases for X_1 to a low of a single case (1.6 percent) for X_6. For the variables with the higher levels of missing data (X_1, X_2, and X_3), the levels are not so excessive that they automatically dictate exclusion of the variable. Given the integral role these variables are expected to play in the various multivariate analyses, all efforts should be made to retain them in the analysis.

One factor that could alleviate some of the high levels of missing data for certain variables is the elimination of cases from the analysis. To determine whether the missing data are concentrated on a select set of cases, Table 2.2 provides a graphical display of the missing data patterns. Except for the six cases already eliminated because of extremely high levels of missing data, we see that no other cases have a disproportionate number of missing values. In fact, only

TABLE 2.2 Graphical Display of Missing Data Patterns

Case Label	Number of Missing Variables	X_1	X_2	X_3	X_4	X_5	X_6	X_7	X_9	X_{10}
202.	2	M		M						
203.	2		M					M		
204.	3	M		M						M
205.	1			M						
207.	3	M		M						M
210.	6	*	*	*	*	*	*	*	*	*
213.	2		M	M						
214.	5	*	*	*	*	*	*	*	*	*
216.	2	M				M				
218.	2	M				M				
219.	2							M	M	
220.	1		M							
221.	3	M		M				M		
222.	2			M		M				
224.	3	M	M						M	
225.	2			M	M					
227.	2		M						M	
228.	2	M			M					
229.	1					M				
231.	1							M		
232.	2	M	M							
233.	5	*	*	*	*	*	*	*	*	*
235.	2						M			M
237.	1		M							
238.	1	M								
240.	1	M								
241.	2			M		M				
244.	1								M	
245.	5	*	*	*	*	*	*	*	*	*
246.	1				M					
248.	2	M	M							
249.	1		M							
250.	2	M		M						
253.	1	M								
255.	2	M		M						
256.	1	M								
257.	2		M	M						
259.	1	M								
260.	1	M								
261.	6	*	*	*	*	*	*	*	*	*
263.	6	*	*	*	*	*	*	*	*	*
267.	2			M	M					
268.	1									M
269.	2	M		M						

Legend:

M = a missing value.

O = a variable that is out of range or missing for all cases.

* = a case with too many missing values.

two cases have more than two missing values. Thus, no case can be eliminated that would markedly improve the missing data problem.

Diagnosing Randomness of the Missing Data

The next step is to examine the patterns of missing data and to determine whether the missing data are distributed randomly across the cases and the variables. The first test for assessing randomness is to compare the observations with and without missing data for each variable on the other variables. For example, the observations with missing data on X_1 are placed in one group and those observations with valid responses for X_1 are placed in another group. Then, these two groups are compared to identify any differences on the remaining variables (X_2 through X_{10}). Once comparisons have been made on the other variables, new groups are formed based on the missing data for the next variable (X_2) and the comparisons are performed again on the remaining variables. This process continues until each variable (X_1 through X_{10}) has been examined for any differences. The objective is to identify any systematic missing data process that would be reflected in patterns of significant differences.

Table 2.3 contains the results for this analysis of the 64 remaining observations from the pretest sample. The first noticeable pattern of significant t values occurs for X_9, where six of the nine comparisons found significant differences between the two groups. However, the impact of these differences is marginal as the number of cases with missing data on X_9 ranged from only three to five. Thus, exclusion of cases with missing data on X_9 is possible because this would not substantially reduce the overall sample size; X_7 showed a pattern of differences similar to X_9 with only four significant differences and small groups of cases with missing data. This analysis indicates that significant differences can be found due to the missing data on two variables (X_7 and X_9), but the small number of cases involved makes this of marginal concern. If later tests of randomness indicate a nonrandom pattern of missing data, these results would then provide a starting point for possible remedies.

A second test for randomness involves the use of correlations between dichotomous variables. The dichotomous variables are formed by replacing valid values with a value of one and missing data with a value of zero. The resulting correlations between the dichotomous variables indicate the extent to which missing data are related in pairs of variables. Low correlations indicate low association between the missing data process for those two variables. Table 2.4 contains the correlations between the nine dichotomized variables. Review of the values indicates that only one correlation is in the moderate range (X_6 and X_{10} have a correlation of .488). This suggests that the missing data process influencing X_{10} corresponds to the missing data process affecting X_6. However, given the absence of any other correlations with even moderate values, the analyst can be assured that no single missing data process is significantly affecting a substantial number of variables.

The final test is an overall test of the missing data for being missing completely at random (MCAR). The test makes a comparison of the missing data with what would be expected if the missing data were totally randomly distributed. In this instance, as shown in Table 2.4, the significance level of the MCAR tests was .2136, indicating that the missing data process can be considered to be MCAR. As a result, the analyst may employ any of the remedies for missing data, because no potential biases exist in the patterns of missing data.

TABLE 2.3 Assessing the Randomness of Missing Data Through Group Comparisons of Observations with Missing Versus Valid Data[1]

Profile Variables for Testing Group Differences

Groups Formed by Missing Data on:	X_1 Delivery Speed	X_2 Price Level	X_3 Price Flexibility	X_4 Mfr. Image	X_5 Overall Service	X_6 Sales Force Image	X_7 Product Quality	X_9 Usage Level	X_{10} Satisfaction Level
X_1	0.00	0.59	−1.78	−2.17	−2.69	−1.41	1.03	−2.34	−2.82
	0	19	14	20	20	23	21	21	20
	45	35	36	40	39	40	39	39	40
X_2	0.40	0.00	−0.10	2.79	4.06	2.08	1.39	1.06	1.27
	9	0	9	12	12	12	12	11	11
	36	54	41	48	47	51	48	49	49
X_3	−0.40	−1.34	0.00	−0.72	−1.11	0.03	0.26	−1.44	−1.35
	8	13	0	15	14	17	16	17	13
	37	41	50	45	45	46	44	43	47
X_4	−0.30	0.13	−0.75	0.00	−0.55	−0.90	−0.74	−0.15	0.65
	4	6	5	0	6	6	6	6	6
	41	48	45	60	53	57	54	54	54
X_5	0.39	1.35	−0.81	−0.18	0.00	1.12	0.80	−0.28	−1.15
	4	6	4	6	0	7	7	7	6
	41	48	46	54	59	56	53	53	54
X_6	−1.69	0.00	−0.72	0.00	−0.08	0.00	0.26	−0.27	0.00
	2	1	2	1	2	0	2	2	0
	43	53	48	59	57	63	58	58	60
X_7	−7.37	0.78	−1.05	8.30	−0.06	2.11	0.00	−0.51	−2.69
	3	4	4	4	5	5	0	3	5
	42	50	46	56	54	58	60	57	55
X_9	−8.68	16.06	−3.04	4.30	1.13	2.38	−1.67	0.00	−7.74
	3	3	5	4	5	5	3	0	5
	42	51	45	56	54	58	57	60	55
X_{10}	−1.36	0.26	−0.03	−0.97	−0.54	−0.84	0.01	−1.03	0.00
	4	5	3	6	6	5	7	7	0
	41	49	47	54	53	58	53	53	60

$t > 2.0$

[1] Each cell contains three values:
  1. *t* value for comparison of the column variable between group A (observations with missing data on the row variable) and group B (observations with valid data on the row variable).
  2. sample size of group A (observations with missing data on row variable).
  3. sample size of group B (observations with valid data for row variable).

Interpretation of the table:
  The upper right cell indicates that the *t* value for the comparison of X_{10} between those observations with missing data on X_1 and those with valid data for X_1 is −2.82. Of the cases with valid values for X_{10}, 20 cases have missing data for X_1 (group A) and 40 cases have valid data for X_1 (group B).

TABLE 2.4 Assessing the Randomness of Missing Data Through Dichotomized Variable Correlations and the Multivariate Test for Missing Completely at Random (MCAR)

	X_1 Delivery Speed	X_2 Price Level	X_3 Price Flexibility	X_4 Mfr. Image	X_5 Overall Service	X_6 Sales Force Image	X_7 Product Quality	X_9 Usage Level	X_{10} Satisfaction Level
X_1	1.000 45								
X_2	0.003 38	1.000 54							
X_3	0.235 38	−0.020 42	1.000 50						
X_4	−0.026 42	−0.111 50	0.176 48	1.000 60					
X_5	0.066 42	−0.125 49	0.128 47	−0.075 55	1.000 59				
X_6	−0.082 44	−0.054 53	−0.067 49	−0.033 59	−0.037 58	1.000 63			
X_7	−0.026 42	0.067 51	0.020 47	−0.067 56	−0.075 55	−0.033 59	1.000 60		
X_9	−0.026 42	0.244 52	−0.137 46	−0.067 56	−0.075 55	−0.033 59	0.200 57	1.000 60	
X_{10}	0.115 43	−0.111 50	0.176 48	−0.067 56	−0.075 55	0.488 60	0.067 56	−0.067 56	1.000 60

Chi-square test for missing completely at Random 172.8784
Probability 0.2136
Degrees of freedom 159

Interpretation:
Values in the table represent the correlation between dichotomized variables where cases with a valid value receive a 1 and missing data receives a 0. The number below the correlation represents the number of cases having missing data on one of the variables in that specific correlation pair.

Remedies for Missing Data

As discussed earlier, numerous remedies are available for dealing with missing data. In this instance, several of the remedies have definite disadvantages. If the complete case approach is taken, the sample size is reduced to 26 observations, barely sufficient for even the simplest univariate analyses, much less multivariate applications. Our earlier examination of the patterns of the missing data demonstrated that there was not a small set of cases that could be deleted and thereby markedly reduce the extent of missing data. The only viable alternative is the elimination of X_1, which has missing data on almost 30 percent of the cases. But even if X_1 was deleted, all the cases with missing data would still have at least one other variable with missing data; thus, this alternative is a less attractive approach as a means of creating more observations with complete data on all variables.

TABLE 2.5 Comparison of Correlations Obtained with the All-Available (Pairwise) and Complete Case (Listwise) Approaches

	X_1 Delivery Speed	X_2 Price Level	X_3 Price Flexibility	X_4 Mfr. Image	X_5 Overall Service	X_6 Sales Force Image	X_7 Product Quality	X_9 Usage Level	X_{10} Satisfaction Level
X_1	1.000 1.000								
X_2	−.456 −.502	1.000 1.000							
X_3	.413 .429	−.359 −.294	1.000 1.000						
X_4	−.107 −.245	.279 .321	−.066 −.061	1.000 1.000					
X_5	.285 .566	.441 .421	.048 .157	.457 .046	1.000 1.000				
X_6	.033 −.094	.241 .356	−.037 −.066	.802 .804	.340 .213	1.000 1.000			
X_7	−.132 −.416	.347 .354	−.347 −.230	.401 .382	.066 −.149	.412 .529	1.000 1.000		
X_9	.329 .599	.147 .048	.601 .648	.231 .191	.711 .683	.271 .301	−.205 −.099	1.000 1.000	
X_{10}	.462 .549	−.197 −.278	.716 .725	.369 .170	.516 .304	.234 .064	−.268 −.405	.694 .566	1.000 1.000

Interpretation: The top value is the correlation obtained with a pairwise or all-available data approach, while the second value is the correlation obtained with a pairwise or complete information approach. A sample size of 26 was used to compute the complete information correlations. Sample sizes for the all-available data approach varied, with the actual sample sizes listed in Table 2.4.

The remaining approach is to employ some form of imputation to estimate replacement values for the missing values. In selecting an imputation technique, the first step is to select between using only observations with complete data or using all available information. The advantage of the complete information approach is that it maintains consistency in the correlation matrix but may also reduce the number of observations used to such a small subset of the sample that the resulting correlations used for imputation differ markedly from those obtained using all available information. For example, Table 2.5 contains the correlations obtained from the all-available and complete information approaches. In most instances the correlations are quite similar, but there are several patterns of substantial differences. Notable differences can be seen in the correlations between X_{10} and X_4, X_5, X_6, and X_7. While the analyst has no proof of greater validity for either approach, these results demonstrate the marked differences sometimes obtained between the two approaches. Whichever approach is chosen, the analyst should examine the correlations obtained by alternative methods to understand the range of possible values.

TABLE 2.6 An Example of the Stepwise Regression Imputation Method

Case	Variable to be Imputed	Imputed Value	R-Squared	Predictor Variables
202.	X_1	4.4300	.606	X_2 X_5 X_4 X_6
202.	X_3	7.9438	.860	X_{10} X_5 X_9 X_4
203.	X_2	2.7849	.644	X_1 X_5 X_{10}
203.	X_7	7.9276	.372	X_6 X_{10} X_4
204.	X_1	3.5701	.606	X_2 X_5 X_4 X_6
204.	X_3	7.7801	.699	X_9 X_5 X_2
204.	X_{10}	4.2322	.700	X_9 X_2 X_4 X_6

Once the approach for selecting information has been made, the analyst can then select the method of imputation for estimating replacement values for the missing data. Table 2.6 contains some illustrative results of a stepwise regression imputation method. In this procedure, a separate stepwise regression equation is developed for each variable that obtains the maximum predictive accuracy possible (interested readers may refer to Chapter 3, which describes stepwise regression in greater detail). The result is a process for estimating replacement values for each variable based on the unique values of each case. For example, the first prediction shown in Table 2.6 involves the missing data of X_1 for case 202. The predicted replacement value is 4.43. The R^2 value represents the predictive accuracy of the predictions for X_1, with an upper limit of 1.0. Sixty percent of the variance in the valid values of X_1 was explained by the four variables (X_2, X_5, X_4, and X_6) in the regression equation. This is a measure of the validity of the regression equation in predicting the valid values, which should also be applicable to missing data that are MCAR. This procedure is repeated for each missing value in the sample. After the imputation process is complete, the data set with replacement values can be saved for further analysis.

Summary

The procedures available for handling missing data are varied in form, complexity, and intent. The analyst must always be prepared to assess and deal with missing data, as it is frequently encountered in multivariate analysis. The decision to use only observations with complete data may always seem to be one of a conservative and "safe" nature, but as the preceding discussion illustrated, there are inherent limitations and biases in this and the other approaches. The analyst has no single method best suited in every situation but instead must make a reasoned judgment of the situation, considering the factors described above.

Outliers

Outliers are observations with a unique combination of characteristics identifiable as distinctly different from the other observations. Outliers cannot be categorically characterized as either beneficial or problematic but instead must be viewed within the context of the analysis and should be evaluated by the types of information they may provide regarding the phenomenon under study. When beneficial, outliers, although different from the majority of the sample, may be indicative of characteristics of the population that would not be discovered in the normal course of analysis. In contrast, problematic outliers are not representative of the population and are counter to the objectives of the analysis. Problematic outliers can seriously distort statistical tests. Owing to the variability in the evaluation of outliers, it is imperative that the analyst examine the data for the presence of outliers to ascertain their type of influence. The reader is also referred to the discussions in Chapter 3 and the appendix to that chapter, which relate to the topic of influential observations. In these discussions, outliers are placed in a framework particularly suited for assessing the influence of individual observations and determining whether this influence is helpful or harmful.

Why do outliers occur? Outliers can be classified into one of four categories. The first category contains those outliers arising from a procedural error, such as a data entry error or a mistake in coding. These types of outliers should be identified in the data cleaning stage, but if overlooked, they should be eliminated or recoded as missing values. The second class of outlier is the observation that occurs as the result of an extraordinary event. In this case, an explanation exists for the uniqueness of the observation. The analyst must decide whether the outlier represents a valid observation in the population. If so, it should be retained; if not, it should be deleted from the analysis. The third class of outlier comprises the extraordinary observations for which the analyst has no explanation. Although these are the outliers most likely to be omitted, they may be retained if the analyst feels they represent a valid segment of the population. The fourth and final class of outlier contains observations that fall within the ordinary range of values on each of the variables but are unique in their combination of values across the variables. In these situations, the analyst should retain the observation unless specific evidence is available discounting the outlier as a valid member of the population.

The following sections detail the methods used in detection of outliers for the univariate, bivariate, and multivariate situations. Once the outliers have been identified, they may be profiled to aid in placing them into one of the four classes described above. Finally, the analyst must decide on the retention or exclusion of each outlier judging not only from the characteristics of the outlier but also from the objectives of the analysis.

Detecting Outliers

Outliers can be identified from a univariate, bivariate, or multivariate perspective. The analyst should utilize as many of these perspectives as possible, looking for a consistency across methods in identifying outliers. The following discussion details the processes involved in each of the three perspectives.

Univariate Detection

The univariate perspective for identifying outliers examines the distribution of observations, selecting as outliers those cases that fall at the outer ranges of the distribution. The primary issue concerns the establishment of the threshold for designation of an outlier. The typical approach first converts the data values to standard scores, which have a mean of zero and a standard deviation of one.

Because the values are expressed in a standardized format, comparisons across variables can be easily made. For small samples (80 or fewer observations), the guidelines suggest identifying those cases with standard scores of 2.5 or greater as outliers. When the sample sizes are larger, the guidelines suggest that the threshold value of standard scores range from 3 to 4. In either case, the analyst must recognize that a certain number of observations may occur normally in these outer ranges of the distribution. The analyst should strive to identify only those truly distinctive observations and designate them as outliers.

Bivariate Detection

In addition to the univariate assessment, pairs of variables can be assessed jointly through a scatterplot. Cases that fall markedly outside the range of the other observations can be noted as isolated points in the scatterplot. To assist in determining the expected range of observations, an ellipse representing a specified confidence interval (varying between 50 and 90 percent of the distribution) for a bivariate normal distribution can be superimposed over the scatterplot. This provides a graphical portrayal of the confidence limits and facilitates identification of the outliers. Another variant of the scatterplot is termed the influence plot. In this scatterplot, each point varies in size in relation to its influence on the relationship. These methods provide some assessment of the influence of each observation to complement the designation of cases as outliers.

Multivariate Detection

The third perspective for identifying outliers involves a multivariate assessment of each observation across a set of variables. Because most multivariate analyses involve more than two variables, the analyst needs a means of objectively measuring the multidimensional position of each observation relative to some common point. The Mahalanobis D^2 measure can be used for this purpose. Mahalanobis D^2 is a measure of the distance in multidimensional space of each observation from the mean center of the observations. While providing a common measure of multidimensional centrality, it also has statistical properties that allow for significance testing. Given the nature of the statistical tests, it is suggested that a very conservative level, perhaps .001, be used as the threshold value for designation as an outlier.

Outlier Designation

When observations that are candidates for designation as an outlier have been identified by the univariate, bivariate, or multivariate methods, the analyst must then select observations that demonstrate real uniqueness in comparison with the remainder of the population. The analyst should refrain from designating too many observations as outliers. Although relatively rare in the general population, valid observations from the population should be retained to maintain generalizability to the population.

Outlier Description and Profiling

Once the potential outliers have been identified, the analyst should generate profiles on each outlier observation and carefully examine the data for the variable(s) responsibile for its being an outlier. In addition to this visual examination, the analyst can also employ multivariate techniques like discriminant analysis or multiple regression to identify the differences between outliers and the other observations. The analyst should continue this analysis until satisfied with the aspects of the data that distinguish the outlier from the other observations. If possible the analyst should assign the outlier to one of the four types described earlier.

Retention or Deletion of the Outlier

After the outliers have been identified, profiled, and categorized, the analyst must decide on the retention or deletion of each one. There are many philosophies among analysts as to how to deal with outliers. Our belief is that they should be retained unless there is demonstrable proof that they are truly aberrant and not representative of any observations in the population. But if they do represent a segment of the population, they should be retained to ensure generalizability to the entire population. As outliers are deleted, the analyst is running the risk of improving the multivariate analysis but limiting its generalizability. If outliers are problematic in a particular technique, many times they can be handled in a particular manner to accommodate them in the analysis but not allow them to seriously distort the analysis.

An Illustrative Example of Analyzing Outliers

As an example of outlier detection, the observations of the HATCO database introduced in Chapter 1 are examined here for outliers. The variables considered in the analysis are the metric variables of X_1, X_2, X_3, X_4, X_5, X_6, X_7, X_9, and X_{10}. The outlier analysis will consider univariate, bivariate, and multivariate diagnoses. If candidates for outlier designation are found, they will be examined and a decision on retention or deletion will be made.

Univariate and Bivariate Detection

The first step is to examine the observations on each of the variables individually. Table 2.7 contains the observations with standardized variable values exceeding ± 2.5. From this univariate perspective, a few observations exceed the threshold

TABLE 2.7 Identification of Univariate and Bivariate Outliers

Univariate Outliers *Cases with Standardized Values (Z scores) Exceeding ± 2.5*		*Bivariate Outliers* *Cases Lying Outside the 90% Confidence Interval Ellipse*	
Variable	Cases	X_9 with	
X_1	39	X_1	1,39,95,96
X_2	71	X_2	3, 49, 57, 7, 96, 97
X_3	none	X_3	11, 57, 96, 100
X_4	82	X_4	5, 22, 42, 50, 72, 82, 93, 96
X_5	96	X_5	3, 22, 39, 57, 71, 96
X_6	5, 42	X_6	5, 7, 42, 82, 96
X_7	none	X_7	57, 58, 95, 96
X_9	none		
X_{10}	none		

on a single variable. For a bivariate perspective, scatterplots are formed for X_1, X_2, X_3, X_4, X_5, X_6, X_7, versus X_9, one of the metric variables used as a dependent variable in many of the multivariate techniques. An ellipse representing the 90 percent confidence interval of a bivariate normal distribution is then superimposed on the scatterplot (see Figure 2.6). The second part of Table 2.7 contains observations falling outside this ellipse. This is a 90 percent confidence interval; thus we would expect some observations normally to fall outside the ellipse. However, several observations (3, 5, 57, and 96) appear several times, perhaps indicating they are bivariate outliers.

Multivariate Detection

The final diagnostic method is to assess multivariate outliers with the Mahalanobis D^2 measure (see Table 2.8). This evaluates the position of each observation compared with the center of all observations on a set of variables. In this case, all the metric variables were used for the evaluation of observations. As noted earlier, the statistical tests for significance with this measure should be very conservative (exceeding .001). With this threshold, two observations (22 and 55) are identified as significantly different. It is interesting that these observations were not seen in earlier univariate and bivariate analyses but appear only in the multivariate tests. This indicates they are not unique on any single variable but instead are unique in combination.

Retention or Deletion of the Outliers

As a result of these diagnostic tests, no observations seem to demonstrate the characteristics of outliers that should be eliminated. Each variable has some observations that are extreme, and they should be considered if that variable is used in an analysis. But no observations are extreme on a sufficient number of variables to be considered unrepresentative of the population. In all instances, the observations designated as outliers, even with the multivariate tests, seem similar enough to the remaining observations to be retained in the multivariate analyses. But the analyst should always examine the results of each specific technique to identify observations that may become outliers in that particular application.

Testing the Assumptions of Multivariate Analysis

The final step in examining the data involves testing the assumptions underlying multivariate analysis. The need to test the statistical assumptions is increased in multivariate applications because of two characteristics of the multivariate analysis. First, the complexity of the relationships, owing to the typical use of a large number of variables, makes the potential distortions and biases more potent when the assumptions are violated. This is particularly true when the violations compound to become even more detrimental than if considered separately. Second, the complexity of the analyses and of the results may mask the "signs" of assumption violations apparent in the simpler univariate analyses. In almost all instances, the multivariate procedures will estimate the multivariate model and

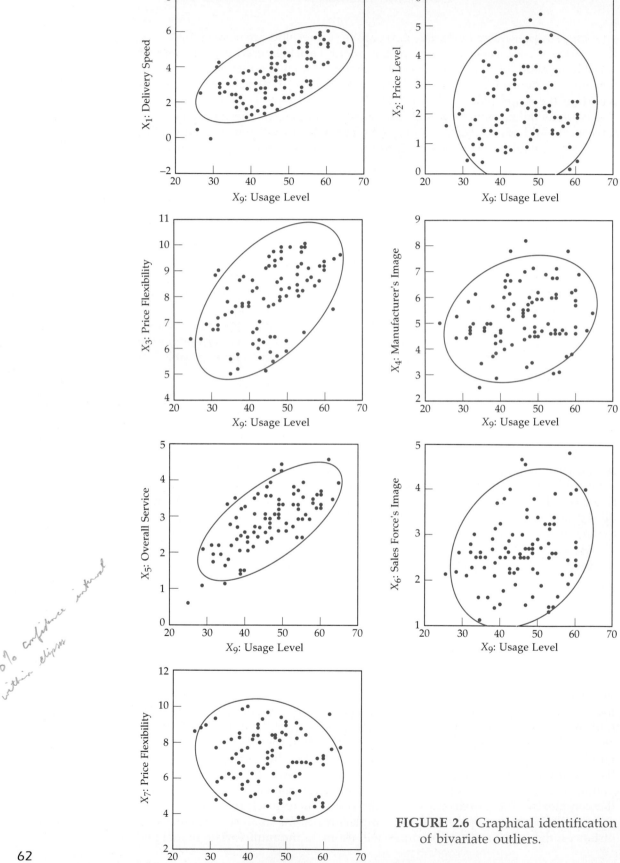

90% confidence interval within ellipse

FIGURE 2.6 Graphical identification of bivariate outliers.

TABLE 2.8 Identifying Multivariate Outliers

Case No	Mahalanobis D^2	D^2/DF	D.F.	Significance	Case No	Mahalanobis D^2	D^2/DF	D.F.	Significance
1	7.031	1.004	7	0.4256	51	6.362	0.909	7	0.4982
2	6.691	0.956	7	0.4617	52	8.467	1.210	7	0.2932
3	7.567	1.081	7	0.3723	53	6.913	0.988	7	0.4380
4	7.103	1.015	7	0.4182	54	3.244	0.463	7	0.8615
5	12.870	1.839	7	0.0753	55	35.197	5.028	7	0.0000
6	.517	0.931	7	0.4809	56	3.082	0.440	7	0.8773
7	8.634	1.233	7	0.2800	57	10.488	1.498	7	0.1626
8	6.563	0.938	7	0.4758	58	5.265	0.752	7	0.6276
9	6.375	0.911	7	0.4967	59	4.348	0.621	7	0.7390
10	3.626	0.518	7	0.8217	60	7.012	1.002	7	0.4276
11	4.237	0.605	7	0.7522	61	13.001	1.857	7	0.0721
12	3.389	0.484	7	0.8468	62	5.798	0.828	7	0.5635
13	3.768	0.538	7	0.8061	63	3.322	0.475	7	0.8537
14	5.030	0.719	7	0.6563	64	6.926	0.989	7	0.4367
15	8.962	1.280	7	0.2554	65	11.683	1.669	7	0.1115
16	6.398	0.914	7	0.4942	66	2.109	0.301	7	0.9536
17	7.212	1.030	7	0.4071	67	4.382	0.626	7	0.7349
18	5.350	0.764	7	0.6173	68	5.925	0.846	7	0.5486
19	5.899	0.843	7	0.5516	69	4.878	0.697	7	0.6749
20	8.962	1.280	7	0.2554	70	5.057	0.722	7	0.6530
21	2.978	0.425	7	0.8870	71	8.294	1.185	7	0.3074
22	35.390	5.056	7	0.0000	72	10.095	1.442	7	0.1833
23	8.333	1.190	7	0.3042	73	5.887	0.841	7	0.5530
24	2.974	0.425	7	0.8874	74	5.363	0.766	7	0.6157
25	4.909	0.701	7	0.6711	75	6.471	0.924	7	0.4859
26	3.463	0.495	7	0.8391	76	4.925	0.704	7	0.6691
27	3.171	0.453	7	0.8687	77	5.847	0.835	7	0.5577
28	5.765	0.824	7	0.5674	78	7.522	1.075	7	0.3766
29	7.601	1.086	7	0.3691	79	12.279	1.754	7	0.0918
30	5.188	0.741	7	0.6370	80	2.270	0.324	7	0.9434
31	2.751	0.393	7	0.9071	81	4.943	0.706	7	0.6669
32	7.024	1.003	7	0.4264	82	14.118	2.017	7	0.0491
33	5.678	0.811	7	0.5778	83	6.837	0.977	7	0.4460
34	3.529	0.504	7	0.8321	84	2.366	0.338	7	0.9369
35	6.539	0.934	7	0.4784	85	3.016	0.431	7	0.8835
36	2.900	0.414	7	0.8941	86	3.493	0.499	7	0.8359
37	6.704	0.958	7	0.4603	87	3.354	0.479	7	0.8504
38	3.030	0.433	7	0.8823	88	2.417	0.345	7	0.9332
39	10.213	1.459	7	0.1768	89	6.011	0.859	7	0.5385
40	3.827	0.547	7	0.7995	90	4.860	0.694	7	0.6771
41	2.898	0.414	7	0.8943	91	3.763	0.538	7	0.8067
42	12.282	1.755	7	0.0917	92	5.841	0.834	7	0.5584
43	7.129	1.018	7	0.4156	93	14.328	2.047	7	0.0456
44	4.819	0.688	7	0.6821	94	5.407	0.772	7	0.6105
45	6.670	0.953	7	0.4640	95	7.391	1.056	7	0.3893
46	7.475	1.068	7	0.3811	96	16.708	2.387	7	0.0194
47	14.094	2.013	7	0.0495	97	8.195	1.171	7	0.3157
48	6.152	0.879	7	0.5221	98	4.990	0.713	7	0.6612
49	7.561	1.080	7	0.3729	99	5.587	0.798	7	0.5888
50	9.029	1.290	7	0.2506	100	4.704	0.672	7	0.6960

produce results even when the assumptions are severely violated. Thus, the analyst must be aware of any assumption violations and the implications they may have for the estimation process or the interpretation of the results.

Assessing Individual Variables Versus the Multivariate Model

Multivariate analysis requires that the assumptions underlying the statistical techniques be tested twice: first for the separate variables, akin to the tests of assumption for univariate analyses, and second for the multivariate model **variate,** which acts collectively for the variables in the analysis and thus must meet the same assumptions as individual variables. This chapter focuses on the examination of individual variables for meeting the assumptions underlying the multivariate procedures. Discussions in each chapter address the methods used to assess the assumptions underlying the variate for each multivariate technique.

Normality

The most fundamental assumption in multivariate analysis is the **normality** of the data, referring to the shape of the data distribution for an individual metric variable and its correspondence to the normal distribution, the benchmark for statistical methods. If the variation from the normal distribution is sufficiently large, all resulting statistical tests are invalid, as normality is required to use the *F* and *t* statistics. Both the univariate and the multivariate statistical methods discussed in this text are based on the assumption of univariate normality, with the multivariate methods also assuming multivariate normality. Univariate normality for a single variable is easily tested, and a number of corrective measures are possible, as we discuss later. In a simple sense, multivariate normality (the combination of two or more variables) means that the individual variables are normal in a univariate sense and that their combinations are also normal. Thus, if a variable is multivariate normal, it is also univariate normal. However, the reverse is not necessarily true (two or more univariate normal variables are not necessarily multivariate normal). Thus a situation in which all variables exhibit univariate normality will help gain multivariate normality, although not guarantee it. Multivariate normality is more difficult to test, but some tests are available for situations in which the multivariate technique is particularly affected by a violation of this assumption. In this text, we focus on assessing and achieving univariate normality for all variables and address multivariate normality when it is especially critical. Even though large sample sizes tend to diminish the detrimental effects of nonnormality, the analyst should assess the normality for all variables included in the analysis.

Graphical Analyses of Normality

The simplest diagnostic test for normality is a visual check of the histogram that compares the observed data values with a distribution approximating the normal distribution (see Figure 2.1). Although appealing because of its simplicity, this method is problematic for smaller samples, where the construction of the histogram (e.g., the number of categories or the width of categories) can distort the visual portrayal to such an extent that the analysis is useless. A more reliable approach is the **normal probability plot,** which compares the cumulative distribution of actual data values with the cumulative distribution of a normal distribution. The normal distribution makes a straight diagonal line, and the plotted data values are compared with the diagonal. If a distribution is normal, the line representing the actual data distribution closely follows the diagonal.

Figure 2.7 shows several normal probability plots and the corresponding univariate distribution of the variable. One characteristic of the distribution's shape, the kurtosis, is reflected in the normal probability plots. **Kurtosis** refers to the "peakedness" or "flatness" of the distribution, compared with the normal distribution. When the line falls below the diagonal, the distribution is flatter than expected. When it goes above the diagonal, the distribution is more peaked than the normal curve. For example, in the normal probability plot of a peaked distribution (Figure 2.7d), we see a distinct S-shaped curve. Initially the distribution is

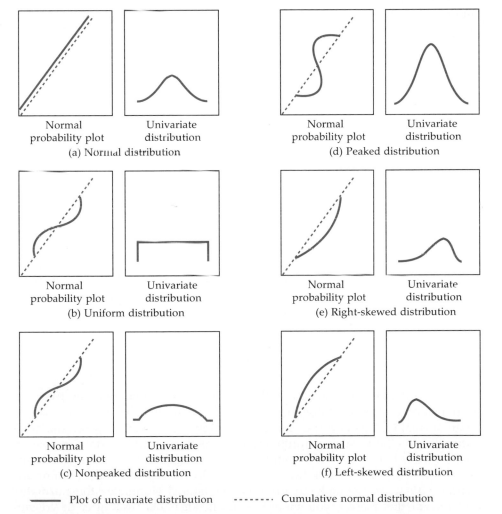

Normal probability plot   Univariate distribution
(a) Normal distribution

Normal probability plot   Univariate distribution
(d) Peaked distribution

Normal probability plot   Univariate distribution
(b) Uniform distribution

Normal probability plot   Univariate distribution
(e) Right-skewed distribution

Normal probability plot   Univariate distribution
(c) Nonpeaked distribution

Normal probability plot   Univariate distribution
(f) Left-skewed distribution

——— Plot of univariate distribution   ------- Cumulative normal distribution

FIGURE 2.7 Normal probability plots and corresponding univariate distributions.

flatter, and the plotted line falls below the diagonal. Then the peaked part of the distribution rapidly moves the plotted line above the diagonal, and eventually the line shifts to below the diagonal again as the distribution flattens. Another common pattern is a simple arc, either above or below the diagonal, indicating the **skewness** of the distribution. A positive skewness (toward the left) is indicated by an arc above the diagonal, while an arc under the diagonal represents a negatively skewed distribution (Figure 2.7e,f). Other patterns corresponding to uniform or heavy-tailed distributions are also shown. An excellent source for interpreting normal probability plots showing the various patterns and interpretations is [6]. These specific patterns not only identify nonnormality but also tell us the form of the original distribution and the appropriate remedy to apply.

Statistical Tests of Normality

In addition to examining the normal probability plot, one can also use statistical tests to assess normality. The simplest test is a rule of thumb based on the skewness value (available as part of the basic descriptive statistics for a variable computed by all statistical programs). The statistic value (z) is calculated as

$$z \text{ value} = \frac{\text{skewness}}{\sqrt{\dfrac{6}{N}}}$$

where N is the sample size. If the calculated z value exceeds a critical value, then the distribution is nonnormal. The critical value is from a z distribution, based on the significance level we desire. For example, a calculated value exceeding ± 2.58 indicates we can reject the assumption about the normality of the distribution at the .01 probability level. Another commonly used critical value is ± 1.96, which corresponds to a .05 error level.

Specific statistical tests are also available in SPSS, SAS, BMDP, and most other programs. The two most common are the Shapiro-Wilks test and a modification of the Kolmogorov-Smirnov test. Each calculates the level of significance for the differences from a normal distribution. The analyst should always remember that tests of significance are less useful in small samples (fewer than 30) and quite sensitive in large samples (exceeding 1,000 observations). Thus, the analyst should always use both the graphical plots and any statistical tests to assess the actual degree of departure from normality.

Remedies for Nonnormality

A number of data transformations are available to accommodate nonnormal distributions and are discussed below. This chapter confines the discussion to univariate normality tests and transformations. However, when we examine other multivariate methods, such as multivariate regression or multivariate analysis of variance, we will discuss tests for multivariate normality as well. Moreover, many times when nonnormality is indicated, it is actually the result of other assumption violations; therefore, remedying the other violations eliminates the normality problem. For this reason, you should perform normality tests after or concurrently with analyses and remedies for other violations. (For those interested in multivariate normality, see [9, 13, and 18].)

Homoscedasticity

Homoscedasticity is an assumption related primarily to dependence relationships between variables. It refers to the assumption that dependent variable(s) exhibit equal levels of variance across the range of predictor variable(s). Homoscedasticity is desirable because the variance of the dependent variable being explained in the dependence relationship should not be concentrated in only a limited range of the independent values. While the dependent variables must be metric, this concept of an equal spread of variance across independent variables can be applied when the independent variables are either metric or nonmetric. With metric independent variables, the concept of homoscedasticity is based on the spread of dependent variable variance across the range of independent variable values, which is encountered in techniques like multiple regression. The same concept also applies when the independent variables are nonmetric. In these instances, such as is found in ANOVA and MANOVA, the focus now becomes the equality of the variance (single dependent variable) or the variance/covariance matrices (multiple independent variables) across the groups formed by the nonmetric independent variables. The equality of variance/covariance matrices is also seen in discriminant analysis, but in this technique the emphasis is on the spread of the independent variables across the groups formed by the nonmetric dependent measure. In each of these instances, the purpose is the same: to ensure that the variance used in explanation and prediction is spread across the range of values, thus allowing for a "fair test" of the relationship across all values of the nonmetric variables.

In most situations, we have many different values of the dependent variable at each value of the independent variable. For this relationship to be fully captured, the dispersion (variance) of the dependent variable values must be equal at each value of the predictor variable. Most problems with unequal variances stem from one of two sources. The first source is the type of variables included in the model. For example, as a variable increases in value (e.g., units ranging from near zero to millions), there is a naturally wider range of possible answers for the larger values. The second source results from a skewed distribution that creates heteroscedasticity. In Figure 2.8, the scatterplots of data points for two variables (X_1 and

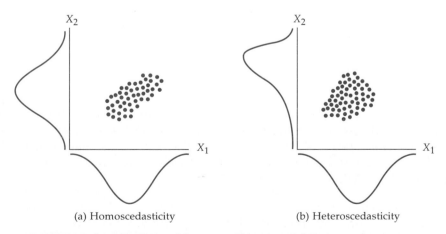

(a) Homoscedasticity   (b) Heteroscedasticity

FIGURE 2.8 Scatterplots of homoscedastic and heteroscedastic relationships.

X_2) with normal distributions exhibit equal dispersion across all data values (i.e., homoscedasticity). However, in Figure 2.8, we also see unequal dispersion **(heteroscedasticity)** caused by skewness of one of the variables (X_2). For the different values of X_1, there are different patterns of dispersion for X_2. This will cause the predictions to be better at some levels of the independent variable than at others. Violating this assumption often makes hypothesis tests either too conservative or too sensitive.

The effect of heteroscedasticity is often also related to sample size, especially when examining the variance dispersion across groups. For example, in ANOVA or MANOVA, the impact of heteroscedasticity on the statistical test depends on the sample sizes associated with the groups of smaller and larger variances. In multiple regression analysis similar effects would occur in highly skewed distributions where there were disproportionate numbers of respondents in certain ranges of the independent variable.

Graphical Tests of Equal Variance Dispersion

The test of homoscedasticity for two metric variables is best examined by graphical means. The most common application of this form of assessment occurs in multiple regression, which is concerned with the dispersion of the dependent variable across the values of the metric independent variables. Because the focus of regression analysis is on the regression variate, the graphical plot of residuals is used to reveal the presence of homoscedasticity (or its opposite, heteroscedasticity, the unequal dispersion of variance). The discussion of residual analysis in Chapter 3 details these procedures.

Statistical Tests for Homoscedasticity

The statistical tests for equal variance dispersion relate to the variances within groups formed by nonmetric variables. The most common test, the Levene test, can be used to assess whether the variances of a single metric variable are equal across any number of groups. If more than one metric variable is being tested so that the comparison involves the equality of variance/covariance matrices, the Box's *M* test is applicable. The Box's *M* test is available in both multivariate analysis of variance and discriminant analysis and is discussed in more detail in later chapters pertaining to these techniques.

Remedies for Heteroscedasticity

Heteroscedastic variables can be remedied through data transformations similar to those used to achieve normality. As mentioned earlier, many times heteroscedasticity is the result of nonnormality of one of the variables, and correction of the nonnormality also remedies the unequal dispersion of variance. A later section discusses data transformations of the variables to make all values have a potentially equal effect in prediction.

Linearity

An implicit assumption of all multivariate techniques based on correlational measures of association, including multiple regression, logistic regression, factor analysis, conjoint analysis, and structural equation modeling is **linearity.** Because correlations represent only the linear association between variables, substantial nonlinear effects that are not represented in the correlation may decrease the correlation value. As a result, it is always prudent to examine all relationships to identify any departures from linearity that may impact the correlation.

Identifying Nonlinear Relationships

The most common way to assess linearity is to examine scatterplots of the variables and to identify any nonlinear patterns in the data. An alternative approach is to run a simple regression analysis (the specifics of this technique are covered in Chapter 3) and to examine the residuals. The residuals reflect the unexplained portion of the dependent variable; thus, any nonlinear portion of the relationship will show up in the residuals. The examination of residuals can also be applied to multiple regression, where the researcher can detect any nonlinear effects not represented in the regression variate. A more detailed discussion of residual analysis is included in Chapter 3.

Remedies for Nonlinearity

If a nonlinear relationship is detected, the most direct approach is to transform one or both variables to achieve linearity. A number of available transformations are discussed below. An alternative to data transformation is the creation of new variables to represent the nonlinear portion of the relationship. The process of creating and interpreting these additional variables, which can be used in all linear relationships, is discussed in Chapter 3.

Absence of Correlated Errors

Predictions in any of the dependence techniques are not perfect, and we will rarely find a situation where they are. However, we do attempt to ensure that any prediction errors are uncorrelated with each other. For example, if we found a pattern that suggests every other error is positive while the alternative error terms are negative, we would know that some unexplained systematic relationship exists in the dependent variable. If such a situation exists, we cannot be confident that our prediction errors are independent of the levels at which we are trying to predict. The most common occurrences of violating this assumption are due to the data collection process. Similar factors may affect one group that do not affect the other. If analyzed separately, the overall effects are constant and do not impact the estimation of the relationship. But if the observations from both groups are combined, then the final estimated relationship must be a "compromise" between the two actual relationships. This causes the results to be biased because an unspecified cause is impacting the estimation of the relationship.

Data Transformations

Data transformations provide a means of modifying variables for one of two reasons: to correct violations of the statistical assumptions underlying the multivariate techniques or to improve the relationship (correlation) between variables.

Data transformations may be based on reasons either "theoretical" (transformations whose appropriateness is based on the nature of the data) or "data derived" (where the transformations are suggested strictly by an examination of the data). Yet in either case the analyst must proceed many times by trial and error, monitoring the improvement versus the need for additional transformations.

All the transformations we describe are easily carried out by simple commands in all the popular statistical packages. We focus on transformations that can be computed in this manner, although more sophisticated and complicated methods of data transformation are available (e.g., see [3]).

Transformations to Achieve Normality and Homoscedasticity

Data transformations provide the principal means of correcting nonnormality and heteroscedasticity. In both instances, patterns of the variables suggest specific transformations. For nonnormal distributions, the two most common patterns are "flat" distributions and skewed distributions. For the flat distribution, the most common transformation is the inverse (e.g., $1/Y$ or $1/X$). Positively skewed distributions are most effectively transformed by taking logarithms of the variable, while negatively skewed distributions are best transformed by employing a square root transformation.

Heteroscedasticity is an associated problem, and in many instances "curing" this problem will deal with normality problems as well. Heteroscedasticity is also due to the distribution of the variable(s). When examining the residuals of regression analysis for heteroscedasticity, we note that an indication of unequal variance is a cone-shaped distribution of the residuals (see Chapter 3 for more specific details of the graphical analysis of residuals). If the cone opens to the right, take the inverse; if the cone opens to the left, take the square root. Some transformations can be associated with certain types of data. For example, frequency counts suggest a square root transformation; proportions are best transformed by the arcsin transformation (new variable = 2 times the arcsin of the square root of the original variable); and proportional change is best handled by taking the logarithm of the variable. In all instances, once the transformations have been performed, the transformed data should be tested to see whether the desired remedy was achieved.

Transformations to Achieve Linearity

There are numerous procedures for achieving linearity between two variables, but most simple nonlinear relationships can be placed in one of four categories (see Figure 2.9). In each quadrant, the potential transformations for both dependent and independent variables are shown. For example, if the relationship looks like that in Figure 2.9a, then either variable can be squared to achieve linearity. When multiple transformation possibilities are shown, start with the top method in each quadrant and move downward until linearity is achieved. An alternative approach is to use additional variables, termed polynomials, to represent the nonlinear components. This method is discussed in more detail in Chapter 3.

General Guidelines for Transformations

There are several points to remember when performing data transformations. They include the following:

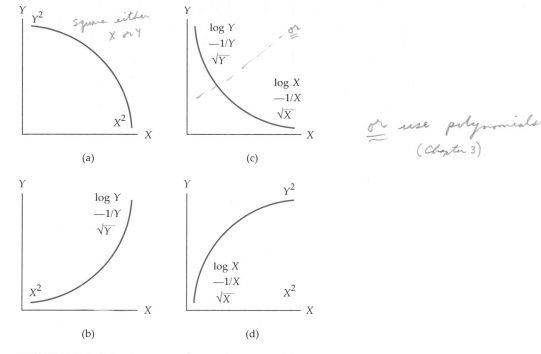

FIGURE 2.9 Selecting transformations to achieve linearity.

Source: Mosteller, F. and J. W. Tukey, *Data Analysis and Regression.* Reading, Mass.: Addison-Wesley, 1977.

1. For a noticeable effect from transformations, the ratio of a variable's mean divided by its standard deviation should be less than 4.0.
2. When the transformation can be performed on either of two variables, select the variable with the smallest ratio from item 1.
3. Transformations should be applied to the independent variables except in the case of heteroscedasticity.
4. Heteroscedasticity can be remedied only by transformation of the dependent variable in a dependence relationship. If a heteroscedasticity relationship is also nonlinear, the dependent and perhaps the independent variables must be transformed.
5. Transformations may change the interpretation of the variables. For example, transforming variables by taking their logarithm translates the relationship into a measure of proportional change (elasticity). Always be sure to explore thoroughly the possible interpretations of the transformed variables.

An Illustration of Testing the Assumptions Underlying Multivariate Analysis

To illustrate the techniques involved in the testing of the data for meeting the assumptions underlying multivariate analysis and to provide a foundation for use of the data in the subsequent chapters, the data set introduced in Chapter 1

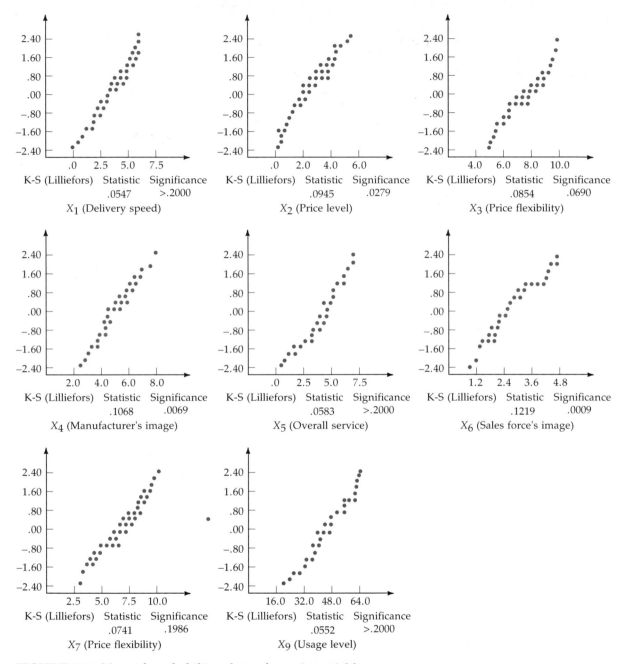

FIGURE 2.10 Normal probability plots of metric variables.

will be examined. In the course of the analysis, the assumptions of normality, homoscedasticity, and linearity will be covered. The fourth basic assumption, the absence of correlated errors, can be addressed only in the context of a specific multivariate model and thus will be covered in later chapters for each multivariate technique. Emphasis will be placed on examining the metric variables, although the nonmetric variables will be assessed where appropriate.

Normality

The first analysis to be conducted in assessing the normality of the metric variables is the derivation of normal probability plots. Figure 2.10 contains the plots for each of the nine variables. In our examination of the graphs, we see some departures from the diagonal, indicative of a departure from normality. Referring to the patterns seen in Figure 2.7, we see that X_2 seems positively skewed, X_3 approximates a uniform distribution, and X_5 seems negatively skewed.

We can complement this visual analysis with statistics reflecting the shape of the distribution (skewness and kurtosis) as well as a statistical test for normality (the modified Kolmogorov-Smirnov test). Table 2.9 shows these values for all the metric variables. Three variables exhibit a statistically significant departure from normality (X_2, X_4, and X_6). Table 2.9 also suggests the appropriate remedy. Each of the variables was transformed by taking its logarithm. In each case, the transformed variable demonstrated normality (see Table 2.9). In situations where the normality of these variables are critical, the transformed variables can be used with the assurance that they meet the assumptions of normality. But the departures from normality were not so extreme that the original variables should never be used in any analysis in their original form. If the technique has a robustness to departures from normality, then the original variables may be preferred for the comparability in the interpretation phase.

TABLE 2.9 Testing for Normality and Possible Remedies

Variable		*Shape* Skewness	Kurtosis	*Normality Test* Statistic	Significance	Description	*Possible Remedies* Transformation	Significance
X_1	Delivery speed	−.0852	−.5112	.0547	>.2000	Normal, slightly flat	None	
X_2	Price level	.4693	−.5094	.0945	.0279	Positive skew, somewhat flat	Log	>.2000
X_3	Price flexibility	−.2891	−1.0731	.0854	.0690	Approaching uniform distribution	None	
X_4	Manufacturer's image	.2179	.0848	.1068	.0069	Slight positive skew	Log	>.2000
X_5	Overall service	−.3726	.1407	.0583	>.2000	Normal distribution	None	
X_6	Sales force's image	.4934	.1071	.1219	.0009	Heavy tails at both ends	Log	>.2000
X_7	Product quality	−.2290	−.8497	.0741	.1986	None		
X_9	Usage level	−.0626	−.7253	.0552	>.2000	Normal distribution	None	

Homoscedasticity

All statistical packages have tests to assess homoscedasticity on a univariate basis (e.g., the Levene test in SPSS) where the variance of a metric variable is compared across levels of a nonmetric variable. For our purposes, we will examine each of the metric variables across the five nonmetric variables in the data set. These are appropriate analyses in preparation for either analysis of variance/multivariate analysis of variance where the nonmetric variables are the independent variables or discriminant analysis where the nonmetric variables are the dependent measures.

Table 2.10 contains the results of the Levene test for each of the nonmetric variables. The nonmetric variables X_8 and X_{11} both show significant heteroscedasticity on the same performance factors (X_4, X_5, X_6 and X_7), while X_{12} and X_{14} have fewer occurrences among the entire set of variables. The implications of these instances of heteroscedasticity must be examined whenever group differences are examinined using these nonmetric variables as independent variables with these metric variables as dependent variables. If the assumption violations are found, variable transformations are available to help rememdy the variance dispersion.

The tests for homoscedasticity of two metric variables, encountered in methods like multiple regression, are best accomplished through graphical analysis, particularly an analysis of the residuals. The interested reader is referred to Chapter 3 for a complete discussion of residual analysis and the patterns of residuals indicative of heteroscedasticity.

Linearity

The final assumption to be examined is the linearity of the relationships. In the case of individual variables, this relates to the patterns of association between

TABLE 2.10 Significance Levels of the Levene Test for Homoscedasticity

Metric Variable	X_8 Size of Firm	X_{11} Specification Buying	X_{12} Structure of Procurement	X_{13} Type of Industry	X_{14} Type of Buying Situation
X_1	.336	.336	.538	.540	.892
X_2	.212	.212	.000	.249	.000
X_3	.277	.277	.031	.662	.000
X_4	.012	.012	.597	.842	.137
X_5	.006	.006	.026	.957	.060
X_6	.024	.024	.205	.609	.182
X_7	.004	.004	.045	.115	.134
X_9	.243	.243	.212	.763	.945
X_{10}	.571	.571	.986	.817	.041

Note: Values represent the statistical significance of the Levene test assessing the variance dispersion of each metric variable across the levels of the nonmetric/categorical variables.

each pair of variables and the ability of the correlation coefficient to adequately represent the relationship. If nonlinear relationships are indicated, then the analyst can either transform one or both of the variables to achieve linearity or create additional variables to represent the nonlinear components. For our purposes, we rely on the visual inspection of the relationships to determine whether nonlinear relationships are present. The reader can refer to Figure 2.3, the scatterplot matrix containing the scatterplot for all the metric variables in the data set. Examination of the scatterplots does not reveal any apparent nonlinear relationships. Thus, transformations are not deemed necessary. The assumption of linearity will also be checked for the entire multivariate model, as is done in the examination of residuals in multiple regression.

Summary

The series of graphical and statistical tests directed toward assessing the assumptions underlying the multivariate techniques found relatively little in terms of violations of the assumptions. Where violations were indicated, they were relatively minor and should not present any serious problems in the course of the data analysis. The analyst is encouraged always to perform these simple, yet revealing, examinations of the data to ensure that potential problems can be identified and resolved before the analysis begins.

Summary

This chapter has provided the analyst with the necessary tools to examine and explore the nature of the data and the relationships among variables before the application of any of the multivariate techniques. While considerable time and effort can be expended in these activities, the prudent analyst wisely invests the necessary resources to thoroughly examine the data to ensure that the multivariate methods are applied in appropriate situations and to assist in a more thorough and insightful interpretation of the results.

Questions

1. List potential underlying causes of outliers. Be sure to include attributions to both the respondent and the researcher.
2. Discuss why outliers might be classified as beneficial and as problematic?
3. Distinguish between data that are missing at random (MAR) and missing completely at random (MCAR). Explain how each type impacts the analysis of missing data.

4. Describe the conditions under which an analyst would delete a case with missing data versus the conditions under which an analyst would use an imputation method.
5. Evaluate the following statement: In order to run most multivariate analyses, it is not necessary to meet all the assumptions of normality, linearity, homoscedasticity, and independence.
6. Discuss the following statement: Multivariate analyses can be run on any data set, as long as the sample size is adequate.

References

1. Anderson, Edgar, "A Semigraphical Method for the Analysis of Complex Problems." *Technometrics,* 2 (August 1969): 387–91.
2. BMDP Statistical Software, Inc. *BMDP / Dynamic User's Guide,* Release 7.0. Los Angeles: 1993.
3. Box, G. E. P., and D. R. Cox., "An Analysis of Transformations." *Journal of the Royal Statistical Society* B (26) (1964): 211–43.
4. Chernoff, Herman. "Graphical Representation as a Discipline," in *Graphical Representation of Multivariate Data*, Peter C. C. Wang, ed. New York: Academic Press, pp. 1–11.
5. Cohen, Jacob, and Patricia Cohen. *Applied Multiple Regression/Correlation Analysis for the Behavioral Sciences*, 2d ed. Hillsdale, N.J.: Lawrence Erlbaum Associates, 1983.
6. Daniel, C., and F. S. Wood. *Fitting Equations to Data*, 2d ed. New York: Wiley-Interscience, 1980.
7. Dempster, A. P., and D. B. Rubin. "Overview," in *Incomplete Data in Sample Surveys: Theory and Annotated Bibliography*, vol. 2. Madow, Olkin and Rubin, eds. New York: Academic Press, 1983.
8. Feinberg, Stephen. "Graphical Methods in Statistics." *American Statistician,* 33 (November 1979): 165–78.
9. Johnson, R. A., and D. W. Wichern. *Applied Multivariate Statistical Analysis.* Englewood Cliffs, N.J.: Prentice-Hall, 1982.
10. Little, Roderick J. A., and Donald B. Rubin. *Statistical Analysis with Missing Data.* Wiley: New York, 1987.
11. SAS Institute, Inc. *SAS User's Guide: Basics, Version 6.* Cary, N.C.: 1990.
12. SAS Institute, Inc. *SAS User's Guide: Statistics, Version 6.* Cary, N.C.: 1990.
13. Seber, G. A. F. *Multivariate Observations.* New York: Wiley, 1984.
14. SPSS, Inc. *SPSS for Windows, Version 6.0.* Chicago: 1993.
15. StatSoft, Inc. *STATISTICA/w.* Tulsa, Okla.: 1993.

16. SYSTAT, Inc. *SYSTAT for Windows, Version 5.* Evanston, Ill.: 1992.
17. Wang, Peter C. C., ed. *Graphical Representation of Multivariate Data.* New York: Academic Press, 1978.
18. Weisberg, S. *Applied Linear Regression.* New York: Wiley, 1985.
19. Wilkinson, L. "An Experimental Evaluation of Multivariate Graphical Point Representations." *Human Factors in Computer Systems: Proceedings,* Gaithersburg, Md.: 202–9.

Multiple Regression Analysis

LEARNING OBJECTIVES

After studying the overview of regression presented in this chapter, you should be able to:

- Determine when regression analysis is the appropriate statistical tool in analyzing a problem.

- Understand how regression helps us make predictions using the least squares concept.

- Be aware of the important assumptions underlying regression analysis and be prepared to provide remedies when violations occur.

- Interpret the results of regression from both a statistical and a managerial viewpoint.

- Apply the diagnostic procedures necessary to assess "influential" observations.

- Explain the difference between stepwise and simultaneous regression.

- Use dummy variables with an understanding of their interpretation.

- Interpret the results of logistic regression.

CHAPTER PREVIEW

This chapter describes multiple regression analysis as it is used to solve important research problems, particularly in business. Regression analysis is by far the most widely used and versatile dependence technique, applicable in every facet of business decision making. Its uses range from the most general problems to the most specific, in each instance relating a factor (or factors) to a specific outcome. For example, regression analysis is the foundation for business forecasting models, ranging from the econometric models predicting the national economy based on certain inputs (income levels, business investment, etc.) to models of a firm's performance in a market if a specific marketing strategy is followed. Regression models are also used to study how consumers make decisions or form impressions and attitudes. Other applications include evaluating the determinants of effectiveness for a program (e.g., what factors aid in maintaining quality) and determining the feasibility of a new product or the expected return for a new stock issue. Even though these examples illustrate only a small subset of all applications, they demonstrate that regression analysis is a powerful analytical tool designed to explore all types of dependence relationships.

Multiple regression analysis is a general statistical technique used to analyze the relationship between a single dependent variable and several independent variables. As noted in Chapter 1, its basic formulation is

$$Y_1 = X_1 + X_2 + \cdots + X_n$$

$$\text{(metric)} \qquad\qquad \text{(metric)}$$

This chapter presents guidelines for judging the appropriateness of multiple regression for various types of problems. Suggestions are provided for interpreting the results of its application from a managerial as well as a statistical viewpoint. Possible transformations of the data to remedy violations of various model assumptions are examined, along with a series of diagnostic procedures that identify observations with particular influence on the results. Finally, a specialized regression procedure is introduced that allows for nonmetric (binary) dependent variables while retaining all the characteristics of regression. Many readers who are already knowledgeable about multiple regression procedures can skim the early portions of the chapter. But for those who are less familiar with the subject, this chapter provides a valuable background for the study of multivariate data analysis.

KEY TERMS

Before beginning the chapter, review the key terms to develop an understanding of the concepts and terminology used. Throughout the chapter the key terms appear in **boldface.** Other points of emphasis in the chapter are *italicized.* Also, cross-references within the key terms are in *italics.*

Adjusted coefficient of determination (adjusted R^2) Modified measure of the *coefficient of determination* that takes into account the number of predictor variables included in the regression equation. While the addition of predictor variables will always cause the coefficient of determination to rise, the adjusted coefficient of determination may fall if the added predictor variables have little explanatory power and are statistically insignificant. This statistic is quite useful for comparison between equations with different numbers of predictor variables.

All-possible subsets regression   Method of selecting the variables for inclusion in the regression model that considers all possible combinations of the independent variables. For example, if the analyst has specified four potential independent variables, this technique would estimate all possible regression models with one, two, three, and four variables. The technique would then select the models with the best predictive accuracy.

Backward elimination   Method of selecting variables for inclusion in the regression model that starts with all independent variables in the model and then eliminates those variables that do not make a significant contribution to prediction.

Beta coefficient (B_n)   Standardized regression coefficient (see *standardization*) that allows for a direct comparison between coefficients as to their relative explanatory power of the dependent variable. While regression coefficients are expressed in terms of the units of the associated variable, thereby making comparisons inappropriate, beta coefficients use standardized data and can be directly compared.

Coefficient of determination (R^2)   Measure of the proportion of the variance of the dependent variable about its mean that is explained by the independent, or predictor, variables. The coefficient can vary between 0 and 1. If the regression model is properly applied and estimated, the analyst can assume that the higher the value of R^2, the greater the explanatory power of the regression equation, and therefore the better the prediction of the criterion variable.

Collinearity   Expression of the relationship between two (collinearity) or more independent variables (*multicollinearity*). Two predictor variables are said to exhibit complete collinearity if their correlation coefficient is 1 and a complete lack of collinearity if their correlation coefficient is 0. *Multicollinearity* occurs when any single predictor variable is highly correlated with a set of other predictor variables.

Correlation coefficient (r)   Indicates the strength of the association between the dependent and the independent variables. The sign ($+$ or $-$) indicates the direction of the relationship. The value can range from -1 to $+1$, with $+1$ indicating a perfect positive relationship, 0 indicating no relationship, and -1 indicating a perfect negative or reverse relationship (as one grows larger, the other grows smaller).

Criterion variable (Y)   See *dependent variable*.

Degrees of freedom (df)   Calculated from the total number of observations minus the number of estimated *parameters*. These parameter estimates are restrictions on the data, because, once made, they define the population from which the data are assumed to have been drawn. For example, in estimating a regression model with a single predictor variable, we estimate two parameters, the constant (b_0) and a regression coefficient for the predictor variable (b_1). In estimating the random error, defined as the sum of the prediction errors (actual minus predicted dependent values) for all cases, we would find ($n - 2$) degrees of freedom. Degrees of freedom provide a measure of how restricted the data are to reach a level of prediction. If the degrees of freedom are small, this suggests that the resulting prediction may be less generalizable, because all but a few observations were incorporated in the prediction. Conversely, a large degrees-of-freedom value indicates that the prediction is fairly "robust" with regard to being representative of the overall sample of respondents.

Dependent variable (Y)   Variable being predicted or explained by the set of independent variables.

Dummy variable   Independent variable used to account for the effect that different levels of a nonmetric variable have in predicting the criterion variable. To account for L levels of an independent variable, $L - 1$ dummy variables are needed. For example, gender is measured as male or female and could be represented by two dummy variables (X_1 and X_2). When the respondent is male, $X_1 = 1$ and $X_2 = 0$. Likewise, when the respondent is female, $X_1 = 0$ and $X_2 = 1$. However, when $X_1 = 1$, we know that X_2 must equal 0. Thus we need only one variable, either X_1 or X_2, to represent the variable gender. We cannot include both variables, because one is perfectly predicted by the other and the regression coefficients cannot be estimated. If a variable has three levels, only two dummy variables are needed. We always have one dummy variable less than the number of levels of the variable used.

Effects coding   Method for specifying the reference category for a set of *dummy variables* where the reference category receives a value of minus one (-1) across the set of dummy variables. With this type of coding, the coefficients for the dummy variables become group deviations from the mean of all groups. This is in contrast to *indicator coding*, where the reference category is given the value of zero across all dummy variables and the coefficients represent group deviations from the reference group.

Heteroscedasticity   See *homoscedasticity*.

Homoscedasticity   When the variance of the error terms (e) appears constant over a range of predictor variables, the data are said to be homoscedastic. The assumption of equal variance of the population error ϵ (where ϵ is estimated from e) is critical to the proper application of linear regression. When the error terms have increasing or modulating variance, the data are said to be *heteroscedastic*. Discussion of residuals in this chapter further illustrate this point.

Independent variable   Variable(s) selected as predictors and potential explanatory variables of the dependent variable.

Indicator coding   Method for specifying the reference category for a set of *dummy variables* where the reference category receives a value of zero across the set of dummy variables. The coefficients represent the group differences from the reference group. Also see *effects coding*.

Influential observation   An observation that has a disproportionate influence on one or more aspects of the regression estimates. This influence may be based on response differences on the predictor variables, extreme observed values for the criterion variable, or a combination of these effects. Influential observations can either be "good" by reinforcing the pattern of the remaining data or "bad," when a single or small set of cases unduly affects the regression estimates. It is not necessary for the observation to be an *outlier*, although many times outliers can be classified as influential observations as well.

Intercept (b_0)   Value on the Y axis (criterion variable axis) where the line defined by the regression equation $Y = b_0 + b_1X_1$ crosses the axis. It is described by the constant term b_0 in the regression equation. In addition to its role in prediction, the intercept may or may not have a managerial interpretation. If the complete absence of the predictor variable has meaning, then the intercept represents that amount. For example, when estimating sales from past advertising expenditures, the intercept represents the level of sales expected if advertising is eliminated. But in many instances the constant has only predictive value because there is no situation in which all predictor variables are absent. An example is predicting product preference based on consumer attitudes. All individuals have an attitude, so the intercept has no managerial use but does aid in a more accurate prediction.

Leverage points   Type of influential observation defined by one aspect of influence termed *leverage*. These observations are substantially different on one or more independent variables so that they affect the estimation of one or more regression coefficients.

Likelihood value   Measure used in *logistic regression* and *logit analysis* to represent the lack of predictive fit. Even though these methods do not use the least squares procedure in model estimation, the likelihood value is similar to the *sum of squared error* in regression analysis.

Linearity   Used to express the concept that the model possesses the properties of additivity and homogeneity. In a simple sense, linear models predict values that fall in a straight line by having a constant unit change of the dependent variable (slope) for a constant unit change of the independent variable. In the population model $Y = b_0 + b_1X_1 + \epsilon$, the effect of a change of 1 in X_1 is to add b_1 (a constant) units of Y. The model $Y = b_0 + b_2(X_1X_2)$ is not additive because a unit change in X_2 does not increase Y by b_2 units; rather, it increases Y by $(X_1)b_2$ units, an amount that varies for different levels of X.

Logistic regression   Special form of regression in which the criterion variable is a nonmetric, dichotomous (binary) variable. While differences exist in some aspects, the general manner of interpretation is quite similar to linear regression.

Logit analysis   See *logistic regression.*

Measurement error   Degree to which the data values do not truly measure the characteristic being represented by the variable. For example, when asking about total family income, there are many sources of measurement error (reluctance to answer full amount, error in estimating total income, etc.) that make the data values imprecise.

Moderator effect   Impact of a third independent variable (the moderator variable) causes the relationship between a dependent/independent variable pair to change depending on the value of the moderator variable. Also known as an interactive effect and similar to the interaction effect seen in analysis of variance methods.

Multicollinearity   See *collinearity.*

Multiple regression   Regression model with two or more independent variables.

Normal probability plot   Graphical comparison of the form of the sample distribution to the normal distribution. In the graph, the normal distribution is represented by a straight line angled at 45 degrees. The actual distribution is plotted against this line, so that any differences are shown as deviations from the straight line, making identification of differences quite simple.

Null plot   Plot of residuals versus the predicted values that exhibits a random pattern. A null plot is indicative of no identifiable violations of the assumptions underlying regression analysis.

Outlier   In strict terms, an observation that has a substantial difference between the actual value for the dependent variable and the predicted value. Cases that are substantially "different," whether in regard to the dependent or independent variables, are often termed outliers as well. In all instances, the objective is to identify observations that represent inappropriate representations of the population from which the sample is drawn, so that they may be discounted or even eliminated from the analysis as unrepresentative.

Parameter   Quantity (measure) characteristic of the population. For example, μ and σ^2 are the symbols used for the population parameters mean (μ) and variance (σ^2). These are typically estimated from sample data where the arithmetic average of the sample is used as a measure of the population average and the variance of the sample is used to estimate the variance of the population.

Part correlation coefficient   Measures the strength of the relationship between a dependent and a single independent variable when the predictive effects of the other independent variables in the regression model are removed. The objective is to portray the *unique* predictive effect due to a single independent variable among a set of independent variables. Differs from the *partial correlation coefficient*, which is concerned with *incremental* predictive effect.

Partial correlation coefficient   Measures the strength of the relationship between the criterion or dependent variable and a single predictor variable when the effects of the other predictor variables in the model are held constant. For example, r_{Y,X_2X_1} measures the variation in Y associated with X_2 when the effect of X_1 on both X_2 and Y is held constant. Used in sequential variable selection methods of regression model estimation to identify the independent variable with the greatest *incremental* predictive power beyond the predictor variables already in the regression model.

Partial F (or t) values   The partial F test is simply an F test for the additional contribution to prediction accuracy of a variable above that of the variables already in the equation. When a variable (X_a) is added to a regression equation after many other variables have already been entered into the equation, its contribution may be very small. The reason is that X_a is highly correlated with the variables already in the equation. A partial F value may be calculated for all variables by simply pretending that each, in turn, is the last to enter the equation. This method gives the additional contribution of each variable above all others in the equation. A t value may be calculated instead of F values in all instances, with the t value being the square root of the F value.

Partial regression plot   Graphical representation of the relationship between the dependent variable and a single independent variable. The scatterplot of points depicts the partial correlation between the two variables, with the effects of other independent variables held constant (see *partial correlation coefficient*). This portrayal is particularly helpful in assessing the form of the relationship (linear versus nonlinear) and the identification of *influential observations*.

Predictor variable (X_n)   See *independent variable*.

PRESS statistic   Validation measure obtained by eliminating each observation one at a time and predicting this dependent value with the regression model estimated from the remaining observations.

Polynomial   Transformation of an independent variable to represent a curvilinear relationship with the dependent variable. By including a squared term (X^2), a single inflection point is estimated. A cubic term estimates a second inflection point. Additional terms of a higher power can also be estimated.

Power   Probability that a significant relationship will be found if it actually exists. Complements the more widely used significance level *alpha* (α).

Prediction error   Difference between the actual and predicted values of the dependent variable for each observation in the sample.

Regression coefficient (b_n)   Numerical value of any parameter estimate directly associated with the independent variables; for example, in the model $Y = b_0 + b_1 X_1$, the value b_1 is the regression coefficient for the variable X_1. In the multiple predictor model (e.g., $Y = b_0 + b_1 X_1 + b_2 X_2$), the regression coefficients are partial because each takes into account not only the relationships between Y and X_1 and between Y and X_2, but also between X_1 and X_2. The coefficient is not limited in range, as it is based on both the degree of association and the scale units of the predictor variable. For instance, two variables with the same association to Y would have different coefficients if one predictor variable was measured on a 7-point scale and another was based on a 100-point scale.

Regression variate   Linear combination of weighted independent variables used collectively to predict the dependent variable.

Residual (e or ϵ)   Error in predicting our sample data is called the residual. Seldom will our predictions be perfect. We assume that random error will occur, but we assume that this error is an estimate of the true random error in the population (ϵ), not just the error in prediction for our sample (e). We assume that the error in the population we are estimating is distributed with a mean of 0 and a constant variance.

Semipartial correlation   See *part correlation coefficient*.

Simple regression   Regression model with a single independent variable.

Specification error   Error in predicting the dependent variable caused by excluding one or more relevant independent variables that can bias the estimated coefficients of the included variables as well as decrease the overall predictive power of the regression model.

Standardization   Process whereby raw data are transformed into new measurement variables with a mean of 0 and a standard deviation of 1. When data are transformed in this manner, the b_0 term (the intercept) assumes a value of 0. When using standardized data, the regression coefficients are known as *beta coefficients*, which allow the researcher to compare the relative effect of each independent variable on the dependent variable.

Standard error of the estimate (SEE)   Measure of the variation in the predicted values that can be used to develop confidence intervals around any predicted value. Similar to the standard deviation of a variable around its mean.

Statistical relationship   Relationship based on the correlation of two or more independent variables with the dependent variable. Measures of association, typically correlations, represent the degree of relationship because there is more than one value of the dependent variable for each value of the independent variable.

Stepwise estimation   Method of selecting variables for inclusion in the regression model that starts with selecting the best predictor of the dependent variable. Additional independent variables are selected in terms of the incremental explanatory power they can add to the regression model. Independent variables are added as long as their *partial correlation coefficients* are statistically significant. Independent variables may also be dropped if their predictive power drops to a nonsignificant level.

Studentized residual Most commonly used form of standardized residual, it differs from other methods in how it calculates the standard deviation used in standardization. To minimize the effect of a single outlier, the residual standard deviation for observation i is computed from regression estimates omitting the ith observation in the calculation of the regression estimates.

Sum of squared errors Sum of the squared *prediction errors (residuals)* across all observations. Used to denote the variance in the dependent variables not yet accounted for by the regression model. If no independent variables are used for prediction, this becomes the squared errors using the mean as the predicted value and thus equals the *total sum of squares*.

Sum of squares regression Sum of the squared differences between the mean and predicted values of the dependent variable for all observations. This represents the amount of improvement in explanation of the independent variable attributable to the independent variable(s).

Tolerance Commonly used measure of collinearity and multicollinearity, the tolerance of variable i (TOL_i) is $1 - R^{2*}_i$, where R^{2*}_i is the coefficient of determination for the prediction of variable i by the other predictor variables. As the tolerance value grows smaller, the variable is more highly predicted (collinear) with the other predictor variables.

Total sum of squares Total amount of variation that exists to be explained by the independent variables. This "baseline" is calculated by summing the squared differences between the mean and actual values for the dependent variable across all observations.

Transformation A variable may have an undesirable characteristic, such as non-normality, that detracts from the ability of the correlation coefficient to represent the relationship between it and another variable. A transformation, such as taking the logarithm or square root of the variable, creates a new variable and eliminates the undesirable characteristic, allowing for a better measure of the relationship. Transformations may be applied to either the dependent or independent variables, or both. The need and specific type of transformation may be based on theoretical reasons (such as transforming a known nonlinear relationship) or empirical reasons (identified through graphical or statistical means).

Variance inflation factor (VIF_i) Indicator of the effect that the other predictor variables have on the variance of a regression coefficient, directly related to the *tolerance* value ($VIF_i = 1/R^{2*}_i$). Large VIF values also indicate a high degree of *collinearity* or *multicollinearity* among the independent variables.

Wald statistic Test used in logistic regression for the significance of estimated coefficients. Its interpretation is like the F or t values used for significance testing of linear regression coefficients.

What Is Multiple Regression Analysis?

Multiple regression analysis is a statistical technique that can be used to analyze the relationship between a single **dependent (criterion) variable** and several **independent (predictor) variables.** The objective of multiple regression analysis is to use the independent variables whose values are known to predict the single dependent value selected by the researcher. Each predictor variable is weighted, the weights denoting their relative contribution to the overall prediction. In calculating the weights, the regression analysis procedure ensures maximal prediction from the set of independent variables in the variate. These weights also facilitate interpretation as to the influence of each variable in making the prediction, although correlation among the independent variables complicates the interpretative process. The set of weighted independent variables is also known as the **regression variate,** a linear combination of the independent variables that best predicts the dependent variable (Chapter 1 contains a more detailed explanation of the variate). The regression equation, also referred to as the regression variate, is the most widely known example of a variate among all the multivariate techniques.

As noted in Chapter 1, multiple regression analysis is a dependence technique. Thus, to use it, you must be able to divide the variables into dependent and independent variables. Regression analysis is also a statistical tool that should be used only when both the dependent and independent variables are metric. However, under certain circumstances, it is possible to include nonmetric data for independent variables (by transforming either ordinal or nominal data with dummy-variable coding) or the dependent variable (by the use of a binary measure in the specialized technique of logistic regression). In summary, to apply multiple regression analysis,

1. The data must be metric or appropriately transformed.
2. Before deriving the regression equation, the researcher must decide which variable is to be dependent and which remaining variables will be independent.

Relating Independent to Dependent Variables with Regression

The objective of regression analysis is to predict a single dependent variable from the knowledge of one or more independent variables. When the problem involves a single independent variable, the statistical technique is called **simple regression.** When the problem involves two or more independent variables, it is called **multiple regression.** The following discussion is divided into three parts to help you understand how regression estimates the relationship of predictor and criterion variables. The three topics covered are (1) prediction without an independent variable, using only a single measure—the average, (2) prediction using a single independent variable—simple regression, and (3) prediction using several independent variables—multiple regression.

Prediction Without an Independent Variable

Let's start with a simple measure. Assume we surveyed eight families and asked how many credit cards were held by the entire family. The data are shown in columns 1 and 2 of Table 3.1. If we were asked to predict how many credit cards a family holds using only these data, we could use any of several descriptive measures—the arithmetic average, the median, or even the mode (the value occurring

TABLE 3.1 HATCO Survey Results for Average Number of Credit Cards

Family Number	Actual Number of Credit Cards	Average Number of Credit Cards*	Error[†]	Errors Squared
1	4	7	+3	9
2	6	7	+1	1
3	6	7	+1	1
4	7	7	0	0
5	8	7	−1	1
6	7	7	0	0
7	8	7	−1	1
8	10	7	−3	9
	56		0	22

*Average number of credit cards = $56 \div 8 = 7$.

[†]*Error* refers to the difference between the actual number of credit cards held by a family and the estimated number of cards held (seven) using the arithmetic average as a predictor.

most often). For each of these statistics, we could make a prediction as to the number of credit cards held by the families. Because we are basing this prediction on only the information about credit card usage, we must make a single prediction and apply it to all families. For example, our prediction with the arithmetic average (the mean) would be seven credit cards. Our prediction would be stated as "The average number of credit cards held by a family is seven."

But the analyst must still answer one question: How accurate is the prediction for each measure? The customary way to assess the adequacy of a predictor variable is to examine the errors in predicting the criterion variable when it is used for prediction. For example, if we predict that family 1 has seven credit cards, we overestimate by three. Thus the error is +3. If this procedure were followed for each family, some estimates would be too high, others would be too low, and still others would give the correct number of credit cards held. By simply adding the errors, we might expect to obtain a measure of prediction accuracy. However, we would not—the errors from using the mean value would always sum to zero. Therefore, we would not have a measure of the adequacy of our prediction, because we know we did not predict perfectly for all families, even though the sum is zero. To overcome this problem, we square each error and then add the results together. The total, referred to as the **sum of squared errors,** provides a better measure of the prediction accuracy. The objective is to obtain the smallest possible sum of squared errors, because this would mean our predictions would be most accurate.

For a single set of observations, the arithmetic average (mean) will produce a smaller sum of squared errors than any other measure of central tendency including the median, mode, any other single value, or any other more sophisticated statistical measure. (Interested readers are encouraged to see if they can find a better predicting value than the mean). Therefore, for our set of eight observations, the average is the best single predictor of the number of credit cards held by families. The sum of squared errors for our example when we use the mean as our single predictor is 22 (see Table 3.1). In our discussion of simple and multiple regression, we will use this prediction by the mean as a baseline for comparison because it represents the best prediction without using any independent variables.

Prediction Using A Single Independent Variable—Simple Regression

As researchers or practitioners, we are always interested in improving our predictions. In the preceding section, we learned that the average is the best predictor if we do not use any independent variables. But in our example survey we also collected information on other measures that could act as independent variables. Let's determine whether knowledge of another measure—the number of people in each family—will help our predictions. This procedure involves a single independent variable and is referred to as *simple regression.*

Simple regression is another procedure for predicting data (just as the average predicts data), and it uses the same rule—minimizing the sum of squared errors of prediction. We know that without using family size we can best describe the number of credit cards held as the mean value, seven. Another way to write the prediction is as follows:

Predicted number of credit cards = Average number of credit cards held

or

$$\hat{Y} = \bar{y}$$

Using our information on family size, we could try to improve our predictions by reducing our prediction errors. To do so, the prediction errors in the number of credit cards held *must* be associated (correlated) with family size. The concept of correlation, represented by the **correlation coefficient (r),** is fundamental to regression analysis and describes the relationship between two variables. Two variables are said to be correlated if changes in one variable are associated with changes in the other variable. In this way, as one variable changes, we would know how the other variable is changing. If family size was correlated with credit card usage, we would then describe the relationship as follows:

Predicted number = Change in number of * (Family size)
of credit cards   credit cards held
                              associated with unit
                              change in family size

or

$$\hat{Y} = b_1 X_1$$

For example, if we find that for each additional member in a family, the number of credit cards increases (on the average) by two, we would predict that families of four would have 8 credit cards and families of five would have 10 credit cards. Thus, the number of credit cards = 2 times (family size). However, we often find that the prediction is improved by adding a constant value because the following pattern may be found:

Family Size	Number of Credit Cards
1	4
2	6
3	8
4	10
5	12

It can be observed that "number of credit cards = 2 times (family size)" is wrong by two credit cards in every case:

Family Size	Number of Credit Cards	Predicted Number of Credit Cards	Error
1	4	2	−2
2	6	4	−2
3	8	6	−2
4	10	8	−2
5	12	10	−2

Therefore, changing our description to

Predicted number of credit cards held = 2 + 2 * (Family size)

gives us perfect predictions in all cases.

We take this approach with our sample of eight families and see how well the description fits our data. The procedure is as follows:

Predicted number   = Constant + Change in number of * (Family size)
of credit cards held   credit cards with
                                    differing family size

or

$$\hat{Y} = b_0 + b_1 X_1 + e$$

If the constant term, also known as the **intercept (b_0),** does not help us predict, the process of minimizing the sum of squared errors will give an estimate of the constant term to be zero. The terms b_0 and b_1 are called **regression coefficients.** The term e is the *residual* or prediction error.

Using a mathematical procedure,* we will find the values of b_0 and b_1 such that the sum of the squared errors of prediction is minimized (see Table 3.2). For this example, the appropriate values are

$$Y = 2.87 + .97 \text{ (Family size)}$$

---

*The mechanics of deriving the regression coefficients for each predictor variable based on the correlations between the independent and dependent variables such that the sum of the squared errors is minimized (also known at least squares estimation) is left to other more technically oriented texts dealing with regression. See [10, 13, 17].

TABLE 3.2 HATCO Survey Results Relating Number of Credit Cards to Family Size

Family Number	Number of Credit Cards* (Y_i)	Family Size† (X_i)	Prediction	Errors Squared
1	4	2	4.81	.66
2	6	2	4.81	1.42
3	6	4	6.75	.56
4	7	4	6.75	.06
5	8	5	7.72	.08
6	7	5	7.72	.52
7	8	6	8.69	.48
8	10	6	8.69	1.72
				5.50

*Average number of credit cards = $(\Sigma\, Y)/n = 56/8 = 7$.
†Average family size = $(\Sigma\, Y)/n = 34/8 = 4.25$.

Because we have used the same criterion (minimizing the sum of squared errors or least squares), we can determine whether our knowledge of family size has helped us better predict credit card holdings when compared with our prediction using only the arithmetic average. The sum of squared errors using the average was 22, and using our new procedure with a single independent variable, we find that the sum of squared errors is 5.50 (see Table 3.2, column 5). Using the least squares procedure and a single independent variable, we see that our new approach, simple regression, is better than using just the average. The equation indicates that for each additional family member, the credit card holdings are higher on average by .97. The constant 2.87 has an interpretation only within the range of values for the independent variable (family size). In this case, a household size of zero is not possible, so the intercept alone does not have practical meaning. However, this fact does not invalidate its use, as it aids in prediction of credit card holdings for the possible family sizes (in our example from 1 to 5). In some instances, the independent variables can take on zero values, and the intercept then has a direct interpretation.*

Because we did not achieve perfect predictions of the dependent variable, we would also like to estimate the range of predicted values that we might expect, rather than relying just on the single (point) estimate. The point estimate is our best estimate of the dependent variable and can be shown to be the average prediction for any given value of the independent variable. Using this point estimate, we can calculate the range of predicted values based on a measure of the prediction errors we expect to make. Known as the **standard error of the estimate,** this measure in simple terms is the standard deviation of the prediction errors. Remember from your basic statistics that we can construct a confidence interval for a variable about its mean value by adding (plus and minus) a certain

*For some special situations where the specific relationship of the regression equation is known to pass through the origin, the constant term may be suppressed (termed regression through the origin). In these cases, the interpretation of the residuals and regression coefficients changes slightly.

number of standard deviations. For example, adding (plus and minus) 1.96 standard deviations to the mean defines a range including 95 percent of the values of a variable.

We can follow a similar method for the predictions we make. Using the point estimate, we can add (plus and minus) a certain number of standard errors of the estimate (depending on the confidence level desired and sample size) to establish the upper and lower bounds for our predictions made with any independent variable(s). The standard error of the estimate (SEE) is calculated by

$$\text{Standard Error of the Estimate (SEE)} = \sqrt{\frac{\text{Sum of Squared Errors}}{\text{Sample Size} - 2}}$$

The number of SEEs to use in deriving the confidence interval is determined by the level of significance (α) and the sample size, which gives a t value. The confidence interval is then calculated as follows:

Lower limit: predicted dependent value − (SEE ∗ t value)

Upper limit: predicted dependent value + (SEE ∗ t value)

For our simple regression example, the standard error of the estimate is .957 (the square root of the value of 5.50 divided by 6). Construct the confidence interval for the predictions by selecting the number of standard errors to add (plus and minus). Then look in a table for the t distribution and select the value for a given confidence level and sample size. In our example, the t value for a 95 percent confidence level with 6 degrees of freedom (sample size (8) − 2 = 6) is 2.447. The amount added (plus and minus) to the predicted value is then .957 × 2.447, or 2.34. If we substitute the average family size (4.25) into the regression equation, the predicted value is 6.99 (it differs from the average of seven only because of rounding). The expected range becomes 4.65 (6.99 − 2.34) to 9.33 (6.99 + 2.34). For a more detailed discussion of these confidence intervals, see [13].

If the sum of squared errors represents a measure of our prediction errors, we should also be able to determine a measure of our prediction success, which we can term the sum of squares regression. Together, these two measures should equal the **total sum of squares.** The total sum of squares is based on the differences of the observations from the mean, the best prediction possible without using any additional variables that provides the baseline prediction. As the analyst adds independent variables, the total sum of squares can now be divided into (1) the sum of squares predicted by the independent variable(s), also known as the **sum of squares regression,** and (2) the sum of squared errors (SSE). The general formula for obtaining the sum of squared errors and the sum of squares regression is

$$\Sigma (y_i - \bar{y})^2 = \Sigma (y_i - \hat{y}_i)^2 + \Sigma (\hat{y}_i - \bar{y})^2$$

$$\text{TSS} = \text{SSE} + \text{SSR}$$

$$\begin{array}{ccc} \text{Total Sum} \\ \text{of Squares} \end{array} = \begin{array}{c} \text{Sum of} \\ \text{Squared Error} \end{array} + \begin{array}{c} \text{Sum of Squares} \\ \text{Regression} \end{array}$$

where

$\bar{y}$ = average of all observations
y_i = value of individual observation
$\hat{y}_i$ = predicted value for observation

We can use this division of the total sum of squares to approximate how well the regression variate we have calculated describes family holdings of credit cards. The average number of credit cards held by our sampled families is our best estimate of the number held by any family. We know that this is not an extremely accurate estimate, but it is the best prediction available without using any other variables. The accuracy of the mean was measured by calculating the squared sum of errors in prediction (sum of squares = 22). Now that we have fitted a regression model using family size, does it explain the variation better than the average? We know it is somewhat better because the sum of squared errors is now 5.50. We can look at how well our model predicts by examining this improvement.

Sum of squared errors in prediction using the average (SS_{Total} or SST)	22.0
−Sum of squared errors in prediction using family size (SS_{Error} or SSE)	5.5
Sum of squared errors explained with family size ($SS_{Regression}$ or SSR)	16.5

Therefore, we explained 16.5 squared errors by changing from the average to a regression model using family size. This is an improvement of 75 percent (16.5 ÷ 22 = .75) over the use of the average. Another way to express this level of prediction accuracy is the **coefficient of determination (R^2)**, the ratio of the regression sum of squares to the total sum of squares as shown in the following equation:

$$\text{Coefficient of determination } (R^2) = \frac{\text{Sum of Squares Regression}}{\text{Total Sum of Squares}}$$

If the regression model using family size perfectly predicted all families holding credit cards, this ratio would equal 1.0. If using family size gave no better predictions than using the average, the ratio (R^2) would be very close to 0. When the regression equation contains more than one independent variable, the R^2 value represents the combined effect of the entire variate in prediction. The R^2 value is simply the squared correlation of the actual values and the predicted values from the variate.

We often use the coefficient of correlation (r) to assess the relationship between Y and X. The sign of the correlation coefficient ($+r$, $-r$) denotes the slope of the regression line. However, the "strength" of the relationship is best represented by R^2. In our example the $R^2 = .75$, thus we know that 75 percent of the variation* in Y is explained by the variable X. The corresponding value of r ($+.86$ or $-.86$) offers the sign as additional information but can mislead analysts to believe a stronger relationship exists.

Prediction Using Several Independent Variables: Multiple Regression Analysis

We previously demonstrated how simple regression helped improve our prediction of credit card holdings. By using data on family size, we predicted the number of credit cards a family would own more accurately than we could by simply using the arithmetic average. This result raises the question of whether we could

---

*The total sum of squares represents the variation about the mean that the analyst attempts to predict with one or more independent variables. When discussions mention the variance of the dependent variable, they are referring to this total sum of squares.

improve our prediction even further by using additional data obtained from the families. Would our prediction of the number of credit cards be improved if we used data on family size in addition to data on another variable, perhaps family income?

The ability of an additional independent variable to improve the prediction of the dependent variable is related not only to the correlation of the additional independent variable to the dependent variable but also to the correlation(s) of the additional independent variable(s) to the independent variable(s) already in the regression equation. **Collinearity** is the association, measured as the correlation, between two independent variables. **Multicollinearity** refers to the correlation among three or more independent variables (evidenced when one is regressed against the others). Although there is a precise distinction in statistical terms, it is rather common practice to use the terms interchangeably.

The impact of multicollinearity is to reduce any individual independent variable's predictive power by the extent to which it is associated with the other independent variables. For example, assume that one independent variable (X_1) has a correlation with the dependent variable of .60 and a second independent variable (X_2) has a correlation of .50. Then X_1 would explain 36 percent (obtained by squaring the correlation of .60) of the variance of the dependent variable, and X_2 would explain 25 percent (correlation of .50 squared). If the two independent variables are not correlated with each other at all, there is no "overlap," or sharing, of their predictive power. The total explanation would be their sum of 51 percent. But as collinearity increases, there is some "sharing" of predictive power and the collective predictive power of the independent variables decreases. Exhibit 3.1 provides further detail on the calculation of unique and shared variance predictions among correlated independent variables

Figure 3.1 portrays the proportions of shared and unique variance for our example of two independent variables above in varying instances of collinearity. As we can see, if their collinearity is zero, then the individual variables predict 36 and

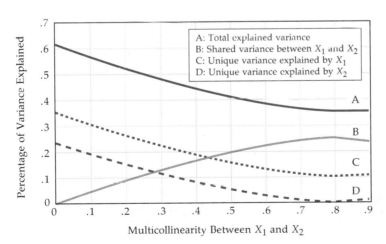

Correlation between dependent and independent variables:
X_1 and dependent (.60), X_2 and dependent (.50)

FIGURE 3.1 Proportions of unique and shared variance by levels of multicollinearity.

EXHIBIT 3.1

Calculating Unique and Shared Variance Among Independent Variables

The basis for estimating all regression relationships is the correlation, which measures the association between two variables. In regression analysis, the correlations between the independent variables and the dependent variable provide the basis for forming the regression variate by estimating regression coefficients (weights) for each independent variable that maximize the prediction (explained variance) of the dependent variable. When the variate contains only a single independent variable, the calculation of regression coefficients is straightforward and based on the direct or univariate correlation between the independent and dependent variable. The percentage of explained variance of the dependent variable is simply the square of the direct correlation.

But as independent variables are added to the variate, the calculations must also consider the intercorrelations among independent variables. If the independent variables are correlated, then they "share" some of their predictive power. Because we use only the prediction of the overall variate, the shared variance must not be "doubled counted" by using the direct correlations. Thus, we calculate some additional forms of the correlation to represent these shared effects. The first is the **partial correlation coefficient,** which is the correlation of an independent (X_i) and dependent (Y) variable when the effects of the other independent variable(s) have been removed from both X_i and Y. A second form of correlation is the **part** or **semipartial correlation,** which reflects the correlation between an independent and dependent variable while controlling for the predictive effects of all other independent variables on X_i. The two forms of correlation differ in that the partial correlation removes the effects of other independent variables from X_i and Y, while the part correlation removes the effects only from X_i. The partial corre-

lation represents the incremental predictive effect of one independent variable from the collective effect of all others and is used to identify independent variables that have the greatest incremental predictive power when a set of independent variables is already in the regression variate. The part correlation represents the unique relationship predicted by an independent variable after the predictions shared with all other independent variables is taken out. Thus, the part correlation is used in apportioning variance among the independent variables. Squaring the part correlation gives the unique variance explained by the independent variable.

The accompanying diagram portrays the shared and unique variance among two correlated independent variables.

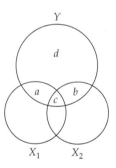

a = variance of Y uniquely explained by X_1
b = variance of Y uniquely explained by X_2
c = variance of Y explained jointly by X_1 and X_2
d = variance of Y not explained by X_1 or X_2

The variance associated with the partial correlation of X_2 controlling for X_1 can be represented as $b \div (d + b)$, where $d + b$ represents the unexplained variance after accounting for X_1. The part correlation of X_2 controlling for X_1 is $b \div (a + b + c + d)$, where $a + b + c + d$ represents the total variance of Y and b in the amount uniquely explained by X_2.

The analyst can also determine the shared and unique variance for independent variables through simple calculations. The part correlation between the dependent variable (Y) and an independent variable (X_1) while controlling for a second independent variable (X_2) is calculated by the following equation:

Part correlation of Y, X_1, given X_2

$$= \frac{\text{Corr of } Y, X_1 - (\text{Corr of } Y, X_2 * \text{Corr of } X_1 X_2)}{\sqrt{1.0 - (\text{Corr of } X_1 X_2)^2}}$$

A simple example with two independent variables $(X_1$ and $X_2)$ illustrates the calculation of both shared and unique variance of the dependent variable (Y). The direct correlations and the correlation between X_1 and X_2 are shown in the following correlation matrix:

	Y	X_1	X_2
Y	1.0		
X_1	.6	1.0	
X_2	.5	.7	1.0

The direct correlations of .60 and .50 represent fairly strong relationships with Y, but the correlation of .70 between X_1 and X_2 means that a substantial portion of this predictive power may be shared. The part correlation of X_1 and Y controlling for X_2 $(r_{Y, X_1(X_2)})$ and the unique variance predicted by X_1 can be calculated as

$$r_{YX_1(X_2)} = \frac{.60 - (.50 \,(.70)}{\sqrt{1.0 - .70^2}} = .35$$

unique variance predicted by $X_1 = .35^2 = .1225$

Because the direct correlation of X_1 and Y is .60, we also know that the total variance predicted by X_1 is $.60^2$, or .36. If the unique variance is .1225, then the shared variance must be .2375 (.36 − .1225).

We can calculate the unique variance explained by X_2 and confirm the amount of shared variance by the following:

$$r_{YX_2(X_1)} = \frac{.50 - (.60)\,(.70)}{\sqrt{1.0 - .70^2}} = .11$$

unique variance predicted by $X_1 = .11^2 = .0125$

With the total variance explained by X_2 being $.50^2$, or .25, the shared variance is calculated as .2375 (.25 − .0125). This confirms the amount found in the calculations for X_1.

Thus, the total variance explained (R^2) by the two independent variables is

Unique variance explained by X_1	.1225
Unique variance explained by X_2	.0125
Shared variance explained by X_1 and X_2	.2375
Total variance explained by X_1 and X_2	.3725

These calculations can be extended to more than two variables, but as the number of independent variables increases, it is easier to allow the statistical programs to perform the calculations.

The calculation of shared and unique variance illustrates the effects of multicollinearity on the ability of the independent variables to predict the dependent variable. Figure 3.1 shows these effects when faced with high to low levels of multicollinearity.

25 percent of the variance in the dependent variable, for an overall prediction (R^2) of 51 percent. But as collinearity increases, the unique variance explained by each variable decreases while the shared prediction percentage rises. Because this shared prediction can count only once, the overall prediction for these two independent variables decreases as multicollinearity increases. To maximize the prediction from a given number of independent variables, the analyst should look for independent variables that have low collinearity with the other independent variables. We will revisit the issues of collinearity and multicollinearity in later sections when we discuss their implications for the selection of independent variables and the interpretation of the regression variate.

To improve further our prediction of credit card holdings, let's use additional data obtained from our eight families. The independent variable we shall add is

TABLE 3.3 HATCO Survey Results Relating Number of Credit Cards to Family Size and Family Income

Family Number	Number of Credit Cards (Y)	Family Size (X₁)	Family Income (X₂)	Prediction	Error Squared
1	4	2	14	4.76	.58
2	6	2	16	5.20	.64
3	6	4	14	6.03	.00
4	7	4	17	6.68	.10
5	8	5	18	7.53	.22
6	7	5	21	8.18	1.39
7	8	6	17	7.95	.00
8	10	6	25	9.67	.11
					3.04

family income (see Table 3.3). We simply expand our simple regression model as follows:

$$\text{Predicted number of credit cards held} = b_0 + b_1 X_1 + b_2 X_2 + e$$

where

b_0 = Constant number of credit cards independent of family size and income
b_1 = Change in credit card holdings associated with unit change in family size
b_2 = Change in credit card holdings associated with unit change in family income
X_1 = Family size
X_2 = Family income

The regression coefficients with two independent variables are

$$Y = .482 + .63X_1 + .216X_2$$

We can again find our residual by predicting Y, subtracting our prediction from the actual value. We then square the resulting prediction error, as in columns 5 and 6 of Table 3.3. The sum of squared errors is 3.04 for our prediction using both family size and family income, compared with 5.50 (Table 3.2) using only family size and 22.0 (Table 3.1) using the arithmetic average. We assume at this point that some improvement has been found.

When family income is added to the regression analysis, the R^2 increases to .86

$$R^2_{\text{(family size+family income)}} = \frac{22.0 - 3.04}{22.0} = \frac{18.96}{22.0} = .86$$

This means that the inclusion of family income in the regression analysis increases the prediction by 11 percent (.85 − .75), all due to the unique incremental predictive power of family income.*

---

*This can also be determined by calculating the part correlation for family income, controlling for family size. The interested reader is encouraged to perform the calculations in a manner similar to those demonstrated in Exhibit 3.1. The correlations are credit card usage and family size (.866); credit card usage and family income (.829); and family size and family income (.673).

A Decision Process for Multiple Regression Analysis

In the previous sections we discussed examples of simple regression and multiple regression. In those discussions, many factors affected our ability to find the best regression model. To this point, however, we have examined these issues only in simple terms, with little regard to how they combine in an overall approach to multiple regression analysis. In the following sections, the six-stage model-building process introduced in Chapter 1 will be used as a framework for discussing the factors that impact the creation, estimation, interpretation, and validation of a regression analysis. The process begins with specifying the objectives of the regression analysis, including the selection of the dependent and independent variables. The analyst then proceeds to design the regression analysis, considering such factors as sample size and the need for variable transformations. With the regression model formulated, the assumptions underlying regression analysis are first tested for the individual variables. If all assumptions are met, then the model is estimated. Once results are obtained, diagnostic analyses are performed to ensure that the overall model meets the regression assumptions and that no observations have undue influence on the results. The next stage is the interpretation of the regression variate; it examines the role played by each independent variable in the prediction of the dependent measure. Finally, the results are validated to ensure generalizability to the population. Figure 3.2 provides a graphical representation of the model-building process for multiple regression, while the following sections discuss each step in detail.

Stage One: Objectives of Multiple Regression

Multiple regression analysis, also known as general linear modeling, is a multivariate statistical technique used to examine the relationship between a single dependent variable and a set of independent variables. The necessary starting point in multiple regression, as with all multivariate statistical techniques, is the research problem. The flexibility and adaptability of multiple regression allows for its use with almost any dependence relationship. In selecting suitable applications of multiple regression, the analyst must consider three primary issues: (1) the appropriateness of the research problem, (2) specification of a statistical relationship, and (3) selection of the dependent and independent variables. Each issue is addressed in the following sections.

Research Problems Appropriate for Multiple Regression

Multiple regression is by far the most widely used multivariate technique of those examined in this text. With its broad applicability, multiple regression has been used for many purposes. The ever-widening applications of multiple regression, however, fall into two broad classes of research problems: prediction and explanation. These research problems are not mutually exclusive, and an application of multiple regression analysis can address either or both types of research problem.

Prediction with Multiple Regression

One fundamental purpose of multiple regression is the prediction of the dependent variable with a set of independent variables. In doing so, multiple regression fulfills one of two objectives. The first objective is to maximize the overall predictive power of the independent variables as represented in the variate. This linear combination of independent variables is formed to be the optimal predictor of the dependent measure. Multiple regression provides an objective means of assessing the predictive power of a set of independent variables. In applications focused on this objective, the researcher is primarily interested in achieving maximum prediction. Multiple regression provides many options in both the form and the specification of the independent variables that may modify the variate to increase its predictive power. Many times prediction is maximized at the expense of interpretation. One example is a variant of regression, time series analysis, in which the sole purpose is prediction and the interpretation of results is useful only as a

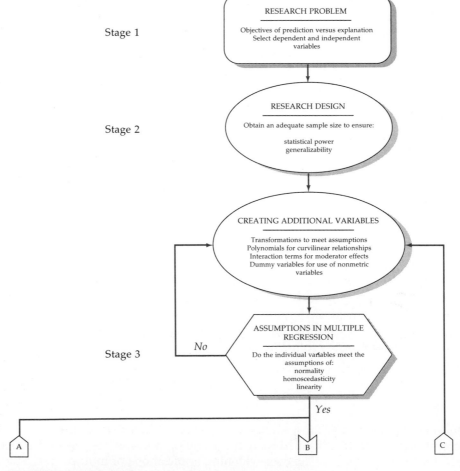

FIGURE 3.2 Multiple regression decision process.

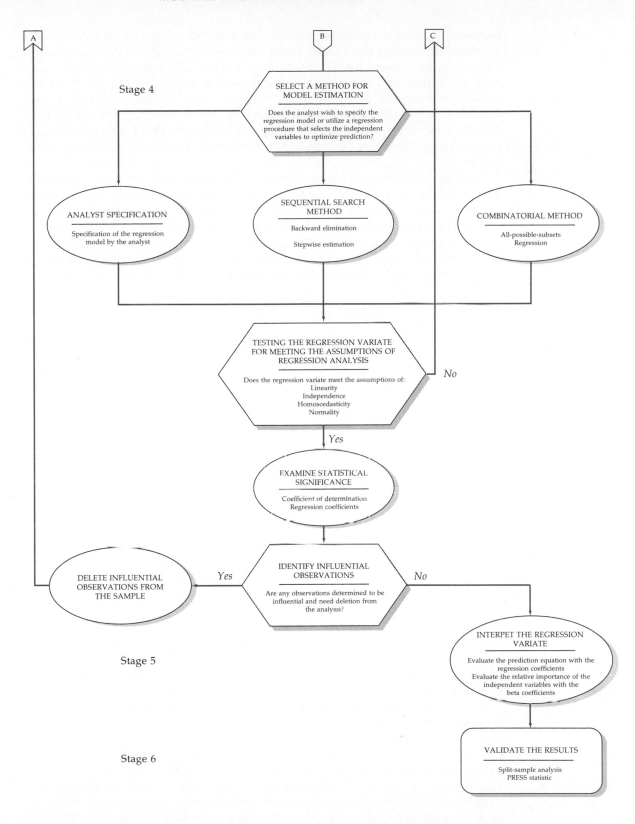

means of increasing predictive accuracy. In other situations, predictive accuracy is crucial to ensuring the validity of the set of independent variables, thus allowing for the subsequent interpretation of the variate. Measures of predictive accuracy are formed and statistical tests regarding the significance of the predictive power can be made. In all instances, whether or not prediction is the primary focus, the regression analysis must achieve acceptable levels of predictive accuracy to justify its application. The analyst must ensure that both statistical and practical significance are considered (see stage four, below, for more discussion of this topic).

Multiple regression can also meet a second objective of comparing two or more sets of independent variables to ascertain the predictive power of each variate. Illustrative of a confirmatory approach to modeling, this use of multiple regression is concerned with the comparison of results across two or more alternative or competing models. The primary focus of this type of analysis is the relative predictive power among models, although in any situation the prediction of the selected model must demonstrate both statistical and practical significance.

Explanation with Multiple Regression

Multiple regression provides a means of objectively assessing the degree and character of the relationship between dependent and independent variables by forming the variate of independent variables. The independent variables, in addition to their collective prediction through the dependent variable, may also be considered for their individual contribution to the variate and its predictions. Interpretation of the variate may rely on any of three perspectives: the importance of the independent variables, the types of relationships found, or the interrelationships among the independent variables.

The most direct interpretation of the regression variate is a determination of the relative importance of each independent variable in the prediction of the dependent measure. In all applications, the selection of independent variables should be based on their theoretical relationships to the dependent variable. Regression analysis then provides a means of objectively assessing the magnitude and direction (positive or negative) of each independent variable's relationship. The multivariate character of multiple regression that differentiates it from its univariate counterparts is the simultaneous assessment of relationships between each independent variable and the dependent measure. In making this simultaneous assessment, the *relative* importance of each predictor is determined.

In addition to assessing the importance of each variable, multiple regression also affords the analyst a means of assessing the nature of the relationships between the predictors and the dependent variable. The assumed relationship is a linear association based on the correlations among the independent variables and the dependent measure. But transformations are also available to assess whether other types of relationships exist, particularly curvilinear relationships. This flexibility ensures that the analyst may examine the true nature of the relationship beyond the assumed linear relationship.

Finally, multiple regression also provides insight into the relationships among independent variables in their prediction of the dependent measure. These interrelationships are important for two reasons. First, correlation among the independent variables may make some variables redundant in the predictive effort. As such, they are not needed to produce the optimal prediction. This does not reflect their individual relationships with the dependent variable but instead indicates that in a multivariate context, they are not needed if another set of independent variables explaining this variance is employed. The analyst must guard against determining the importance of independent variables based solely on the derived variate, because relationships among the independent variables may "mask" relationships that are not needed for predictive purposes but that represent substantive findings nonetheless. The interrelationships among variables can extend not only to their predictive power but also to interrelationships among their estimated effects. This is best seen when one independent variable's effect is contingent on another independent variable. Multiple regression provides diagnostic analyses that can determine whether such effects exist based on empirical or theoretical rationale.

Specifying a Statistical Relationship

Multiple regression is appropriate for use when the analyst is interested in a statistical, not a functional, relationship. For example, examine the following relationship:

$$\text{Total cost} = \text{Variable cost} + \text{Fixed cost}$$

If the variable cost is \$2 per unit, the fixed cost is \$500, and we produce 100 units, we assume that the total cost will be exactly \$700 and that any deviation from \$700 is caused by our inability to measure cost since the relationship between costs is fixed. This is called a *functional relationship* because we expect there will be no error in our prediction.

But in our earlier example dealing with sample data representing human behavior, we were assuming that our description of credit card usage was only approximate and not a perfect prediction. It was thought to be a **statistical relationship** because there will always be some random component to the relationship being examined. We found two families with two members, two with four members, and so on, who had different numbers of credit cards. *More than one value of the dependent value will usually be observed for any value of a predictor variable in a statistical relationship.* The dependent variable is assumed to be a random variable, and for a given predictor we can only hope to estimate the average value of the dependent variable associated with it. In our simple regression example, the two families with four members held an average of 6.5 credit cards, and our prediction was 6.75. Our prediction is not as accurate as we would like, but it is better than just using the average of 7 credit cards. The error is assumed to be the result of random behavior among credit card holders.

In summary, a functional relationship calculates an exact value, while a statistical relationship estimates an average value. Throughout this book, we will be concerned with statistical relationships. Both of these relationships are displayed in Figure 3.3.

Selection of Dependent and Independent Variables

The ultimate "success" of any multivariate technique, including multiple regression, starts with the selection of the variables to be used in the analysis. Because multiple regression is a dependence technique, the analyst must specify which variable is the dependent variable and which variables are used as predictor variables. The selection of both types of variables should be based principally on conceptual or theoretical grounds. Chapters 1 and 11 discuss the role of theory in multivariate analysis, and those issues strongly apply to multiple regression. The

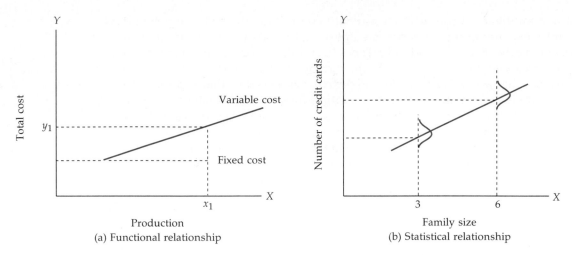

FIGURE 3.3 Comparison of functional and statistical relationships.

fundamental task of variable selection must be performed by the analyst, even though the analyst has many options and program features to assist in model estimation. If the analyst does not exert judgment during variable selection but instead (1) selects variables indiscriminately or (2) allows for the selection of an independent variable to be based solely on empirical bases, several of the basic tenets of model development will be violated.

The selection of a dependent variable is many times dictated by the research problem. But in all instances, the analyst must be aware of the **measurement error,** especially in the dependent variable. Measurement error refers to the degree that the variable is an accurate and consistent measure of the concept being studied. If the variable used as the dependent measure has substantial measurement error, then even the best independent variables may be unable to achieve acceptable levels of predictive accuracy. Measurement error can come from several sources (see Chapter 1 for a more detailed discussion). The analyst must always be concerned with obtaining the best dependent and independent measures, based both on conceptual and empirical factors.

The most problematic issue in independent variable selection is **specification error,** which concerns the inclusion of irrelevant variables or the omission of relevant variables from the set of independent variables. Although the inclusion of irrelevant variables does not bias the results for the other independent variables, it does have some impact on them. First, it reduces model parsimony, which may be critical in the interpretation of the results. Second, the additional variables may mask or replace the effects of more useful variables, especially if some hierarchical form of model estimation is used (this is discussed in more detail in stage four). Finally, the additional variables may make the testing of statistical significance of the independent variables less precise and reduce the statistical and practical significance of the analysis.

Given the problems associated with adding irrelevant variables, should the analyst be concerned with excluding relevant variables? The answer is definitely yes, because the exclusion of relevant variables can seriously bias the results and negatively affect any interpretation of them. In the simplest case, the omitted variables are uncorrelated with the included variables, and the only effect is to reduce the overall predictive accuracy of the analysis. But when correlation exists between the included and omitted variables, the effects of the included variables become biased to the extent that they are correlated with the omitted variables. The greater the correlation, the greater the bias. The estimated effects for the included variables now represent not only their actual effects but also the effects that the included variables share with the omitted variables. This can lead to serious problems in model interpretation and the assessment of statistical and managerial significance.

The analyst must be careful in the selection of the variables to avoid both types of specification error. Perhaps most troublesome is the omission of relevant variables, as the variables' effect cannot be assessed without their inclusion. This heightens the need for theoretical and practical support for all variables included or excluded in a multiple regression analysis.

Measurement error also affects the independent variables by reducing the predictive power as their measurement error increases. Multiple regression has no direct means of correcting for known levels of measurement error for the independent variables. If the analyst suspects that measurement error may be problematic in the independent variables, structural equation modeling (Chapter 11) should be examined as a means of accommodating measurement error in estimating the effects of independent variables.

Stage Two: Research Design of a Multiple Regression Analysis

In the design of a multiple regression analysis, the analyst must consider issues like sample size, the nature of the independent variables, and the possible creation of new variables to represent special relationships between the dependent and independent variables. In doing so, the criteria of statistical and practical significance must always be maintained. The ability of multiple regression to address many types of research questions is greatly impacted by the research design issues to be discussed.

Sample Size

The sample size used in multiple regression is perhaps the most influential single element under the control of the analyst in designing the analysis. The effects of sample size are seen most directly in the statistical power of the significance testing and the generalizability of the result. Both issues are addressed in the following sections.

Statistical Power and Sample Size

The size of the sample has a direct impact on the appropriateness and the statistical power of multiple regression. Small samples, usually characterized as having fewer than 20 observations, are appropriate only for analysis by simple regression with a single independent variable. Even in these situations, only very strong relationships can be detected with any degree of certainty. Likewise, very large samples of 1,000 observations or more make the statistical significance tests overly sensitive, indicating that almost any relationship is statistically significant. With very large samples the analyst must ensure that the criteria of practical significance are met along with statistical significance.

Power in multiple regression refers to the probability of detecting as statistically significant a specific level of R^2 or a regression coefficient at a specified significance level and a specific sample size (see Chapter 1 for a more detailed discussion). Sample size has a direct and sizable impact on power. Table 3.4 illustrates the interplay among the sample size, the significance level (α) chosen, and the number of independent variables in detecting a significant R^2. The table values are the minimum R^2 that the specified sample size will detect as statistically significant at the specified alpha level with a probability (power) of .80. For example, if the researcher employs five independent variables, specifies the .05 significance level, and is satisfied with detecting the R^2 80 percent of the time it occurs (corresponding to a power of .80), a sample of 50 respondents will detect R^2 values of 23 percent and greater. If the sample is increased to 100 respondents, then R^2 values of 12 percent and above will be detected. But if the 50 respondents are all that are available and the analyst wants a .01 significance level, the analysis will detect R^2 values only in excess of 29 percent. The analyst should always consider the role of sample size in significance testing before the actual collection of the data. If weaker relationships are expected, the analyst can make informed judgments as to the necessary sample size to reasonably detect the relationships, if they exist. For example, Table 3.4 demonstrates that sample sizes of 100 will detect fairly small R^2 values (10 percent to 15 percent) with up to ten independent variables and a significance level of .05. However, if the sample size falls to 50

TABLE 3.4 Minimum R^2 That Can Be Found Statistically Significant with a Power of .80 for Varying Numbers of Independent Variables and Sample Sizes

	Significance Level (α) = .01 *No. of Independent Variables*				*Significance Level (α) = .05* *No. of Independent Variables*			
Sample Size	*2*	*5*	*10*	*20*	*2*	*5*	*10*	*20*
20	45	56	71	NA	39	48	64	NA
50	23	29	36	49	19	23	29	42
100	13	16	20	26	10	12	15	21
250	5	7	8	11	4	5	6	8
500	3	3	4	6	3	4	5	9
1000	1	2	2	3	1	1	2	2

Source: Based on calculations contained in Cohen and Cohen [6].

NA: Not applicable because number of independent variables exceeds sample size

Note: Values in the table indicate the minimum R^2 value that can be found statistically significant at a power of .80 for a given significance level, sample size, and number of independent variables. For example, the leftmost value (45) is interpreted as meaning that a regression equation based on 20 observations with two independent variables requires a R^2 value of at least 45% for the relationship to be deemed statistically significant with a power of .80 if the significance level is set at .01. Relationships lower than 45% will not be deemed statistically significant. Lower table values indicate the ability of the regression model to identify a wider range of significant relationships with an acceptable level of power.

observations in these situations, the minimum R^2 that can be detected doubles. The researcher must always be aware of the anticipated power of any proposed multiple regression analysis and understand the elements of the research design that can be changed to meet the requirements for an acceptable analysis [11].

The analyst can also determine the sample size needed to detect effects for individual independent variables given the expected effect size (correlation), the alpha level, and the power desired. The possible computations are too numerous for presentation in this discussion, but the interested reader is referred to several texts dealing with power analyses [5, 6] or to a computer program to calculate sample size or power for a given situation [3].

Generalizability and Sample Size

In addition to sample size's role in determining statistical power, it also affects the generalizability of the results by the ratio of observations to independent variables. A general rule is that the ratio should never fall below five, meaning that there should be five observations for each independent variable in the variate. As this ratio falls below five, the analyst encounters the risk of "overfitting" the variate to the sample, making the results too specific to the sample and thus lacking generalizability. While the minimum ratio is 5 to 1, the desired level is between 15 to 20 observations for each independent variable. When this level is reached, the results should be generalizable if the sample is representative. However, if a stepwise procedure is employed (discussed in stage four under model estimation approaches), the recommended level increases to 50 to 1. In cases when the available sample does not meet these criteria, the analyst should be certain to validate the generalizability of the results.

Fixed Versus Random Effects Predictors

The examples of regression models we have discussed to this point have assumed that the levels of the predictor variables are fixed. For example, if we wish to know the impact on preference of three levels of sweetener in a cola drink, we make up three batches of cola and have a number of people sample each. We then predict the preference rating of each cola, using level of sweetener as the predictor. We have fixed the level of sweetener and are interested in its effect at these levels. We do not assume the three levels to be a random sample for a large number of possible levels of sweetener. A random predictor variable is one in which the levels of the predictor are selected at random. When using a random predictor variable, the interest is not just in the levels examined but rather in the larger population of possible predictor levels from which we selected a sample.

Most regression models based on survey data are random effects models. As an illustration, a survey was conducted to help assess the relationship between age of the respondent and frequency of visits to physicians. The predictor variable "age of respondent" was randomly selected from the population, and the inference regarding the population is of concern, not just knowledge of the individuals in the sample.

The estimation procedures for models using both types of predictor variables are the same except for the error terms. In the random effects models, a portion of the random error comes from the sampling of the predictors. However, the statistical procedures based on the fixed model are quite robust, so using the statistical analysis as if you were dealing with a fixed model (as most analysis packages assume) may still be appropriate as a reasonable approximation.

Creating Additional Variables

The basic relationship represented in multiple regression is the *linear* association between *metric* dependent and independent variable(s) based on the product-moment correlation. The next section detailing stage three's testing of regression assumptions examines the issue of linearity in more detail. However, there are situations in which linear relationships are inadequate or inappropriate. For example, the relationships between dependent and independent variables may be nonlinear (e.g., U-shaped). Or the analyst may wish to include nonmetric independent variablees.

In these situations, new variables must be created by **transformations,** as multiple regression is totally reliant on the types of variables in the model to represent any effects other than linear relationships. Transforming the data provides the researcher with a means of modifying either the dependent or independent variables for one of two reasons: to improve or modify the relationship between independent and dependent variables or to allow the use of nonmetric variables in the regression variate. Data transformations may be based on reasons either "theoretical" (transformations whose appropriateness is based on the nature of the data) or "data derived" (transformations that are suggested strictly by an examination of the data). In either case the analyst must proceed many times by trial and error, constantly assessing the improvement versus the need for additional transformations. We explore these issues with discussions of data transformations that allow the regression analysis to best represent the actual data and a discussion of the creation of variables to supplement the original variables.

All the transformations we describe are easily carried out by simple commands in all the popular statistical packages. We focus on transformations that can be computed in this manner, although other, more sophisticated and complicated methods of data transformation are available (e.g., see [4]).

Representing Curvilinear Effects with Polynomials

Several types of data transformations are appropriate for linearizing a curvilinear relationship. Direct approaches, discussed in Chapter 2, involve modifying the values through some arithmetic transformation (e.g., taking the square root or logarithm of the variable). However, such transformations have several limitations. First, they are helpful only in a simple curvilinear relationship (a relationship with only one turning or inflection point). Second, they do not provide any statistical means for assessing whether the curvilinear or linear model is the more appropriate. Finally, they accommodate only univariate relationships and not the interaction between variables when more than one independent variable is involved. We now discuss a means of creating variables to explicitly model the curvilinear components of the relationship and address each of the limitations inherent in data transformations.

Polynomials are power transformations of an independent variable that add a nonlinear component for each additional power of the independent variable. The power of 1 (X^1) represents the linear component and is the form we have discussed throughout this chapter. The power of 2, the variable squared (X^2), represents the quadratic component. In graphical terms, X^2 represents the first inflection point. A cubic component, represented by the variable cubed (X^3), adds a second infection point. With these variables, and even higher powers, more complex relationships can be accommodated than are possible with only transformations. For example, in a simple regression model, a curvilinear model with one turning point can be modeled with the equation

$$Y = b_0 + b_1 X_1 + b_2 X_1^2$$

where

b_0 = Intercept
X_1 = Linear effect of X_1
X_1^2 = Curvilinear effect of X_1

Although any number of nonlinear components may be added, the cubic term is usually the highest power used. As each new variable is entered into the regression equation, we can also perform a direct statistical test of the nonlinear components that is not available with data transformations. For interpretation purposes, the positive quadratic term indicates a ∪-shaped upward curve, while a negative coefficient indicates a ∩-shaped downward relationship.

Multivariate polynomials are created when the regression equation contains two or more independent variables. We follow the same procedure for creating the polynomial terms as before but must also create an additional term, the interaction term ($X_1 * X_2$), which is needed for each variable combination to represent fully the multivariate effects. In graphical terms, a two-variable multivariate polynomial is portrayed by a surface with one peak or valley. For higher-order polynomials, interpretation is best made by plotting the surface from the predicted values.

How many terms should be added? Common practice is to start with the linear component and then sequentially add higher-order polynomials until nonsignificance is achieved. The use of polynomials, however, is not without potential problems. First, each additional term requires a degree of freedom, which may be particularly restrictive with small sample sizes. This limitation is not found with data transformations. Also, multicollinearity is introduced by the additional terms. Care must be taken to assess the impact of multicollinearity when polynomials are introduced.

Representing Interaction or Moderator Effects

The nonlinear relationships discussed above require the creation of an additional variable (for example, the squared term) to represent the changing slope of the relationship over the range of the independent variable. This focuses on the relationship between a single independent variable and the dependent variable. But what if an independent/dependent variable relationship is affected by another independent variable? This is termed a **moderator effect,** which occurs when the moderator variable, a second independent variable, changes the *form* of the relationship between another independent variable and the dependent variable. This is also known as an *interaction effect* and is similar to the interaction term found in analysis of variance and multivariate analysis of variance (see Chapter 5 for more detail on interaction terms).

The most common moderator effect employed in multiple regression is the *quasi-* or *bilinear moderator,* where the slope of the relationship of one independent variable (X_1) changes across values of the moderator variable (X_2)[9, 16]. In our earlier example of credit card usage, assume that family income (X_2) was found to be a positive moderator of the relationship between family size (X_1) and credit card usage (Y_1). This would mean that the expected change in credit card usage based on family size (b_1, the regression coefficient for X_1) would be lower for families with low income and higher for families with higher incomes. Without the moderator effect, we assumed that family size had a "constant" effect on the number of credit cards used. But the interaction term tells us that this relationship changes depending on family income level. Note that this does not necessarily mean the effects of family size or family income by themselves are unimportant, but instead the interaction term complements their explanation of credit card usage.

The moderator effect is represented in multiple regression by a term quite similar to the polynomials described earlier to represent nonlinear effects.* The moderator term is a compound variable formed by multiplying X_1 by the moderator X_2, which is entered into the regression equation. The moderated relationship is represented as

$$Y = b_0 + b_1X_1 + b_2X_2 + b_3X_1X_2$$

where

$$b_0 = \text{Intercept}$$
$$X_1 = \text{Linear effect of } X_1$$
$$X_2 = \text{Linear effect of } X_2$$
$$X_1X_2 = \text{Moderator effect of } X_2 \text{ on } X_1$$

To determine whether the moderator effect is significant, the analyst first estimates the original (unmoderated) equation and then estimates the moderated relationship. If the change in R^2 is statistically significant, then a significant moderator effect is present.

The interpretation of the regression coefficients changes slightly in moderated relationships. The b_3 coefficient, the moderator effect, indicates the unit change in the effect of X_1 as X_2 changes. The b_1 and b_2 coefficients now represent the effects of X_1 and X_2, respectively, when the other independent variable is zero. In the unmoderated relationship, the b_1 coefficient represents the effect of X_1 across all levels of X_2, and vice versa for b_2. Thus, in unmoderated regression, the regression coefficients b_1 and b_2 are "averaged" across levels of the other independent variables, while in a moderated relationship they are separate from the other independent variables. To determine the total effect of an independent variable, the separate and moderated effects must be combined. The overall effect of X_1 for any value of X_2 can be found by substituting the X_2 value into the following:

$$b_{(\text{total})} = b_1 + b_3X_2$$

For example, assume a moderated regression resulted in the following coefficients: $b_1 = 2.0$ and $b_3 = .5$. If the value of X_2 ranges from one to seven, the analyst

---

*The reader will note the similarity of the nonlinear and interaction terms. In fact, the nonlinear term can be viewed as a form of interaction, where the independent variable "moderates" itself, thus the squared term ($X_1 * X_1$).

can calculate the total effect of X_1 at any value of X_2. When X_2 equals 3, the total effect of X_1 is 3.5 (2.0 + .5(3)). When X_2 increases to 7, the total effect of X_1 is now 5.5 (2.0 + .5(7)). We can see the moderator effect at work, making the relationship of X_1 and the dependent variable change, given the level of X_2. Excellent discussion of moderated relationships in multiple regression are available in a number of sources [6, 9, 16].

Incorporating Nonmetric Data with Dummy Variables

To this point, all our illustrations have assumed metric measurement for both predictor and criterion variables. When the criterion variable is measured as a dichotomous (0, 1) variable, either discriminant analysis (discussed in Chapter 3) or a specialized form of regression discussed in the next section is appropriate. But what can we do when the independent variables are nonmetric, with two or more categories? We use a set of dichotomous variables, known as **dummy variables,** that act as replacement predictor variables. A dummy variable is a dichotomous variable that represents one category of a nonmetric independent variable. Any nonmetric variable with k categories can be represented as k-1 dummy variables. The following example will help clarify this concept.

Assume we want to predict the number of credit cards held by families according to the age of the head of household. Moreover, we have measured the age of the head of the household with a nonmetric variable with two categories—those above or below 40 years of age.* To represent the nonmetric variable, we would create two new dummy variables (X_1 and X_2) as follows:

Dummy Variable	Age of Head of Household
$X_1 = 1$	if <40 years else $X_1 = 0$
$X_2 = 1$	if ≥40 years else $X_2 = 0$

Both variables (X_1 and X_2) are not necessary, however, because when $X_1 = 0$, age must be ≥40 by definition. Thus we need include only one of the variables (X_1 or X_2) to test the effect of the head of household's being above or below age 40 on the number of credit cards held.

Correspondingly, if we had also measured household income with three levels, as shown in the following table, we would need only include any two of the variables to represent the effects of household income.

Dummy Variable	Household Income Level	
$X_3 = 1$	if < \$15,000,	else $X_3 = 0$
$X_4 = 1$	if ≥ \$15,000 and <\$25,000	else $X_4 = 0$
$X_5 = 1$	if ≥ \$25,000,	else $X_5 = 0$

---

*Dummy variables can also represent divisions of the sample based on forming categories from metric variables. For example, we could have measured the actual age of the head of the household and then formed the two categories from the metric variable.

There are three ways to represent the income levels with two dummy variables:

Method 1		Method 2		Method 3		
X_3	X_4	X_3	X_5	X_4	X_5	Income Levels
1	0	1	0	0	0	<$15,000
0	1	0	0	1	0	≥$15,000 and < $25,000
0	0	0	1	0	1	≥$25,000

This form of dummy-variable coding is known as **indicator coding.** The regression coefficients for the dummy variables represent deviations from the comparison group (i.e., the omitted group that received all zeros) on the criterion variable. The deviations represent the differences between means for each group of respondents formed by a dummy variable and the comparison group. This form is most appropriate when there is a logical comparison group, such as in an experiment. In an experiment with a control group acting as comparison group, the coefficients are the mean differences on the dependent variable for each treatment group from the control group. Any time dummy-variable coding is used, we must be aware of the comparison group and remember that the coefficients represent the differences in group means from this group. Also, the coefficients are in the same units as the dependent variable.

An alternative method of dummy-variable coding is termed **effects coding.** It is exactly the same as indicator coding except that the comparison or omitted group (the group that got all zeros) is now given the value of −1 instead of 0 for the dummy variables. Now the coefficients represent differences for any group from the mean of all groups, rather than from the omitted group. Both forms of dummy-variable coding will give exactly the same predictive results, coefficient of determination, and regression coefficients for the continuous variables. The only differences will be in the interpretation of the dummy-variable coefficients.

Stage Three: Assumptions in Multiple Regression Analysis

We have shown how improvements in prediction of the dependent variable are possible by adding independent variables and even transforming them to represent aspects of the relationship that are not linear. But to do so we must make several assumptions about the relationships between the dependent and independent variables that affect the statistical procedure (least squares) used for multiple regression. In the following sections we discuss testing for the assumptions and corrective actions to take if violations occur.

Assessing Individual Variables Versus the Variate

The assumptions underlying multiple regression analysis apply both to the individual variables (dependent and independent) and to the relationship as a whole. Chapter 2 examined the available methods for assessing the assumptions for individual variables. But in multiple regression, once the variate has been derived, it acts collectively in predicting the dependent variable. This necessitates assessing the assumptions not only for individual variables but also for the variate itself. This section focuses on examining the variate and its relationship with the dependent variable for meeting the assumptions of multiple regression. These analyses actually must be performed *after* the regression model has been estimated in stage four. Thus, the testing for assumptions must occur not only in the initial phases of the regression but also after the model has been estimated.

The basic issue is whether, in the course of calculating the regression coefficients and predicting the dependent variable, the assumptions of regression analysis have been met. Are the errors in prediction a result of an actual absence of a relationship among the variables, or are they caused by some characteristics of the data not accommodated by the regression model? The assumptions to be examined are as follows:

1. The linearity of the phenomenon measured
2. The constant variance of the error terms
3. The independence of the error terms
4. The normality of the error term distribution

The principal measure of prediction error for the variate is the **residual**—the difference between the observed and predicted values for the dependent variable. Plots of the residuals versus the independent or predicted variables are a basic method of identifying assumption violations for the overall relationship. When examining residuals, some form of standardization is recommended, as it makes the residuals directly comparable. (In their original form, larger predicted values naturally have larger residuals). The most widely used is the **studentized residual**. Its values correspond to *t* values; we will see that this correspondence makes it quite easy to assess the statistical significance of particularly large residuals. We will also examine a series of statistical tests that can complement the visual examination of the residual plots.

The most common residual plot involves the residuals (r_1) versus the predicted dependent values (Y_i).* Violations of each assumption can be identified by specific patterns of the residuals. Figure 3.4 contains a number of residual plots addressing the basic assumptions discussed in the following sections. One plot of special interest is the **null plot** (Figure 3.4a), the plot of residuals when all assumptions are met. The null plot shows the residuals falling randomly, with relatively equal dispersion about zero and no strong tendency to be either greater or less than zero. Likewise, no pattern is found for large versus small values of the independent variable.

Linearity of the Phenomenon

The **linearity** of the relationship between dependent and independent variables represents the degree to which the change in the dependent variable associated with the predictor variable (the regression coefficient) is constant across the range

---

*In a simple regression model, the residuals may be plotted against either the dependent or independent variable, because they are directly related. In multiple regression, however, only the predicted dependent values represent the entire effect of the independent variable set (variate). Thus, unless the residual analysis intends to concentrate on a single variable, the predicted dependent values (Y_i) are used.

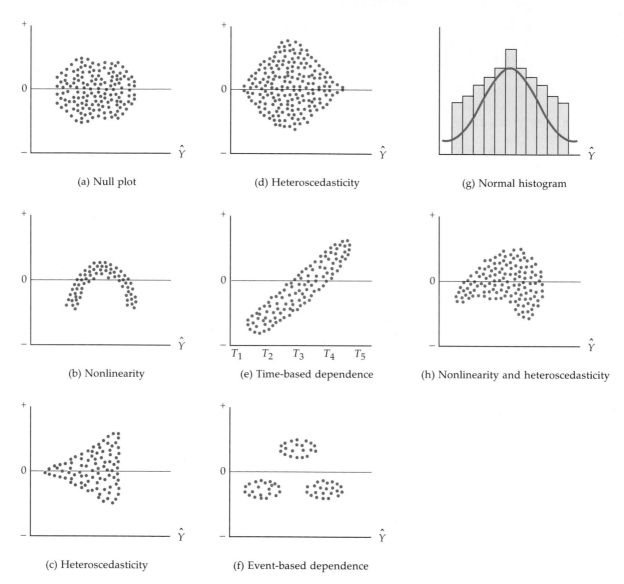

(a) Null plot

(d) Heteroscedasticity

(g) Normal histogram

(b) Nonlinearity

(e) Time-based dependence

(h) Nonlinearity and heteroscedasticity

(c) Heteroscedasticity

(f) Event-based dependence

FIGURE 3.4 Graphical analysis of residuals.

of values for the independent variable. The concept of correlation is based on a linear relationship, thus making it a critical issue in regression analysis. Linearity is easily examined by residual plots. Figure 3.4b shows a typical pattern of residuals indicating the existence of a nonlinear relationship not represented in the current model. Any consistent curvilinear pattern in the residuals indicates that corrective action will increase both the predictive accuracy of the model and the validity of the estimated coefficients. Remedies through transforming the data are discussed in Chapter 2 [12]. We might also wish to include many nonlinear relationships in our regression models. Methods have been developed (data transformations like polynomial regression, and specific methods like nonlinear regression) that can accommodate the curvilinear effects of one or more independent variables.

In multiple regression with more than one independent variable, an examination of the residuals shows the combined effects of all predictor variables, but we cannot examine any predictor variable separately in a residual plot. To do so, we use what are called **partial regression plots,** which show the relationship of a single predictor variable to the criterion variable. They differ from the residual plots we just discussed in that the line running through the center of the points, which was horizontal in the earlier plots (refer to Figure 3.4), will now slope up or down depending on whether the regression coefficient for that predictor variable is positive or negative. Examining the residuals around this line is done exactly as before.

We look for the curvilinear pattern of residuals indicating a nonlinear relationship between a specific predictor variable and the criterion variable. This is the more useful method when we have several predictor variables, as we call tell which specific variables violate the assumption of linearity and apply the needed remedies only to them. Also, the identification of outliers or influential observations is facilitated on the basis of one predictor variable at a time.

Constant Variance of the Error Term

The presence of unequal variances **(heteroscedasticity)** is one of the most common assumption violations. Diagnosis is made with residual plots or simple statistical tests. Plotting the residuals (studentized) against the predicted criterion values and comparing them to the null plot (see Figure 3.4a) shows a consistent pattern if the variance is not constant. Perhaps the most common pattern is triangle-shaped in either direction (Figure 3.4c). A diamond-shaped pattern (Figure 3.4d) can be expected in the case of percentages where more variation is expected in the midrange than at the tails. Many times, a number of violations occur simultaneously, such as the nonlinearity and heteroscedasticity shown in Figure 3.4h. Remedies for one of the violations often corrects problems in other areas as well.

Each statistical computer program has statistical tests for heteroscedasticity. For example, SPSS provides the Levane test for homogeneity of variance, which measures the equality of variances for a single pair of variables. Its use is particularly recommended because it is less affected by departures from normality, another frequently occurring problem in regression.

If heteroscedasticity is present, two remedies are available. If the violation can be attributed to a single independent variable, the procedure of weighted least squares can be employed. More direct and easier, however, are a number of variance-stabilizing transformations discussed in Chapter 2 that allow the transformed variables to be used directly in our regression model.

Independence of the Error Terms

We assume in regression that each predicted value is independent. By this we mean that the predicted value is not related to any other prediction; that is, they are not sequenced by any variable. We can best identify such an occurrence by plotting the residuals against any possible sequencing variable. If the residuals are independent, the pattern should appear random and similar to the null plot of residuals. Violations will be identified by a consistent pattern in the residuals. Figure 3.4e displays a residual plot that exhibits an association between the residuals and time, a common sequencing variable. Another frequent pattern is shown in Figure 3.4f. This pattern occurs when basic model conditions change but are not included in the model. For example, swimsuit sales are measured monthly for 12 months, with two winter seasons versus a single summer season, yet no seasonal indicator is estimated. The residual pattern will show negative residuals for the winter months versus positive residuals for the summer months. Data transformations, such as first differences in a time series model, inclusion of indicator variables, or specially formulated regression models can address this violation if it occurs.

Normality of the Error Term Distribution

Perhaps the most frequently encountered assumption violation is nonnormality of the independent or dependent variables, or both [15]. The simplest diagnostic for the set of predictor variables in the equation is a histogram of residuals, with a visual check for a distribution approximating the normal distribution (see Figure 3.4g). Although attractive because of its simplicity, this method is particularly difficult in smaller samples, where the distribution is ill-formed. A better method is the use of **normal probability plots.** They differ from residual plots in that the standardized residuals are compared with the normal distribution. The normal distribution makes a straight diagonal line, and the plotted residuals are compared with the diagonal. If a distribution is normal, the residual line closely follows the diagonal. The same procedure can compare the dependent or independent variables separately to the normal distribution [7]. Refer to Chapter 2 for a more detailed discussion of the interpretation of normal probability plots.

Summary

Analysis of residuals, whether with the residual plots or statistical tests provides a simple yet powerful set of analytical tools for examining the appropriateness of our regression model. Too often, however, these analyses are not made, and the violations of assumptions are left intact. Thus users of the results are unaware of the potential inaccuracies that may be present. These range from inappropriate tests of the significance of coefficients (either showing significance when it is not present or vice versa) to the biased and inaccurate predictions of the dependent variable. We strongly recommend that these methods be applied for each set of data and regression model. Application of the remedies, particularly transformations of the data, will increase confidence in the interpretations and predictions from multiple regression.

Stage Four: Estimating the Regression Model and Assessing Overall Fit

Having specified the objectives of the regression analysis, selected the independent and dependent variables, addressed the issues of research design, and assessed the variables for meeting the assumptions of regression, the analyst is now ready to estimate the regression model and assess the overall predictive accuracy of the independent variables. In this stage, the analyst must accomplish three basic tasks: (1) select a method for specifying the regression model to be estimated, (2) assess the statistical significance of the overall model in predicting the dependent variable, and (3) determine whether any of the observations exert an undue influence on the results. Each of these tasks is addressed in the following sections.

General Approaches to Variable Selection

In most instances of multiple regression, the researcher has a number of possible independent variables from which to choose for inclusion in the regression equation. Sometimes the set of independent variables may be closely specified and the regression model is essentially used in a confirmatory approach. In other instances, the analyst may wish to pick and choose among the set of independent variables. We have discussed the impact of collinearity on the "value" of additional predictors. There are several approaches (sequential search methods and combinatorial processes) to assist the researcher in finding the "best" regression model. Each of these models for specifying the regression model is discussed next.

Confirmatory Specification

The simplest yet perhaps most demanding approach for specifying the regression model is to employ a confirmatory perspective wherein the analyst completely specifies the set of independent variables to be included. As compared with the approaches to be discussed next, the analyst has total control over the variable selection. Although confirmatory specification is simple in concept, the analyst must be assured that the set of variables achieves the maximum prediction while maintaining a parsimonious model. Guidelines for model development are discussed in Chapters 1 and 11.

Sequential Search Approaches

Sequential search methods have in common the general approach of estimating the regression equation with a set of variables and then selectively adding or deleting variables until some overall criterion measure is achieved. This approach provides an objective method for selecting variables that maximizes the prediction with the smallest number of variables employed. While this approach to variable selection seems ideal, the analyst must be aware of certain caveats. First, the multicollinearity among independent variables can have substantial impact on the final model specification. Let's examine this situation with two highly correlated independent variables that have almost equal correlations with the dependent variable. Because these variables are highly correlated, there is little unique variance for each variable separately. If one of these variables enters the regression model, it is highly unlikely that the other variable will also enter. The criterion for inclusion or deletion in these approaches is to maximize the incremental predictive power of the additional variable. For this reason, the analyst must assess the effects of multicollinearity in model interpretation and examine the direct correlations of all potential independent variables to avoid concluding that the independent variables that do not enter the model are inconsequential.

Thus, although the sequential search approaches will maximize the predictive ability of the regression model, the analyst must be quite careful in model interpretation.

Backward Elimination   **Backward elimination** is largely a trial-and-error procedure for finding the best regression estimates. It involves computing a regression equation with all the predictor variables, then going back and deleting independent variables that do not contribute significantly. The steps are as follows:

1. Compute a single regression equation using all the predictor variables that interest you.
2. Calculate a **partial F value** for each variable that tests its unique variance explained after the variance accounted for by all other predictor variables is removed.
3. Eliminate predictor variables with partial F tests that indicate they are not making a statistically significant contribution.
4. After eliminating variables, reestimate the regression model using only the remaining predictor variables.
5. Return to step 2 and continue the process until you identify all variables of interest and determine their contributions. This process is time-consuming, but with adequate computer facilities it is a satisfactory process for many researchers.

Stepwise Estimation   **Stepwise estimation** is perhaps the most popular sequential approach to variable selection. This approach allows you to examine the contribution of each predictor variable to the regression model. However, rather than deleting variables, as in the backward elimination procedure, each variable is considered for inclusion prior to developing the equation. The primary distinction of this approach is the ability to add or delete variables at each stage. The stepwise procedure is illustrated in Figure 3.5. The specific issues at each stage are as follows:

1. Start with the simple regression model in which only the one predictor that is the most highly correlated with the criterion variable is used. The equation would be $Y = b_0 + b_1X_1$.
2. Examine the **partial correlation coefficients** to find an additional predictor variable that explains both a significant portion and the largest portion of the error remaining from the first regression equation.
3. Recompute the regression equation using the two predictor variables, and examine the partial F value for the original variable in the model to see whether it still makes a significant contribution, given the presence of the new predictor variable. If it does not, eliminate the variable. This ability to eliminate variables already in the model distinguishes the stepwise model from simple forward addition models. If the original variable still makes a significant contribution, the equation would be $Y = b_0 + b_1X_1 + b_2X_2$.
4. Continue this procedure by examining all predictors not in the model to determine whether one should be included in the equation. If a new predictor is included, examine all predictors previously in the model to judge whether they should be kept. A potential bias in the stepwise procedure results from considering only one variable for selection at a time. Suppose variables X_3 and X_4 together would explain a significant portion of the variance (each given the

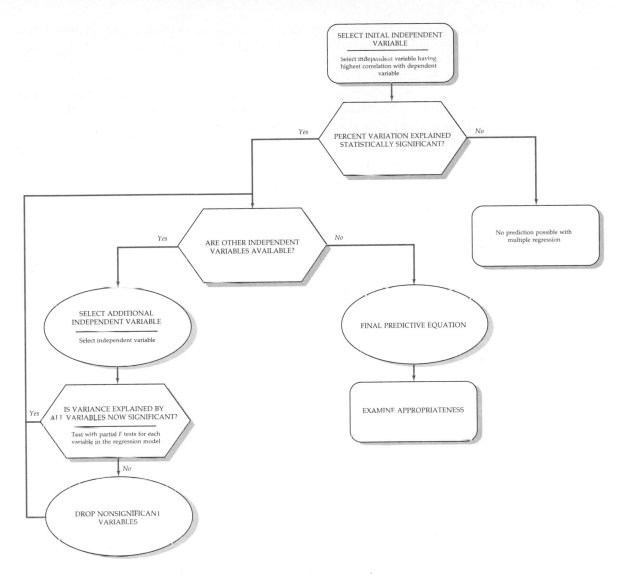

FIGURE 3.5 Flowchart of the stepwise estimation procedure.

presence of the other), but neither is significant by itself. In this situation, neither would be considered for the final model.

A second caveat pertains primarily to the stepwise procedure. In this approach, multiple significance tests are preformed in the model estimation process. To ensure that the overall error rate across all significance tests is reasonable, the analyst should employ quite conservative thresholds (e.g., .01) in adding or deleting variables.

Combinatorial Approach

The combinatorial approach is primarily a generalized search process across all possible combinations of independent variables. The best-known procedure is **all-possible-subsets regression,** which is exactly as the name suggests. All possible combinations of the independent variables are examined and the best-fitting set of variables are identified. For example, for a model with ten independent variables, there exist 1,024 possible regressions (one equation with only the constant, 10 equations with a single independent variable, 45 equations with all combinations of two variables, and so on). With computerized estimation procedures, this process can be managed today for even rather large problems, identifying the best overall regression equation for any number of measures of predictive fit. The researcher must remember that issues like multicollinearity, the identification of outliers and influentials, and the interpretability of the results are not addressed in selecting the final model. When these issues are considered, the "best" equation may have serious problems that affect its appropriateness, and another model may ultimately be selected.

Overview of the Model Selection Approaches

Whether a confirmatory, sequential search, or combinatorial method is chosen, the most important criterion is the researcher's substantive knowledge of the situation that determines the variables to be included as well as the expected signs and magnitude of their coefficients. Without this knowledge, the regression results can have high predictive accuracy without any managerial or theoretical relevance. One should never be totally guided by these methods but instead should use them after careful consideration of the alternative approaches, and then accept the results only after careful scrutiny.

Testing the Regression Variate for Meeting the Regression Assumptions

With the independent variables selected and the regression coefficients estimated, the analyst must now assess the estimated model for meeting the assumptions underlying multiple regression. As discussed in stage three, the individual variables must meet the assumptions of linearity, constant variance, independence, and normality. In addition to individual variables, the regression variate must also meet these assumptions. The diagnostic tests discussed in stage three can be applied to assessing the collective effect of the variate through examination of the residuals. If substantial violations are found, the analyst must take corrective actions and then reestimate the regression model.

Examining the Statistical Significance of Our Model

If we were to take repeated samples of eight families and ask them how many family members and credit cards they have, we would seldom get exactly the same values for $Y = b_0 + b_1 X_1$ from all the samples. We would expect chance variation to cause differences among many samples. Usually we take only one sample and base our predictive model on it. We can test certain hypotheses concerning our predictive model to ensure that it represents the population of all families having credit cards rather than just our one sample of eight people. These tests may take one or two basic forms: a test of the variation explained (coefficient of determination) and a test of coefficients.

Significance of the Overall Model:
The Coefficient of Determination

To test the hypothesis that the amount of variation explained by the regression model is more than the variation explained by the average (i.e., that R^2 is greater than zero), the F ratio is used. The test statistic F is defined as

$$F \text{ ratio} = \frac{\dfrac{\text{Sum of squared error}_{\text{regression}}}{\text{Degrees of freedom}_{\text{regression}}}}{\dfrac{\text{Sum of squared error}_{\text{Total}}}{\text{Degrees of freedom}_{\text{residual}}}}$$

where

$\text{Degrees of freedom}_{\text{regression}}$ = Number of estimated coefficients (including the constant) $- 1$

$\text{Degrees of freedom}_{\text{residual}}$ = Sample size $-$ the number of estimated coefficients (including the constant)

Two important features of this ratio should be noted:

1. Each sum of squares divided by its appropriate **degrees of freedom** is simply the variance of the prediction errors.
2. Intuitively, one knows that if the ratio of the explained variance to the variance about the mean is high, the use of family size must be of significant value in explaining the number of credit cards held by families.

For our example, the F ratio for the simple regression model discussed earlier in the chapter is $(16.5 \div 1)/(5.50 \div 6) = 18.0$. When compared with the tabled F statistic of 1 and six degrees of freedom of 5.99 (which would occur with a probability of .95), it leads us to reject the hypothesis that the reduction in error we obtained by using family size to predict credit card holdings was a chance occurrence. This outcome means that, considering the sample used for estimation, we can explain 18 times more variation than when using the average, and that this is not very likely to happen by chance (less than 5 percent of the time). Likewise, the F ratio for the multiple regression model with two independent variables is $(18.96 \div 2)/(3.04 \div 5) = 15.59$. The multiple regression model is also statistically significant, indicating that the additional independent variable was substantial in adding to the regression model's predictive ability.

We also know that R^2 is influenced by the number of predictor variables relative to the sample size. Several rules of thumb have been proposed, ranging from 10 to 15 observations per predictor to an absolute minimum of 4 observations per predictor. As we approach or fall below these limits, we need to adjust for the inflation in R^2 from "overfitting" the data. As part of all regression programs, an **adjusted coefficient of determination (adjusted R^2)** is given along with the coefficient of determination. Interpreted the same as the unadjusted coefficient of determination, the adjusted R^2 becomes smaller as we have fewer observations per predictor variable. The adjusted R^2 value is particularly useful in comparing across regression equations involving different numbers of predictors or different sample sizes, because it makes allowances for the specific number of predictors and the sample size upon which each model is based.

Significance Tests of Regression Coefficients

In the simple regression model, we said that the number of credit cards equaled 2.87 + .971 (family size). We would test two hypotheses for this regression model:

Hypothesis 1. The intercept (constant term) value of 2.87 is due to sampling error, and the real constant term appropriate to the population is zero.

With this hypothesis, we would simply be testing whether the constant term should be considered appropriate for our predictive model. If it is found not to differ significantly from zero, we would assume that the constant term should not be used for predictive purposes. The appropriate test is the *t* test, which is commonly available on computerized regression analysis programs. From a practical point of view, this test is seldom necessary. If the data used to develop the model did not include some observations with all the predictors measured at zero, the constant term is "outside" the data and acts only to position the model. It is then not necessary to test the constant term.

Hypothesis 2. The coefficient .971 indicates that an increase of one unit in family size is associated with an increase in the average number of credit cards held by .971 and that this coefficient also differs significantly from zero.

If it occurred because of sampling error, we would conclude that family size has no impact on the number of credit cards held. Note that this is not a test of any exact value of the coefficient but rather of whether it should be used at all. Again, the appropriate test is the *t* test. The expected variation of the estimated coefficients (both the constant and the regression coefficients) is termed the *sampling error* of the coefficients. The analyst should remember that the statistical test of the regression coefficient is to ensure that across all the possible samples that could be drawn, the regression coefficient should be different than zero. One important consideration is the sample size used to estimate the regression model. The larger the sample size, the more generalizable (and less variable) the estimated coefficient. To illustrate the possible variation in estimated coefficients, a prior research project for HATCO that had over 1,000 respondents was used to draw 20 random samples for sample sizes of 10, 25, 50, and 100 respondents. Exhibit 3.2 contains the estimated regression coefficient for a simple regression model for the 20 samples at each sample size. As expected, the estimated coefficients for the smaller sample sizes are much more variable, ranging from 2.19 to 6.06 for sample sizes of 10 respondents. The coefficients become more consistent as the sample sizes increase. But even for the largest sample size, the range of coefficients was from 2.79 to 4.90. Thus, statistical tests for the estimated coefficients is essential to ensure the validity of the estimated model.

Identifying Influential Observations

Up to now, we have focused on identifying general patterns within the entire set of observations. Here we shift our attention to individual observations, with the objective of finding the observations that lie outside the general patterns of the data set or that strongly influence the regression results. We should remember that these observations are not necessarily bad in the sense that they must be

EXHIBIT 3.2

How the Variation in Estimated Regression Coefficients Due to Sampling Variability Provides the Basis for Statistical Significance Testing

Statistical significance testing for the estimated coefficients in regression analysis is appropriate and necessary when the analysis is based on a sample of the population rather than a census. When using a sample for estimating the regression model, the analyst is not interested in the regression estimate for just that sample but is really interested in how generalizable the results are to the population. For each sample drawn from a population, a different value for the coefficient will be obtained. For small sample sizes, the estimated coefficients will most likely vary widely from sample to sample. But as the size of the sample increases, the samples become more representative of the population and the variation in the estimated coefficients for these large samples will be expected to become smaller. This is true until the analysis is estimated using the population. Then, there is no need for significance testing because the "sample" is equal to and thus perfectly representative of the population.

Significance testing of regression coefficients provides a statistically based probability estimate of whether the estimated coefficients across a large number of samples of a certain size will indeed be different than zero. If the sample size is small, the variation may be too great to say with a needed degree of certainty (what we refer to as the significance level) that the coefficient is not equal to zero. However, if the sample size is larger, the test has greater precision because the variation in the coefficients becomes less. Larger samples do not guarantee that the coefficients will not equal zero, but instead make the test more precise.

To illustrate this point, 20 random samples for each of four sample sizes (10, 25, 50, and 100 respondents) were drawn from a large database. A simple regression was performed for each sample and the estimated regression coefficient recorded in the accompanying table below.

	Estimated Coefficient for Samples of Size:			
Sample	*10*	*25*	*50*	*100*
1	2.5820	2.5671	2.9700	3.5968
2	2.4519	2.8134	2.9081	3.6955
3	2.1965	3.7337	3.5762	3.8800
4	6.0634	5.6444	4.9984	4.2007
5	2.5866	4.0091	4.0800	3.1592
6	5.0585	3.0825	3.8894	3.6815
7	4.6778	2.6610	3.0735	2.7984
8	5.5996	4.1157	3.6539	4.5845
9	3.9122	4.0546	4.6204	3.3416
10	3.0413	3.0368	3.6805	3.3184
11	3.7412	3.4548	4.0412	3.4773
12	5.1964	4.1892	4.4263	3.2336
13	5.8241	4.6824	5.1972	3.6833
14	2.2270	3.7651	3.9938	4.3018
15	5.1717	4.8842	4.7634	4.9013
16	3.6944	3.0855	4.0186	3.7453
17	3.1682	3.1372	2.9052	3.1680
18	2.6261	3.5546	3.7211	3.4412
19	3.4851	5.0245	5.8530	4.3093
20	4.5711	3.6140	5.1233	4.2122
Minimum	2.1965	2.5171	2.9052	2.7984
Maximum	6.0634	5.6444	5.8530	4.9013
Range	3.8669	3.1273	2.9478	2.1029
Standard Deviation	1.28	.85	.83	.54

As we can see, the variation in the estimated coefficients is greatest for samples of 10 respondents, varying from a low coefficient of 2.1965 to a high of 6.0634. As the sample sizes increase to 25 and 50 respondents, the variation in coefficients lowers considerably. Finally, the samples of 100 respondents have a range of almost one-half that for the samples of 10 respondents (2.10 versus 3.86). From this we can see that the ability of the statistical test to determine whether the coefficient is actually greater than zero is made more precise with the larger sample sizes.

One final note concerning the variation in regression coefficients. Many times analysts forget that the estimated coefficients in their regression analysis are specific to the sample used in estimation. They are the best estimates for that sample of observations, but as the results above point out, the coefficients can vary quite markedly from sample to sample. This points to the need for concerted efforts to validate any regression analysis on a different sample(s). While in doing so the analyst must expect the coefficients to vary, the attempt is to demonstrate that the relationship generally holds in other samples so that the results can be assumed to be generalizable to any sample drawn from the population.

---

omitted. In many instances they represent the distinctive elements of the data set. However, we must first identify them and assess their impact before we can proceed. The following section introduces the concept of influential observations and their potential impact on the regression results, while Appendix 3A contains a more detail discussion of the procedures for identifying influential observations.

Influential observations contain three basic types: outliers, leverage points, and influentials. **Outliers** are observations that have large residual values and can be identified only with respect to a specific regression model. Outliers have traditionally been the only form of influential observation considered in regression models, and specialized regression methods (robust regression) have even been developed to deal specifically with outliers' impact on the regression results [1, 14]. Chapter 2 provides additional procedures for identifying outliers. **Leverage points** are observations that are distinctive from the remaining observations based on their independent variable values. Their impact is particularly noticeable in the estimated coefficients for one or more predictor variables. Finally, **influential observations** is the broadest category, including all observations that have a disproportionate effect on the regression results. Influential observations potentially include outliers and leverage points but may include other observations as well. Also, not all outliers and leverage points are necessarily influential observations.

Influential observations can exhibit many patterns. Figure 3.6 illustrates several forms of influential observations and their correspondence to residuals. In each instance, the residual for the influential points (the perpendicular distance from the point of the estimated regression line) would not be expected to be so large as to be classified as an outlier. Thus focusing only on large residuals would generally ignore these additional influential observations. In Figure 3.6a, the influential point is a "good" one, reinforcing the general pattern of the data and lowering the standard error of the prediction and coefficients. It is a leverage point but has a small or zero residual value, as it is predicted well by the regression model. However, influential points can also have an effect that is contrary to the general pattern of the remaining data but still have small residuals (see Figure 3.6b and 3.6c). In Figure 3.6b, two influential observations almost totally account for the observed relationship, because without them no real pattern emerges from the

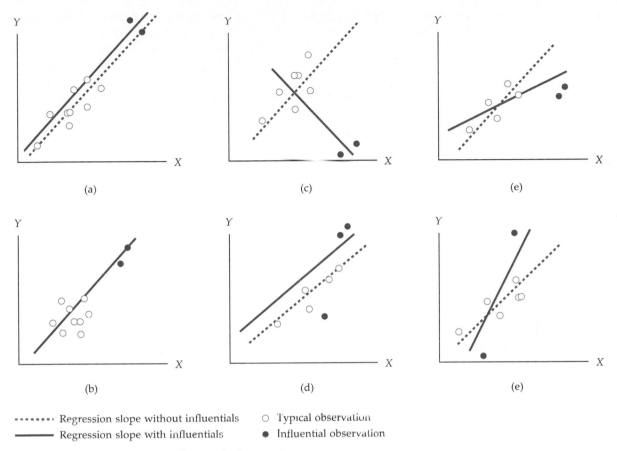

········ Regression slope without influentials   ○ Typical observation
———— Regression slope with influentials   ● Influential observation

FIGURE 3.6 Patterns of influential observations.

Adapted from [2, 9].

other data points. They also would not be identified if only large residuals were considered, because their residual value would be small. In Figure 3.6c, an even more profound effect is seen where the influential observations counteract the general pattern of all the remaining data. In this case, the "real" data would have larger residuals than the "bad" influential points. The influential observations may affect only a portion of the results, as in Figure 3.6d, where the slope remains constant but the intercept is shifted. Finally, multiple influential points may work toward the same result. In Figure 3.6e, two influential points have the same relative position, making the detection somewhat harder. And in Figure 3.6f, influentials have quite different positions but a similar effect on the results. These examples illustrate that we must develop a bigger tool kit of methods for identifying these influential cases.

Procedures for identifying all types of influential observations are quite numerous and still less well defined than many other aspects of regression analysis. All computer programs provide an analysis of residuals from which those with large values (particularly standardized residuals greater than 2.0) can be easily identified. Moreover, most computer programs now provide at least some of the diagnostic measures for identifying leverage points and other influential observations.

The need for additional study of leverage points and influentials is highlighted when we see the substantial extent to which the generalizability of the results and the substantive conclusions (the importance of variables, level of fit, etc.) can be changed by only a small number of observations. Whether "good" (accentuating the results) or "bad" (substantially changing the results), these observations must be identified to assess their impact. Influentials, outliers, and leverage points are based on one of four conditions:

1. An error in observations or data entry
2. A valid but exceptional observation that is explainable by an extraordinary situation
3. An exceptional observation with no likely explanation
4. An ordinary observation in its individual characteristics but exceptional in its combination of characteristics

Courses of action can be recommended for dealing with influentials from each condition. For an error in observation, correct the data or delete the case. With the valid but exceptional observation (condition 2), deletion of the case is warranted unless variables reflecting the extraordinary situation are included in the regression equation. The unexplained observation (condition 3) presents a special problem because there is no reason for deleting the case, but its inclusion cannot be justified either. Finally, the observation that is ordinary on each variable separately yet exceptional in its combination of characteristics (condition 4) indicates modifications to the conceptual basis of the regression model and should be retained.

In all situations, the analyst is encouraged to delete truly exceptional observations but still guard against deleting observations that, while different, are representative of the population. Remember that the objective is to ensure the most representative model for the sample data so that it will best reflect the population from which it was drawn. This extends beyond achieving the highest predictive fit, because some outliers may be valid cases that the model should attempt to predict, even if poorly. The analyst should also be aware of instances where the results would be changed substantially by deleting just a single observation or a very small number of observations.

Stage Five: Interpreting the Regression Variate

The analyst's next task is to interpret the regression variate by evaluating the estimated regression coefficients for their explanation of the dependent variable. As we will see in our discussion, the analyst must evaluate not only the regression model that was estimated but also the potential independent variables that were omitted if a sequential search or combinatorial approach was employed. In those approaches, multicollinearity may substantially affect the variables ultimately included in the regression variate. Thus, in addition to assessing the estimated coefficients, the analyst must also evaluate the potential impact of omitted variables to ensure that the managerial significance is evaluated along with statistical significance.

Using the Regression Coefficients

The estimated regression coefficients are used to calculate the predicted values for each observation and to express the expected change in the dependent variable for each unit change in the independent variables. In addition to making the prediction, we would also like to know which independent variable is the most helpful in predicting the dependent variable. In the multiple regression example discussed earlier, we would like to know which variable—family size or family income—is more helpful in predicting the number of credit cards held by a family. Unfortunately, the regression coefficients (b_0, b_1, and b_2) do not give us this information. To illustrate why, we can use a rather obvious case. Suppose we wanted to predict teenagers' monthly expenditures on CDs (Y), using two predictor variables; X_1 is parents' income in thousands of dollars and X_2 is the teenager's monthly allowance measured in dollars. We found the following model by a least squares procedure:

$$Y = -.01 + X_1 + .001X_2$$

You might assume that X_1 is more important because its coefficient is 1,000 times larger than the coefficient for X_2. This assumption is not true, however. A $10 increase in the parents' income produces a $1 \times \$10 \div \$1,000$ change in average CD purchases (we divide $10 by 1000 because the X_1 value is measured in thousands of dollars). This change is .01 in the average number of CDs. A change of $10 in the teenager's monthly allowance produces a (.001)($10) change in average CD expenditures or a .01 change in the average number of CDs (because the teenager's allowance was measured in dollars).

A $10 change in the parents' income produced the same effect as a $10 change in the teenager's allowance. Both variables are equally important, but the regression coefficients do not directly reveal this fact. We can resolve this problem by using a modified regression coefficient called the *beta coefficient*.

Standardizing the Regression Coefficients: Beta Coefficients

If each of our predictor variables had been **standardized** before we estimated the regression equation, we would have found different regression coefficients. The coefficients resulting from standardized data are called **beta coefficients.** Their value is that they eliminate the problem of dealing with different units of measurement (as illustrated previously), and they reflect the relative impact on the criterion variable of a change in one standard deviation in either variable. Now that we have a common unit of measurement, we can determine which variable is the most influential.

Three cautions must be observed when using beta coefficients. First, they should be used as a guide to the relative importance of individual independent variables only when collinearity is minimal. Second, the beta values can be interpreted only in the context of the other variables in the equation. For example, a beta value for family size reflects its importance only in relation to family income, not in any absolute sense. If another predictor variable was added to the equation, the beta coefficient for family size would probably change, because there would likely be some relationship between family size and the new predictor variable.

The third caution is that the levels (e.g., families of size 5, 6, and 7) affect the beta value. Had we found families of size 8, 9, and 10, the value of beta would likely change. In summary, use beta only as a guide to the relative importance of the predictor variables included in your equation, and only over the range of values for which you actually have sample data.

Assessing Multicollinearity

A key issue in interpreting the regression variate is the correlation among the predictor variables. This is a data problem, not a problem of model specification. But it has substantial effects on the results of the regression procedure. First, it limits the size of the coefficient of determination and makes it increasingly more difficult to add unique explanatory prediction from additional variables. Second, and just as important, it makes determining the contribution of each independent variable difficult because the effects of the independent variables are "mixed" or confounded, owing to collinearity. As shown earlier, high multicollinearity results in larger portions of shared variance and lower levels of unique variance from which the effects of the individual predictor variables can be determined. Thus regression coefficients may be incorrectly estimated and have the wrong signs. The following example illustrates this point:

Variables in the Regression Analysis

	Dependent	Independent	
Respondent	D	A	B
1	5	6	13
2	3	8	13
3	9	8	11
4	9	10	11
5	13	10	9
6	11	12	9
7	17	12	7
8	15	14	7

The two regressions using A and B separately are

$$D = -5 + 1.5(A)$$
$$D = 30 - 2.0(B)$$

It is clear that the relationship between A and D is positive, while the relationship between B and D is negative. The multiple regression equation is

$$D = 50 - 1.0(A) - 3.0(B)$$

It would now appear to the casual observer that the relationship between A and D is negative when in fact we know it is not. The sign of A is wrong in an intuitive sense but reflects the strong negative correlation between A and B.

We have seen that the effects of multicollinearity can be substantial. In any regression analysis, the assessment of multicollinearity should be undertaken in two steps: (1) identification of the extent of collinearity and (2) assessment of the degree to which the estimated coefficients are affected. If corrective action is dictated, several options exist. We first discuss the identification and assessment procedures, and then examine some possible remedies.

The simplest and most obvious means of identifying collinearity is an examination of the correlation matrix for the independent variables. The presence of high correlations (generally those of .90 and above) is the first indication of substantial collinearity. Lack of any high correlation values does not ensure a lack of collinearity. Collinearity may be due to the combined effect of two or more other independent variables.

Two of the more common measures for assessing both pairwise and multiple variable collinearity are (1) the **tolerance** value and (2) its inverse—the **variance inflation factor (VIF).** These measures tell us the degree to which each independent variable is explained by the other independent variables. In simple terms, each independent variable becomes a dependent variable and is regressed against the remaining independent variables. Tolerance is the amount of variability of the selected independent variable not explained by the other independent variables. Thus very small tolerance values (and large VIF values) denote high collinearity. A common cutoff threshold is a tolerance value of .10, which corresponds to VIF values above 10. Each analyst must determine the degree of collinearity he or she will accept, as most defaults or recommended thresholds still allow for substantial collinearity. For example, the suggested cutoff for the tolerance value of .10 corresponds to a multiple correlation of .95. Moreover, a multiple correlation of .9 between one independent variable and all others (similar to the rule we applied in the pairwise correlation matrix) would result in a tolerance value of .19. Thus any variables with tolerance values below .19 (or above a VIF of 5.3) would be correlated more than .90.

We strongly suggest that the analyst always specify the tolerance values in regression programs, as the default values for excluding collinear variables allow for an extremely high degree of collinearity. For example, the default tolerance value in SPSS for excluding a variable is .0001, which means that until more than 99.99 percent of variance is predicted by the other independent variables, the variable could be included in the regression equation. Estimates of the actual effects of high collinearity on the estimated coefficients are possible but beyond the scope of this text (see [13]).

Even with diagnoses using VIF or tolerance values, we still do not necessarily know which variables are intercorrelated. A procedure developed by Belsley et al. [2] allows for the intercorrelated variables to be identified, even if we have correlation among several variables. It provides the analyst greater diagnostic power in assessing the extent and impact of multicollinearity and is discussed in the appendix to this chapter.

Once the degree of collinearity has been determined, the researcher has a number of options:

- Omit one or more highly correlated predictor variables and identify other predictor variables to help the prediction.
- Use the model with the highly correlated predictors for prediction only (i.e., make no attempt to interpret the partial regression coefficients).
- Use the simple correlations between each predictor and the dependent variable to understand the predictor–dependent variable relationship.

• Use a more sophisticated method of analysis such as Bayesian regression (or a special case—ridge regression) or regression on principal components to obtain a model that more clearly reflects the simple effects of the predictors. These procedures are discussed in more detail in several texts [2, 13].

Stage Six: Validation of the Results

After identifying our best regression model, our final step is to ensure that it represents the general population (generalizability) and is appropriate for the situations in which it will be used (transferability). Our best guideline is how well it matches an existing theoretical model or set of previously validated results on the same topic. In many instances, however, prior results or theory are not available. Thus we will also discuss empirical approaches to model validation.

Additional or Split Samples

The most appropriate empirical validation approach is to test the regression model on a new sample drawn from the general population. A new sample will ensure representativeness and can be used in several ways. First, the original model can predict values in the new sample, and predictive fit can be calculated. Second, a separate model can be estimated with the new sample and then compared with the original equation on such characteristics as the significant variables included; sign, size, and relative importance of variables; and predictive accuracy. In both instances, the researcher determines the validity of the original model by comparing it to regression models estimated with the new sample.

Many times the ability to collect new data is limited or precluded by such factors as cost, time pressures, or availability of respondents. When this is the case, the researcher may then divide the sample into two parts: an estimation subsample for creating the regression model and the holdout/validation subsample used to "test" the equation. Many procedures, both random and systematic, are available for splitting the data, each drawing two independent samples from the single data set. All the popular statistical packages have specific options to allow for estimation and validation on separate subsamples. See Chapter 4 for a discussion of the use of estimation and validation subsamples.

Whether a new sample is drawn or not, it is likely that differences will occur between the original model and other validation efforts. The researcher's role now shifts to being a mediator among the varying results, looking for the best model across all samples. The need for continued validation efforts and model refinements reminds us that no regression model, unless estimated from the entire population, is the final and absolute model.

Calculating the PRESS Statistic

An alternative approach to obtaining additional samples for validation purposes is to employ the original sample in a specialized manner by calculating the **PRESS statistic**, a measure similar to R^2 used to assess the predictive accuracy of the estimated regression model. It differs from the prior approaches in that not one, but $n - 1$ regression models, are estimated. The procedure, similar to bootstrapping techniques, omits one observation in the estimation of the regression model and then predicts the omitted observation with the estimated model. In this way, the observation cannot affect the coefficients of the model used to calculate its predicted value. The procedure is applied again, omitting another observation, estimating a new model, and making the prediction. The residuals for the observations can then be summed to provide an overall measure of predictive fit.

Comparing Regression Models

When comparing regression models, the most common standard used is overall predictive fit. We discussed earlier that the coefficient of determination (R^2) provides us with this information, but it has one drawback: As more variables are added, it can never decrease. Thus, by including all independent variables, we will never find a higher R^2, although we may have achieved the same R^2 with a smaller number of variables or find that a smaller number of predictors results in an almost identical value. To compare between models with different numbers of predictors, we use the adjusted R^2, which is also useful in comparing models between different data sets, as it will compensate for the different sample sizes.

Predicting with the Model

Model predictions can always be made by applying the estimated model to a new set of independent variable values and calculating the criterion variable values. However, in doing so, we must consider several factors that can have a serious impact on the quality of the new predictions:

1. When applying the model to a new sample, remember that the predictions now have not only the sampling variations from the original sample but also those of the newly drawn sample as well. Thus we should always calculate the confidence intervals of our predictions in addition to the point estimate to see the expected range of criterion values.
2. Make sure that the conditions and relationships measured at the time the original sample was taken have not changed materially. For instance, in our credit card example, if most companies started charging higher fees for their cards, actual credit card holdings might change substantially, yet this information would not be included in the model.
3. Finally, do not use the model to estimate beyond the range of independent variables found in the sample. For instance, if the largest family had six members, it might be unwise to predict credit card holdings for families with ten members. One cannot assume that the relationships are the same for values of the independent variables substantially greater or less than those in the original estimation sample.

Regression with a Binary Dependent Variable

In the discussions of multiple regression, the one element that never changed was the metric nature of the dependent variable. When multivariate techniques were classified in Chapter 1, the single metric dependent variable distinguished regression from other dependence methods. However, an alternative form of regression, the linear probability model, now relaxes this assumption and provides a wide range of diagnostic and explanatory techniques for nonmetric dependent variables, especially binary measures. In this section, we discuss one of the most widely used linear probability models, **logistic regression,** or logit analysis, highlighting both its similarities and differences from traditional regression analysis.

As we will see in Chapter 4, discriminant analysis is also appropriate when the dependent variable is nonmetric. However, logit analysis may be preferred for several reasons. First, discriminant analysis relies on strictly meeting the assumptions of multivariate normality and equal variance-covariance matrices across groups, features not found in all situations. Logit analysis does not face these strict assumptions, thus making its application appropriate in many more situations. Second, even if the assumptions are met, many researchers prefer logit analysis because it is similar to regression with its straightforward statistical tests, ability to incorporate nonlinear effects, and wide range of diagnostics. For these and more technical reasons, logit analysis is equivalent to discriminant analysis and may be more appropriate in certain situations.

Unique Characteristics of Logit Analysis

Logit analysis presents a unique complement to multiple regression in its ability to utilize a binary dependent variable. In many respects, logit analysis or logistic regression is very similar to multiple regression. But the nonmetric nature of the dependent variable requires us to rethink our notions of multiple regression in several regards. The following sections detail areas that require attention when applying logit analysis.

Use of a Binary Dependent Variable

The primary difference from the user's viewpoint between logit analysis and multiple regression is the use of a dichotomous dependent variable. Many phenomena are dichotomous in nature, such as whether a person is in good versus poor health or an acceptable credit risk or not, whether a firm has failed or not, or whether the product was purchased or not. But such situations cannot be studied with ordinary regression, because doing so would violate several assumptions. Most critical is that the error term of a discrete variable follows the binomial distribution instead of the normal distribution, thus invalidating all statistical testing performed in regression.

To understand the effects of the independent variables more fully, logit analysis does not predict just whether an event occurred or not (one or zero), but instead predicts the probability of an event. In this manner, the dependent variable can be any value between zero and one. This also means that the predicted value must be bounded to fall within the range of zero and one. This is one aspect of logit analysis that makes regression analysis invalid in these situations. To define a relationship bounded by zero and one, logit analysis uses an assumed relationship between the independent and dependent variable that resembles an S-shaped curve (see Figure 3.7). At very low levels of the independent variable, the probability approaches zero. As the independent variable increases, the probability increases up the curve. But then the slope starts decreasing so that at any level of the independent variable, the probability will approach one but never exceed it. As we can see, the traditional linear models of regression cannot accommodate

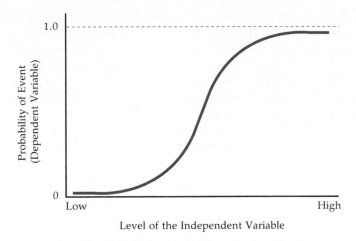

FIGURE 3.7 Form of the logistic relationship between dependent and independent variables.

such a relationship. Thus, logit analysis was developed. But this unique feature of the assumed relationship between dependent and independent variables requires a somewhat different approach in estimating the model and interpreting the coefficients.

Interpreting the Coefficients

One of the advantages of logit analysis is that we need only know whether an event (purchase or not, credit risk or not, firm failure or success) occurred to then use a dichotomous value as our dependent variable. From this dichotomous value, the procedure predicts its estimate of the probability that the event will or will not occur. If the predicted probability is greater than .50, then the prediction is yes, otherwise no. Logit analysis derives its name from the logit transformation used with the dependent variable. When this transformation is used, however, the logit coefficients take on a somewhat different meaning from that found in regression with a metric dependent variable.

The procedure that calculates the logit coefficient, similar to the least squares criterion discussed earlier, compares the probability of an event occurring with the probability of its not occurring. This *odds ratio* can be expressed as

$$\frac{\text{Prob (event)}}{\text{Prob (no event)}} = e^{B_0 + B_1 X_1 + \cdots + B_n X_n}$$

The estimated coefficients (B_0, B_1, B_2, . . . , B_n) thus actually are measures of the changes in the ratio of the probabilities, termed the odds ratio. Moreover, they are expressed in logarithms, such that they need to be transformed back (the antilog of the value has to be taken) so that their relative effect on the probabilities is assessed more easily. (The computer programs perform this procedure automatically and give both the actual coefficient and the transformed coefficient.) Use of this procedure *does not change* in any manner the way we interpret the sign of the coefficient. A positive coefficient increases the probability, while a negative value decreases the predicted probability.

Let's look at a simple example to see what we mean. If B_i is positive, its transformation (antilog) will be greater than 1, and the odds ratio will increase. This increase occurs when the predicted probability of the event's occurring increases and the predicted probability of its not occurring is reduced. Thus the model has a higher predicted probability of occurrence. Likewise, if B_i is negative, the antilog is less than one and the odds will be decreased. A coefficient of zero equates to a value of 1.0, resulting in no change in the odds.*

In our earlier discussion of the assumed distribution of possible dependent variables, we described an **S**-shaped or logistic curve. To represent that relationship between the dependent and independent variables, the coefficients must actually represent nonlinear relationships among the dependent and independent variables. While the transformation process of taking logarithms provides a linearization of the relationship, the analyst must remember that the coefficients actually represent different slopes in the relationship across the values of the independent variable. In this way, the **S**-shaped distribution can be estimated. If the analyst is interested in the slope of the relationship at various values of the independent variable, the coefficients can be calculated and the relationship assessed [5].

Assessing the Goodness-of-Fit of the Estimated Model

Logistic regression is similar to multiple regression in many of its results, but it is different in the method of estimating coefficients. Instead of minimizing the squared deviations (least squares), logit analysis maximizes the "likelihood" that an event will occur. Using this alternative estimation technique also requires that we assess model fit in different ways.

The overall measure of how well the model fits, similar to the residual or error sums of squares value for multiple regression, is given by the **likelihood value.** (It is actually -2 times the log of the likelihood value and is referred to as $-2LL$ or -2 log likelihood.) A well-fitting model will have a small value for $-2LL$. The minimum value for $-2LL$ is zero. (A perfect fit has a likelihood of 1, and $-2LL$ is then 0.) The likelihood value can be compared between equations as well, with the difference representing the change in predictive fit from one equation to another. The statistical programs have automatic tests for the significance of these differences.

The statistical test for the reduction in the log likelihood value is based on Chi-square and is provided in all statistical programs. A null model, which is similar to calculating the total sum of squares using only the mean, provides the baseline for comparison. In addition to the statistical Chi-square tests, the analyst can also construct a "pseudo R^2" value for logit analysis similar to the R^2 value in regression analysis [5]. The R^2 for a logit model (R^2_{logit}) can be calculated as

$$R^2_{\text{Logit}} = \frac{-2 \log L_{\text{null}} - (-2 \log L_{\text{model}})}{-2 \log L_{\text{null}}}$$

We can also assess how well the model predicts by creating a classification table where we compare the actual events (occurring or not) versus the predicted val-

---

*A more detailed discussion of interpretation of the coefficients, logistic transformation, and the estimation procedure can be found in numerous texts [8].

ues (occurrence or not). We can then easily see how many events were correctly predicted and where our mispredictions occurred. The reader may wish to refer to Chapter 4 for a more detailed discussion of evaluating classification tables for predictive accuracy.

Testing for Significance of the Coefficients

Logit analysis also can test the hypothesis that a coefficient is different from zero (remember, the odds ratio does not change, thus the probability is not affected) as is done in multiple regression, where we use the *t* value to assess the significance of each coefficient. Although the programs use a different statistic, the **Wald statistic,** they provide the statistical significance for each estimated coefficient so that hypothesis testing can occur just as it did in multiple regression.

Other Similarities

Despite the differences we have described, the format of logit analysis is much like that of multiple regression. Just as in regression, nominal and categorical data may be included through some form of dummy-variable coding. In addition to a similar method of testing the fit of the model and the significance of the coefficients, model selection procedures such as those found in multiple regression (forward and backward stepwise) are also available. Finally, to examine the results more clearly, many of the diagnostic measures, such as residuals, residual plots, and measures of influence, are also available.

The researcher faced with a dichotomous variable need not resort to methods designed to accommodate the limitations of multiple regression. Logit analysis addresses these problems and provides a method developed to deal directly with this situation in the most efficient manner possible.

Illustration of a Regression Analysis

The issues concerning the application and interpretation of regression analysis have been discussed in the preceding sections by following the six-stage model-building framework introduced in Chapter 1 and discussed earlier in this chapter. To provide an illustration of the important questions at each stage, an illustrative example is presented in the following sections detailing the application of multiple regression to a research problem specified by HATCO. Chapter 1 introduced a research setting in which HATCO had obtained a number of measures in a survey of customers. To demonstrate the use of multiple regression, we will show the procedures used by analysts to attempt to predict the product usage levels of the individuals in the sample with a set of seven independent variables.

Stage One: Objectives of the Multiple Regression

HATCO management has long been interested in more accurately predicting the level of business obtained from its customers in the attempt to provide a better basis for production controls and marketing efforts. To this end, analysts at HATCO proposed that a multiple regression analysis should be attempted to predict the product usage levels of the customers based on their perceptions of HATCO's performance. In addition to finding a way to accurately predict usage levels, the researchers were also interested in identifying the factors that led to increased product usage for application in differentiated marketing campaigns.

To apply the regression procedure, researchers selected product usage (X_9) as the dependent variable (Y) to be predicted by independent variables representing perceptions of HATCO's performance. The following seven variables were included as predictor variables:

X_1	Delivery speed
X_2	Price level
X_3	Price flexibility
X_4	Manufacturer's image
X_5	Service
X_6	Sales force image
X_7	Product quality

The relationship among the seven predictor variables and product usage was assumed to be statistical, not functional, because it involved perceptions of performance and may have had levels of measurement error.

Stage Two: Research Design of the Multiple Regression Analysis

The HATCO survey obtained 100 respondents from their customer base. All 100 respondents provided complete responses, resulting in 100 observations available for analysis. The first question to be answered concerning sample size is the level of relationship (R^2) that can be reliably detected with the proposed regression analysis. Table 3.4 indicates that the sample of 100, with seven potential independent variables, is able to detect relationships with R^2 values of approximately 30 percent at a power of .80 with the significance level set at .01. If the significance level is relaxed to .05, then the analysis will identify relationships explaining about 25 percent of the variance. The proposed regression analysis was deemed sufficient to identify not only statistically significant relationships but also relationships that had managerial significance.

The sample of 100 observations also meets the proposed guideline for the ratio of observations to independent variables with a ratio of 15 to 1. While the analysts can be assured that they will not be in danger of overfitting the sample, they must still validate the results to ensure the generalizability of the findings to the entire customer base.

Stage Three: Assumptions of the Multiple Regression Analysis

Meeting the assumptions of regression analysis is essential to ensure both that the results obtained were truly representative of the sample and that we have obtained the best results possible. Any serious violations of the assumptions must be detected and corrected if at all possible. Analysis to ensure that the research is meeting the basic assumptions of regression analysis involves two steps: testing the individual dependent and independent variables and testing the overall relationship after model estimation. This section addresses the assessment of individual variables, while the examination of the overall relationship will occur after the model has been estimated.

The three assumptions to be addressed for the individual variables are linearity, constant variance, and normality. For purposes of the regression analysis, we will summarize the results found in Chapter 2 detailing the examination of the dependent and independent variables. First, scatterplots of the individual variables did not indicate any nonlinear relationships between the dependent variable and the independent variables. Tests for heteroscedasticity found that only one of the variables (X_2) violated this assumption. Finally, in the tests of normality, three of the variables $(X_2, X_4,$ and $X_6)$ were found to violate the statistical tests. In each case, transformations by taking logarithms were indicated. The series of tests for these three assumptions underlying regression analysis indicated that the concerns should center on the normality of three independent variables. While regression analysis has been shown to be quite robust even when the normality assumption is violated, analysts should estimate the regression analysis with both the original and transformed variables to assess the consequences of nonnormality of the independent variables on the interpretation of the results.

Stage Four: Estimating the Regression Model and Assessing Overall Model Fit

With the regression analysis specified in terms of dependent and independent variables, the sample deemed adequate for the objectives of the study and the assumptions assessed for the individual variables, the process now proceeds to estimation of the regression model and assessing the overall model fit. For purposes of illustration, the stepwise procedure is employed to select variables for inclusion in the regression variate. After the regression model has been estimated, the variate will be assessed for meeting the assumptions of regression analysis. Finally, the observations will be examined to determine whether any observation should be deemed influential. Each of these issues will be discussed in the following sections.

Stepwise Estimation: Selecting the First Variable

Table 3.5 displays all the correlations among the seven independent variables and their correlations with the dependent variable (Y). Examination of the correlation matrix indicates that predictor 5 (X_5) is most closely correlated with the dependent variable (.70). Our first step is to build a regression equation using this best predictor. Note that the correlation of predictor 1 with the dependent variable is .68. However, X_1 is correlated (.61) with X_5. This is our first clue that use of both predictors $(X_1$ and $X_5)$ might not be appropriate because they are as highly correlated with each other as they are with the dependent variable. The results of this first step appear as shown in Table 3.6. The concepts from Table 3.6 with which you should be familiar follow.

Multiple R Multiple R is the correlation coefficient (at this step) for the simple regression of X_5 and the dependent variable. It has no plus or minus sign because in multiple regression the signs of the individual variables may vary, so this coefficient reflects only the degree of association.

TABLE 3.5 Correlation Matrix: HATCO Data

	Predictors						
Variables	X_1	X_2	X_3	X_4	X_5	X_6	X_7
Predictors							
X_1 Delivery speed	1.00						
X_2 Price level	−.35	1.00					
X_3 Price flexibility	.51	−.49	1.00				
X_4 Manufacturer's image	.05	.27	−.12	1.00			
X_5 Overall service	.61	.51	.07	.30	1.00		
X_6 Sales force's image	.08	.19	−.03	.79	.24	1.00	
X_7 Product quality	−.48	.47	−.45	.20	−.06	.18	1.00
Dependent							
$Y(X_9)$ Usage level	.68	.08	.56	.22	.70	.26	−.19

TABLE 3.6 Example Output: Step 1 of HATCO Multiple Regression Example

Variable entered: X_5 Overall service

Multiple R	.701
Multiple R^2	.491
Adjusted R^2	.486
Standard error of estimate	6.446

Analysis of Variance

	Sum of Squares	df	Mean Square	F Ratio
Regression	3,927.31	1	3,927.31	94.52
Residual	4,071.69	98	41.65	

	Variables in Equation				Variables Not in the Equation	
Variables	Coefficient	Standard Error of Coefficient	Standardized Regression Coefficient (beta)	Partial t Value	Partial Correlation	t Value
Y-intercept	21.65					
X_5 Overall service	8.38	.86	.70	9.722		
X_1 Delivery speed					.438	4.812
X_2 Price level					−.453	−5.007
X_3 Price flexibility					.720	10.210
X_4 Manufacturer's image					.022	.216
X_6 Sales force's image					.126	1.252
X_7 Product quality					−.216	−2.178

R Square   *R* square (R^2) is the correlation coefficient squared, also referred to as the coefficient of determination. This value indicates the percentage of total variation of *Y* explained by X_5. The total sum of squares (3,927.31 + 4,071.69 = 7,999.0) is the squared error that would occur if we used only the mean of *Y* to predict the dependent variable. Using the values of X_5 reduces this error by 49.1 percent (3,927.31 ÷ 7,999.0 = 49.1%).

Standard Error of the Estimate   The standard error of the estimate is another measure of the accuracy of our predictions. It is the square root of the sum of the squared errors divided by the degrees of freedom. It represents an estimate of the standard deviation of the actual dependent values around the regression line; that is, it is a measure of variation around the regression line. The standard error of the estimate can also be viewed as the standard deviation of the prediction errors and thus becomes a measure to assess the absolute size of the prediction error. It is also used in estimating the size of the confidence interval for the predictions. See Neeter, Wasserman, and Kunter [13] for details regarding this procedure.

Variables in the Equation (Step 1)   In Step 1, a single predictor variable (X_5) is used to calculate the regression equation for predicting the dependent variable. For each variable in the equation, several measures need to be defined: the regression coefficient, the standard error of the coefficient, and the *t* value of variables in the equation.

- *Regression Coefficient.* The value 8.38 is the regression coefficient (b_5) of the predictor variable (X_5). Thus the predicted value for each value of X_5 is the intercept plus the regression coefficient times the value of the predictor variable (21.65 + 8.38X_5). The standardized regression coefficient, or beta value, of .70 is the value calculated from standardized data. With only one independent variable, the squared beta coefficient equals the coefficient of determination. The beta value allows you to compare the effect of X_5 on *Y* to the effect on *Y* of other predictor variables at each stage, because this value reduces the regression coefficient to a comparable unit, the number of standard deviations. (Note that at this time we have no other variables available for comparison.)
- *Standard Error of the Coefficient.* The standard error of the coefficient is the standard error of the estimate of b_5. The value of b_5 divided by the standard error (8.38 ÷ .86 = 9.74) is the calculated *t* value for a *t* test of the hypothesis $b_5 = 0$. A smaller standard error implies more reliable prediction. Thus we would like to have small standard errors and therefore smaller confidence intervals. This coefficient is also referred to as the standard error of the regression coefficient; it is an estimate of how much the regression coefficient will vary between samples of the same size taken from the same population; that is, if one were to take multiple samples of the same size from the same population and use them to calculate the regression equation, this would be an estimate of how much the regression coefficient would vary from sample to sample (see Exhibit 3.2 for a more detailed discussion of statistical significance testing of regression coefficients).
- *t Value of Variables in the Equation.* The *t* value of variables in the equation measures the significance of the partial correlation of the variable reflected in the regression coefficient. It is particularly useful in step 5 of Figure 3.5 in helping to determine whether a variable should be dropped from the equation once a variable has been added. Also given in the table is the level of significance, which is compared to the threshold level set by the researcher for dropping the

variable. In our example, we have set a .10 level for dropping variables from the equation. The critical value for a significance level of .10 with 98 degrees of freedom is 1.658. Therefore, X_5 meets our requirements for inclusion in the regression equation. *F* values are often given at this stage rather than *t* values. Remember that they are directly comparable, as the *t* value is the square root of the *F* value.

Variables Not in the Equation   Although X_5 has been included in the regression equation, six other potential independent variables remain for inclusion to improve the prediction of the criterion variable. For those values, two measures are available to assess their potential contribution: partial correlations and *t* values.

- *Partial Correlation.* The partial correlation is a measure of the variation in Y not accounted for by the variables in the equation (only X_5 in step 1) that can be accounted for by each of these additional variables. For example, the value .720 represents the partial correlation of X_3 given that X_5 is the equation. Remember, the partial correlation can be misinterpreted. It does not mean that we explain 72.0 percent of the previously unexplained variance. It means that 51.8 percent $(72.0^2 = 51.8\%$, the partial coefficient of determination) of the unexplained (not the total) variance can now be accounted for by X_3. Because 49.1 percent was already explained by X_5, $(1 - 49.1) \times 51.8\% = 26.4$ percent of the total variance could be explained by adding variable X_3. A Venn diagram illustrates this concept.

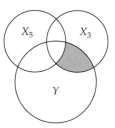

The shaded area of X_3 as a proportion of the shaded area of Y represents the *partial correlation* of X_3 with Y given X_5. This represents the variance in X_3 (after removal of the effects of X_5 on X_3) in common with the remaining variance in Y (after removing the effects of X_5 on Y). The calculation of the unique variance associated with adding X_3 can also be determined through the *part correlation*, as described in Exhibit 3.1.

- *t Values of Variables Not in the Equation.* The column of *t* values measures the significance of the partial correlations for variables not in the equation. These are calculated as a ratio of the additional sum of squares explained by including a particular variable and the sum of squares left after adding that same variable. If this *t* value does not exceed a specified significance level, the variable will not be allowed to enter the equation. The tabled *t* value for a significance level of .05 with 97 degrees of freedom is 1.98. Looking at the column of *t* values, note that four variables (X_1, X_2, X_3, and X_7) exceed this value and are candidates for inclusion.

Recall that the simple correlation of X_1 with the dependent variable was .68, but it was only .56 for X_3. Therefore, you may have thought variable X_1 would be included in the model next. But in deciding which additional variables to include in the equation, we would first select the predictor variable that exhibits the highest partial correlation with the dependent variable (not the highest correlation with Y). The partial correlation of X_3 is the largest (.720), and therefore X_3 (and even X_2) will be considered for addition to the model before X_1.

We now know that a significant portion of the variance in the dependent variable is explained by predictor variable 5. We can also see that predictor variable 3 has the highest partial correlation coefficient with the dependent variable, and that the t value is significant at the .05 level. (It is significant at the .01 level as well.) We can now look at the new model using both variables 5 and 3.

Stepwise Estimation: Adding X_3

The multiple R and R squared values have both increased with the addition of X_3 (see Table 3.7). The R^2 has increased by the 26.4 percent, the amount we predicted when we examined the partial correlation coefficient from X_3 of .720. The increase

TABLE 3.7 Example Output: Step 2 of HATCO Multiple Regression Example

Variable entered: X_3 Price flexibility

Multiple R	.869
Multiple R^2	.755
Adjusted R^2	.750
Standard error of estimate	4.498

Analysis of Variance

	Sum of Squares	df	Mean Square	F Ratio
Regression	6,036.5	2	3,018.25	149.18
Residual	1,962.5	97	20.23	

	Variables in Equation				*Variables Not in the Equation*	
Variables	Coefficient	Standard Error of Coefficient	Standardized Regression Coefficient (beta)	Partial t Value	Partial Correlation	t Value
Y-intercept	−3.489					
X_3 Price flexibility	3.336	.327	.514	10.210		
X_5 Overall service	7.974	.603	.666	13.221		
X_1 Delivery speed					.021	.205
X_2 Price level					−.027	−.267
X_4 Manufacturer's image					.181	1.303
X_6 Sales force's image					.236	2.378
X_7 Product quality					.169	1.683

in R^2 of 26.4 percent is derived by multiplying the 50.9 percent of variation that was not explained after step 1 by the partial correlation squared: $50.9 \times (.720)^2 = 26.4$; that is, of the 50.9 percent unexplained with X_5, $(.720)^2$ of this variance was explained by adding X_3, yielding a total variance explained of .755—that is, $.491 + (.509 \times (.720)^2)$.

The value of b_5 has changed very little. This is a further clue that variables X_5 and X_3 are relatively independent (the simple correlation between the two variables is .07). If the effect of X_3 on Y were totally independent of the effect of X_5, the b_5 coefficient would not change at all.

The partial t values indicate that both X_5 and X_3 are statistically significant predictors of Y. The t value for X_5 is now 13.221, where it was 9.722 in step 1. The t value for X_3 examines the contribution of this variable given that X_5 is already in the equation. Note that the t value for X_3 (10.210) is the same value shown for X_3 in step 1 under the heading "Variables Not in the Equation" (see Table 3.6).

Because predictors 3 and 5 both make significant contributions to the explanation of variation in the dependent variable, we can ask, Are other predictors available? Looking at the partial correlations for the variables not in the equation in Table 3.7, we see that X_6 has the highest partial correlation (.236). This variable would explain 5.6 percent of the heretofore unexplained variance ($.236^2 = .056$) or 1.4 percent of the total variance [$(1 - .755) \times .056 = 0.14$]. This is a very modest contribution of explanatory power of our prediction, even though the partial correlation is significant at the .05 significance level. (Note: The tabled t value for 96 degrees of freedom at a .05 level is 1.98, while the t value for X_6 is 2.378.)

Stepwise Estimation: A Third Variable Is Added—X_6

With X_6 entered into the regression equation, the results are shown in Table 3.8. As we predicted, the value of R^2 increases by 1.4 percent. In addition, examination of the partial correlations for X_1, X_2, X_4, and X_7 indicates that no additional value will be gained by adding them to the predictive equation. These partial correlations are all very small and have partial t values associated with them that would not be statistically significant at the level (.05) chosen for this model.

Evaluating the Variate for the Assumptions of Regression Analysis

In evaluating the estimated equation, we have considered statistical significance. We must also address two other basic issues: (1) meeting the assumptions underlying regression, and (2) identifying the influential data points. We consider each of these issues in the following sections.

The assumptions to examine are linearity, constant variance, independence of the residuals, and normality. The principal measure used in evaluating the regression variate is the residual—the difference between the actual dependent variable value and its predicted value. For comparison purposes, we use the studentized residuals. The most basic type of residual plot is shown in Figure 3.8, the studentized residuals versus the predicted values. As we can see, the residuals fall within a generally random pattern, very similar to the null plot in Figure 3.4a. However, we must make specific tests for each assumption to check for violations.

TABLE 3.8 Example Output: Step 3 of HATCO Multiple Regression Example

Variable entered: X_6 Sale force's image

Multiple R	.877
Multiple R^2	.768
Adjusted R^2	.761
Standard error of estimate	4.394

Analysis of Variance

	Sum of Squares	df	Mean Square	F Ratio
Regression	6,145.7	3	2,048.6	106.11
Residual	1,853.3	96	19.3	

	Variables in Equation					*Variables Not in the Equation*	
Variables	*Coefficient*	*Standard Error of Coefficient*	*Standardized Regression Coefficient (beta)*	*Partial t Value*		*Partial Correlation*	*t Value*
Y-intercept	−6.520						
X_3 Price flexibility	3.376	.320	.521	10.562			
X_5 Overall service	7.621	.607	.637	12.547			
X_6 Sales force's image	1.406	.591	.121	2.378			
X_1 Delivery speed						.040	.389
X_2 Price level						−.041	−.405
X_4 Manufacturer's image						−.002	−.021
X_7 Product quality						.130	1.273

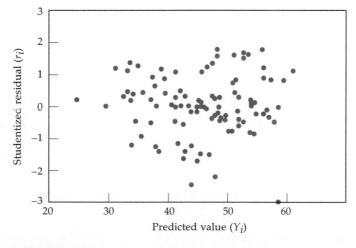

FIGURE 3.8 Analysis of studentized residuals.

Linearity   The first assumption, linearity, will be assessed through the analysis of residuals and partial regression plots. Figure 3.8 does not exhibit any nonlinear pattern to the residuals, thus ensuring that the overall equation is linear. But we must also be certain, when using more than one predictor variable, that each predictor variable's relationship is linear as well to ensure its best representation in the equation. To do so, we use the partial regression plot for each predictor in the equation. In Figure 3.9 we see that the relationships for X_3 and X_5 are quite well defined; thus they have strong and significant effects in the regression equation. The variable X_6 is less well defined, both in slope and scatter of the points, thus explaining its lesser effect in the equation (evidenced by the smaller coefficient, beta value, and significance level). For all three variables, no nonlinear pattern is shown, thus meeting the assumption of linearity for each predictor variable.

Homoscedasticity   The next assumption deals with the constancy of the residuals across values of the predictor variables. Our analysis is again through examination of the residuals (Figure 3.8), which shows no pattern of increasing or decreasing residuals. This finding indicates homoscedasticity in the multivariate (the set of predictor variables) case.

Independence of the Residuals   The third assumption deals with the effect of carry-over from one observation to another, thus making the residual not independent. When carry-over is found in such instances as time-series data, the analyst must identify the potential sequencing variables (such as time in a time-series problem) and plot the residuals by this variable. For example, assume that the identification number represented the order in which we collected our responses. We could plot the residuals and see whether a pattern emerges. In our example, several variables, including the identification number and each predictor variable, were tried and no consistent pattern was found. Remember, use the residuals in this analysis, not the original dependent variable values, because the focus is on the prediction errors, not the relationship captured in the regression equation.

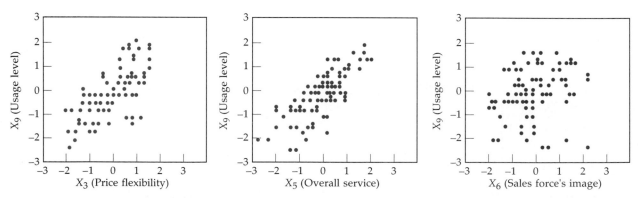

Axes are expressed in standardized scores.

FIGURE 3.9 Standardized partial regression plots.

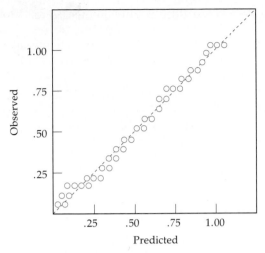

Modified K-S (Lilliefors): .0688 ($p > .2000$)

FIGURE 3.10 Normal probability plot: standardized residuals.

Normality   The final assumption we will check is normality of the error term of the variate with a visual examination of the normal probability plots of the residuals. As shown in Figure 3.10, the values fall along the diagonal with no substantial or systematic departures; thus, the residuals are considered to represent a normal distribution. The regression variate is found to meet the assumption of normality.

Applying Remedies for Assumption Violations   After testing for violations of the four basic assumptions of multivariate regression for both individual variables and the regression variate, the researcher should assess the impact of any remedies on the results. In our example, the only remedies needed were the transformations of X_2, X_4, and X_6 to achieve normality. If we substitute these variables for their original values and reestimate the regression equation, we achieve almost identical results (see Table 3.9). The same variables enter the equation, with the only substantive difference being a slightly stronger coefficient for the transformed X_6 variable and a slight improvement in the R^2 value. (.771 versus .768). The predictor variables not in the equation still show nonsignificant levels for entry, even those that were transformed. Thus, in this case, the remedies for violating the assumptions improved the prediction slightly but did not alter the substantive findings.

Identifying Outliers as Influential Observations

For our final analysis, we attempt to identify any observations that are influential (having a disproportionate impact on the regression results) and determine whether they should be excluded from the analysis. Although more detailed procedures are available for identifying outliers as influential observations, we address in the following section the use of residuals in identifying outliers.

TABLE 3.9 Example Output: Multiple Regression Results after Remedies for Violation of Assumptions

Multiple R	.87806
Multiple R^2	.77099
Standard error of estimate	4.36828

Analysis of Variance

	Sum of Squares	df	Mean Square	F Ratio
Regression	6,167.1	3	2,055.71	107.73
Residual	1,831.8	96	19.08	

		Variables in Equation				Variables Not in the Equation	
Variables		Coefficient	Standard Error of Coefficient	Standardized Regression Coefficient (beta)	Partial t Value	Partial Correlation	t Value
Y-intercept		−6.792	3.225		−2.11		
X_3	Price flexibility	3.409	.319	.5258	10.70		
X_5	Overall service	7.640	.589	.6385	12.75		
log X_6	Sales force's image	3.952	1.511	.1311	2.62		
X_1	Delivery speed					.048	.469
log X_2	Price level					−.075	−.737
log X_4	Manufacturer's image					−.047	−.463
X_7	Product quality					.118	1.163

The most basic diagnostic tool involves the residuals and identification of any outliers—that is, observations not predicted well by the regression equation which have large residuals. Figure 3.11 shows the studentized residuals for each observation. Because the values correspond to t values, upper and lower limits can be set once the desired confidence interval has been established. Perhaps the most widely used level is 95 percent confidence ($\propto$ =.05). The corresponding t value is 1.96, thus identifying statistically significant residuals as those with residuals greater than this value. Four observations (7, 11, 14, 100) have significant residuals and can be classified as outliers. Outliers are important because they are observations not represented by the regression equation for one or more reasons—one of which may be an influential effect on the equation that requires a remedy.

Examination of the residuals also can be done through the partial regression plots (see Figure 3.9). These plots help identify influential observations for each predictor–criterion variable relationship. As noted in Figure 3.9, a set of separate and distinct points (observations 7, 11, 14, 100) can be identified for variables X_3

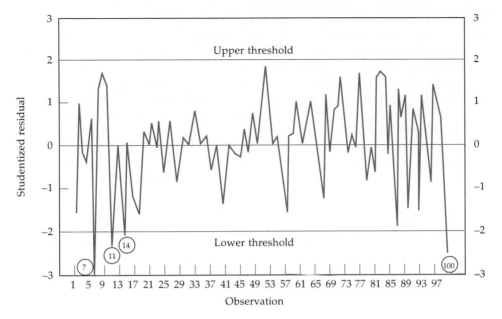

FIGURE 3.11 Plot of studentized residuals.

and X_6. These points are not well represented by the relationship and thus could affect the partial correlation as well. More detailed analyses to ascertain whether any of the observations can be classified as influential observations, as well as what may be the possible remedies, are discussed in Appendix 3A.

Stage Five: Interpreting the Variate

With the model estimation completed, the regression variate specified, and the diagnostic tests administered that confirm the appropriateness of the results, we can now examine our predictive equation, which includes X_3, X_5, and X_6. The section of Table 3.8 headed "Variables in Equation" yields the prediction equation from the column labeled "Coefficient." From this column, we read the constant term (-6.520) and the coefficients (3.376, 7.621, and 1.406) for variables 3, 5, and 6, respectively. The predictive equation would be written

$$Y = -6.520 + 3.376\, X_3 + 7.621\, X_5 + 1.406\, X_6$$

With this equation, the expected usage level for any customer could be calculated if his or her evaluations of HATCO were known. For illustration purposes, assume that a customer rated HATCO with a value of 4.0 for each of these three measures. The predicted product usage level for that customer would be

Predicted Level of
Product Usage $= -6.520 + 3.376(4.0) + 7.621(4.0) + 1.406(4.0)$
                   $= -6.520 + 13.504 + 30.484 + 5.624$
                   $= 43.902$

In addition to providing a basis for predicting product usage levels, the regression coefficients also provide a means of assessing the relative importance of the individual variables in the overall prediction of product usage. In this situation, all the variables are expressed on the same scale and thus direct comparisons can be made. But in most instances the beta coefficients are used for comparison between independent variables. In Table 3.8, the beta coefficients are listed in the column headed "Standardized Regression Coefficient." The researcher can make direct comparisons among the variables to ascertain their relative importance in the regression variate. For our example, X_5 (overall service) was the most important, followed closely by X_3 (price flexibility). The third independent variable, X_6 (sales force's image) was notably lower in importance. This supports its lower incremental amount of variance explained and the lower univariate correlation with product usage. Although significant, X_6 does not merit the attention that should be accorded to the other two independent variables.

Measuring the Degree and Impact of Multicollinearity

In any interpretation of the regression variate, the researcher must be aware of the impact of multicollinearity. As discussed earlier, highly collinear variables can distort the results substantially or make them quite unstable and thus not generalizable. Two measures are available for testing the impact of collinearity: (1) calculating the tolerance and VIF values, and (2) using the condition indices and decomposing the regression coefficient variance. The tolerance value is 1 minus the proportion of the variable's variance explained by the other predictors. Thus a high tolerance value indicates little collinearity, and tolerance values approaching zero indicate that the variable is almost totally accounted for by the other variables. The variance inflation factor (VIF) is the reciprocal of the tolerance value; thus we look for small VIF values as indicative of low intercorrelation among variables. In our example, tolerance values all exceed .93, indicating very low levels of collinearity (see Table 3.10). Likewise, the VIF values are all quite close to 1.0. These results indicate that interpretation of the regression variate coefficients should not be affected adversely by multicollinearity.

A second approach to identifying multicollinearity and its effects is through the decomposition of the coefficient variance. Analysts are encouraged to explore this technique and the additional insights it offers into the interpretation of the regression equation. Details of this method are discussed in the Appendix to this chapter.

While multicollinearity does not have a substantial impact on the estimated regression variate, it does have an impact on the estimation of the variate. After X_5 (the first variable added to the regression variate), the second-highest correlation with the dependent variable was X_1. Yet X_1 also had a fairly high level of collinearity (.61) with X_5. Because X_5 entered the regression variate first in the stepwise procedure, there was not enough unique variance in X_1 to justify its

TABLE 3.10 Testing for Multicollinearity: Assessing Tolerance and VIF Values

	Variable	Tolerance	Variance Inflation
X_3	Price flexibility	0.99287009	1.00718111
X_5	Overall services	0.93639766	1.06792236
X_6	Sales force's image	0.93946418	1.06443654

inclusion. Therefore, only X_5 entered the regression variate. However, it would be substantively incorrect to interpret from these results that X_1 had no impact on product usage when in fact it was the independent variable with the second highest bivariate correlation with the dependent variable. The correct interpretation would be that X_5 or X_1 demonstrates high impact, but that the similarity of their effect on product usage (high collinearity) dictates that only one of them is needed in the prediction process. The analyst must never allow an estimation procedure to dictate the interpretation of the results but instead must understand the issues of interpretation accompanying each estimation procedure. For example, if all seven independent variables had been entered into the regression variate, the analyst would still have to contend with the effects of collinearity on the interpretation of the coefficients for X_5 and X_1 but in a different manner than if stepwise were used.

Stage Six: Validating the Results

The final task facing the analyst involves the validation process of the regression model. The primary concern of this process is to ensure that the results are generalizable to the population and not specific to the sample used in estimation. The most direct approach to validation is to obtain another sample from the population and assess the correspondence of the results from the two samples. In the absence of an additional sample, the analyst can assess the validity of the results in several approaches. The first involves examination of the adjusted R^2 value. In this situation, the adjusted R^2 value is .761 (as compared with a R^2 value of .768), which indicates that the estimated model is not overfitted to the sample.

A second approach is to divide the sample into two subsamples, estimate the regression model for each subsample, and compare the results. Table 3.11 con-

TABLE 3.11 Split-Sample Validation of the Stepwise Estimation

Model Component	Overall (n = 100)	Sample 1 (n = 53)	Sample 2 (n = 47)
Independent Variables			
X_3			
Regression coefficient	3.376	3.491	3.425
Beta coefficient	.521	.545	.520
t-value	10.56	7.32	7.66
X_5			
Regression coefficient	7.621	8.694	6.805
Beta coefficient	.637	.714	.579
t value	12.54	9.59	8.27
X_6			
Regression coefficient	1.406	Not Entered	1.883
Beta coefficient	.121		.170
t value	2.38		2.50
Model Fit			
R^2	.768	.725	.816
Adjusted R^2	.761	.715	.803
Standard error of the estimate	4.39	4.66	4.15

tains the overall stepwise results plus the results from stepwise models estimated for two subsamples of 53 and 47 observations. Comparison of the overall model fit demonstrates a high level of similarity of the results in terms of R^2, adjusted R^2, and the standard error of the estimate. But in comparing the individual coefficients, one difference does appear. In sample one, X_6 did not enter in the stepwise results as it did in sample two and the overall sample. The omission of X_6 in one of the subsamples confirms that it was a marginal predictor, as indicated by the low beta and t values in the overall model.

Summary

This chapter presents a simplified introduction to the rationale and fundamental concepts underlying multiple regression analysis. It emphasizes that multiple regression analysis can describe and predict the relationship between two or more intervally scaled variables. Also, multiple regression analysis, which can be used to examine the incremental and total explanatory power of many variables, is a great improvement over the sequential analysis approach necessary with univariate techniques. Both stepwise and simultaneous techniques can be used to estimate a regression equation, and under certain circumstances nonmetric dummy-coded variables can be included in the regression equation. Finally, we have seen that numerous diagnostic techniques exist for testing both the assumptions underlying regression analysis and the existence of cases exerting an undue influence on the resulting equation or predictions. This chapter has provided a fundamental presentation of how regression works and what it can achieve. Familiarity with the concepts presented in this chapter will help you better understand the more complex and detailed technical presentations in other textbooks.

Questions

1. How would you explain the relative importance of the predictor variables used in a regression equation?
2. Why is it important to examine the assumption of linearity when using regression? How can nonlinearity be corrected or accounted for in the regression equation?
3. Could you find a regression equation that would be acceptable as statistically significant and yet offer no acceptable interpretational value to management? How could such an equation exist?
4. What is the difference in interpretation between regression coefficients associated with interval-scale predictor variables and dummy-coded (0, 1) predictor variables?
5. What are the differences between interactive and correlated predictor variables? Do any of these differences affect your interpretation of the regression equation?
6. Are influential cases always to be omitted? Give examples of occasions when they should or should not be omitted.

References

1. Barnett, V., and T. Lewis. *Outliers in Statistical Data*, 2d ed. New York: Wiley, 1984.
2. Belsley, D. A., E. Kuh, and R. E. Welsch. *Regression Diagnostics: Identifying, Influential Data and Sources of Collinearity*. New York: Wiley, 1980.
3. BMDP Statistical Software, Inc. *SOLO Power Analysis*, Los Angeles: 1991.
4. Box, G. E. P., and D. R. Cox. "An Analysis of Transformations." *Journal of the Royal Statistical Society* B (26) (1964): 211–43.
5. Cohen, J. *Statistical Power Analysis for the Behavioral Sciences*, rev. ed. New York: Academic Press, 1977.
6. Cohen, J., and P. Cohen. *Applied Multiple Regression/Correlation Analysis for the Behavioral Sciences*, 2d ed. Hillsdale, N.J.: Lawrence Erlbaum, 1983.
7. Daniel, C., and F. S. Wood. *Fitting Equations to Data*, 2d ed. New York: Wiley-Interscience, 1980.
8. Hosmer, D. W., and S. Lemeshow. *Applied Logistic Regression*. New York: Wiley, 1989.
9. Jaccard, J., R. Turrisi, and C. K. Wan. *Interaction Effects in Multiple Regression*. Beverly Hills, Calif.: Sage Publications, 1990.
10. Johnson, R. A., and D. W. Wichern. *Applied Multivariate Statistical Analysis*. Englewood Cliffs, N.J.: Prentice-Hall, 1982.
11. Mason, C. H., and W. D. Perreault, Jr. "Collinearity, Power, and Interpretation of Multiple Regression Analysis." *Journal of Marketing Research* XXVIII (August 1991): 268–80.
12. Mosteller, F., and J. W. Tukey. *Data Analysis and Regression*. Reading, Mass.: Addison-Wesley, 1977.
13. Neter, J., W. Wassermann, and M. H. Kutner. *Applied Linear Regression Models*. Homewood, Ill.: Irwin, 1989.
14. Rousseeuw, P. J., and A. M. Leroy. *Robust Regression and Outlier Detection*. New York: Wiley, 1987.
15. Seer, G. A. F. *Multivariate Observations*. New York: Wiley, 1984.
16. Sharma, S., R. M. Durand, and O. Gur-Arie. "Identification and Analysis of Moderator Variables." *Journal of Marketing Research* XVIII (August 1981): 291–300.
17. Weisberg, S. *Applied Linear Regression*. New York: Wiley, 1985.
18. Wilkinson, L. "Tests of Significance in Stepwise Regression." *Psychological Bulletin* 86(1979): 168–74.

APPENDIX **3A**

Advanced Diagnostics for Multiple Regression Analysis

LEARNING OBJECTIVES

After reading our discussion of these techniques, you should be able to:

- Understand how the condition index and regression coefficient variance–decomposition matrix isolate the effects, if any, of multicollinearity on the estimated regression coefficients.

- Identify those variables with unacceptable levels of collinearity or multicollinearity.

- Identify the observations with a disproportionate impact on the multiple regression results.

- Isolate the influential observations and assess the relationships when the influential observations are deleted.

PREVIEW

Perhaps the most widely used statistical technique is regression analysis, and multiple regression has led the movement toward increased usage of multivariate techniques. In moving from simple to multiple regression, the increased analytical power of the multivariate form also requires added diagnostics to deal with the correlations between variables and the observations with substantial impact on the results. This appendix describes two advanced diagnostic techniques for assessing (1) the impact of multicollinearity and (2) the identity of influential observations and their impact on multiple regression analysis. While Chapter 3 dealt with the basic diagnoses for these issues, here we discuss more sensitive procedures that have recently been proposed specifically for multivariate situations. These procedures are not refinements to the estimation procedures but instead address questions in interpreting the results that occur with multicollinearity and influential observations.

KEY TERMS

Before reading the appendix, review the key terms to develop an understanding of the concepts and terminology used. Throughout the appendix the key terms appear in boldface. Other points of emphasis in the appendix are italicized. Also, cross-references in the key terms are in italics.

Collinearity Relationship between two (collinearity) or more (multicollinearity) variables. Variables exhibit complete collinearity if their correlation coefficient is 1 and a complete lack of collinearity if their correlation coefficient is 0.

Condition index Measure of the relative amount of variance associated with an *eigenvalue* so that a large condition index indicates a high degree of collinearity.

Cook's distance (D_i) Summary measure of the influence of a single case (observation) based on the total changes in all other residuals when the case is deleted from the estimation process. Large values (usually greater than 1) indicate substantial influence by the case in affecting the estimated regression coefficients.

COVRATIO Measure of the influence of a single observation on the entire set of estimated regression coefficients. A value close to 1 indicates little influence. If the COVRATIO value minus 1 is greater than $\pm 3p \div n$ (where p is the number of predictors and n is the sample size), the observation is deemed influential based on this measure.

DFBETA Measure of the change in a regression coefficient when an observation is omitted from the regression analysis. The value of DFBETA is in terms of the coefficient itself and a standardized form (SDFBETA) is also available. No threshold limit can be established, although the researcher can look for values substantially different from the remaining observations to assess potential influence.

DFFIT Measure of the impact of an observation on the overall model fit or its standardized version (SDFFIT). The best rule of thumb is to identify as influential any standardized values (SDFFIT) that exceed $2 \div \sqrt{(p \div n)}$, where p is the number of predictors and n is the sample size. There is no threshold value for the DFFIT measure.

Eigenvalue Also known as the latent root or characteristic root, a measure of the amount of variance contained in the correlation matrix so that the sum of the eigenvalues is equal to the number of variables.

Hat matrix Represents the impact of the observed dependent variable on its predicted value. If all cases had equal influence, each would have a value of $(p \div n)$, where p equals the number of predictor variables and n is the number of cases. If a case had no influence, its value would be $-1 \div n$, while if a single case dominated totally, the value would be $(n - 1) \div n$. Generally, points exceeding twice the average $(2p \div n)$ are classified as influential.

Influential observation Observation with a disproportionate influence on one or more aspects of the regression estimates. This influence may have as its basis (1) substantial differences on the set of predictor values from other cases (2) extreme (either high or low) observed values for the criterion variables, or (3) a combination of these effects. Influential observations can either be "good," by reinforcing the pattern of the remaining data, or "bad," when a single or small set of cases unduly affects (biases) the regression estimates.

Leverage point   Observation is termed a leverage point if it has substantial impact on the regression results due to its differences from other observations on one or more of the independent variables. Values exceeding $2p \div n$ for larger samples (where p is the number of predictors and n is the sample size) or $3p \div n$ for smaller samples (less than $n = 30$) are designated as leverage points and become candidates for influential point classification.

Mahalanobis distance   Measure of the impact of a single case based on differences between the case's value and the mean value for all other cases across all the independent variables. The source of influence on regression results is for the case to be quite different on one or more predictor variables, thus causing a shift of the entire regression equation.

Multicollinearity   See *collinearity*.

Regression coefficient variance–decomposition matrix   Method of determining the relative contribution of each *eigenvalue* to each estimated coefficient. If two or more eigenvalues are highly associated with a single coefficient, an unacceptable level of multicollinearity is indicated.

Studentized residual   Most commonly used form of standardized residual, which differs from other methods in how it calculates the standard deviation used in standardization. To minimize the effect of a single outlying observation, the residual's standard deviation for an observation, used to standardize the residual, is computed from regression estimates omitting this observation in the calculation of the regression estimates and residuals. This is done repeatedly for each observation, each time omitting one observation from the calculation of the regression estimates.

Tolerance   Commonly used measure of *collinearity* and multicollinearity. The tolerance of variable i (TOL_i) is $1 - R_i^{*2}$, where R_i^{*2} is the coefficient of determination for the prediction of variable i by the other predictor variables. Tolerance values approaching zero indicate that the variable is highly predicted (collinear) with the other predictor variables.

Variance inflation factor (VIF_i)   Measure of the effect of other predictor variables on a regression coefficient. Directly related to the *tolerance* value ($VIF_i = 1 \div R_i^{*2}$). Large VIF values (a usual threshold is 10.0) also indicate a high degree of *collinearity* or multicollinearity among the independent variables.

Assessing Multicollinearity

As discussed in Chapter 3, **collinearity** and **multicollinearity** can have several harmful effects on multiple regression, both in the interpretation of the results and in how they are obtained, such as stepwise regression. The use of several variables as predictors makes the assessment of multiple correlation between the independent variables necessary to identify multicollinearity. But this is not possible by examining only the correlation matrix (which shows only simple correlations between two variables). We are now going to discuss a method developed specifically to diagnose the amount of multicollinearity present and the variables exhibiting the high multicollinearity. All major statistical programs (SPSS, SAS, and BMDP) have optional analyses providing the collinearity diagnostics.

TABLE 3A.1 Hypothetical Coefficient Variance Decomposition Analysis with Condition Indices

Condition Index (u_i)		b_1	b_2	b_3	b_4	b_5	b_6
				Proportion of variance of coefficient:			
1.0	u_1	.003	.001	.000	.003	.000	.000
4.0	u_2	.000	.021	.005	.003	.000	.000
16.5	u_3	.000	.012	.003	.010	.000	.001
45.0	u_4	.001	.963	.003	.972	.983	.000
87.0	u_5	.003	.002	.000	.009	.015	.988
122.0	u_6	.991	.001	.987	.003	.002	.011

A Two-Part Process

The method has two components. First is the **condition index,** which represents the collinearity of combinations of variables in the data set (actually the relative size of the **eigenvalues** of the matrix). The second is the **regression coefficient variance–decomposition matrix,** which shows the proportion of variance for each regression coefficient (and its associated variable) attributable to each eigenvalue (condition index). When we combine these two, in simple terms, the procedure is

1. Identify all condition indices above a threshold value. The threshold value usually is in a range of 15 to 30, with 30 the most commonly used value.
2. For all condition indices exceeding the threshold, identify variables with variance proportions above .50 percent.
3. A collinearity problem is indicated when a condition index identified in step 1 accounts for a substantial proportion of variance (.90 or above) for two or more coefficients.

The example shown in Table 3A.1 illustrates the basic procedure and shows both the condition indices and variance decomposition values. Using the condition index threshold of 30, we select three condition indices (u_4, u_5, and u_6). Coefficients exceeding the .90 threshold for these three condition indices are b_1 and b_3 with u_6, b_2, b_4, and b_5 with u_4 and b_6 with u_5 (see the underlined values in Table 3A.1). However, u_5 has only a single value (b_6) associated with it; thus no collinearity is shown for this coefficient. As a result, we would attempt to remedy the significant correlations among two sets of variables: (V_1, V_3) and (V_2, V_4, V_5).

An Illustration of Assessing Multicollinearity

In Chapter 3, we discussed the use of multiple regression in predicting the usage level for HATCO customers. The stepwise procedure identified three statistically significant predictors: X_3, X_5 and X_6. However, before we accept these regreession results as valid, we must examine the degree of multicollinearity and its effect on the results. To do so, we employ the condition indices and the decomposition of the coefficient variance and make comparisons with the **variance inflation factor (VIF)** and **tolerance** values (see Table 3A.2). As discussed in Chapter 3 and also presented in Table 3A.2, the tolerance/VIF values indicate inconsequential collin-

TABLE 3A.2 Testing for Multicollinearity in Multiple Regression

Assessing Tolerance and VIF Values

Variable	Tolerance	Variance Inflation
X_3	0.99287009	1.00718111
X_5	0.93639766	1.06792236
X_6	0.93946418	1.06443654

(handwritten: >10% above Tolerance; >10 above Variance Inflation)

Using the Condition Indices and Decomposition of Coefficient Variance Matrix

Number	Eigenvalue	Condition Index	Proportion of Coefficient Variance			
			Intercep	X_3	X_5	X_6
1	3.88226	1.00000	0.0012	0.0019	0.0037	0.0046
2	0.05997	8.04622	0.0138	0.1097	0.0213	0.8500
3	0.04541	9.24595	0.0197	0.1357	0.9089	0.0422
4	0.01237	17.71863	0.9653	0.7527	0.0661	0.1032

(handwritten: >30 above Condition Index; >.9 above X_3)

earity, since no VIF value exceeds 10.0 and the tolerance values show that in no case does collinearity explain more than 10 percent of any predictor variable's variance. This conclusion is supported when we examine the condition indices. We fail to pass the first step, as no condition index is greater than 30.0. Even if we were to use a threshold value of 15 for the condition index, we would select only u_4 and only one coefficient (the intercept) loads highly. Thus, we can find no support for the existence of multicollinearity.

Identifying Influential Observations

In Chapter 3, we examined only one approach to identifying **influential observations,** that being the examination of studentized residuals to find outliers. As noted then, however, cases may be classified as influential while not being recognized as an outlier. Thus, we need to examine more specific procedures to measure an observation's influence in several aspects of multiple regression [2]. In the following discussion, we discuss a four-step process of identifying outliers, leverage points, and influential observations. As noted before, an observation may fall into one or more of these classes, and it is the judgment of the researcher, based on the best available evidence, as to the course of action to be taken.

Step 1: Examining Residuals

While residuals were instrumental in detecting violations of model assumptions, they also play a role in identifying observations that are outliers on the dependent variable. We employ three methods of detection: studentized residuals (discussed in Chapter 3), dummy variable regression, and partial regression plots.

The **studentized residual** is the primary indicator of an observation that is an outlier on the dependent variable. With a fairly large sample size (50 or above), we may use a rule of thumb that studentized residuals greater than ± 2.0 are substantial, if not statistically significant. Most computer programs have the option of examining the residuals with a case-by-case listing. Observations falling outside the range can be considered outliers. A stricter test of significance has been proposed, which accounts for the multiple comparisons being made across various sample sizes [4].

As a complement to examining the studentized residual, a form of dummy-variable regression may be used. In this method, a dummy variable is added for each observation.* The model is then reestimated and those dummy variables with significant coefficients are the observations with significant residuals. You must be aware of the potential problems with overfitting the data when using such a method. To use such a method, you must have at least $N + 1$ (where N is the sample size) degrees of freedom left as you will be adding one variable for each observation. If the sample size is small or too few degrees of freedom are left, the process may be done in stages. At each stage, a set of dummy variables is added for selected observations and the regression model reestimated. This is repeated until each observation has been included.

To graphically portray the impact of individual cases, the partial regression plot is most effective. Because the slope of the regression line of the partial regression plot is equal to the variable's coefficient in the regression equation, an outlying case's impact on the regression slope (and the corresponding regression equation coefficient) can be readily seen. The effects of outlying cases on individual regression coefficients are portrayed visually. Again, most computer packages have the option of plotting the partial regression plot for you, so you need look only for outlying cases separated from the main body of observations.

Step 2: Identifying Leverage Points from the Predictors

Our next step is finding those observations separated from the remaining observations on one or more independent variables. These cases are termed **leverage points** in that they may "lever" the relationship in their direction because of their difference from the other observations (see Chapter 3 for a general description of leverage points).

When only two predictor variables are involved, plotting each variable on an axis of a two-dimensional plot will show those observations substantially different from the others. Yet, when a larger number of predictor variables are included in the regression equation, the task quickly becomes impossible through univariate methods. However, we are able to use a special matrix, the **hat matrix,** to identify multivariate leverage points. The hat matrix represents the combined effects of all independent variables for each case.

The diagonal of the hat matrix measures two aspects of influence. First, for each observation, it is a measure of the distance of the observation from the mean center of all other observations on the independent variables. (This is similar to the **Mahalanobis distance,** see below.) Second, large diagonal values also indicate that the observation carries a disproportionate weight in determining its predicted dependent variable value, thus minimizing its residual. This is an indica-

---

*For those unfamiliar with dummy-variable coding, see Chapter 3.

tion of influence, because the regression line must be closer to this observation, (i.e., strongly influenced) for the small residual to occur. This is not necessarily "bad," as illustrated in Chapter 3, when the influential observations fall in the general pattern of the remaining observations. So on both aspects, the observation has a substantial influence on the regression results.

What is a large leverage value? The average value is $p \div n$, where p is the number of predictors (the number of coefficients plus one for the constant) and n is the sample size. The rule of thumb for situations where p is greater than 10 and the sample size exceeds 50 is to select observations with a leverage value greater than twice the average ($2p \div n$). When the number of predictors or the sample size is less, use of three times the average ($3p \div n$) is suggested. The more widely used computer programs (SPSS, SAS, BMDP) all have options for calculating and printing the leverage values for each observation. The analyst must then select the appropriate threshold value ($2p \div n$ or $3p \div n$) and identify observations with values larger than the threshold.

A comparable measure is the Mahalanobis distance, which considers only the distance of an observation from the mean values of the independent variables and not the impact on the predicted value. The Mahalanobis distance as a means of identifying outliers is discussed in Chapter 2. It is limited in this case because threshold values depend on a number of factors, and a rule-of-thumb threshold value is not possible. It is possible to determine statistical significance from published tables [1]. Yet even without the published tables, the researcher can look at the values and identify any observations with substantially higher values than the remaining observations. For example, if a small set of observations with the highest Mahalanobis values are two to three times the next-highest value, this would be a substantial break in the distribution and another indication of influence in support of other measures.

Step 3: Single-Case Diagnostics
Identifying Influential Observations

Up to now we have found outlying points on the predictor and criterion variables but have not formally estimated the influence of a single observation on the results. In this third step, all the methods rely on a common proposition: deleting one or more observations and observing the changes in the regression results in terms of the residuals, individual coefficients, the matrix of coefficients, or overall model fit. Most computer programs with residual analysis options will print any of these single-case diagnostic values for each observation. The researcher then need only examine the values and select those observations that exceed the specified value.

The first diagnostic is the studentized deleted residual, which is the studentized residual for observation i when it is deleted from calculation of the regression equation. The rationale is that if an observation is extremely influential in an equation, it may not be identified by the normal studentized residuals because of its impact on the estimated regression model. The studentized deleted residual eliminates the case's impact on the regression estimates, however, and offers a "less influenced" residual measure. A large value (greater than ± 2.0), as with other residuals, indicates that the case differs substantially from the other cases.

The impact of a single observation on each regression coefficient is shown by the **DFBETA.** Calculated as the change in the coefficient when the observation is deleted, it is the relative effect of an observation on each coefficient. Guidelines

DFBeta (Degree of Fit of the Beta = considers each coefficient separately

for identifying particularly high values suggest that a lower threshold of 1.0 be applied to small and medium-sized samples, with $2 \div \sqrt{n}$ used for larger data sets.

A similar measure is the **COVRATIO,** which estimates the effect of the observation on the efficiency of the estimation process. It differs from the DFBETA in that it considers all coefficients collectively rather than each individual coefficient. Large values, where the value of the ratio minus 1 is greater than $\pm 3p \div n$, are another indicator that the observation has a substantial influence on the set of coefficients.

The final measure, **Cook's distance** (D_i), is considered the single most representative measure of influence. It captures the impact of an observation from two sources: the size of changes in the predicted values when the case is omitted (outlying studentized residuals) as well as the observation's distance from the other observations (leverage). A rule of thumb is to identify observations with a Cook's distance of 1.0 or greater. However, even if no observations exceed this threshold, additional attention is dictated if a small set of observations has substantially higher values than all of the remaining observations.

A comparable measure is **DFFIT,** the degree to which the fitted values change when the case is deleted. A cutoff value of $2*\sqrt{(p \div n)}$ has been suggested to detect substantial influence. Even though both Cook's distance and DFFIT are measures of overall fit, they must be complemented by the measures of steps 1 and 2 to enable us to determine whether influence arises from the residuals, leverage, or both.

Step 4: Selecting and Accommodating Influential Observations

The identification of influential observations is more a process of convergence by multiple methods than a reliance on a single measure. Because no single measure totally represents all dimensions of influence, it is a matter of interpretation, although these measures typically identify a small set of observations. In selecting the observations with large values on the diagnostic measures, you should first identify all observations exceeding the threshold values. Then, examine the range of values in the data set being analyzed and look for large gaps between the highest values and the remaining data. Many times additional observations will be detected that should be classified as influential.

After identification, several courses of action are possible. First, if the number of observations is small and a justifiable argument can be made for their exclusion, the cases should be deleted and the new regression equation estimated. If, however, deletion cannot be justified, several more "robust" estimation techniques are available, among them robust regression [3]. Whatever action is taken, it should meet our original objective of making the data set most representative of the actual population, to ensure validity and generalizability.

Example from the HATCO Database

To illustrate the diagnostic procedures for influential observations, we examine the regression example from Chapter 3. In identifying any observations that are "influential" (having a disproportionate impact of the regression results), we will also determine whether they should be excluded from the analysis. The four-step process described above is used in the following sections.

Step 1: Examining the Residuals

Our first diagnosis involves examination of the residuals and identification of any outliers (i.e., observations with large residuals which are not predicted well by the regression equation). Figure 3A.1 shows the studentized residuals for each observation. Because the residuals correspond to t values, upper and lower limits have been set at the 95 percent confidence interval (t value = 1.96). Statistically significant residuals are those falling outside these limits. Four observations (7, 11, 14, 100) have significant residuals and can be classified as outliers. An outlier is not necessarily an influential point, nor do all influential points have to be outliers. But outliers are important to note as they represent observations not represented by the regression equation because of one or more reasons, one of which may have an influential effect on the equation and may require a remedy.

Examination of the residuals can also be done though the partial regression plots (see Figure 3A.2). These plots help identify influential observations for each predictor–criterion variable relationship. As noted in Figure 3A.2, a set of separate and distinct points (observation 7, 11, 14, 100) can be identified for variables X_3 and X_6. These points are not well represented by the relationship and thus could affect the partial correlation as well.

Step 2: Identifying Leverage Points

Although the residual analysis could identify outliers, we were unable to see the magnitude of each observation's impact on the predictions. To do so, we use the measure of leverage (see Table 3A.3 and Figure 3A.3). We use the threshold limits of $2p \div N$, because the sample size exceeds 50. The calculated value is $2 \times 4 \div 100 = .08$. Using this threshold limit, we identify seven cases (5, 7, 42, 71, 82, 93, 96) as leverage points. These cases do have an influential effect on the results, but

Sequence

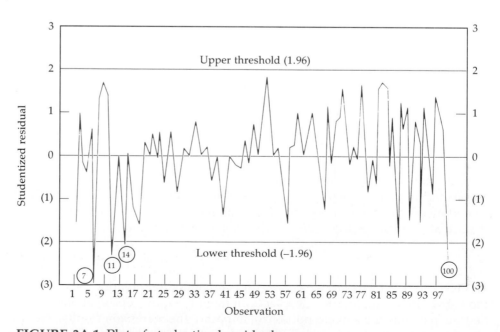

FIGURE 3A.1 Plot of studentized residuals.

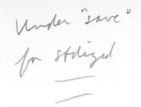

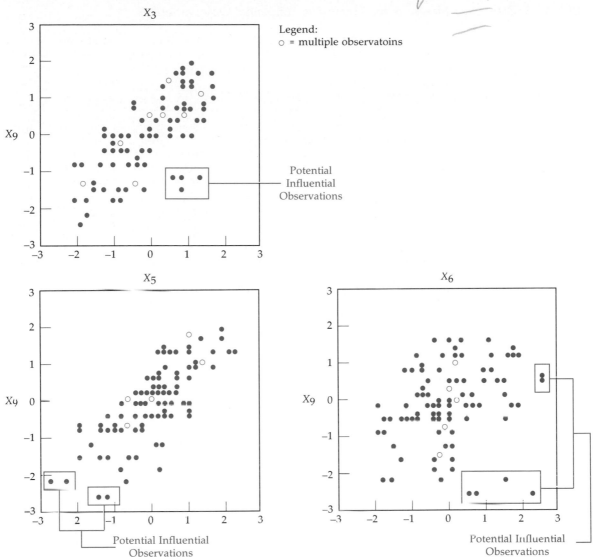

Note: Axes are expressed in standardized scores. *Y* axes are dependent variable.

FIGURE 3A.2 Standardized partial regression plots and influential observations.

as discussed earlier, all influential points are not bad (i.e., they do not distort or detract from the relationship). We must examine the observations further to asses what part(s) of the results they affect.

Step 3: Single-Case Diagnostics

While we have identified the potentially influential cases, we now complement this analysis with measures reflecting the impact on specific portions of the regression results. In each instance, we will delete a case from the regression estimation and observe the changes in the results. Our first measure is the studentized deleted residual (see Table 3A.3 and Figure 3A.4). As with the residual analysis in step 1, four cases are again significant but are joined by an additional case (case

TABLE 3A.3 Diagnostic Measures for Identifying Influential Observations

										DFBETA		
Obser-vation	Predicted Y value	Residual	Studentized Residual	Cook's Distance	Deleted Studentized Residual	Leverage	COVRATIO	DFFITS	Intercept	X_3	X_5	X_6
1	38.2986	−6.2986	−1.449	0.011	−1.4570	0.0206	0.9746	−0.2113	−0.1678	0.1021	0.0792	0.0522
2	39.4246	3.5754	0.840	0.011	0.8384	0.0608	1.0781	0.213	0.0453	−0.0884	−0.0799	0.1623
3	49.2903	−1.2903	−0.306	0.002	−0.3041	0.0763	1.1245	−0.0874	−0.0199	0.0551	−0.0641	0.0159
4	34.4009	−2.4009	−0.556	0.003	−0.5542	0.0348	1.0665	−0.1052	−0.0752	0.0273	0.0778	0.0082
5	58.2680	−0.2680	−0.064	0.000	−0.00637	0.0912	1.1472	−0.0202	0.0139	−0.0088	0.0004	−0.0167
6	42.6366	2.3634	0.543	0.002	0.5414	0.0203	1.0513	0.077	0.0381	−0.0008	−0.0102	−0.0505
7	58.5520	−12.5520	−2.983	0.201	−3.1155	0.0829	0.7705	−0.9368	0.6566	−0.4016	−0.0376	−0.7588
8	38.8434	5.1566	1.191	0.011	1.1939	0.0294	1.0122	0.207	0.1668	−0.1518	0.0106	−0.0788
9	56.1060	6.8940	1.591	0.018	1.6044	0.0279	0.9639	0.271	−0.1906	0.1720	0.0997	0.0506
11	42.3672	−10.3672	−2.400	0.050	−2.4622	0.0332	0.8422	−0.4566	−0.0070	−0.1919	0.3409	−0.1314
12	50.5797	−3.5797	−0.822	0.003	−0.8208	0.0181	1.0324	−0.1114	0.0404	−0.0710	−0.0085	0.0170
13	38.7983	0.2017	0.047	0.003	0.0467	0.0425	1.0890	0.009	0.0053	0.0007	−0.0035	−0.0068
14	48.2920	−10.2920	−2.386	0.053	−2.4468	0.0359	0.8471	−0.4723	0.1724	−0.1518	0.1689	−0.3717
15	53.4211	0.5789	0.134	0.000	0.1332	0.0312	1.0755	0.023	−0.0133	0.0196	0.0003	−0.0005
16	49.1944	−0.1944	−0.045	0.000	−0.0450	0.0419	1.0883	−0.0094	0.0011	−0.0058	0.0004	0.005
17	44.2364	−6.2364	−1.453	0.025	−1.4616	0.0458	0.9998	−0.3201	−0.1313	0.2473	−0.1484	−0.0004
18	47.4963	−7.4963	−1.746	0.036	−1.7647	0.0447	0.9595	−0.3818	−0.1105	0.0495	−0.1918	0.3137
19	56.8596	−2.8596	−0.669	0.006	−0.6674	0.0546	1.0825	−0.1603	0.1174	−0.0933	−0.0017	−0.1098
20	53.4211	1.5789	0.365	0.001	0.3634	0.0312	1.0704	0.065	−0.0363	0.0534	0.0009	−0.0015
21	41.0485	−0.0485	−0.011	0.000	−0.0112	0.0328	1.0781	−0.0021	−0.0007	−0.0006	0.0010	0.0010
22	33.0359	1.9641	0.463	0.004	0.4616	0.0697	1.1109	0.126	0.0618	0.0197	−0.0995	−0.0337
23	55.6555	−0.6555	−0.151	0.000	−0.1507	0.0299	1.0739	−0.0265	0.0186	−0.0133	−0.0076	−0.0128
24	34.0938	1.9062	0.442	0.002	0.4397	0.0346	1.0714	0.083	0.0626	−0.0345	−0.0574	0.0034
25	51.6563	−2.6563	−0.609	0.002	−0.6075	0.0160	1.0435	−0.0775	0.0289	−0.0331	−0.0310	0.0114
26	49.9988	−0.9988	−0.229	0.000	−0.2278	0.0142	1.0555	−0.0274	0.0042	0.0009	−0.0143	−0.0004
27	33.7562	2.2438	0.520	0.003	0.5181	0.0357	1.0693	0.099	0.0767	−0.0446	−0.0674	0.0038
28	56.0974	−2.0974	−0.491	0.003	−0.4890	0.0547	1.0921	−0.1176	0.0835	−0.0689	0.0057	−0.0822
29	52.9182	−3.9182	−0.914	0.010	−0.9131	0.0479	1.0576	−0.2047	0.0522	−0.1284	−0.0429	0.1202
30	47.3386	−1.3386	−0.313	0.001	−0.3113	0.0510	1.0943	−0.0721	−0.0179	0.0490	−0.0435	−0.0014
31	41.5791	1.4209	0.329	0.001	0.3270	0.0310	1.0713	0.058	0.0303	−0.0456	0.0083	0.010
32	53.1749	−0.1749	−0.040	0.000	−0.0400	0.0214	1.0655	−0.0059	0.0036	−0.0029	−0.0012	−0.0025
33	56.4494	3.5506	0.822	0.006	0.8205	0.0333	1.0486	0.152	−0.0751	0.0768	0.0911	−0.0474
34	46.5403	0.4597	0.106	0.000	0.1059	0.0345	1.0795	0.020	0.0030	−0.0118	0.0070	0.007
35	33.9777	1.0223	0.241	0.001	0.2395	0.0655	1.1131	0.063	0.0418	−0.0030	−0.0287	−0.0422
36	41.2269	−2.2269	−0.510	0.001	−0.5085	0.0143	1.0464	−0.0611	−0.0361	0.0134	0.0239	0.011
37	44.7739	−0.7739	−0.181	0.000	−0.1798	0.0495	1.0956	−0.0410	−0.0181	0.0298	−0.0232	0.0104
38	46.1474	−0.1474	−0.034	0.000	−0.0338	0.0234	1.0678	−0.0052	0.0008	−0.0032	0.0026	−0.0005
39	28.8124	0.1876	0.044	0.000	0.0442	0.0745	1.1267	0.012	0.0075	−0.0025	−0.0112	0.0022
40	34.1836	−6.1836	−1.432	0.018	−1.4398	0.0338	0.9900	−0.2694	−0.2375	0.1512	0.1306	0.059
41	42.4672	−2.4672	−0.565	0.001	−0.5630	0.0123	1.0418	−0.0628	−0.0309	0.0109	0.0207	0.007
42	58.2680	−0.2680	−0.064	0.000	−0.0637	0.0912	1.1472	−0.0202	0.0139	−0.0088	0.0004	−0.0167
43	54.1411	−1.1411	−0.269	0.001	−0.2677	0.0681	1.1156	−0.0724	0.0075	−0.0237	−0.0365	0.0558
44	49.6539	−1.6539	−0.379	0.001	−0.3773	0.0137	1.0510	−0.0445	0.0099	−0.0183	−0.0102	0.0098
45	36.1048	1.8952	0.440	0.002	0.4379	0.0380	1.0752	0.087	0.0749	−0.0458	−0.0142	−0.0525
46	55.3814	−1.3814	−0.326	0.002	−0.3241	0.0679	1.1138	−0.0875	0.0144	−0.0310	−0.0475	0.0641
47	51.5535	3.4465	0.802	0.007	0.8009	0.0443	1.0621	0.172	−0.0943	0.1313	−0.0598	0.0742
48	43.0555	−0.0555	−0.013	0.000	−0.0129	0.0548	1.1032	−0.0031	−0.0017	0.0023	−0.0016	0.0011
49	49.4264	7.5736	1.737	0.012	1.7561	0.0156	0.9322	0.221	−0.0475	0.1118	0.0269	−0.0620
50	51.4171	1.5829	0.372	0.002	0.3706	0.0638	1.1074	0.096	0.0113	−0.0040	0.625	−0.0764
51	40.6820	0.3180	0.073	0.000	0.0728	0.0228	1.0669	0.011	0.0034	0.0022	−0.0080	0.000

TABLE 3A.3 Diagnostic Measures for Identifying Influential Observations

Obser-vation	Predicted Y value	Residual	Studentized Residual	Cook's Distance	Deleted Studentized Residual	Leverage	COVRATIO	DFFITS	DFBETA Intercept	X_3	X_5	X_6
52	52.1142	0.8858	0.204	0.000	0.2027	0.0206	1.0628	0.029	−0.0060	0.0028	0.0201	−0.0090
53	52.2998	−2.2998	−0.541	0.005	−0.5387	0.0628	1.0992	−0.1395	0.0041	0.0699	−0.0980	−0.0257
54	38.8072	−6.8072	−1.565	0.012	−1.5771	0.0200	0.9595	−0.2252	−0.1659	0.1320	0.0775	−0.0012
55	37.9971	1.0029	0.234	0.001	0.2326	0.0468	1.0914	0.051	0.0156	0.0164	−0.0400	−0.0068
56	45.8752	1.1248	0.260	0.000	0.2583	0.0270	1.0687	0.043	−0.0050	−0.0017	−0.0119	0.0338
57	59.1432	2.8568	0.677	0.010	0.6752	0.0780	1.1095	0.196	−0.0810	−0.0313	0.1349	0.0851
58	60.5919	4.4081	1.026	0.012	1.0265	0.0443	1.0440	0.221	−0.1612	0.1206	0.1358	0.015
59	46.2779	−0.2779	−0.064	0.000	−0.0633	0.0120	1.0553	−0.0070	−0.0019	0.0016	−0.0021	0.0019
60	48.7309	1.2691	0.292	0.000	0.2901	0.0183	1.0585	0.039	0.0037	0.0001	0.0198	−0.022
61	51.0753	2.9247	0.679	0.005	0.6774	0.0397	1.0652	0.137	−0.0721	0.1053	−0.0478	0.0531
62	55.2091	4.7909	1.108	0.010	1.1089	0.0308	1.0219	0.197	−0.0835	0.0959	0.1117	−0.0757
63	47.7253	−0.7253	−0.167	0.000	−0.1663	0.0249	1.0681	−0.0266	0.0076	−0.0180	0.0112	−0.0041
64	41.6153	−5.6153	−1.309	0.021	−1.3142	0.0471	1.0183	−0.2923	−0.1680	0.2420	−0.1111	0.0459
65	34.4792	5.5208	1.293	0.025	1.2974	0.0554	1.0291	0.314	0.1208	−0.0437	−0.2611	0.1518
66	46.9124	−1.9124	−0.440	0.001	−0.4378	0.0199	1.0554	−0.0623	0.0054	−0.0359	0.0168	0.015
67	55.1716	3.8284	0.892	0.010	0.8911	0.0460	1.0573	0.195	−0.1308	0.0908	0.0048	0.144
68	41.4645	4.5355	1.044	0.006	1.0448	0.0230	1.0197	0.160	0.1064	−0.1171	0.0269	−0.0324
69	51.5215	6.4785	1.499	0.019	1.5086	0.0321	0.9800	0.274	−0.0142	0.0498	0.1571	−0.1845
70	47.5978	1.4022	0.321	0.000	0.3198	0.0135	1.0525	0.037	0.0020	0.0084	0.0069	−0.016
71	51.7709	−1.7709	−0.420	0.004	−0.4186	0.0810	1.1264	−0.1243	−0.0135	0.0700	−0.0972	0.0128
72	53.8977	1.1023	0.259	0.001	0.2579	0.0631	1.1099	0.066	−0.0006	0.0009	0.0488	−0.0476
73	51.6750	−0.6750	−0.157	0.000	−0.1560	0.0403	1.0854	−0.0320	0.0090	−0.0224	−0.0018	0.0151
74	53.1473	6.8527	1.575	0.012	1.5877	0.0197	0.9580	0.225	−0.1195	0.1345	0.0724	−0.0059
75	37.8480	3.1520	0.726	0.003	0.7240	0.0231	1.0441	0.111	0.0802	−0.0647	−0.0478	0.0126
76	53.3979	−4.3979	−1.013	0.006	−1.0127	0.0229	1.0223	−0.1549	0.0881	−0.1094	−0.0304	−0.0009
77	41.9223	0.0777	0.018	0.000	0.0178	0.0207	1.0649	0.002	0.0006	0.0007	−0.0018	0.000
78	50.4376	−3.4376	−0.803	0.008	−0.8010	0.0496	1.0680	−0.1829	0.0197	−0.1014	−0.0212	0.1227
79	33.2389	5.7611	1.351	0.028	1.3573	0.0586	1.0257	0.338	0.1488	−0.0554	−0.2876	0.1444
80	49.4292	6.5708	1.511	0.012	1.5210	0.0201	0.9665	0.217	−0.0162	0.0810	0.0606	−0.1216
81	53.2400	5.7600	1.331	0.014	1.3368	0.0303	0.9982	0.236	−0.0488	0.0643	0.1474	−0.1303
82	48.7182	−1.7182	−0.412	0.005	−0.4104	0.1000	1.1505	−0.1368	0.0003	0.0751	−0.0460	−0.0812
83	36.9440	4.0560	0.952	0.015	0.9517	0.0601	1.0682	0.240	0.0826	−0.1139	−0.1103	0.1636
84	45.5157	−8.5157	−1.949	0.011	−1.9786	0.0113	0.8974	−0.2115	−0.0681	0.0466	−0.0372	0.0520
85	47.4515	5.5485	1.271	0.005	1.2751	0.0129	0.9870	0.145	−0.0272	0.0594	−0.0306	0.0313
86	41.5442	1.4558	0.334	0.001	0.3328	0.0182	1.0571	0.045	0.0314	−0.0225	0.0014	−0.0208
87	46.2112	4.7888	1.097	0.004	1.0980	0.0125	1.0041	0.123	−0.0059	0.0435	−0.0374	0.0154
88	43.0351	−7.0351	−1.612	0.009	−1.6260	0.0137	0.9473	−0.1914	−0.1081	0.0610	0.0021	0.077
89	36.8815	−2.8815	−0.665	0.003	−0.6627	0.0263	1.0513	−0.1089	−0.0682	0.0232	0.0792	−0.0043
90	56.4119	3.5881	0.839	0.010	0.8375	0.0522	1.0683	0.196	−0.1370	0.0916	0.0131	0.145
91	48.0182	0.9818	0.227	0.000	0.2261	0.0325	1.0755	0.041	−0.0104	0.0000	−0.0058	0.0344
92	46.2560	−7.2560	−1.692	0.035	−1.7087	0.0472	0.9696	−0.3803	−0.1349	0.0601	−0.1684	0.3225
93	43.7570	−0.7570	−0.180	0.001	−0.1790	0.0834	1.1361	−0.0540	−0.0118	0.0376	−0.0124	0.0273
94	30.6410	5.3590	1.259	0.026	1.2624	0.0608	1.0388	0.321	0.2848	−0.2658	−0.1007	−0.0117
95	35.8180	−4.8180	−1.113	0.009	−1.1141	0.0288	1.0194	−0.1919	−0.1648	0.0940	0.0834	0.063
96	23.3730	1.6270	0.392	0.005	0.3899	0.1058	1.1587	0.134	0.0983	−0.0367	−0.1152	−0.0029
97	53.6254	6.3746	1.466	0.012	1.4753	0.0211	0.9730	0.216	−0.1272	0.1370	0.0616	0.015
98	33.1216	4.8784	1.140	0.018	1.1418	0.0514	1.0410	0.265	0.2191	−0.2230	−0.0672	0.0140
99	39.9910	2.0090	0.462	0.001	0.4600	0.0203	1.0549	0.066	0.0387	−0.0380	−0.0208	0.0175
100	44.8478	−11.8478	−2.741	0.062	−2.8399	0.0321	0.7777	−0.5170	0.0836	−0.2611	0.3354	−0.2118

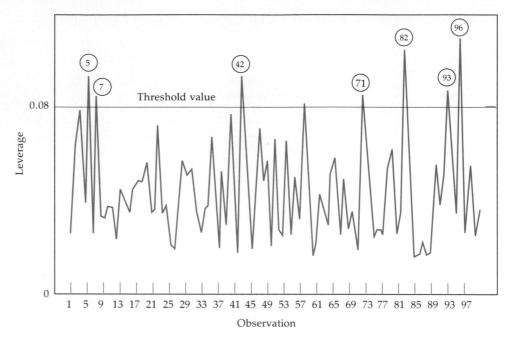

FIGURE 3A.3 Leverage values.

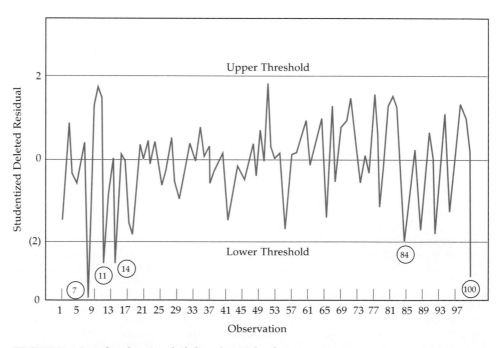

FIGURE 3A.4 Studentized deleted residuals.

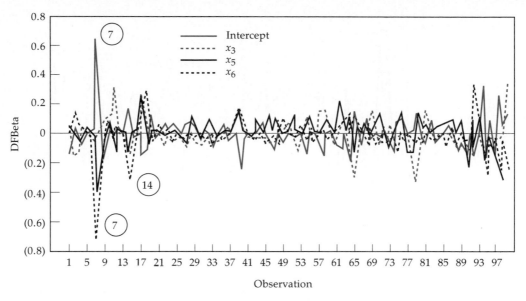

FIGURE 3A.5 Plot of DFBETAs.

84). In each instance, these are outliers from the regression equation estimated *without* their influence.

Our next step is to identify the impact on the estimated regression coefficients by use of the DFBETA values (see Figure 3A.5). With no exact limits available, one must look for substantial impacts, as noted in Figure 3A.3. As we see, case 7 substantially impacts three of the four coefficients (intercept, X_3, and X_6). The only other impact of note is for case 14 and X_6. If we consider all coefficients simultaneously by examining the COVRATIO (see Figure 3A.6), we identify 11 additional points (3, 5, 11, 14, 39, 42, 71, 82, 93, 96, 100) as having some disproportionate impact on the *set* of coefficients.

Thus far in our analysis, several points (7, 11, 14, 100) have emerged as potentially bad influential points owing to their substantial influence *and* differences from the remaining observations. We have one remaining measure, Cook's distance, which simultaneously captures both the leverage effects along with the change in residuals (see Table 3A.3). In this case, only case 7 is substantially higher than the other observations, thus making it a potential candidate for remedy.

Step 4: Selecting and Accommodating Influential Cases

Although there is no single procedure for identifying influential cases and then deciding on the course of action, the basic premise is quite simple. In the absence of data entry error or other correctable reasons, influential cases that are substantially different from the remaining data on one or more variables should be closely examined. If it is ascertained that a case is unrepresentative of the general population, it should be eliminated. Our objective is to estimate the regression equation on a representative sample to obtain generalizable results. If the sample

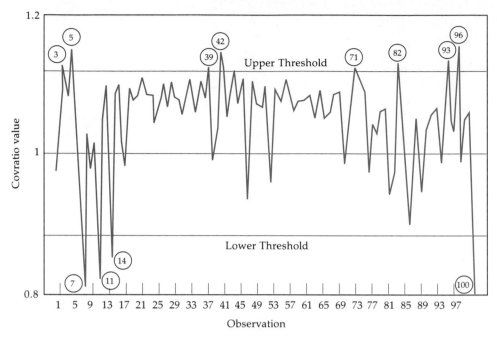

FIGURE 3A.6 COVRATIO values.

contains one or more unrepresentative observations, it hinders us in achieving this objective.

For purposes of illustration, we will select four cases (7, 11, 14, 100) for elimination. These cases were consistently identified by the diagnostic analyses and are deemed the cases with the most impact on improving the regression equation. Table 3A.4 shows the final regression model with these four cases eliminated. Comparing these results with those in Table 3.8 in Chapter 3, we see substantial improvement in every area. Overall prediction was improved, with the R^2 changing from .768 to .833, more than the effect we obtained by adding the third variable (X_6) to the equation. Also, the standard error decreased from 4.39 to 3.69, a 16 percent improvement. Moreover, each coefficient improved in statistical significance, indicating a strengthening of the relationships by removing these influential outliers.

Summary

The identification of influential cases is an essential step in interpreting the results of regression analysis. The analyst must be careful, however, to use discretion in the elimination of cases identified as influential. There are always outliers in any population, and the researcher must be careful not to trim the data set so that good results are almost guaranteed. Yet, one must also attempt to best represent the relationships in the sample, and the influence of just a few cases may distort or completely inhibit achieving this objective. Thus, we recommend that one use these techniques wisely and with care, as they represent both potential benefit and harm.

TABLE 3A.4 Multiple Regression Results After Eliminating Four Influential
Observations

Overall Regression Model Results

Multiple R	.913
Multiple R-square	.833
Standard error of estimate	3.698

Analysis of Variance

	Sum of Squares	Degrees of Freedom	Mean Square	F ratio
Regression	6291.8	3	2097.3	153.37
Residual	1258.1	92	13.7	

Variables in the Equation

Variables	Coefficient	Standard Error of Coefficient	Standardized Regression Coefficient	Partial t value
Intercept	−9.645	2.803		−3.440
X_3 Price flexibility	3.719	.274	.582	13.566
X_5 Overall service	7.093	.521	.600	13.605
X_6 Sales force image	2.337	.521	.198	4.482

Variables Not in the Equation

	Partial Correlation	t value
X_1 Delivery speed	−.004	−.036
X_2 Price level	−.019	−.177
X_4 Manufacturing image	−.116	−1.117
X_7 Product quality	.136	1.311

Summary

As the applications of regression analysis increase in both scope and complexity, it becomes essential to explore the issues addressed in this appendix. Both multicol-linearity and influential observations can have substantial impact on the results and their interpretation. However, recent advances in diagnostic techniques such as those described earlier now provide the analyst a simplified method of performing analyses that will identify problems in each of these areas. Whenever the regression analysis encounters either of these problem areas, the analyst is encouraged to investigate the issues raised here and make the appropriate remedies if needed.

Questions

1. Describe the reasons for not relying solely on the univariate correlation matrix for diagnosing multicollinearity.
2. In what instances does the detection of outliers potentially miss other influential observations?
3. Describe the differences in using residuals (including the studentized residual) versus the single-case diagnostics of DFBETA and DIFFITS.
4. What criteria would you suggest for determining whether an observation was to be deleted from the analysis?

References

1. Barnett, V., and T. Lewis. *Outliers in Statistical Data, 2d ed.* New York: Wiley, 1984.
2. Belsley, D. A., E. Kuh, and R. E. Welsch. *Regression Diagnostics: Identifying Influential Data and Sources of Collinearity.* New York: Wiley, 1980.
3. Rousseeuw, P. J., and A. M. Leroy. *Robust Regression and Outlier Detection.* New York: Wiley, 1987.
4. Weisberg, S. *Applied Linear Regression.* New York: Wiley, 1985.

Factors Affecting the Performance of Individual Chain Store Units: An Empirical Analysis

Richard T. Hise
Myron Gable
J. Patrick Kelly
James B. McDonald

Retail chain store executives are constantly faced with the problem of achieving success through the choice of a combination of decision variables. Although many executives are increasingly using more sophisticated tools for decision making, there still appear to be instances of decisions by intuition, hunch, or untested rules of thumb.

There are a number of variables or factors which can have an impact on the success of individual chain store units. They include such variables as product offerings; store location; strength, number, and strategies of competitors; promotional efforts; store factors, such as store size, inventory levels, and number of employees; store manager characteristics, including such factors as the store manager's experience, age, marital status, and educational level; and market factors, such as disposable income and population.

While nonretailing industries have had large-scale studies conducted on the factors that affect their performance, especially in the area of return on investment (Schoeffler, Buzzell, and Heany 1974; Buzzell, Gale, and Sultan 1975), few such studies have been done in the retailing sector. Those that have are relatively small-scale in nature and are now somewhat dated. One early study in retailing focused upon the variety of merchandise offered by a store and its impact on sales, cost, and profits (Baumol and Ide, 1962). This study indicated that inventory level and va-

riety of merchandise had a positive effect on sales and profits. A later study focused upon how profit was influenced by a number of internally controllable operating factors, such as markdowns, rent, publicity, sales volume, stock turnover, and average sale (Dalrymple 1966). In that study, sales volume was found to explain the greatest variance in profit levels.

None of these prior studies has dealt with the importance of relatively uncontrollable, long-run, and irreversible variables on store performance. The present study reports the results of an analysis of the impact of 18 independent variables—controllable/uncontrollable, short-run/long-run, and reversible/irreversible—on three performance factors of 132 retail chain store units. The study's objectives, all relating to individual retail chain store units, were:

1. To determine the aggregate impact of these 18 independent variables on the three performance measures.
2. To identify the effect of four major groups of predictor variables on the three performance measures.
3. To identify those individual independent variables which had the greatest impact on the three measures of performance.
4. To develop some conclusions as to the value of using these independent variables to predict unit performance.
5. To formulate some recommendations regarding a marketing strategy that chain store executives should consider in efforts to improve the performance of their retail stores.

From "Factors Affecting the Performance of Individual Chain Store Units: An Empirical Analysis." *Journal of Retailing*, Vol. 59, No. 2 (Summer 1983), pp. 22–39. Reprinted with permission.

Methodology

The 18 independent variables used to predict the chain stores' performance can be grouped into four major areas. These variables are identified with their mean, minimum, maximum, and standard deviations in Table 1. The performance factors (or dependent variables) were sales volume, contribution income (gross margin less direct expenses), and return on assets. The chain's top management considered return on assets to be the most important of the three performance variables.

The executives of this large retailing corporation agreed to provide data on all its units. These data consisted of the three performance measures and three of the four groups of independent variables: store, competition, and location. The store manager factors were obtained from a questionnaire administered to each store manager.

A total of 179 out of 180 units responded to the questionnaire. Some of the secondary data on the units contained missing or unusable data. Because of missing and unusable data, 37 units were dropped from the analysis; the usable sample was thus 132 units.

The corporation sells nonclothing items in stores located in malls throughout the United States. (Company executives asked the authors not to reveal the name of the company of the nature of its products.) Although the product is essentially a shopping good, customers frequently make purchase decisions before visiting the retail outlet, so the product sold tends to take on some of the characteristics of a specialty good.

Average annual sales volume for these stores was approximately $565,000; their average size was slightly over 3,300 square feet, and the average number of employees, including the store manager, was 7.4. The average number of years the stores had been open was 5.7, and the average mall size in which the stores operated was 778,100 square feet. Because the product assortment was similar for each store, and their promotional budgets, copy, and media were also consistent and basically the same, these factors were not included as independent variables.

The nine store manager variables were included as independent variables because of previous research on retail employees. While previous studies did not focus on store managers, their results are perhaps indicative of what might be found if store managers were analyzed. Weaver (1969) found that older, better-educated, married, or divorced salespersons were more productive. Cotham (1969) found that retail employees who were old, had prior retailing experience, and had worked for another retail firm were more productive. Paul and Bell (1968) identified older, more experienced, and slightly better educated retail employees as being more productive. They also found retail employees who "worked harder" were able to generate higher sales. These results suggest that retail store managers who are older, married, and have more children might be more serious about their work and therefore more productive.

While some of the other nine variables have been used in previous research (Baumol and Ide 1962: Dalrymple 1966), several were used because of their availability from company records and the intuitive belief that they would likely be an important determinant of chain store units' performance. For example, because the products sold are basically shopping goods, the presence of secondary, rather than primary, competitive stores should increase sales and profits. Larger malls with more traffic and larger SMSAs would tend to suggest high sales volume and possibly better profit results. Stores with more inventory and more sales help should result in more acceptable products and greater sales rates. Stores that were in business longer would have overcome the start-up problems that initially would adversely affect sales and profits in newer locations.

The above factors can be viewed in terms of their reversibility, the control that management can exert, and their time duration. Store location and store size, especially in mall locations, are examples of somewhat irreversible factors, whereas, for example, inventory levels and number of employees are relatively reversible. Although such factors as store manager characteristics and number of employees are factors over which management can exert a good deal of control, market dimensions and various actions by competitors are largely beyond their control.

TABLE 1 Grouping of 18 Variables Used to Predict Retail Store Performance

Category of Variables	Mean	Min	Max	Standard Deviation	Units of Measure
Store manager variables					
Age	25.43	18.00	48.00	3.60	Years
Annual income[a]	134.39	70.00	612.00	47.758	Dollars (00)
Marital status[b]	2.67	1.00	4.00	1.47	1 = single, 2 = separated & divorced, 3 = widow, 4 = married
Number of children	.34	0.00	3.00	.62	Number
Educational level	14.02	9.00	18.00	1.71	Years: high school graduate = 12
Hours worked per week	51.19	35.00	80.00	7.23	Hours
Experience in retailing	5.58	1.00	22.00	2.56	Years
Experience in present position	2.38	0.00	10.00	2.036	Years
Experience with present employer	4.09	0.00	10.00	1.92	Years
Store variables					
Store size	33.24	6.00	52.00	8.26	Sq. feet (000)
Fixed assets	569.31	102.00	1,825.00	277.53	Dollars (000)
Inventory level	670.18	362.00	1,051.00	136.50	Dollars (000)
Years store opened	5.74	2.00	19.00	2.99	Years
Employees per store	7.41	4.00	20.00	3.39	Number
Competitive variables					
Primary competitors— Similar retail stores in the same mall location	.45	0.00	2.00	.60	Number
Secondary competitors— Similar departments in stores in the mall location	2.24	0.00	5.00	1.02	Number
Location variables					
Mall size	778.10	75.00	2,267.00	412.63	Sq. feet (000)
Market size	2.10	1.00	3.00	.80	Classified by SMSA size 1 = small, 2 = medium, 3 = large
Performance variables					
Sales	565.21	217.00	1,243.00	200.28	Dollars (000)
Contribution income	89.26	2.97	234.33	52.40	Dollars (000)
Return assets	74.36	3.50	232.00	44.31	Percentage

[a] Annual income is a salary plus bonus. The bonus can add an additional 7 to 10 percent to the base salary.
[b] Marital status codes represent a movement toward stability.

TABLE 2 Regression Estimates of Sales Relationship

Variable	Order Entered	Regression Coefficient	Standard Error	t-Statistic $(H_0 b_i = 0)$	Partial Correlation Coefficient	Standardized Regression Coefficients
Number of employees	1	25.752	3.964	6.496[b]	.509	.436
Inventory level	2	.685	.106	6.483[b]	.508	.467
Hours worked per week	3	−1.894	1.748	−1.084	−.098	−.068
Store size	4	−3.764	1.744	−2.159[b]	−.193	−.155
Fixed assets	5	.123	.057	2.169[b]	.193	.171
Years experience with present employer	6	24.240	9.311	2.603[b]	.230	.231
Years experience in present position	7	−15.898	8.362	−1.910[a]	−.170	−.162
Number of primary competitors	8	−36.369	22.736	−1.600	−.144	−.108
Mall size	9	.057	.041	1.388	.125	.117
Market size	10	18.388	16.829	1.093	.099	.073
Intercept		−60.39	118.64	−.509		

[a] $P < .10$
[b] $P < .05$

Regression equation characteristics:

$R^2 = .598$ $s = 132.11$ $D.W. = 2.167$
$R^2 = .565$ $F(10.121) = 18.01$ $N = 132$

Chow test of collective explanatory power of seven excluded variables:
$F(8.113) = .31$
Chow test of stability of estimated coefficients:
$F(11.121) = 1.11$

Data Evaluation

Regression analysis was used to estimate the relationship between the independent factors (X_1) and the performance factor (PF). For each performance factor, the technique of least-squares was used to estimate the regression coefficients (b_i) in an equation of the form:

$$PF = b_0 + b_1 X_1 + b_2 X_2 + \ldots + b_n X_n + u$$

where u denotes a random disturbance term. The regression coefficient (b_i) represents the expected change in the performance indicator associated with a one-unit change in the ith independent variable. Forward and backward stepwise regression was used to determine which of the 18 independent variables to include in the final regression equation for each of the three performance factors. Forward stepwise regression is based upon adding variables one at a time, whereas backward stepwise regression starts with including all the variables and then deleting variables one at a time. The criterion for adding (deleting) variables is to select the variable which increases (reduces) the sum of squared errors the least (most). The two procedures do not necessarily yield the same equations with the same number of corresponding variables. For both the forward and backward regressions considered, the agreement was very similar and the equation with the largest adjusted R^2 was selected. The stepwise regressions run on the three performance factors contained 10, 10, and 12 independent variables; results are reported in Tables 2, 3, and 4. Tables 2, 3, and 4 report the t statistics and regression results, along with some additional diagnostic statistics, standardized regression coefficients, and the partial correlation coefficients.

One problem which can occur with this type of data is multicollinearity—the situation where two or more of the independent variables are intercorrelated. The multicollinearity problem is a

TABLE 3 Regression Estimates of Contribution Income Relationship

Variable	Order Entered	Regression Coefficient	Standard Error	t-Statistic $(H_0 b_i = 0)$	Partial Correlation Coefficient	Standardized Regression Coefficients
Inventory level	1	.182	.031	5.871[b]	.474	.474
Number of employees	2	5.963	1.154	5.168[b]	.428	.386
Store size	3	−1.032	.467	−2.208[b]	−.198	−.163
Number of primary competitors	4	−16.185	6.690	−2.419[b]	−.216	−.184
Hours worked per week	5	−.684	.513	−1.332	−.121	−.094
Market size	6	7.707	4.881	1.579	.143	.118
Educational level	7	2.864	2.016	1.420	.129	.093
Marital status	8	3.612	2.426	1.489	.135	.102
Mall size	9	.012	.012	1.055	.096	.096
Years experience with present employer	10	5.982	2.720	2.199[b]	.198	.219
Years experience in present position	11	−5.716	2.492	−2.294[b]	−.206	−.222
Number of secondary competitors	12	3.962	3.606	1.099	−.100	−.077
Intercept		−95.528	43.250	−2.209		

[b] $P < .05$

Regression equation characteristics:

$R^2 = .510$ $s = 38.48$ D.W. = 2.19
$R^2 = .461$ $F(12.119) = 10.32$ N = 132

Chow test of collective explanatory power of eight excluded variables:
$F(6.113) = .31$

Chow test of stability of estimated coefficients:
$F(13.106) = .93$

question of degree. It can result in making it difficult to obtain accurate estimates of the individual effects of variables, i.e., b_i. This is frequently associated with variables that appear individually to lack significant explanatory power, while collectively being associated with significant explanatory power. A Chow test[1] was used to test for the collective explanatory power of the independent variables deleted from each of the performance factor equations. In each case, the deleted independent variables lacked significant explanatory power. Another characteristic of multicollinearity is that the estimates may be very sensitive to the addition of more data. As a check on the stability of estimates, the data were divided into two equal groups, and separate regressions run on the two subgroups. The two sets of estimates were compared using a Chow test, and in neither case were the differences found to be statistically significant.

[1] The Chow test can be used to test the hypothesis that the coefficients satisfy certain restrictions. It consists of estimating the model without the constraints imposed to yield the sum of squared errors (SSE) with associated degrees of freedom equal to $N - K$ (sample size minus the number of estimated coefficients). The model is then reestimated with the parameter constraints imposed with the corresponding sum of squared errors and degree of freedom denoted SSE* and $(N - K)^*$

The statistic defined by

$$\frac{\dfrac{SSE^* - SSE}{(N - K)^* - (N - K)}}{\dfrac{SSE}{N - K}}$$

is distributed as an $F[(N - K)^* - (N - K), N - K]$ with "large" values providing the basis for rejecting the hypotheses that the constraints are valid.

TABLE 4 Regression Estimates of Rate of Return on Assets Relationship

Variable	Order Entered	Regression Coefficient	Standard Error	t-Statistic $(H_0:b_i = 0)$	Partial Correlation Coefficient	Standardized Regression Coefficients
Fixed assets	1	−.048	.013	−3.624[b]	−.313	−.300
Number of employees	2	4.550	1.008	4.514[b]	.380	.348
Number of primary competitors	3	−12.554	5.406	−2.322[b]	−.206	−.169
Inventory	4	.085	.027	3.141[b]	.274	.261
Store size	5	−1.099	.446	−2.465[b]	−.219	−.205
Marital status	6	3.910	2.206	1.773[a]	.159	.130
Educational level	7	2.943	1.829	1.610	.145	.113
Number of secondary competitors	8	5.633	3.104	1.815[a]	.163	.130
Years' experience in present position	9	−7.122	2.200	−3.238[b]	−.282	−.327
Years' experience with present employer	10	6.730	2.340	−2.876[b]	.253	.291
Intercept		−21.579	31.558	−.684		

[a] $P < .10$
[b] $P < .05$

Regression equation characteristics:
$$R^2 = .427 \qquad s = 34.89 \qquad D.W. = 2.26$$
$$\bar{R}^2 = .380 \qquad F(10.121) = 9.03 \qquad N = 132$$
Chow test of collective explanatory power of eight excluded variables:
$$F(8.113) = .40$$
Chow test of stability of estimated coefficients:
$$F(11.110) = 1.12$$

Finally, the estimated residuals associated with each of the equations were analyzed for systematic behavior. The results do not provide a basis for rejecting the assumption of independently and identically distributed random disturbances.

The coefficients (b_i) depend upon the units of measurement for PF and X_i. The standardized beta or standardized regression coefficients do not depend upon the units of measurement and facilitate a comparison of the relative impact of different variables. They are obtained from b_i by multiplying b_i by (S_{Xi}/S_{PF}) where s denotes the standard deviation of the indicated variable. The standardized beta coefficient can then be interpreted as the number of standard deviations that PF is expected to change in response to a one-standard deviation change in X_i.

Partial correlation coefficients are frequently used in comparing the impact of variables in multiple-regression equations. These coefficients are

independent of the units of measurement and show the correlation between the performance factor and each of the independent variables when the influences of the other variables in the equation are held constant. The regression coefficients and standardized beta coefficients also do this, but the partial correlation coefficients must be between 1 and −1.

Findings

Various combinations of the 18 predictor variables could account for 43 percent to 60 percent of the variation in the three performance variables analyzed. Of the four groups of independent factors employed—store manager variables, store variables, competitive variables, and location variables—store variables were the most important in explaining the variability in the three performance measures. If we look at the first three independent variables entered into each of the three regression models, we find inventory levels and number of employees as being the two most important variables in explaining variation in sales and contribution income, while the fixed assets item was first in explaining return on assets with number of employees, number of primary competitors, and inventory levels entering in that order. The managers' years of experience in the present position (a store manager variable) also was important in explaining variation in all three performance measures; however, there appears to be an inverse relationship for each of the performance measures and managers' years of experience in the present position.

Sales Volume

The results for predicting sales volume appear in Table 2. The estimated regression relationship for sales volume is:

Sales =

$$
\begin{aligned}
&- 60.385 + 25.752 \text{ (employees)} \\
&+ .685 \text{ (inventory)} - 1.894 \text{ (hours)} \\
&- 3.765 \text{ (store size)} + .123 \text{ (fixed assets)} \\
&+ 24.240 \text{ (experience with present employer)} \\
&- 15.898 \text{ (experience in present position)} \\
&- 36.369 \text{ (primary competitors)} \\
&+ .057 \text{ (mall size)} \\
&+ 18.388 \text{ (market size)}
\end{aligned}
$$

The interpretation of this equation is that some of the independent variables have a positive effect upon sales volume, while others have negative effects. The variables with a positive effect are number of employees, inventory levels, fixed assets, managers' experience with present employer, mall size, and market size. An increase in any of these variables is expected to increase the sales volume. The amount of increase expected would differ for each variable on the basis of the regression coefficient. For example, a $100 increase in the inventory would produce an expected increase in sales of $685 (see Table 1 for units of measure for each independent variable).

Of the four variables having a positive impact on sales volume, one is a store manager variable and three are store variables. This suggests that selection of the store manager possessing certain characteristics can have a positive effect on a retail store's sales volume, thus further suggesting that proper attention should be given to the re-

cruiting and employee selection process. While store variables, such as inventory levels, number of employees, and fixed assets, have a positive impact on sales, these are often the same areas where controls are established to minimize the investment in inventory, employees' work schedules, and dollars invested in fixed assets. Also, future site selection considerations should be made in larger malls that have a minimized number of primary competitors and that are located in large market areas.

Those variables having a negative effect on sales volume in this regression equation are managers' years of experience in their present position, hours worked, store size, and primary competitors. Two of these are store manager variables, one is a competitor variable, and one is a store variable.

The regression equation characteristics of sales volume indicate an R^2 of .60. This indicates that 60 percent of the variation in sales volume is explained by this equation. The order of the variables' entry into the regression equation is presented in Tables 2, 3, and 4. When the partial correlation coefficient is used to indicate impact, the variable with the greatest effect is number of employees (.509), followed closely by inventory levels (.508), managers' years of experience with present employer (.230), fixed assets (.193), and so on.

Contribution Income

Contribution income is the gross margin less the directly identifiable expenses of that store. The results for predicting contribution income appear in Table 3. The estimated regression relationship for contributions income is:

Contribution income =

$$
\begin{aligned}
&-95.528 + .182 \text{ (inventory)} \\
&+5.963 \text{ (number of employees)} \\
&-1.032 \text{ (store size)} \\
&-16.185 \text{ (primary competitors)} \\
&-.684 \text{ (hours worked)} \\
&+7.707 \text{ (market size)} \\
&+2.864 \text{ (educational level)} \\
&+3.612 \text{ (marital status)} \\
&+.012 \text{ (mall size)} \\
&+5.982 \text{ (experience with present employer)} \\
&-5.716 \text{ (experience in present position)} \\
&+3.962 \text{ (secondary competitors)}
\end{aligned}
$$

Those variables having a positive effect on contribution income are inventory levels, number of employees, market size, managers' educational level, managers' marital status, mall size, managers' years of experience with present employer, and number of secondary competitors. Three of the positive variables are store manager variables, two are store variables, two are location variables, and one is a competitive variable.

The positive variables on this list are very similar to those influencing total sales volume. The similarity is further supported by the high pairwise correlation of sales with contribution income shown in Table 5. This correlation is $R = .93$.

The variables having a negative impact are store size, primary competitors, hours worked, and the managers' years of experience in their present position. These represent two of the store manager variables, one store, and one competitive variable described in Table 1.

The regression equation characteristics of contribution income indicate an R^2 of .51. This equation then explains 51 percent of the variation of contribution income among stores in this study. Using the same impact measures of partial correlation, the variables with the greatest effect are inventory level (.474), number of employees (.428), primary competitors (−.216), managers' years of experience in present position (−.206), and so on through the partial correlation coefficients in a descending order.

Return on Assets

The result for predicting return on assets appear in Table 4. The estimated regression relationship for return on assets (ROA) is:

Return on assets =
$$-21.579 - .048 \text{ (fixed assets)}$$
$$+4.550 \text{ (number of employees)}$$
$$-12.554 \text{ (primary competitors)}$$
$$+.0847 \text{ (inventory)} - 1.099 \text{ (store size)}$$
$$+3.910 \text{ (marital status)}$$
$$+2.943 \text{ (educational level)}$$
$$+5.633 \text{ (secondary competitors)}$$
$$-7.122 \text{ (experience in present position)}$$
$$+6.730 \text{ (experience with present employer)}$$

The variables having a positive impact on return on assets are fixed assets, number of employees, inventories, marital status, manager's education, secondary competitors, and manager's years of experience with present employer. Table 5 indi-

TABLE 5 Pairwise Correlation Coefficients Between the Three Performance Factors

	Contribution Income	Return on Assets
Sales	.93	.70
Contribution income		.86

cates a pairwise correlation between ROA and sales of .70 and ROA with contribution income at .86. This again suggests that the same variables have a positive impact on sales, contribution income, and ROA.

Of the ten variables included in this equation, there are four store manager variables, four store variables, and both of the competitive variables.

The variables having a negative effect are fixed assets, primary competitors, store size, and managers' years of experience in the present position. (These also had a negative effect in the contribution income and sales equations.)

The regression equation characteristics of the return on assets indicated an $R^2 = .428$. This indicates that the equation presented above explains 43 percent of the variance in ROA for the stores in this study. The partial correlations indicate the variables with the greatest effect in the equation: number of employees (.380), fixed assets (−.313), managers' years of experience in the present position (−.282), and so on.

Unexplained Variance and Limitations

While the results presented thus far have focused on the level of performance variation explained by the regression equations, it is also helpful to indicate the amount of dependent variable variation not explained. In this study, the independent variables were not able to account for 40 percent of the variation in sales volume, 49 percent of contribution income, and 57 percent of the variation in return on assets. Because management believes that product assortments, prices, and promotional budgets, advertising copy, and media are essentially the same for all stores in this study, the authors believe that the most important independent variables excluded in this analysis were the various marketing strategies of competitive firms, including their prices, product assortment, promotion mixes, and expenditures, as well as their various store and store manager characteristics.

This analysis dealt with only one type of store and was cross-sectional in nature. Additional research on factors affecting performance of chain store units should be expanded to different types of stores, such as supermarkets, department stores, drug stores, etc., and should be done longitudinally in order to assess the impact of time. The results of the present study have identified a number of controllable variables as the most significant predictors of chain store units' performance. It will be important to find out, in future studies, the effect of less controllable, more longrun, and less reversible factors—such as location and competition—on the performances of retail chain store units. These findings point to the competitive variable as being potentially important, but substantiation will have to await further research.

Another important limitation is that the use of linear regression equations suggests that the addition of one more unit of the independent variables will continue to produce a positive or negative effect on a continual basis. Nonlinear specifications might provide useful insight on an optimal level of inventory, number of employees, store size, etc.

A final limitation is that the results presented here are drawn from one retail chain at one point in time. Generalizing the results to other retail firms should be done with caution. The value or contribution of this study's findings exists in the methods and techniques used to identify the factors influencing selected performance variables. The authors strongly urge other chain store retailers to pursue similar analysis for their own operations.

Executive Summary

The results of this study have provided insight into some prediction factors that have an important impact in explaining the variation in total sales, contribution income, and return on assets. They were identified as the three key performance measures of 132 retail chain store units. From the perspective of the chain's executives, these findings should assist in developing a set of marketing strategies that can potentially help their stores in the aggregate to maintain acceptable levels of performance. What is most significant in this regard is that those factors found to influence performance the most strongly were number of employees, inventory levels, years the store manager had spent in the same position, fixed assets, and the manager's years of experience with the present employer.

These independent variables had three important characteristics in common:

1. They are short-run in their scope.
2. They are controllable by corporate management.
3. They are reversible.

These characteristics thus suggest that marketing strategies can be instituted and manipulated by the corporation's executives, and that the results of changes in these strategies can be noted relatively soon. These advantages obviously result from the fact that, based upon partial correlations, store or store manager factors were the more powerful in predicting store performance compared with location (relatively long run and irreversible) and the competitive (largely uncontrollable) factors.

The impact that level of inventory has on these stores' performance suggests that management needs to ensure that individual units are stocking an adequate product assortment, that stockouts are minimized, and that any unbalanced inventory situations are handled as quickly as possible. Checkout systems with automatic inventory maintenance capabilities are likely to be helpful in stores where extensive product assortments are carried and stockouts occur frequently. Efficient and rapid delivery systems are also important in reducing the number of stockouts.

The fact that this company's products take on the character of specialty goods probably does much to explain the significance of the inventory factor in explaining performance levels. That is, customers are predisposed to purchase certain products and will do so if they are available in this company's outlets. If these items are not available, it is highly unlikely that a substitute product will be purchased. Rather, other stores will be visited in search of the desired item.

This study's findings indicate number of employees as having important positive influence on the performance of retail stores. This finding suggests that management needs to consider whether:

1. The store is employing, in general, a sufficient number of employees.
2. It is employing sufficient numbers of clerical personnel to handle customer checkout.
3. It has enough salespersons available to accommodate peak periods of customer traffic.
4. It has allocated enough clerical personnel to maintain adequate inventory levels. This recommendation seems especially important in view of the significant impact of level of inventory on the performance of these chain store units.

The relationship between the store managers' length of time in that position and their stores' performance should be considered when management formulates promotion, transfer, and retention policies. Since an inverse relationship between length of tenure and stores' performance is suggested, corporate executives may want to investigate whether store manager enthusiasm and motivation wane as their tenure in that position increases. Another explanation for this negative effect is that those managers with the best performance are promoted, while those remaining continue to perform at constant and satisfactory levels. The cross-sectional rather than longitudinal nature of the data on each manager makes it difficult to interpret exactly this finding. If a negative relationship does exist, company executives will have to decide what should be done with managers of individual stores who have protracted stays in their position. Assuming these individuals are not promotable to higher-level positions that are more challenging, dismissal or movement to new store locations may have to be considered.

Hiring policies should also be implemented to focus upon prospective employees with higher levels of education and those who are married, since these two conditions were found to be associated with higher store performance factors.

A final set of variables needs to be considered in assessing the potential performance of site locations. Although this decision, once it is made, is fairly long-run and irreversible in nature, management can have an influence over location decisions before they are made. Results of this study indicated that locating in a mall with secondary competitors is advantageous, but that locating in malls with primary competitors should be avoided. Thus it appears that departments in other stores selling similar products may draw shoppers to the mall but, because of inadequate assortments in stores not specializing in these products, customers make purchases in the store offering a wider assortment. But when primary competitors are added to the mall, they may divide the market into pieces that are too small.

Retail managers should attempt to understand the factors that have the greatest effect on the performance levels of their stores, and then to make appropriate adjustments to maximize each store's performance. This study has attempted to identify such performance factors.

References

Baumol, William J., and Edward A. Ide (1962), "Variety in Retailing," in *Mathematical Models and Methods in Marketing.* Frank M. Bass et al. (eds), Homewood, Ill.: Richard D. Irwin, 128–144.

Buzzell, Robert D., Bradley T. Gale, and Ralph G. M. Sultan (1975), "Market Share—A Key to Profitability," *Harvard Business Review,* 53 (January–February), 97–106.

Cotham, James C., III (1969), "Using Personal History Information in Retail Salesmen Selection." *Journal of Retailing.* 45 (Summer), 31–39+.

Dalrymple, Douglas J. (1966). *Merchandising Decision Models for Department Stores,* East Lansing: Michigan State University.

Paul, Robert J., and Robert W. Bell (1968), "Evaluating the Retail Salesman," *Journal of Retailing* 44 (Summer), 17–26.

Schoeffler, Sidney, Robert D. Buzzell, and Donald F. Heany (1974), "Impact of Strategic Planning on Profit Performance," *Harvard Business Review,* 52 (March–April), 137–145.

Weaver, Charles N. (1969), "An Empirical Study to Aid in the Selection of Retail Salesclerks," *Journal of Retailing.* 45 (Fall), 22–26+.

CHAPTER 4

Multiple Discriminant Analysis

LEARNING OBJECTIVES

Upon completing this chapter, you should be able to do the following:

- State the circumstances under which a linear discriminant function rather than multiple regression should be used.

- Understand the assumptions underlying discriminant analysis in assessing the appropriateness of its use for a particular problem.

- Identify the major issues in the application of discriminant analysis.

- Describe the two computation approaches for discriminant analysis and state when each should be used.

- Tell how to interpret the nature of the linear discriminant function, that is, to identify independent variables with significant discriminatory power.

- Explain the usefulness of the classification matrix methodology and tell how to develop a classification matrix.

- Describe the various approaches to evaluating the classificatory power of the discriminant function, including the distinction between the hit ratio and multiple regression's R^2.

- Justify the use of a split-sample approach in validating the discriminant function.

Much has been written on the multivariate statistical technique of multiple discriminant analysis. This chapter discusses this complex and sophisticated technique without resorting to statistical jargon and mathematical formulas or glossing over important concepts. The chapter has two major objectives: (1) to introduce the underlying nature, philosophy, and conditions of discriminant analysis, and (2) to demonstrate its application and interpretation with an illustrative example.

If you recall from Chapter 1, the basic purpose of discriminant analysis is to estimate the relationship between a single nonmetric (categorical) dependent variable and a set of metric independent variables, in this general form:

$$Y_1 \quad = X_1 + X_2 + X_3 + \ldots + X_n$$
$$\text{(nonmetric)} \qquad \text{(metric)}$$

Multiple discriminant analysis has widespread application in situations where the primary objective is identifying the group to which an object (e.g., person, firm, or product) belongs. Potential applications include predicting the success or failure of a new product, deciding whether a student should be admitted to graduate school, classifying students as to vocational interests, determining what category of credit risk a person falls into, or predicting whether a firm will be successful or not. In each instance, the objects fall into groups and group membership for each object, it is hoped, can be predicted or explained by a set of independent variables selected by the analyst.

KEY TERMS

Before starting the chapter, review the key terms to develop an understanding of the concepts and terminology to be used. Throughout the chapter the key terms appear in **boldface**. Other points of emphasis in the chapter are *italicized*. Also, cross-references within the key terms are in *italics*.

Analysis sample When constructing classification matrices, the original sample should be divided randomly into two groups, one for computing the discriminant function and the other for validating it. The group used to compute the discriminant function is called the analysis sample.

Categorical variable Also referred to as a nonmetric, nominal, binary, qualitative, or taxonomic variable, a categorical variable uses values that serve merely as a label or means of identification. The number on a football jersey is an example.

Centroid Mean value for the discriminant Z scores of a particular category or group. For example, a two-group discriminant analysis has two centroids, one for each of the groups.

Classification matrix Also called a confusion, assignment, or prediction matrix, it is a matrix containing numbers that reveal the predictive ability of the discriminant function. The numbers on the diagonal of the matrix represent correct classifications, and the off-diagonal numbers are incorrect classifications.

Cross-validation Procedure of dividing the sample into two parts: the *analysis sample* used to estimate the discriminant function(s) and the *holdout sample* used to validate the discriminant function(s). Cross-validation avoids the "overfitting" of the discriminant function by allowing its validation on a totally separate sample.

Cutting score   Criterion (score) against which each individual's discriminant score is judged to determine into which group the individual should be classified. When the analysis involves two groups, the *hit ratio* is determined by computing a single "cutting" score. Entities whose Z scores are below this score are assigned to one group, while those whose scores are above it are classified in the other group.

Discriminant coefficient   See *discriminant weight*.

Discriminant function   Linear equation in the following form:

$$Z = W_1 X_1 + W_2 X_2 + \ldots + W_n X_n$$

where

Z = Discriminant score
W_i = Discriminant weight for independent variable i
X_i = Independent variable i

Discriminant loadings   Also called structure correlations, they measure the simple linear correlation between the independent variables and the discriminant function.

Discriminant score   Referred to as the Z score, defined by the *discriminant function*.

Discriminant weight   Also called *discriminant coefficient*, its size is determined by the variance structure of the original variables. Independent variables with large discriminatory power usually have large weights, and those with little discriminatory power usually have small weights; however, multicollinearity among the independent variables will cause an exception to this rule.

Hit ratio   Percentage of statistical units (individuals, respondents, objects, etc.) correctly classified by the discriminant function.

Holdout sample   Group of subjects held out of the total sample when the function is computed. This group is then used to validate the discriminant function on another sample of respondents. Also called validation sample.

Metric variable   Variable with a constant unit of measurement. If a variable is scaled from 1 to 9, the difference between 1 and 2 is the same as that between 8 and 9.

Maximum chance criterion   Measure of predictive accuracy in the *classification matrix* that compares the percentage correctly classified (also known as the *hit ratio*) with the percentage of respondents in the largest group. The rationale is that the best uninformed choice is to classify every observation into the largest group.

Optimum cutting score   *Discriminant score* value that best separates the groups.

Polar extremes approach   The method of constructing a categorical dependent variable from a metric variable. First, the metric variable is divided into categories. Then the extreme categories are used in the discriminant analysis, with the middle category not included in the analysis.

Potency index   Composite measure of the discriminatory power of a predictor variable when more than one discriminant function is estimated. Based on discriminant loadings, it is a relative measure on which predictors can be compared.

Press's *Q* statistic Measure of the classificatory power of the discriminant function when compared with the results expected from a chance model. The calculated value is compared to a critical value based on the chi-square distribution. If the calculated value exceeds the critical value, the classification results are significantly better than would be expected by chance.

Proportional chance criterion Another criterion for assessing the *hit ratio,* in which the "average" probability of classification is calculated considering all group sizes.

Simultaneous estimation Estimation of the discriminant function(s) in a single step where weights for all independent variables are calculated simultaneously, in contrast to *stepwise estimation,* where independent variables are entered sequentially according to discriminating power.

Split-sample validation See *cross-validation.*

Stepwise estimation Process of estimating the discriminant function(s) whereby independent variables are entered sequentially according to the discriminatory power they add to the discriminant function(s).

Tolerance Proportion of the variation in the independent variables not explained by the variables already in the model (function). It can be used to protect against multicollinearity. A tolerance of 0 means that a predictor (independent variable) under consideration is a perfect linear combination of variables already in the model (equation). A tolerance of 1 means that a predictor is totally independent of other predictors already in the model. The default option in most computer packages sets the minimum acceptable tolerance at .01. This default allows quite a bit of redundancy or multicollinearity in the predictors. In short, if at least 1 percent of the variation in the response variable remains unexplained by the predictors already included in the function, the predictor variable under consideration will be allowed to enter the function.

Variate Also called linear combination, linear compound, and *discriminant function,* it represents the weighted sum of two or more independent variables.

Vector Representation of the direction and magnitude of a variable's role as portrayed in a graphical interpretation of discriminant analysis results.

What Is Discriminant Analysis?

In attempting to choose an appropriate analytical technique, we sometimes encounter a problem that involves a categorical dependent variable and several metric independent variables. For example, we may wish to distinguish good from bad credit risks. If we had a metric measure of credit risk, then we could use multivariate regression. But we may be able to ascertain only if someone is in the good or bad risk category. This is not the metric type measure required by multivariate regression analysis.

Discriminant analysis is the appropriate statistical technique when the dependent variable is categorical (nominal or nonmetric) and the independent variables are metric. In many cases, the dependent variable consists of two groups or classifications, for example, male versus female or high versus low. In other instances, more than two groups are involved, such as a three-group classification involving low, medium, and high classifications. Discriminant analysis is capable of handling either two groups or multiple groups (three or more). When two classifications are involved, the technique is referred to as two-group discriminant analysis. When three or more classifications are identified, the technique is referred to as *multiple discriminant analysis (MDA)*.

Discriminant analysis involves deriving a **variate,** the linear combination of the two (or more) independent variables that will discriminate best between a priori defined groups. Discrimination is achieved by setting the variate's weights for each variable to maximize the between-group variance relative to the within-group variance. The linear combination for a discriminant analysis, also known as the **discriminant function,** is derived from an equation that takes the following form:

$$Z = W_1X_1 + W_2X_2 + W_3X_3 + \ldots + W_nX_n$$

where

Z = Discriminant score
W_i = Discriminant weight for variable i
X_i = Independent variable i

Discriminant analysis is the appropriate statistical technique for testing the hypothesis that the group means of a set of independent variables for two or more groups are equal. To do so, discriminant analysis multiplies each independent variable by its corresponding weight and adds these products together (see the preceding equation). The result is a single composite discriminant score for each individual in the analysis. By averaging the discriminant scores for all the individuals within a particular group, we arrive at the group mean. This group mean is referred to as a **centroid.** When the analysis involves two groups, there are two centroids; with three groups, there are three centroids, and so forth. The centroids indicate the most typical location of any individual from a particular group, and a comparison of the group centroids shows how far apart the groups are along the dimension being tested.

The test for the statistical significance of the discriminant function is a generalized measure of the distance between the group centroids. It is computed by comparing the distribution of the discriminant scores for the two or more groups. If the overlap in the distribution is small, the discriminant function separates the groups well. If the overlap is large, the function is a poor discriminator between the groups. The distributions of discriminant scores shown in Figure 4.1 further illustrate this concept. For example, the top diagram represents the distribution of discriminant scores for a function that separates the groups well, whereas the lower diagram shows the distribution of discriminant scores on a function that is a relatively poor discriminator between groups A and B. Note that the shaded areas represent probabilities of misclassifying statistical units from A and B.

Multiple discriminant analysis is unique in one characteristic among the dependence relationships we will study: If there are more than two groups in the dependent variable, discriminant analysis will calculate more than one discriminant function. As a matter of fact, it will calculate NG −1 functions, where NG is the number of groups. Each discriminant function will calculate a discriminant score. In the case of a three-group dependent variable, each object will have a score for discriminant functions one and two, allowing the objects to be plotted in two dimensions, with each dimension representing a discriminant function. Thus, dis-

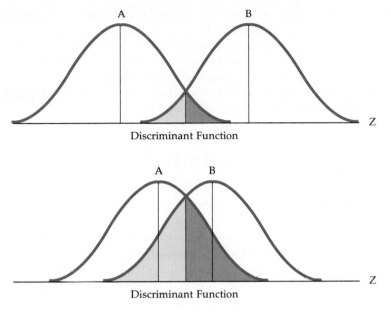

A B

Discriminant Function

A B

Discriminant Function

FIGURE 4.1 Univariate representation of discriminant Z scores.

criminant analysis is not limited to a single variate, as is multiple regression, but creates multiple variates representing dimensions of discrimination among the groups.

Analogy with Regression and ANOVA

The application and interpretation of discriminant analysis is much the same as in regression analysis; that is, the discriminant function is a linear combination (variate) of metric measurements for two or more independent variables and is used to describe or predict a single dependent variable. The key difference is that discriminant analysis is appropriate for research problems in which the dependent variable is categorical (nominal or nonmetric), whereas regression is utilized when the dependent variable is metric.

In Chapter 3 we introduced a specialized form of regression known as logistic regression, which has the same basic properties as regression. The form of the logistic regression variate is similar to multiple regression. The variate represents a single multivariate relationship with regressionlike coefficients indicating the relative impact of each predictor variable. Logistic regression's differences from discriminant analysis will become more apparent as you become familiar with discriminant analysis. When the basic assumptions of both methods are met, both methods give comparable predictive and classificatory results. Logistic regression, however, has the advantage of being less affected than discriminant analysis when the basic assumptions, particularly normality of the variables, are not met. It also can accommodate nonmetric variables through dummy-variable coding, just as regression can. It is limited, however, to prediction of only a two-group dependent measure. Thus, in cases when three or more groups form the dependent measure, discriminant analysis is better suited.

Discriminant analysis is also comparable to "reversing" multivariate analysis of variance (MANOVA), which we discuss in Chapter 5. In discriminant analysis, the single dependent variable is categorical, and the independent variables are metric. The opposite is true of MANOVA, which involves metric dependent variables and categorical independent variable(s).

Hypothetical Example of Discriminant Analysis for Two Groups

First we will examine a HATCO research problem to demonstrate the underlying logic of MDA. Suppose HATCO wants to find out whether one of its new products—a new and improved food mixer—will be commercially successful. In carrying out the investigation, HATCO is primarily interested in identifying (if possible) those consumers who would purchase the new product and those who would not purchase it. In statistical terminology, then, HATCO would like to minimize the number of errors it would make in predicting which consumers would buy the new food mixer and which would not. To assist in identifying potential purchasers, HATCO has devised three rating scales to be used by consumers to evaluate the new product. Consumers would evaluate the durability, performance, and style of the food mixer with the three scales. Rather than relying on each scale as a separate measure, HATCO believes that a weighted combination of all three would better predict whether a consumer is likely to purchase the new product.

MDA can obtain a weighted combination of the three scales. This weighted combination can predict the likelihood that a consumer will purchase the product. In addition to determining whether persons who are likely to purchase the new product can be distinguished from those who are not, HATCO would like to know which characteristics of its new product are useful in differentiating purchasers from nonpurchasers; that is, which of the three characteristics of the new product best separates purchasers from nonpurchasers? For example, if the response "would purchase" is always associated with a high durability rating and the response "would not purchase" is always associated with a low durability rating, HATCO could conclude that the characteristic durability distinguishes purchasers from nonpurchasers. In contrast, if HATCO found that about as many persons with a high rating on style said they would purchase the food mixer as those who said they would not, then style is a characteristic that discriminates poorly between purchasers and nonpurchasers.

Table 4.1 lists the ratings on these three characteristics of the new mixer with a specified price by a panel of ten housewives who are potential purchasers. (In rating the food mixer, each housewife would be implicitly comparing it with products already on the market.) After the product was evaluated, the housewives were asked to state their buying intentions ("would purchase" or "would not purchase"). Five housewives stated that they would purchase the new mixer and five said they would not.

TABLE 4.1 HATCO Survey Results for the Evaluation of a New Consumer Product

Purchase Intention	Subject Number	Evaluation of New Product*		
		X_1 Durability	X_2 Performance	X_3 Style
Group 1				
Would purchase	1	8	9	6
	2	6	7	5
	3	10	6	3
	4	9	4	4
	5	4	8	2
Group Mean		7.4	6.8	4.0
Group 2				
Would not purchase	6	5	4	7
	7	3	7	2
	8	4	5	5
	9	2	4	3
	10	2	2	2
Group Mean		3.2	4.4	3.8
Difference between group means		4.2	2.4	0.2

*Evaluations made on a 0 (very poor) to 10 (excellent) rating scale.

After examining Table 4.1, we can identify several discriminating variables. First, there is a substantial difference between the mean ratings for the "would purchase" and "would not purchase" groups on the characteristic of durability (7.4 − 3.2 = 4.2). Thus durability appears to discriminate well between the "would purchase" and "would not purchase" groups and is likely to be an important characteristic to potential purchasers.* On the other hand, the characteristic of style has a much smaller difference between mean ratings [only 0.2 (4.0 − 3.8 = 0.2)] for the "would purchase" and "would not purchase" groups. Therefore, we would expect this characteristic to be less discriminating in terms of a decision to purchase or not to purchase.

Because we have only ten respondents in two groups and three independent variables, we will also look at the data graphically to determine what discriminant analysis is trying to accomplish. Figure 4.2 shows the ten respondents on each of the three variables. The "would purchase" group is represented by circles and the "would not purchase" group by the squares. Respondent identification numbers are inside the shapes. Looking first at X_1, durability, which had a substantial difference in mean scores, we see that we could almost perfectly discriminate between the groups using only this variable. If we established the value of

*This conclusion is based only on differences in the means and could possibly change after consideration of the standard deviations of the two sets of data; that is, with large standard deviations the difference between the means may not be statistically significant.

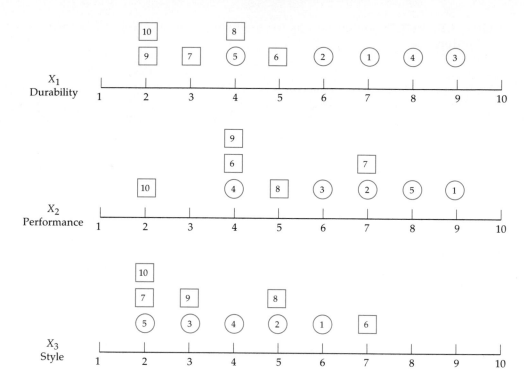

Shapes indicate purchase intention group (◯ = Would Purchase;   ☐ = Would Not Purchase), and valves represent respondent identification numbers.

FIGURE 4.2 Graphical representation of ten potential purchasers on three possible discriminating variables.

5.5 as our cutoff point to discriminate between the two groups, then we would "misclassify" only respondent five, one of the "would purchase" group members. This is indicative of the large difference in the means for the two groups on durability. Examining X_2's performance, we see there is a less clearcut distinction between the two groups. However, this variable does provide high discrimination for respondent five, who was misclassified if we used X_1. In addition, the respondents who would be misclassified using X_2 are well separated on X_1. Thus X_1 and X_2 might be used quite effectively *in combination* to predict group membership. Finally, X_3 shows little differentiation between the groups. Thus, by forming a variate of X_1, X_2, and X_3 and weighting X_1 and X_2 highly and X_3 lower, if at all, a discriminant function can be formed that maximizes the separation of the groups on the discriminant score.

The MDA technique follows a procedure very similar to that shown in the hypothetical example. It identifies the variables with the greatest differences between the groups and derives a discriminant weighting coefficient for each variable to reflect these differences. MDA then uses the weights and each individual's ratings on the characteristics to develop the discriminant score for each respondent and finally assigns each respondent to a group according to the discriminant score.

A Geometric Representation of the Discriminant Function

A graphic illustration of another two-group analysis will help to further explain the nature of discriminant analysis [6]. Figure 4.3 is a scatter diagram and projection that shows what happens when a two-group discriminant function is computed. Assume we have two groups, A and B, and two measurements, X_1 and X_2, on each member of the two groups. We can plot in a scatter diagram the association of variable X_1 with variable X_2 for each member of the two groups. Group membership is identified by large and small dots. In Figure 4.3 the small dots represent the variable measurements for the members of group B and the large dots for group A. The ellipses drawn around the large and small dots would enclose some prespecified proportion of the points, usually 95 percent or more in each group. If we draw a straight line through the two points where the ellipses intersect and then project the line to a new Z axis, we can say that the overlap between the univariate distributions A' and B' (represented by the shaded area) is smaller than would be obtained by any other line drawn through the ellipses formed by the scatterplots [6].

The important thing to note about Figure 4.3 is that the Z axis expresses the two-variable profiles of groups A and B as single numbers (discriminant scores). By finding a linear combination of the original variables X_1 and X_2, we can project the result as a discriminant function. For example, if the dots and circles are projected onto the new Z axis as discriminant Z scores, the result condenses the information about group differences (shown in the X_1X_2 plot) into a set of points (Z scores) on a single axis.

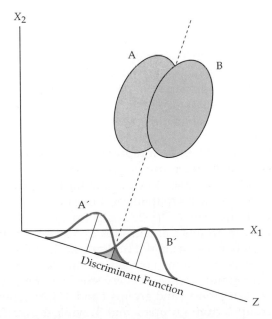

FIGURE 4.3 Graphic illustration of two-group discriminant analysis.

To summarize, for a given discriminant analysis problem, a linear combination of the independent variables is derived, resulting in a series of discriminant scores for each individual in each group. The discriminant scores are computed according to the statistical rule of maximizing the variance between the groups and minimizing the variance within them. If the variance between the groups is large relative to the variance within the groups, we say that the discriminant function separates the groups well.

A Hypothetical Example of a Three-group Discriminant Analysis

After examining the two-group example above, we should grasp the rationale and benefit of combining independent variables into a variate for purposes of discriminating between groups. But discriminant analysis also has another means of discrimination: the estimation and use of *multiple* variates in instances where there are three or more groups. These discriminant functions now become dimensions of discrimination, each dimension separate and distinct from the other. Thus, besides improving the explanation of group membership, these additional discriminant functions add insight into the various combinations of independent variables that discriminate between groups.

As an illustration of a three-group application of discriminant analysis, we will examine research conducted by HATCO concerning the possibility of a competitor's customers switching suppliers. A small-scale pretest interviewed 15 customers of a major competitor. In the course of the interview, the customers were asked their probability of switching suppliers on a three-category scale. The three possible responses were "Definitely Switch", "Undecided," and "Definitely Not Switch." Customers were assigned to groups 1, 2, or 3, respectively, according to their response. The customers also rated the competitor on the two characteristics of price competitiveness and service level. The research issue is now to determine whether the customers' ratings of their current supplier can predict their probability of leaving that supplier. Because the dependent variable of probability of switching was measured as a categorical (nonmetric) variable and the ratings of price and service are metric, discriminant analysis is appropriate.

With three categories of the dependent variable, discriminant analysis can estimate two discriminant functions, each representing a different dimension of discrimination. Table 4.2 contains the survey results for the 15 customers, 5 in each category of the dependent variable. As we did in the two-group example, we can look at the mean scores for each group to see if one of the variables discriminates well among all the groups. For X_1, price competitiveness, we see a rather large mean difference between groups 1 and groups 2 or 3 (0.0 versus 1.6 and 2.25). X_1 may discriminate well between groups 1 and groups 2 or 3 but probably will be much less effective in discriminating between groups 2 and 3. For X_2, we see that the difference between groups 1 and 2 is very small (0.0 versus .2), while a large difference exists between group 3 and groups 1 and 2 (4.2 versus 0.0 and .2). Thus, X_1 distinguishes group 1 from groups 2 and 3, while X_2 distinguishes group 3 from groups 1 and 2. As a result, we see that X_1 and X_2 provide different "dimensions" of discrimination between the groups.

TABLE 4.2 HATCO Survey Results of Potential Customers

		Evaluation of Current Supplier*	
Probability of Switching Subject			
Suppliers	Number	X_1 Price Competitiveness	X_2 Service Level
Group 1			
Definitely Switch	1	0	0
	2	−1	0
	3	1	0
	4	0	−1
	5	0	1
Group Mean		0	0
Group 2			
Undecided	6	2	0
	7	2	1
	8	3	−1
	9	3	0
	10	3	1
Group Mean		1.6	0.2
Group 3			
Definitely Not Switch	11	0	4
	12	1	4
	13	2	4
	14	3	4
	15	3	5
Group Mean		2.25	4.2

*Evaluations made on a −5 (Very Poor) to +5 (Very Good) rating scale.

To illustrate this graphically, Figure 4.4 portrays the three groups on each of the independent variables separately. Viewing the group members on either variable, we can see that no variable discriminates well among all the groups. But if we now construct two simple discriminant functions, the results become much clearer. For illustration purposes, we calculate two discriminant functions with weights of 0.0 or 1.0 for the variables. Discriminant function one will be comprised of X_1 by giving it a weight of 1.0, while X_2 has a weight of 0.0. Likewise, discriminant function two is comprised of X_2 by giving it a weight of 1.0, and X_1 receives a weight of 0.0. The functions can be stated mathematically as

$$\text{Discriminant Function 1} = 1.0 * X_1 + 0.0 * X_2$$
$$\text{Discriminant Function 2} = 0.0 * X_1 + 1.0 * X_2$$

This is in simple terms how the discriminant analysis procedure estimates weights to maximize discrimination.

With the two functions, we can now calculate two discriminant scores for each respondent. Figure 4.4 also contains a plot of each respondent in a two-dimensional representation. The separation between groups now becomes quite apparent, and each group can be easily distinguished. We can establish values on each

A. Individual Variables

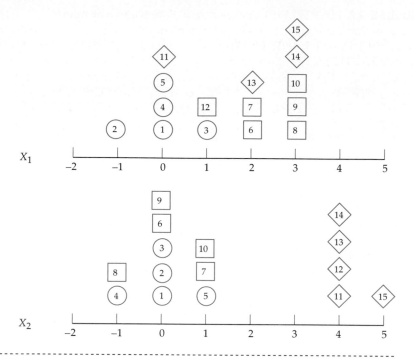

B. Two-Dimensional
Representation of
Discriminant Functions

Discriminant
Function 1 = 1.0 * X_1 + 0 * X_2

Discriminant
Function 2 = 0 * X_1 + 1.0 * X_2

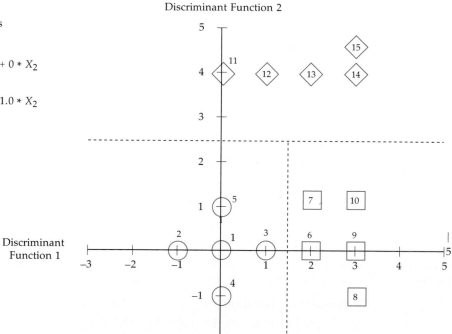

FIGURE 4.4 Graphical representation of potential discriminating variables for a three-group discriminant analysis.
  *Shapes indicate probability of switching suppliers (○ = Definitely Switch, □ = Undecided, ◇ = Definitely Not Switch) and values represent respondent identification numbers.

dimension (e.g., 1.5 for dimension 1 and 2.5 for dimension 2) that will define regions containing each group. All members of group 1 are in the region less than 1.5 on dimension 1 and less than 2.5 on dimension 2. Each of the other groups can be similarly defined in terms of the ranges of their discriminant function scores.

Moreover, the two discriminant functions provide the dimensions of discrimination. The first discriminant function, comprised of price competitiveness, distinguishes between undecided customers and those customers that have decided to switch. But price competitiveness does not distinguish those who have decided not to switch. Instead, the perception of service level, defining the second discriminant function, predicts whether a customer will decide not to switch versus whether a customer is undecided or determined to switch suppliers. The analyst can present to management the separate impacts of both price competitiveness and service level in making this decision.

The estimation of more than one discriminant function, when possible, provides the analyst with both improved discrimination and additional perspectives on the features and the combinations that best discriminate among the groups. The following sections detail the necessary steps for performing a discriminant analysis, for assessing its level of predictive fit, and then for interpreting the influence of independent variables in making that prediction.

The Decision Process for Discriminant Analysis

The application of discriminant analysis can be viewed from the six-stage model-building perspective introduced in Chapter 1 and portrayed in Figure 4.5. As with all multivariate applications, setting the objectives is the first step in the analysis. Then the analyst must address specific design issues and make sure the underlying assumptions are met. The analysis proceeds with the derivation of the discriminant function and the determination whether or not a statistically significant function can be derived to separate the two (or more) groups. The discriminant results are then assessed for predictive accuracy by developing a classification matrix. Next, interpretation of the discriminant function determines which of the independent variables contribute the most to discriminating between the groups. Finally, the discriminant function can be validated with a holdout sample. Each of these stages is discussed in the following sections.

Stage One: Objectives of Discriminant Analysis

A review of the objectives for applying discriminant analysis should further clarify its nature. Discriminant analysis can address any of the following research questions:

1. Determining whether statistically significant differences exist between the average score profiles on a set of variables for two (or more) a priori defined groups.

2. Determining which of the independent variables account the most for the differences in the average score profiles of the two or more groups.
3. Establishing procedures for classifying statistical units (individuals or objects) into groups on the basis of their scores on a set of independent variables.
4. Establishing the number and composition of the dimensions of discrimination between groups formed from the set of independent variables.

As can be noted from these objectives, discriminant analysis is useful when the analyst is interested either in understanding group differences or in correctly classifying statistical units into groups or classes. Discriminant analysis, therefore, can be considered either a type of profile analysis or an analytical predictive

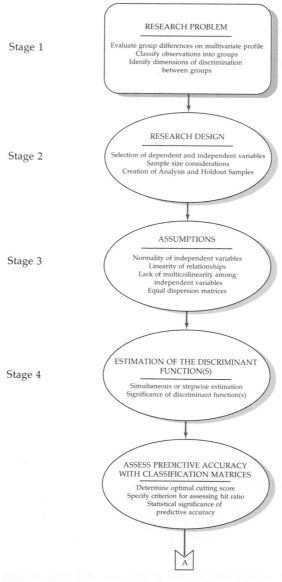

FIGURE 4.5 Discriminant analysis decision process.

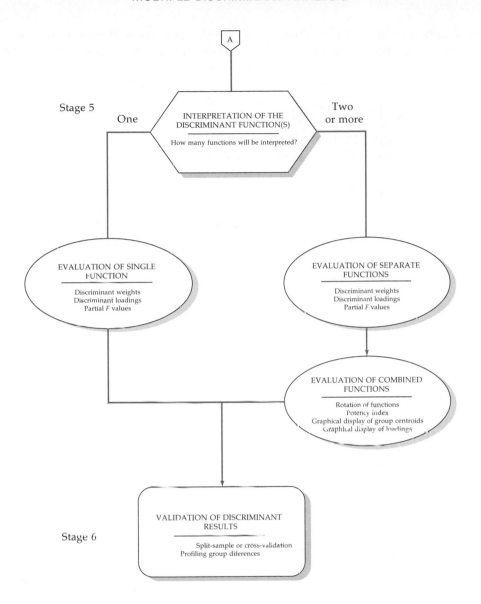

technique. In either case, the technique is most appropriate where there is a single categorical dependent variable and several metrically scaled independent variables. As a profile analysis, MDA provides an objective assessment of group differences between groups on a set of independent variables. In this situation, MDA is quite similar to multivariate analysis of variance (see Chapter 5 for a more detailed discussion of multivariate analysis of variance). For understanding group differences, discriminant analysis lends insight into the role of individual variables as well as defining combinations of these variables that represent dimensions of discrimination between groups. These dimensions are the collective effects of several variables that work jointly to distinguish between the groups. For classification purposes, discriminant analysis provides a basis for classifying not only the sample used to estimate the discriminant function but also any other observations that can have values for all the independent variables. In this way, the discriminant analysis can be used to classify other observations into the defined groups.

Stage Two: Research Design for Discriminant Analysis

The successful application of discriminant analysis requires consideration of several issues. These issues include the selection of both the dependent and the independent variables, the sample size needed for estimation of the discriminant functions, and the division of the sample for validation purposes.

Selection of Dependent and Independent Variables

To apply discriminant analysis, the analyst must first specify which variables are to be independent and which variable is to be dependent. Recall that the dependent variable is categorical and the independent variables are metric.

The analyst should focus on the dependent variable first. The number of dependent variable groups (categories) can be two or more, but these groups must be mutually exclusive and exhaustive. By this we mean that each observation can be placed into only one group. In some cases, the dependent variable may be two groups (dichotomous), such as good versus bad. In other cases, the dependent variable may involve several groups (multichotomous), such as occupation—for example, physician, attorney, or professor.

The preceding examples of categorical variables were true dichotomies (or multichotomies). There are some situations, however, where discriminant analysis is appropriate even if the dependent variable is not a true categorical variable. We may have a dependent variable that is of ordinal or interval measurement that we wish to use as a categorical dependent variable. In such cases, we would have to create a categorical variable. For example, if we had a variable that measured the average number of cola drinks consumed per day, and the individuals responded on a scale from zero to eight or more per day, we could create an artificial trichotomy (three groups) by simply designating those individuals who consumed no, one, or two cola drinks per day as light users, those who consumed three, four, or five per day as medium users, and those who consumed six, seven, eight, or more as heavy users. Such a procedure would create a three-group categorical variable in which the objective would be to discriminate among light, medium, and heavy users of colas.

Any number of artificial categorical groups can be developed. Most frequently the approach would involve creating two, three, or four categories. But a larger number of categories could be established if the need arose. When three or more categories are created, the possibility arises of examining only the extreme groups in a two-group discriminant analysis. This procedure is called the polar-extremes approach.

The **polar extremes approach** involves comparing only the extreme two groups and excluding the middle group from the discriminant analysis. For example, the analyst could examine the light and heavy users of cola drinks and exclude the medium users. This approach can be used any time the analyst wishes to examine only the extreme groups. However, the analyst may also want to try this approach when the results of a regression analysis are not as good as anticipated. Such a procedure may be helpful because it is possible that group differences may appear even though regression results are poor; that is, the polar extremes approach with discriminant analysis can reveal differences that are not as prominent in a regression analysis of the full data set [6]. Such manipulation of the data naturally would necessitate caution in interpreting one's findings.

After a decision has been made on the dependent variable, the analyst must decide which independent variables to include in the analysis. Independent variables usually are selected in two ways. The first approach involves identifying variables either from previous research or from the theoretical model that is the underlying basis of the research question. The second approach is intuition—extending the researcher's knowledge and intuitively selecting variables for which no previous research or theory exists but that logically might be related to predicting the groups for the dependent variable.

Sample Size

Discriminant analysis is quite sensitive to the ratio of sample size to the number of predictor variables. Many studies suggest a ratio of 20 observations for each predictor variable. Although this ratio may be hard to maintain in practice, the analyst must note that the results become unstable as the sample size decreases relative to the number of independent variables.

In addition to the overall sample size, the analyst must also consider the sample size of each group. At a minimum, the smallest group size must exceed the number of independent variables. As a practical guideline, each group should have at least 20 observations. But even if all groups exceed 20 observations, the analyst must also consider the relative sizes of the groups. If the groups vary widely in size, this may impact the estimation of the discriminant function and the classification of observations. In the classification stage, larger groups have a disproportionately higher chance of classification. If the group sizes do vary markedly, the analyst may wish to randomly sample from the larger group(s), thereby reducing their size to a level comparable to the smaller group(s).

Division of the Sample

One final note about the impact of sample size in discriminant analysis. As will be discussed later, many times the sample is divided into two subsamples, one used for estimation of the discriminant function and another for validation purposes. It is essential that each subsample be of adequate size to support conclusions from the results.

A number of procedures have been suggested for dividing the sample, but the most popular one involves developing the discriminant function on one group and then testing it on a second group. The usual procedure is to divide the total sample of respondents randomly into two groups. One of these groups, the **analysis sample,** is used to develop the discriminant function. The second group, the **holdout sample,** is used to test the discriminant function. This method of validating the function is referred to as the split-sample or **cross-validation approach** [4, 8, 14].

No definite guidelines have been established for dividing the sample into analysis and holdout groups. The most popular procedure is to divide the total group so that one-half of the respondents are placed in that analysis sample and the other half are placed in the holdout sample. However, no hard-and-fast rule has been established, and some researchers prefer a 60–40 or 75–25 split between the analysis and the holdout groups.

When selecting the individuals for the analysis and holdout groups, one usually follows a proportionately stratified sampling procedure. If the categorical groups for the discriminant analysis are equally represented in the total sample, an equal number of individuals is selected. If the categorical groups are unequal, the sizes of the groups selected for the holdout sample should be proportionate to the total sample distribution. For instance, if a sample consists of 50 males and 50 females, the holdout sample would have 25 males and 25 females. If the sample contained 70 females and 30 males, then the holdout samples would consist of 35 females and 15 males.

Several additional comments need to be made regarding the division of the total sample into analysis and holdout groups. One is that if the analyst is going to divide the sample into analysis and holdout groups, the sample must be sufficiently large to do so. Again, no hard-and-fast rules have been established, but it seems logical that the analyst would want at least 100 in the total sample to justify dividing it into the two groups. One compromise procedure the analyst can select if the sample size is too small to justify a division into analysis and holdout groups is to develop the function on the entire sample and then use the function to classify the same group used to develop the function. This procedure results in an upward bias in the predictive accuracy of the function, but is certainly better than not testing the function at all.

Stage Three: Assumptions of Discriminant Analysis

It is desirable to meet certain conditions for proper application of discriminant analysis. The key assumptions for deriving the discriminant function are multivariate normality of the independent variables and unknown (but equal) dispersion and covariance structures (matrices) for the groups as defined by the dependent variable [7, 9]. While there is mixed evidence that discriminant analysis is very sensitive to violations of these assumptions, any analyst should examine the data and if assumptions are violated, the analyst should examine the alternative methods available and the impacts on the results that can be expected. Data not meeting the multivariate normality assumption can cause problems in the estimation of the discriminant function. Therefore, it is suggested that logistic regression be used as an alternative technique, if possible.

Unequal covariance matrices can adversely affect the classification process. If the sample sizes are small and the covariance matrices are unequal, then the statistical significance of the estimation process is adversely affected. But more likely is the case of unequal covariances among groups of adequate sample size, whereby observations are "overclassified" into the groups with larger covariance matrices. This effect can be minimized by increasing the sample size and also by using the group-specific covariance matrices for classification purposes, but this approach mandates cross-validation of the discriminant results. Finally, quadratic classification techniques are available in many of the statistical programs if large differences exist between the covariance matrices of the groups and the remedies do not minimize the effect [5, 10, 12].

Another characteristic of the data that can affect the results is multicollinearity among the independent variables. Multicollinearity denotes that two or more independent variables are highly correlated, so that one variable can be highly explained or predicted by the other variables and thus adds little to the explanatory power of the entire set. This consideration becomes especially critical when stepwise procedures are employed. The analyst, in interpreting the discriminant function, must be aware of the level of multicollinearity and its impact on determining which variables enter the stepwise solution. For a more detailed discussion of multicollinearity and its impact on stepwise solutions, see Chapter 3. If you are not familiar with the procedures for detecting the violations of assumptions or the presence of multicollinearity, see Chapter 2.

As with any of the multivariate techniques employing a variate, an implicit assumption is that all relationships are linear. Nonlinear relationships are not reflected in the discriminant function unless specific variable transformations are made to represent nonlinear effects. Finally, outliers can have a substantial impact on the classification accuracy of any discriminant analysis results. The analyst is encouraged to examine all results for the presence of outliers and to eliminate true outliers if needed. For a discussion of some of the techniques for outlier detection, the reader is again referred to Chapter 2.

Stage Four: Estimation of the Discriminant Model and Assessing Overall Fit

To derive the discriminant function, the analyst must decide on the method of estimation and then determine the number of functions to be retained. The predictive accuracy is assessed by the number of observations classified into the correct groups. A number of criteria are available to assess whether the classification process achieves practical and/or statistical significance.

Computational Method

Two computational methods can be utilized in deriving a discriminant function: the simultaneous (direct) method and the stepwise method. **Simultaneous estimation** involves computing the discriminant function so that all of the independent variables are considered concurrently. Thus the discriminant function(s) is computed based upon the entire set of independent variables, regardless of the discriminating power of each independent variable. The simultaneous method is appropriate when, for theoretical reasons, the analyst wants to include all the independent variables in the analysis and is not interested in seeing intermediate results based only on the most discriminating variables.

Stepwise estimation is an alternative to the simultaneous approach. It involves entering the independent variables into the discriminant function one at a time on the basis of their discriminating power. The stepwise approach begins by choosing the single best discriminating variable. The initial variable is then paired with each of the other independent variables one at a time, and the variable that is best able to improve the discriminating power of the function in combination with the first variable is chosen. The third and any subsequent variables are selected in a similar manner. As additional variables are included, some previously selected variables may be removed if the information they contain about group differences is available in some combination of the other variables included at later stages. Eventually, either all independent variables will have been included in the function or the excluded variables will have been judged as not contributing significantly to further discrimination.

The stepwise method is useful when the analyst wants to consider a relatively large number of independent variables for inclusion in the function. By sequentially selecting the next best discriminating variable at each step, variables that are not useful in discriminating between the groups are eliminated and a reduced set of variables is identified. The reduced set typically is almost as good as, and sometimes better than, the complete set of variables.

Statistical Significance

After the discriminant function has been computed, the analyst must assess its level of significance. A number of different statistical criteria are available. The measures of Wilks' lambda, Hotelling's trace, and Pilliai's criteria all evaluate the statistical significance of the discriminatory power of the discriminant function(s). Roy's greatest characteristic root evaluates only the first discriminant function. For a more detailed discussion of the advantages and disadvantages of each criterion, the reader is referred to the discussion of significance testing in multivariate analysis of variance in Chapter 5.

If a stepwise method is used for estimating the discriminant function, the Mahalanobis D^2 and Rao's V measures are most appropriate. The Mahalanobis procedure is based on generalized squared Euclidean distance that adjusts for unequal variances. The major advantage of this procedure is that it is computed in the original space of the predictor variables rather than as a collapsed version used in other measures. The Mahalanobis procedure becomes particularly critical as the number of predictor variables increases, because it does not result in any reduction in dimensionality. The loss in dimensionality causes a loss of information, because it decreases predictor variability. In general, Mahalanobis is the preferred procedure when one is interested in the maximal use of available information. The Mahalanobis D^2 procedure performs a stepwise discriminant analysis similar to a stepwise regression analysis. This stepwise procedure is designed to develop the best one-variable model, followed by the best two-variable model, and so forth, until no other variables meet the desired selection rule. The selection rule in this procedure is to maximize Mahalanobis distance (D^2) between groups. Both stepwise and simultaneous methods are available in the major statistical programs [1, 15, 16, 17, 18].

The conventional criterion of .05 or beyond is often used. Many researchers believe that if the function is not significant at or beyond the .05 level, there is little justification for going further. Some social scientists and business analysts, however, disagree. Their decision rule for continuing is the cost versus the value of the information, and higher levels of risk (e.g., significance levels > .05) may be acceptable for these purposes. For example, they may decide to examine discriminant functions that are significant at the .2 or even the .3 level if the circumstances justify it.

If the number of groups is three or more, then the analyst must decide not only if the discrimination between groups overall is statistically significant but also if each of the estimated discriminant functions is statistically significant. As discussed earlier, discriminant analysis estimates one fewer discriminant function than there are groups. If three groups are analyzed, then two discriminant functions will be estimated. For four groups, three functions will be estimated, and so on. The computer programs all provide the analyst the information necessary to ascertain the number of functions needed to obtain statistical significance, without including discriminant functions that do not increase the discriminatory power significantly. If one or more functions are deemed not statistically significant, the discriminant model should be reestimated with the number of functions to be derived limited to the number of significant functions. In this manner, the assessment of predictive accuracy and the interpretation of the discriminant functions will be based only on significant functions.

Assessing Overall Fit

Once the significant discriminant functions have been identified, the attention shifts to ascertaining the overall fit of the retained discriminant function(s). This assessment involves several major considerations: the reason for developing classification matrices, the cutting score determination, construction of the classification matrices, and standards for assessing classification accuracy.

Why Classification Matrices Are Developed

The statistical tests for assessing the significance of the discriminant functions do not tell how well the function predicts. For example, suppose the two groups are significantly different beyond the .01 level. With sufficiently large sample sizes, the group means (centroids) could be virtually identical, and we still would have statistical significance. In short, these statistics suffer the same drawbacks as the classical tests of hypotheses. Thus the level of significance of these statistics is a very poor indication of the function's ability to discriminate between the two groups. To determine the predictive ability of a discriminant function, the analyst must construct classification matrices.

To clarify further the usefulness of the classification matrix procedure, we shall relate it to the concept of an R^2 in regression analysis. Most of us have probably read academic articles in which the author has found statistically significant relationships and yet has explained only 10 percent (or less) of the variance (i.e., $R^2 = 0.10$). Usually this R^2 is significantly different from zero simply because the sample size is large. With multiple discriminant analysis, the hit ratio (percentage correctly classified) is analogous to regression's R^2. The hit ratio reveals how well the discriminant function classified the statistical units; the R^2 indicates how much variance the regression equation explained. The F test for statistical significance of the R^2 is, therefore, analogous to the Chi-square (or D^2) test of significance in discriminant analysis. Clearly, with a sufficiently large sample size in discriminant analysis, we could have a statistically significant difference between the two (or more) groups and yet correctly classify only 53 percent (when chance is 50 percent, with equal group sizes) [13].

Hit ratio! ratio of how many I "hit" correctly

Cutting Score Determination

If the statistical test indicates that the function discriminates significantly, it is customary to develop classification matrices to provide a more accurate assessment of the discriminating power of the function. Before a classification matrix can be constructed, however, the analyst must determine the cutting score. The **cutting score** is the criterion (score) against which each individual's discriminant score is judged to determine into which group the individual should be classified.

In constructing classification matrices, the analyst will want to determine the **optimum cutting score** (also called a critical Z value). The optimal cutting score will differ depending on whether the sizes of the groups are equal or unequal. If the groups are of equal size, the optimal cutting score will be halfway between the two group centroids. The cutting score for two groups of equal size is therefore defined as

$$Z_{CE} = \frac{Z_A + Z_B}{2}$$

where

Z_{CE} = Critical cutting score value for equal group sizes
Z_A = Centroid for group A
Z_B = Centroid for group B

Specifying Probabilities of Classification

The analyst must also determine whether to specify if the observed group sizes reflect the actual population proportions or whether the population group sizes should be assumed to be equal. The default assumption is equal probabilities: in other words, each group is assumed to have an equal chance of occurring. If the analyst is unsure if the observed proportions in the sample are representative of the population proportions, then equal probabilities should be employed. However, if the sample is randomly drawn from the population so that the groups do estimate the population proportions in each group, then the best estimates of actual group sizes and the prior probabilities are not equal values but, instead, the sample proportions. The impact of specifying the prior probabilities varies by the discrepancy of the sample proportions from the population proportions. But the analyst should specify the probabilities in all analyses to ensure that the correct assumptions underlay the classification process.

If the groups are not of equal size and are assumed to be representative of the population, a weighted average of the group centroids will provide a weighted optimal cutting score, calculated as follows:

$$Z_{CU} = \frac{N_A Z_A + N_B Z_B}{N_A + N_B}$$

where

Z_{CU} = Critical cutting score value for unequal group sizes
N_A = Number in group A
N_B = Number in group B
Z_A = Centroid for group A
Z_B = Centroid for group B

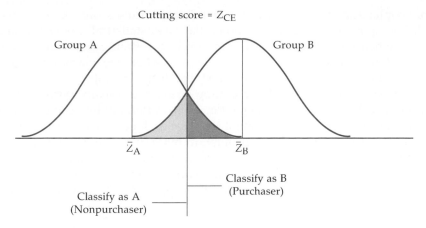

FIGURE 4.6 Optimal cutting score with equal sample sizes.

Both of the formulas for calculating the optimal cutting score assume that the distributions are normally distributed and the group dispersion structures are known.

The concept of an optimal cutting score for equal and unequal groups is illustrated in Figures 4.6 and 4.7. The optimal cutting score for equal groups is shown in Figure 4.6. The effect of one group's being larger than the other is illustrated in Figure 4.7. Both the weighted and unweighted cutting scores are shown. It is apparent that if group A is much smaller than group B, the optimal cutting score will be closer to the centroid of group A than to the centroid of group B. Also, note that if the unweighted cutting score were used, none of the individuals in group A would be misclassified, but a substantial portion of those in group B would be misclassified.

The optimal cutting score also must consider the cost of misclassifying an individual into the wrong group. If the costs of misclassifying an individual are approximately equal, the optimal cutting score will be the one that will misclassify the fewest number of individuals in all groups. If the misclassification costs are unequal, the optimum cutting score will be the one that minimizes the costs of misclassification.

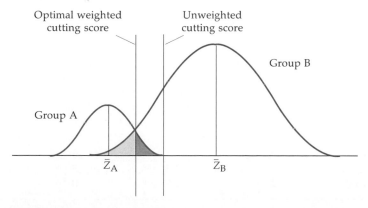

FIGURE 4.7 Optimal cutting score with unequal sample sizes.

More sophisticated approaches to determining cutting scores are discussed in [3, 11]. The approaches are based upon a Bayesian statistical model and are appropriate when the costs of misclassification into certain groups are very high, when the groups are of grossly different sizes, or when one wants to take advantage of a priori knowledge of group membership probabilities.

In practice, when calculating the cutting score, it is usually not necessary to insert the raw variable measurements for every individual into the discriminant function and to obtain the discriminant score for each person to use in computing the Z_A and Z_B (group A and B centroids). In many instances the computer program will provide the discriminant scores, as well as the Z_A and Z_B as regular output. When the analyst has the group centroids and sample sizes, he or she must merely substitute the values in the appropriate formula to obtain the optimal cutting score.

Constructing Classification Matrices

To validate the discriminant function through the use of classification matrices, the sample should be randomly divided into two groups. One of the groups (the analysis sample) is used to compute the discriminant function. The other group (the holdout or validation sample) is retained for use in developing the classification matrix. The procedure involves multiplying the weights generated by the analysis sample by the raw variable measurements of the holdout sample. Then the individual discriminant scores for the holdout sample are compared with the critical cutting score value and classified as follows:

1. Classify an individual into group A if $Z_n < Z_{ct}$.
2. Classify an individual into group B if $Z_n > Z_{ct}$.

where

$$Z_n = \text{Discriminant } Z \text{ score for the } n\text{th individual}$$
$$Z_{ct} = \text{Critical cutting score value}$$

The results of the classification procedure are presented in matrix form, as shown in Table 4.3. The entries on the diagonal of the matrix represent the number of individuals correctly classified. The numbers off the diagonal represent the incorrect classifications. The entries under the column labeled ''Actual Total'' represent the number of individuals actually in each of the two groups. The entries at

TABLE 4.3 Classification Matrix for HATCO's New Consumer Product

| | Predicted Group | | | |
	Would Purchase (1)	Would Not Purchase (2)	Actual Total	Group Classification Percentage
Actual group 1	22	3	25	88
Actual group 2	5	20	25	80
Predicted total	27	23	50	

Percent correctly classified (hit ratio) = $100 \times [(22 + 20)/50] = 84\%$.

the bottom of the columns represent the number of individuals assigned to the groups by the discriminant function. The percentage correctly classified for each group is shown at the right side of the matrix, and the overall percentage correctly classified, also known as the **hit ratio,** is shown at the bottom. For example, in our HATCO problem that attempted to predict which housewives would purchase a new consumer product, the number of individuals correctly assigned to actual group 1, "Would Purchase," was 22. The number incorrectly assigned to group 2, "Would Not Purchase," was 3. Similarly, the number of correct classifications to actual group 2 was 20, and the number of incorrect assignments to group 1 was 5. Thus the classification accuracy percentage of the discriminant function for actual groups 1 and 2 would be 88 and 80 percent, respectively. The overall classification accuracy (hit ratio) would be 84 percent.

One final classification-procedures topic should be discussed. A *t* test is available to determine the level of significance for the classification accuracy. The formula for a two-group analysis (equal sample size) is

$$t = \frac{p - .5}{\sqrt{\dfrac{.5 * (1.0 - .5)}{N}}}$$

where

p = Proportion correctly classified
N = Sample size

This formula can be adapted for use with more groups and unequal sample sizes.

Measures of Predictive Accuracy

As noted earlier, the predictive accuracy of the discriminant function is measured by the hit ratio, which is obtained from the classification matrix. The analyst may ask, What is considered an acceptable level of predictive accuracy for a discriminant function, and what is not considered acceptable? For example, is 60 percent an acceptable level, or should one expect to obtain 80 to 90 percent predictive accuracy? To answer this question, the analyst must first determine the percentage that could be classified correctly by chance (without the aid of the discriminant function).

When the sample sizes of the groups are equal, the determination of the chance classification is rather simple, obtained by dividing 1 by the number of groups. The formula is $C = 1 \div$ number of groups. For instance, in a two-group function the chance probability would be .50; for a three-group function the chance probability would be .33, and so forth.

The determination of the chance classification for situations where the group sizes are unequal is somewhat more involved. Let's assume that we have a sample in which 75 subjects belong to one group and 25 to the other. We could arbitrarily assign all the subjects to the larger group and achieve a 75 percent classification accuracy without the aid of a discriminant function. It could be concluded that unless the discriminant function achieves a classification accuracy higher than 75 percent, it should be disregarded because it has not helped us improve our prediction accuracy.

Determining the chance classification based on the sample size of the largest group is referred to as the **maximum chance criterion.** It is determined by computing the percentage of the total sample represented by the largest of the two (or more) groups. For example, if the group sizes are 65 and 35, the maximum chance criterion is 65 percent correct classifications. Therefore, if the hit ratio for the discriminant function does not exceed 65 percent, it has not helped us predict, based on this criterion.

The maximum chance criterion should be used when the sole objective of the discriminant analysis is to maximize the percentage correctly classified [13]. But situations in which we are concerned only about maximizing the percentage correctly classified are rare. Usually the analyst uses discriminant analysis to correctly identify members of all groups. In cases where the sample sizes are unequal and the analyst wants to classify members of all groups, the discriminant function defies the odds by classifying a subject in the smaller group(s). But the chance criterion doesn't take this fact into account [13]. Therefore, another chance model— the **proportional chance criterion**—should be used in most situations.

The proportional chance criterion should be used when group sizes are unequal and the analyst wishes to correctly identify members of the two (or more) groups. The formula for this criterion is

$$C_{\text{PRO}} = p^2 + (1 - p)^2$$

where

$$p = \text{Proportion of individuals in group 1}$$
$$1 - p = \text{Proportion of individuals in group 2}$$

Using the group sizes from our earlier example (75 and 25), we see that the proportional chance criterion would be 62.5 percent compared with 75 percent. Therefore, in this instance, a prediction accuracy of 75 percent would be acceptable because it is above the 62.5 percent proportional chance criterion.

These chance model criteria are useful only when computed with holdout samples (split-sample approach). If the individuals used in calculating the discriminant function are the ones being classified, the result will be an upward bias in the prediction accuracy. In such cases, both of these criteria would have to be adjusted upward to account for this bias.

The question of classification accuracy is crucial. If the percentage of correct classifications is significantly larger than would be expected by chance, an attempt can be made to interpret the discriminant functions in the hope of developing group profiles. However, if the classification accuracy is no greater than can be expected by chance, whatever structural differences appear to exist merit little or no interpretation; that is, differences in score profiles would provide no meaningful information for identifying group membership.

The question, then, is how high should the classification accuracy be relative to chance? For example, if chance is 50 percent (two-group, equal sample sizes), does a classification (predictive) accuracy of 60 percent justify moving to the interpretation stage? No general guidelines have been developed to answer this question. Ultimately the decision depends on the cost in relation to the value of the information. If the costs associated with a 60 percent predictive accuracy (relative to 50 percent by chance) are greater than the value to be derived from the findings, there is no justification for interpretation. If the value is high relative to the costs, 60 percent accuracy would justify moving on to interpretation.

The cost-versus-value argument offers little assistance to the neophyte data analyst. Therefore, the authors suggest the following criterion: the classification accuracy should be at least one-fourth greater than that achieved by chance. For example, if chance accuracy is 50 percent, the classification accuracy should be 62.5 percent (50% + $\frac{1}{4}$*50%). If chance accuracy is 30 percent, the classification accuracy should be 37.5 percent. This criterion provides only a rough estimate of the acceptable level of predictive accuracy. The criterion is easy to apply with groups of equal size. With groups of unequal size, an upper limit is reached when the maximum chance model is used to determine chance accuracy. This does not present too great a problem, however, because under most circumstances, the maximum chance model would not be used with unequal group sizes (see the section on chance models).

Statistically-Based Measures of Classification Accuracy Relative to Chance

A statistical test for the discriminatory power of the classification matrix when compared with a chance model is **Press's Q statistic.** This simple measure compares the number of correct classifications with the total sample size and the number of groups. The calculated value is then compared with a critical value (the Chi-square value for 1 degree of freedom at the desired confidence level). If it exceeds this critical value, then the classification matrix can be deemed statistically better than chance. The Q statistic is calculated by the following formula:

$$\text{Press's } Q = \frac{[N - (n * K)]^2}{N(K - 1)}$$

where

N = Total sample size
n = Number of observations correctly classified
K = Number of groups

For example, in Table 4.3, the Q statistic would be based on a total sample of 50, 42 correctly classified observations and 2 groups. The calculated statistic would be

$$\text{Press's } Q = \frac{[50 - (42 * 2)]^2}{50(2 - 1)} = 23.12$$

The critical value at a significance level of .01 is 6.63. Thus we would conclude that in the example the predictions were significantly better than chance, which would be a correct classification rate of 50 percent. This simple test is sensitive to sample size, where large samples are more likely to show significance than small sample sizes of the same classification rate. For example, if the sample size is increased to 100 in the example and the classification rate remains at 84 percent, the Q statistic increases to 46.24. One must be careful in drawing conclusions based solely on this statistic, however, because as the sample sizes become larger, a lower classification rate will still be deemed significant.

Stage Five: Interpretation of the Results

If the discriminant function is statistically significant and the classification accuracy is acceptable, the analyst should focus on making substantive interpretations of the findings. This process involves examining the discriminant functions to determine the relative importance of each independent variable in discriminating between the groups. Three methods of determining the relative importance have been proposed: (1) standardized discriminant weights, (2) discriminant loadings (structure correlations), and (3) partial *F* values.

Discriminant Weights

The traditional approach to interpreting discriminant functions examines the sign and magnitude of the standardized **discriminant weight** (sometimes referred to as a discriminant coefficient) assigned to each variable in computing the discriminant functions. When the sign is ignored, each weight represents the relative contribution of its associated variable to that function. Independent variables with relatively larger weights contribute more to the discriminating power of the function than do variables with smaller weights. The sign only denotes that the variable makes either a positive or a negative contribution [3].

The interpretation of discriminant weights is analogous to the interpretation of beta weights in regression analysis and is therefore subject to the same criticisms. For example, a small weight may indicate either that its corresponding variable is irrelevant in determining a relationship or that it has been partialed out of the relationship because of a high degree of multicollinearity. Another problem with the use of discriminant weights is that they are subject to considerable instability. These problems suggest caution in using weights to interpret the results of discriminant analysis.

Discriminant Loadings

In recent years, loadings have increasingly been used as a basis for interpretation because of the deficiencies in utilizing weights. **Discriminant loadings,** referred to sometimes as structure correlations, measure the simple linear correlation between each independent variable and the discriminant function. The discriminant loadings reflect the variance that the independent variables share with the discriminant function and can be interpreted like factor loadings in assessing the relative contribution of each independent variable to the discriminant function. (Chapter 7 further discusses factor-loading interpretation.)

Discriminant loadings (like weights) may be subject to instability. Loadings are considered relatively more valid than weights as a means of interpreting the discriminating power of independent variables because of their correlational nature. The analyst still must be cautious when using loadings for interpreting discriminant functions.

Partial F Values

As discussed earlier, two computational approaches—simultaneous and stepwise—can be utilized in deriving discriminant functions. When the stepwise method is selected, an additional means of interpreting the relative discriminating power of the independent variables is available through the use of the partial *F* values. This is accomplished by examining the absolute sizes of the significant *F* values and ranking them. Large *F* values indicate greater discriminating power. In practice, rankings using the *F*-values approach are the same as the ranking derived from using the weights, but the *F* values indicate the associated level of significance for each variable.

Interpretation of Two or More Functions

When there are two or more significant discriminant functions, we are faced with additional problems of interpretation. First, can we simplify the discriminant weights or loadings to facilitate the profiling of each function? Second, how do we represent the impact of each variable across the functions? This problem is found both in measuring the total discriminating effects across functions and in assessing the role of each variable in profiling each function separately. We address these two questions in the following paragraphs by introducing the concepts of rotation of the functions, potency index, and stretched attribute vectors in graphical representations.

Rotation of the Discriminant Functions

After the discriminant functions have been developed, they can be "rotated" to redistribute the variance. (The concept is more fully explained in Chapter 7.) Basically, rotation preserves the original structure and the reliability of the discriminant solution while at the same time making the functions easier to interpret substantively. In most instances, the varimax rotation is employed as the basis for rotation.

Potency Index

In the previous sections, we discussed using the standardized weights or discriminant loadings as measures of a variable's contribution to a discriminant function. When two or more functions are derived, however, a composite or summary measure is useful in describing the contributions of a variable across *all* significant functions. The **potency index** is a relative measure among all variables that is indicative of each variable's discriminating power [14]. It includes both the contribution of a variable to a discriminant function (its discriminant loading) and the relative contribution of the function to the overall solution (a relative measure among the eigenvalues of the functions). The composite is simply the sum of the individual potency indices across all significant discriminant functions. Interpretation of the composite measure is limited, however, by the fact that it is useful only in depicting the relative position (such as the rank order) of each variable, and the absolute value has no real meaning. The potency index is calculated by the following two-step process:

> *Step 1: Calculate a potency value for each significant function.* In the first step, the discriminating power of a variable, represented by the squared value of the discriminant loading, is "weighted" by the relative contribution of the discriminant function to the overall solution. First, the relative eigenvalue measure for each significant discriminant function is calculated simply as

$$\text{Relative eigenvalue of discriminant function} = \frac{\text{Eigenvalue of discriminant function}}{\text{Sum of eigenvalues across all significant functions}}$$

The potency value of each variable on a discriminant function is then

$$\text{Potency value of variable } i \text{ on single function} = (\text{Discriminant loading})^2 \times \text{Relative eigenvalue of discriminant function}$$

> *Step 2: Calculate a composite potency index across all significant functions.* Once a potency value has been calculated for each function, the composite po-

tency index is calculated as the sum of potency values on each significant discriminant function. Stated in equation form,

Potency index of variable i = Sum of potency values for all significant discriminant functions

The potency index now represents the *total* discriminating effect of the variable across all of the significant discriminant functions. Remember that it is only a relative measure, and its absolute value has no substantive meaning.

Graphical Display of Group Centroids

Group centroids (means of the discriminant function scores for observations in each discriminant group) on each discriminant function can be plotted to demonstrate the results from a global prospective. Plots are usually prepared for the first two or three discriminant functions (assuming they are statistically significant and valid predictive functions). The values for each group show its position in reduced discriminant space (so called because not all the functions and thus not all the variance is plotted). The analyst can see the differences between the groups on each function; however, visual inspection does not totally explain what these differences are. Circles enclosing the distribution of observations around their respective centroids can be drawn to clarify group differences further, but this procedure is beyond the scope of this text (see [3]).

Recall that these centroids are reported in the form of Z scores. These aggregate Z scores (group centroids) can be interpreted as the number of standard deviations each group is away from the average of all groups (with standardization, the average of all groups is zero).

Graphical Display of Discriminant Loadings

To depict differences in the groups on the predictor variables, the analyst can plot the discriminant loadings. The simplest approach is to plot actual rotated or unrotated loadings on a graph. The preferred approach would be to plot the rotated loadings. An even more accurate approach, however, involves what is called *stretching the vectors*.

Before explaining the process of stretching, we must first define a vector in this context. A **vector** is merely a straight line drawn from the origin (center) of a graph to the coordinates of a particular variable. The length of each vector is indicative of the relative importance of each variable in discriminating among the groups. To stretch a vector, the analyst multiplies the discriminant loading (preferably after rotation) by its respective univariate F value.

The plotting process always involves all the variables included in the model as significant. But the analyst may also plot the other variables with significant univariate F ratios that were not significant in the discriminant function. This procedure shows the importance of collinear variables that are not included, such as in a stepwise solution. By using this procedure, we note that vectors point to the groups having the highest mean on the respective predictor and away from the groups having the lowest mean scores. The group centroids are also stretched in this procedure by multiplying them by the approximate F value associated with each discriminant function. If the loadings are stretched, the centroids must be stretched as well to plot them accurately on the same graph. The approximate F values for each discriminant function are obtained by the following formula:

$$F \text{ value}_{\text{function}_i} = \text{Eigenvalue}_{\text{function}_i} * \frac{\text{Sample size used for estimation} - \text{Number of groups}}{\text{Number of groups} - 1}$$

As an example, assume that the sample of 50 observations was divided into three groups. The multiplier of each eigenvalue would be $(50 - 3) \div (3 - 1) = 23.5$. For more details on this procedure, see [3].

For those who do not wish to stretch the attribute vectors and centroids, there is an alternative in the "territorial maps" provided by most programs. It does not include the vectors, but it does plot the centroids and the boundaries for each group.

Which Interpretive Method to Use?

Several methods for interpreting the nature of discriminant functions have been discussed, both for single- and multiple-function solutions. Which methods should be used? The loadings approach is somewhat more valid than the use of weights and should be utilized whenever possible. The use of univariate and partial F values allows the analyst to use several measures and look for some consistency in evaluations of the variables. If two or more functions are estimated, then the analyst can employ several graphical techniques and the potency index, which aid in interpreting the multidimensional solution. The most basic point is that the analyst should employ all available methods to arrive at the most accurate interpretation.

Stage Six: Validation of the Results

The final stage of a discriminant analysis involves validating the discriminant results to provide assurances that the results have external as well as internal validity. With discriminant analysis's propensity to inflate the hit ratio if evaluated only on the analysis sample, cross-validation is an essential step. Most often the cross-validation is done with the original sample, but it is possible to employ an additional sample as the holdout sample. In addition to cross-validation, the analyst should use group profiling to ensure that the group means are valid indicators of the conceptual model used in selecting the independent variables. Both these approaches will now be discussed.

Split-Sample or Cross-Validation Procedures

Recall that the most frequently utilized procedure in validating the discriminant function is to divide the groups randomly into analysis and holdout samples once. This involves developing a discriminant function with the analysis sample and then applying it to the holdout sample. The justification for dividing the total sample into two groups is that an upward bias will occur in the prediction accuracy of the discriminant function if the individuals used in developing the classification matrix are the same as those used in computing the function; that is, the classification accuracy will be higher than is valid for the discriminant function, if it was used to classify a separate sample. The implications of this upward bias are particularly important when the analyst is concerned with the external validity of the findings.

Other researchers have suggested, however, that greater confidence could be placed in the validity of the function by following this procedure several times [14]. Instead of randomly dividing the total sample into analysis and holdout groups once, the analyst would randomly divide the total sample into analysis and holdout samples several times, each time testing the validity of the function through the development of a classification matrix and a hit ratio. Then the several hit ratios would be averaged to obtain a single measure.

More sophisticated methods based on estimation with multiple subsets of the sample have been suggested for validating discriminant functions [2, 3]. The two most widely used approaches are the U-method and the jackknife method. Both methods are based on the "leave-one-out" principle, where the discriminant function is fitted to repeatedly drawn samples of the original sample. The most prevalent use of this method has been to estimate $k - 1$ samples, eliminating one observation at a time from a sample of k cases. The primary difference in the two approaches is that the U-method focuses on the classification accuracy, whereas the jackknife approach addresses the stability of the discriminant coefficients. Both approaches are quite sensitive to small sample sizes. Guidelines suggest that either of these two approaches be used only when the smallest group size is at least three times as great as the number of predictor variables, and most analysts suggest a ratio of five to one [11]. In spite of these limitations, both methods provide the most valid and consistent estimate of the classification accuracy rate. The use of the U-method and jackknife methods has been limited because only one of the major computer packages [1] provides them as a program option.

Profiling Group Differences

Another validation technique is to profile the groups on the independent variables to ensure their correspondence with the conceptual bases used in the original model formulation. When the analyst has identified the independent variables that make the greatest contribution in discriminating between the groups, the next step is to profile the characteristics of the groups based on the group means. This profile enables the analyst to understand the character of each group according to the predictor variables. For example, referring to the HATCO survey data presented in Table 4.1, we see that the mean rating on "Durability" for the "Would Purchase" group is 8.0, while the comparable mean rating on "Durability" for the "Would Not Purchase" group is 3.2. Thus a profile of these two groups shows that the "Would Purchase" group rates the perceived durability of the new product substantially higher than the "Would Not Purchase" group.

Another approach is to profile the groups on a separate set of variables that should mirror the observed group differences. This separate profile provides an assessment of external validity in that the groups vary on both the independent variable(s) and the set of associated variables. This is similar in character to the validation of derived clusters described in Chapter 8.

A Two-group Illustrative Example

To illustrate the application of a two-group discriminant analysis, we shall use variables drawn from the HATCO database introduced in Chapter 1. This example examines each of the six stages of the model-building process to a research problem particularly suited to multiple discriminant analysis.

Stage One: Objectives of the Discriminant Analysis

You will recall that one of the customer characteristics obtained by HATCO in its survey was a categorical variable indicating which purchasing approach a firm used: total value analysis versus specification buying. Firms that employ total value analysis evaluate each aspect of the purchase, including both the product and the services being purchased. Specification buying, on the other hand, defines all product and service characteristics desired, and the seller then makes a bid to fill the specifications. Both approaches have merit in certain situations, but HATCO's management team expects that firms using these two approaches would emphasize different characteristics of suppliers in their selection decision. The objective is to identify the perceptions of HATCO that differ significantly between firms using these two purchasing methods. The company would then be able to tailor sales presentations and benefits offered to best match the buyer's perceptions. To do so, discriminant analysis was selected to identify those perceptions of HATCO that best distinguish firms using each buying approach.

Stage Two: Research Design of the Discriminant Analysis

The research design stage focuses on three key issues: selecting dependent and independent variables, assessing the adequacy of the sample size for the planned analysis, and dividing the sample for validation purposes. Each of these issues is discussed in the following sections.

Selection of Dependent and Independent Variables

Because the dependent variable, the purchasing approach employed by a firm, is a two-group categorical variable, discriminant analysis is the appropriate technique. The survey also collected perceptions of HATCO that can now be used to differentiate between the two groups of firms. The discriminant analysis uses as independent variables the first seven variables from the database (X_1 to X_7) to discriminate between firms applying each purchasing method (X_{11}).

Sample Size

The sample of 100 observations meets the suggested minimum size for application of discriminant analysis. Moreover, it provides a 15-to-1 ratio of observations to independent variables (100 observations for 7 potential independent variables) that is quite close to the suggested ratio of 20 to 1. Finally, the two groups of firms contain 60 and 40 observations, making them comparable enough in size not to impact either the estimation or the classification processes.

Division of the Sample

Previous discussion has emphasized the need for validating the discriminant function with a split sample or holdout sample. Any time a holdout sample is used, the analyst must ensure that the resulting sample sizes are sufficient to support the number of predictors included in the analysis. The HATCO database has 100 observations, and it was decided that a holdout sample of 40 observations would be sufficient for validation purposes and would still leave 60 observations for estimation of the discriminant function. It is important to ensure randomness in the selection of the holdout sample so that any ordering of the observations does not affect the processes of estimation and validation. The control cards necessary for both selection of the holdout sample and performance of the two-group discriminant analysis are shown in Appendix A.

Stage Three: Assumption of Discriminant Analysis

The principal assumption underlying discriminant analysis involves the formation of the variate or discriminant function (normality, linearity, and multicollinearity) and the estimation of the discriminant function (equal variance/covariance matrices). The examination of the independent variables for normality, linearity, and multicollinearity is performed in Chapter 2. The interested reader is encouraged to review these results. For purposes of our illustration of discriminant analysis, these assumptions are met at acceptable levels.

The assumption of equal covariance or dispersion matrices is also addressed in Chapter 2. In most statistical programs, one or more statistical tests for this assumption are provided. The most common test is *Box's M*. Again, the reader is referred to Chapter 2 for a more complete discussion. In the two-group example, the significance of differences in the covariance matrices between the two groups is .0320. Even though the significance is less than .05 (remember that in this test the analyst looks for values above the desired significance level), the sensitivity of the test to factors other than just covariance differences (e.g., normality of the variables and increasing sample size) make this an acceptable level. No additional remedies are needed before estimation of the discriminant function can be performed.

Stage Four: Estimation of the Discriminant Function and Assessing Overall Fit

Let us begin our analysis of the two-group discriminant analysis by examining Table 4.4, which shows the unweighted group means for each of the independent variables, based on the 60 observations constituting the analysis sample. Table 4.5 shows the univariate analysis of variance which is used to assess the significance between means of the independent variables for the two groups.

TABLE 4.4 Group Descriptive Statistics for the Two-group Discriminant Analysis Sample

Dependent Variable**		Independent Variables*							Sample
Group Means	X_{11}	X_1	X_2	X_3	X_4	X_5	X_6	X_7	Size
Specification buying	0	2.227	2.973	6.873	5.155	2.578	2.555	8.464	22
Total value analysis	1	4.258	2.082	8.568	5.437	3.179	2.832	6.013	38
Total		3.513	2.408	7.947	5.333	2.958	2.730	6.912	60

Dependent Variable		Independent Variables*							Sample
Standard Deviations	X_{11}	X_1	X_2	X_3	X_4	X_5	X_6	X_7	Size
Specification buying	0	1.053	1.187	.763	.815	.936	.580	.945	22
Total value analysis	1	1.099	1.119	1.280	1.319	.501	.919	1.322	38
Total		1.458	1.214	1.384	1.160	.745	.817	1.683	60

*X_1 = Delivery speed; X_2 = Price level; X_3 = Price flexibility; X_4 = Manufacturer's image; X_5 = Service; X_6 = Sales force's image; X_7 = Product quality.
**X_{11} = Specification buying

Estimation of the Discriminant Function

Because the objective of this analysis was to determine which variables are the most efficient in discriminating between firms using the two purchasing approaches, a stepwise procedure was used. If the objective had simply been to determine the discriminating capabilities of the entire set of benefits, with no regard to the impact of any individual benefit sought, all variables would have been entered into the model simultaneously. The Mahalanobis D^2 measure will be used in the stepwise procedure.

The stepwise procedure begins with all of the variables excluded from the model and selects the variable that maximizes the Mahalanobis distance between the groups. In this example, a minimum F value of 1.00 (the default value) was required for entry. This limitation eliminated variable X_4 from consideration for

TABLE 4.5 Test for Equality of Group Means Between Firms Using Different Buying Approaches

Independent Variables		Wilks' Lambda	Univariate F Ratio	Significance
X_1	Delivery speed	.54209	48.99	.0000
X_2	Price level	.87279	8.45	.0052
X_3	Price flexibility	.64530	31.88	.0000
X_4	Manufacturer's image	.98602	.82	.3682
X_5	Overall service	.84578	10.58	.0019
X_6	Sales force's image	.97284	1.62	.2082
X_7	Product quality	.49924	58.18	.0000

Wilks' lambda (U statistic) and univariate F ratio with 1 and 58 degrees of freedom.

TABLE 4.6A Results from Step 1 of Stepwise Two-group Discriminant Analysis Model

STEP 1: X_7 (Product Quality) Included in the Analysis.
Summary Statistics

		Degrees of Freedom		Significance	Between Groups
Wilks' lambda	.4992	1	1	58.0	
Equivalent F	58.1760		1	58.0	.0000
Minimum D^2	4.1753				0 and 1
Equivalent F	58.1760		1	58.0	.0000

Variables in the Analysis after Step 1

Variable	Tolerance	F to Remove
X_7 Product quality	1.000	58.176

Variables Not in the Analysis after Step 1

Variables	Tolerance	Minimum Tolerance	F to Enter	D^2	Between Groups
X_1 Delivery speed	.9733	.9733	16.680	6.615	0 and 1
X_2 Price level	.9327	.9327	.454		
X_3 Price flexibility	.9968	.9968	18.196	6.837	0 and 1
X_4 Manufacturer's image	.9629	.9629	2.873	4.596	0 and 1
X_5 Overall service	.9944	.9944	7.203	5.229	0 and 1
X_6 Sales force's image	.9618	.9618	3.895	4.745	0 and 1

Significance Testing of Group Differences after Step 1*

	Group 0: Specification Buying
Group 1: Total Value Analysis	58.176
	(.0000)

*F statistic and significance level (in parentheses) between groups after Step 1. Each F statistic has 1 and 58 degrees of freedom.

possible entry into the discriminant function. The maximum Mahalanobis distance (D^2) is associated with X_7 (see Table 4.6A). After X_7 entered the model, the remaining variables were evaluated on the basis of distance between their means after the variance associated with X_7 was removed. Again, variables with F values less than 1.00 (only X_2) were eliminated from consideration for entry at the next step. Variable X_3 was the next variable to enter the model because it had the highest F value (18.196). Given X_1's large F value (16.68), it is very likely that it will enter the model at a later step if it is not highly correlated with variables previously selected. The F value must be calculated after the effect of the variable in the models are removed. For instance, high multicollinearity of X_1 with variables in the model could substantially reduce the F value. Also, note that in cases

TABLE 4.6B Results from Step 2 of Stepwise Two-group Discriminant Analysis Model

STEP 2: X_3 **(Price Flexibility) Included in the Analysis.**
Summary Statistics

		Degrees of Freedom		Significance	Between Groups
Wilks' lambda	.3784	2	1	58.0	
Equivalent F	46.8104		2	57.0	.0000
Minimum D^2	6.8371				0 and 1
Equivalent F	46.8104		2	57.0	.0000

Variables in the Analysis after Step 2

Variables		Tolerance	F to Remove
X_3	Price flexibility	.9968	18.196
X_7	Product quality	.9968	40.195

Variables Not in the Analysis after Step 2

Variables		Tolerance	Minimum Tolerance	F to Enter	D^2	Between Groups
X_1	Delivery speed	.9316	.9316	7.974	8.403	0 and 1
X_2	Price level	.8093	.8093	.661		0 and 1
X_4	Manufacturer's image	.9459	.9459	3.885	7.600	0 and 1
X_5	Overall service	.9796	.9796	7.769	8.363	0 and 1
X_6	Sales force's image	.9595	.9577	3.557	7.536	0 and 1

Significance Testing of Group Differences after Step 2*

	Group 0: Specification Buying
Group 1: Total Value Analysis	46.810
	(.0000)

*F statistic and significance level (in parentheses) between groups after Step 2. Each F statistic has 2 and 57 degrees of freedom.

where two or more variables are entered into the model, the variables already in the model are evaluated for possible removal. A variable may be removed if high multicollinearity exists between it and the other included independent variables.

In step 2 (see Table 4.6B), X_3 enters the model as expected. As in step 1, the overall model is significant ($F = 46.81$), as is the discriminating power of both variables included to this point (X_3 and X_7). As noted earlier, X_1 is the next candidate for inclusion, but the F-to-enter value has been reduced substantially because of the multicollinearity of X_1 with X_3 and X_7.

Table 4.7 provides the overall stepwise discriminant analysis results after all the significant discriminators have been included in the estimation of the discriminant function. The summary table indicates that four variables (X_1, X_3, X_6, and X_7) entered the model and were significant discriminators based on their Wilks'

TABLE 4.7 Summary of Two-group Stepwise Discriminant Analysis Results

Summary Table

	Action		Wilks' Lambda		Minimum D^2		
Steps	Entered	Removed	Value	Significance	Value	Significance	Between Groups
1	X_7	Product quality	.49924	.0000	4.1753	.0000	0 and 1
2	X_3	Price flexibility	.37843	.0000	6.8371	.0000	0 and 1
3	X_1	Delivery speed	.33126	.0000	8.4034	.0000	0 and 1
4	X_6	Sales force's image	.31601	.0000	9.0101	.0000	0 and 1

Canonical Discriminant Functions

		Percent of Variance		Canonical	After	Wilks'			
Function	Eigenvalue	Function	Cumulative	Correlation	Function	Lambda	Chi-square	D.F.	Significance
					0	.3160	64.512	4	.0000
1*	2.1645	100.00	100.00	.8270					

Standardized Canonical Discriminant Function Coefficients

Independent Variables		Discriminant Function Coefficients: Function 1
X_1	Delivery speed	.41954
X_3	Price flexibility	.47140
X_6	Sales force's image	.26519
X_7	Product quality	−.69078

Structure Matrix†

Independent Variables		Discriminant Function Loadings: Function 1
X_7	Product quality	−.68074
X_1	Delivery speed	.62470
X_3	Price flexibility	.50394
X_2	Price level	−.40281
X_5	Overall service	.17118
X_6	Sales force's image	.11358
X_4	Manufacturer's image	.03390

Group Means (Centroids) of Canonical Discriminant Functions

Group	Group Centroids: Function 1
Specification buying	−1.90106
Total value analysis	1.10062

*Marks the 1 canonical discriminant function remaining in the analysis.

†Pooled-within-groups correlations between discriminating variables and canonical discriminant functions. (Variables ordered by size of correlation within function.)

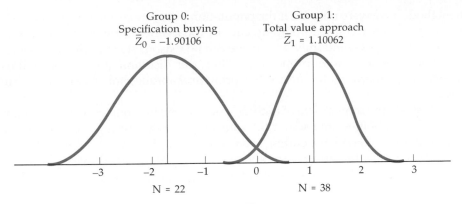

Group 0:
Specification buying
$\overline{Z}_0 = -1.90106$

Group 1:
Total value approach
$\overline{Z}_1 = 1.10062$

N = 22

N = 38

FIGURE 4.8 Plot of group centroids ($\overline{Z}$).

lambda and minimum D^2 values. The multivariate aspects of the model are reported under the heading "Canonical Discriminant Functions." Note that the discriminant function is highly significant (.000) and displays a canonical correlation of .8270. One interprets this correlation by squaring it $(.8270)^2 = .6839$ and concluding that 68.39 percent of the variance in the dependent variable (X_{11}) can be accounted for (explained) by this model, which includes only four independent variables. The standardized canonical discriminant function coefficients are the weights that will be used in the validation phase. The loadings are reported under the heading "Structure Matrix" and are ordered from highest to lowest by the size of the loading. The loadings are discussed later under the interpretation phase. Group centroids are also reported, and they represent the mean of the individual discriminant function scores for each group.

Group centroids can be used to interpret the discriminant function results from a global or an overall perspective. Table 4.7 reveals that the group centroid for the firms using specification buying (group 0) is -1.90106, while the group centroid for the firms using the total value analysis approach (group 1) is 1.10062. Looking at Figure 4.8, we can see a plot of the centroids showing each group's deviation from the overall mean of the two groups. The centroids are measured by Z scores. To show that the overall mean is zero, multiply the number in each group by its centroid and add the result (e.g., $-1.90106 \times 22 + 1.10062 \times 38 = 0.0$).

Assessing Overall Fit

The second step in the estimation stage is to assess the predictive accuracy of the discriminant function. To accomplish this, we must develop a classification matrix. Classification matrices for both the analysis and the holdout samples are calculated. While examination of the holdout sample and its predictive accuracy is actually performed in the validation stage, the results will be discussed now for ease of comparison between estimation and holdout samples. Before developing the classification matrices, we must determine the cutting score. Recall that the cutting score is the criterion against which each individual's discriminant Z score is judged to determine into which group the individual should be classified.

In this situation, we know that the dependent variable consists of two groups, 40 firms following the specification buying approach (40 percent of the sample) and the remaining 60 firms (60 percent of the sample) using the total value analysis method. If we were not sure if the population proportions were represented by the sample, then we should employ equal probabilities. However, because our sample of firms was randomly drawn, we can be reasonably sure that this sample does reflect the population proportions. Thus, this discriminant analysis will use the sample proportions to specify the prior probabilities for classification purposes.

To illustrate the importance of cutting score determination, let us focus on how the prior probabilities are used in the calculation of the cutting score. If the two groups are of equal size, the cutting score would simply be the average of the two centroids. Because the groups are unequal in size, a weighted average must be used to account for the difference in variance. The weighted average is calculated as follows:

$$Z_{CU} = \frac{N_A Z_A + N_B Z_B}{N_A + N_B}$$

where

$$Z_{CU} = \text{Critical cutting score for unequal group sizes}$$
$$N_A = \text{Number in group A}$$
$$N_B = \text{Number in group B}$$
$$Z_A = \text{Centroid for group A}$$
$$Z_B = \text{Centroid for group B}$$

By substituting the appropriate values in the formula, we can obtain the critical cutting score (assuming equal costs of misclassification):

$$Z_{CU} = \frac{[(22) * (-1.90106)] + [(38) * (1.10062)]}{60} = 0.0$$

Group sizes used in the preceding calculation are based on the data set used in the analysis sample and do not include the holdout sample. By using this approach, the cutting score will always be zero (0) when the data are standardized. Because the critical cutting score is 0, the procedure for classifying firms is as follows:

1. Classify a firm as using specification buying if its discriminant score is negative.
2. Classify a firm as using the total value analysis approach if its discriminant score is positive.

Using this criteria, the computer program developed classification matrices for the observations in both the analysis and the holdout samples. The results are shown in Table 4.8. The 93.33 percent accuracy of the analysis sample is slightly higher than the 92.50 percent accuracy of the holdout sample, as anticipated.

The 92.5 percent classification accuracy is quite high. For illustration purposes, however, let us compare it with the a priori chance of classifying individuals correctly without the discriminant function. The proportional chance criterion is the appropriate chance model to use for our HATCO example. We have unequal group sizes and we want to identify members of both groups correctly. The formula is

$$C_{PRO} = p^2 + (1 - p)^2$$

where

$$C_{PRO} = \text{The proportional chance criterion}$$
$$p = \text{Proportion of firms in group 1}$$
$$1 - p = \text{Proportion of firms in group 2}$$

Substituting the appropriate values, we obtain

$$
\begin{aligned}
C_{PRO} &= (.367)^2 + (.633)^2 \\
&= .135 + .401 \\
&= .536
\end{aligned}
$$

The maximum chance criterion can also be calculated. It is simply the percentage correctly classified if all observations were placed in the group with the greatest probability of occurrence. Because one group occurs 63.3 percent of the time, we could be correct 63.3 percent of the time if we assigned all observations to this group. Because the maximum chance criterion is larger than the proportional test criterion, our model should outperform the 63.3 percent level.

The classification accuracy of 92.50 percent is substantially higher than the proportional chance criterion of 53.6 percent and the maximum chance criterion of 63.3 percent. Note that the proportional chance criterion is compared with the

TABLE 4.8 Classification Matrices for Two-group Discriminant Analysis for Both Analysis and Holdout Samples

Classification Results: Analysis Sample*

Actual Group	Number of Cases	Predicted Group Membership	
		Specification Buying	Total Value Analysis
Specification buying	22	21 95.5%	1 4.5%
Total Value analysis	38	3 7.9%	35 92.1%

Classification Results: Holdout Sample†

Actual Group	Number of Cases	Predicted Group Membership	
		Specification Buying	Total Value Analysis
Specification buying	18	16 88.9%	2 11.1%
Total Value analysis	22	1 4.5%	21 95.5%

*Percent of "grouped" cases correctly classified: 93.33% [(21 + 35)/60 = 93.33%].
†Percent of "grouped" cases correctly classified: 92.50% [(16 + 21)/40 = 92.50%].

percentage correctly classified in the holdout sample, reducing the upward bias seen in the classification of the analysis sample.

The final measure of classification accuracy is Press's Q. From the discussion earlier, the calculation for the analysis sample is

$$\text{Press's } Q = \frac{[60 - (56 * 2)]^2}{60(2 - 1)} = 45.1$$

And the calculation for the holdout sample is

$$\text{Press's } Q = \frac{[40 - (37 * 2)]^2}{40(2 - 1)} = 28.9$$

In both instances, the calculated values exceed the critical value of 6.63. Thus the classification accuracy for the analysis and, more important, the holdout sample exceed at a statistically significant level the classification accuracy expected by chance. Remember always to use caution in the application of a holdout sample with small data sets. In this case the small sample size of 40 was adequate, but larger sizes are always more desirable.

Stage Five: Interpretation of the Discriminant Function

After estimating the function, the next phase is interpretation. This stage involves examining the function to determine the relative importance of each independent variable in discriminating between the groups. Table 4.9 contains, among the interpretive measures, the discriminant weights and loadings for the function. The independent variables were screened by the stepwise procedure, and four were significant enough to be included in the function—X_1, X_3, X_6, and X_7. For interpretation purposes, we rank the independent variables in terms of both their weights and loadings—indicators of their discriminating power. Signs do not affect the rankings; they indicate a positive or negative relationship with the dependent variable. Because the loadings are considered more valid than the weights, we shall use them in our example.

When using the discriminant-loadings approach, we need to know which variables are significant discriminators. With stepwise procedures, this determination

TABLE 4.9 Summary of Interpretive Measures for Two-group Discriminant Analysis

Variable	Standardized Weights Value	Discriminant Loadings Value	Rank	Univariate F Ratio Value	Rank
X_1 Delivery speed	.41954	.62470	2	48.99	2
X_2 Price level	NI	−.40281	4	8.45	5
X_3 Price flexibility	.47140	.50394	3	31.88	3
X_4 Manufacturer's image	NI	.03390	7	.82	7
X_5 Overall service	NI	.17118	5	10.58	4
X_6 Sales force's image	.26519	.11358	6	1.62	6
X_7 Product quality	−.69078	−.68074	1	58.18	1

NI: not included in the stepwise solution.

is made easier because the criteria specified for the technique prevent nonsignificant variables from entering the function. In simultaneous discriminant analysis, all variables are entered in the function, and generally any variables exhibiting loadings ± .30 or higher are considered significant.

The analyst is usually interested in substantive interpretations of the individual variables. Such interpretations are accomplished by identifying the variables that are significant in the discriminant function and understanding what the differing group means on each variable indicate. For example, for all the variables in this model, higher scores indicate more favorable perceptions of HATCO on that attribute (for more detail, see Chapter 1). From Table 4.9 we can use the structure matrix information (loadings) and the univariate F values to determine the ranking of these variables in terms of their discriminating value. Both measures exhibit a high degree of correspondence. Of the four variables in the function, X_7 discriminates the most and X_6 discriminates the least. One will note that several variables not included in the model have higher loadings than X_6. They were not included, however, because their collinearity with the variables already included in the model reduced the additional discriminating power they could provide. Thus, although X_6 is lower, it provides a unique and statistically significant source of discrimination not found in the other variables. Referring back to Table 4.4, we note that on three of the four variables (X_1 delivery speed, X_3 price flexibility, and X_6 sales force's image), the means for those firms employing the total-value approach are higher, meaning that they have more favorable perceptions of HATCO than do firms using specification buying. Only on product quality (X_7) is the mean for firms using specification buying higher. One can conclude that firms using the total-value analysis approach employ a wider range of factors, while specification buying focuses on product quality.

Stage Six: Validation of the Discriminant Results

The final stage addresses the internal and external validity of the discriminant function. The primary means of validation is through the use of the holdout sample and the assessment of its predictive accuracy. In this manner, validity is established if the discriminant function performs at an acceptable level in classifying observations that were not used in the estimation process. If the holdout sample is formed from the original sample, then this approach establishes internal validity. If another separate sample, perhaps from another population or segment of the population, forms the holdout sample, then this addresses the external validity of the discriminant results.

In our example, the holdout sample came from the original sample. Thus, the acceptable levels on all measures of predictive accuracy found in the holdout sample do establish internal validity. The analyst is encouraged to extend the validation process through expanded profiling of the groups and the possible use of additional samples to establish external validity.

A Three-group Illustrative Example

To illustrate the application of a three-group discriminant analysis, we once again use the HATCO database. In the previous example, we were concerned with discriminating between only two groups, so we were able to develop a single discriminant function and a cutting score to divide the two groups. In this example, it is necessary to develop two separate discriminant functions to distinguish among three groups. The first function separates one group from the other two, and the second separates the remaining two groups. As with the prior example, each of the six stages of the model-building process will be discussed.

Stage One: Objectives of the Discriminant Analysis

HATCO's objective in this research is to determine the relationship between the firms' perceptions of HATCO and the type of purchasing situation most often faced. Firms that predominantly deal with HATCO in different purchasing situations may view and evaluate HATCO differently. The resulting discriminant model, like the two-group model discussed earlier, allows for a precise determination of the perceptions uniquely held by firms for each type of purchase situation. From this information, HATCO can develop targeted strategies in each purchasing situation that accentuate the perceived strengths of HATCO.

Stage Two: Research Design of the Discriminant Analysis

To test this relationship, a discriminant analysis is performed using X_{14} as the dependent variable and the perceptions of HATCO by these firms (X_1 to X_7) as the independent variables. Note that X_{14} differs from the dependent variable in the prior example in that it has three categories in which to classify a firm by the purchasing situation (new task, modified rebuy, or straight rebuy) most often engaged in with HATCO. The sample size of 100 is adequate to support inclusion of the seven independent variables. Examination of the three category sizes also finds adequate sizes in each group as well. For purposes of validation, a holdout sample of 40 observations will be drawn. This will leave 60 observations to be used in the estimation of the discriminant functions.

Stage Three: Assumptions in Discriminant Analysis

As was the case in the two-group example, the assumptions of normality, linearity, and collinearity of the independent variables have already been discussed at length in Chapter 2. The interested reader is referred to these discussions for more details. The analyses performed in Chapter 2 indicated that the independent variables met these assumptions at adequate levels to allow for the analysis to continue without additional remedies. The remaining assumption, the equality of the variance/covariance or dispersion matrices, is also addressed in Chapter 2. The Box's M test assesses the similarity of the dispersion matrices of the independent variables between the three groups (categories). The test statistic indicated that differences were indicated at the .01 significance level. While in many instances this significance level would necessitate remedial action, the sensitivity of the Box's M test to sample size and other characteristics of the independent variables makes it a very liberal test. Thus, the statistical test is judged to provide inadequate evidence that the dispersion matrices are sufficiently different to require corrective action, and the analysis can proceed.

TABLE 4.10 Group Descriptive Statistics for the Three-group Discriminant Analysis Sample

Dependent Variable *		*Independent Variables†*							*Sample*
Group Means	X_{14}	X_1	X_2	X_3	X_4	X_5	X_6	X_7	*Size*
New task	1	2.429	2.157	7.233	5.067	2.281	2.686	7.762	21
Modified rebuy	2	3.227	3.520	6.980	5.587	3.353	2.687	7.307	15
Straight rebuy	3	4.642	1.933	9.175	5.408	3.305	2.796	5.921	24
Total		3.513	2.408	7.947	5.333	2.958	2.730	6.912	60

Dependent Variable *		*Independent Variables†*							*Sample*
Standard Deviations	X_{14}	X_1	X_2	X_3	X_4	X_5	X_6	X_7	*Size*
New task	1	1.162	.915	.881	.867	.670	.731	1.373	21
Modified rebuy	2	1.130	1.379	1.343	1.120	.638	.766	1.750	15
Straight rebuy	3	1.023	.893	.700	1.387	.372	.939	1.406	24
Total		1.458	1.214	1.384	1.160	.745	.817	1.683	60

* X_{14} = Type of buying situation.
† X_1 = Delivery speed; X_2 = Price level; X_3 = Price flexibility; X_4 = Manufacturer's image; X_5 = Overall service; X_6 = Sales force's image; X_7 = Product quality.

Stage Four: Estimation of the Discriminant Function and Assessment of Overall Fit

As in the previous example, we begin our analysis by reviewing the group means and standard deviations to see if the groups are significantly different on any single variable. Table 4.10 gives the group means and standard deviations, and Table 4.11 displays the Wilks' lambda and univariate F ratios (simple ANOVAs) for each independent variable. Review of the significance levels of the individual variables in Table 4.11 reveals that on a univariate basis, all the variables except X_4 and X_6 display significant differences between the group means. We do not know whether the differences are between groups 1 and 2, 2 and 3, or 1 and 3. But we do know that significant differences exist.

TABLE 4.11 Test for Equality of Group Means Between Different Buying Situations

Independent Variables		*Wilks' Lambda* *	*Univariate F Ratio* *	*Significance*
X_1	Delivery speed	.54970	23.35	.0000
X_2	Price level	.70941	11.67	.0001
X_3	Price flexibility	.46070	33.36	.0000
X_4	Manufacturer's image	.96737	.9612	.3885
X_5	Overall service	.54606	23.69	.0000
X_6	Sales force's image	.99560	.1261	.8818
X_7	Product quality	.75411	9.293	.0003

*Wilks' lambda (U statistic) and univariate F ratio with 2 and 57 degrees of freedom.

TABLE 4.12A Results from Step 1 of Stepwise Three-group Discriminant Analysis Model

STEP 1: X_1 (Delivery Speed) Included in the Analysis.
Summary Statistics

			Degrees of Freedom		Significance	Between Groups
Wilks' lambda	.5497	1	2	57.0		
Equivalent F	23.3464		2	57.0	.0000	
Minimum D^2	.5264					1 and 2
Equivalent F	4.6060		1	57.0	.0361	

Variables in the Analysis after Step 1

Variable	Tolerance	F to Remove	D^2	Between Groups
X_1	1.0000	23.346		

Variables Not in the Analysis after Step 1

Variables		Tolerance	Minimum Tolerance	F to Enter	D^2	Between Groups
X_2	Price Level	.7657	.7657	14.609	2.718	2 and 3
X_3	Price flexibility	.9822	.9822	17.273	.659	1 and 2
X_4	Manufacturer's image	.9977	.9977	.690		
X_5	Overall service	.7036	.7036	13.243	2.537	2 and 3
X_6	Sales force's image	.9909	.9909	.020		
X_7	Product quality	.8689	.8689	1.142	.529	1 and 2

Significance Testing of Group Differences after Step 1*

	Group 1: New Task	Group 2: Modified Rebuy
Group 2: Modified Rebuy	4.6061 (.0361)	
Group 3: Straight Rebuy	45.335 (.0000)	15.274 (.0002)

*F statistics and significance level (in parentheses) between pairs of groups after step 1. Each F statistic has 1 and 57 degrees of freedom.

Estimation of the Discriminant Functions

The stepwise procedure is performed in the same manner as in the two-group example. The data in Table 4.12A show that the first variable to enter the model is X_1. Review of the F to enter reveals that of the variables not included in the model after step 1 (see Table 4.12B), all but X_4 and X_6 have values large enough to be considered for inclusion in the model at later steps. (Recall that we set cutoff for F to enter at 1.0 or larger).

TABLE 4.12B Results from Step 2 of Stepwise Three-group Discriminant Analysis Model

STEP 2: X_2 (Price Level) Included in the Analysis.
Summary Statistics

		Degrees of Freedom		Significance	Between Groups
Wilks' lambda	.3612	2	2	57.0	
Equivalent F	18.5870		4	112.0	.0000
Minimum D^2	2.7180				2 and 3
Equivalent F	12.3245		2	56.0	.0000

Variables in the Analysis after Step 2

	Variable	Tolerance	F to Remove	D^2	Between Groups
X_1	Delivery speed	.7657	26.988		
X_2	Price level	.7657	14.609		

Variables Not in the Analysis after Step 2

	Variables	Tolerance	Minimum Tolerance	F to Enter	D^2	Between Groups
X_3	Price flexibility	.8776	.6843	17.162	4.201	1 and 2
X_4	Manufacturer's image	.9628	.7389	.009		
X_5	Overall service	.0171	.0166	.846		
X_6	Sales force's image	.9519	.7356	.533		
X_7	Product quality	.7860	.6927	2.140	2.755	2 and 3

Significance Testing of Group Differences after Step 2†

	Group 1: New Task	Group 2: Modified Rebuy
Group 2: Modified Rebuy	17.749 (.0000)	
Group 3: Straight Rebuy	26.405 (.0000)	12.325 (.0000)

†F statistics and significance level (in parentheses) between pairs of groups after step 2. Each F statistic has 2 and 58 degrees of freedom.

The information provided in Table 4.13 summarizes the four steps of the three-group discriminant analysis. Note that variables X_1, X_2, X_3, and X_7 were all entered into the discriminant function. By comparing these results with the univariate results in Tables 4.10 and 4.11, one can see that it is not always possible to predict solely from univariate results which variables will be included in a discriminant function.

Because this is a three-group discriminant analysis model, two canonical discriminant functions are calculated to discriminate between the three groups. The variables are now entered into the canonical discriminant procedure, and linear

TABLE 4.13 Summary of Three-group Discriminant Analysis Results

| | Action | | | Wilks' Lambda | | Minimum D^2 | | Between |
Steps	Entered	Removed	Value	Significance	Value	Significance	Groups
1	X_1 Delivery speed		.54970	.0000	.52641	.0361	1 and 2
2	X_2 Price level		.36123	.0000	2.71800	.0000	2 and 3
3	X_3 Price flexibility		.22243	.0000	4.20140	.0000	1 and 2
4	X_7 Product quality		.20569	.0000	4.67826	.0000	1 and 2

composites are formulated. Note that the discriminant functions are based only on the variables included in the discriminant model (X_1, X_2, X_3, and X_7).

However, after the linear composites are calculated, the procedure correlates all seven independent variables with the canonical discriminant functions to develop a structure (loadings) matrix. This procedure enables us to see where the discrimination would occur if all seven variables were included in the model (that is, if none were excluded by multicollinearity or lack of statistical significance). Much of this discussion is based upon concepts presented in the chapters on canonical correlation (Chapter 6) and factor analysis (Chapter 7), so you may wish either to skim these chapters now or come back to this topic after you have read them.

The linear composites are similar to a regression line (i.e., they are a linear combination of variables). Just as a regression line tries to explain the maximum amount of variation in its dependent variable, these linear composites attempt to explain the variations or differences in discriminant dependent categorical variables. The first linear composite is developed to explain (account for) the largest amount of variation (difference) in the discriminant groups. The second linear composite, which is orthogonally independent of the first, explains the largest percentage of the remaining (residual) variance after the variance for the first composite function is removed.

Table 4.14 contains the results for the canonical discriminant functions. Note that the functions are statistically significant, as measured by the Chi-square statistic, and that the first function accounts for 79.44 percent of the variance. Below the summary information are the discriminant function coefficients (weights) for the predictive models and the unrotated structure matrix. Because we will rotate the linear composites to facilitate interpretation, there is no need to interpret the structure loadings at this point. After the first function is extracted, the Chi-square is recalculated. The results show that significant differences are present in the remaining variance. If more groups are used in the model (e.g., a four-group discriminant analysis), additional canonical discriminant functions would be possible, and the Chi-square statistic would be continually recalculated on the residual variance to test for significant differences until the maximum number of canonical discriminant functions were extracted (maximum number of discriminant functions = number of groups − 1).

Assessing Overall Fit of the Discriminant Functions

Before the analyst can move to the interpretation of the functions, he or she must determine that the functions are valid predictors. This determination is accomplished in the same fashion as with the two-group discriminant model, by exami-

TABLE 4.14 Multivariate Results for Three-group Discriminant Analysis

Canonical Discriminant Functions

Function*	Eigenvalue	Percent of Variance		Canonical Correlation	After Function	Wilks' Lambda	Chi-square	df	Significance
		Function	Cumulative						
1	2.1327	79.44	79.44	.8251	0	.2057	87.766	8	.0000
2	.5519	20.56	100.00	.5963	1	.6444	24.390	3	.0000

Standardized Canonical Discriminant Function Coefficients

Independent Variables	Discriminant Function Coefficients	
	Function 1	Function 2
X_1 Delivery speed	.68934	.44566
X_2 Price level	.61626	1.00929
X_3 Price flexibility	.77181	−.32567
X_7 Product quality	−.34675	−.20219

Structure Matrix†

Independent Variables	Discriminant Function Loadings	
	Function 1	Function 2
X_3 Price flexibility	.67451*	−.60246
X_1 Delivery speed	.61972*	−.01317
X_7 Product quality	−.38826*	.09097
X_2 Price level	−.13374	.82044*
X_5 Overall service	.50853	.76481*
X_4 Manufacturer's image	.02097	.16564*
X_6 Sales force's image	.14643	.14853*

*Marks the two canonical discriminant functions remaining in the analysis.

†Pooled-within-groups correlations between discriminating variables and canonical discriminant functions (variables ordered by size of correlation within function).

nation of the classification matrices. Table 4.15 shows that the discriminant functions achieve a high degree of classification accuracy. The hit ratio for the analysis sample is 83.33 percent, whereas that for the holdout sample is 67.50 percent. These results indicate the upward bias that is likely without a holdout sample. Although both these hit ratios are high, they must be compared with the maximum chance and the proportional chance criteria to assess their "true" effectiveness.

The maximum chance criterion is simply the hit ratio obtained if we assign all the observations to the group with the highest probability of occurrence. In the present sample of 100 observations, 34 were in group 1, 32 in group 2, and 34 in group 3. From this information, we can see that the highest probability would be

TABLE 4.15 Classification Matrices for Three-group Discriminant Analysis for Both Analysis and Holdout Samples

*Classification Results: Analysis Sample**

| | | Predicted Group Membership | | |
Actual Group	Number of Cases	New Task	Modified Rebuy	Straight Rebuy
New task	21	16	3	2
		76.2%	14.3%	9.5%
Modified rebuy	15	2	10	3
		13.3%	66.7%	20.0%
Straight rebuy	24	0	0	24
		0%	0%	100.0%

Classification Results: Holdout Sample†

| | | Predicted Group Membership | | |
Actual Group	Number of Cases	New Task	Modified Rebuy	Straight Rebuy
New task	13	10	1	2
		76.9%	7.7%	15.4%
Modified rebuy	17	4	7	6
		23.5%	41.2%	35.3%
Straight rebuy	10	0	0	10
		0%	0%	100.0%

*Percent of "grouped" cases correctly classified: 83.33 percent [(16 + 10 + 24)/60 = 83.33%].
†Percent of "grouped" cases correctly classified: 67.50 percent [(10 + 7 + 10)/40 = 67.50%].

34 percent (groups 1 or 3). Based on the maximum chance criterion, therefore, our model is very good.

The proportional chance criterion is calculated by squaring the proportions of each group, as shown in Table 4.16. The calculated value is 33.36 percent. Because the C_{MAX} is greater than the C_{PRO}, the maximum chance criterion is the measure to outperform. The hit ratios of both 83.33 (analysis sample) and 67.50 (holdout sample) both exceed this criterion substantially, so we again conclude that the discriminant model is valid based on these measures.

The final measure of classification accuracy is Press's Q, calculated for both analysis and holdout samples. It tests the statistical significance that the classification accuracy is better than chance. The calculated value for the analysis sample is

$$\text{Press's } Q = \frac{[60 - (50 * 3)]^2}{60(3 - 1)} = 67.5$$

The calculated value for the holdout sample is

$$\text{Press's } Q = \frac{[40 - (27 * 3)]^2}{40(3 - 1)} = 21.0$$

TABLE 4.16 Calculation of Chance Criteria for Classification

Maximum chance criteria
    Group 1: 34/100 = 34%
    Group 2: 32/100 = 32%
    Group 3: 34/100 = 34%
      C_{MAX} = 34 percent

Proportional chance criteria
    $C_{PRO} = p_1^2 + p_2^2 + \frac{2}{3}$
    $C_{PRO} = .34^2 + .32^2 + .34^2$
    $C_{PRO} = .3336$ or 33.36%

Because the critical value at a .01 significance level is 6.63, the discriminant analysis can confidently be described as predicting group membership better than chance, as indicated by this and the other classification accuracy measures.

Table 4.17 contains additional classification data for the three-group discriminant analysis. The observation number is shown on the left side of the table. The "Analysis Sample" column indicates whether an observation was selected to be included in the analysis or the holdout group. "Yes" indicates that the observation is in the analysis sample, and "No" indicates that it is in the holdout sample. In the "Actual Group" column, a 1 indicates group 1 (new task), a 2 indicates group 2 (modified rebuy), and a 3 indicates group 3 (straight rebuy). The asterisk beside the numbers indicates that a particular observation was misclassified by the discriminant function. The "Highest Probability" column shows the group assignment of an observation by the model that is most likely using the discriminant function, and the "Second-Highest" column shows the second–most likely assignment using the discriminant function. The discriminant scores for each observation on each function are shown on the right side of the table. (*Note:* When there is one discriminant function [two groups], classification of cases is based on the values for the single function. When there are three or more groups, a case's values on all functions are considered simultaneously.)

Stage Five: Interpretation of Three-group Discriminant Analysis Results

The next stage of the discriminant analysis is interpretation of the discriminant functions. The first step is to examine the contributions of the predictor variables to each function separately (i.e., discriminant loadings), then to examine their cumulative effect (potency index). Graphical display can also benefit the analyst in this two-dimensional solution to understand the relative position of each group and the interpretation of the relevant variables in determining this position. Each of these questions is addressed in the following sections.

Rotation

After the discriminant functions are developed, they can be "rotated" to redistribute the variance (this concept is more fully explained in Chapter 7). Basically, rotation preserves the original structure and reliability of the discriminant models

TABLE 4.17 Classification Data for Three-group Discriminant Analysis

Case Number	Analysis Sample	Actual Group	Group Membership Prediction					Discriminant Scores	
			Highest Probability:			Second-Highest Probability:			
			Group	P(D/G)	P(G/D)	Group	P(G/D)	Function 1	Function 2
1	Yes	1	1	.8422	.9232	2	.0565	−.5869	−1.3552
2	Yes	1	1	.5308	.9012	2	.0984	−2.2363	−.9047
3	No	2	2	.1301	.9936	1	.0043	−1.9840	2.6321
4	No	1	1	.5106	.9939	2	.0056	−1.2684	−2.3917
5	Yes	3	3	.6071	.9977	2	.0020	2.5906	.5794
6	Yes	2*	1	.8395	.7906	2	.1894	−.9661	−.6788
7	Yes	1*	3	.9880	.9759	2	.0214	1.6561	.6540
8	Yes	2	2	.3995	.6761	1	.3219	−2.2039	.4671
9	No	3	3	.9533	.9851	2	.0119	1.8398	.3244
10	No	2	2	.9373	.8797	1	.0981	−1.2616	1.0386
11	Yes	1	1	.2777	.6780	3	.2654	.4375	−1.3232
12	Yes	2*	3	.6169	.8413	2	.0798	.9351	−.2175
13	Yes	1	1	.5795	.9086	3	.0486	−.1568	−1.5247
14	Yes	1*	3	.2547	.5449	1	.3610	.6028	−.8122
15	Yes	3	3	.6318	.9901	2	.0051	2.0760	−.3160
16	No	3	3	.8963	.9813	2	.0130	1.7629	.0757
17	Yes	2	2	.4729	.9666	1	.0291	−1.8514	1.7120
18	No	2*	3	.4845	.6186	2	.2974	.3930	.4247
19	No	3	3	.8183	.9928	2	.0055	2.1285	.1727
20	Yes	3	3	.6318	.9901	2	.0051	2.0760	−.3160
21	No	2*	1	.2419	.7828	3	.1853	.4695	−1.6612
22	No	1	1	.2449	.8832	3	.0973	.3650	−1.9338
23	Yes	3	3	.5342	.7685	2	.2210	.7231	1.2161
24	Yes	1	1	.7216	.9852	2	.0143	−1.6052	−1.9119
25	Yes	2*	3	.9896	.9731	2	.0239	1.6115	.6552
26	Yes	3	3	.6041	.7522	2	.2243	.6517	.8592
27	No	1	1	.6881	.9858	2	.0138	−1.6925	−1.9192
28	Yes	3	3	.7788	.9920	2	.0055	2.0989	.0171
29	Yes	3	3	.8625	.9876	2	.0116	1.9470	.9255
30	No	2	2	.6758	.9345	1	.0585	−1.6915	1.3548
31	Yes	1*	2	.4323	.5377	1	.4592	−1.9866	.1716
32	Yes	3	3	.5057	.6662	2	.3031	.4890	.8903
33	Yes	3	3	.6635	.9907	2	.0090	2.1017	−1.2620
34	No	1*	2	.8417	.8611	1	.1268	−1.5112	.9367
35	No	1	1	.5572	.9851	2	.0102	−.6249	−2.1764
36	Yes	1	1	.6364	.6339	2	.3412	−1.0513	−.2929
37	No	2	2	.5043	.9026	1	.0945	−2.0690	−1.1568
38	No	3	3	.1368	.6122	1	.3586	1.0196	−1.3981
39	Yes	1	1	.0627	.9992	2	.0008	−2.5095	−3.1656
40	No	1	1	.5905	.9810	2	.0188	−2.0598	−1.7329
41	No	1	1	.5280	.5383	2	.4228	−.9343	−.1299
42	Yes	3	3	.6307	.9975	2	.0022	2.5539	.5144
43	Yes	3	3	.8537	.9809	2	.0182	1.7845	1.0412
44	No	2*	3	.9949	.9647	2	.0297	1.4944	.4923
45	Yes	1	1	.6843	.9464	2	.0531	−2.0320	−1.2365
46	No	3	3	.6879	.9847	2	.0149	1.9089	1.3175
47	Yes	3	3	.7973	.9909	2	.0062	2.0510	.0181
48	Yes	2	2	.3904	.8501	1	.1483	−2.2957	.9444
49	Yes	3	3	.3618	.9132	1	.0689	1.4517	−.9068
50	Yes	3	3	.6130	.8561	2	.1384	.9606	1.2722

*Indicates misclassification.

Case Number	Analysis Sample	Actual Group	Group	Highest Probability: P(D/G)	P(G/D)	Second-Highest Probability: Group	P(G/D)	Function 1	Function 2
				Group Membership Prediction				*Discriminant Scores*	
51	Yes	2*	1	.2872	.7230	3	.2268	.4096	−1.4059
52	Yes	2	2	.5973	.7797	3	.2023	−.2370	1.6598
53	Yes	2	2	.5258	.9684	3	.0182	−1.2234	2.0065
54	Yes	1	1	.8648	.9212	2	.0772	−1.6791	−1.0886
55	No	1*	3	.1497	.5321	1	.4253	.8331	−1.2821
56	No	2*	1	.5707	.6089	2	.3068	−.5392	−.3799
57	No	2	2	.2179	.8316	3	.1653	−.2084	2.5049
58	Yes	3	3	.6813	.9888	2	.0108	2.0296	1.2720
59	Yes	3	3	.4291	.5599	2	.3323	.3007	.3705
60	No	2	2	.8824	.8124	3	.1316	−.5008	1.1784
61	Yes	3	3	.8236	.9886	2	.0077	1.9363	.0108
62	No	3	3	.7798	.9882	2	.0113	1.9847	1.0991
63	No	3	3	.2495	.8127	1	.1617	1.2313	−1.1144
64	No	2	2	.4737	.9218	1	.0756	−2.0928	1.2728
65	Yes	1	1	.2079	.9980	2	.0020	−2.0093	−2.7935
66	No	2*	3	.2883	.7162	1	.2319	.9306	−.9190
67	Yes	3	3	.9563	.9427	2	.0489	1.2977	.5540
68	Yes	2	2	.6047	.8653	1	.1305	−1.9259	.9714
69	No	3	3	.9145	.9515	2	.0448	1.3888	.8820
70	Yes	2	2	.6602	.5290	1	.2879	−.3547	.2017
71	Yes	2	2	.0879	.9946	3	.0034	−1.7499	2.9579
72	Yes	3	3	.5642	.9099	2	.0881	1.2017	1.5075
73	Yes	3	3	.9130	.9904	2	.0081	2.0112	.4244
74	No	3	3	.7911	.9591	2	.0231	1.4887	−.1644
75	No	1	1	.9153	.9339	2	.0644	−1.5792	−1.1909
76	No	2*	3	.9048	.9411	2	.0425	1.2931	.1808
77	No	2*	1	.2407	.5962	3	.3433	.5266	−1.2430
78	No	3	3	.9891	.9792	2	.0182	1.7129	.5997
79	Yes	1	1	.1432	.9986	2	.0014	−2.1263	−2.9564
80	Yes	3	3	.6440	.8571	2	.0740	.9764	−.1943
81	Yes	3	3	.8321	.9713	2	.0274	1.6298	1.1173
82	Yes	2	2	.2081	.9899	1	.0074	−1.9305	2.3715
83	No	1	1	.4243	.9511	2	.0488	−2.4704	−1.2305
84	Yes	1*	2	.9137	.7107	1	.2043	−.7227	.5390
85	No	2*	3	.3578	.4870	2	.3332	.2081	.1437
86	Yes	1	1	.6885	.7236	2	.2688	−1.4469	−.4218
87	No	2*	3	.2560	.3624	2	.3332	.0545	−.0842
88	Yes	1*	2	.7825	.5925	1	.3627	−.9567	.2131
89	Yes	1	1	.7037	.9864	2	.0118	−1.0344	−2.0658
90	Yes	3	3	.9787	.9558	2	.0384	1.4074	.6035
91	No	2*	1	.4128	.4579	2	.4084	−.4291	−.1264
92	Yes	2*	3	.4067	.5394	2	.3262	.2760	.2618
93	Yes	2	2	.2434	.9706	1	.0282	−2.3620	1.7848
94	No	1	1	.0954	.9598	2	.0402	−3.3288	−1.2301
95	Yes	1	1	.8967	.9499	2	.0405	−.7770	−1.5026
96	Yes	1	1	.0397	.9994	2	.0006	−2.6915	−3.2652
97	Yes	3	3	.6146	.9819	1	.0101	1.8758	−.4336
98	No	1	1	.1398	.9432	2	.0567	−3.1387	−1.0826
99	Yes	1	1	.9117	.8881	2	.1083	−1.4692	−.9374
100	No	1*	3	.1828	.4799	1	.4509	.6349	−1.0624

TABLE 4.18 Results for Varimax Rotated Three-group Discriminant Analysis

Varimax Rotation Transformation Matrix

	Function 1	Function 2
Function 1	.88536	.46491
Function 2	−.46491	.88536
Percent of variance	66.72	33.28

Rotated Standardized Discriminant Function Coefficients*

		Discriminant Function Coefficients	
Independent Variables		Function 1	Function 2
X_3	Price flexibility	.83473*	.07049
X_2	Price level	.07639	1.18009*
X_1	Delivery speed	.40312	.71505*
X_7	Product quality	−.21304	−.34014*

Correlations Between Rotated Canonical Discriminant Functions and Discriminating Variables*

		Discriminant Function Loadings	
Independent Variables		Function 1	Function 2
X_1	Delivery speed	.55480	.27645
X_2	Price level	−.49984	.66420
X_3	Price flexibility	.87727	−.21980
X_7	Product quality	.06059	.19958
X_4	Manufacturer's image	−.38604	−.09996
X_5	Overall service	−.05844	.15640
X_6	Sales force's image	.09466	.91355

Group Means (Centroids) of Canonical Discriminant Functions

	Discriminant Function Centroids	
Group	Function 1	Function 2
New task	−1.16101	−1.23718
Modified rebuy	−.92457	.91279
Straight rebuy	1.59374	.51204

*Variables ordered by size of coefficient within function.

while at the same time making them easier to interpret substantively. In the present application, we chose the most widely used procedure of varimax rotation. The results are portrayed in Table 4.18.

Assessing the Contribution of Predictor Variables

The first task is to plot the group centroids to present an overall perspective of the differences in the three groups. The plots (see Figure 4.9) depict the centroids for the first two functions. This graphic presentation shows that there appear to be differences in the groups on the eight predictor variables. But it does not provide a basis for explaining these differences.

To assess the contributions of the seven predictors, the analyst has a number of measures to employ—discriminant loadings, univariate F ratios, and the potency index. Of particular note in a multifunction solution is the potency index. Table 4.19 illustrates the calculation of the potency index for each of the predictor variables, representing the total discriminating effect across both discriminant functions. Table 4.20 presents the three preferred interpretive measures for each variable. The results generally support the stepwise analysis, although X_5 has a substantial univariate F and potency index but was not included because of collinearity. The other variables not included (X_2, X_4, and X_6) all have very low loadings, nonsignificant F values, and low potency index values.

To depict the differences in terms of the predictor variables, the loadings and the group centroids can be plotted in reduced discriminant space. As noted earlier, the most valid representation is the use of stretched attribute vectors and

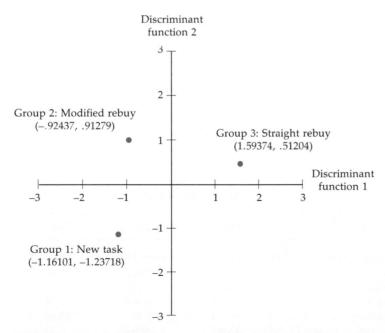

FIGURE 4.9 Plot of group centroids in reduced discriminant space.

TABLE 4.19 Calculation of the Potency Indices for the Three-group Discriminant Analysis

		Discriminant Function 1				Discriminant Function 2				
Variables		Loading (L)	Squared Loading	Relative Eigenvalue*	Potency Value†	Loading (L)	Squared Loading	Relative Eigenvalue*	Potency Value†	Potency Index
X_1	Delivery speed	.61972	.38405	.7944	.30509	.01317	.00017	.2056	.00003	.30512
X_2	Price level	−.13374	.01789	.7944	.01421	.82044	.67312	.2056	.13839	.15260
X_3	Price flexibility	.67451	.45496	.7944	.36142	−.60246	.36296	.2056	.07462	.43604
X_4	Manufacturer's image	.02097	.00044	.7944	.00035	.16564	.02744	.2056	.00564	.00599
X_5	Overall service	.50853	.25860	.7944	.20543	.76481	.58493	.2056	.12026	.32569
X_6	Sales force's image	.14693	.02144	.7944	.01703	.14853	.02206	.2056	.00453	.02156
X_7	Product quality	−.38826	.15075	.7944	.11976	.09097	.00828	.0256	.0017	.12146

*Relative eigenvalue = eigenvalue of the discriminant function divided by the sum of the eigenvalues for all significant discriminant functions. In our example, the relative eigenvalue for function 1 is [2.1327/(2.1327 + .5519)] = .7944.
†Potency value = squared loading × relative eigenvalue.

group centroids. Table 4.21 shows the calculations for stretching both the discriminant loadings (used for attribute vectors) and the group centroids. The plotting process always involves all the variables included in the model by the stepwise procedure (in our example, variables X_1, X_2, X_3, and X_7). But the analyst frequently plots the variables not included in the discriminant function if their respective univariate F ratios are significant. This procedure shows the importance of collinear variables that were not included in the final stepwise model. The significance of including these variables in the plotting process is demonstrated in our example. Variable X_7 is included in the final model because it is significant with a portion of the residual variance. Variable X_5 exhibits much stronger differences between the groups but was not included because of collinearity.

The plots of the stretched attribute vectors for the rotated discriminant loadings are shown in Figure 4.10. By plotting the vectors using this procedure, one causes them to point to the groups having the highest mean on the respective predictor and away from the groups having the lowest mean scores. Thus the interpretation

TABLE 4.20 Summary of Interpretive Measures for Three-group Discriminant Analysis

		Rotated Discriminant Loading		Univariate F Ratio	Potency Index
Variables		Function 1	Function 2		
X_1	Delivery speed	.61972	−.01317	23.35	.30512
X_2	Price level	−.13374	.82044	11.67	.15260
X_3	Price flexibility	.67451	−.60246	33.36	.43604
X_4	Manufacturer's image	.02097	.16564	.96	.00599
X_5	Overall service	.50853	.76481	23.69	.32569
X_6	Sales force's image	.14643	.14853	.13	.02156
X_7	Product quality	−.38826	.09097	9.29	.12146

TABLE 4.21 Calculation of the Stretched Attribute Vectors and Group Centroids in Reduced Discriminant Space

		Rotated Discriminant Loading		Univariate F ratio	Reduced Space Coordinates	
Variables		Function 1	Function 2		Function 1	Function 2
X_1*	Delivery speed	.55480	.27645	23.35	12.95	6.45
X_2*	Price level	−.49984	.66420	11.67	−5.83	7.75
X_3*	Price flexibility	.87727	−.21980	33.36	29.26	−7.33
X_4	Manufacturer's image	−.38604	−.09996	.96	−.37†	−.09†
X_5	Overall service	−.05844	.15640	23.69	−1.38	3.71
X_6	Sales force's image	.09466	.91355	.13	.01†	.11†
X_7*	Product quality	.06059	.19958	9.29	.56	1.85

	Group Centroids		Approximate F Value		Reduced Space Coordinates	
Group	Function 1	Function 2	Function 1	Function 2	Function 1	Function 2
Group 1: New task	−1.16101	−1.23718	60.78	17.72	−70.56	−19.46
Group 2: Modified rebuy	−.92457	.91279	60.78	17.72	−56.20	14.36
Group 3: Straight rebuy	1.59374	.51204	60.78	17.72	96.87	8.05

*Denotes variables entered in the stepwise solution.
†Vectors not plotted because of nonsignificant F ratio.

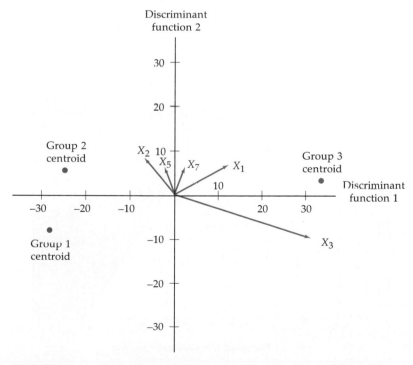

FIGURE 4.10 Plot of stretched attribute vectors (variables) in reduced discriminant analysis.

of the plot in Figure 4.10 indicates that the first discriminant function (1) is the primary source of difference between groups 1 and 2 versus group 3. Moreover, the first function corresponds most closely to variable X_3 and somewhat with variables X_1 and X_2. Thus the distinguishing characteristics of firms primarily making straight rebuy decisions (group 3) are more favorable perceptions of delivery speed and price flexibility. Similar profiles can be made for the two remaining groups for the first discriminant function.

The second discriminant function (2) provides the distinction between group 1 versus groups 2 and 3. The close correspondence between the X_5 and X_7 vectors and the second function signifies that the perceptions of service and product quality are most descriptive of the second discriminant function.

An alternative approach is the use of the territorial map, which contains the group centroids and each observation's discriminant scores. The map is used to develop group boundaries that will enable us to classify the holdout observations in much the same way as the cutting score is applied in the two-group case. Such a map is shown in Figure 4.11.

At this point, you may feel that all this plotting is too complicated. What other procedure is simpler, yet effective? The easiest approach is to use the rotated correlations (loadings) provided in Table 4.18 and the group centroids. The asterisks beside the loadings indicate which are significant for each function (e.g., X_3 for function 1 versus X_2, X_1, and X_7 for function 2). To determine which groups each function discriminates between, simply look at the group centroids and see

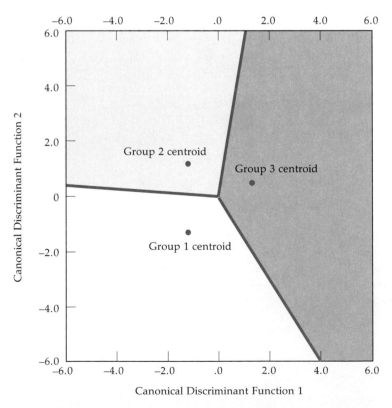

FIGURE 4.11 Territorial map for three-group discriminant analysis.

where differences lie. This is merely a distance assessment, not a statistical measure, but it is usually sufficient. For example, looking at function 1, we see that the centroid for group 1 is -1.16101, for group 2 it is $-.92457$, and for group 3 it is 1.59374. From this we conclude that the primary source of differences for this function is between groups 1 and 2 versus group 3. A similar approach can be used for function 2. However, because function 1 represents substantially more variance than function 2, one must be cautious in determining the impact of variables based on their loadings on this function.

Stage Six: Validation of the Discriminant Results

The internal validity of the discriminant results are supported by the levels of predictive accuracy found in the holdout sample. For each measure of predictive accuracy, the discriminant functions applied to the holdout sample exceeded the criterion values. The researcher is also encouraged to extend the validation process through profiling the groups on additional sets of variables and/or applying the discriminant function to other sample(s) representative of the overall population or segments within the population.

Summary

The underlying nature, concepts, and approach to multiple discriminant analysis have been presented. Basic guidelines for its application and interpretation were included to clarify further the methodological concepts. Illustrative examples for both two-group and three-group solutions were presented based on the HATCO database. These applications demonstrated the major points you need to be familiar with in applying discriminant analysis.

Multiple discriminant analysis helps you to understand and explain research problems that involve a single categorical dependent variable and several metric independent variables. A mixed data set (both metric and nonmetric) is also possible for the independent variables if the nonmetric variables are dummy coded (0–1). The result of a discriminant analysis can assist you in profiling the intergroup characteristics of the subjects and in assigning them to their appropriate groups. Potential applications of discriminant analysis to both business and nonbusiness problems are numerous.

Some of the concepts presented in this chapter are based on material discussed in Chapters 6 and 7. Thus it is recommended that these three chapters be studied together.

Questions

1. How would you differentiate among multiple discriminant analysis, regression analysis, and analysis of variance?
2. What criteria could you use in deciding whether or not to stop a discriminant analysis after estimating the discriminant function(s)? After the interpretation stage?

3. What procedure would you follow in dividing your sample into analysis and holdout groups? How would you change this procedure if your sample consisted of fewer than 100 individuals or objects?
4. How would you determine the optimum cutting score?
5. How would you determine whether or not the classification accuracy of the discriminant function is sufficiently high relative to chance classification?
6. How does a two-group discriminant analysis differ from a three-group analysis?
7. Why would an analyst stretch the loadings and centroid data in plotting a discriminant analysis solution?

References

1. BMDP Statistical Software, Inc., *BMDP Statistical Software Manual, Release 7,* vols. 1 and 2. Los Angeles: 1992.
2. Crask, M., and W. Perreault. "Validation of Discriminant Analysis in Marketing Research." *Journal of Marketing Research* 14 (February 1977): 60–68.
3. Dillon, W. R., and M. Goldstein. *Multivariate Analysis: Methods and Applications.* New York: Wiley, 1984.
4. Frank, R. E., W. E. Massey, and D. G. Morrison. "Bias in Multiple Discriminant Analysis." *Journal of Marketing Research* 2, no. 3 (1965): 250–58.
5. Gessner, Guy, N. K. Maholtra, W. A. Kamakura, and M. E. Zmijewski. "Estimating Models with Binary Dependent Variables: Some Theoretical and Empirical Observations." *Journal of Business Research* 16, no. 1 (1988): 49–65.
6. Green, P. E., D. Tull, and G. Albaum. *Research for Marketing Decisions.* Englewood Cliffs, N.J.: Prentice-Hall, 1988.
7. Green, P. E. *Analyzing Multivariate Data.* Hinsdale, Ill.: Holt, Rinehart, and Winston, 1978.
8. Green, P. E., and J. D. Carroll. *Mathematical Tools for Applied Multivariate Analysis.* New York: Academic Press, 1978.
9. Harris, R. J. *A Primer of Multivariate Statistics.* New York: Academic Press, 1975.
10. Huberty, C. J. "Issues in the Use and Interpretation of Discriminant Analysis." *Psychological Bulletin* 95 (1984): 156–71.
11. Huberty, C. J., J. W. Wisenbaker, and J. C. Smith. "Assessing Predictive Accuracy in Discriminant Analysis." *Multivariate Behavioral Research* 22 (July 1987): 307–29.
12. Johnson, N., and D. Wichern. *Applied Multivariate Statistical Analysis.* Englewood Cliffs, N.J.: Prentice-Hall, 1982.
13. Morrison, D. G. "On the Interpretation of Discriminant Analysis." *Journal of Marketing Research* 6, no. 2 (1969): 156–63.
14. Perreault, W. D., D. N. Behrman, and G. M. Armstrong. "Alternative Approaches for Interpretation of Multiple Discriminant Analysis in Marketing Research." *Journal of Business Research* 7 (1979): 151–73.
15. SAS Institute, Inc. *SAS User's Guide: Basic, Version 6.* Cary, N.C.: 1990.
16. SAS Institute, Inc. *SAS User's Guide; Statistics, Version 6.* Cary, N.C.: 1990.
17. SPSS, Inc. *SPSS User's Guide,* 4th ed. Chicago: 1990.
18. SPSS, Inc. *SPSS Advanced Statistics Guide,* 4th ed. Chicago: 1990.

Private Physicians or Walk-In Clinics: Do the Patients Differ?

Rajiv P. Dant
James R. Lumpkin
Robert P. Bush

Over the past decade, a one-third increase in the number of physicians has been observed in the health care industry. More than half a million physicians are estimated to be in practice in the United States today and the number is expected to grow well into the 1990s (U.S. Department of Commerce 1987, p. 202). In response to both increasing competition and pressures from insurance companies and Medicare to reduce medical care costs, many physicians are setting up walk-in clinics as a way to provide low-cost medical care.

Physicians as well as potential investors are beginning to monitor walk-in clinics to assess how customer patronage patterns will affect the future structure of the health care industry. We empirically compare walk-in clinics and traditional private practices in terms of (1) consumer expectations about these two delivery systems, (2) perceived consumer performance evaluation for both systems, and (3) the influence of demographic characteristics and the nature of medical needs on patronage behavior.

Prior Research

Review of the health care literature indicates that most empirical research has addressed marketing issues and problems from the health care provider's perspective rather than the consumer's. Recently, however, a growing number of studies have begun to concentrate on understanding the

From "Private Physicians or Walk-In Clinics: Do the Patients Differ?" Rajiv P. Dant, James R. Lumpkin, and Robert P. Bush. *Journal of Health Care Marketing*, vol. 10, no. 2 (June 1990), pp. 25–35. Reprinted with permission.

role of consumers in the consumption of health care offerings (Harrell and Fors 1985; Ortinau 1986).

Strategic Issues for Health Care Providers

With this recent surge of consumer-oriented interest in the field of health care marketing, issues of consumer choice behavior among competing hospitals (Hisrich and Peters 1982; Lane and Lindquist 1988; Malhotra 1983; Woodside, Sertich, and Chakalas 1987), as well as other types of health care offerings, have been investigated (Crane and Lynch 1988). Much of the work has concentrated on understanding and predicting how the preference structures of consumers are formulated, the correlates of patient satisfaction, and their strategic implications.

Hisrich and Peters (1982) identified significant differences in the importance patients attach to the different elements of a health care delivery system. They found that the most significant reasons for hospital choice for inpatient care were being on the hospital staff, location, and prior family experience, in that order. Recommendations of physicians and reputation were found to be low in significance. From a strategy perspective, their findings suggest that hospitals must pay close attention to public relations and, perhaps more importantly, that they should be cognizant of differences in attribute significance.

Malhotra (1983) proposed a stochastic modeling approach for predicting consumer preferences, based on factors considered significant in selecting a facility, that would be useful in the development of positioning strategies by health care providers. Fundamentally, such an approach presumes consumers will choose probabilistically between alternate health care facilities, and choice probabilities can be modeled as a function of the attributes of the alternatives. In his illustrative study, Malhotra found proximity (traditionally, a strategic advantage associated with walk-in clinics) to be the most significant attribute influencing hospital choice. The extension of stochastic modeling for predicting choices between *competing health care delivery systems*, however, requires an understanding of salient factors that cause switching behavior—an issue of significant managerial and theoretical interest of which little is currently known.

Woodside, Sertich, and Chakalas (1987) found that affects of internal versus external attributions of hospital choice and satisfaction may be different for male and female patients, implying that different marketing strategies may be needed for the two groups. Lane and Lindquist (1988) concluded that a large variety of choice influencers and correlates remain to be systematically investigated, especially in light of the increasing proactive health care movement. Some of their major conclusions relate to (1) the significance of seriousness of illness in hospital choice, (2) the dominant role of patients in hospital choice as deciders in accident and illness situations, (3) the dominant role of physicians as deciders in surgery situations, and (4) the impact of a variety of hospital attributes (ranging from staff and quality to physical facilities, convenience, and reputation of the hospital) on choice behavior. In sum, though the number of studies is limited, consumer-oriented research demonstrates that preconceptions about consumer choice determinants must be investigated carefully if consumer orientation is to guide strategic responses and positioning decisions.

Intratype and Intertype Competition

Adding urgency to the need for consumer-oriented strategic thinking is the increasing competition among offerings from a *variety* of health care delivery systems. Today's competition has taken two forms: (1) intratype competition between the *same* type and providers (e.g., between hospitals) and (2) intertype competition across *different* types of providers (e.g., private physi-

cians vs. walk-in clinics). The impact of this expanded competition on hospitals is well documented. For instance, Malhotra (1983) reports lowered hospital bed utilization rates: 60% in the 1980s in contrast to 80% in the early 1970s. The repercussions of increased intertype competition can be expected to influence other delivery systems as well. Yet, though intertype competition has received some scrutiny from a consumer perspective, virtually no investigations have been conducted on how consumers react to increased intertype competition. Strategic responses to intertype competition must build on an understanding of consumer expectations, performance evaluations, and their correlates (e.g., demographic characteristics, medical need types).

Several authors (Bessom and Jackson 1975; James 1987; Kotler and Clarke 1987) have pointed out that the core concepts associated with strategic marketing will remain essentially unchanged in the context of health care marketing, but will differ in their implementation in keeping with the unique aspects of health care as a service. Needless to say, one must understand consumer behavior patterns to comprehend fully what these unique aspects might be. In the context of consumer reaction to intertype competition, two central consumer behavior issues are the assessment of (1) consumer needs satisfaction patterns and (2) perceived differentials in rival delivery systems that could provide clues for potential segmentation strategies.

Consumer needs are a complex phenomenon to understand and evaluate. It is important to recognize that motivation to satisfy a particular need is but one of many influences that govern a consumer's action (Robertson 1972). However, by attempting to isolate this one intervening variable, health care providers may be able to gain greater insights into various aspects of consumer choice. Health care providers must determine what specific needs their service is capable of satisfying, and which attributes of their service are deemed most important by consumers (James 1987), especially in the context of emerging intertype competition. Needs assessment must be accomplished in light of the fact that needs in health care markets are extremely heterogeneous, levels of services sought vary considerably, and consumers may perceive similar services very differently in terms of price, quality, and convenience (Lamb 1988).

Market segmentation is viewed as essential in the planning process of health care providers (Gregory, Kingstrom, and Reardon 1982; Kotler and Bloom 1984; Kotler and Clarke 1987). The task of health care providers is not only to identify their particular market segments, but also to understand specific attitudes, benefits, and expectations of the segments in relation to services offered and quality of services offered (Carroll and Gagon 1983; Harrell and Fors 1985; Neuhaus 1982). These insights would allow the structuring of marketing and/or promotional appeals for the market segments that are most attractive to the health care provider's offering.

Past research provides useful guidelines for strategy development in health care marketing. Yet, solid information from which to assess the adaptability of such guidelines across *alternative* health care delivery systems has not been developed. The important question that remains is whether these guidelines can be generalized to *various types* of health care providers. The purpose of our study is to assess empirically the adaptability of need satisfaction assessment and market segmentation to private physicians and walk-in clinic facilities.

Strategy-Related Issues of Walk-In Clinics

Because of external pressures from insurance companies as well as governmental agencies, health care providers are seeking ways of controlling costs (Lim and Zallocco 1988). Many large medical organizations like Hospital Corporation of America, Humana, and Medstop have emerged to offer a complete range of health care services of good quality at a lower cost within a more flexible time frame. One form of alternative delivery system, commonly referred to as walk-in clinics, has begun to broaden the type of care provided (hence contributing to intertype competition) to include many nonemergency services typically limited to private physician offices (i.e., preventive health care, physical examinations, obstetrics-gynecology services). Researchers have

speculated that this continued-care service could begin to blur the distinction between walk-in clinics and private physicians, thus creating an increasingly competitive marketplace (Phillips and Reeder 1987; Ross et al. 1983).

Despite the recent expansion of walk-in clinics' services, research on health care delivery systems suggests that the administrators of walk-in clinics are still positioning them to attract the price-conscious, convenience-oriented, emergency-need consumer (Phillips and Reeder 1987). Tucker and Tucker (1985) found that customers of walk-in facilities are seeking convenience, lower costs, physician quality, and flexible hours. Similarly, Ortinau (1986) found the major need of walk-in clinic patients to be specific treatment of emergency situations. Despite the changes occurring in service offerings among walk-in clinics, administrators of these facilities are maintaining their past market orientation.

Hypotheses

Trends associated with walk-in clinics suggest several research questions. In the wake of increased intertype competition and the broadening of services offered by walk-in clinics, are the administrators of those facilities ignoring potential segments once served only by private physicians? Further, is the expansion of continued-care services offered by walk-in clinics blurring the distinction between those facilities and private physician offices? Finally, if overlap is occurring among market segments of health care delivery systems, is it based on consumer expectations or do demographic variables influence the trend?

These questions, derived from gaps in the literature, suggest a three-part research agenda for understanding consumer reactions to increasing intertype competition. First, we need to understand the decision process, expectations, and performance criteria used by the purchasers of health care services (Lamb 1988; Lane and Lindquist 1988). A review of medical facility choice criteria suggests that the users of walk-in clinics emphasize convenience of location, low cost of the services, and ability to obtain a quick appointment. Ortinau (1986) suggests that short waiting time and reputation of physicians and staff are important attributes for walk-in clinic users. Hence we investigate two hypotheses about consumer expectations and performance evaluations.

H_1: Users of walk-in clinics have higher *expectation determinants* for such attributes as convenience of location, lower cost of services, ease of gaining appointments, and flexible hours of operation than do users of private physicians.

H_2: Users of walk-in clinics have higher *performance evaluations* for such attributes as convenience of location, lower cost of services, ease of gaining appointments, and flexible hours of operation than do users of private physicians.

The second component of the research agenda is to understand the relationship between expectations and performance criteria as they relate to demographic characteristics of health care service users (Parasuraman, Zeithaml, and Berry 1985; Ross et al. 1983). Results of several studies indicate that walk-in clinic users are more likely to be employed in blue collar occupations, lower in education and income, and somewhat younger than individuals who use private physicians (Carroll and Gagon 1983; Woodside et al. 1988).

Because of the lack of adequate *comparative* descriptive evidence in the literature, we do not propose formal demographic hypotheses. Instead, the influence of demographic variables is examined in an exploratory vein.

The third and final dimension of the research agenda is to develop market segmentation techniques whereby different types of health care providers can direct their marketing efforts to heavy users of their services (Kotler and Bloom 1984). Malhotra (1983) and Ortinau (1986) suggest heavy users of walk-in facilities are primarily seeking treatment for emergencies or for illnesses occurring beyond the normal hours of private physicians. Heavy users of private physicians, as suggested by Hisrich and Peters (1982) as well as Lane and Lindquist (1988), tend to emphasize routine physical examinations, preventive health services, and consistent family medical care (i.e., obstetrics-gynecology, pediatrics, internal medicine). Hence,

H_3: Need and treatment requirements of walk-in clinic users center on emergency care proce-

dures, whereas for users of private physicians they center on nonemergency procedures, physical examinations, and preventive health programs.

Our study, focusing on comparative consumer expectations, performance evaluations, and some of the correlates of their patronage behavior, provides information that can be used to develop appropriate strategic responses.

Method

Sampling Frame and Sampling Procedures

The data for our study were obtained in telephone interviews by trained interviewers. To provide some generalizability, the sample was drawn randomly from the residential sections of current telephone directories of 15 different cities in five contiguous southeastern states. A total of 2777 phone calls were made during the evening hours on weekdays to maximize the probability of adult contact. The male/female head of household was asked (after being told the purpose of the call and that the study was on attitudes about health care services) two screening questions to determine eligibility for participation. Specifically, only adults who had personally been to *or* had taken someone else to a medical doctor *within the past six months* were qualified. Though somewhat arbitrary, this screening question was used to ensure that the respondents could recall details about specific medical facilities. Precedence for qualifying respondents on the basis of time of the last visit is found in the literature. Woodside and Shinn (1988), for example, used respondents who had been hospital inpatients two to three months prior to the start of their study.

Because of our interest in comparing walk-in clinics and private practitioners, persons who had visited *other* health care delivery facilities (such as hospitals or hospital emergency rooms) were disqualified. We defined a "walk-in clinic" as a medical office where no appointment is necessary (though possible in some cases) and general medical services are provided during extended hours of operation in a freestanding facility as opposed to a collection of private practices housed in the same building/office (Phillips and Reeder 1987). The definition of walk-in clinics was explained to potential respondents. In addition, examples of walk-in clinics prepared in advance for each city were used to illustrate the type of clinics being investigated. If the head of the household answering the phone did not quality, the spouse (if any) was asked the screening questions. If neither adult qualified, the interview was terminated. The interviews lasted 12 to 20 minutes on the average. The following results were obtained from the 2777 calls.

Completed interviews	670
Not contacted (no answer/busy— no callback attempted)	1125
Refused—eligibility not determined	528
Ineligible	
Adult head of household not available	42
Not been to a medical facility in 6 months	104
Not visited types of facilities under study	308
Total ineligible	454
	2777

Of the 670 completed interviews, 602 were sufficiently complete to be used in the study. On the basis of the approach outlined by Wiseman and Billington (1984), a response rate of 41% was obtained.

Though the sample was chosen randomly from telephone directories, such a method (as opposed to random digit dialing, for instance) clearly introduces biases due to exclusion of households with unlisted numbers or without residential telephones. However, given the relatively small proportions of unlisted and nontelephone homes in the general population, the representation biases are presumably minimal. Further, such exclusion may not be theoretically very significant. If one assumes nontelephone households typically belong to the relatively lower income category, and the unlisted ones (which pay more) belong to the relatively higher income category, some eligible respondents from both ends of the income continuum may have been omitted by our method. Theoretically (in terms of H_3), some eli-

gible patrons of clinics *as well as* of private physicians were excluded (though not necessarily in the same proportion).

The Survey Instrument

The survey included questions about the respondent's last medical visit as well as their demographic characteristics. Anticipating that some respondents would have patronized both private physicians and walk-in clinics, we requested the respondents at the very beginning to answer *all* the questions as they pertained to the facility they had *visited last*. Again, the aim of such anchoring was to obtain specific answers about their most recent experience. Consequently, the categorization of patrons into clinic users and private physician users is not meant to imply they utilize a particular format exclusively. However, 89% of the respondents indicated that they had a *regular* clinic or physician and for the large majority (72%) the last visit had been to their regular facility. A comparison between those who did and those who *did not* have a regular facility revealed no statistically significant differences.

Following the approach of Hawes and Rao (1985), we first asked the respondents to rate the *importance* of 10 characteristics/attributes that clinics/private practices might have and that influence the choice of health care providers. These characteristics were measured on a 3-point scale (unimportant 1, somewhat important 2, very important 3). Lehmann and Hulbert (1972) concluded that when the focus of the research is on averages across people or group behavior (as in our study), a 3-point scale is sufficient. Such a scaling approach also reduces the complexity (thereby increasing the accuracy) of obtaining responses to scaled questions over the phone.

The respondents next were asked to indicate the degree to which their health care provider (either the walk-in clinic or the private practitioner) had each of the same 10 characteristics/attributes. A 3-point scale (not at all 1, only somewhat 2, very much so 3) was used for this set of questions.

The 10 characteristics/attributes measured (see Table 2) encompass the characteristics identified by Crane and Lynch (1988) and Lim and Zallocco (1988) as influencing the health care provider de-

TABLE 1 Summary of Key Demographic Characteristics of the Usable Sample

Characteristics	Frequency	Percent of Sample[a] (n = 602)	Characteristics	Frequency	Percent of Sample[a] (n = 602)
Marital Status			Employed?		
Married	375	62.3	Yes	412	68.4
Divorced	55	9.1	No	147	24.4
Widowed	29	4.8	Other answers	30	5.0
Never married	140	23.3		589	97.8
	599	99.5			
			Age		
Sex of Respondent			35 or younger	252	41.9
Female	346	57.5	35–50 years	212	35.2
Male	253	42.0	51–64 years	88	14.6
	599	99.5	65 or older	33	5.5
				585	97.2
Children at Home?					
No	307	51.0	Income ($)		
Yes	288	47.8	20,000 or less	143	23.8
	595	98.8	20,001–30,000	148	24.6
			30,001–40,000	99	16.4
Number of Children at Home[b]			40,001–50,000	81	13.5
1	137	47.6	50,001 or more	117	19.4
2	103	35.7		588	97.7
3 or more	48	16.7			
	288	100.0	Education		
			High school or less	113	18.8
Age of Youngest Child at Home[b]			Some college	177	29.4
Up to 10 years	231	80.2	College degree	237	39.4
11–20 years	43	14.9	Graduate degree	57	9.5
21 years or older	14	4.9		584	97.1
	288	100.0			

[a]Some of the percentages do not add up to 100 because of rounding errors and missing values.
[b]Of those who had children living at home.

cision. Further, these characteristics also had emerged as significant issues in preliminary focus group interviews conducted separately with walk-in clinic and private practitioner users.

Statistical Analysis

Multivariate analysis of variance (MANOVA) and multiple discriminant analysis (MDA) were the principal inferential techniques. MANOVA is useful when there are multiple intervally scaled criterion variables and one categorical predictor variable (Green 1978). In principle, MANOVA is similar to the univariate analysis of variance (ANOVA). However, MANOVA is concerned with the differences among the population (group) centroids. The overall test for differences across the predictor groups, a generalization of the univariate F-ratio, is based on statistics such as Wilks' lambda that are convertible into equivalent multivariate F-ratios. If MANOVA indicates overall group differences, further analysis to determine the source of these group differences becomes appropriate, traditionally by ANOVA for each criterion variable (Cooley and Lohnes 1971).

Darden and Perreault (1975) have demonstrated the advantages of using MDA in conjunction with MANOVA to help determine the direction and intensity of each criterion variable's impact on the overall group differences. Whereas MANOVA (and ANOVA) test for significant departure from the null hypothesis of no differences across groups, MDA determines the weights of the combination of criterion variables that maximize the departure from the null hypothesis. These weights are best measured by discriminant loadings, which are the correlations between the discriminating variables and the canonical discriminant function (Hair et al. 1979). Hence, evaluating the contribution of each criterion variable to the discriminant function by discriminant loadings can increase our understanding of each variable's impact on the group separation.

Results

Sample Patronage Characteristics

Of the 602 respondents, 397 (66%) reported using private physicians for their medical needs on a regular basis. The remaining 205 (34%) reported using walk-in clinics. In the six months prior to the data collection, 54% of the respondents had visited their medical service provider once, another 27% had sought medical help twice, and 19% had sought it three or more times.

A large majority (72%) indicated they had visited their *regular* physician or clinic in *their last visit.* Another 17% had a regular physician or clinic but went to an *alternate* facility *in the last visit,* and only 11% of the respondents reported not having a regular physician or clinic. Further, results show that respondents had been patronizing a particular physician or clinic for more than 4 years on the average. In other words, patronage behavior for medical services (whether private physicians or clinics) shows a high degree of "brand loyalty" as well as "repeat purchase."

The most common reason for the *last medical visit* was to get help "for themselves" (81%), followed by "for children" (13%) and "for others" (6%). Fifty-seven percent of the last visits were routine or followup visits, 24% were necessitated by major illnesses or injuries, 9% were caused by emergencies, and the remaining 10% were for other reasons. Most visits were made to general practitioners (55%), though a fairly large percentage (45%) were to specialists.

A large majority of the respondents expressed satisfaction with their service providers (89%), 7% were neutral on the satisfaction-dissatisfaction scale, and only 4% expressed dissatisfaction. Moreover, 55% rated their service providers as better than others, 43% rated them of comparable quality, and only 2% rated them worse than others. Thus, respondents generally tended to be satisfied with the medical service providers they patronize, which accounts for the "loyalty."

Sample Demographic Characteristics

Most respondents were married (62%), 14% were divorced or widowed, and 23% were single. Women (58%) outnumbered men (42%) in the sample. Approximately half (48%) of the respondents had children living with them. Most frequently, only one child was in residence (48%), though the mean number of children at home was 1.7. Most children at home were 10 years old or younger (80%), though 15% of the respondents reported children between the ages of 11 to 20 living at home.

Sixty-eight percent of the sample reported being employed outside the home and 33% had an employed spouse. Most respondents belonged to two age categories, 35 years or younger (42%) and 36 to 50 years (35%). In income, the majority were in two categories, $20,000 or less (24%) and $20,001 to $30,000 (25%). Most had college degrees (39%) and 10% held graduate degrees. Only 19% had a high school education or less and the remainder (29%) had attended college or technical school. Table 1 presents the complete profile.

Direct comparisons between this demographic profile and population characteristics are difficult because the sample was prescreened and restricted to persons who had visited a medical facility within the last six months that was either a walk-in clinic or a private physician office. As noted, 26% of the prospective respondents (412 adults) were excluded because of these criteria.

Nevertheless, a comparison with the *1980 Census of Population* (U.S. Bureau of the Census 1982) figures for the five contiguous states indicates that the sample includes a disproportionately larger number of married respondents (sample 62%, population 38%), a larger number of female respondents (sample 58%, population 51%), a smaller number of people in the 35 years or younger group (sample 42%, population 50%), a larger number of people in the $50,001 or more income bracket (sample 19%, population 3%), and a larger number of college degree holders (sample 39%, population 13%). If we assume that the visit to a doctor within the past six months indi-

TABLE 2 MANOVA and MDA Results for Differences in Importance of Expected Attributes Across Patrons of Clinics Versus Patrons of Private Physicians

Attributes	F-Ratio	Significance Level	R^2	Discriminant Loadings[b]	Group Means[a]	
					Clinics[c]	Private Physicians[d]
Convenient location	6.14	.01	.01	.554	2.360(9)	2.214(9)[d]
Friendly staff	2.07	.15	.00	−.322	2.510(8)	2.584(5)
Good reputation	3.56	.06	.00	−.422	2.795(3)	2.860(3)
Different specialists in same building	.03	.86	.00	−.038	1.815(10)	1.827(10)
Ability to get appointment quickly	2.72	.09	.00	.369	2.660(5)	2.582(6)[d]
Short waiting time	5.75	.01	.01	.536	2.595(6)	2.472(8)[d]
Reasonable costs	1.44	.23	.00	.268	2.575(7)	2.513(7)
Friendly doctors	.14	.70	.00	−.085	2.680(4)	2.696(4)
Competence of doctors	.31	.57	.00	−.125	2.905(1)	2.921(1)
Physicians' willingness to spend time and explain	.09	.76	.00	−.068	2.855(2)	2.865(2)
Multivariate significance level						
Percent correctly classified (hit ratio)						
Analysis sample	64.58					
Holdout sample	65.79					
Maximum chance criterion	66.22					
Proportional chance criterion	55.26					

[a] 1 = unimportant, 2 = somewhat important, 3 = very important.

[b] Correlation between discriminating variables and canonical discriminant function.

[c] The numbers in parentheses denote summary ranks of attributes *within each group.* For ranks of *all 10 attributes* taken together *across both groups,* the coefficient of concordance given by Kendall's tau is .8444 ($p = .005$).

[d] Kruskal-Wallis approximate chi square tests on ranks show these pairs of mean ratings as significantly different at $p < .05$ across the two groups.

cates a greater propensity to obtain medical help, then to the extent that better educated, older individuals with higher income tend to see physicians more often, the sample could be from the population. Statistics that incorporate propensity to obtain medical care as a variable are needed for meaningful comparison with the sample. Unfortunately, this type of information at the population level is not available. In a subsequent section we discuss the relationship between demographic categories and the type of medical service provider selected by the respondents.

H[1]: Expectations from Service Providers

The first hypothesis posits that the expectations of walk-in clinic patrons differ from those of persons who use private physicians in terms of the

characteristics or attributes sought from the medical facilities. Specifically, clinic users are predicted to be more likely to be concerned about such attributes as convenient location, costs, ability to get an appointment quickly, and shorter waiting time. To test this hypothesis, respondents were grouped into clinic users and patrons of private physicians on the basis of the facility last utilized, then the 10 characteristics/attributes were contrasted across the two groups.

As reported in Table 2, the multivariate F-ratio of MANOVA is significant at the .03 level, indicating overall differences between the two groups. However, the subsequent univariate ANOVA procedures and MDA reveal that at the .05 level, the groups are statistically different on *only two* of the 10 attributes, convenient location and short waiting time. The group means and the discriminant loadings for these two attributes are supportive of the hypothesis in terms of *direction*. However, the differences in means of the two groups are small. Further, effect sizes are no larger than 1% for the significant results, suggesting that the effects are largely an artifact of the sample size. As an additional check, the predictive validity of the discriminant function was assessed by comparing the *overall* hit ratio (66.05%) with the proportional chance criterion (55.26%). Hair et al. (1979) suggest that the classification accuracy reflected in the overall hit ratio should be at least 25% higher than the proportional chance criterion (i.e., should be 69.08% or more) before one can have confidence in the predictive validity of the MDA function. Because this criterion is *not* met, only two attribute differences are statistically significant, and the effect sizes are small, the hypothesis appears *not to be supported*.

However, because the spirit of the hypothesis suggests that the level of importance attached to the 10 attributes differs across users of clinics and private physicians (and also because the mean differences are relatively small), additional analyses were carried out using ranks. As is evident in Table 2, the ranks (of the importance ratings) also indicate that the patrons of both clinics and private physicians have a high degree of concordance in their expectations about the service attributes in terms of importance.

Two inferential tests were carried out to test this concordance. First, we compared the ranks for the 10 attributes *across* clinic and private physician patrons *for each individual attribute* using the Kruskal-Wallis approximate chi square test on ranks. Only three of 10 chi squares are significant at the .05 level. The finding that clinic patrons ranked convenient location, ability to get an appointment quickly, and short waiting time *higher* in importance than did private physician patrons partially supports the multivariate analysis.

Because most attribute ranks were not rated statistically differently by the two groups, Kendall's coefficient of concordance (tau) was computed to investigate whether the rankings *across all 10 attributes* were similar for the groups. This global test on ranks reveals a strong degree of concordance at tau = .8444; the probability of getting such concordance by chance alone is as low as .005.

In summary, though clinic patrons did *rank* three attributes as significantly more important than did their private physician counterparts, when all 10 attributes are considered simultaneously, the two subgroups show a high degree of agreement. Thus, H_1 appears to be generally rejected.

Another noteworthy feature in Table 2 is the absolute magnitude of the importance ratings of the attributes. With the exception of "having different specialists in the same building" (rated between unimportant and somewhat important), all attributes were rated as important, signaling further commonality of expectations between the two groups. This finding also indicates that the attributes chosen for the questionnaire were successful in addressing concerns significant to the patrons of medical services.

H_2: Performance of Service Providers

H_2 predicts that clinics would receive higher *performance* ratings by their patients on such attributes as convenient location, costs, ability to get an appointment quickly, and shorter waiting time than those accorded to private physicians by their patients. MANOVA and MDA were used to test this hypothesis. Comparisons of performance ratings across the same 10 attributes were made between the two user groups. The respondents were asked whether the 10 attributes were exhibited by the physician or the clinic *they last visited*. The results are reported in Table 3.

Though the multivariate test is significant, the MDA results indicate that the performance judgments are generally not statistically different across the two groups for a majority of the attributes. Also, the classification accuracy achieved by the MDA (*overall* hit ratio) is 67.12%, which falls short of the requisite 69.26% accuracy (125% of the proportional chance criterion; Hair et al. 1979), suggesting that the predictive validity of the discriminant function may be low. Only three means are significantly different across the two subgroups: clinic patrons rated the convenience of location of their service provider higher than did the private physician patrons, whereas the latter rated the reputation and competence of the physicians higher than did the clinic patrons. However, in each case the effect size is small.

TABLE 3 MANOVA and MDA Results for Differences in Performance on Attributes Across Patrons of Clinics Versus Patrons of Private Physicians

Attributes	F-Ratio	Significance Level	R^2	Discriminant Loadings[b]	Group Means[a]	
					Clinics[c]	Private Physicians[c]
Convenient location	6.03	.01	.01	.523	2.429	2.286
Friendly staff	.09	.76	.00	.064	2.587	2.572
Good reputation	9.50	.00	.03	−.657	2.673	2.802
Different specialists in same building	.00	.96	.00	−.010	1.837	1.840
Ability to get an appointment quickly	2.82	.09	.00	.358	2.582	2.495
Short waiting time	1.13	.28	.00	.226	2.429	2.369
Reasonable costs	.58	.44	.00	.163	2.418	2.376
Friendly doctors	.03	.85	.00	−.038	2.724	2.732
Competence of doctors	4.29	.03	.01	−.442	2.801	2.871
Physicians' willingness to spend time and explain	.69	.40	.00	−.176	2.735	2.771
Multivariate significance level		.01				

Percent correctly classified (hit ratio)
Analysis sample	64.18
Holdout sample	67.55
Maximum chance criterion	66.44
Proportional chance criterion	55.41

[a] 1 = not at all, 2 = only somewhat, 3 = very much so.
[b] Correlation between discriminating variables and canonical discriminant function.

Hence H_2 does not receive much support. As with the expectation variables, both groups generally gave high marks to their respective service providers on nine of the 10 attributes, the exception being "availability of different specialists in the same building," which both groups rated between not at all and only somewhat important.

Effect of Demographic Characteristics on Choice of Service Provider

Though no formal hypothesis was proposed for the impact of demographic characteristics on choice behavior, limited prior research suggests that clinic users are likely to be younger and to have lower income and educational background. We tested this proposition using MANOVA and stepwise MDA where the demographic char-

acteristics (when obtained in nonmetric form) were dummy coded. The results are reported in Table 4.

The multivariate test is highly significant ($p = .00001$), suggesting that demographic variables do, in fact, discriminate in the classification of patrons as users of clinics and users of private physicians. Using stepwise MDA, we find that the *overall* hit ratio of 72% and the stability of percent correctly classified across the analysis and holdout samples further support this conclusion. Moreover, the overall hit ratio exceeds the proportional chance criterion (55%) by the requisite 25% cutoff (i.e., exceeds 69%), allowing some confidence in the predictive validity of the discriminant function and thereby further legitimizing inferences based on the univariate results. Additional cues come from the statistical significance

TABLE 4 MANOVA and Stepwise MDA Results for Differences in Patronage Preference Across Demographic Variables and the Type of Medical Need

Demographics/Medical Need[a]	F-Ratio	Significance Level	Discriminant Loadings[b]	R^2	% Frequencies[c] Clinics	Private Physicians
Marital Status						
Divorced	.717	.39	.082	.00	29.09	70.91
Married	4.573	.03	.209	.01	30.48	69.52
Widowed	.3693	.06	.188	.00	17.24	82.76
Never married[d]					48.20	51.80
Education						
Some college	2.206	.13	−.145	.00	38.42	61.58
Graduate degree[d]					29.82	70.18
Sex of Respondent[e]						
Male	4.826	.02	−.215	.01	28.65	61.35
Female					30.06	69.94
Income ($)						
20,000 or less	7.780	.00	−.273	.03	44.26	55.74
20,001–30,000	3.109	.07	−.173	.00	40.94	59.06
50,000 or more[d]					27.00	73.00
Number of Children at Home						
One	14.350	.00	−.371	.03	24.79	74.21
Two					36.36	63.64
Three or more					39.02	60.98
Age of Youngest Child at Home						
10 or younger	44.270	.00	.652	.03	35.80	64.20
11–20 years					26.97	73.03
21 or older					20.00	80.00
Type of Medical Need[f]						
Major illness/injury	6.021	.01	.236	.01	42.86	57.14
Routine/followup	24.130	.00	−.473	.03	25.83	74.17
Emergency	16.430	.00	.390	.03	58.44	41.51
Other[d]					37.70	62.30
Multivariate significance level		.00				
Percent correctly classified (hit ratio)						
Analysis sample	70.10					
Holdout sample	69.93					
Maximum chance criterion	66.33					
Proportional chance criterion	55.33					

[a] Only demographic variables and their levels retained by the stepwise MDA with the minimum tolerance level of .001 are presented in descending order. Responses to "number of children at home" and "age of youngest child at home" were obtained in metric data. All other demographic variables (and types of medical need) were measured by categorical methods and hence were dummy coded.
[b] Correlation between discriminating variables and canonical discriminant function.
[c] For ease of interpretation, frequency percentages are reported so that they add to 100% rowwise.
[d] Comparison group for dummy codes.
[e] 0 = female and 1 = male
[f] Results of regular MDA (not stepwise).

and the magnitude of effect (R^2) of the demographic measures.

Stepwise MDA procedures show that all demographic variables measured (see Table 1) were not significant to the discriminant function. Notably, employment status, presence of children at home, and age of the respondent were eliminated from the model. Also eliminated were *certain levels* of other demographic measures (e.g., "high school or less" and "college degree" levels were *not* retained in the education variable though "some college" was).

Not all retained variables are statistically significant. Specifically, education is not significant. In other cases, certain levels of the retained variables fail to achieve significance (the divorced and widowed categories within marital status and the $20,001–$30,000 income category). These nonsignificant variables also have effect sizes of less than 1%.

EFFECT OF INCOME A positive relationship is found between increasing income and the use of private physicians. At income levels up to $20,000, clinics appear to be preferred. The trend is reversed at income levels of more than $20,000. In the $50,000 or more category, almost three-fourths (73%) of the respondents chose private physicians.

Though age by itself is not significant, the possibility of an income effect on age was assessed by using the chi square test on joint frequencies. The results affirm such an effect and show that 62% of the respondents in the 35 years or younger age group had income *less than* $30,000 whereas 61% of the 36 years plus respondents had income *exceeding* $30,000 ($p = .000$). For the joint effect, the magnitude of effect is .38 in terms of phi.

EFFECT OF MARITAL STATUS AND SEX The married respondents appear to prefer private physicians over clinics. Additional tests on joint frequencies reveal that the preference of married respondents also can be explained by an income effect ($\chi^2 p = .000$; $\phi = .49$), which is consistent with the affordability argument favoring private physicians. Sixty-eight percent of the married respondents had income greater than $30,000; analogous percentages for divorced and widowed respondents are 31% and 22%, respectively. By itself, the sex variable shows that both men and women appear

to prefer private physicians. Joint frequencies of sex and employment status reveal that 64% of the female and 79% of the male respondents were employed ($\chi^2 p = .000$; $\phi = .24$). Relatedly, 53% of male respondents reported income of more than $30,000 versus 49% of female respondents ($\chi^2 p = .005$). In other words, income may be the underlying dimension here as well.

INFLUENCE OF CHILDREN AT HOME The presence of children in the home does not appear to discriminate in the facility choice. However, the *number* and *age* of the children significantly affect the facility choice (each individually yielding a phi value of .11). Clinic users tend to have more children at home (mean = 1.9) than users of private physicians (mean = 1.7). Clinic users also appear to have younger children at home (mean = 9.1 years) than the users of private physicians (mean = 11.7 years). In light of the noted income effect, this finding is not surprising because having more children usually means having more expenditures for medical care, and younger children often require more frequent medical visits.

The frequency distributions corroborate such an interpretation. We find use of clinics increasing with the number of children (correspondingly, the use of private physicians declines) presumably because of the lower costs of walk-in clinics. Relatedly, respondents with younger children favor clinics more than do those with older children, undoubtedly for the same economic reasons.

H$_3$: Type of Medical Need and Service Chosen

The third hypothesis posits the walk-in clinics are more likely to be chosen when immediate attention is required, as in the case of emergencies, whereas private physicians are preferred for non-emergency visits. The results reported in Table 4 indicate that this hypothesis is supported, as the nature of medical need has a significant effect on the type of service chosen by the respondents. In the case of emergencies, clinics are significantly preferred—presumably because immediate attention is required. However, private physicians are preferred when the visit involves a routine checkup or followup. For major illness/injury, private physicians are preferred over clinics, but not at the same level as for routine followup.

Discussion

In response to the need for more empirical research in order to understand health care delivery systems from the perspective of the consumers (especially in light of the increasing intertype competition), we used a trait-behavior model to investigate consumer perceptions of two health care delivery systems in terms of the significant expectation attributes, performance evaluation, and patronage motives that may underlie the selection of a health care provider.

H_1 and H_2

The assumption that consumers have different expectations for clinics and private physicians (H_1) is generally rejected. Even though expectations about convenient location and shorter waiting time are statistically different and consistent with the hypothesis, the magnitude of effects is small, the predictive ability of the discriminant function is low, and overall the rank orders of the attributes are not significantly different. Much more noteworthy is the lack of differences in the remaining eight attributes. For instance, reasonable costs and ease of obtaining appointments (traditional differential advantages of clinics) fail to discriminate between the two types of patrons. Equally significant is the lack of differences on what may be called the "quality" dimension. Clinic users rated their expectations about competence, reputation, friendliness of physicians, and in-office features (such as friendly staff) just as highly as did the patrons of private physicians.

These findings suggest that one or several of the following changes are taking place in the marketplace:

1. Clinic patrons have become demanding in that, regardless of how the positioning was traditionally conceived by the clinics, the expectations are consistently high on all attributes but one.

2. Conversely, patrons of private physicians have developed high expectations about conveniences traditionally associated with walk-in clinics.
3. The clinics have succeeded in broadening their appeals by providing services and a level of quality traditionally not associated with them.
4. Private physicians have failed in communicating/conveying the differential advantages consistent with their specific positioning.

Regardless of which of these interpretations are valid, they all essentially suggest the demise of the traditional positioning from the consumer's perspective.

We perhaps could have viewed the findings on consumer expectations as idealized representations of what the consumers hope to receive from their health care providers *if* they had rated the actual performance of the two facilities *differently*. However, H_2 pertaining to perceived performance is also generally rejected for reasons similar to the ones stated for H_1. Both groups gave high marks on most attributes to their respective health care providers, which suggests one or more of the following conclusions:

1. Clinics are doing a superb job on all expectation variables.
2. Private physicians are *not* doing a good enough job to be perceived as distinct from clinics.
3. Both are doing such a good job that it is difficult to differentiate which has the better performance.

Again, regardless of which interpretation is closer to the "truth," performance differentials as perceived by the consumers appear to be minimal, once more suggesting the erosion of the *de facto* segmentation and unique positioning strategies.

Two rival interpretations can be offered for the rejection of the first two hypotheses. One could argue that a between-subjects design (such as was used in our study) is not appropriate for making such comparative statements. Instead, a within-subjects design is required whereby one would sample from the population of consumers who ostensibly patronize both clinics and private physicians interchangeably. Though such an argument has merit, 89% of the respondents indicated that they have a regular clinic or physician, and a large majority (72%) of consumers had gone to their regular facility on their last visit. Only 11% denied having a regular facility and only 17% had visited an alternate (rather than their regular) facility. Moreover, consumers had been patronizing a particular facility for four years on the average. In sum, though it would be worthwhile to replicate our study with a within-subjects design, the practice or switching between facilities does not appear to be prevalent.

A second rival hypothesis could be drawn from the arguments based on cognitive dissonance theory (cf. Festinger 1957), self-perception theory (cf. Bem 1967), and attitude theory (cf. Osgood and Tannenbaum 1955; Ross et al. 1987). This body of literature suggests that the links between attitudes and choice/decision are not always unidirectional (i.e., an attitude structure does not necessarily precede choice/decision). Bem (1967), for instance, suggests that people may determine their own attitudes, beliefs, and expectations in the same way as they assess others' beliefs, attitudes, and expectations: by observing their own actions and *post hoc* rationalizing that their actions must be consistent with their cognitive structures. In our context, one could argue that people may initially begin to patronize a particular type of facility for a variety of reasons (e.g., referrals, family traditions, or urgent needs) and then formulate their expectations on the basis of what they have experienced. In other words, consumers may not indulge in comparative shopping *consciously*. Hence, their subsequent performance ratings would also be influenced by their experiences because they may be "locked on" to a particular facility without any comparative yardsticks to evaluate performance. The "brand loyalty" alluded to before would tend to reinforce such a thesis and hence warrants the attention of investigators.

H_3 and Demographic Variables

H_3 is generally supported. Exploratory predictions related to demographic variables receive partial support. A strong effect of economic differentials is particularly notable with several demographic variables. Though income, marital status, sex, number of children at home, and the age of the youngest child at home contribute significantly to the discriminant function, income is the underlying correlate in most cases. Further investigation is needed in which consumers are asked directly to divulge their reasons for selecting a particular facility. Validation of the income effect as the dominant construct could simplify the formulation of future positioning strategies. In conjunction with the findings discussed for H_1 and H_2, such an income effect would suggest that segmentation should be based not on the attributes offered by health care providers, but on the economic differences.

The support found for H_3 suggests that future research may fruitfully draw on the rationale of benefit segmentation (as opposed to demographic segmentation) for determining workable positionings for alternate health care delivery systems. Differential uniqueness can be communicated successfully to the potential consumers on the basis of the benefits sought and economic status. Another theoretical framework suggested by the H_3 results appears to be risk theory (cf. Bauer 1967; Cox 1967; Cunningham 1967; Jacoby and Kaplan 1972). Specifically, perceived temporal risk (one of the four types of risk) appears to be a strong predictor.

Implications

Our research shows that though there are differences in criteria used for the selection of health care providers, the significance of those differences is not as dramatic as previous research has suggested. However, upon examination of consumers' perceived characteristics of private physicians versus walk-in clinics, some basic observations emerge. Consumers clearly perceive walk-in clinics as more convenient in terms of location and securing appointments for medical services. Yet, it is interesting to note that both private physicians and walk-in clinic users ranked these criteria relatively low as health care service patronage determinants. Another basic observation pertains to the competency and reputation of physicians. Though users of private physicians perceived these characteristics as more important, the difference between their perceptions and those of walk-in clinic users are not as pronounced as previous research would indicate. Additionally, both private physician users and users of walk-in clinics rank these criteria relatively high as a patronage determinant for health care providers.

These findings, from a market segmentation perspective, suggest similarities between users of private physicians and users of walk-in clinics. Whereas conventional wisdom would indicate two distinct market segments, our study suggests that the two segments may be dissipating and that users of the two delivery systems may not belong to two distinct groups, thus defying segmentation except on an economic basis. This basic finding lends credence to the marketing strategies currently being used by walk-in clinics and provides an empirical explanation for their success and growth rates. On a broader note, health care professionals would do well to go beyond intratype competition considerations and conceptualize the competitive threat in light of the emergent intertype competition. Stated differently, strategic responses and positioning decisions that are made without accounting for the intertype competition may lead to costly fratricid rivalry while competing delivery systems encroach on traditional domains.

The results show that certain demographic variables seem to be better predictors of choice than others. Health care marketers should note that, though intuitively appealing, employment status, age, and the presence/absence of children at home may not be useful bases for segmentation or strategic response to intertype competition between private physicians and walk-in clinics. However, income, the number and age of children at home, sex, and educational status may be useful for both segmentation and strategic positioning. Especially noteworthy is the influence of income, which appears to be the underlying dimension for several demographic variables.

The demographic findings also provide some initial clues for the profiles of the users of private physicians and walk-in clinics. They help to assess the direction of changes/encroachments when compared with the traditional preconceptions about these health care delivery systems. Consistent with other findings (e.g., Lane and Linquist 1988) is our picture of savvy health care consumers who want value for their money.

The final implication that can be drawn from our findings pertains to medical service needs. In general, users of walk-in facilities are found to have treatment needs centering on emergency care, whereas users of private physicians predominantly seek routine/followup medical care. This difference provides clues to benefit segmentation as perceived by the consumers.

In conclusion, on the basis of information relating to opinion-behavior-demographic characteristics, our study suggests that intertype rivalry and competitive parity may be emerging in the two health care delivery systems, at least in the eyes of the consumers. The major competitive advantage once held by private physicians (i.e., consumer perceptions of high competence and good reputation) is being challenged by walk-in clinics. Private practice physicians may need to enhance their marketing programs and strategies by defining their market segments more precisely on the basis of demographic data and treatment needs of the patient/consumer. Though perceptual differences relating to service and treatment needs of consumers may persist, walk-in clinic facilities seem to be employing better tactics to reinforce the consumers' perceptions of their specialized skill associated with convenient emergency medical care provision.

References

Bauer, Raymond A. (1967), "Consumer Behavior as Risk Taking," in *Risk-Taking and Information-Handling in Consumer Behavior*, Donald F. Cox, ed. Boston: Division of Research, Graduate School of Business Administration, Harvard University, 22–33.

Bem, Daryl J. (1967), "Self-Perception: An Alternative Interpretation of Cognitive Dissonance Phenomenon," *Psychological Review*, 74 (May), 83–7.

Bessom, Richard M. and Donald W. Jackson (1975), "Service Retailing: A Strategic Marketing Approach," *Journal of Retailing*, 51 (Summer), 75–86.

Carroll, Norman V. and Jean Paul Gagon (1983), "Identifying Consumer Segments in Health Service Markets: An Application of Conjoint and Cluster Analyses to the Ambulatory Care Pharmacy Market," *Journal of Health Care Marketing*, 3 (Summer), 22–34.

Cooley, William W. and Paul R. Lohnes (1971), *Multivariate Data Analysis*. New York: John Wiley & Sons, Inc.

Cox, Donald F. (1967), "Risk-Taking and Information-Handling in Consumer Behavior—An Intensive Study of Two Cases," in *Risk-Taking and Information-Handling in Consumer Behavior*, Donald F. Cox, ed. Boston: Division of Research, Graduate School of Business Administration, Harvard University, 34–81.

Crane, F. G. and J. E. Lynch (1988), "Consumer Selection of Physicians and Dentists: An Examination of Choice Criteria and Cue Usage," *Journal of Health Care Marketing*, 8 (September), 16–19.

Cunningham Scott M. (1967), "Major Dimensions of Perceived Risk," in *Risk-Taking and Information-Handling in Consumer Behavior*, Donald F. Cox, ed. Boston: Division of Research, Graduate School of Business Administration, Harvard University.

Darden, William R. and William D. Perreault (1975), "A Multivariate Analysis of Media Exposure and Vacation Behavior With Life Style Covariates," *Journal of Consumer Research*, 2 (September), 93–103.

Festinger, Leon (1957), *A Theory of Cognitive Dissonance*. Stanford, CA: Stanford University Press.

Green, Paul E. (1978), *Analyzing Multivariate Data*. Hinsdale, IL: Dryden Press.

Gregory, Douglas, Paul Kingstrom, and Timothy Rearden (1982), "Planning Models for Outpatient Care: A Marketing Approach," *Journal of Health Care Marketing*, 2 (Winter), 21–30.

Hair, Joseph F., Ralph E. Anderson, Ronald L. Tatham, and Bernie J. Grablowsky (1979), *Multivariate Data Analysis*. Tulsa, OK: Petroleum Publishing Company.

Harrell, Gilbert D. and Matthew F. Fors (1985), "Marketing Ambulatory Care to Women: A Segmentation Approach," *Journal of Health Care Marketing*, 5 (Spring), 19–28.

Hawes, Jon M. and C. P. Rao (1985), "Using Importance-Performance Analysis to Develop Health Care Marketing Strategies," *Journal of Health Care Marketing*, 5 (Fall), 19–25.

Hisrich, Robert D. and Michael J. Peters (1982), "Comparison of Perceived Hospital Affiliation and Selection Criteria by Primary Market Segments," *Journal of Health Care Marketing*, 2 (Summer), 24–30.

Jacoby, Jacob and Leon B. Kaplan (1972), "The Components of Perceived Risk," in *Proceedings of the Third Annual Conference of the Association for Consumer Research*, M. Venkatesan, ed. Chicago: University of Chicago, 382–93.

James, Frank E. (1987), "Controversy Mounts Over Efforts to Measure Quality of Health Care," *Wall Street Journal* (December 17), 20.

Kotler, Philip and Paul N. Bloom (1984), *Marketing Professional Services*. Englewood Cliffs, NJ: Prentice-Hall, Inc.

——— and Roberta N. Clarke (1987), *Marketing for Healthcare Organizations*. Englewood Cliffs, NJ: Prentice-Hall, Inc.

Lamb, Charles W. (1988), "An Important New Challenge Facing Health Care Marketers," *Journal of Health Care Marketing*, 8 (September), 2–4.

Lane, Paul M. and Jay D. Lindquist (1988), "Hospital Choice: A Summary of the Key Empirical and Hypothetical Findings of the 1980s," *Journal of Health Care Marketing*, 8 (December), 5–20.

Lehmann, Donald R. and James Hulbert (1972), "Are Three-Point Scales Always Good Enough?" *Journal of Marketing Research*, 9 (November), 444–6.

Lim, Jeen-Su and Ron Zallocco (1988), "Determinant Attributes in Formulation of Attitudes Toward Four Health Care Systems," *Journal of Health Care Marketing*, 8 (June), 25–30.

Malhotra, Naresh K. (1983), "Stochastic Modeling of Consumer Preferences for Health Care Institutions," *Journal of Health Care Marketing*, 3 (Fall), 18–26.

Neuhaus, Evelyn R. (1982), "A Methodology for Determining the Market Feasibility of a Satellite Clinic," *Journal of Health Care Marketing*, 2 (Summer), 31–8.

Ortinau, David J. (1986), "Discriminating Users and Nonusers of Preventive Health Care Practices and Emergency Medical Walk-In Clinics," *Journal of Health Care Marketing*, 6 (June), 26–37.

Osgood, Charles E. and Percy H. Tannenbaum (1955), "The Principle of Congruity in the Prediction of Attitude Change," *Psychological Review*, 62 (January), 42–55.

Parasuraman, A., Valerie A. Zeithaml, and Leonard L. Berry (1985), "A Conceptual Model of Service Quality and Its Implications for Future Research," *Journal of Marketing*, 49 (Fall), 41–50.

Phillips, Jan Hirsch and C. E. Reeder (1987), "Ambulatory Care Centers: Structure, Services, and Marketing Techniques," *Journal of Health Care Marketing*, 7 (December), 27–32.

Robertson, Thomas (1972), "Low-Commitment Consumer Behavior," *Journal of Advertising Research*, 16 (April), 19–24.

Ross, Caroline K., Gayle Frommelt, Lisa Hazelwood, and Rowland W. Chang (1987), "The Role of Expectations in Patient Satisfaction With Medical Care," *Journal of Health Care Marketing*, 7 (December), 16–26.

Ross, Michael, Cathy McFarland, Michael Conway, and Mark P. Zanna (1983), "Reciprocal Relation Between Attitudes and Behavior Recall: Committing People to Newly Formed Attitudes," *Journal of Personality and Social Psychology*, 45 (August), 257–67.

Tucker, Frances Gaither and James B. Tucker (1985), "An Evaluation of Patient Satisfaction and Level of Physician Training," *Journal of Health Care Marketing*, 5 (Summer), 31–8.

U.S. Bureau of the Census (1982), *1980 Census of Population*. Washington, DC: U.S. Government Printing Office.

U.S. Department of Commerce (1987), *Statistical Abstracts of the United States*. Washington, DC: U.S. Government Printing Office.

Wiseman, Frederick and Maryann Billington (1984), "Comment on a Standard Definition of Response Rates," *Journal of Marketing Research*, 21 (August), 336–8.

Woodside, Arch G., Robert L. Nielsen, Fred Walters, and Gale D. Muller (1988), "Preference Segmentation of Health Care Services: The Old-Fashioneds, Value Conscious, Affluents, and Professional Want-It-Alls," *Journal of Health Care Marketing*, 8 (June), 14–24.

———, Chris M. Sertich, and James M. Chakalas (1987), "Hospital Choice: Patient Attribution of the Decision and Satisfaction With the Services," *Journal of Health Care Marketing*, 7 (March), 61–8.

——— and Raymond Shinn (1988), "Consumer Awareness and Preferences Toward Competing Hospital Services," *Journal of Health Care Marketing*, 8 (March), 39–47.

5

Multivariate Analysis of Variance

LEARNING OBJECTIVES

Upon completing this chapter, you should be able to:

- Explain the difference between the univariate null hypothesis of ANOVA and the multivariate null hypothesis of MANOVA.

- Discuss the advantages of a multivariate approach to significance testing compared to the more traditional univariate approaches.

- State the assumptions for the use of MANOVA.

- Discuss the different types of test statistics that are available for significance testing in MANOVA.

- Describe the purpose of post hoc tests in ANOVA and MANOVA.

- Interpret interaction results when more than one independent variable is used in MANOVA.

- Describe the purpose of multivariate analysis of covariance (MANCOVA).

As a theoretical construct, multivariate analysis of variance (MANOVA) was introduced several decades ago by Wilks' original formulation [26]. However, it was not until the development of appropriate test statistics with tabled distributions and the wide availability of programs to compute these statistics on high-speed computers that MANOVA became a practical tool for applied researchers.

Multivariate analysis of variance is an extension of analysis of variance (ANOVA) to accommodate more than one criterion variable. It is a dependence technique that measures the differences for two or more metric variables based on a set of categorical variables acting as predictors. MANOVA and ANOVA can be stated in the following general forms:

$$\text{Multivariate Analysis of Variance}$$
$$\underset{\text{(metric)}}{Y_1 + Y_2 + Y_3 + \cdots + Y_n} = \underset{\text{(nonmetric)}}{X_1 + X_2 + X_3 + \cdots + X_n} \quad \text{MANOVA}$$

$$\text{Analysis of Variance}$$
$$\underset{\text{(metric)}}{Y_1} = \underset{\text{(nonmetric)}}{X_1 + X_2 + X_3 + \cdots + X_n} \quad \text{ANOVA}$$

Like ANOVA, MANOVA is concerned with differences between groups (or experimental treatments). However, ANOVA is termed a univariate procedure because we use it to assess group differences on a single metric dependent variable. MANOVA is termed a multivariate procedure because we use it to assess group differences across multiple metric-dependent variables simultaneously. That is, in MANOVA, each treatment group is observed on two or more dependent variables.

Both ANOVA and MANOVA are particularly useful when used in conjunction with **experimental designs**—that is, research designs in which the researcher directly controls or manipulates one or more independent variables to determine the effect on one (ANOVA) or more (MANOVA) dependent variables. ANOVA and MANOVA provide the tools necessary to judge the reliability of any observed effects (i.e., whether an observed difference is due to a treatment effect or to random sampling variability). See Chapter 1 for a discussion of how these techniques relate to other multivariate procedures.

Before proceeding to the discussion of MANOVA, you should familiarize yourself with the key terms used in multivariate analysis. Review the list of terms below and refer to them whenever necessary during chapter discussions.

KEY TERMS

Before starting the chapter, review the key terms to develop an understanding of the concepts and terminology used. Throughout the chapter the key terms appear in **boldface**. Other points of emphasis in the chapter are *italicized*. Also, cross-references within the key terms are in *italics*.

Alpha (α) Significance level associated with the statistical testing of the differences between two or more groups. Typically small values, such as .05 or .01, are specified to minimize the possibility of making a *Type I error*.

A priori test See *planned comparison*.

Analysis of variance (ANOVA) Statistical technique used to determine whether samples come from populations with equal means. Univariate analysis of variance employs one dependent measure, whereas multivariate analysis of variance compares samples based on two or more dependent variables.

Beta (β) See *Type II error.*

Blocking factor Characteristic of respondents in the ANOVA/MANOVA that is used to reduce within-group variability. This characteristic becomes an additional treatment in the analysis. In doing so, additional groups are formed that are more homogeneous. As an example, assume that customers were asked buying intentions for a product and the independent measure used was age. Examination of the data found that substantial variation was due to gender. Then gender could be added as an additional treatment so that each age category was split into male and female groups with greater within-group homogeneity.

Bonferroni inequality Approach for adjusting the selected *alpha* level to control for the overall *Type I error* rate. The procedure involves (1) computing the adjusted rate as α divided by the number of statistical tests to be performed, and (2) then using the adjusted rate as the critical value in each separate test.

Box test Statistical test for the equality of the variance/covariance matrices of the dependent variables among the groups. The test is very sensitive, especially to the presence of a nonnormal variable(s). A significance level of .01 or less is used as an adjustment for the sensitivity of the statistic.

Contrast Procedure for investigating specific group differences of interest in conjunction with ANOVA and MANOVA—for example, comparing group mean differences for a specified pair of groups.

Covariates, or **covariate analysis** Use of regressionlike procedures to remove extraneous (nuisance) variation in the dependent variables due to one or more uncontrolled metric independent variables (covariates). The covariates are assumed to be linearly related to the dependent variables. After adjusting for the influence of covariates, a standard ANOVA (or MANOVA) is carried out. This adjustment process (known as ANCOVA and MANCOVA) usually allows for more sensitive tests of treatment effects.

Discriminant function "Dimension" of difference or discrimination between the groups in the MANOVA analysis. The discriminant function is a variate of the dependent variables.

Disordinal interaction Form of *interaction effect* among independent variables that invalidates interpretation of the *main effects* of the treatments. A disordinal interaction is exhibited graphically by plotting the means for each group and having the lines intersect and cross. In this type of interaction the mean differences not only vary, given the unique combinations of independent variable levels, but the relative ordering of groups changes as well.

Effect size Standardized measure of group differences used in the calculation of statistical power.

Experimental design Research plan in which the researcher directly manipulates or controls one or more predictor variables (see *treatment*) and assesses their effect on the dependent variables. Common in the physical sciences, it is gaining in popularity in business and the social sciences. For example, respondents are shown separate advertisements that vary systematically on a characteristic, such as different appeals (emotional versus rational) or types of presentation (color versus black-and-white) and are then asked their attitudes, evaluations, or feelings toward the different advertisements.

Factor   Nonmetric independent variable, also referred to as a *treatment* or experimental variable.

Factorial design   Design with more than one *factor* (treatment). In factorial designs, we examine the effects of several factors simultaneously by forming groups based on all possible combinations of the levels (values) of the various treatment variables.

Greatest characteristic root (gcr)   Statistic for testing the null hypothesis in MANOVA. It measures the linear combination of the dependent variables that minimizes group differences.

Hotelling's T^2   Test to assess the statistical significance of the difference between two sets of sample means. It is a special case of MANOVA used with two groups or levels of a treatment variable.

Independence   Critical assumption of ANOVA/MANOVA that requires that the dependent measures for each respondent be totally uncorrelated with the responses from other respondents in the sample. A lack of independence severely affects the statistical validity of the analysis unless corrective action is taken.

Interaction effect   In factorial designs, the joint effects of two treatment variables in addition to the individual *main effects*. This means that the differences between groups on one treatment variable varies depending on the level on the second treatment variable. For example, assume that respondents were classified by income (three levels) and gender (males versus female). A significant interaction would be found when the differences on the independent variable(s) between males and females varied substantially across the three income levels.

Main effect   In factorial designs, the individual effect of each *treatment* variable on the dependent variable.

Multivariate normal distribution   Generalization of the univariate normal distribution to the case of p variables. A multivariate normal distribution of sample groups is a basic assumption required for the validity of the significance tests in MANOVA. Also, see Chapter 2 for more discussion of this topic.

Null hypothesis   Hypothesis that samples come from populations with equal means for either a dependent variable (univariate test) or a set of dependent variables (multivariate test). The null hypothesis can be accepted or rejected depending on the results of a test of statistical significance.

Ordinal interaction   Acceptable type of *interaction effect* in which the magnitudes of differences between groups vary but the groups' relative positions remain constant. Graphically represented by plotting mean values and observing nonparallel lines which do not intersect.

Orthogonal   Statistical independence or an absence of association. Orthogonal *variates* explain unique variance, with no variance explanation shared between them. Orthogonal contrasts are planned comparisons that are statistically independent and represent unique comparisons of group means.

Planned comparison   *A priori test* that tests a specific comparison of group mean differences. These tests are performed in conjunction with the tests for *main* and *interaction effects.*

Post hoc test   Statistical test of mean differences performed after the statistical tests for *main effects* have been performed.

Power   Probability of identifying a treatment effect when it actually exists in the sample. Defined as $1 - \beta$ (see *beta*).

Repeated measures   Use of two or more responses from a single individual in an ANOVA or MANOVA analysis. The purpose of a repeated measures design is to control for individual-level differences that may affect the within-group variance.

Replication   Readministration of an experiment with the intent of validating the results in another sample of respondents.

Significance level   See *alpha*.

Standard error   Measure of the dispersion of the means or mean differences expected due to sampling variation. The standard error is used in the calculation of the *t statistic*.

Stepdown analysis   Test for the incremental discriminatory power of a dependent variable after the effects of other dependent variables have been accounted for. Similar to stepwise regression or discriminant analysis, this procedure, relying on a specified order of entry, determines how much an additional dependent variable adds to the explanation of the differences between the groups in the MANOVA analysis.

t **statistic**   Test statistic that assesses the statistical significance between two groups on a single dependent variable (see *t test*).

t **test**   Test to assess the statistical significance of the difference between two sample means for a single dependent variable. A special case of ANOVA for two groups or levels of a treatment variable.

Treatment   Independent variable that a researcher manipulates to see the effect (if any) on the dependent variables. The treatment variable can have several levels. For example, different intensities of advertising appeals might be manipulated to see the effect on consumer believability.

Type I error   Probability of rejecting the null hypothesis when it should be accepted, that is, concluding that two means are significantly different when in fact they are the same. Small values of alpha (e.g., .05 or .01), also denoted as α (see *alpha*), lead to rejection of the null hypothesis as untenable and acceptance of the alternative hypothesis that population means are unequal.

Type II error   Probability of failing to reject the null hypothesis when it should be rejected, that is, concluding that two means are not significantly different when in fact they are different. Also known as the *beta* (β) *error*.

U **statistic**   See *Wilks' lambda*.

Variate   Linear combination of variables. In MANOVA, the dependent variables are formed into *variates* in the discriminant function(s).

Vector   Set of real numbers (e.g., $X_1 \ldots X_N$). Vectors can be written either for columns or rows. Column vectors are considered conventional and row vectors are considered transposed. Column vectors and row vectors are shown as follows:

$$X = \begin{bmatrix} X_1 \\ X_2 \\ \vdots \\ X_N \end{bmatrix} \qquad X^T = [X_1 X_2 \cdots X_N]$$

Column Vector   Row Vector

The T on the row vector indicates that it is the transpose of the column vector.

Wilks' lambda   One of the four principal statistics for testing the null hypothesis in MANOVA. Also referred to as the maximum likelihood criterion or *U statistic*.

What Is Multivariate Analysis of Variance?

Multivariate analysis of variance is the multivariate extension of the univariate techniques for assessing the differences between group means. The univariate procedures include the *t* test for two-group situations and analysis of variance (ANOVA) for situations with three or more groups defined by two or more independent variables. Before proceeding with our discussion of the unique aspects of multivariate analysis of variance, we will review the basic principles of the univariate techniques.

The t Test

The *t* **test** assesses the statistical significance of the difference between two independent sample means. For example, a researcher may expose two groups of respondents to different advertisements reflecting different advertising messages, one informational and one emotional, and subsequently ask each group about the appeal of the message on a ten-point scale, with one being poor and ten being excellent. The two different advertising messages represent a **treatment** with two levels (informational versus emotional). A treatment is a categorical independent variable, experimentally manipulated or observed, that can be represented in various levels. In our example, the treatment is the effect of emotional versus informational appeals. To determine whether the two messages are viewed differently (meaning that the treatment has an effect), a *t* statistic is calculated. The *t* **statistic** is the ratio of the difference between the sample means ($\mu_1 - \mu_2$) to its standard error. The **standard error** is an estimate of the difference between means to be expected because of sampling error, rather than real differences between means. By forming the ratio of the actual difference between the means to the difference expected due to sampling error, we quantify the amount of the actual impact of the treatment that is due to random sampling error. Absolute values of the *t* statistic that exceed the critical value of the *t* statistic (t_{crit}) lead to rejection of the **null hypothesis** of no difference in the appeals of the advertising messages between groups. This means that the actual difference due to the appeals is statistically larger than the difference expected from sampling error.

We determine a critical value (t_{crit}) for our *t* statistic and test the statistical significance of the observed differences by the following:

1. Compute the *t* statistic as described above.
2. Specify a **Type I error** level (denoted as α, or **significance level**), which indicates the probability level the analyst will accept in concluding that the group means are different when in fact they are not.
3. Refer to the *t* distribution with $N_1 + N_2 - 2$ degrees of freedom and a specified α to identify the critical *t* value. If the value of the computed *t* statistic exceeds t_{crit}, we conclude that the two advertising messages have different levels of appeal (i.e., $\mu_1 \neq \mu_2$), with a Type I error probability of α.

[handwritten margin note:] $\left| \dfrac{(\mu_1 - \mu_2)}{\sigma} \right| = t_{\text{crit}}$

[handwritten margin note:] $H_0: \mu_1 \neq \mu_2 \,/\, \mu_1 - \mu_2 = 0$
$H_1: (\mu_1 - \mu_2) > 0$

4. If a statistically significant difference is found, the analyst can then examine the actual mean values to determine which group is higher on the dependent value.

Analysis of Variance (ANOVA)

In our discussion of the t test, a researcher exposed two groups of respondents to different advertising messages and subsequently asked them to rate the appeal of the advertisements on a ten-point scale. Suppose we were interested in evaluating three advertising messages rather than two (i.e., $k = 3$). Respondents would be randomly assigned to one of three groups, and we would have three sample means to compare. To analyze these data, we might be tempted to conduct separate t tests for the difference between each pair of means (i.e., group 1 versus group 2; group 1 versus group 3; and group 2 versus group 3).

However, multiple t tests inflate the overall type 1 error rate (we discuss this in more detail in the next section). **ANOVA** avoids this Type I error inflation across comparisons of a number of treatment groups by determining whether the entire set of sample means suggests that the samples were drawn from the same general population. That is, ANOVA is used to determine the probability that differences in means across several groups are due solely to sampling error.

The logic of an ANOVA test is fairly straightforward. As the name "analysis of variance" implies, the following two independent estimates of the variance for the dependent variable are compared, one that is sensitive to treatment effects and one that is not:

1. Within-groups estimate of variance (MS_W: mean square within groups): This is an estimate of the random respondent variability on the dependent variable within a treatment group and is based on deviations of individual scores from their respective group means, but not the differences between group means. This is comparable to the standard error between two means calculated in the t test. The value MS_W is sometimes referred to as error variance.
2. Between-groups estimate of variance (MS_B: mean square between groups): The second estimate of variance is the variability of the treatment group means on the dependent variable. It is based on deviations of *group means* from the overall grand mean of all scores. Under the null hypothesis of no treatment effects (i.e., $\mu_1 = \mu_2 = \mu_3 = \ldots = \mu_k$), this variance, like MS_W, is a simple estimate of the sampling variance of scores. However, this variance estimate, unlike MS_W, reflects any treatment effects that exist; that is, differences in treatment means increase the expected value of MS_B.

Given that the null hypothesis of no group differences is true, MS_W and MS_B represent independent estimates of population variance. Therefore, the ratio of MS_B to MS_W is a measure of how much variance is attributable to the different treatments versus the variance expected from random sampling. The ratio of MS_B to MS_W gives us a value for an F statistic. Because group differences tend to inflate MS_B, large values of the F statistic lead to rejection of the null hypothesis of no difference in means across groups. If the analysis has several different treatments, then estimates of MS_B are calculated for each treatment and F statistics calculated for each treatment.

To determine if the F statistic is sufficiently large to support rejection of the null hypothesis, follow a process similar to the t test. First, determine the critical value for the F statistic (F_{crit}) by referring to the F distribution with $(k - 1)$ and $(N - k)$ degrees of freedom for a specified level of α (where $N = N_1 + \ldots + N_k$ and k = number of groups). If the value of the calculated F statistic exceeds F_{crit}, conclude that the means across all groups are not all equal.

Examination of the group means then allows the analyst to assess the relative standing of each group on the dependent measure. While the F statistic test assesses the null hypothesis of equal means, it does not address the question of which means are different. For example, in a three-group situation, all three groups may differ significantly, or two may be equal but differ from the third. To assess these differences, the analyst can employ either planned comparisons or post hoc tests. We examine each of these methods in a later section.

Multivariate Analysis of Variance (MANOVA)

As statistical inference procedures, both the univariate techniques (t test and ANOVA) and MANOVA are used to assess the statistical significance of differences between groups. In the t test and ANOVA, the null hypothesis tested is the equality of dependent variable means across groups. In MANOVA, the null hypothesis tested is the equality of **vectors** of means on multiple dependent variables across groups. The distinction between the hypotheses tested in ANOVA and MANOVA is illustrated in Figure 5.1. In the univariate case, a single dependent measure is tested for equality across the groups. In the multivariate case, a **variate** is tested for equality. The concept of a variate has been instrumental in our discussions of the previous multivariate techniques and is covered in detail in Chapter 1. In MANOVA, the analyst actually has two variates, one made up of the

ANOVA

$$H_0 : \mu_1 = \mu_2 = \ldots \mu_k$$

Null hypothesis (H_0) = all the group means are equal, that is, they come from the same population.

MANOVA

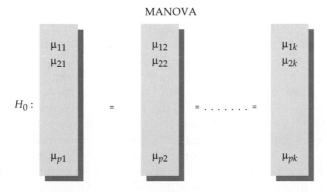

μ_{pk} = mean of variable p, group k

Null hypothesis (H_0) = all the group mean vectors are equal, that is, they come from the same population.

FIGURE 5.1 Null hypothesis testing of ANOVA and MANOVA.

dependent variables and another from the independent variables. The dependent variable variate is of more interest, because the metric-dependent measures can be combined in a linear combination as we have already seen in multiple regression and discriminant analysis. The unique aspect of MANOVA is that the variate optimally combines the multiple dependent measures into a single value that maximizes the differences across groups.

The Two-Group Case: Hotelling's T^2

In our earlier univariate example, researchers were interested in the appeal of two advertising messages. But what if they also wanted to know about the purchase intent generated by the two messages? If only univariate analyses were used, the researchers would perform separate t tests on the ratings of both the appeal of the messages and the purchase intent generated by the messages. Yet the two measures are interrelated; thus, what is really desired is a test of the differences between the messages on both variables collectively. This is where Hotelling's T^2, a specialized form of MANOVA that is a direct extension of the univariate t test, can be used.

Hotelling's T^2 provides a statistical test of the variate formed from the dependent variables that produces the greatest group difference. It also addresses the problem of "inflating" the Type I error rate that arises when making a series of t tests of the comparisons between group means on the dependent measures. It controls this inflation of the Type I error rate by providing a single overall test of group differences across all dependent variables at a specified α level.

How does Hotelling's T^2 achieve these goals? Consider the following equation for a variate of the dependent variables:

$$C = W_1Y_1 + W_2Y_2 + \cdots + W_nY_n$$

where

$$C = \text{Composite or variate score for a respondent}$$
$$W_i = \text{Weight for dependent variable } i$$
$$Y_i = \text{Dependent variable } i$$

In our example, the ratings of message appeal are combined with the purchase intentions to form the composite. For any set of weights, we could compute composite scores for each respondent and then calculate an ordinary t statistic for the difference between groups on the composite scores. However, if we can find a set of weights that gives the maximum value for the t statistic for this set of data, these weights would be the same as the discriminant function between the two groups (which should not be surprising to those who have read Chapter 4). The maximum t statistic that results from the composite scores produced by the discriminant function can be squared to produce the value of Hotelling's T^2 [11]. The computational formula for Hotelling's T^2 represents the results of mathematical derivations used to solve for a maximum t statistic (and, implicitly, the most discriminating linear combination of the dependent variables). This is equivalent to saying that if we can find a discriminant function for the two groups that produces a significant T^2, the two groups are considered different across the mean vectors.

How does Hotelling's T^2 provide a test of the hypothesis of no group difference on the vectors of mean scores? Just as the t statistic follows a known distribution under the null hypothesis of no treatment effect on a single dependent variable, Hotelling's T^2 follows a known distribution under the null hypothesis of no treatment effect on any of a set of dependent measures. This distribution turns out to be an F distribution with p and $N_1 + N_2 - 2 - 1$ degrees of freedom after adjustment (where p = the number of dependent variables). To get the critical value for the Hotelling's T^2, we find the tabled value for F_{crit} at a specified α level and compute T^2_{crit} as follows:

$$T^2_{crit} = \frac{p(N_1 + N_2 - 2)}{N_1 + N_2 - p - 1} \times F_{crit}$$

The k-Group Case: MANOVA

MANOVA can be considered a simple extension of Hotelling's T^2 procedure; that is, we devise dependent variable weights to produce a variate score for each respondent, as described earlier. If we wanted to evaluate three advertising messages both for their appeal and for the purchase intentions they generate, we would use MANOVA. In MANOVA we now want to find the set of weights that maximizes the ANOVA F value computed on the variate scores for all the groups. But MANOVA can also be considered an extension of discriminant analysis (see Chapter 4) in that multiple variates of the dependent measures can be formed if the number of groups is three or more. The first variate, termed a **discriminant function,** specifies a set of weights that maximize the differences between groups, thereby maximizing the F value. The maximum F value itself allows us to compute directly what is called the **greatest characteristic root statistic.** The greatest characteristic root statistic can be calculated as: gcr = $(k - 1) F_{max} \div (N - k)$ [11].

To obtain a single test of the hypothesis of no group differences on the vectors of mean scores, we could refer to tables of the gcr distribution. Just as the F distribution follows a known distribution under the null hypothesis of equivalent group means on a single dependent variable, the gcr statistic follows a known distribution under the null hypothesis of equivalent group mean vectors (i.e., group means are equivalent on a set of dependent measures). A comparison of the observed gcr to gcr$_{crit}$ gives us a basis for rejecting the overall null hypothesis of equivalent group mean vectors.

Any subsequent discriminant functions are **orthogonal;** they maximize the differences among groups based on the remaining variance not explained by the prior function(s). Thus, in many instances, the test for differences between groups involves not just a single variate score but a set of variate scores that are evaluated simultaneously. A range of multivariate tests is available, each best suited to specific situations of testing these multiple variates.

When Should We Use MANOVA?

With the ability to examine several dependent measures simultaneously, the analyst can gain in several ways from the use of MANOVA. We discuss the issues in using MANOVA from the perspectives of controlling statistical accuracy and efficiency while still providing the appropriate forum for testing multivariate questions.

Control of Experimentwide Error Rate

The use of separate univariate ANOVAs or *t* tests can create a problem when trying to control the overall, or experimentwide, error rate. For example, assume that we evaluate a series of five dependent variables by separate ANOVAs, each time using .05 as the significance level. Given no real differences in the dependent variables, we would expect to observe a significant effect on any given dependent variable 5 percent of the time. However, across our five separate tests, the probability of a Type I error will lie somewhere between 5 percent (if all dependent variables are perfectly correlated) and $(1 - .95^5)$ 23 percent if all dependent variables are uncorrelated. Thus a series of separate statistical tests leaves us without control of our effective overall or experimentwide Type I error rate. If the researcher desires to maintain control over the experimentwide error rate and there is at least some degree of intercorrelation among the dependent variables, then MANOVA is appropriate.

Differences Among A Combination of Dependent Variables

A series of univariate ANOVA tests also ignores the possibility that some composite (linear combination) of the dependent variables may provide evidence of an overall group difference that may go undetected by examining each dependent variable separately. Individual tests ignore the correlations among the dependent variables and thus use less than the total information available for assessing overall group differences. In the presence of multicollinearity among the dependent variables, MANOVA will be more powerful than the separate univariate tests. In this manner, the MANOVA may detect *combined* differences not found in the univariate tests. Moreover, if multiple variates are formed, then they may provide *dimensions* of differences that can distinguish among the groups better than single variables. However, in some instances of a large number of dependent variables, the statistical power of the ANOVA tests exceeds that obtained with a single MANOVA. The considerations involving sample size, number of dependent variables, and statistical power will be considered in a subsequent section.

A Decision Process for MANOVA

The process of performing a multivariate analysis of variance is similar to that found in many other multivariate techniques, so that it can be described through the six-stage model-building process described in Chapter 1. The process begins with the specification of research objectives. It then proceeds to a number of design issues facing a multivariate analysis and then an analysis of the assumptions underlying MANOVA. With these issues addressed, the process proceeds to estimation of the MANOVA model and the assessment of overall model fit. When an acceptable MANOVA model is found, then the results can be interpreted in more detail. The final step involves efforts to validate the results to ensure generalizability to the population. Figure 5.2 provides a graphical portrayal of the process, which is discussed in detail in the following sections.

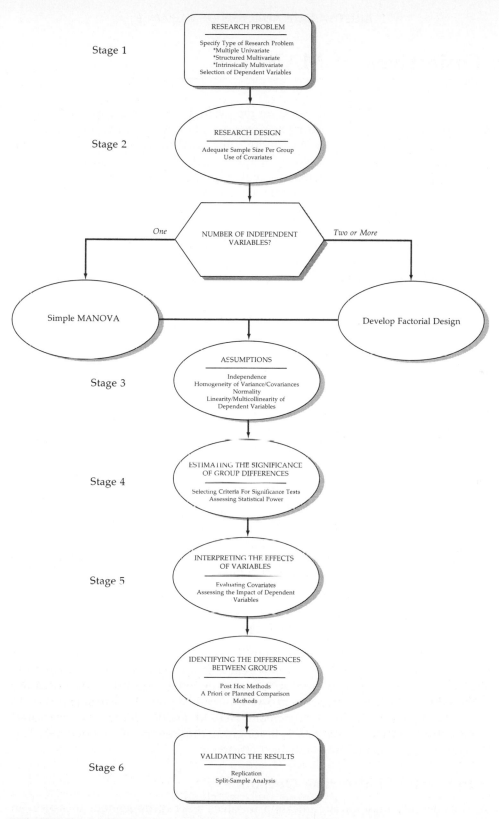

FIGURE 5.2 Multivariate analysis of variance decision process.

Stage One: Objectives of MANOVA

The selection of MANOVA is based on the desire to analyze a dependence relationship represented as the differences in a *set* of dependent measures across a series of groups formed by one or more categorical independent measures. As such, MANOVA represents a powerful analytical tool suitable to a wide array of research questions. Whether used in actual or quasi-experimental situations (such as field settings or survey research where the independent measures are categorical), MANOVA can provide insights into not only the nature and predictive power of the independent measures but also the interrelationships and differences seen in the set of dependent measures.

Types of Multivariate Questions Suitable for MANOVA

The advantages of MANOVA versus a series of univariate ANOVAs extend past the statistical domain discussed earlier and are also found in its ability to provide a single method of testing a wide range of differing multivariate questions. Throughout the text, we emphasize the interdependent nature of multivariate analysis. MANOVA has the flexibility to allow the researcher to select the test statistics most appropriate for the question of concern. Hand and Taylor have classified multivariate problems into three categories, each of which employs different aspects of MANOVA in its resolution [10]. The three categories and their use of MANOVA are described below.

Multiple Univariate Questions

A researcher studying multiple univariate questions identifies a number of separate dependent variables (e.g., age, income, education of consumers) that are to be analyzed separately but needs some control over the experimentwide error rate. In this instance, MANOVA is used to assess whether an overall difference is found between groups, and then the separate univariate tests are employed to address the individual issues for each dependent variable.

Structured Multivariate Questions

A researcher dealing with structured multivariate questions gathers two or more dependent measures that have specific relationships between them. A common situation in this category is repeated measures, where multiple responses are gathered from each subject, perhaps over time or in a pretest-posttest exposure to some stimulus, such as an advertisement. Here MANOVA provides a structured method for specifying the comparisons of group differences on a set of dependent measures while maintaining statistical efficiency.

Intrinsically Multivariate Questions

An intrinsically multivariate question involves a set of dependent measures in which the principal concern is how they differ *as a whole* across the groups. Differences on individual dependent measures are of less interest than their collective effect. One example is the testing of multiple measures of response that should be consistent, such as attitudes, preference, and intention to purchase, all of which relate to differing advertising campaigns. The full power of MANOVA is utilized in this case by assessing not only the overall differences but also the differences among combinations of dependent measures that would not otherwise be apparent. This type of question is served well by MANOVA's ability to detect multivariate differences, even when no single univariate test shows differences.

Selecting the Dependent Measures

In identifying the questions appropriate for MANOVA, it is also important to discuss briefly the development of the research question, specifically the selection of the dependent measures. A common problem encountered with MANOVA is the tendency of researchers to misuse one of its strengths—the ability to handle multiple dependent measures—by including variables without a sound conceptual or theoretical basis. The problem occurs when the results indicate that a subset of the dependent variables has the ability to influence the overall differences among groups. If some of the dependent measures with the strong differences are not really appropriate for the research question, then "false" differences may lead the analyst to draw incorrect conclusions about the *set as a whole*. Thus, the analyst should always scrutinize the dependent measures and make sure there is a solid rationale for including them. Any ordering of the variables, such as possible sequential effects, should also be noted. MANOVA provides a special test, stepdown analysis, to assess the statistical differences in a sequential manner, much like the addition of variables to a regression analysis.

In summary, the researcher should assess all aspects of the research question carefully and ensure that MANOVA is applied in the correct and most powerful way. In the following sections, we address many issues that have an impact on the validity and accuracy of MANOVA; however, it is ultimately the responsibility of the analyst to employ the technique properly.

Stage Two: Issues in the Research Design of MANOVA

Although MANOVA tests for assumptions in the same way as ANOVA and follows the same basic principles, several issues are unique in the application of MANOVA. The issues concern both the design and statistical testing of the MANOVA model.

Sample Size Requirements—Overall and by Group

As might be expected, MANOVA requires greater sample sizes than univariate ANOVAs and the sample size must exceed specific thresholds in each cell (group) of the analysis. Most important, the sample in each cell must be greater than the number of dependent variables included. Although this concern may seem minor, the inclusion of just a small number of dependent variables (from five to

[handwritten marginalia: must not be overspecified]

ten) in the analysis places a sometimes bothersome constraint on data collection. This is particularly a problem in field experimentation or survey research, where the researcher has less control over the achieved sample.

Factorial Designs—Two or More Treatments

Up to this point, we have discussed only situations in which there was a single treatment in either the ANOVA or MANOVA tests. But many times, the researcher wishes to examine the effects of several independent variables or treatments. An analysis with more than two treatments is called a **factorial design.** (In general, a design with n treatments is called an n-way factorial design.)

Selecting Treatments

The most common use of factorial designs are those research questions that relate two or more nonmetric independent variables to a set of dependent variables. In these instances, the independent variables have been specified in the design of the experiment or included in the design of the field experiment or survey questionnaire. But in some instances, treatments are added after the analysis is designed. The most common use of additional treatments is as a **blocking factor,** which is a nonmetric characteristic used post hoc to segment the respondents to obtain greater within-group homogeneity and reduce the MS_W source of variance. By doing so, the ability of the statistical tests to identify differences is enhanced. As an example, assume that in our earlier advertising example, it was discovered that males reacted differently to the advertisements than females. If gender is then used as a blocking factor, the differences between the messages may become more apparent, whereas the differences were obscured when males and females were assumed to react similarly and were not separated. The effects of message type and gender are then evaluated separately, providing a more precise test of their separate effects.

A Hypothetical Example

As an example of a simple two-treatment factorial design, assume that a cereal manufacturer wishes to examine the impact of three different color possibilities (red, blue, and green) and three different shapes (stars, cubes, and balls) on the overall consumer evaluation of a new cereal. We could examine the impact of both of these independent variables simultaneously by employing a 3 × 3 factorial design. Respondents would be randomly assigned to evaluate one of the nine possible combinations of color and shape (using, for example, a ten-point overall evaluation scale). In analyzing this design, three different overall effects can be tested with ANOVA:

1. The **main effect** of color: Are there any differences between the mean ratings given to red (i.e., including all ratings of red stars, red cubes, and red balls), blue, and green?
2. The main effect of shape: Are there any differences between the mean ratings given to stars (i.e., including all ratings of red stars, blue stars, and green stars), cubes, and balls?

3. The **interaction effect** of color and shape: As to the overall difference between colors, is this difference the same when we examine it separately for stars, cubes, and balls? For example, if red was rated very high overall but received a very low rating when it was rated as a ball (relative to blue and green), this outcome would be evidence of an interaction effect; that is, the effect of color depends on what shape we are considering. We could pose this interaction question in an equivalent fashion by asking whether the effect of shape depends on what color we are considering.

In ANOVA for factorial designs, each of these three effects would be tested with an F statistic. The MANOVA for factorial designs is a straightforward extension of ANOVA; that is, for every F statistic in ANOVA that evaluates an effect on a single dependent variable, there is a corrresponding multivariate statistic (e.g., gcr or Wilks' lambda) that evaluates the same effect on a set (vector) of dependent variable means.

Interpreting Interaction Terms

The interaction term represents the joint effect of two treatments and is the effect that must be examined first. If the interaction effect is not statistically significant, then the effects of the treatments are independent. Independence in factorial designs means that the effect of one treatment is the same for each level of the other treatment(s) and that the main effects can be interpreted directly. If the interaction term is significant, then the type of interaction must be determined. Interactions can be termed **ordinal** or **disordinal.** An ordinal interaction occurs when the effects of a treatment are not equal across all levels of another treatment, but the magnitude is always the same direction. In a disordinal interaction, the effects of one treatment are positive for some levels and negative for other levels of the other treatment.

The differences between interactions are best portrayed graphically. In Figure 5.3, we use the example of cereal shapes and colors described above. The vertical axis represents the mean evaluations of each group of respondents across the combinations of levels. In case A, there is no interaction. This is shown by the parallel lines representing the differences of the various shapes across the levels of color (the same effect would be seen if the differences in color were graphed across the three types of shape). In the case of no interaction, the effects of each treatment are constant at each level and the lines are roughly parallel. In case B, we see that the effects of each treatment are not constant and thus the lines are not parallel. The differences for red are large, but they decline slightly for blue cereal and even more for green cereal. Thus, the differences by color vary across the shapes. But the relative ordering among levels of shape are the same, with stars always highest, followed by the cubes and then the ball shapes. Finally, in case C, the differences in color not only vary in magnitude but also in direction. This is shown by lines that are not parallel and that cross between levels. For example, cubes have a higher evaluation than stars when the color is red, but the evaluation is lower for the colors of blue and green.

If the significant interactions are ordinal, the analyst must interpret the interaction term and ensure that its results are acceptable conceptually. If so, then the effects of each treatment can be described. But if the significant interaction is disordinal, then the main effects of the treatments cannot be interpreted and the

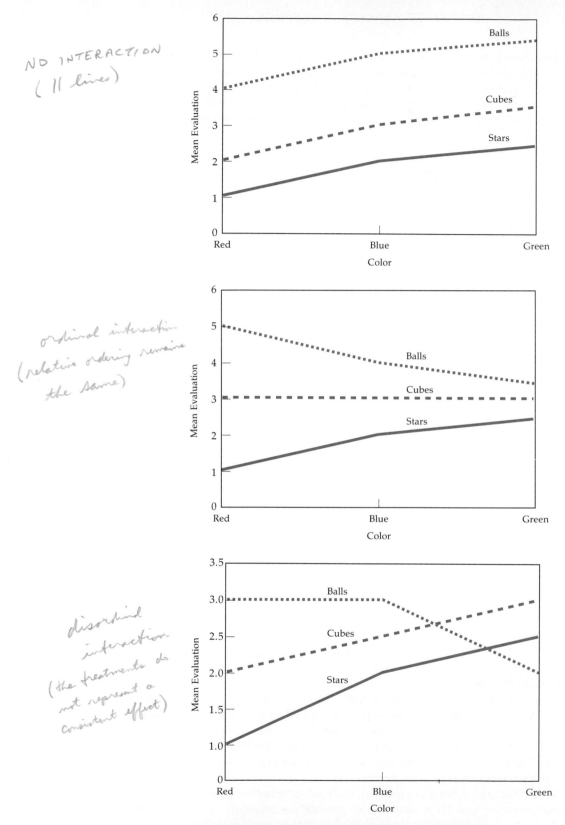

NO INTERACTION
(11 lines)

ordinal interaction
(relative ordering remains
the same)

disordinal
interaction
(the treatments do
not represent a
consistent effect)

FIGURE 5.3 Interaction effects in factorial designs.

study must be redesigned. This stems from the fact that with disordinal interactions, the effects vary not only across treatment levels but also in direction (positive or negative) as well. Thus, the treatments do not represent a consistent effect.

MANOVA Counterparts of Other ANOVA Designs

As the reader may be aware, many types of ANOVA designs are discussed in standard experimental design texts [13, 15, 16]. For every ANOVA design, there is a multivariate counterpart; that is, any ANOVA on a single dependent variable can be extended to MANOVA designs. To illustrate, we would have to discuss each ANOVA design in detail. Clearly, this is not possible in a single chapter, because entire books are devoted to the subject of ANOVA designs. For more information, the reader is referred to more statistically oriented texts [1, 4, 5, 7, 8, 9, 11, 16, 25].

Using Covariates—ANCOVA and MANCOVA

In any univariate ANOVA design, metric independent variables, referred to as **covariates,** can be included. The design is then termed an analysis of covariance (ANCOVA) design. Metric covariates are typically included in an experimental design to remove extraneous influences from the dependent variable, thus increasing the within-group variance (MS_W). Procedures similar to linear regression are employed to remove variation in the dependent variable associated with one or more covariates. Then a conventional ANOVA is carried out on the adjusted dependent variable.

Objectives of Covariance Analysis

A covariate analysis is appropriate to achieve two specific purposes: (1) to eliminate some systematic error outside the control of the researcher that can bias the results, and (2) to account for differences in the responses due to unique characteristics of the respondents. A systematic bias can be eliminated by the random assignment of respondents to various treatments. However, in nonexperimental research, such controls are not possible. For example, in testing advertising, effects may differ depending on the time of day or the composition of the audience and their reactions. The purpose of the covariate is to eliminate any effects that affect only a portion of the respondents. For instance, personal differences, such as attitudes or opinions, may affect responses, but the experiment does not include them as a treatment factor. The researcher uses a covariate to take out any differences due to these factors before the effects of the experiment are calculated. This is the second role of covariance analysis.

Selecting Covariates

An effective covariate in ANCOVA is one that is highly correlated with the dependent variable but not correlated with the independent variables; that is, variance in the dependent variable forms the basis of our error term in ANOVA. If our covariate is correlated with the dependent variable, we can explain some of the variance (through linear regression), and we are left with only residual variance in the dependent variable. This residual variance provides a smaller error term (MS_W) for the F statistic and thus a more efficient test of treatment effects.

A common question is how many covariates to add to the analysis. While the researcher wants to account for as many extraneous effects as possible, too large a number will reduce the statistical efficiency of the procedures. A rule of thumb is that the number of covariates should be less than (.10 × sample size) − (number of groups − 1). For example, for a sample size of 100 respondents and 5 groups, the number of covariates should be less than 6, that is, .10 × 100 − (5 − 1) [12]. The researcher should always attempt to minimize the number of covariates, while also ensuring that effective covariates are not eliminated, because in many cases, particularly with small sample sizes, they can markedly improve the sensitivity of the statistical tests.

There are two requirements for use of an analysis of covariance: (1) the covariates must have some relationship with the dependent measures, and (2) the covariates must have a homogeneity of regression effect, meaning that the covariate(s) have equal effects on the dependent variable across the groups. In regression terms, this implies equal coefficients for all groups. Statistical tests are available to assess whether this assumption holds true for each covariate used. If either of these requirements is not met, then the use of covariates is inappropriate.

Multivariate analysis of covariance (MANCOVA) is a simple extension of the principles of ANCOVA to multivariate (multiple dependent variables) analysis; that is, MANCOVA can be viewed as MANOVA of the regression residuals, i.e., variance in the dependent variables not explained by the covariates.

A Special Case of MANOVA: Repeated Measures

We have discussed a number of situations in which we wish to examine differences on several dependent measures. A special situation of this type occurs when the same respondent provides several measures, such as test scores over time, and we wish to examine them to see whether any trend emerges. Without special treatment, however, we would be violating the most important assumption, independence. There are special MANOVA models, termed **repeated measures,** that can account for this dependence and still ascertain whether any differences occurred across individuals for the set of dependent variables. The within-person perspective is important so that each person is placed on "equal footing." For example, assume we were assessing improvement on test scores over the semester. We must account for the earlier test scores and how they relate to later scores, as we might expect to see different trends for those with low versus high initial scores. Thus we must "match" each respondent's scores when performing the analysis. We will not address the details of repeated measures models in this text, because it is a specialized form of MANOVA. The interested reader is referred to any number of excellent treatments on the subject and the available statistical packages [2, 19, 22, 25, 27].

Stage Three: Assumptions of ANOVA and MANOVA

The univariate test procedures of ANOVA described in this chapter are valid (in a formal sense) only if it is assumed that the dependent variable is normally distributed and that variances are equal for all treatment groups. There is evidence [15,

27], however, that F tests in ANOVA are robust with regard to these assumptions except in extreme cases. For the multivariate test procedures of MANOVA to be valid, three assumptions must be met: The observations must be independent, the variance-covariance matrices must be equal for all treatment groups, and the set of p-dependent variables must follow a multivariate normal distribution (i.e., any linear combination of the dependent variables must follow a normal distribution) [11]. In addition to the strict statistical assumptions, the researcher must also consider several issues that influence the possible effects—namely, the linearity and multicollinearity of the variate of dependent variables. Each of these issues is addressed in the following sections.

Independence

The most basic, yet most serious, violation of an assumption occurs when there is a lack of **independence** among observations. There are a number of both experimental and nonexperimental situations in which this assumption can easily be violated. For example, a time-ordered effect (serial correlation) may occur if measures are taken over time, even from different respondents. Another common problem is gathering information in group settings, so that a common experience (such as a noisy room or confusing set of instructions) would cause a subset of individuals (those with the common experience) to have answers that are somewhat correlated. Finally, extraneous and unmeasured effects can affect the results by creating dependence among the respondents. Although there are no tests with an absolute certainty of detecting all forms of dependence, the researcher should explore all possible effects and correct for them if found. If dependence is found among groups of respondents, then a possible solution is to combine those within the groups and analyze the group's average score instead of the scores of the separate respondents. Another approach is to employ some form of covariate analysis to account for the dependence. In either case, or when dependence is suspected, the researcher should use a lower level of significance (.01 or even lower).

Equality of Variance-Covariance Matrices

The second assumption of MANOVA is the equivalence of covariance matrices across the groups. Here, as with the problem of heteroscedasticity we addressed in multiple regression, we are concerned with substantial differences in the amount of variance of one group versus another for the same variables. In MANOVA, however, the interest is in the variance/covariance matrices of the dependent measures for each group. The requirement of equivalence is a strict test, because instead of equal variances for a single variable in ANOVA, the MANOVA test examines all elements of the covariance matrix. For example, for five dependent variables, the five correlations and ten covariances are all tested for equality across the groups. Fortunately, a violation of this assumption has minimal impact if the groups are of approximately equal size (if the largest group size divided by the smallest group size is less than 1.5). If the sizes differ more than this, then the analyst should test and correct for unequal variances, if possible. MANOVA programs provide the test for equality of covariance matrices— typically the **Box test**—and provide significance levels for the test statistic. If the analyst encounters a significant difference that requires a remedy, try one of the many variance-stabilizing transformations available. The reader is referred to Chapter 2 for a discussion of these approaches. One note on the use of the Box test [11, 23]: It is very sensitive to departures from normality. Thus one should always check for univariate normality of all dependent measures before performing this test.

If the unequal variances persist after transformation and the group sizes differ markedly, the analyst should make adjustments for their effects. First, one has to ascertain which group has the largest variance. This determination is easily made either by examining the variance-covariance matrix or by using the determinant of the variance-covariance matrix, which is provided by all statistical programs. If the larger variances are found with the larger group sizes, the alpha level is overstated. This result means that differences should actually be assessed using a somewhat lower value (e.g., use .03 instead of .05). If the larger variance is found in the smaller group sizes, then the reverse is true. The power of the test has been reduced, and the analyst should increase the significance level.

Normality

Finally, the last assumption concerns normality of the dependent measures. In the strictest sense, the assumption is that all the variables are **multivariate normal.** Multivariate normality assumes that the joint effect of two variables is normally distributed. While this assumption underlies most multivariate techniques, there is no direct test for multivariate normality. Therefore, most analysts test for univariate normality of each variable. While univariate normality does not guarantee multivariate normality, if all variables meet this requirement, then any departures from multivariate normality are usually inconsequential. Violations of this assumption have little impact with larger sample sizes, just as is found with ANOVA. Violating this assumption primarily creates problems in applying the Box test, but transformations can correct these problems in most situations. If you are unfamiliar with transforming variables, please refer to Chapter 2. With moderate sample sizes, modest violations can be accommodated as long as the differences are due to skewness and not outliers.

Linearity and Multicollinearity Among the Dependent Variables

While MANOVA assesses the differences across combinations of dependent measures, it can construct a linear relationship only between the dependent measures (and any covariates, if included). The researcher is again encouraged first to examine the data, this time assessing the presence of any nonlinear relationships. If these exist, then the decision can be made whether they need to be incorporated into the dependent variable set, at the expense of increased complexity but greater representativeness. Again, if you are unfamiliar with such tests, refer to Chapter 2.

In addition to the linearity requirement, the dependent variables should not have high multicollinearity (discussed in Chapter 3), because this indicates only redundant dependent measures and decreases statistical efficiency.

Sensitivity to Outliers

In addition to the impact of heteroscedasticity discussed earlier, MANOVA (and ANOVA) are especially sensitive to outliers and their impact on Type I error. The researcher is strongly encouraged first to examine the data for outliers and eliminate them from the analysis if at all possible, because their impact will be disproportionate in the overall results.

Stage Four: Estimation of the MANOVA Model and Assessing Overall Fit

Once the MANOVA analysis has been formulated and the assumptions tested for compliance, the assessment of significant differences among the groups formed by the treatment(s) can proceed. In making this assessment, the analyst must select the test statistics most appropriate for the study objectives. Moreover, in any situation, but especially as the analysis becomes more complex, the analyst must evaluate the power of the statistical tests to provide the most informed perspective on the results obtained. Each of these issues is addressed in the following sections.

Criteria for Significance Testing

MANOVA presents the analyst with several criteria with which to assess multivariate differences across groups. The four most popular are Roy's greatest characteristic root; Wilks' lambda (also known as the U statistic); Hotelling's trace; and Pillai's criterion. As you may remember from the discussions of discriminant analysis, these criteria assess the differences across "dimensions" of the dependent variables (see Chapter 4 for a more detailed discussion). Roy's greatest characteristic root, as the name implies, measures the differences on only the first canonical root (or discriminant function) among the dependent variables. This criterion provides some advantages in power and specificity of the test but makes it less useful in certain situations where all dimensions should be considered. Roy's gcr test is most appropriate when the dependent variables are strongly interrelated on a single dimension, but it is also the measure most likely to be severely affected by violations of the assumptions.

The other three measures assess all sources of difference among the groups. Readers who have some familiarity with MANOVA from other texts or statistical programs probably have encountered a more commonly used test statistic for overall significance in MANOVA called **Wilks' lambda.** We have referred to the greatest characteristic root and the first discriminant function, and these terms imply that there may be additional characteristic roots and discriminant functions. Actually, where p is the number of dependent variables and k is the number of groups, there are p or $(k - 1)$ (whichever is the smaller) characteristic roots or discriminant functions. Unlike the gcr statistic, which is based on the first (greatest) characteristic root, Wilks' lambda considers all the characteristic roots; that is, it examines whether groups are somehow different without being concerned with whether they differ on at least one linear combination of the dependent variables. As it turns out, Wilks' lambda is much easier to calculate than the gcr statistic. Its formulation is $|W| \div |W + A|$ where $|W|$ is the determinant (a single number) of the within-groups multivariate dispersion matrix and $|W + A|$ is the determinant of the sum of W and A where A is the between-groups multivariate dispersion matrix. The larger the between-groups dispersion, the smaller the value of Wilks' lambda and the greater the implied significance. Although the distribution of Wilks' lambda is complex, good approximations for significance testing are available by transforming it into an F statistic [17].

Which statistic is preferred? Researchers have these two statistics plus a number of possible measures to choose from. Other widely used measures include

Pillai's criterion and Hotelling's trace, both of which are similar to Wilks' lambda, because they consider all the characteristic roots and can be approximated by an *F* statistic. The measure to use is the one most immune to violations of the assumptions underlying MANOVA and yet maintains the greatest power. There is agreement that either Pillai's criterion or Wilks' lambda best meets these needs, although evidence suggests Pillai's criterion is more robust and should be used if sample size decreases, unequal cell sizes appear, or homogeneity of covariances is violated. However, if the researcher is confident that all assumptions are strictly met and the dependent measures are representative of a single dimension of effects, then Roy's gcr is the most powerful test statistic. The major statistical packages provide all these measures, and a comparison among them can be made.

Statistical Power of the Multivariate Tests

In simple terms, **power** is the probability that the statistical test will identify a treatment's effect if it actually exists. Power can be defined as one minus the probability of a **Type II error** (β). As such, power is also related to the significance or alpha (α) level, which defines the acceptable Type I error. For a more detailed discussion of the relationships between Type I and Type II errors and power, the reader is referred to Chapter 1.

The level of power for any of the four statistical criteria discussed above is based on three considerations: the alpha level, the effect size of the treatment, and the sample size of the groups. Power is inversely related to the alpha level selected. As alpha increases (becomes more conservative, such as moving from .05 to .01), the power decreases. Therefore, if the analyst reduces the alpha level to reduce the Type I error, such as in the case of a possible dependence among observations or an adjustment for multiple comparisons, power decreases. The analyst must always be aware of the implications of adjusting the alpha level, because the overriding objective of the analysis is not only avoiding Type I errors but also identifying the treatment effects if they do indeed exist. If the alpha level is set too stringently, then the power may be too low to identify valid results. We suggest that the researcher consider not only the alpha level but also the resulting power, and try to maintain an acceptable alpha level with power in the range of .80.

But how does the analyst increase power once an alpha level has been specified? The primary "tool" at the analyst's disposal is the sample size of the groups. But before we assess the role of sample size, we need to understand the impact of **effect size,** which is a standardized measure of group differences, typically expressed as the differences in group means divided by their standard deviation. The magnitude of the effect size has a direct impact on the power of the statistical test. For any given sample size, the power of the statistical test will be higher the larger the effect size. Conversely, if a treatment has a small expected effect size, it is going to take a much larger sample size to achieve the same power as a treatment with a large effect size.

With the alpha level specified and the effect size identified, the final element affecting power is the sample size. In many instances, this is the element most under the control of the researcher. As discussed before, increased sample size generally reduces sampling error and increases the sensitivity (power) of the test. In analyses with group sizes of fewer than 50 members, obtaining desired power levels can be quite problematic. Increasing sample sizes in each group has marked effects until group sizes of approximately 150 are reached, and then the increases in power slow markedly. One note of caution must be made with large sample sizes as well. For many statistical tests, large sample sizes reduce the sampling error component to such a small level that any small difference is regarded as statistically significant. When the sample sizes do become large and statistical significance is indicated, the analyst must examine the power and effect sizes to ensure not only statistical significance but practical significance as well.

The estimation of power should be used both in planning the analysis and in assessing the results. In the planning stage, the analyst determines the sample size needed to identify the estimated effect size. In many instances, the effect size can be estimated from prior research or reasoned judgments, or even set at a minimum level of practical significance. In each case, the sample size needed to achieve a given level of power with a specified alpha level can be determined.

By assessing the power of the test criteria after the analysis has been completed, the analyst is providing a context for interpreting the results, especially if significant differences were not found. The analyst must first determine whether the achieved power was sufficient (.80 or above). If not, can the analysis be reformulated to provide more power? A possibility includes some form of blocking treatment or covariant analysis that will make the test more efficient by accentuating the effect size. If the power was adequate and statistical significance was not found for a treatment effect, then most likely the effect size for the treatment was too small to be of statistical or practical significance.

To calculate power for ANOVA analyses, both published sources [6, 24] and computer programs [3] are now available. The methods of computing the power of MANOVA, however, are much more limited. Fortunately, most computer programs provide an assessment of power for the significance tests and allow the researcher to determine whether power should play a role in the interpretation of the results. In terms of published material for planning purposes, little exists for MANOVA because many elements affect the power of a MANOVA analysis. One source [14] of published tables presents the power in a number of common situations where MANOVA is applied. Table 5.1 provides an overview of the sample sizes needed for various levels of analysis complexity. A review of the table leads to several general points. First, increasing the number of dependent variables requires increased sample sizes to maintain a given level of power. The additional

TABLE 5.1 Sample Size Requirements per Group for Achieving Statistical Power of .80 in MANOVA

	Number of Groups											
	3				4				5			
	Number of Dependent Variables				Number of Dependent Variables				Number of Dependent Variables			
Effect Size	2	4	6	8	2	4	6	8	2	4	6	8
Very large	13	16	18	21	14	18	21	23	16	21	24	27
Large	26	33	38	42	29	37	44	48	34	44	52	58
Medium	44	56	66	72	50	64	74	84	60	76	90	100
Small	98	125	145	160	115	145	165	185	135	170	200	230

Source: [14]

sample size needed is more pronounced for the smaller effect sizes. Second, if the effect sizes are expected to be small, the analyst must be prepared to engage in a substantial research effort to achieve acceptable levels of power. For example, to achieve the suggested power of .80 when assessing small effect sizes in a four-group design, 115 subjects per group are required if two dependent measures are used. The required sample size increases to 185 per group if eight dependent variables are considered. The benefits of parsimony in the dependent variable set occur not only in interpretation but in the statistical tests for group differences as well.

Stage Five: Interpretation of the MANOVA Results

Once the statistical significance of the treatments has been assessed, the analyst may wish to examine the results through any one of three methods: (1) interpreting the effects of covariates if employed, (2) assessing which dependent variable(s) exhibited differences across the groups, or (3) identifying which groups differ on a single dependent variable or the entire dependent variate. The following sections will first examine the methods by which the significant covariates and dependent variables are identified and then address the methods by which differences among individual groups can be measured.

Evaluating Covariates

Having met the assumptions for applying covariates, the analyst may wish to interpret the actual effect of the covariates on the dependent variate and their impact on the actual statistical tests of the treatments. Because ANCOVA/MANCOVA are an application of regression procedures within the analysis of variance method, assessing the impact of the covariates on the dependent variables is quite similar to examining regression equations. For each covariate, a regression equation that details the strength of the predictive relationship is formed. If the covariates represent theoretically based effects, then these results provide an objective basis for accepting or rejecting the proposed relationships. In a practical vein, the analyst can examine the impact of the covariates and eliminate those with little or no effect.

The analyst should also examine the overall impact of adding the covariate(s) in the statistical tests for the treatments. The most direct approach is to run the analysis with and without the covariates. Effective covariates will improve the statistical power of the tests and reduce within-group variance. If the analyst does not see any substantial improvement, then the covariates may be eliminated, because they reduce the degrees of freedom available for the tests of treatment effects. This approach also can identify those instances in which the covariate is "too powerful" and reduces the variance to such an extent that the treatments are all nonsignificant. This occurs many times when a covariate is included that is correlated with one of the independent variables and thus "removes" this variance, thereby reducing the explanatory power of the independent variable.

Assessing the Dependent Variate

The next step is an analysis of the dependent variate to assess which of the dependent variables contribute to the overall differences indicated by the statistical tests. This is essential because a set of variables may be identified that either accentuates the differences while the other variables are nonsignificant or masks the significant effects of the remainder. The procedures described in this section are termed **post hoc tests**—tests that are decided on after examining the pattern of the data. Another approach is the use of **a priori tests**—tests planned prior to looking at the data from a theoretical or practical decision-making viewpoint. From a pragmatic standpoint, situations arise wherein a key dependent variable must be isolated and tested with maximum power. We recommend that an a priori test be performed in such situations. The most common approach is to perform univariate tests for the selected variables. For example, in a two-group case, an ordinary t test is an a priori test for a given dependent variable. However, researchers should be aware that as the number of these a priori tests increases, one of the major benefits of the multivariate approach to significance testing — control of the Type I error rate—is negated unless specific adjustments are made that control for the inflation of the Type I error. The next section describes how these adjustments can be made.

For a two-group MANOVA, this involves an adjustment of the T^2 statistic. Given that the T^2 statistic exceeds T^2_{crit} for a specified α level, we concluded that the vectors of the mean scores are different. The discriminant function (if computed) tells us what linear combination of the dependent variables produces the most reliable group difference, but other group comparisons may also be of interest. If we wished to test the group differences individually for each of the dependent variables, we could compute a standard t statistic and compare it to the square root of T^2_{crit} (i.e., T_{crit}) to judge its significance. This procedure would ensure that the probability of any Type I error across all the tests would be held to α (where α was specified in the calculation of T^2_{crit}) [11]. We could make similar tests for k-group situations by adjusting the α level by the **Bonferroni inequality,** which states that the alpha level should be adjusted for the number of tests being made. The adjusted alpha level used in any separate test is defined as the overall alpha level divided by the number of tests (adjusted alpha = overall alpha ÷ number of tests).

A procedure known as **stepdown analysis** [23] may also be used to assess individually the differences of the dependent variables. This procedure involves computing a univariate F statistic for a dependent variable after eliminating the effects of other dependent variables preceding it in the analysis. The procedure is somewhat similar to stepwise regression, but here we examine whether a particular dependent variable contributes unique (uncorrelated) information on group differences. The stepdown results would be exactly the same as performing a covariate analysis with the other preceding dependent variables used as the covariates. A critical assumption of stepdown analysis is that the analyst knows the order in which the dependent variables should be entered, because the interpretations can vary dramatically given different entry orders. If the ordering has theoretical support, then the stepdown test is valid. Variables indicated to be nonsignificant are "redundant" with the earlier significant variables, as they add no further information concerning differences about the groups.

Other procedures involve further analysis of the discriminant functions, in particular the first discriminant function, to gain additional information about which variables best differentiate between the groups. All these analyses are directed toward assisting the analyst in understanding which of the dependent variables contribute to the differences in the dependent variate across the treatment(s).

Identifying Differences Between Individual Groups

Although the univariate and multivariate tests of ANOVA and MANOVA allow us to reject the null hypothesis that the groups' means are all equal, they do not pinpoint where the significant differences lie. Multiple t tests are not appropriate for testing the significance of differences between the means of paired groups because the probability of a Type I error increases with the number of intergroup comparisons made (similar to the problem of using multiple univariate ANOVAs versus MANOVA). Many procedures are available for further investigation of specific group mean differences of interest, all of which can be classified as either a priori or post hoc. These procedures use different approaches to control Type I error rates across multiple tests.

Post Hoc Methods

Among the more common post hoc procedures are (1) Scheffe's test, (2) Tukey's honestly significant difference (HSD) method, (3) Tukey's extension of the Fisher least significant difference (LSD) approach, (4) Duncan's multiple-range test, and (5) the Newman-Kuels test. Each method identifies which comparisons among groups (e.g., group 1 versus groups 2 and 3) have significant differences. They provide the analyst with tests of each combination of groups, thus simplifying the interpretative process.

While they simplify the identification of group differences, these methods all share the problem of having quite low levels of power. Because the post hoc tests must examine all possible combinations, the power of any individual test is rather low. These five post hoc or multiple-comparison tests of significance have been contrasted for power [23]. The conclusions are that Scheffe's test is the most conservative with respect to Type I error. The remaining tests are ranked in this order: Tukey HSD, Tukey LSD, Newman-Kuels, and Duncan. If the effect sizes are large or the number of groups is small, the post hoc methods may identify the group differences. But the analyst must also recognize the limitations of these methods and employ other methods if more specific comparisons can be identified. A discussion of the options available with each method is beyond the scope of this chapter. Excellent discussions and explanations of these procedures can be found in several texts [12, 27].

A Priori or Planned Comparisons

The analyst can also make specific comparisons between groups by using a priori or **planned comparisons.** This method is similar to the post hoc tests described above but differs in that the analyst specifies which group comparisons are to be made in the planned comparisons versus testing the entire set, as done in the post hoc tests. The planned comparisons are more powerful because the number of comparisons are fewer, but more power is of little use if the analyst does not specifically test for "correct" group comparisons. Planned comparisons are most appropriate when conceptual bases can support the specific comparisons to be made. Planned comparisons should not be used in an exploratory manner because they do not have effective controls against inflating the overall Type I error levels.

The analyst specifies the groups to be compared through a **contrast,** which is just a combination of group means that represent a specific planned comparison. Contrasts can be stated generally as

$$C = W_1G_1 + W_2G_2 + \cdots + W_kG_k$$

where

$$C = \text{Contrast value}$$
$$W = \text{Weights}$$
$$G = \text{Group means}$$

The contrast is formulated by assigning positive and negative weights to specify the groups to be compared while ensuring that the weights sum to zero. For example, assume we have three group means. To test for a difference between G_1 and G_2, $C = (1)G_1 + (-1)G_2 + (0)G_3$. To test whether the average of G_1 and G_2 differs from G_3, the contrast is specified as $C = (.5)G_1 + (.5)G_2 + (-1)G_3$. A separate F statistic is computed for each contrast. In this manner, the analyst can create any comparisons desired and test them directly, but the probability of a Type I error for each a priori comparison is equal to α. Thus, several planned comparisons will inflate the overall Type I error level. All the statistical packages can perform either a priori or post hoc tests for single dependent variables.

If the analyst wishes to perform comparisons of the entire dependent variate, extensions of these methods are available. After concluding that the group mean vectors are not equivalent, the analyst might be interested in whether there are any group differences on the composite dependent variate. A standard ANOVA F statistic can be calculated and compared to $F_{crit} = (N - k)\text{gcr}_{crit}/(k - 1)$, where the value of gcr_{crit} is taken from the gcr distribution with appropriate degrees of freedom.

Stage Six: Validation of the Results

Analysis of variance techniques (ANOVA and MANOVA) were developed in the tradition of experimentation, with **replication** as the primary means of validation. The specificity of experimental treatments allows for a widespread use of the same experiment in multiple populations to assess the generalizability of the results. This is a principal tenet of the scientific method. In social science and business research, however, true experimentation is many times replaced with statistical tests in nonexperimental situations like survey research. The ability to validate the results in these situations is based on the replicability of the treatments. In many instances, demographic characteristics such as age, gender, income, and the like are used as treatments. These treatments may seem to meet the requirement of comparability, but the analyst must ensure that the additional element of randomized assignment to cell is also met; however, many times in survey research this is not true. For example, using age and gender as the independent variables is a common example of the use of ANOVA/MANOVA in survey research. But in terms of validation, the analyst must be wary of analyzing multiple populations and comparing results as the sole proof of validity. Because respondents in a simple sense "select themselves," the treatments in this case cannot be assigned by the researcher and thus randomized assignment is impossible. So the analyst should strongly consider the use of covariates to control for other characteristics that might be characteristic of the age/gender groups that could affect the dependent variables but are not included in the analysis.

Another issue is the claim of causation when experimental methods or techniques are employed. The principles of causation are examined in more detail in Chapter 11. For our purposes here, the analyst must remember that in all research settings, including experiments, certain conceptual criteria (e.g., temporal ordering of effects and outcomes) must be established before causation may be supported. The single application of a particular technique used in an experimental setting does not ensure causation.

Summary

We have discussed the appropriate applications and important considerations of MANOVA in addressing multivariate analyses with multiple dependent measures. Although there are considerable benefits from its use, MANOVA must be carefully and appropriately applied to the question at hand. When doing so, analysts have at their disposal a technique with flexibility and statistical power. We will now illustrate the application of MANOVA (and its univariate counterpart ANOVA) in a series of examples.

Example 1: Difference Between Two Independent Groups

To introduce the practical benefits of a multivariate analysis of group differences, we begin our discussion with one of the best-known experimental designs: the two-group randomized design, in which each respondent is randomly assigned to only one of the two levels (groups) of the treatment (independent variable). In the univariate case, a single metric dependent variable is measured, and the null hypothesis is that the two groups have equal means. In the multivariate case, multiple metric dependent variables are measured, and the null hypothesis is that the two groups have equal vectors of means. For a two-group univariate analysis, the appropriate test statistic is the t statistic (a special case of ANOVA); for a multivariate analysis, the appropriate test statistic is Hotelling's T^2. Because both analyses employ the same context, they do not differ in the appropriate approaches for validation, which essentially involve replication in other samples or through a split sample. Therefore, validation issues (stage six) will not be discussed for each of the separate analyses.

A Univariate Approach: The t *Test*

Stage One: Objectives of the Analysis

HATCO decided to randomly survey some of their customers (firms that they supply) and to assess various measures of HATCO's performance. Of particular interest was whether HATCO performed better when specification buying or total-value analysis was used by the customers (refer to Chapter 1 for a complete description of the variables and the entire study). As a measure of performance, HATCO decided to use X_9, the percentage of the customer's business that comes from HATCO. The higher the value, the more HATCO supplies, on a percentage basis, to that firm. The null hypothesis that HATCO wishes to test is that they serve both types of firms equally well; that is, the type of buying methodology has no impact on usage level (H_0: $\mu_1 = \mu_2$). The alternative hypotheses is that the buying methodology does define groups who buy in significantly different proportions from HATCO (i.e., H_A: $\mu_1 < \mu_2$ or H_A: $\mu_1 > \mu_2$).

Stage Two: Research Design of the ANOVA

The principal consideration in the design of the two-group ANOVA is the sample size in each of the cells. As is the case in most survey research, the cells sizes are unequal. Upon completion of the survey, 40 firms indicated that they used specification buying, and 60 firms used total-value analysis. Unequal cell sizes make the statistical tests more sensitive to violations of the assumptions, especially the test for homogeneity of variance of the dependent variable. HATCO researchers did not identify any variables appropriate for inclusion as covariates. Finally, additional independent variables that would create a factorial design were deemed unsuitable at this time.

Stage Three: Assumptions in ANOVA

The independence of the respondents was ensured as much as possible by the random sampling plan. The assumption of normality and the presence of outliers for the dependent variable, X_9, were examined in Chapter 2 and found to be acceptable. The assumption particularly important to ANOVA is the homogeneity of the variance of the dependent variable between groups. Several tests are available for testing this assumption. The Levene statistic indicates no difference (significance = .2434), as do the Cochran's C (significance = .396) and the Bartlett-Box (significance = .411) tests. Thus, the unequal cell sizes should not impact the sensitivity of the statistical tests of group differences.

Stage Four: Estimation of the ANOVA Model and Assessing Overall Fit

Usage, measured in percentage terms (0 to 100%), is shown for all firms in Table 5.2. The boxplots in Figure 5.4 portray the usage levels for respondents in two groups—customers using specification buying and customers using total-value analysis. As can be seen, the customers using specification buying conducted an average of 42.1 percent of their business with HATCO, while those using

TABLE 5.2 Firm Usage Level by Buying Method (Basic Data for Univariate *t* Test or Two-Group ANOVA)

Buying Method

Group 1: Specification Buying		Group 2: Total-Value Analysis	
Firm	X_9 (Usage Level)	Firm	X_9 (Usage Level)
2	43.0	1	32.0
3	48.0	5	58.0
4	32.0	7	46.0
6	45.0	9	63.0
8	44.0	11	32.0
10	54.0	12	47.0
13	39.0	14	38.0
24	36.0	15	54.0
27	36.0	16	49.0
30	46.0	17	38.0
31	43.0	18	40.0
34	47.0	19	54.0
35	35.0	20	55.0
36	39.0	21	41.0
37	44.0	22	35.0
39	29.0	23	55.0
40	28.0	25	49.0
41	40.0	26	49.0
45	38.0	28	54.0
48	43.0	29	49.0
52	53.0	32	53.0
53	50.0	33	60.0
54	32.0	38	46.0
57	62.0	42	58.0
60	50.0	43	53.0
65	40.0	44	48.0
68	46.0	46	54.0
70	49.0	47	55.0
71	50.0	49	57.0
75	41.0	50	53.0
79	39.0	51	41.0
83	41.0	55	39.0
85	53.0	56	47.0
86	43.0	58	65.0
87	51.0	59	46.0
89	34.0	61	54.0
94	36.0	62	60.0
96	25.0	63	47.0
98	38.0	64	36.0
99	42.0	66	45.0
		67	59.0
		69	58.0
		72	55.0
		73	51.0
		74	60.0
		76	49.0
		77	42.0

TABLE 5.2 *Continued*

Group 1: Specification Buying		*Group 2: Total-Value Analysis*	
Firm	X_9 *(Usage Level)*	*Firm*	X_9 *(Usage Level)*
		78	47.0
		80	56.0
		81	59.0
		82	47.0
		84	37.0
		88	36.0
		90	60.0
		91	49.0
		92	39.0
		93	43.0
		95	31.0
		97	60.0
		100	33.00
Means	42.10		48.77
Variances	60.66		77.40
Sample sizes	$N_1 = 40$		$N_2 = 60$

Calculating the t *statistic*

Standard error:* $\sqrt{\dfrac{77.40}{60} + \dfrac{60.66}{40}} = 1.675$

t statistic: $\dfrac{48.77 - 42.10}{1.675} = 3.98$ $\dfrac{\mu_1 - \mu_2}{\sigma}$

---

*This formula for standard error is appropriate for equal cell sizes or where all cells have greater than 30 observations. For situations with unequal cell sizes or small cell sizes (fewer than 30), see [21].

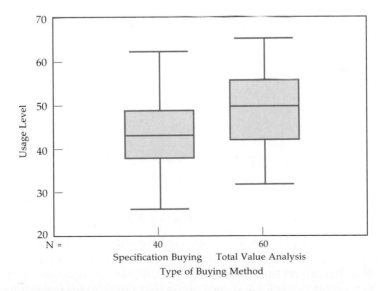

FIGURE 5.4 Boxplots of usage level for two-group ANOVA.

total-value analysis gave HATCO 48.77 percent. The t test analysis will examine the difference between groups and statistically test for the equality of the two group means.

To conduct the test, we first choose the significance level .05 (the maximum allowable Type I error rate). Thus, before we conduct the study, we know that 5 times out of 100 we might conclude that the buying methodology had an impact on the firm's usage rate when in fact it did not. Although all statistical programs will automatically calculate the significance levels of the differences, we will illustrate how the calculations are performed. To determine the value for t_{crit}, we refer to the t distribution with $40 + 60 - 2 = 98$ degrees of freedom and $\alpha = .05$. We find that $t_{crit} = 1.66$. Next, we compute the value of our t statistic. As shown at the bottom of Table 4.1, $t = 3.98$. Because this exceeds t_{crit}, we conclude that the buying methodology does affect a firm's usage rate of HATCO products. The statistical power of the test is .97, ensuring that the difference found is statistically significant.

Stage Five: Interpretation of the Results

The single dependent variable and the presence of only two groups eliminates the need to examine either the dependent variate or the differences between the groups in addition to the overall tests described in stage four. The analyst can report statistically significant percentages of business from the HATCO customers using the two buying methods. Firms using total-value analysis buy a significantly greater percentage of their products from HATCO than do firms using specification buying. The analyst must assess, however, whether the difference of approximately 6 percent has practical significance for managerial decision making.

A Multivariate Approach: Hotelling's T^2

It is probably unrealistic to assume that a difference between any two experimental groups will be manifested only in a single dependent variable. For example, two advertising messages not only may produce different levels of purchase intent but also may affect a number of other (potentially correlated) aspects of the response to advertising (e.g., overall product evaluation, message credibility, interest, attention). Many researchers handle this multiple-criterion situation by repeated application of individual univariate t tests until all the dependent variables have been analyzed. This approach has serious deficiencies. As discussed earlier, consider what might happen to the Type I error rate (inflation over multiple t tests) and the inability of paired t tests to detect differences among combinations of the dependent variables that are not apparent in univariate tests.

Stage One: Objectives of the MANOVA

In our univariate example, HATCO compared the usage level of firms (X_9 is the dependent variable) utilizing different buying methodologies (X_{11} is the independent variable). To convert this example into a multivariate example, we require at least two dependent variables. Let us assume that HATCO was also interested in the satisfaction levels of firms with the two buying approaches. We would now select X_{10} (satisfaction with HATCO) as the second dependent variable (see Table

5.3). The null hypothesis HATCO is now testing is that the vectors of the mean scores for each group are equivalent (i.e., that buying method has no effect on either usage or satisfaction).

Stage Two: Research Design of MANOVA

The primary consideration in a two-group MANOVA is still the sample size for each of the groups. As discussed in the univariate example, the group sizes are 60 and 40, exceeding both the minimum and recommended sizes. These sample sizes should be adequate to provide the recommended power of .80 for at least medium effect sizes.

Stage Three: Assumptions in MANOVA

Before calculating the test statistics for mean differences across the groups, the analyst must first determine if the dependent measures are significantly correlated. The most widely used test for this purpose is Bartlett's test for sphericity [19, 22]. It examines the correlations among all dependent variables and assesses whether, collectively, significant intercorrelation exists. In our example, a significant degree of intercorrelation does exist (.657) (see Table 5.4).

The other critical assumption concerns the homogeneity of the variance/covariance matrices among the two groups. The first analysis assesses the univariate homogeneity of variance across the two groups. As shown in Table 5.4, univariate tests for both variables are nonsignificant. The next step is to assess the dependent variables collectively by testing the equality of the entire variance/covariance matrices between the groups. As shown in Table 5.4, a difference is seen in the correlation among the two dependent variables in the two groups (.823 for the customers using specification buying and .559 for those using total-value analysis). This illustrates the differing levels of covariance. The test for overall equivalence of the variance/covariance matrices is the Box's M test, which in this example has a significance level of .01. Given the sensitivity of this test, the significance level is deemed acceptable and the analysis proceeds.

Stage Four: Estimation of the MANOVA Model and Assessing Overall Fit

The means of each deendent variable (usage level and satisfaction) for the two groups of firms were presented in Table 5.3. To conduct the test, we again specify our significance level (.05 in this example) as the maximum allowable Type I error. To determine the value for T_{crit}^2, we refer to the F distribution with 2 and 97 degrees of freedom. With an F_{crit} of 3.09, the T_{crit}^2 can be calculated as follows:

$$T_{crit}^2 = \frac{p(N_1 + N_2 - 2)}{N_1 + N_2 - p - 1} \times F_{crit}$$

$$= \frac{2(60 + 40 - 2)}{60 + 40 - 2 - 1} \times 3.09$$

$$= 6.24$$

As shown at the bottom of Table 5.3, the computed value of Hotelling's T^2 is 26.33. Because this exceeds T_{crit}^2, we reject the null hypothesis and conclude that buying method has had some impact on the set of dependent measures. Moreover, the

TABLE 5.3 Firm Usage and Satisfaction by Buying Method (Basic Data for Hotelling's T^2* or Two-Group MANOVA)

			Buying Method		
	Group 1: Specification Buying			*Group 2: Total-Value Approach*	
Firm	X_9 (Usage Level)	X_{10} (Satisfaction)	Firm	X_9 (Usage Level)	X_{10} (Satisfaction)
2	43.0	4.3	1	32.0	4.2
3	48.0	5.2	5	58.0	6.8
4	32.0	3.9	7	46.0	5.8
6	45.0	4.4	9	63.0	5.4
8	44.0	4.3	11	32.0	4.3
10	54.0	5.4	12	47.0	5.0
13	39.0	4.4	14	38.0	5.0
24	36.0	3.7	15	54.0	5.9
27	36.0	3.7	16	49.0	4.7
30	36.0	5.1	17	38.0	4.4
31	43.0	3.3	18	40.0	5.6
34	47.0	3.8	19	54.0	5.9
35	35.0	4.1	20	55.0	6.0
36	39.0	3.6	21	41.0	4.5
37	44.0	4.8	22	35.0	3.3
39	29.0	3.9	23	55.0	5.2
40	28.0	3.3	25	49.0	4.9
41	40.0	3.7	26	49.0	5.9
45	38.0	3.2	28	54.0	5.8
48	43.0	4.7	29	49.0	5.4
52	53.0	5.2	32	53.0	5.0
53	50.0	5.5	33	60.0	6.1
54	32.0	3.7	38	46.0	5.1
57	62.0	6.2	42	58.0	6.7
60	50.0	5.0	43	53.0	5.9
65	40.0	3.4	44	48.0	4.8
68	46.0	4.5	46	54.0	6.0
70	49.0	4.8	47	55.0	4.9
71	50.0	5.4	49	57.0	4.9
75	41.0	4.4	50	53.0	3.8
79	39.0	3.3	51	41.0	5.0
83	41.0	4.1	55	39.0	3.7
85	53.0	5.6	56	47.0	4.2
86	43.0	3.7	58	65.0	6.0
87	51.0	5.5	59	46.0	5.6
89	34.0	4.0	61	54.0	4.8
94	36.0	3.6	62	60.0	6.1
96	25.0	3.4	63	47.0	5.3
98	38.0	3.7	64	36.0	4.2
99	42.0	4.3	66	45.0	4.9
			67	59.0	6.0
			69	58.0	4.3
			72	55.0	3.9
			73	51.0	4.9
			74	60.0	5.1
			76	49.0	5.2
			77	42.0	5.1

TABLE 5.3 *Continued*

	Group 1: Specification Buying			Group 2: Total-Value Approach	
Firm	X_9 (Usage Level)	X_{10} (Satisfaction)	Firm	X_9 (Usage Level)	X_{10} (Satisfaction)
			78	47.0	5.1
			80	56.0	5.1
			81	59.0	4.5
			82	47.0	5.6
			84	37.0	4.4
			88	36.0	4.3
			90	60.0	6.1
			91	49.0	4.4
			92	39.0	5.5
			93	43.0	5.2
			95	31.0	4.0
			97	60.0	5.2
			100	33.0	4.4
Means	42.10	4.30		48.77	5.09
Variances	60.66	.612		77.40	.569
Sample size	$N_1 = 40$			$N_2 = 60$	

*Hotelling's $T^2 = 26.33$.

TABLE 5.4 Diagnostic Information for Two-Group MANOVA

Test of Assumptions: Homogeneity of Variance

Variance/Covariance Matrices
(values in parentheses are correlations)

	Group 1: Specification Buying		Group 2: Total Value Analysis	
	X_9	X_{10}	X_9	X_{10}
X_9 Usage Level	60.656		77.402	
X_{10} Satisfaction	5.011	.611	3.707	.569
	(.823)		(.559)	

	X_9: Usage Level		X_{10}: Satisfaction		Overall	
	Statistic	Signif.	Statistic	Signif.	Statistic	Signif.
Univariate Tests						
Cochran's C	.561	.396	.518	.803		
Bartlett-box	.517	.411	.060	.807		
Levene test	1.38	.243	.323	.571		
Multivariate Test						
Box's M					11.68	.010

Test of Assumption: Correlation of Dependent Variables

	Statistic	Signif.
Bartlett test of sphericity	54.47	.000

power for the multivariate test was almost 1.0, indicating that the sample sizes and the effect size were sufficient to ensure that the significant differences would be detected if they existed beyond the differences due to sampling error.

Stage Five: Interpretation of the Results

Given the significance of the multivariate test indicating group differences on the dependent variate (vector of means), the analyst must examine the results to assess their logical consistency. The group using total-value analysis not only has a higher usage level (refer to Figure 5.4) but also has a higher level of satisfaction (Figure 5.5). The question the analyst must now assess is whether both dependent variables are significantly different or whether the results are derived mainly from differences of only one of the two dependent variables.

One post hoc test of obvious interest is whether the buying method had an impact on usage level (X_9) or on satisfaction (X_{10}), each considered separately. The group means and univariate tests of mean differences are as follows:

	Dependent variables	
	X_9 *Usage Level*	X_{10} *Satisfaction*
Group Means		
Total-value analysis	48.77	5.09
Specification buying	42.10	4.30
Difference	+6.67	+0.79
Univariate Test of Group Differences		
t statistic	3.96	5.03
Significance level	.000	.000

The *t* statistic was previously computed as 3.96 for the difference in usage level, and as shown above, the *t* statistic for the difference in satisfaction levels is 5.03. Both of the *t* statistics exceed the square root of T^2_{crit} ($\sqrt{6.24} = 2.50$). Thus we can conclude that the buying method had a positive impact on usage level and satisfaction. In our use of T^2, we also are confident that the probability of a Type I error is held to 5 percent across both of the post hoc tests.

A second analysis of the dependent variate is the stepdown test, which examines the significance of group differences while allowing for dependent variable intercorrelation. Table 5.5 shows that both X_9 and X_{10} are significantly different, even when controlling for their intercorrelation. Thus, after examining the results and the tests described above, the analyst can safely conclude that the two groups, both collectively and individually, differed significantly on both variables.

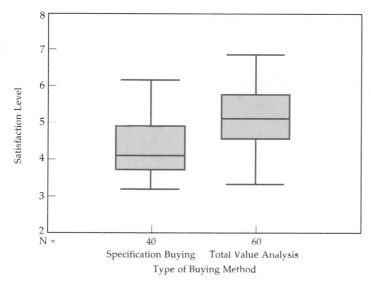

FIGURE 5.5 Boxplots of second dependent variable, satisfaction, for two-group MANOVA.

Example 2: Difference Between k Independent Groups

The two-group randomized design (example 1) is a special case of the more general k-group randomized design. In the general case, each respondent is randomly assigned to one of k levels (groups) of the treatment (independent variable). In the univariate case, a single metric dependent variable is measured, and the null hypothesis is that all group means are equal (i.e., $\mu_1 = \mu_2 = \mu_3 = \ldots = \mu_k$). In the multivariate case, multiple metric dependent variables are measured, and the null hypothesis is that all group vectors of mean scores are equal (i.e., $\mu_1 = \mu_2 = \mu_3 = \ldots = \mu_k$, where μ refers to a vector or set of mean scores). For a univariate analysis, the appropriate test statistic is the F statistic resulting from ANOVA. For a multivariate analysis, we will examine two of the more widely used test statis-

TABLE 5.5 Stepdown Tests for Two-Group MANOVA

Roy-Bargman Stepdown F tests

Variable	Between-Groups Mean Square	Within-Groups Mean Square	Stepdown F	Degrees of Freedom Between	Within	Significance of Stepdown F
X_9	1066.667	70.738	15.08	1	98	.000
X_{10}	3.246	.336	9.65	1	97	.002

tics including the greatest characteristic root (gcr) statistic (also known as Roy's largest root) and Wilks' lambda (also referred to as Wilks' likelihood ratio criterion or the U statistic).

A Univariate Approach: k-Groups ANOVA

Stage One: Objectives of the ANOVA

The HATCO survey also asked the customers to classify the type of purchases being made with HATCO (X_{14}) as primarily a new-task situation, a modified rebuy, or a straight rebuy (see Chapter 1 for a more detailed discussion of these data). HATCO is also interested in knowing whether usage varies across the type of buying situation. The overall null hypothesis HATCO now wishes to test is that $\mu_1 = \mu_2 = \mu_3$ (i.e., all three groups are equivalent in their usage level).

Stage Two: Research Design of ANOVA

In the sample, 34 firms indicated that the new-task buying situation best characterized their relationship with HATCO, 32 firms responded with modified rebuy, and 34 firms said straight rebuy. These sample sizes are adequate to obtain sufficient power with medium or larger effect sizes (refer to Table 5.1). If the effect size was small or the samples sizes were to decrease owing to missing data or other factors, the power would fall below recommended levels. The analyst would then need to carefully assess the statistical tests for statistical power and the practical significance of the differences.

Stage Three: Assumptions in ANOVA

The univariate tests for homogeneity of variance for X_9, usage level, across the three groups shows no significant differences with any of the three tasks (see Table 5.6). Thus, the analyst again can eliminate, as was done in the two-group example, the varying group sizes as impacts on the statistical tests of group differences.

Stage Four: Estimation of the ANOVA
Model and Assessing Overall Fit

The ANOVA model tests for the differences in group means among those HATCO customers using one of the three buying methods. Figure 5.6 contains a graphical depiction of the responses by group, while the usage levels of all 100 firms by group are displayed in Table 5.6. To conduct the test by hand, we specify .05 as the Type I error rate. To determine the value for F_{crit}, we refer to the F distribution with $(3 - 1) = 2$ and $(100 - 3) = 97$ degrees of freedom with $\alpha = .05$. We find that $F_{crit} = 3.09$. The calculation of the F statistic from ANOVA is usually summarized in an ANOVA table similar to that shown in Table 5.7. The mean square values for both between-groups and within-groups variances are calculated as the sum of squares (the sum of squared deviations) divided by the appropriate degrees of freedom. As shown in Table 5.7, the resulting F statistic = 106.66 (2749.833 ÷ 25.776). Because this exceeds F_{crit}, we can conclude that all group means are not equal.

TABLE 5.6 Firm Usage Rate by Type of Buying Situation (Basic Data for Three-Group ANOVA)

Buying Situation

Group 1: New Task		Group 2: Modified Rebuy		Group 3: Straight Rebuy	
Firm	X_9 (Usage Level)	Firm	X_9 (Usage Level)	Firm	X_9 (Usage Level)
1	32.0	3	48.0	5	58.0
2	43.0	6	45.0	9	63.0
4	32.0	8	44.0	15	54.0
7	46.0	10	54.0	16	49.0
11	32.0	12	47.0	19	54.0
13	39.0	17	38.0	20	55.0
14	38.0	18	40.0	23	55.0
22	35.0	21	41.0	26	49.0
24	36.0	25	49.0	28	54.0
27	36.0	30	46.0	29	49.0
31	43.0	37	44.0	32	53.0
34	47.0	44	48.0	33	60.0
35	35.0	48	43.0	38	46.0
36	39.0	51	41.0	42	58.0
39	29.0	52	53.0	43	53.0
40	28.0	53	50.0	46	54.0
41	40.0	56	47.0	47	55.0
45	38.0	57	62.0	49	57.0
54	32.0	60	50.0	50	53.0
55	39.0	64	36.0	58	65.0
65	40.0	66	45.0	59	46.0
75	41.0	68	46.0	61	54.0
79	39.0	70	49.0	62	60.0
83	41.0	71	50.0	63	47.0
84	37.0	76	49.0	67	59.0
86	43.0	77	42.0	69	58.0
88	36.0	82	47.0	72	55.0
89	34.0	85	53.0	73	51.0
94	36.0	87	51.0	74	60.0
95	31.0	91	49.0	78	47.0
96	25.0	92	39.0	80	56.0
98	38.0	93	43.0	81	59.0
99	42.0			90	60.0
100	33.0			97	60.0
Means	36.91		46.53		54.88
Variances	25.60		28.13		23.74
Sample sizes	$N_1 = 34$		$N_2 = 32$		$N_3 = 34$

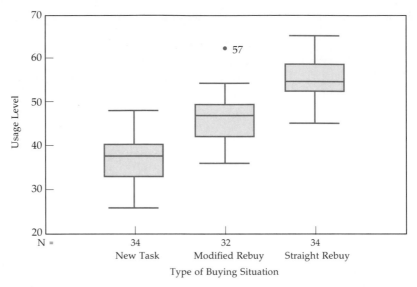

FIGURE 5.6 Boxplots of usage level for three-group ANOVA.

Stage Five: Interpretation of the Results

As shown in Table 5.6, the group means are as follows:

	Usage Level
New task	36.91
Modified rebuy	46.53
Straight rebuy	54.88

Examining these means, we note that HATCO usage increases as we proceed from new task to modified rebuy to straight rebuy. This type of question can be tested with one of the post hoc procedures. One hypothesis of interest is whether there is a significant difference between the new-task or modified-rebuy versus straight-rebuy situations [i.e., (36.91 + 46.53)/2 versus 54.88]. The contrast is significant (assume $\alpha = .05$ as the criterion; formulas for calculation can be found in texts oriented more to the statistician). Thus we can conclude that the straight-

TABLE 5.7 Three-Group ANOVA Results: Usage Level by Type of Buying Situation

Source of Variance	Sum of Squares	Mean Square	Degrees of Freedom	F Ratio
Between groups	5,498.767	2,749.383	2	106.667
Within-groups error	2,500.234	25.776	97	

rebuy situation had higher levels of usage than the other two buying situations. Another approach is to use one of the post hoc procedures that tests all group differences and identifies those differences that are statistically significant. One such statistical test is the Scheffe method, perhaps the most widely used of these post hoc methods. In this example, while controlling the overall error rate so as not to exceed .05, the Scheffe test still identifies that all the groups are significantly different from each other. From this, the analyst will know that the significant differences are due to each group comparison and not specific to differences only among certain groups.

In summary, univariate ANOVA suggests that the type of buying situation leads to higher usage levels. Post hoc tests enable the analyst to identify these significant differences quite easily and to maintain statistical control on the overall significance level.

A Multivariate Approach: k-*Groups MANOVA*

In *k*-group designs where multiple dependent variables are measured, many researchers proceed with a series of individual *F* tests (ANOVAs) until all the dependent variables have been analyzed. As the reader should suspect, this approach suffers from the same deficiencies as a series of *t* tests across multiple dependent variables; that is, a series of *F* tests with ANOVA (1) results in an inflated Type I error rate, and (2) ignores the possibility that some composite of the dependent variables may provide reliable evidence of overall group differences. In addition, because individual *F* tests ignore the correlations among the independent variables, they use less than the total information available for assessing overall group differences.

MANOVA again provides a solution to these problems. MANOVA solves our Type I error rate problem by providing a single overall test of group differences at a specified α level. It solves our composite variable problem by implicitly forming and testing the linear combination(s) of the dependent variables that provides the strongest evidence of overall group differences.

Stage One: Objectives of the MANOVA

In our earlier univariate example, HATCO assessed its performance across firms having one of three types of buying situations when dealing with HATCO (new-task buying, modified rebuy, or straight rebuy). To convert this example to a multivariate example, we require at least two dependent variables. As in our earlier multivariate extension of a univariate example, let us assume that HATCO also wished to examine differences in satisfaction with HATCO across the three groups. The values on both dependent measures (X_9 and X_{10} from our database) are presented in Table 5.8. The null hypothesis HATCO now wishes to test is that the three sample vectors of the mean scores are equivalent.

Stage Two: Research Design of the MANOVA

As discussed in the univariate example of the three-group analysis, the sample sizes are adequate based on the number of dependent variables. A more important consideration is the effect of the group sample sizes on the statistical power

TABLE 5.8 Usage Level and Satisfaction Level by Buying Situation (Basic Data for Three-Group MANOVA)

				Buying Situation				
Group 1: New-Task Scores			*Group 2: Modified-Rebuy Scores*			*Group 3: Straight-Rebuy Scores*		
Firm	$X_9{}^*$	$X_{10}\dagger$	*Firm*	$X_9{}^*$	$X_{10}\dagger$	*Firm*	$X_9{}^*$	$X_{10}\dagger$
1	32.0	4.2	3	48.0	5.2	5	58.0	6.8
2	43.0	4.3	6	45.0	4.4	9	63.0	5.4
4	32.0	3.9	8	44.0	4.3	15	54.0	5.9
7	46.0	5.8	10	54.0	5.4	16	49.0	4.7
11	32.0	4.3	12	47.0	5.0	19	54.0	5.9
13	39.0	4.4	17	38.0	4.4	20	55.0	6.0
14	38.0	5.0	18	40.0	5.6	23	55.0	5.2
22	35.0	3.3	21	41.0	4.5	26	49.0	5.9
24	36.0	3.7	25	49.0	4.9	28	54.0	5.8
27	36.0	3.7	30	46.0	5.1	29	49.0	5.4
31	43.0	3.3	37	44.0	4.8	32	53.0	5.0
34	47.0	3.8	44	48.0	4.8	33	60.0	6.1
35	35.0	4.1	48	43.0	4.7	38	46.0	5.1
36	39.0	3.6	51	41.0	5.0	42	58.0	6.7
39	29.0	3.9	52	53.0	5.2	43	53.0	5.9
40	28.0	3.3	53	50.0	5.5	46	54.0	6.0
41	40.0	3.7	56	47.0	4.2	47	55.0	4.9
45	38.0	3.2	57	62.0	6.2	49	57.0	4.9
54	32.0	3.7	60	50.0	5.0	50	53.0	3.8
55	39.0	3.7	64	36.0	4.2	58	65.0	6.0
65	40.0	3.4	66	45.0	4.9	59	46.0	5.6
75	41.0	4.1	68	46.0	4.5	61	54.0	4.8
79	39.0	3.3	70	49.0	4.8	62	60.0	6.1
83	41.0	4.1	71	50.0	5.4	63	47.0	5.3
84	37.0	4.4	76	49.0	5.2	67	59.0	6.0
86	43.0	3.7	77	42.0	5.1	69	58.0	4.3
88	36.0	4.3	82	47.0	5.6	72	55.0	3.9
89	34.0	4.0	85	53.0	5.6	73	51.0	4.9
94	36.0	3.6	87	51.0	5.5	74	60.0	5.1
95	31.0	4.0	91	49.0	4.4	78	47.0	5.1
96	25.0	3.4	92	39.0	5.5	80	56.0	5.1
98	38.0	3.7	93	43.0	5.2	81	59.0	4.5
99	42.0	4.3				90	60.0	6.1
100	33.0	4.4				97	60.0	5.2
Means:	36.91	3.93		46.53	5.00		54.88	5.39

$^*X_9$ = Usage level.
$\dagger X_{10}$ = Satisfaction level.

of the tests of group differences. In referring to Table 5.1, sample sizes of 30 and above will provide adequate power for large effect sizes and somewhat lower levels of power for medium effect sizes. These sample sizes, however, are not adequate to provide the recommended power of .80 for small effect sizes. The required sample size for small effect sizes in this situation would be 98 respondents per group. Thus, any nonsignificant results should be examined closely to evaluate whether the effect size has managerial significance, because the low statistical power precluded designating it as statistically significant.

Stage Three: Assumptions in MANOVA

The two univariate tests for homogeneity of variance indicate a nonsignificant difference for X_9, the usage level, but mixed results for X_{10}, the satisfaction with HATCO (Table 5.9). In the case of satisfaction with HATCO, combining the two tests (.033 and .070) provides a sufficient level of nonsignificance for the test of homogeneity of variance to proceed to the multivariate test. Using the Box's M test for homogeneity of the variance/covariance matrices, we find that the groups have no significant differences.

TABLE 5.9 Diagnostic Information for Three-Group MANOVA

Test of Assumptions: Homogeneity of Variance

Variance/Covariance Matrices
(values in parentheses are correlations)

	Group 1: New Task		Group 2: Modified Rebuy		Group 3: Straight Rebuy	
	X_9	X_{10}	X_9	X_{10}	X_9	X_{10}
X_9 Usage Level	25.598		28.128		23.743	
X_{10} Satisfaction	.648	.282	1.366	.237	.763	.509
	(.241)		(.529)		(.219)	

	X_9: Usage Level		X_{10}: Satisfaction		Overall	
	Statistic	*Signif.*	*Statistic*	*Signif.*	*Statistic*	*Signif.*
Univariate Tests						
Cochran's C	.363	.965	.495	.033		
Bartlett-box	.114	.892	2.67	.070		
Levene test	.056	.945	3.30	.041		
Multivariate Test						
Box's M					9.79	.147

Test of Assumption: Correlation of Dependent Variables

	Statistic	*Signif.*
Bartlett test of sphericity	9.47	.002

The second assumption to test is the correlation among the dependent variables. In this case the Bartlett's test of sphericity has a significance level of .002, satisfying the necessary level of intercorrelation to justify MANOVA. See Table 5.9 for more details.

Stage Four: Estimation of the MANOVA Model and Assessing Overall Fit

From examination of the boxplots of responses in each group for usage level (refer to Figure 5.6) and satisfaction (Figure 5.7), the indications are that both variables may differ across the three groups. The purpose of the multivariate test is to assess these differences collectively rather than singularly with univariate tests. Table 5.10 provides summary output from the MANOVA performed on the data of Table 5.8. The Pillai's criterion has a significance level (.0000) well below our prespecified level of .05. The value of the greatest characteristic root is .723. Referring to the gcr distribution with appropriate degrees of freedom and setting $\alpha =$ .05, we see that $gcr_{crit} = .310$. Because .723 exceeds this value, we again conclude that the mean vectors of the three groups are not equal. As we see in Table 5.10, the statistical program estimates a significance level of .000 for this measure as well. The value of Wilks' lambda, also shown in Table 5.10, is .264. An approximate F statistic associated with this value of Wilk's lambda is 45.4. With 4 and 192 degrees of freedom and an α level of .05, $F_{crit} = 2.41$. Because 45.4 well exceeds this value, we again reach the same conclusion that the mean vectors of the three groups are not equal. Using any of the measures of multivariate differences results in the same conclusion: Both dependent variables, usage and satisfaction, vary across the three buying situations.

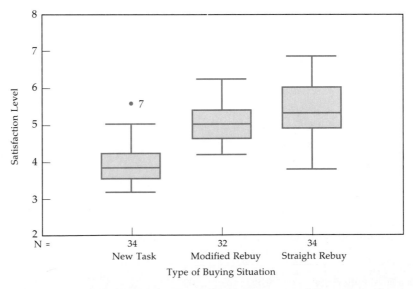

FIGURE 5.7 Boxplots of second dependent variable, satisfaction, for three-group MANOVA

TABLE 5.10 Three-Group MANOVA Summary Table

Multivariate Tests of Significance

Test Name	Value	Approximate F	Degrees of Freedom Between Group	Degrees of Freedom Within Group	Significance of F Statistic
Pillai's criterion	.77089	30.41887	4	194	.000
Hotelling's trace	2.65480	63.05157	4	190	.000
Wilks' lambda	.26405	45.41066	4	192	.000
Roy's gcr	.72253				

Statistical Power of MANOVA Tests

	Effect Size	Power
Pillai's criterion	.385	1.00
Hotelling's trace	.570	1.00
Wilks' lambda	.486	1.00

Univariate F tests

Variable	Between-Groups Sum of Squares	Within-Groups Sum of Squares	Degrees of Freedom	Between-Groups Mean Square	Within-Groups Mean Square	F Statistic	Significance
X_9	5498.76654	2500.23346	2 and 97	2749.38327	25.77560	106.7	.000
X_{10}	39.00680	33.45910	2 and 97	19.50340	.34494	56.5	.000

Roy-Bargman Stepdown F tests

Variable	Between-Groups Mean Square	Within-Groups Mean Square	Stepdown F	Degrees of Freedom Between	Degrees of Freedom Within	Significance of Stepdown F
X_9	2749.38327	25.77560	106.7	2	97	.000
X_{10}	2.78337	.31560	8.8	2	96	.000

Stage Five: Interpretation of the Results

As shown in Table 5.8, the group means are as follows:

	Usage Level	Satisfaction Level
New task	36.91	3.93
Modified rebuy	46.53	5.00
Straight rebuy	54.88	5.39

A number of post hoc tests may be of interest. For example, are group mean differences statistically significant for each dependent variable considered alone? We can examine this question with the individual *F* tests for the two dependent variables. Consistent with the multivariate results, both variables show significant differences across the groups. A stepdown analysis, as shown in Table 5.10, shows that both variables have unique differences across the groups; that is, they are not so highly correlated that there were no unique differences in satisfaction after the effects of usage were accounted for. This result suggests that buying situation has significant separate effects on satisfaction that are unrelated to a firm's usage level. Finally, are the differences in usage and satisfaction found between all buying situations? For example, do firms with the modified rebuy situation show a difference in satisfaction when compared with those in the straight rebuy? All such questions can be answered with the multivariate extension of contrast procedure outlined earlier. Examination of discriminant functions (which have to be obtained by using a discriminant analysis program, as described in Chapter 3) are generally more useful first steps in post hoc analysis as the number of dependent variables increases.

Example 3: A Factorial Design for MANOVA with Two Independent Variables

In the prior two cases, the MANOVA analyses have been extensions of univariate two- and three-group analyses. In this example, we will explore a multivariate factorial design—two independent variables used as treatments to analyze differences of two dependent variables. In the course of our discussion, we will assess the interactive or joint effects between the two treatments on the dependent variables separately and collectively.

Stage One: Objectives of the MANOVA

In the previous multivariate research questions, HATCO has considered only the effect of a single-treatment variable on the dependent variables. But the possibility of joint effects among two or more independent variables must also be considered. After deliberation, one proposed analysis focused on extending the prior three-group analysis of the differences in the two dependent variables, usage level (X_9) and satisfaction (X_{10}), by considering not only the effects of buying situations (X_{14}) but also industry type (X_{13}). The objective is to reduce the within-group variance of the buying situation groups by adding the second treatment—industry type. This will form a separate group for each industry type within each buying situation. The boxplots in Figure 5.8 show that the two industry types do not vary substantially on either of the dependent variables. But this does not preclude its value as a means of reducing within-group variance. The data for each group (combination of buying situation and industry type) are shown in Table 5.11. In this manner, the impact of industry type can be evaluated simultaneously with buying situation along with examining for the possible occurrence of interaction effects.

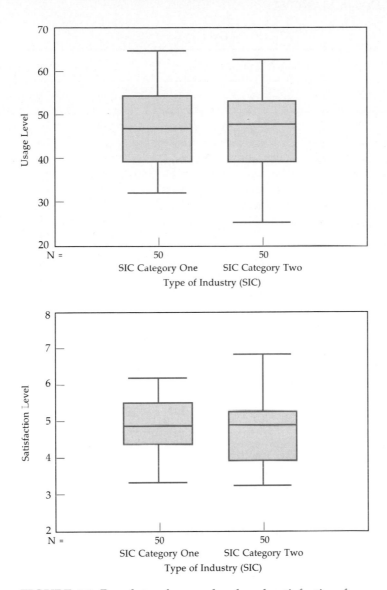

FIGURE 5.8 Boxplots of usage level and satisfaction for second independent variable, industry type, in 3 × 2 factorial design MANOVA.

Stage Two: Research Design of the MANOVA

The factorial design of two independent variables—X_{14} and X_{13}—raises the issue of adequate sample size in the various groups. Because there are three levels of X_{14} (new task, modified rebuy, and straight rebuy) and two levels of X_{13} (SIC category one and SIC category two), this is a 3 × 2 design with six groups. The analyst must ensure in creating the factorial design that each group has sufficient sample size to (1) meet the minimum requirements of group sizes exceeding the number of dependent variables, as well as (2) provide the statistical power to

TABLE 5.11 Firm Usage and Satisfaction Level by Buying Situation and Industry Type (Basic Data for 3 × 2 Factorial Design MANOVA)

SIC Category One

| | Group 1: New Task | | | Group 3: Modified Rebuy | | | Group 5: Straight Rebuy | |
ID	X_9	X_{10}	ID	X_9	X_{10}	ID	X_9	X_{10}
2	43	4.3	8	44	4.3	15	54	5.9
11	32	4.3	10	54	5.4	16	49	4.7
13	39	4.4	18	40	5.6	20	55	6.0
22	35	3.3	21	41	4.5	23	55	5.2
24	36	3.7	25	49	4.9	32	53	5.0
27	36	3.7	30	46	5.1	33	60	6.1
31	43	3.3	44	48	4.8	43	53	5.9
34	47	3.8	51	41	5.0	46	54	6.0
35	35	4.1	53	50	5.5	58	65	6.0
55	39	3.7	57	62	6.2	62	60	6.1
75	41	4.1	66	45	4.9	67	59	6.0
83	41	4.1	68	46	4.5	69	58	4.3
84	37	4.4	77	42	5.1	73	51	4.9
88	36	4.3	85	53	5.6	80	56	5.1
94	36	3.6	87	51	5.5	81	59	4.5
98	38	3.7	92	39	5.5	90	60	6.1
99	42	4.3						
100	33	4.4						
Mean	38.3	3.97		46.94	5.15		56.31	5.49

SIC Category Two

| | Group 2: New Task | | | Group 4: Modified Rebuy | | | Group 6: Straight Rebuy | |
ID	X_9	X_{10}	ID	X_9	X_{10}	ID	X_9	X_{10}
1	32	4.2	3	48	5.2	5	58	6.8
4	32	3.9	6	45	4.4	9	63	5.4
7	46	5.8	12	47	5.0	19	54	5.9
14	38	5.0	17	38	4.4	26	49	5.9
36	39	3.6	37	44	4.8	28	54	5.8
39	29	3.9	48	43	4.7	29	49	5.4
40	28	3.3	52	53	5.2	38	46	5.1
41	40	3.7	56	47	4.2	42	58	6.7
45	38	3.2	60	50	5.0	47	55	4.9
54	32	3.7	64	36	4.2	49	57	4.9
65	40	3.4	70	49	4.8	50	53	3.8
79	39	3.3	71	50	5.4	59	46	5.6
86	43	3.7	76	49	5.2	61	54	4.8
89	34	4.0	82	47	5.6	63	47	5.3
95	31	4.0	91	49	4.4	72	55	3.9
96	25	3.4	93	43	5.2	74	60	5.1
						78	47	5.1
						97	60	5.2
Mean	35.37	3.88		46.12	4.86		53.61	5.31

assess differences deemed practically significant. In this case, the sample sizes range from 16 to 18 respondents per group. This exceeds the number of dependent variables (two) in each group, but the statistical power is quite low. While tabled values do not deal with a factorial design, values for a six-group MANOVA indicate that this sample size will detect only moderately large or better effect sizes with a power of .70 [14]. Thus, the analyst must recognize that unless the effect sizes are substantial, the limited sample sizes in each group preclude the identification of significant differences.

TABLE 5.12 Diagnostic Information for 3 × 2 Factorial Design MANOVA

Test of Assumptions: Homogeneity of Variance

Variance/Covariance Matrices
(values in parentheses are correlations)

SIC Category One

	Group 1: New Task		Group 3: Modified Rebuy		Group 5: Straight Rebuy	
	X_9	X_{10}	X_9	X_{10}	X_9	X_{10}
X_9	15.389		37.662		16.762	
X_{10}	−.162	.141	1.643	.252	.984	.423
	(−.110)		(.533)		(.370)	

SIC Category Two

	Group 2: New Task		Group 4: Modified Rebuy		Group 6: Straight Rebuy	
	X_9	X_{10}	X_9	X_{10}	X_9	X_{10}
X_9	34.117		20.117		27.663	
X_{10}	1.461	.456	1.053	.192	.375	.600
	(.370)		(.536)		(.092)	

	X_2: Usage Level		X_{10}: Satisfaction		Overall	
	Statistic	*Signif.*	*Statistic*	*Signif.*	*Statistic*	*Signif.*
Univariate Tests						
Cochran's C	.248	.447	.291	.115		
Bartlett-box	1.06	.380	2.36	.038		
Levene test	1.33	.257	1.51	.191		
Multivariate Test						
Box's M					24.04	.090

Test of Assumption: Correlation of Dependent Variables

	Statistic	Signif.
Bartlett test of sphericity	8.22	.004

Stage Three: Assumptions in MANOVA

As with the prior MANOVA analyses, the assumption of greatest importance is the homogeneity of variance/covariance matrices across the groups. In this instance, there are six groups involved in testing the assumption. The univariate tests for usage level and satisfaction are both nonsignificant, except for the Bartlett-Box test of satisfaction, which had a significance level of .038 (see Table 5.12). With the univariate tests showing nonsignificance, the analyst can proceed to the multivariate test. The Box's M test has a significance level of .09, thus allowing us to accept the null hypothesis of homogeneity of variance/covariance matrices at the .05 level. Meeting this assumption allows for direct interpretation of the results without having to consider group sizes, level of covariances in the group, and so forth.

The second assumption is the correlation of the dependent measures, which is assessed with Bartlett's test of sphericity. In this example, the significance is .004, indicative of a significant level of correlation between the two dependent measures (See Table 5.12).

Stage Four: Estimation of the MANOVA Model and Assessing Overall Fit

The boxplots for each dependent variable across the six groups (Figure 5.9) show that differences do seem to exist between them. The differences are the most pronounced across buying situations, but differences within each buying situation for the two industry types can also be seen. The MANOVA model tests not only for the main effects of both independent variables but also their interaction or joint effect on the two dependent variables. The first step is to examine the interaction effect and determine whether it is statistically significant. Table 5.13 contains the MANOVA results for testing the interaction effect. All four multivariate tests indicate that the interaction effect is not significant. This means that the differences between industry types are roughly equal across the three buying situations for both dependent variables collectively. The univariate tests confirm that this finding holds for each variable separately. Figure 5.10 documents the lack of interaction effect for each dependent variable. In the graphs for each dependent variable, the differences between the two industry types are relatively equal across the three buying situations. With a nonsignificant interaction effect, the direct effects can be interpreted directly without adjustment.

Table 5.14 contains the MANOVA results for the main effects of buying situation and industry type. Industry type (X_{13}) has a significance level of .069 for the multivariate tests, indicating a nonsignificant difference attributable to industry type. The analyst should consider, however, raising the required α level, because the power of the multivariate tests is reduced owing to the rather small sample sizes per group. If this was done, then industry type would be considered a significant effect. The second independent variable, satisfaction (X_{14}), shows highly significant effects for all the multivariate tests. In each instance, the significance level exceeds .000. Moreover, the statistical power is 1.0, indicating that the very large effect sizes ensured high levels of power even with the small sample sizes per group. The impact of the two independent variables can be compared by

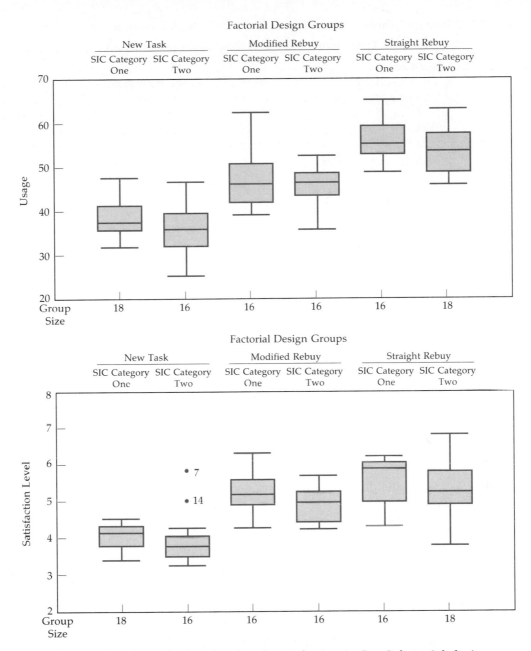

FIGURE 5.9 Boxplots of usage level and satisfaction in 3×2 factorial design MANOVA.

examining the relative effect sizes. The effect sizes for buying situation are eight to ten times larger than those associated with usage level. This comparison gives the analyst an evaluation of practical significance separate from the statistical significance tests. In this example, buying situation is the dominant effect, with industry type having a slight effect. Moreover, the interaction or joint effects between the two treatments are nonsignificant for both dependent variables.

TABLE 5.13 3×2 Factorial Design MANOVA Summary Table: Interaction Effect

Interaction Effect: Industry Type (X_{13}) by Buying Situation (X_{14})

Multivariate Tests of Significance

Test Name	Value	Approximate F	Degrees of Freedom Between Group	Degrees of Freedom Within Group	Significance of F statistic
Pillai's criterion	.01954	.46383	4	188	.762
Hotelling's trace	.01992	.45819	4	184	.766
Wilks' lambda	.98406	.61036	4	186	.764
Roy's gcr	.01923				

Statistical Power of MANOVA Tests

	Effect Size	Power
Pillai's criterion	.010	.16
Hotelling's trace	.010	.16
Wilks' lambda	.010	.16

Univariate F tests

Variable	Between-Groups Sum of Squares	Within-Groups Sum of Squares	Degrees of Freedom	Between-Groups Mean Square	Within-Groups Mean Square	F Statistic	Significance
X_9	21.681	2361.763	2 and 94	10.841	25.125	.431	.651
X_{10}	.17013	32.435	2 and 94	.08507	.34505	.246	.782

Roy-Bargman Stepdown F tests

Variable	Between-Groups Mean Square	Within-Groups Mean Square	Stepdown F	Degrees of Freedom Between	Degrees of Freedom Within	Significance of Stepdown F
X_9	10.841	25.125	.651	2	94	.651
X_{10}	.15802	.31909	.495	2	93	.611

Stage Five: Interpretation of the Results

A comparison among the six groups should not be attempted with the post hoc tests, such as the Scheffe method, which make all possible comparisons while controlling overall Type I error. In this example, the relatively small sample sizes and the large number of tests required for all comparisons of six groups result in such low levels of statistical power that only very large effect sizes can be detected reliably. The analyst thus should examine the differences for practical significance in addition to statistical significance. If specific comparisons among the groups can be formulated, then planned comparisons can be specified and tested directly in the analysis.

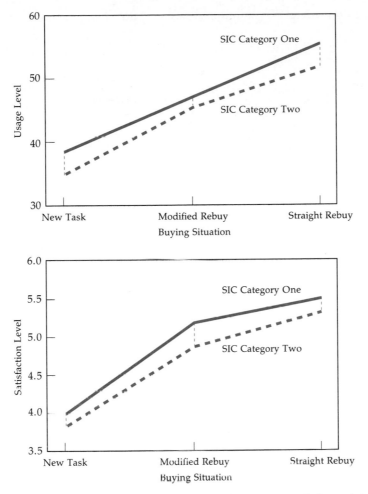

FIGURE 5.10 Plots of interaction effects in 3 × 2 factorial design MANOVA for usage level and satisfaction.

While the group comparisons are limited by the small sample sizes, examination of each dependent variable is still warranted. For example, the industry type (X_{13}) was judged to be nonsignificant at the .05 alpha level when the set of dependent variables is evaluated. But if the analyst examines the univariate tests, some interesting points emerge. First, the industry type does affect the usage level but not satisfaction. Thus, when evaluated collectively, the set of dependent variables is found to be nonsignificant. The practical significance of demonstrating an effect on the usage level may lead to use of the univariate results in addition to the multivariate results. This can be contrasted to the second independent variable—buying situation (X_{14})—where univariate tests for both dependent variables support its multivariate effect. This also holds in the stepdown analysis, which can be interpreted that buying situation affects not only the set of dependent variables but also these dependent variables separately, even after the impact of other dependent variables has been considered. These results confirm the differences between the impacts of the two independent variables discussed above.

TABLE 5.14 3×2 Factorial Design MANOVA Summary Table: Main Effects of Industry Type and Buying Situation

Main Effect: X_{13}—Industry Type

Multivariate Tests of Significance

Test Name	Value	Approximate F	Between Group	Within Group	Significance of F Statistic
			Degrees of Freedom		
Pillai's criterion	.05587	2.752	2	93	.069
Hotelling's trace	.05918	2.752	2	93	.069
Wilks' lambda	.94413	2.752	2	93	.069
Roy's gcr	.05587				

Statistical Power of MANOVA Tests

	Effect Size	Power
Pillai's criterion	.056	.53
Hotelling's trace	.056	.53
Wilks' lambda	.056	.53

Univariate F tests

Variable	Between-Groups Sum of Squares	Within-Groups Sum of Squares	Degrees of Freedom	Between-Groups Mean Square	Within-Groups Mean Square	F Statistic	Significance
X_9	114.019	2351.764	1 and 94	114.019	25.125	4.54	.036
X_{10}	.87188	32.435	1 and 94	.87188	.34504	2.53	.115

Roy-Bargman Stepdown F tests

Variable	Between-Groups Mean Square	Within-Groups Mean Square	Stepdown F	Between	Within	Significance of Stepdown F
				Degrees of Freedom		
X_9	114.019	25.125	4.54	1	94	.036
X_{10}	.30858	.31909	.97	1	93	.328

Summary

It may be unrealistic to assume that a difference between experimental treatments will be manifested only in a single measured dependent variable. Unfortunately, many researchers handle multiple-criterion situations by repeated application of individual univariate tests until all the dependent variables have been analyzed. This approach can seriously inflate Type I error rates and ignores the possibility that some composite of the dependent variables may provide the strongest evidence of reliable group differences. Appropriate use of MANOVA provides solutions to both these problems.

TABLE 5.14 *Continued*

Main Effect: X_{14}—Buying Situation

Multivariate Tests of Significance

			Degrees of Freedom		
Test Name	Value	Approximate F	Between Group	Within Group	Significance of F Statistic
Pillai's criterion	.78570	30.411	4	188	.000
Hotelling's trace	2.85220	65.600	4	184	.000
Wilks' lambda	.25026	45.452	4	186	.000
Roy's gcr	.73690				

Statistical Power of MANOVA Tests

	Effect Size	Power
Pillai's criterion	.393	1.00
Hotelling's trace	.588	1.00
Wilks' lambda	.500	1.00

Univariate F tests

Variable	Between-Groups Sum of Squares	Within-Groups Sum of Squares	Degrees of Freedom	Between-Groups Mean Square	Within-Groups Mean Square	F Statistic	Significance
X_9	5580.664	2361.764	2 and 94	2790.332	25.125	111.06	.000
X_{10}	39.250	32.435	2 and 94	19.652	.34505	56.88	.000

Roy-Bargman Stepdown F tests

				Degrees of Freedom		
Variable	Between-Groups Mean Square	Within-Groups Mean Square	Stepdown F	Between	Within	Significance of Stepdown F
X_9	2790.332	25.125	111.06	2	94	.000
X_{10}	2.7928	.31909	8.75	2	93	.000

This chapter has not covered all the types of experimental designs for MANOVA, nor would such an undertaking be practical in a book oriented to the nonstatistician. We hope the reader has obtained sufficient stimulation, understanding, and confidence to tackle some of the more sophisticated designs of MANOVA described in more statistically oriented texts.

Questions

1. Design a two-way factorial MANOVA experiment. What are the different sources of variance in your experiment? What would a significant interaction tell you?

2. Locate and compare at least two "canned" MANOVA computer programs. What are the essential differences between them, especially with regard to the printout of results?

3. Besides the overall, or global, significance, there are at least three approaches to doing follow-up tests: (a) use of Scheffe contrast procedures: (b) stepdown analysis, which is similar to stepwise regression in that each successive F statistic is computed after eliminating the effects of the previous dependent variables; and (c) examination of the discriminant functions. Name the practical advantages and disadvantages of each of these approaches.

4. Describe some data analysis situations in which MANOVA and MANCOVA would be appropriate in your areas of interest. What types of uncontrolled variables or covariates might be operating in each of these situations?

References

1. Anderson, T. W. *Introduction to Multivariate Statistical Analysis*. New York: Wiley, 1958.
2. BMDP Statistical Software, Inc. *BMDP Statistical Software Manual, Release 7*, vols. 1 and 2. Los Angeles: 1992.
3. BMDP Statistical Software, Inc. *SOLO Power Analysis*. Los Angeles: 1991.
4. Cattell, R. B., ed., *Handbook of Multivariate Experimental Psychology*. Chicago: Rand McNally, 1966.
5. Cooley, W. W., and P. R. Lohnes. *Multivariate Data Analysis*. New York: Wiley, 1971.
6. Cohen, J. *Statistical Power Analysis for the Behavioral Sciences*. New York: Academic Press, 1977.
7. Green, P. E. *Analyzing Multivariate Data*. Hinsdale, Ill.: Holt, Rinehart, & Winston, 1978.
8. Green, P. E., and J. Douglas Carroll. *Mathematical Tools for Applied Multivariate Analysis*. New York: Academic Press, 1978.
9. Green, P. E., and D. S. Tull. *Research for Marketing Decisions*, 3d ed. Englewood Cliffs, N.J.: Prentice-Hall, 1979.
10. Hand, D. J., and C. C. Taylor. *Multivariate Analysis of Variance and Repeated Measures*. London: Chapman and Hall, 1987.
11. Harris, R. J. *A Primer of Multivariate Statistics*. New York: Academic Press, 1975.
12. Huitema, B. *The Analysis of Covariance and Alternatives*. New York: Wiley, 1980.
13. Kirk, R. E. *Experimental Design: Procedures for the Behavioral Sciences*, 2d ed. Belmont, Calif.: Brooks/Cole, 1982.
14. Läuter, J. "Sample Size Requirements for the T^2 Test of MANOVA (Tables for One-Way Classification)." *Biometrical Journal* 20 (1978): 389–406.
15. Meyers, J. L. *Fundamentals of Experimental Design*. Boston: Allyn & Bacon, 1975.
16. Morrison, D. F. *Multivariate Statistical Methods*. New York: McGraw-Hill, 1967.
17. Rao, C. R. *Linear Statistical Inference and Its Application*, 2d ed. New York: Wiley, 1978.
18. SAS Institute, Inc. *SAS User's Guide: Basics, Version 6*. Cary, N.C.: 1990.
19. SAS Institute, Inc. *SAS User's Guide: Statistics, Version 6*. Cary: N.C., 1990.
20. Sheth, J. H. *Multivariate Methods for Market and Survey Research*. Chicago: American Marketing Association, 1977, pp. 83–96.

21. SPSS, Inc. *SPSS User's Guide,* 4th ed. Chicago: 1990.
22. SPSS, Inc. *SPSS Advanced Statistics Guide,* 4th ed. Chicago: 1990.
23. Stevens, J. P. "Four Methods of Analyzing Between Variations for the k-Group MANOVA Problem." *Multivariate Behavioral Research* 7 (October 1972): 442–54.
24. Stevens, J. H. "Power of the Multivariate Analysis of Variance Tests." *Psychological Bulletin* 88 (1980): 728–37.
25. Tatsuoka, M. M. *Multivariate Analysis: Techniques for Education and Psychological Research.* New York: Wiley, 1971.
26. Wilks, S. S. "Certain Generalizations in the Analysis of Variance." *Biometrika* 24 (1932): 471–94.
27. Winer, B. J. *Statistical Principles in Experimental Design.* New York: McGraw-Hill, 1962.

The American Export Trading Company: Designing a New International Marketing Institution

Daniel C. Bello
Nicholas C. Williamson

As marketing intermediaries, Japanese general trading companies have been extraordinarily successful in developing world markets for Japan's domestic industries (Young 1979). Japanese trading companies (JTCs) are enormous global marketing firms with extensive investments in foreign offices, communication centers, and logistical facilities. With these numerous resources, JTCs achieve significant economies of scale in international trade and provide a full range of low cost export services to Japanese producers.

However, Japanese style trading companies have never developed in the U.S., and the closest entity to American trading companies (ATCs) have been export management companies (EMCs) whose scale and business diversity are much smaller (Maker 1982, Terpstra 1983). Although EMCs are independent exporters and perform many of the functions associated with large trading companies, individual EMCs do not provide complete one-stop export services, since they tend to be undercapitalized and operate with few employees (Bello and Williamson 1985, Brasch 1978). Regulatory barriers developed over the years to promote domestic competition have prevented the development of large scale American trading companies (Scouton 1982). The 1982 Export Trading Company Act (Public Law 97-290) removes many of the legal roadblocks to their development. Essentially the Act enables domestic competitors to participate in a cooperative trading company venture by providing com-

plete immunity from U.S. antitrust laws. In addition, the Act removes a long-standing ban on bank ownership of commercial enterprises by permitting banks to hold equity participation in trading companies.

While the 1982 Act provides the legal basis for banks, manufacturers, and others to jointly form large scale, highly capitalized ATCs, the operating characteristics and export services of these ATCs are not specified in the Act. Rather, the law leaves it to the participating firms to define whatever type of trading company best fits its particular needs (Barovick 1983). The type of firms forming an ATC, the range of products exported, and the specific intermediary functions performed by the ATC are all to be determined by the private sector. Since the Act's passage 45 ATCs have been established, but these few ventures ". . . have mostly failed to fulfill their promise. Almost no U.S. trading company is making money, and most have managed to pull off only a handful of small deals" (Weiner and Johnson 1984). However, analysts recognize that the benefits of the Act will not be realized immediately since the legislation provides a great deal of flexibility and not every ATC design will be successful (Hurt 1982).

This study evaluates the type of export services appropriate to ATCs by analyzing the services that export management companies currently provide domestic producers. As will be shown, the particular mix of export services provided by international intermediaries such as EMCs and JTCs tend to be associated with basic operating characteristics. Characteristics such as the type of product exported, the export role adopted (agent or merchant), and the export volume of the suppliers represented are particularly likely to influ-

From "The American Export Trading Company: Designing a New International Marketing Institution," Daniel C. Bello and Nicholas C. Williamson, *Journal of Marketing*, Vol. 49 (Fall 1985), pp. 60–69. Reprinted by permission.

ence the marketing services exporters provide. A better understanding of these relationships will enable those forming ATCs to identify the export services that are most compatible with their product, role, and supplier decisions.

1982 Export Trading Company Act

Prior to the 1982 ATC legislation, U.S. firms had been handicapped in forming trading companies and other joint exporting efforts by fear of antitrust prosecution and inadequate capitalization. Oddly, the antitrust threat stemmed from the Webb-Pomerene Export Trade Act of 1918 which permitted U.S. firms to combine exporting efforts in order to compete against foreign cartels. While ostensibly permitting firms to jointly export as long as domestic competition was unharmed, the Act actually discouraged joint exporting since the courts had severely limited the antitrust exemption (Scouton 1982). Another legal roadblock was the 1934 Glass-Steagall Act which banned bank ownership of commercial enterprises. This led to chronic capital shortages for independent exporters since conventional collateral requirements limited the amount of credit they could obtain. The size of international transactions that U.S. EMCs could finance was limited to three or four times their net worth (Hay 1977). A marked contrast existed in Japan where banks held equity in trading companies and played a central role in their success. Greater access to capital enabled JTCs to finance transactions up to 20 times their net worth and support operations at a dramatically larger scale (Kunio 1982).

The 1982 Act permits U.S. bank equity participation and provides protection from antitrust laws for export trading companies. By allowing banks to own up to 100% of the stock, the Act ensures ATCs will have access to sufficient financial resources. In addition, the new legislation grants complete immunity from U.S. antitrust laws. Through a preclearance procedure, firms entering into a cooperative export venture with a competitor or bank can get prior antitrust approval. By removing the antitrust threat, the 1982 Act encourages many firms to participate in ATCs and thus allows these ventures to develop efficiencies through large scale operations.

Designing American Export Trading Companies

EMCs and JTCs, as well as European trading houses, provide important insights into ATC design. While these institutions share a need to provide useful export services, they differ in terms of their product, geographic market, and functional diversification (Cho 1984). EMCs are the least diversified in product and market and usually restrict their activities to exporting; large general JTCs possess the greatest product and market diversification, and import as well as export (Dooley 1984). European trading houses, which were the nineteenth century models for JTCs, operate today by importing and exporting a wide variety of items but restrict their markets to a limited number of former colonies.

Analysts expect ATCs to be much more diversified than EMCs but less than foreign trading companies (Cho 1984). ATCs are likely to be less diversified than their foreign counterparts because the purpose of the 1982 Act is to promote U.S. exports; while ATCs may import and trade between foreign countries, such activity is to be secondary and done only when necessary to promote exports (Williams and Baliga 1983). By concentrating on exports, ATCs will be less able to develop the highly efficient infrastructure associated with two-way trade. Thus, ATCs will have a much broader financial and operational base than EMCs but will have a somewhat similar export focus.

To aid in ATC design, the influence of basic operating characteristics on the export services provided by existing institutions can be examined. For purposes of analysis export services can be divided into two broad categories: a flow of transaction creating efforts and a flow of physical fulfillment efforts (Bowersox 1978). Transactioncreating export services include foreign market advertising, personal selling, and all the other activities that stimulate foreign demand. Physical fulfillment services include the activities necessary to supply foreign demand, such as export documentation, international transportation, and foreign market warehousing. The way basic operating characteristics are likely to affect the provision of these two broad categories of export services is discussed below.

Type of Product Exported

Although JTCs provide an extensive range of export services, these large scale firms are most effective for undifferentiated products, such as food commodities, raw materials, and standardized industrial goods (Roehl 1982). Since the major purchase criterion for undifferentiated products is the delivered cost of the goods, the huge economies of scale achievable by JTCs in physical fulfillment activities provide these firms with a low price differential advantage (Hay 1977). Other services of JTCs, like information gathering, also are most effective for standardized commodities. JTCs are not as effective in procuring market information for brand specific products; rather, information networks are most useful in reducing speculative risks for undifferentiated items (Daido 1976). As a result, JTCs are not large traders of consumer durables and industrial equipment, since these require transaction creating effort and are not subject to great scale economies in physical fulfillment (Kunio 1982).

In contrast, EMCs often handle differentiated products that require aggressive transaction creating efforts (Hay 1977). For many EMCs, export services are concentrated in sales-related activities (Cao 1981). In addition, many EMCs do not invest in specialized logistical facilities, since their ability to compete in foreign markets does not depend on economies in physical fulfillment. In Japan differentiated consumer and industrial products are usually exported through specialty trading companies called *senmon shosha*, or directly by Japanese manufacturers (Shimizu 1980). Similar to EMCs, specialty trading companies are small entrepreneurial firms focusing on a product group or geographic area (Roehl 1982). Because of their specialization, these trading companies excel in transaction creating activities, aggressively promoting and servicing differentiated products in foreign markets.

Role of the Export Intermediary

Services provided by export intermediaries also vary, depending upon whether title is taken for the products exported (Brasch 1978, Kunio 1982). These firms may act as commissioned agents or as merchant distributors, depending upon the marketing situation and the needs of the supplier. When acting as an agent, the trading firm does not take title and is compensated solely by generating foreign market sales. After obtaining an order the agent may not have any further involvement with the transaction. By retaining title, the domestic supplier bears the financial risk of collecting from foreign buyers and retains control over the physical fulfillment activities associated with exporting its brand.

When acting as a merchant, the trading firm purchases the product and resells it abroad on its own account. To the supplier the transaction is simply another domestic sale, and he/she often has little further responsibility or interest. The trading firm bears all foreign financial and commercial risks and is responsible for physical fulfillment. The exporter's compensation, like that of any merchant wholesaler, depends on its ability to maintain its margin by minimizing distribution costs.

Supplier's Export Sales Volume

Indirect intermediaries, like export trading and management companies, face a common problem of being bypassed by their suppliers. Once a supplier gains experience in exporting and its foreign sales volume reaches a level sufficient to support an in-house export department, a supplier may decide to drop the indirect exporter and absorb the export function. Suppliers are motivated to export directly by a desire to gain control over their foreign marketing program and to reduce costs. Even the highly efficient JTCs which offer low cost, comprehensive export services increasingly lose suppliers that decide to export directly (Abegglen and Stalk 1983).

Similarly, EMCs face the chronic problem of developing foreign demand for products only to be dropped once a high level of sales is achieved. Suppliers often test international marketing with minimum risk and cost initially through an EMC, then "go it alone" once foreign demand is established (Cao 1981). Some analysts suggest that at export volumes above $1 million, American producers can economically duplicate the export services performed by EMCs with an in-house department (Miller 1982).

Hypotheses

A basic decision for those forming American trading companies under the 1982 Act is the particular mix of export services that should be provided. To gain insight into ATC design, the following hypotheses examine whether the provision of export services is associated with basic characteristics of export intermediaries.

H1: The importance of services provided by export intermediaries is influenced by the type of product exported, export role of the intermediary, and supplier's export sales volume.

The first hypothesis states each characteristic—product, role, and supplier volume—affects the importance of export services. If these factors are associated with service provision, inferences can be drawn for the types of services ATCs should be capable of providing.

H2: Type of product influences the importance of services provided by export intermediaries in the following manner:
  a. Transaction creating services are more important for differentiating products.
  b. Physical fulfilling services are more important for undifferentiated products.

The second hypothesis posits the specific way product influences the provision of export services. Since differentiated products require brand specific demand stimulation in foreign markets, the transaction creating export services are expected to be most important for intermediaries handling these products. For undifferentiated products, physical fulfilling export services are likely to be most important. Since these products are sold on a price basis, competency and cost effectiveness in fulfillment activities are expected to be a competitive necessity.

H3: Export role influences the importance of services provided by export intermediaries in the following manner:
  a. Transaction creating services are more important for agents.
  b. Physical fulfilling services are more important for merchants.

The third hypothesis states how service provision varies for agent and merchant export inter-

mediaries. Since agents are compensated on the basis of sales volume generated, transaction creating services are expected to be most important for such intermediaries. While transaction creation is necessary for merchants, the size of their profit margin in international trade is directly affected by the speed and cost of fulfillment. Thus, physical fulfillment services are expected to be most important.

H4: Supplier's export sales volume influences the importance of services provided by export intermediaries in the following manner:
  a. Transaction creating services are more important for intermediaries with suppliers exporting over $1 million.
  b. Physical fulfilling services are more important for intermediaries with suppliers exporting over $1 million.

The final hypothesis posits that both service categories are more important for intermediaries with high volume suppliers compared to intermediaries with low volume suppliers. Since suppliers exporting over $1 million could support an in-house export effort (Miller 1982), intermediaries continuing to hold clients beyond this point are expected to provide more important export services.

Methodology

The hypotheses were tested on a national sample of American EMCs. The population for the study was the approximate 1,100 EMCs listed in the U.S. Department of Commerce *Directory* of EMCs (1981). In the summer of 1982 a mail survey addressed to the president was sent to all the EMCs listed in the *Directory*. Approximately 200 were undeliverable by the post office, reducing the population to 900 firms.

To ensure an adequate response rate, three mailing waves were used. The first contained only a prenotification letter. The second wave contained the questionnaire and a cover letter explaining the academic nature of the project. Three weeks later a follow-up questionnaire and cover letter were sent. Questionnaires were to be returned anonymously to a university address, and the only incentive to participants was the promise of a copy of the results. Of the 308 questionnaires returned, 258 were sufficiently complete to be used in the analysis, yielding a response rate of 29%.

The data were considered adequate for the research because the survey respondents were the presidents/owners of the exporting firms. These key informants possess the ability to provide reliable information because they make virtually all the decisions and there is no alternative informant within the firm. Further, as a check for nonresponse bias, the present sample was compared with two earlier national studies of EMCs (Brasch 1978, Hay 1977) on several key descriptive parameters. The present sample showed close correspondence to the earlier studies on marketing procedures, export volume, and geographic markets.

Measures Used: Export Services

Based on prelininary field interviews with EMC principals and a review of the export literature, a large list of export services was developed. A second round of field interviews reduced the list to the 20 items which most represent the mix of transaction creating and physical fulfillment activities performed for suppliers. In the final mail questionnaire, respondents rated the importance of each service to their major supplier. A 6-point scale ranging from (1) *not a provided service* to (6) *a most important provided service* was used. Major supplier was defined as the firm whose products account for the largest dollar share of EMC sales.

The 20 items were factor analyzed using a varimax rotation which revealed four underlying export service dimensions. Items loading above .4 on each factor were combined to form an export service scale (see Appendix). The first scale, labeled promotion services, consisted of seven items dealing with various foreign market promotional activities, such as trade shows, advertising, selecting distributors, and tariff-related pricing. The second scale, technical export services, included six items dealing with foreign price quotes, documentation, export marking, and other export issues. The third scale, market contact services, consisted of four items referring to the EMC's physical presence in foreign markets. The final scale, consolidation services, included two items referring to the ability to consolidate overseas orders and shipments.

Since the promotion and market contact scales contained activities that stimulate foreign demand, these scales were treated as transaction creating export activities for hypothesis testing. Since the technical export and consolidation scales contained activities necessary to supply foreign demand, these scales were treated as physical fulfillment export activities.

Product Exported

Each EMC respondent was asked to describe the largest-selling product line that it obtained from its major supplier. A total of 130 different types of products were reported. Three judges independently classified the major product handled by the respondents into differentiated and undifferentiated product categories. An undifferentiated product was defined (Roehl 1982) for the judges as a standardized commondity item lacking brand identifiable characteristics and tending to be sold on the basis of price (i.e., basic chemicals, unprocessed food, steel). A differentiated product was defined as a unique, brandable item possessing distinctive characteristics and tending to be sold on the basis of nonprice factors (i.e., refrigerators, clothing, tools).

Interjudge reliability (Kassarjian 1977) is commonly assessed by the ratio of coding agreements between each pair of judges to the total number of coding decisions, in this case the 130 different product descriptions. The composite reliability score for the three judges was 86%, which is above the minimum reliability of 80% suggested by Kassarjian (1977). For the products without a three-judge consensus, classification decisions were based on the two-judge agreement. This procedure placed 126 respondents in the undifferentiated category and 132 respondents in the differentiated category.

Export Role

In the relationship with his/her major supplier, each respondent was asked to indicate whether he/she acted as a commercial agent or as a title taking merchant distributor. A total of 78 respondents were agents and 180 were merchants.

Suppliers Export Sales Volume

Each EMC respondent was asked to indicate his/her firm's annual export sales volume and the percentage of this figure due to the products obtained from the major supplier. These figures were combined and respondents were classified into groups representing suppliers exporting over or under $1 million. A total of 163 respondents were in the under $1 million category and 95 were in the over $1 million category.

Statistical Procedure

Multivariate analysis of variance (MANOVA) was used to test the first hypothesis. The remaining hypotheses were tested directly by a univariate ANOVA for each export service. The MANOVA used the four export service scales as dependent variables and product, role, and volume as independent variables, resulting in a $2 \times 2 \times 2$ design. Since a requirement of MANOVA is that the dependent variables be correlated, the appropriateness of the multivariate technique was tested by Bartlett's test of sphericity (Cooley and Lohnes 1971). The test (Bartlett's = 235.96 with 6df, $p < .001$) indicated that MANOVA is appropriate for analyzing the data.

Although all eight cells in the design contained responses, the Ns in the cells were unequal. As a consequence, a model comparison approach was used to test effects in the MANOVA model. Model comparison is recommended for the nonorthogonal case since each effect is tested eliminating the confounding influence of other factors (Appelbaum and Cramer 1974, Perreault and Darden 1975).

Results

Table 1 shows the results of the MANOVA for the first hypothesis. Since a significant overall main effect is found for the product, role, and volume factors ($p < .001, .007, .070$, respectively), H1 can be accepted. None of the possible two- or three-way interactions among the factors is significant, indicating the overall main effects can be interpreted directly. Thus, the results show the dependent variables, the four export service scales, dif-

fer, depending upon the type of product exported, the intermediaries' export roles, and the supplier's export volume (significant at the $p < .10$ level), as hypothesized.

For the second hypothesis, Table 1 shows the univariate results for product exported. Both of the transaction creating services, promotion and market contact, differ significantly ($p < .001, .006$) for the product groups. Of the two physical fulfilling services, only consolidation ($p < .063$) is marginally significant. The export service means (Table 2) for both promotion and market contact are in the direction hypothesized in H2a. However, the means for consolidation services are in a direction opposite to that hypothesized in H2b. Thus, only H2a is supported. Apart from technical export services, all of the export services examined are more important for differentiated products.

For the third hypothesis, Table 1 shows the univariate results for export role. Of the four export services, only consolidation ($p < .002$) significantly differs for the agent and merchant groups. No support is found for H3a since neither transaction-creating service is significant, and only partial support is found for H3b since only one physical fulfilling service is significant.

The only support for H3 is found in the importance of consolidation services. As hypothesized, this physical fulfilling activity receives a higher importance rating (Table 2) for merchants, $X = 2.84$, than for agents, $X = 2.31$. By taking domestic title to products, merchants assume full responsibility for completing the international exchange and, of course, capture the profits generated. Surprisingly, operating in this manner compared to merely acting as a commissioned agent does not affect the importance of most export activities. This suggests that role per se is not a particularly critical operating characteristic in terms of its affect on the export services the intermediary must be capable of providing client suppliers.

For the fourth hypothesis, Table 1 shows the univariate results for supplier export sales volume. Of the two transaction-creating services, only market contact ($p < .015$) is significant. Of the two physical fulfilling services, only consolidation ($p < .082$) is marginally significant. The ex-

TABLE 1 Summary for Importance of Export Services, Multi- and Univariate Analysis of Variance

Source of Variation[a]	Multivariate F-Ratio	Univariate F Ratio	Degrees of Freedom	P Less Than
Product (p)	5.04		4;247	.001
Promotion		14.55	1;250	.001
Technical export		.01	1;250	.963
Market contact		7.64	1;250	.006
Consolidation		3.49	1;250	.063
Export role (r)	3.65		4;247	.007
Promotion		.70	1;250	.340
Technical export		.91	1;250	.340
Market contact		.41	1;250	.520
Consolidation		10.00	1;250	.002
Supplier volume (v)	2.20		4;247	.070
Promotion		.65	1;250	.423
Technical export		.01	1;250	.936
Market contact		6.05	1;250	.015
Consolidation		3.05	1;250	.082
Interactions				
$p \times r$	1.07		4;247	.372
$p \times v$	1.33		4;247	.260
$v \times r$	.73		4;247	.575
$p \times r \times v$	1.30		4;247	.272

[a] Using a model comparison approach to test main effects, the significance test for product was computed eliminating confounding by role and volume, the test for role was computed eliminating confounding by product and volume, and the test for volume as computed eliminating confounding by product and role.

TABLE 2 Dependent Variable Means: Importance of Export Services[a]

Export services	For entire sample
Promotion	3.48
Technical export	4.06
Market contact	4.42
Consolidation	2.68

	For product		
Export services	Undifferentiated	Differentiated	p<
Promotion	3.21	3.73	.001
Technical export	4.06	4.06	NS.
Market contact	4.23	4.60	.006
Consolidation	2.55	2.81	.063

TABLE 2 *Continued*

Export services	For export role		
	Merchant	*Agent*	*p<*
Promotion	3.43	3.59	NS.
Technical export	4.10	3.96	NS.
Market contact	4.44	4.37	NS.
Consolidation	2.84	2.31	.002

Export services	For major supplier's export sales volume		
	Under $1 Million	*Over $1 Million*	*p<*
Promotion	3.44	3.55	NS.
Technical export	4.06	4.05	NS.
Market contact	4.29	4.64	.015
Consolidation	2.57	2.87	.082

[a] Higher means indicate greater importance for export services.

port service means (Table 2) for both promotion (not significant) and market contact are in the direction hypothesized in H4a. The means for consolidation are as hypothesized in H4b. Thus, the fourth hypothesis is partially supported.

The results of H4 indicate market contact and consolidation services are more important for EMCs whose major suppliers export over $1 million. When representing high volume suppliers, it is critical ($X = 4.64$) for intermediaries to have market contact capabilities, such as sales personnel in foreign markets and personal contact with foreign buyers. Since market contacts require years to develop and continuous foreign travel to maintain (Hay 1977), these capabilities are probably the most difficult for domestic suppliers to develop in-house. As a consequence, intermediaries retaining high volume suppliers possess this operating characteristic to a greater degree than those representing low volume clients. In addition, since the less important consolidation service is the only other export service differentiating high and low volume intermediaries, market contact services appear to be the critical ingredient in retaining suppliers.

Implications for Designing an American Export Trading Company

Several limitations must be considered before implications can be drawn about ATCs based on data collected from EMCs. First, ownership of the firms may differ, since EMCs tend to be owned by one or two principals, while ATCs may be jointly owned by manufacturers, banks, and other entities. Since complex ownership increases the number of decision makers and external stakeholders, ATC decision making may be less flexible and less opportunistic (Wind and Perlmutter 1977). However, the effect of ownership may be moderate; since banks and manufacturers generally lack essential export knowledge, it is expected that EMCs will participate and play a central role in many ATCs (Dooley 1984).

Second, a difference in scale is likely, since ATCs are expected to be more diversified in product categories and geographic markets (Cho 1984). Unlike EMCs, which tend to be small operations with few employees, ATCs have the potential to be much larger enterprises. With greater size and a more diversified product/market portfolio (Wind and Douglas 1981), ATCs face more complex resource allocation decisions. In making decisions about export services, ATCs must explicitly consider the costs, potential economies, and risks associated with joint operations of multiple products in multiple markets.

While these differences may make conclusions about EMCs only roughly applicable to ATCs, relationships found in the present study may have broad application. Since both EMCs and ATCs are indirect export intermediaries, each must be organized to provide services consistent with basic operating characteristics. For example, the data indicate that, regardless of role, the importance of export services remains largely the same for EMCs. Even considering differences in ownership and scale, this result suggests that the decision to operate an ATC as an agent or merchant intermediary does not have major implications for service provision.

However, decisions about the type of product to export and the type of supplier to represent appear to fundamentally affect the export services EMCs as well as ATCs must be capable of supplying.

A decision to export differentiated products implies the ATC must provide very effective transaction creation and physical fulfillment services. Compared to intermediaries handling undifferentiated products, these ATCs must have a greater physical presence in foreign markets. ATC sales personnel should personally call on foreign buyers, attend overseas trade shows, prepare foreign advertising, and otherwise engage in activities to stimulate foreign demand. In addition, these firms must have a superior shipment and order consolidation capability. To provide these services, the ATC would probably need to maintain foreign-based sales offices. Unlike EMCs, however, ATCs may experience greater synergy in providing these services, since deploying a foreign salesforce across multiple products in multile markets may be more efficient and efective (Wind and Douglas 1981).

The data also suggest an ATC's decision to represent suppliers with a high or low export potential has implications for the services it must provide. ATCs intending to establish relationships with suppliers capable of generating export volumes in excess of $1 million will require very strong market contact and consolidation capabilities. High volume suppliers are likely to look toward the ATC's intimate contact with foreign decision makers as the major reason to continue the indirect export relationship. This is a crucial point, since the major question facing ATCs, according to some analysts, is ". . . whether or not they will find and keep a sufficient regular customer base to make their services worth the effort" (Maker 1982, p. 66).

Product and Supplier Decisions

Separate discussions of product and supplier decisions fail to account for the joint effect these decisions have on the importance of services. Since any given ATC must select both products and suppliers, the export services provided ought to reflect both decisions. Fortunately, the individual effects of each operating characteristic shown in Table 2 can be combined since no interactions were found in the MANOVA (Cooley and Lohnes 1971). For example, the net requirement for exporting the differentiated goods of low volume suppliers is approximately an average capability for market contact. This occurs because the means for market contact across this type of product and supplier balance out and approximate the total sample mean.

Figure 1 summarizes the requirements for each export service in terms of the two critical operating decisions, product and supplier selection. These decisions have such a substantial impact on services that each product supplier configuration requires a unique trading organization. The only common trait across these firms is a need to possess an expertise in technical export services.

Undifferentiated, Low Volume

An ATC exporting the undifferentiated products of low volume suppliers requires a less than average capability in the transaction creating and consolidation services. When representing these suppliers, ATCs that specialize in standardized commodity items such as unprocessed food or basic chemicals do not need particularly aggres-

FIGURE 1 Service requirements for American export trading companies.

Products Exported

Suppliers represented	*Undifferentiated*	*Differentiated*
Low export volume	Requires a less than average capability in promotion, market contact, and consolidation.	Requires an above average capability in promotion, but an average capability in market contact and consolidation.
High export volume	Requires a less than average capability in promotion, but an average capability in market contact and consolidation.	Requires an above average capability in promotion, market contact, and consolidation.

sive promotional and market contact capabilities in foreign markets. Rather, the undifferentiated products of suppliers with a modest export potential probably can be handled by corresponding with foreign distributors. Product shipment can be delegated to an independent freight forwarder. Thus, these ATCs can function effectively without making investments in foreign sales offices and specialized logistic facilities.

Differentiated, Low Volume

When representing low volume suppliers, ATCs specializing in undifferentiated products require an above average capability in promotion. These firms need to participate in foreign trade shows, advertise in foreign media, and be able to select very competent foreign distributors. However, only an average capability in personal contacts with foreign customers is required. This suggests the transaction creating function can be handled through overseas travel by ATC personnel. The salesforce should visit foreign markets to coordinate promotional activities but need not maintain extremely close personal contact with foreign buyers. Similarly, only an average physical fulfillment capability is necessary.

Undifferentiated, High Volume

An ATC representing suppliers with a high export volume of undifferentiated products requires a less than average capability in promotion but an average ability in market contact and consolidation. These firms must maintain personal contact with foreign customers and have a good capability in physical fulfillment. However, because they specialize in undifferentiated products, promotion at foreign trade shows and through foreign media is not a high requirement. The organization needed to supply this service mix includes some form of foreign market presence. At the minimum, extended visits to overseas markets by ATC personnel is necessary to maintain customer contact and ensure proper physical fulfillment. In some cases it may be necessary to staff overseas sales offices.

Differentiated, High Volume

ATCs exporting the differentiated products of high volume suppliers require the capacity to provide above average export services. These firms must be organized for aggressive transaction creating activities. An experienced and highly qualified field salesforce must be maintained. Further, foreign based sales offices are probably necessary to coordinate promotional programs and ensure a high level of personal contact with foreign customers. Since superior physical fulfillment is required, foreign logistic facilities also may be needed.

Conclusions

While the analysis suggests that product exported and supplier volume have fundamental implications for service provision, future research should assess the effect of product and market diversification on ATCs. Since a superior financial base may allow ATCs to export a mix of differentiated and/or undifferentiated products to many foreign markets, the effect of interdependencies among these actions must be analyzed (Wind and Douglas 1981). For example, ATCs need to evaluate whether export services can be standardized across their entire portfolio, or whether services

APPENDIX Questionnaire Items Used For Export Service Scales

Scale Name	Questionnaire Items[a]	Factor Loading	Coefficient Alpha
Promotion services	1. Promotion of supplier's product at foreign trade shows.	.68	.85
	2. Ability to pioneer new foreign markets for supplier's product.	.66	
	3. Preparation of advertising and sales literature for use in foreign markets.	.60	
	4. Ability to select competent foreign distributors.	.58	
	5. Ability to develop a marketing strategy for foreign markets.	.57	
	6. Ability to train and instruct foreign sales network in product.	.56	
	7. Ability to inform supplier on impending tariff increases.	.51	
Technical export services	1. Arranging for cost, insurance, and freight quotes in response to inquiries from abroad.	.73	.80
	2. Understanding of export documentation requirements.	.65	
	3. Advising on or arranging for all export packaging and marking.	.64	
	4. Knowledge of international transportation practices.	.57	
	5. Assumption of responsibility for physical delivery of product to foreign buyer.	.44	
	6. Evaluation of credit risk associated with foreign buyers.	.43	
Market contact services	1. Sales personnel calling on foreign customers in person.	.63	.76
	2. Understanding of foreign market competitive conditions.	.59	
	3. Personal contacts with potential foreign buyers.	.59	
	4. Ability to provide or arrange after-sale support to foreign buyers.	.48	
Consolidation services	1. Ability to consolidate overseas shipments with products of other suppliers to lower freight costs.	.59	.63
	2. Ability to consolidate orders to the supplier from several overseas customers.	.59	

[a] A six-point response scale was used for each service item ranging from (1) *not a provided service,* to (6) *a most important provided service.*

must be individualized to specific foreign markets and products. Effective guidance will enable ATCs to function more effectively and to compete better in the global market.

References

Abegglen, James C. and George Stalk (1983). "Japanese Trading Companies: A Dying Industry?," *The Wall Street Journal,* 202 (July 18), 19.

Appelbaum, Mark I. and Elliot M. Cramer (1974). "Some Problems in the Nonorthogonal Analysis of Variance," *Psychological Bulletin,* 81 (no. 6), 335–343.

Barovick, Richard L. (1983), "New Trading Company Law Heightens Interest in Exporting," *Business America,* 6 (March), 6–10.

Bello, Daniel C. and Nicholas C. Williamson (1985). "Contractual Arrangement and Marketing Practices in the Indirect Export Channel," *Journal of International Business Studies,* forthcoming.

Bowersox, Donald J. (1978), *Logistical Management.* New York: Macmillan.

Brasch, John J. (1978). "Export Management Companies," *Journal of International Business Studies,* 9 (Spring/Summer), 59–71.

Cao, A. D. (1981). "U.S. Export Trading Company: A Model of Export Promotion in the 1980's," *Business,* 31 (September/October), 32–38.

Cho, Dong Sung (1984), "The Anatomy of the Korean General Trading Company," *Journal of Business Research*, 12 (June), 241–255.

Cooley, William W. and Paul R. Lohnes (1971), *Multivariate Data Analysis*, New York: Wiley.

Daido, Eisuko (1976), "Why Are They General Trading Firms?," *Japanese Economic Studies*, 4 (Summer), 44–62.

Dooley, Brian (1984), "Export Management Companies," *American Import Export Management*, 99 (May), 48–59.

Hay Associates (1977), "A Study to Determine the Feasibility of the Export Trading Company Concept as a Viable Vehicle for Expansion of United States Exports," U.S. Department of Commerce, (March), 1–130.

Hurt, William E. (1982), "Export Trading Companies: Forging A New Sales Tool," *Business America*, 5 (May), 2–5.

Kassarjian, Harold H. (1977), "Content Analysis in Consumer Research," *Journal of Consumer Research*, 4 (June), 8–18.

Kunio, Yoshihara (1982). *Sogo Shosha: The Vanguard of the Japanese Economy*, New York: Oxford University Press.

Maker, Phillip (1982), "Trading Companies: A U.S. Export Panacea?," *Industrial Marketing*, 67 (October), 59–68.

Miller, Larry (1982), "Pricing EMC's," *American Import Export Management*, 97 (July), 16–19.

Perreault, William D. and William R. Darden (1975), "Unequal Cell Sizes in Marketing Experiments: Use of the General Linear Hypothesis," *Journal of Marketing Research*, 12 (August), 333–342.

Roehl, Thomas (1982), "The General Trading Companies: A Transactions Cost Analysis of Their Function in the Japa

nese Economy," in *Marketing Channels: Domestic and International Perspectives*, M. G. Harvey and R. F. Lusch., eds., Normal, OK: Center for Economic Research, The University of Oklahoma, 86–100.

Scouton, William (1982), "Exporting Trading Companies: A New Tool for American Business," *Business America*, 5 (October), 3–6.

Shimizu, Ryuei (1980), *The Growth of the Firm in Japan*, Tokyo: Keio Tsushin, Ltd.

Terpstra, Vern (1983), "Critical Mass and International Marketing Strategy," *Journal of the Academy of Marketing Science*, 11 (Summer), 269–282.

U.S. Department of Commerce (1981), *Directory of U.S. Export Management Companies*, Washington, DC: International Trade Administration, 1–188.

Weiner, Steve and Robert Johnson (1984), "Export Trading Firms in U.S. Are Failing to Fulfill Promise," *The Wall Street Journal*, 203 (May 24), 1.

Williams, Harold R. and G. M. Baliga (1983), "The U.S. Export Trading Company Act of 1982," *Journal of World Trade Law*, 17 (May), 224–235.

Wind, Yoram and Susan Douglas (1981), "International Portfolio Analysis and Strategy: The Challenge of the 80s," *Journal of International Business Studies*, 12 (Fall), 69–82.

——— and Howard Perlmutter (1977), "On the Identification of Frontier Issues in Multinational Marketing," *Columbia Journal of World Business*, 12 (Winter), 131–139.

Young, Alexander (1979), *The Sogo Shosha: Japan's Multinational Marketing Companies*, Boulder, CO: Westview Press.

CHAPTER 6

Canonical Correlation Analysis

LEARNING OBJECTIVES

An understanding of the most important concepts in canonical correlation analysis should enable you to do the following:

- State the similarities and differences between multiple regression, factor analysis, discriminant analysis, and canonical correlation.

- Summarize the conditions that must be met for application of canonical correlation analysis.

- State what the canonical root measures and point out its limitations.

- State how many independent canonical functions can be defined between the two sets of original variables.

- Compare the advantages and disadvantages of the three methods for interpreting the nature of canonical functions.

- Define redundancy and compare it with multiple regression's R^2.

CHAPTER PREVIEW

Until recent years, canonical correlation analysis was a relatively unknown statistical technique. The availability of computer programs has facilitated its increased application to research problems. It is particularly useful in situations where multiple output measures like satisfaction, purchase, or sales volume are available. If the predictor variables were only categorical, multivariate analysis of variance could be used. But what if the predictor variables are metric? Canonical correlation is the answer, allowing for the assessment of the relationship between metric predictor variables and multiple dependent measures. As discussed in Chapter 1, canonical correlation is considered to be the general model on which many other multivariate techniques are based because it can use both metric and nonmetric data for either the dependent or independent variables. We express the general form of canonical analysis as

$$Y_1 + Y_2 + Y_3 + \ldots + Y_n = X_1 + X_2 + X_3 + \ldots + X_n$$

(metric, nonmetric)   (metric, nonmetric)

This chapter introduces the data analyst to the multivariate statistical technique of canonical correlation analysis. Specifically, we (1) describe the nature of canonical correlation analysis, (2) illustrate its application, and (3) discuss its potential advantages and limitations. Before reading the chapter, you should familiarize yourself with the key terms.

KEY TERMS

Before starting the chapter, review the key terms to develop an understanding of the concepts and terminology used. Throughout the chapter the key terms appear in **boldface.** Other points of emphasis in the chapter are *italicized.* Also, cross-references within the key terms are in *italics.*

Canonical correlation Measure of the strength of the overall relationships between the linear composites of the predictor and the criterion sets of variables. In effect, it represents the bivariate correlation between the two linear composites.

Canonical function Relationship (correlational) between two linear composites. Each canonical function has two separate linear composites (*canonical variates*), one for the set of criterion variables and one for the set of predictor variables. The strength of the relationship is given by the *canonical correlation*.

Canonical loadings Also known as canonical structure correlations, they measure the simple linear correlation between the independent variables and their respective canonical variates and can be interpreted like factor loadings.

Canonical roots Squared *canonical correlations*, which provide an estimate of the amount of shared variance between the respective optimally weighted linear composites (*canonical variate*) of criterion and predictor variables.

Canonical variates Also referred to as linear combinations, linear compounds, and linear composites, they represent the weighted sum of two or more variables and can be defined for either criterion or predictor variables.

Criterion variables Dependent variables.

Eigenvalues See *canonical roots.*

Linear composites See *canonical variates.*

Orthogonal   Mathematical constraint specifying that the canonical functions are independent of each other. In other words, the canonical functions are derived so that each is at a right angle to all other functions when plotted in multivariate space, thus ensuring statistical independence between the canonical functions.

Predictor variables   Independent variables.

Redundancy index   Amount of variance in a canonical variate (dependent or independent) explained by the other canonical variate in the canonical function. It can be computed for both the dependent and the independent canonical variates in each canonical function. For example, a redundancy index of the dependent variate represents the amount of variance in the dependent variables explained by the independent canonical variate.

What Is Canonical Correlation?

In Chapter 3 you studied multiple regression analysis, which can predict the value of a single (metric) criterion variable from a linear function of a set of predictor (independent) variables. For some research problems, interest may not center on a single criterion (dependent) variable; rather, the analyst may be interested in relationships between sets of multiple criterion and multiple predictor variables. **Canonical correlation analysis** is a multivariate statistical model that facilitates the study of interrelationships among sets of multiple criterion (dependent) variables and multiple predictor (independent) variables [6, 7]; that is, whereas multiple regression predicts a single dependent variable from a set of multiple independent variables, canonical correlation simultaneously predicts multiple dependent variables from multiple independent variables.

Canonical correlation places the fewest restrictions on the types of data on which it operates. Because the other techniques impose more rigid restrictions, it is generally believed that the information obtained from them is of higher quality and may be presented in a more interpretable manner. For this reason, many researchers view canonical correlation as a last-ditch effort, to be used when all other higher-level techniques have been exhausted. But in situations with multiple dependent and independent variables, canonical correlation is the most appropriate and powerful multivariate technique.

Hypothetical Example of Canonical Correlation

To clarify further the nature of canonical correlation, let us consider an extension of the example used in Chapter 3. Recall that the HATCO survey results used family size and income as predictors of the number of credit cards a family would hold. The problem involved examining the relationship between two independent variables and a single dependent variable.

Suppose HATCO was interested in the broader concept of credit usage by consumers. To measure this concept, it seems logical that HATCO should consider

TABLE 6.1 Canonical Correlation of Credit Usage (Number of Credit Cards and Usage Rate) with Customer Characteristics (Family Size and Family Income)

Measures of Credit Usage		*Measures of Customer Characteristics*
Number of credit cards held by the family		Family size
Average monthly dollar expenditures on all credit cards		Family Income
Composite of Dependent (Criterion) Variables	*Canonical Correlation*	*Composite of Independent (Predictor) Variables*
Multiple dependent variables	}R_c{	Multiple independent variables

not only the number of credit cards held by the family but also the family's average monthly dollar charges on all credit cards. The problem involves predicting two dependent measures simultaneously (number of credit cards and average dollar charges), and multiple regression is capable of handling only a single dependent variable. Thus, canonical correlation should be used because it is able to examine the relationship with multiple dependent variables.

The problem of predicting credit usage is illustrated in Table 6.1. The two dependent variables used to measure credit usage—number of credit cards held by the family and average monthly dollar expenditures on all credit cards—are listed at the left. The two independent variables selected to predict credit usage—family size and family income—are shown on the right. By using canonical correlation analysis, HATCO can predict a composite measure of credit usage that consists of both dependent variables, rather than having to compute a separate regression equation for each of the dependent variables. The result of applying canonical correlation is a measure of the strength of the relationship between two sets of multiple variables (variates). This measure is expressed as a canonical correlation coefficient (R_c) between the two variates.

Analyzing Relationships with Canonical Correlation

Canonical correlation analysis is the most generalized member of the family of multivariate statistical techniques. It is directly related to several dependence methods. Similar to regression, canonical correlation's goal is to quantify the strength of the relationship, in this case between the two sets of variables (independent and dependent). It also resembles discriminant analysis in its ability to determine independent dimensions (similar to discriminant functions) for *each* variable set that produces the maximum correlation between the dimensions. Thus, canonical correlation identifies the optimum structure or dimensionality of each variable set that maximizes the relationship between independent and dependent variable sets.

Our discussion of canonical correlation analysis is organized around the model-building process described in Chapter 1. Figure 6.1 depicts the stages for canonical correlation analysis, which include (1) specifying the objectives of canonical correlation, (2) developing the analysis plan, (3) assessing the assumptions underlying canonical correlation, (4) estimating the canonical model and assessing overall model fit, (5) interpreting the canonical variates, and (6) validating the model.

Stage One: Objectives of Canonical Correlation Analysis

As noted in the preceding description, canonical analysis is a method that deals with a composite association between sets of multiple criterion and predictor variables. This technique makes it possible to develop a number of independent **canonical functions** that maximize the correlation between the **linear composites** of sets of **criterion** and **predictor** variables. Each canonical function is actually based on the correlation between two **canonical variates,** one variate for the dependent variables and one for the independent variables. The unique feature of canonical correlation is that the variates are derived to maximize their correlation. Moreover, canonical correlation does not stop with the derivation of a single relationship between the sets of variables. Instead, a number of canonical functions (pairs of canonical variates) may be derived.

The basic input data for canonical correlation analysis are two sets of variables. We assume that each set can be given some theoretical meaning, at least to the extent that one set could be defined as the independent variables and the other as the dependent variables. Once this distinction has been made, canonical correlation can address a wide range of objectives, which may be any or all of the following:

1. Determining whether two sets of variables (measurements made on the same objects) are independent of one another or, conversely, determining the magnitude of the relationships that may exist between the two sets.
2. Deriving a set of weights for each set of criterion and predictor variables so that the linear combinations of each set are maximally correlated. Additional linear functions that maximize the remaining correlation are independent of the preceding set(s) of linear compounds.
3. Explaining the nature of whatever relationships exist between the sets of criterion and predictor variables, generally by measuring the relative contribution of each variable to the canonical functions (relationships) that are extracted.

Stage Two: Designing a Canonical Correlation Analysis

As the most general form of multivariate analysis, canonical correlation analysis shares basic implementation issues common to all multivariate techniques. Discussions in other chapters (particularly multiple regression, discriminant analysis, and factor analysis) on the impact of sample size (both small and large),

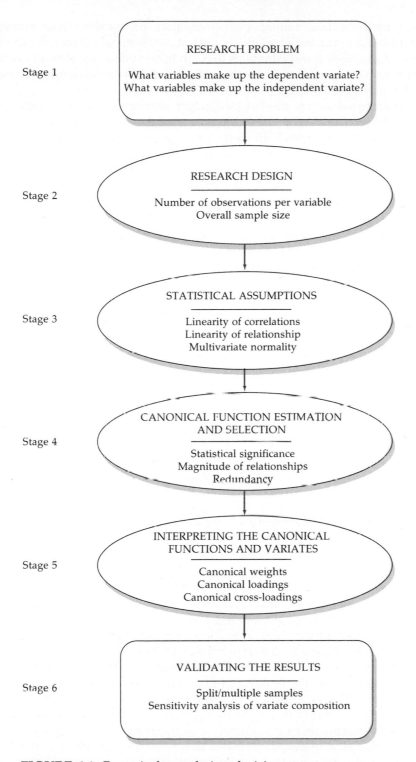

FIGURE 6.1 Canonical correlation decision process.

the necessity for sufficient number of observations per variable (a recommended level is 10 or more per variable), the impact of measurement error, and the types of variables and their transformations that can be included are relevant to canonical correlation analysis as well. We suggest you review those chapters if you are unfamiliar with the issues.

The classification of variables as dependent or independent is of little importance for the statistical estimation of the canonical functions, because canonical correlation analysis weights both variates to maximize the correlation and places no particular emphasis on either variate. Yet because the technique produces variates to maximize the correlation between them, a variable in either set relates to all other variables in both sets. This allows the addition or deletion of a single variable to affect the entire solution, particularly the other variate. The composition of each variate, either independent or dependent, becomes critical. A researcher must have conceptually linked sets of variables before applying canonical correlation analysis. This makes the specification of dependent versus independent variates essential to establishing a strong conceptual foundation for the variables.

Stage Three: Assumptions in Canonical Correlation

The generality of canonical correlation analysis also extends to its underlying statistical assumptions. The assumption of linearity affects two aspects of canonical correlation results. First, the correlation coefficient between any two variables is based on a linear relationship. If the relationship is nonlinear, then one or both variables should be transformed if possible. Second, the canonical correlation is the linear relationship between the variates. If the variates relate in a nonlinear manner, the relationship will not be captured by canonical correlation. Thus, while canonical correlation analysis is the most generalized multivariate method, it is still constrained to identifying linear relationships.

Canonical correlation analysis can accommodate any metric variable without the strict assumption of normality. Normality is desirable because it standardizes a distribution to allow for a higher correlation among the variables. But in the strictest sense, canonical correlation analysis can accommodate even nonnormal variables if the distributional form (e.g., highly skewed) does not decrease the correlation with other variables. This allows for transformed nonmetric data (in the form of dummy variables) to be used as well. However, multivariate normality is required for the statistical inference test of the significance of each canonical function. Because tests for multivariate normality are not readily available, the prevailing guideline is to ensure that each variable has univariate normality. Thus, although normality is not strictly required, it is highly recommended that all variables be evaluated for normality and transformed if necessary.

Homoscedasticity, to the extent that it decreases the correlation between variables, should also be remedied. Finally, multicollinearity among either variable set will confound the ability of the technique to isolate the impact of any single variable, making interpretation less reliable. Readers unfamiliar with these statistical assumptions, the tests for their diagnosis, or the alternative remedies when the assumptions are not met should refer to Chapter 2.

Stage Four: Deriving the Canonical Functions and Assessing Overall Fit

The first step of canonical correlation analysis is to derive one or more canonical functions. Each function consists of a pair of variates, one representing the independent variables and the other representing the dependent variables. The maximum number of canonical variates (functions) that can be extracted from the sets of variables equals the number of variables in the smallest data set, independent or dependent. For example, when the research problem involves five independent (predictor) variables and three dependent (criterion) variables, the maximum number of canonical functions that can be extracted is three.

The derivation of successive canonical variates is similar to the procedure used with unrotated factor analysis (see Chapter 7); that is, the first factor extracted accounts for the maximum amount of variance in the set of variables. Then the second factor is computed so that it accounts for as much as possible of the variance not accounted for by the first factor, and so forth, until all factors have been extracted. Therefore, successive factors are derived from residual or leftover variance from earlier factors. Canonical correlation analysis follows a similar procedure but focuses on accounting for the maximum amount of the relationship between the two sets of variables, rather than within a single set. The result is that the first pair of canonical variates is derived so as to have the highest intercorrelation possible between the two sets of variables. The second pair of canonical variates is then derived so that it exhibits the maximum relationship between the two sets of variables (variates) not accounted for by the first pair of variates. In short, successive pairs of canonical variates are based on residual variance, and their respective canonical correlations (which reflect the interrelationships between the variates) become smaller as each additional function is extracted; that is, the first pair of canonical variates exhibits the highest intercorrelation, the next pair the second-largest correlation, and so forth.

One additional point about the derivation of canonical variates: as noted, successive pairs of canonical variates are based on residual variance. Therefore, each of the pairs of variates is **orthogonal** and independent of all other variates derived from the same set of data.

The strength of the relationship between the pairs of variates is reflected by the canonical correlation. When squared, the canonical correlation represents the amount of variance in one canonical variate accounted for by the other canonical variate. This also may be called the amount of shared variance between the two canonical variates. Squared canonical correlations are called **canonical roots** or **eigenvalues.**

Which Canonical Functions Should Be Interpreted?

As with research using other statistical techniques, the most common practice is to analyze functions whose canonical correlation coefficients are statistically significant beyond some level, typically .05 or above. If other independent functions are deemed insignificant, these relationships among the variables are not interpreted. Interpretation of the canonical variates in a significant function is based on the premise that variables in each set that contribute heavily to shared variances for these functions are considered to be related to each other.

The authors believe that the use of a single criterion such as the level of significance is too superficial. Instead, they recommend that three criteria be used in conjunction with one another to decide which canonical functions should be interpreted. The three criteria are (1) level of statistical significance of the function, (2) magnitude of the canonical correlation, and (3) redundancy measure for the percentage of variance accounted for from the two data sets.

Level of Significance

The level of significance of a canonical correlation generally considered to be the minimum acceptable for interpretation is the .05 level, which (along with the .01 level) has become the generally accepted level for considering a correlation coefficient statistically significant. This consensus has developed largely because of the availability of tables for these levels. These levels are not necessarily required in all situations, however, and researchers from various disciplines frequently must rely on results based on lower levels of significance. The most widely used test, and the one normally provided by computer packages, is the *F* statistic, based on Rao's approximation [3].

In addition to separate tests of each canonical function, a multivariate test of all canonical roots can also be used for evaluating the significance of canonical roots. Many of the measures for assessing the significance of discriminant functions, including Wilks' lambda, Hotelling's trace, Pillai's trace, and Roy's greatest root, are also provided. See Chapter 4 for a discussion of these measures.

Magnitude of the Canonical Relationships

The practical significance of the canonical functions, represented by the size of the canonical correlations, also should be considered in deciding which functions to interpret. No generally accepted guidelines have been established regarding acceptable sizes for canonical correlations. Rather, the decision is usually based on the contribution of the findings to better understanding of the research problem being studied. It seems logical that the guidelines suggested for significant factor loadings (see Chapter 7) might be useful with canonical correlations, particularly when one considers that canonical correlations refer to the variance explained in the canonical variates (linear composites), not the original variables.

Redundancy Measure of Shared Variance

Recall that squared canonical correlations (roots) provide an estimate of the shared variance between the canonical variates. Although this is a simple and appealing measure of the shared variance, it may lead to some misinterpretation, because the squared canonical correlations represent the variance shared by the linear composites of the sets of criterion and predictor variables, and not the variance extracted from the sets of variables [1]. Thus, a relatively strong canonical correlation may be obtained between two linear composites (canonical variates), even though these linear composites may not extract significant portions of variance from their respective sets of variables [9].

Because canonical correlations may be obtained that are considerably larger than previously reported bivariate and multiple correlation coefficients, there may be a temptation to assume that canonical analysis has uncovered substantial relationships of conceptual and practical significance. Before such conclusions are warranted, however, further analysis involving measures other than canonical correlations must be undertaken to determine the amount of the dependent variable variance accounted for or shared with the independent variables [8].

To overcome the inherent bias and uncertainty in using canonical roots (squared canonical correlations) as a measure of shared variance, a **redundancy index** has been proposed [13]. It is the equivalent of computing the squared multiple correlation coefficient between the total predictor set and each variable in the criterion set, and then averaging these squared coefficients to arrive at an average R^2. It provides a summary measure of the ability of a set of predictor variables (taken as a set) to explain variation in the criterion variables (taken one at a time). As such, the redundancy measure is perfectly analogous to multiple regression's R^2 statistic, and its value as an index is similar.

The Stewart-Love index of redundancy calculates the amount of variance in one set of variables that can be explained by the variance in the other set. This index serves as a measure of accounted-for variance, similar to the R^2 calculation used in multiple regression. The R^2 represents the amount of variance in the dependent (criterion) variable explained by the regression function of the independent (predictor) variables. In regression, the total variance in the dependent variable is equal to 1, or 100 percent. Remember that canonical correlation is different from multiple regression in that it does not deal with a single criterion variable but has a criterion set that is a composite of several variables, and this composite has only a portion of each dependent variable's total variance. For this reason, we cannot assume that 100 percent of the variance in the criterion set is available to be explained by the predictor set. The predictor set of variables can be expected to account only for the shared variance in the criterion canonical variate. For this reason, the calculation of the redundancy index is a two-step process. The first step involves calculating the amount of shared variance from the set of criterion variables included in the criterion canonical variate. The second step involves calculating the amount of variance in the criterion canonical variate that can be explained by the predictor canonical variate. The redundancy index is then found by multiplying these two components.

>   **Step 1.** To calculate the amount of shared variance in the criterion set included in the criterion canonical variate, let us first consider how the regression R^2 statistic is calculated. The R^2 is simply the square of the correlation coefficient R, which represents the correlation between the actual dependent variable and the predicted value. In the canonical case, we are concerned with the correlation between the criterion canonical variate and each of the criterion variables. Such information can be obtained from the canonical loadings (L_1), which represent the correlation between each input variable and its own canonical variate (discussed in more detail in the following section). By squaring each of the criterion loadings (L_i^2), one may obtain a measure of the amount of variation in each of the criterion variables explained by the criterion canonical variate. To calculate the amount of shared variance explained by the canonical variate, a simple average of the squared loadings is used.
>
>   **Step 2.** The second step of the redundancy process involves the percentage of variance in the criterion canonical variate that can be explained by the predictor canonical variate. This is simply the squared correlation be-

tween the predictor canonical variate and the criterion canonical variate, which is otherwise known as the canonical correlation. The squared canonical correlation is commonly called the canonical R^2.

Step 3. The redundancy index of a variate is then derived by multiplying the two components (shared variance of the variate by the squared canonical correlation) to find the amount of shared variance that can be explained by each canonical function. To have a high redundancy index, one must have a high canonical correlation and a high degree of shared variance explained by the criterion variate. A high canonical correlation alone does not ensure a valuable canonical function. Redundancy indices are calculated for both the dependent and the independent variates, although in most instances the researcher is concerned only with the variance extracted from the dependent variable set, which provides a much more realistic measure of the predictive ability of canonical relationships.

What is the minimum acceptable redundancy index needed to justify the interpretation of canonical functions? Just as with canonical correlations, no general accepted guidelines have been established. The analyst must judge each canonical function in light of its theoretical and practical significance to the research problem being investigated to determine whether the redundancy index is sufficient to justify interpretation. Also, a test for the significance of the redundancy index has been developed [2], although it has not been widely utilized.

Stage Five: Interpreting the Canonical Variate

If the canonical relationship is statistically significant and the magnitudes of the canonical root and the redundancy index are acceptable, the analyst still needs to make substantive interpretations of the results. Making these interpretations involves examining the canonical functions to determine the relative importance of each of the original variables in the canonical relationships. Three methods have been proposed: (1) canonical weights (standardized coefficients), (2) canonical loadings (structure correlations), and (3) canonical cross-loadings.

Canonical Weights

The traditional approach to interpreting canonical functions involves examining the sign and the magnitude of the canonical weight assigned to each variable in its canonical variate. Variables with relatively larger weights contribute more to the variates, and vice versa. Similarly, variables whose weights have opposite signs exhibit an inverse relationship with each other, and variables with weights of the same sign exhibit a direct relationship. However, interpreting the relative importance or contribution of a variable by its canonical weight is subject to the same criticisms associated with the interpretation of beta weights in regression techniques. For example, a small weight may mean either that its corresponding variable is irrelevant in determining a relationship or that it has been partialed out of the relationship because of a high degree of multicollinearity. Another problem with the use of canonical weights is that these weights are subject to considerable instability (variability) from one sample to another. This instability occurs because the computational procedure for canonical analysis yields weights that maximize the canonical correlations for a particular sample of observed dependent and independent variable sets [8]. These problems suggest considerable caution in using canonical weights to interpret the results of a canonical analysis.

Canonical Loadings

Canonical loadings have been increasingly used as a basis for interpretation because of the deficiencies inherent in canonical weights. **Canonical loadings,** also called canonical structure correlations,* measure the simple linear correlation between an original observed variable in the dependent or independent set and the set's canonical variate. The canonical loading reflects the variance that the observed variable shares with the canonical variate and can be interpreted like a factor loading in assessing the relative contribution of each variable to each canonical function. The methodology considers each independent canonical function separately and computes the within-set variable-to-variate correlation [10]. The larger the coefficient, the more important it is in deriving the canonical variate. Also, the criteria for determining the significance of canonical structure correlations are the same as with factor loadings (see Chapter 7).

Canonical loadings, like weights, may be subject to considerable variability from one sample to another. This variability suggests that loadings, and hence the relationships ascribed to them, may be sample-specific, resulting from chance or extraneous factors [8]. Although canonical loadings are considered relatively more valid than weights as a means of interpreting the nature of canonical relationships, the analyst still must be cautious when using loadings for interpreting canonical relationships, particularly with regard to the external validity of the findings.

Canonical Cross-Loadings

The computation of canonical cross-loadings has been suggested as an alternative to conventional loadings [5]. This procedure involves correlating each of the original observed dependent variables directly with the independent canonical variate, and vice versa. Recall that conventional loadings correlate the original observed variables with their respective variates after the two canonical variates (dependent and independent) are maximally correlated with each other. This may also seem similar to multiple regression but differs in that each independent variable, for example, is correlated with the dependent variate instead of a single dependent variable. Thus cross-loadings provide a more direct measure of the dependent-independent variable relationships by eliminating an intermediate step involved in conventional loadings.

---

*Some canonical analyses do not compute correlations between the variables and the variates. In such cases the canonical weights are considered comparable but not equivalent for purposes of our discussion.

Which Interpretation Approach to Use

Several different methods for interpreting the nature of canonical relationships were discussed. The question remains, however: Which method should the analyst use? Because most canonical problems require a computer, the analyst frequently must use whichever method is available in the standard statistical packages. Cross-loadings is the preferred approach and is provided by many computer programs, such as SAS [9, 10]. If the cross-loadings are not available, the analyst is forced either to compute the cross-loadings by hand or to select another method of interpretation. The SPSS statistical [11, 12] package does provide canonical loadings, while the BMDP package [4] provides canonical weights. The canonical loadings approach is somewhat more valid than the use of weights. Therefore, whenever possible the loadings approach is recommended as a second alternative to the canonical cross-loadings method.

Stage Six: Validation and Diagnosis

As with any other multivariate technique, canonical correlation analysis should be subjected to validation methods to ensure that the results are not specific only to the sample data and can be generalized to the population. The most direct procedure is to create two subsamples of the data (if sample size allows) and perform the analysis on each subsample separately. Then the results can be compared for similarity of canonical functions, variate loadings, and the like. If marked differences are found, the researcher should consider additional investigation to ensure that the final results are representative of the population values, not those solely of a single sample. Another approach is to assess the sensitivity of the results to the removal of a dependent and/or independent variable. Because the canonical correlation procedure maximizes the correlation and does not optimize the interpretability, the canonical weight and loadings may vary substantially if one variable is removed from either variate. To ensure the stability of the canonical weights and loading, the researcher should estimate multiple canonical correlations, each removing a different independent or dependent variable.

Although there are few diagnostic procedures developed specifically for canonical correlation analysis, the researcher should view the results within the limitations of the technique. Among the limitations that can have the greatest impact on the results and their interpretation are the following:

1. The canonical correlation reflects the variance shared by the linear composites of the sets of variables, not the variance extracted from the variables.
2. Canonical weights derived in computing canonical functions are subject to a great deal of instability.
3. Canonical weights are derived to maximize the correlation between linear composites, not the variance extracted.
4. The interpretation of the canonical variates may be difficult because they are calculated to maximize the relationship, and there are no aids for interpretation such as rotation of variates as seen in factor analysis.

5. It is difficult to identify meaningful relationships between the subsets of independent and dependent variables because precise statistics have not yet been developed to interpret canonical analysis, and we must rely on inadequate measures like loadings or cross-loadings [8].

These limitations are not meant to discourage the use of canonical correlation. Rather, they are pointed out to enhance the effectiveness of canonical correlation as a research tool.

An Illustrative Example

To illustrate the application of canonical correlation, we use variables drawn from the database introduced in Chapter 1. Recall that the data consisted of a series of measures obtained on a sample of 100 HATCO customers. The variables included ratings of HATCO on seven attributes (X_1 to X_7) and two measures reflecting the effects of HATCO's efforts (X_9, usage of HATCO products; and X_{10}, customer satisfaction with HATCO).

As with previous chapters, the discussion of this application of canonical correlation analysis follows the six-stage process discussed earlier in the chapter. At each stage the results illustrating the decisions in that stage are examined.

Stage 1: Objectives of Canonical Correlation Analysis

In demonstrating the application of canonical correlation, we use all nine variables as input data. The HATCO ratings (X_1 through X_7) are designated as the set of multiple independent variables or the predictor variables. The measures of usage level and level of satisfaction (variables X_9 and X_{10}) are specified as the set of multiple dependent variables or the criterion variables. The statistical problem involves identifying any latent relationships between a customer's perceptions about HATCO and the customer's level of usage and satisfaction.

Stages 2 and 3: Designing a Canonical Correlation Analysis and Testing the Assumptions

The designation of the variables includes two metric-dependent and seven metric-independent variables. The conceptual basis of both sets is well established, so there is no need for alternative model formulations testing different sets of variables. The seven variables resulted in a 13-to-1 ratio of observations to variables, well exceeding the guideline of 10 observations per variable. Finally, both dependent and independent variables were assessed in Chapter 2 for meeting the basic distributional assumptions underlying multivariate analyses and passed all statistical tests. Interested readers are referred to Chapter 2 for more detail.

TABLE 6.2 Canonical Correlation Analysis Relating Level of Usage and Satisfaction with HATCO to Perceptions of HATCO

Measures of Overall Model Fit for Canonical Correlation Analysis

Canonical Function	Canonical Correlation	Adjusted Can. Corr.	Approx. Std. Err.	Canonical R^2	F Statistic	Prob.
1	.9369	.9328	.0123	.8778	30.2353	0.0001
2	.5100	.4760	.0744	.2601	5.3908	0.0001

Multivariate Test Statistics

Statistic	Value	Approximate F statistic	Prob.
Wilks' lambda	0.0904	30.2353	.0001
Pillai's trace	1.1379	17.3485	.0001
Hotelling's trace	7.5353	48.4410	.0001
Roy's greatest root	7.1837	94.4142	.0001

Stage 4: Deriving the Canonical Functions and Assessing Overall Fit

The canonical correlation analysis was restricted to deriving two canonical functions, because the dependent variable set contained only two variables. To determine the number of canonical functions to include in the interpretation stage, analysis focused on the level of statistical significance, the practical significance of the canonical correlation, and the redundancy indices for each variate.

Statistical and Practical Significance

The first statistical significance test is for the canonical correlations of each of the two canonical functions. In this example, both canonical correlations are statistically significant (see Table 6.2). In addition to tests of each canonical function separately, multivariate tests of both functions simultaneously are also performed. The test statistics employed are Wilks' lambda, Pillai's criterion, Hotelling's trace, and Roy's greatest root. Table 6.2 also details the multivariate test statistics, which all indicate that the canonical functions, taken collectively, are statistically significant at the .01 level.

In addition to statistical significance, the canonical correlations were both of sufficient size to be deemed practically significant. The final step was to perform redundancy analysis on both canonical functions.

Redundancy Analysis

A redundancy index is calculated for the independent and dependent variates of the first function in Table 6.3. As can be seen, the redundancy index for the criterion variate is substantial (.7503). The predictor variate, however, has a sub-

TABLE 6.3 Calculation of the Redundancy Indices for the First Canonical Function

Variate/Variable	Canonical Loading	Canonical Loading^2	Average Loading^2	Canonical R^2	Redundancy Index[a]
Dependent Variables					
X_9 Usage level	.9129	.8334			
X_{10} Satisfaction	.9358	.8759			
Dependent Variate		1.7093	.8547	.8778	.7503
Independent Variables					
X_1 Delivery speed	.7643	.5842			
X_2 Price level	.0614	.0038			
X_3 Price flexibility	.6237	.3890			
X_4 Mfr. image	.4145	.1718			
X_5 Overall service	.7653	.5857			
X_6 Sales image	.3479	.1210			
X_7 Product quality	−.2783	.0775			
Independent Variate		1.9330	.2761	.8778	.2424

[a]The redundancy index is calculated as the average loading^2 times the canonical R^2.

TABLE 6.4 Redundancy Analysis of Dependent and Independent Variates for Both Canonical Functions

Standardized Variance of the Dependent Variables Explained By

Canonical Function	Their Own Canonical Variate (Shared Variance)			The Opposite Canonical Variate (Redundancy)	
	Percentage	Cumulative Percentage	Canonical R^2	Percentage	Cumulative Percentage
1	.8547	.8547	.8778	.7503	.7503
2	.1453	1.0000	.2601	.0378	.7881

Standardized Variance of the Independent Variables Explained By

Canonical Function	Their Own Canonical Variate (Shared Variance)			The Opposite Canonical Variate (Redundancy)	
	Percentage	Cumulative Percentage	Canonical R^2	Percentage	Cumulative Percentage
1	.2761	.2761	.8778	.2424	.2424
2	.0823	.3584	.2601	.0214	.2638

stantially lower redundancy index (.2424), although in this case, because there is a clear delineation between dependent and independent variables, this lower value is not unexpected or problematic. The low redundancy of the predictor variate results from the relatively low shared variance in the predictor variate (.2761), not the canonical R^2. From the redundancy analysis and the statistical significance tests, the first function should be accepted.

The redundancy analysis for the second function produces quite different results (see Table 6.4). First, the canonical R^2 is substantially lower (.2601). Moreover, both variable sets have low shared variance in the second function (.1453 for the dependent variate and .0823 for the independent variate). Their combination with the canonical root in the redundancy index produces values of .0378 for the dependent variate and .0214 for the independent variate. Thus, while the second function is statistically significant, it has little practical significance. With such a small percentage, one must question the value of the function. This is an excellent example of a statistically significant canonical function that does not significantly explain a large proportion of the criterion variance.

Stage 5: Interpreting the Canonical Variates

With the canonical relationship deemed statistically significant and the magnitude of the canonical root and the redundancy index acceptable, the analyst proceeds to making substantive interpretations of the results. Although the second function could be considered practically nonsignificant, owing to the low redundancy value, it is included in the interpretation phase for illustrative reasons. These interpretations involve examining the canonical functions to determine the relative importance of each of the original variables in deriving the canonical relationships. The three methods for interpretation are (1) canonical weights (standardized coefficients), (2) canonical loadings (structure correlations), and (3) canonical cross-loadings.

TABLE 6.5 Canonical Weights for the Two Canonical Functions

	Function 1	*Function 2*
Standardized Canonical Coefficients for the Independent Variables		
X_1 Delivery speed	.2249	−.9648
X_2 Price level	.1029	−.8680
X_3 Price flexibility	.5686	.1596
X_4 Manufacturer image	.3480	−1.4557
X_5 Overall Service	.4453	1.5304
X_6 Salesforce's image	−.0508	.7362
X_7 Product quality	.0007	.4776
Standardized Canonical Coefficients for the Dependent Variables		
X_9 Usage level	.5007	1.3304
X_{10} Satisfaction level	.5801	−1.2977

Canonical Weights

Table 6.5 contains the standardized canonical weights for each canonical variate for both dependent and independent variables. As discussed earlier, the magnitude of the weights represents their relative contribution to the variate. Based on the size of the weights, the order of contribution of independent variables to the first variate is X_3, X_5, X_4, X_1, X_2, X_6, and X_7, while the dependent variable order on the first variate is X_{10}, then X_9. Similar rankings can be found for the variates of the second canonical function. Because canonical weights are typically unstable, particularly in instances of multicollinearity, owing to their calculation solely to optimize the canonical correlation, the canonical loading and cross-loadings are considered more appropriate.

Canonical Loadings

Table 6.6 contains the canonical loadings for the dependent and independent variates for both canonical functions. In the first dependent variate, both variables have loadings exceeding .90, resulting in the high shared variance (.8547). The first independent variate has a quite different pattern, with loadings ranging from .0614 to .7653, with one independent variable (X_7) even having a negative loading.

TABLE 6.6 Canonical Structure of the Two Canonical Functions

Canonical Loadings

Correlations Between the Independent Variables and Their Canonical Variates

X_1	Delivery speed	.7643	.1091
X_2	Price level	.0614	.1414
X_3	Price flexibility	.6237	.1229
X_4	Manufacturer's image	.4145	−.6262
X_5	Overall service	.7653	.2216
X_6	Sales force's image	.3479	−.1995
X_7	Product quality	−.2783	.2189

Correlations Between the Dependent Variables and Their Canonical Variates

X_9	Usage level	.9129	.4081
X_{10}	Satisfaction level	.9359	−.3522

Canonical Cross-Loadings

Correlations Between the Independent Variables and the Dependent Canonical Variates

X_1	Delivery speed	.7161	.0556
X_2	Price level	.0575	.0721
X_3	Price flexibility	.5843	.0627
X_4	Manufacturer's image	.3883	−.3194
X_5	Overall service	.7170	.1130
X_6	Sales force's image	.3260	−.1017
X_7	Product quality	−.2607	.1116

Correlations Between the Dependent Variables and the Independent Canonical Variates

X_9	Usage level	.8553	.2081
X_{10}	Satisfaction level	.8769	−.1796

The result is a variate that, while maximized for correlation with the dependent variate, is not "optimized" for interpretation. This makes identification of relationships more difficult. The researcher should also perform a sensitivity analysis of the independent variate in this case to see whether the loadings change when an independent variable is deleted.

The second variate's poor redundancy values are exhibited in the substantially lower loadings for both variates on the second function. Thus, the poorer interpretability as reflected in the lower loadings, coupled with the low redundancy values, reinforce the low practical significance of the second function.

Canonical Cross-Loadings

Table 6.6 also includes the cross-loadings for the two canonical functions. In studying the first canonical function, we see that both independent variables (X_9 and X_{10}) exhibit high correlations with the predictor canonical variate (P1): .8553 and .8769, respectively. By squaring these terms, we find the percentage of the variance for each of the variables explained by P1. The results show that 73 percent of the variance in X_9 and 77 percent of the variance in X_{10} is explained by P1. Looking at the predictor variables' cross-loadings, we see that variables X_1 and X_5 both have high correlations of roughly .72 with the criterion canonical variate. From this information, we see that approximately 52 percent of the variance in each of these two variables is explained by the criterion variate (the 52 percent is obtained by squaring the correlation coefficient; $.72 * .72 = .52$). One should note the correlation of X_3 (.5843). Although this may appear high, one must realize that after squaring this correlation, only 34 percent of the variation is included in the canonical variate.

The final issue of interpretation is examining the signs of the cross-loadings. All independent variables except X_7 (product quality) have a positive, direct relationship. For the second function, two predictor variables (X_4 and X_6), plus a criterion variable (X_{10}), are negative. Thus all the relationships are direct except for one inverse relationship in the first function.

Stage 6: Validation and Diagnosis

The last stage should involve a validation of the canonical correlation analyses through one of several procedures. Among the available approaches would be (1) splitting the sample into estimation and validation samples, or (2) sensitivity analysis of the independent variable set. Table 6.7 contains the result of such a sensitivity analysis where the canonical loadings are examined for stability when individual independent variables are deleted from the analysis. As seen, the canonical loadings in our example are remarkably stable and consistent in each of the three cases where an independent variable (X_1, X_2, or X_7) is deleted. The overall canonical correlations also remain stable. But if the reader examined the canonical weights (not presented in the table), there would be widely varying results dependent on which variable was deleted. This reinforces the procedure of using the canonical loading and cross-loading for interpretation purposes.

TABLE 6.7 Sensitivity Analysis of the Canonical Correlation Results to Removal of an Independent Variable

	Complete Variate	Results After Deletion of		
		X_1	X_2	X_7
Canonical Correlation (R)	.937	.936	.937	.937
Canonical Root (R^2)	.878	.877	.878	.878
Independent Variate				
Canonical Loadings				
X_1 Delivery speed	.764	omitted	.765	.764
X_2 Price level	.061	.062	omitted	.061
X_3 Price flexibility	.624	.624	.624	.624
X_4 Mfr. image	.414	.413	.414	.415
X_5 Overall service	.765	.766	.766	.765
X_6 Sales force's image	.348	.348	.348	.348
X_7 Product quality	−.278	−.278	−.278	omitted
Shared variance	.2761	.2250	.3216	.3092
Redundancy	.2424	.1972	.2822	.2714
Dependent Variate				
Canonical Loadings				
X_9 Usage level	.913	.915	.914	.913
X_{10} Satisfaction	.936	.934	.935	.936
Shared variance	.8547	.8549	.8548	.8547
Redundancy	.7503	.7494	.7501	.7503

Summary

Canonical correlation analysis is a useful and powerful technique for exploring the relationships among multiple criterion and predictor variables. The technique is primarily descriptive, although it may be used for predictive purposes. Results obtained from a canonical analysis should suggest answers to questions concerning the number of ways in which the two sets of multiple variables are related, the strengths of the relationships, and the nature of the relationships defined.

Canonical analysis enables the data analyst to combine into a composite measure what otherwise might be an unmanageably large number of bivariate correlations between sets of variables. It is useful for identifying overall relationships between multiple independent and dependent variables, particularly when the data analyst has little a priori knowledge about relationships among the sets of variables. Essentially, the analyst can apply canonical correlation analysis to a set of variables, select those variables (both independent and dependent) that appear to be significantly related, and run subsequent canonical correlations with the more significant variables remaining, or perform individual regressions with these variables.

Questions

1. Under what circumstances would you select canonical correlation analysis instead of multiple regression as the appropriate statistical technique?
2. What three criteria should you use in deciding which canonical functions should be interpreted? Explain the role of each.
3. How would you interpret a canonical correlation analysis?
4. What is the relationship among the canonical root, the redundancy index, and multiple regression's R^2?
5. What are the limitations associated with canonical correlation analysis?
6. Why has canonical correlation analysis been used much less frequently than the other multivariate techniques?

References

1. Alpert, Mark I., and Robert A. Peterson. "On the Interpretation of Canonical Analysis." *Journal of Marketing Research* 9, (May 1972): 187.
2. Alpert, Mark I., Robert A. Peterson, and Warren S. Martin. "Testing the Significance of Canonical Correlations." *Proceedings, American Marketing Association* 37 (1975): 117–119.
3. Bartlett M. S. "The Statistical Significance of Canonical Correlations." *Biometrika* 32 (1941): 29.
4. BMDP Statistical Software, Inc. *BMDP Statistical Software Manual, Release 7*, vol. 1 and 2, Los Angeles: 1992.
5. Dillon, W. R., and M. Goldstein. *Multivariate Analysis: Methods and Applications.* New York: Wiley, 1984.
6. Green, P. E. *Analyzing Multivariate Data.* Hinsdale, Ill.: Holt, Rinehart, & Winston, 1978.
7. Green, P. E., and J. Douglas Carroll. *Mathematical Tools for Applied Multivariate Analysis.* New York: Academic Press, 1978.
8. Lambert, Z., and R. Durand. "Some Precautions in Using Canonical Analysis." *Journal of Marketing Research* 12 (November 1975): 468–75.
9. SAS Institute, Inc. *SAS User's Guide: Basics, Version 6.* Cary, N.C.: 1990.
10. SAS Institute, Inc. *SAS User's Guide: Statistics, Version 6.* Cary, N.C.: 1990.
11. SPSS, Inc. *SPSS User's Guide*, 4th ed. Chicago: 1990.
12. SPSS, Inc. *SPSS Advanced Statistics Guide*, 4th ed. Chicago: 1990.
13. Stewart, Douglas, and William Love. "A General Canonical Correlation Index." *Psychological Bulletin* 70 (1968): 160–63.

The Impact of Channel Leadership Behavior on Intrachannel Conflict

Patrick L. Schul
William M. Pride
Taylor L. Little

Introduction

Distribution channel researchers and marketing practitioners have shown considerable interest in channel leadership behavior. As defined here, channel leadership behavior includes activities performed by a distribution channel member to influence the marketing policies and strategies of other channel members for the purpose of controlling various aspects of channel operations (El-Ansary and Robicheaux 1974, Little 1970, Stern 1967). A channel organization's survival and effectiveness are influenced by channel leadership (Bucklin 1968, 1973; El-Ansary and Robicheaux 1974; Etgar 1977; Little 1970; Mallen 1964; Robicheaux and El-Ansary 1975; Speh and Bonfield 1978; Stern 1967; Sturdivant and Granbois 1968).

Though the leadership construct has been discussed in the marketing literature, little is known about the impact of channel leadership behavior on the attitudes and behaviors of other channel members. Research efforts have focused primarily on identifying factors associated with the emergence of channel leaders (Etgar 1977, 1978) and exploring power bases for channel leader control (El-Ansary and Stern 1972, Etgar 1976, Hunt and Nevin 1974, Lusch 1976, Wilkinson 1972).

Two general conditions characterize research in channel leadership behavior. The definition of the construct lacks precision; distinct processes associated with power, authority and control have been used synonymously in defining channel

From "The Impact of Channel Leadership Behavior on Intrachannel Conflict," Patrick L. Schul, William M. Pride, and Taylor L. Little, *Journal of Marketing*, vol. 47 (Summer 1983), pp. 21–34. Reprinted by permission.

leadership. Second, adequate measures of the channel leadership construct are lacking; studies attempting to analyze channel leadership have relied on general measures of control or power (Etgar 1977, Speh and Bonfield 1978) and have not always incorporated applicable theory and research on leadership developed in intraorganizational studies.

This article addresses both deficiencies by defining and measuring channel leadership behavior and its relation to intrachannel conflict, the extent of disagreements between firms occupying different levels in the same channel of distribution (Rosenberg and Stern 1970). Given the nature of conflict in distribution channels (Etgar 1979; Firat, Tybout and Stern 1974; Rosenberg 1971; Rosenberg and Stern 1971; Stern and El-Ansary 1977), one expects attempts to assume channel leadership to affect the level of conflict. When strategies and policies implemented by a channel leader to influence another channel member's marketing policies are perceived as impeding the other channel member's organizational goals, vertical channel conflict should arise (Rosenberg and Stern 1971).

There have been several studies that have shown the level of intrachannel conflict to be significantly affected by the type of power resources (noncoercive or coercive) utilized by the channel leader in the process of administering the channel (see Walker and Shooshtari 1979 for a review of these studies). These power resources stem from the channel leader's access to economic, social and psychological resources (Etgar 1977). In effect, these studies show that the use of noncoercive power resources (e.g., reward, referent, expert and legitimate) can reduce the frequency of intrachannel conflict. Conversely, the use of coercive power resources increases the frequency of conflict.

The level of intrachannel conflict may, however depend in part on *how* the channel leader utilizes power resources; it may depend on the style of leadership. Research studies in intraorganizational behavior show that an effective leader utilizes a particular style (e.g., participative, supportive, or achievement oriented) to secure subordinates' compliance (Ivancevich, Szilagyi and Wallace 1977). Channel leaders exhibit a variety of leadership styles in channel administration; some increase while others reduce the level of intrachannel conflict. By identifying the effects of different leadership styles on the level of intrachannel conflict, a channel leader could design a channel management approach that would reduce or at least contain the level of vertical conflict.

The research reported here examined relationships between each of several identifiable types of channel leadership behavior and intrachannel conflict in a distribution channel that has a well-defined leadership structure—the franchise-franchisor distribution channel. Given the franchisor's defined leadership role in the franchise agreement, the franchise channel provides an appropriate channel structure for exploring selected effects of leadership behavior on intrachannel conflict.

The franchise distribution channel for real estate brokerage services was selected for this study. Like many traditional distribution channels, this channel has a relatively informal leader-subordinate relationship; some of the structural aspects of the channel (e.g., rules, operating policies, control procedures and compensation arrangements) are not formally organized (Davis 1979). Unlike traditional channels, however, this franchise channel has a well-defined authority structure (Baird, Hay and Bailey 1977) that permits a more careful identification and assessment of different styles of leadership behavior. Finally, the distribution channel for real estate brokerage services experiences frequent and visible conflicts (Baird, Hay and Bailey 1977) resulting in franchisors using a variety of approaches to channel operations (Davis 1979). These characteristics make the real estate brokerage distribution channel an acceptable choice for analyzing selected effects of different channel leadership styles.

Conceptual Perspectives and Hypotheses

Intrachannel Conflict

Intrachannel conflict is said to arise when a "component (channel member) perceives the behavior of another to be impeding the attainment of its goals and effective performance of its instrumental behavior patterns" (Stern and Gorman 1969, p. 156). The effects of intrachannel conflict can be functional or dysfunctional to the operation of the channel (see Rosenbloom 1978, pp. 67–84). While functional conflict produces channel efficacy (Assael 1969, Bower 1965, Deutsch 1971, Lusch 1976, Rosenbloom 1973, Stern and Heskett 1969), dysfunctional conflict can impede the channel's performance and may eventually destroy the channel as a competitive entity (Alderson 1965, Reve 1977, Rosenberg and Stern 1970, Stern and Gorman 1969).

Using Rosenberg and Stern's process model of intrachannel conflict, Rosenberg (1971), Firat, Tybout and Stern (1974), Brown and Day (1981) and Etgar (1979) suggest that channel members who experience intrachannel conflict initially enter a cognitive or affective stage of conflict before exhibiting any conflict-related behavior. It is in this early stage that the channel member's feelings of stress, tension and animosity form toward the offending party. As Rosenberg and Stern indicate, "It would be myopic to identify conflict only when it manifest because in its more subdued stages (affective conflict) it is potentially disruptive, since it may be capable of being escalated" (Rosenberg and Stern 1970, pp. 43–44). Based on this rationale, a perceptual measure of affective conflict has been used to represent intrachannel conflict in this study, as it has in previous studies (Lusch 1976; Pearson and Monoky 1976; Rosenberg 1971; Stern, Sternthal and Craig 1973).

Channel Leadership Behavior

Channel leadership behavior can be defined as activities carried out by a channel member to influence the marketing policies and strategies of other channel members for the purpose of controlling various aspects of channel operations. Style of leadership focuses on actual leader behavior, what the leader does and how it is done (Katz and Kahn 1953; Katz, Maccoby and Morse 1950; Stogdill and Coons 1957). In recent years, empirical efforts have isolated styles of leadership (e.g., instrumental, supportive, achievement oriented, participative) that influence leadership effectiveness. (For a review of these empirical studies, see Ivancevich, Szilagyi and Wallace 1977.) Researchers have assessed the effects of leadership styles on attitudinal and behavioral outcomes such as role clarity, conflict, performance and satisfaction. In this study, channel leadership behavior was operationalized by isolating leadership styles exhibited by franchise channel leaders in real estate brokerage services and developing and testing separate measures for each leadership style. Exploratory interviews with franchisees, franchisors and others familiar with the franchised real estate brokerage industry aided in identifying three types of leadership behaviors—participative, supportive and directive.

A participative leader consults with subordinates, solicits their suggestions and considers these suggestions before making a decision (House and Mitchell 1974). The importance of the subordinate's contributing input has received considerable attention in the organizational literature. While some studies indicate subordinate involvement in decision making provides more occasions for conflict and makes the participants more aware of latent conflicts (Coleman 1957, Corwin 1969, Gamson 1966), much of the evidence suggests that participative leadership behavior facilitates cooperation and thus reduces conflict (House and Mitchell 1974, House and Rizzo 1972, Seashore and Bowers 1970, Strauss and Rosenstein 1970, Tannebaum and Cooke 1974).

Applied to channels, collaboration between channel members appears conducive to cooperation (Bucklin 1973, El-Ansary and Robicheaux 1974. Mallen 1964, Stern 1967, Walker 1970), and thus could be successful in reducing intrachannel conflict. More specifically, it seems reasonable that the level of intrachannel conflict is lower if the channel leader solicits channel members' inputs regarding channel-related decisions, and intrachannel conflict is higher if the leader does not solicit such inputs. This contention has not been adequately tested in a marketing channel context.

Accordingly, the following hypothesis was tested:

H_1: The higher the level of participative leadership behavior exhibited by the channel leader, the lower the level of intrachannel conflict.

Supportive leadership considers subordinates' needs, displays concern for subordinates' wellbeing and creates a friendly and pleasant task environment (Ivancevich, Szilagyi and Wallace 1977). The support dimension is often described as consideration, socioemotional or expressive (House and Mitchell 1974, Parsons 1951). House, Filley and Kerr (1971) maintain that supportive activities reflecting subordinate consideration allow for social and normative integration of group members. By providing a more pleasant task environment, this integration can directly influence the level of conflict.

Empirical evidence regarding the relationship between supportive leadership and conflict in the channels context is nonexistent. However, studies of other organizations indicate supportive leadership behavior, emphasizing a concern for subordinates' welfare and personal growth, facilitates coordination and thus helps reduce conflict (House and Mitchell 1974, House and Rizzo 1972, Seashore and Bowers 1970). It appears reasonable that channel members who receive supportive leadership are more likely to be motivated to cooperate with channel management and experience less intrachannel conflict. The hypothesized relationship between supportive leadership behavior and intrachannel conflict suggested the following:

H_2: The higher the level of supportive leadership behavior exhibited by the channel leader, the lower the level of intrachannel conflict.

A directive leader organizes and defines the task environment, assigns the necessary functions to be performed, establishes communication networks and evaluates work group performance (Ivancevich, Szilagyi and Wallace 1977). Applied to channels, directive leadership refers to the instrumental behavior of the channel leader. According to Parsons (1951), instrumental activities, which are highly directive and task oriented, are necessary for any social system to solve basic functional problems of adapting to its environment and attaining goals through allocation and mobilization of resources. As applied to the distribution channel, instrumental leadership behavior includes planning, organizing, coordinating and controlling channel-related activities. A channel leader using a directive leadership style would exhibit instrumental behavior that involves establishing and communicating to channel members consistent, channel-wide objectives, policies and operating procedures.

Considerable research suggests that organizational conflict varies within and across organizational subunits as a function of subordinates' perceptual evaluations of their leaders' directive behavior. (For a review, see Ivancevich, Szilagyi and Wallace 1977.) In general, directive leadership (i.e., standardization, emphasis on rules, and close supervision) appears to create tension and pressure (conflict antecedents) when subordinates engage in highly structured tasks and reduces conflict when subordinates engage in ambiguous or unstructured tasks (Burke 1966, Corwin 1969, Pondy 1967). Findings from intraorganizational research efforts, when applied to the distribution channel organization with its inherent lack of traditional structure, suggest that channel members who perceive their channel leader to be high on directiveness are more likely to understand and appreciate channel policies and procedures. Consequently, they are likely to be more satisfied with the channel arrangement and experience less conflict with channel management. The following hypothesized relationship between directive leadership behavior and intrachannel conflict was tested:

H_3: The higher the level of directive leadership behavior exhibited by the channel leader, the lower the level of intrachannel conflict.

Methodology

Sample and Research Procedure

The data used to test the hypotheses were collected as part of a general study of franchisee-franchisor relations in the real estate brokerage industry. Franchised real estate brokers representing six major real estate franchise organizations doing business in three south central states participated in a mail survey. Prior to conducting the survey, a pilot study was administered to pre-

test the validity and reliability of variable measures, to make needed modifications in the research design and to pretest the analysis procedure. Eighty-five randomly selected franchisees were asked to complete the questionnaire. These franchisees were, in turn, deleted from the list of franchisees participating in the final study. Thirty-one franchisees returned questionnaires. All variable measures contained in the research instrument were inspected, and scale items not stable or consistent were modified or deleted. Also, response rate and sample representation were assessed. Finally, format adjustments were made.

To initiate the main study, an eight-page questionnaire was mailed to 1,052 franchised real estate brokers representing the six franchise organizations. The questionnaire consisted of four parts and gathered information about the franchisee's (1) general organization characteristics, (2) perceptions of conflict, (3) satisfaction with the franchise arrangement, and (4) feelings toward the franchise organization. Of the initial questionnaires mailed in the first wave, 265 were returned. After three weeks follow-up questionnaires were sent to participants who had not responded to the first mailing. Of the 787 questionnaires in the second wave, 126 were returned. For both waves, 349 participants responded with usable questionnaires, resulting in a 33% response rate.

To assess nonresponse bias, 36 randomly selected nonrespondents were contacted and asked several of the more important descriptive questions contained in the original test instrument. As the results provided in Table 1 show, a series of χ^2 tests indicated no significant differences between respondents and nonrespondents on any of the measures analyzed. These results suggest that the sample was representative of the overall population of franchised real estate brokers.

Measurement of Variables

A review of the organizational behavior literature and interviews with franchisees preceded the development of a measure of channel leadership behavior. From this review, 19 scale items were developed to represent the three types of leadership behavior. Based on the pilot study results,

TABLE 1 Comparison of Responses Across Respondent and Nonrespondent Groups

Variable	Respondents		Nonrespondents		χ^2	d.f.	Sig. Level
	n	Percentage	n	Percentage			
Brokerage experience					2.97	2	.063
Less than 4 years	81	23.2	10	26.7			
4–6 years	107	30.7	12	34.8			
Over 6 years	161	46.1	14	38.5			
Franchisee experience					2.24	2	.328
Less than 3 years	109	31.2	10	28.5			
3–4 years	147	42.1	15	41.1			
Over 4 years	93	26.6	11	30.4			
Franchise affiliation					5.51	2	.662
Century 21	179	51.3	17	46.6			
Realty World	28	8.0	1	3.8			
Red Carpet	62	17.8	6	15.7			
Gallery of Homes	26	7.4	4	11.2			
First Mark	21	6.0	3	8.8			
E.R.A.	33	9.5	5	13.9			
Geographical location					3.41	2	.772
Texas	284	81.4	28	78.3			
Oklahoma	46	13.2	6	15.6			
Arkansas	19	5.4	2	6.1			
Salesforce size					4.65	2	.721
Less than 4 salespeople	35	10.0	5	12.7			
4–6 salespeople	104	29.8	9	26.2			
7–10 salespeople	95	27.2	12	34.3			
Over 10 salespeople	115	33.0	10	26.7			
Frequency of disagreement					4.22	3	.261
Very frequently	23	6.6	1	3.9			
Frequently	50	14.3	5	13.8			
Infrequently	232	66.5	23	65.0			
Never	44	12.6	6	17.3			
General Satisfaction					1.77	2	.226
Yes	250	71.6	28	77.9			
No	46	13.2	4	10.1			
Questionable	53	15.2	4	12.0			

the list was reduced to nine items (see Table 2). Using a 5-point Likert-type scale that ranged from "completely agree" to "completely disagree," franchisees rated their franchisors on each of these nine items.

Thus, leadership behavior was assessed on the basis of a self-reported perceptual measure. While perceptual measures such as these represent time-collapsed perceptions of behavior, and thus suffer limitations stemming from response bias (Churchill 1979), they are frequently used by researchers investigating organizational constructs such as leadership. This is because perceptual processes, not objective properties, affect organizational behavior. "To the extent a researcher has a strong interest in understanding and anticipating the human component within organizations, it is probably desirable to employ perceptual measures" (Hellriegel and Slocum 1974, p. 260).

TABLE 2 Factor Analysis of Channel Leadership Items[a]

	Derived Factors[b]		
Leadership Statements	(1) Participative Leadership	(2)[c] Supportive Leadership	(3) Directive Leadership
In my franchise arrangement:			
1. Franchisees have major influence in the determination of policies and standards for this franchise organization.	(.932)[d]	.067	.008
2. Good ideas from franchisees often do not get passed along to franchise management.	(.976)	−.153	.044
3. Franchisees are not allowed to provide input into the determination of standards and promotional allowances.	(.869)	−.040	−.007
4. There is a definite lack of support, coaching and feedback.	.165	(−.747)	.006
5. Once they've sold you the franchise, they forget all about you . . . except when your fees are due again.	−.049	(−1.00)	−.059
6. This franchise organization is highly interested in the welfare of its franchisees.	−.018	(−.821)	.109
7. I am provided sufficient guidelines and careful instruction on how to manage my franchise operations.	−.108	.052	(−.753)
8. The rights and obligations of all parties concerned are *clearly* spelled out in the franchise contract.	.162	−.078	(.733)
9. I am encouraged to use uniform procedures.	.156	−.123	(.635)
Eigenvalues	5.34	.86	1.43
Percent of total variation	59.30	7.43	15.90

[a] An oblique rotation was performed on the initial factor matrix. The use of an oblique rotation with its relaxed requirements of orthogonality was deemed more appropriate, given the nature of the leadership construct, the content of the items themselves, and the resulting potential for intercorrelated factors.

[b] Interfactor correlations: $r_{12} = .523$, $r_{13} = .352$, $r_{23} = .458$

[c] While the eigenvalue for supportive leadership was not overly significant (>1.00), this factor was maintained given the conceptual nature of the leadership construct, the strength of the coefficients for items loading on the factor, and the sizable amount of explained variation accounted for by the factor (7.43%).

[d] Loadings above .60 in parentheses.

To establish the content or face validity of the leadership behavior scale, a series of interviews with franchisees and others familiar with the franchised real estate brokerage industry was conducted. This aided the development of a channel leadership measure that adequately represented the domain of leadership behavior in the distribution channel under study. Approximately 50 unstructured interviews were conducted to obtain information about franchise operations and other related topics.

The unidimensionality of the three leadership measures was assessed using factor analysis. Table 2 provides the results of an oblique rotation of the initial factor matrix underlying the leadership items. Three independent factors extracted from the data accounted for 82.63% of the variation in the leadership scale items. A cutoff of .60 was used for item scale selection. These results provide evidence of unidimensionality, since those items intended to measure each of the three leadership dimensions loaded appropriately on a corresponding factor. Aggregate measures were created for each of the three leadership variables by summing the raw scores for the items loading on the representative factor.

The reliability of the channel leadership scales was assessed using coefficient alpha (Cronbach 1960). The coefficients ranged from a low of .80 for the directive behavior scale to a high of .92 for the supportive behavior scale. Thus, all three scales exhibited well over .50 reliability levels suggested by Nunnally (1967) as a minimum level for acceptable reliability.

Intrachannel conflict was defined operationally as the intensity of affective conflict over a set of derived channel specific issues. Initially, a large set of conflict issues was obtained through interviews with franchisees and a search of the trade literature. As shown in Table 3, the overall list was subsequently reduced to 14 key issues following the results of the pilot study. Intensity of

conflict was measured on a four-point scale ranging from high to none.

Factor analysis was used to identify latent factors within the overall conflict intensity measure. Table 3 provides the results of an oblique rotation of the factor matrix underlying the conflict items. Two independent factors with eigenvalues greater than one were extracted from the data. These factors accounted for over 53% of the variation in the conflict scale items. A cutoff of .60 was used for item-scale selection. Following an inspection of the items loading on each factor, the two factors were subsequently labeled "administrative conflict" and "product-service conflict," each representing a separate class of conflict issues. Aggregate measures were then created for each dimension of conflict by summing the raw scores for the items loading on each of the two representative factors. An examination of Table 3 shows that all but two of the items loaded signifi-

TABLE 3 Factor Analysis of Intrachannel Conflict Issues[a]

	Derived Factors[b]	
Conflict Issue	(1) Administrative Conflict	(2) Product-Service Conflict
1. Bureaucratic red tape	(.664)[c]	.064
2. Quality of sales training programs	−.018	(.769)
3. General responsiveness of franchisor to franchise needs	.226	(.613)
4. Quality of meetings and conventions	−.118	(.807)
5. Quality of intercity referral program	.012	(.654)
6. Local advertising assistance[d]	.337	.413
7. Service and/or advertising fees	(.641)	.223
8. Contract terms or arrangements	(.672)	.298
9. Accounting information requirements	(.639)	.186
10. Initial franchise fees (or fees on additional new offices)	(.653)	.090
11. Sales promotion assistance[d]	.526	.286
12. Quality of national advertising program	.083	(.708)
13. Quality of management training program	.111	(.691)
14. Exclusive territory arrangements	(.733	−2.38
Eigenvalues	6.24	1.16
Percent of total variation	44.6	8.3

[a] An oblique rotation was performed on the initial factor matrix.
[b] Interfactor correlation, r_{12} = .520.
[c] Loadings above .60 are in parentheses.
[d] Denotes item that failed to load significantly on either factor.

cantly on an appropriate factor. Thus, the unidimensionality of the intrachannel conflict scales was established.

The internal reliability of the conflict scale was assessed using coefficient alpha. The alpha coefficient for the administrative conflict scale was .78, while the coefficient for the product-service scale was .83, thus indicating an acceptable level of internal reliability.

Data Analysis

The data were analyzed using canonical correlation to test for relationships between the channel leadership variables as sets of predictors and the intrachannel conflict variables as a set of criterion variables. The strength of the association in this study between each set of predictor variables (channel leadership styles) and the criterion variable set (intrachannel conflict) was assessed by inspecting the magnitudes of both the canonical correlation coefficients and the redundancy index for each pair of linear composites derived from the data. By inspecting the canonical correlation coefficients, a rough estimate of the strength of the relationship between each set of variables was derived. Specifically, the canonical coefficient indicates the correlation between the canonical scores for each linear combination of variables (Green 1978). An analysis of the canonical correlation coefficient does not, however, reveal the amount of variance shared by the two sets of variables. Consequently, it necessitates inspection of the magnitude of the redundancy index, an asymmetric index measuring how much variance in one set of variables is shared by the variability in the other set (Stewart and Love 1968). Lambert and Durand (1975) recommend the redundancy index as a more indicative measure of the explanatory capability of canonical analysis in accounting for criterion variance.

The relative importance of a variable in each set of variables was indicated by the canonical weights extracted for the variable, their canonical loadings (within set, variable-variate correlations) and canonical cross-loadings (between set, variable-variate correlations). These statistics computed for the most significant linear composite provided a basis for subset interpretation. The canonical loading reflects the variance that an observed variable in one set of variables shares with the canonical score for that set. Conversely, the cross-loading value reflects the variable's correlation with the canonical score for the other set of variables.[1]

A key consideration influencing the objective interpretation of the canonical correlation analysis concerns weight instability. The values of canonical weights can be unstable and can vary across repeated samples from the same population (Alpert and Peterson 1972, Lambert and Durand 1975). As a result, a statistically significant canonical correlation can occur even though the criterion and predictor sets are not strongly related.

To evaluate weight instability, a diagnostic test suggested by Lambert and Durand (1975) was employed. Initially, the total sample was randomly split into two groups—analysis and validation. Canonical analysis was then applied to each subsample, thus deriving separate weights for each group. The structural coefficients for the first linear composite for each group are presented in Table 4.

The diagnostic test for weight instability applied the estimated canonical weights from the analysis group to the validation group and vice versa. Canonical scores were then calculated for both predictor and criterion sets within each subsample. The resulting canonical correlations were .52 for the analysis group and .59 for the validation group. Both correlations were significant at the .001 level.

The high degree of association between the two sets of canonical correlations and the observed consistency in the weights themselves supports the proposition that weight instability does not affect the objective interpretation of the results of the canonical correlation analysis from the overall sample.

Findings and Implications

The findings and implications should be approached with caution due to several limitations of the study. The franchise distribution channel

[1] The significance of loading and cross-loading statistics was assessed using a subjective cutoff criterion developed by Lambert and Durand (1975). This was used due to the lack of an objective test for assessing the statistical significance of canonical weights.

TABLE 4 Results of Canonical Analysis Showing the Effects of Channel Leadership Variables on Intrachannel Conflict: Total and Split Sample Results

Variables	Canonical Weights[b]			Canonical Loadings	Canonical Cross-Loading
	1st Split Sample (n = 174)	2nd Split Sample (n = 166)	Total Sample (n = 340)		
Predictor Set–Leadership Variables[a]					
1. Participative leadership behavior	.1380	.1356	.1628	.7869	.4380
2. Directive leadership behavior	.0949	.1040	.0700	.4985	.3210
3. Supportive leadership behavior	.8500	.8518	.8386	.9922	.5547
				Redundancy coefficient = .1934	
Criterion Set–Intrachannel Conflict Variables[a]					
1. Administrative conflict	.6605	.6606	.6618	.9441	.5334
2. Product-service conflict	.4104	.4328	.4204	.8915	.4915
				Redundancy coefficient = .2640	
Canonical correlation coefficient	.5997	.5221	.5596		
Canonical root (eigenvalue)	.3540	.2726	.3132		
χ^2	74.809	50.7103	123.9497		
d.f.	6	6	6		
$P(\chi^2)$	.000	.000	.000		

[a] Scoring for the separate dimensions of leadership behavior and intrachannel conflict was handled by summing the items corresponding to each dimension.

[b] Canonical loading and cross-loading statistics are provided for only the total sample group. Canonical weights are provided for all three groups (total group and both subsample groups) strictly for sample diagnostic purposes.

for real estate brokerage services is predominantly service oriented. Consequently, the results of this study may not be comparable to results of other studies analyzing channels for tangible goods. The variance among channel operations may limit the generalizability of the results. Additionally, factors other than channel leadership behavior (e.g., social, environmental influences and perceptual differences) may affect the level of intrachannel conflict. Finally, although the hypotheses were expressed as proposed unidimensional or monotonic relationships between channel leadership behavior and intrachannel conflict, the actual relationships may be more accurately expressed as curvilinear relationships or as mathematical relationships other than monotonic relationships.

The results of separate canonical correlation analyses of relationships between each of the three predictor sets (leadership styles) and the criterion set (intrachannel conflict) appear in Tables 5, 6, and 7. Since omitted variables could have introduced bias, the results of these separate

analyses should be interpreted with caution. However, certain insights can be gained by analyzing the individual effects of different leadership styles on dimensions of intrachannel conflict. The results of an aggregate analysis, including all three leadership styles in the predictor set, are provided in Table 4.

Separate Canonical Correlation Analyses

Table 5 shows the results of correlating the three participative leadership items with the two dimensions of intrachannel conflict. The canonical correlation is .490, which is significantly different than zero at the .000 level by the chi-square test. The redundancy index for the canonical function indicates that 19.64% of the variance in the intrachannel conflict dimensions is accounted for by the variability in the participative leadership items. Conversely, 18.47% of the variation in the participative leadership items is accounted for by the variability in the intrachannel conflict dimen-

TABLE 5 Participative Leadership Behavior Correlated with Intrachannel Conflict

	Canonical Weights	Canonical Loadings	Canonical Cross-Loadings
Predictor Behavior Variables Set–Participative Leadership			
1.			
policies and standards	.0212	.7884	.4008
2.[a] Good ideas from franchisees often don't get passed			
along to management	.6616	.9432	.4746
3.[a] Not allowed to provide input into determination of			
standards and promotional allowances	.4118	.8926	.4478
		Redundancy coefficient = .18479	
Criterion Set–Intrachannel Conflict Variables			
1. Administrative conflict	.8139	.9803	.4456
2. Product-service conflict	.2460	.8212	.3655
		Redundancy coefficient = .19641	

Canonical correlation coefficient =.4901
Canonical root (eigenvalue) = .2402
   χ^2 = 87.5912
  d.f. = 6
$P(\chi^2)$ = .000

[a]Scoring revised before analysis.

TABLE 6 Supportive Leadership Behavior Correlated with Intrachannel Conflict

	Canonical Weights	Canonical Loadings	Canonical Cross-Loadings
Predictor Set Supportive Leadership Behavior Variables			
1.[a] Lack of support, coaching and feedback	.5108	.9376	.5425
2.[a] Only interested in selling franchises	.2872	.9268	.5286
3. Franchisor highly interested in franchisee welfare	.2758	.8663	.5008
		Redundancy coefficient = .26918	
Criterion Set Intrachannel Conflict Variables			
1. Administrative conflict	.6311	.9354	.5294
2. Product-service conflict	.4533	.9028	.4936
		Redundancy coefficient = .27421	

Canonical correlation coefficient = .5696
Canonical root (eigenvalue) = .3245
   χ^2 = 125.0152
  d.f. = 6
$P(\chi^2)$ = .000

[a]Scoring revised before analysis.

TABLE 7 Directive Leadership Behavior Correlated with Intrachannel Conflict

	Canonical Weights	Canonical Loadings	Canonical Cross-Loadings
Predictor Set–Directive Leadership Behavior Variables			
1. Am provided sufficient guidelines and careful operational instructions on how to manage my franchise.	.4450	.6937	.1037
2. Rights and obligations of all parties concerned are clearly spelled out by my franchisor.	.7752	.8607	.3934
3. Encouraged to use uniform procedures.	.5088	.7926	.3681
		Redundancy coefficient = .1229	
Criterion Set–Intrachannel Conflict Variables			
1. Administrative conflict	.6020	.9267	.2822
2. Product-service conflict	.4090	.9128	.3848
		Redundancy coefficient = .2226	

Canonical correlation coefficient = .4463
Canonical root (eigenvalue) = .1992
$\quad \chi^2 = 77.7988$
$\quad$ d.f. = 6
$P(\chi^2) = .000$

sions. Thus, participative leadership behavior and intrachannel conflict are significantly related.

An examination of the structural coefficients for the canonical function indicated that the composite score for the participative leadership items is significantly related to both dimensions of intrachannel conflict (the cross-loading values for both conflict dimensions exceed the .30 level suggested by Lambert and Durand (1975) as an acceptable minimum loading value). In other words, channel members who perceive their channel leader to encourage member participation in channel-wide decision making tend to experience less conflict over both administrative and product-service issues. A closer inspection of the structural coefficients reveals, however, that participative leadership behavior is more significantly related to conflict over administrative issues (e.g., bureaucratic red tape, franchise fees, territory restrictions) than conflict arising from product-service disagreements (e.g., national advertising programs, training programs, sales promotional assistances).

Because of channel members' sensitivity to administrative issues stemming from elements such as service and advertising fees, accounting information requirements and contract terms,

they prefer more participation in policymaking. Thus, when the channel leader uses a participative leadership style, the channel member tends to experience less conflict over administrative issues.

One explanation for this association is that the relationship between participative style and subordinate behavior may be moderated by personality characteristics or situational demands. Studies by Tannenbaum and Allport (1956) and Vroom (1959), for instance, showed that subordinates who prefer autonomy and self-control respond more positively to participative leadership, and thus experience less conflict (especially conflict over administrative issues) than subordinates who do not have such preferences. Given the autonomous nature of the real estate brokerage franchise channel, such could be the case in the findings presented here.

These findings suggest that if channel management wants to minimize potential dysfunctional effects of conflict, a climate of trust, responsibility and participation is needed. The franchisor should consult with franchisees, solicit their suggestions and consider these suggestions when designing and introducing channel-wide policies and procedures. If the franchisor does these things, the formal task of administering the channel organization should become more efficient and effective as a result of reduced conflict.

The canonical correlation results for the supportive leadership dimension (presented in Table 6) indicate that the three supportive leadership items, when correlated with dimensions of intrachannel conflict, produce a significant canonical correlation coefficient of .569. According to the redundancy index, over 27% of the variation in the intrachannel conflict dimensions is shared with variance in the supportive leadership items. Conversely, 26% of the variance in the supportive leadership items is accounted for by the variability in the intrachannel conflict dimensions.

The cross-loading values for both dimensions of conflict exceed the .30 minimum value, indicating a significant association between supportive leadership behavior and both dimensions of intrachannel conflict. Thus, channel members who perceive their channel leader to be exhibiting leadership behavior emphasizing support and consideration tend to experience less intrachannel conflict. As with participative leadership, it appears that leadership behavior reflecting support and consideration tends to be more significantly associated with conflict over administrative issues rather than with product-service issues.

These findings suggest that to reduce intrachannel conflict, the franchisor should create a leadership climate of support or consideration. Emphasis should be placed on (1) giving support and consideration to franchisee needs and behavior, (2) exhibiting confidence in franchisee effort, (3) emphasizing the prestige of the franchise organization, (4) working to make the franchisee-franchisor relationship a friendly and pleasant arrangement, and (5) accentuating overall franchise group accomplishments. A franchisor should realize that both social and economic considerations motivate franchisees.

Table 7 shows the results of the canonical correlation between directive leadership behavior and intrachannel conflict. The .446 canonical correlation value indicates a significant relationship between the two sets of variables. The redundancy index shows that 16.26% of the variation in intrachannel conflict is associated with the variation in the directive leadership behavior items, while 12.29% of the variance in the conflict dimensions is explained by the variation in directive leadership.

An examination of the structural coefficients for the canonical function shows the composite canonical score for directive leadership behavior to be inversely related to both administrative and product-service conflict. Thus, when the channel member perceives a high degree of task specificity or instrumental behavior being exhibited by the channel leader, the member tends to experience less intrachannel conflict in general. However, in the case of directive leadership behavior, the composite canonical score for the leadership items is significantly related to the level of product-service conflict than to conflict arising over administrative issues.

This result is converse to the previously described findings for the participative and supportive leadership dimensions. In both of the previous analyses, leadership behavior was more significantly related to conflict over administrative rather than product-service issues. This difference in results may be attributed to the specific nature of each of the three leadership styles, and more importantly, to how important channel members view conflicts over administrative versus product-service issues. In this particular channel, members view product-service issues as important because they are complex and are closely related to extrinsic compensation (Davis 1979, Kilborn 1979). Consequently, to minimize ambiguity surrounding the administration of these complex programs, the channel member may prefer more structured or directive channel leadership. On the other hand, given the inherent desire for autonomy typically observed among channel members, in franchised channel systems (Davidson 1970), channel members may prefer to participate actively and/or receive individual consideration or support. Accordingly, one might expect conflict over administrative issues to be highly associated with participative and supportive leadership behavior.

These directive leadership findings suggest that when a franchisee feels that the franchisor communicates what is expected, gives special guidance in franchise operations, maintains definite performance standards, and asks franchisees to follow standard rules and regulations, there is less conflict with the franchisor. The franchisees may perceive the situations to be less ambiguous when franchisors provide close supervision and a task-oriented franchise arrangement. As such, franchisees understand and more fully accept franchise-related policies and procedures and thus experience less confusion, resulting in less conflict with their franchisors. By designing and communicating a directive channel policy to franchisees, the possibility for conflict may be reduced.

Aggregate Canonical Analysis

The results of the three separate canonical correlation analyses suggest that intrachannel conflict is significantly related to all three types of channel leadership behavior. To obtain a better understanding of the interactive statistical relationships between the different types of leadership behavior and intrachannel conflict, a combined analysis was performed using the three types of leadership behavior as individual components in one predictor set. These results are reported in Table 4 and suggest that the three types of channel leadership behavior explain a significant share of the variance in intrachannel conflict. The combined data produced a canonical correlation value of .559, which is significantly different from zero at the .000 level by the chi-square test. The findings show that about 27% of the variation in intrachannel conflict is accounted for by the variability in the type of channel leadership behavior by the channel leader. Conversely, over 19% of the variation in leadership behavior is accounted for by the variability in the intrachannel conflict dimensions. Thus, it is apparent from both the individual and overall analyses that channel leadership behavior and intrachannel conflict are related.

An inspection of the structural coefficients for the canonical function (Table 4) indicates that the composite canonical score for channel leadership behavior is significantly related to both dimensions of intrachannel conflict. The cross-loadings for both dimensions of intrachannel conflict exceed the .30 level suggested by Lambert and Durand (1975) as an acceptable minimum loading value. Table 4 shows that all three types of leader-

ship behavior had cross-loading values that exceed .30. Thus, all three leadership styles were significantly related to the composite canonical score for intrachannel conflict. Supportive leadership behavior does, however, appear to have a stronger relationship with conflict than do the other two types of leader behavior. This is evident in the larger cross-loading value derived for supportive leadership.

These data indicate that, as predicted in the rationale for H_1, participative leadership is inversely related to channel members' perceptions of intrachannel conflict. This finding supports the underlying theory and research regarding participative leadership style (Strauss and Rosenstein 1970, Tannebaum and Cooke 1974). Franchisees experience less conflict with their franchisor when they have some influence in determining channel-related policies and procedures. These findings are consistent with previous studies' results showing participation to be inversely related to conflict (House and Mitchell 1974, House and Rizzo 1972, Seashore and Bowers 1970).

The franchisee's perceptions of the level of supportive leadership behavior exhibited by the franchisor appear to have a significant, inverse effect on intrachannel conflict. Thus, H_2 is not rejected. This finding supports the belief that when a franchisee is participating in a supportive franchise arrangement in which the franchisor is perceived as giving support consideration to the needs of franchisees, the franchisee better understands and more fully accepts franchise-related policies and procedures, and thus less conflict is likely to occur.

Consistent with the expectations stated in support of H_3, directive leadership behavior has a significant, inverse effect on channel members' perceptions of intrachannel conflict. Channel members who perceive a high degree of directiveness being exhibited by channel management tend to experience less conflict with channel management. One explanation for this association is that directive leadership reduces a channel member's role ambiguity (or its consequences) and allows for a more cooperative, less conflictual relationship. This explanation is particularly relevant when one considers the high degree of ambiguity typically found in a channel organization.

Conclusion

From a managerial perspective, it is important to determine how channel leadership can improve relations with channel members to reduce intrachannel conflict. This study demonstrates that perceived intrachannel conflict, as described by Pondy (1967), Rosenberg and Stern (1970) and Rosenbloom (1973), varies on the basis of channel members' evaluations of the leadership style exhibited by their channel leader. The findings suggest that if a channel leader wants to minimize the potential detrimental effects of intrachannel conflict, the leader should implement measures to ensure the orchestration of an effective leadership climate necessary to adjust the channel environment to fit the needs and predispositions of franchisees better.

Several research issues merit further study regarding channel leadership behavior and its impact on behavioral outcomes such as intrachannel conflict. First, it has been noted in this study that channel leadership is a topic that has not achieved a sufficient level of research rigor. The construct of channel leadership behavior must be further refined and operationalized. Specifically, attempts should be made to further integrate concepts from traditional leadership theory with existing knowledge about channel behavior. In doing so, additional insights may be gained for developing a contingency theory of channel leadership. Second, the construct of intrachannel conflict needs further development. Separate measures are needed for assessing intrachannel conflict at different stages (e.g., cognitive, affective, behavioral) in the distribution channel. One could then examine leadership patterns as they develop over time in response to conflict in a channel of distribution. Third, this study found significant relationships between channel leadership behavior and channel conflict. Leadership behavior could also have a significant impact on other attitudinal and behavioral outcomes such as channel performance and satisfaction. Future research could investigate possible relationships between leadership behavior, intrachannel conflict and channel performance. Fourth, a limitation noted earlier was that the hypothesized relationships were unidimensional or monotonic in nature. Future research needs to explore the possibility of higher order relationships, such as curvilinear or exponential, between leadership behavior and intrachannel conflict. Fifth, increasing attention should focus on the mechanisms or conditions that facilitate or hinder channel leadership and determine the precise effects of leadership behavior on outcomes such as conflict, performance and satisfaction as they occur over time. Finally, additional interactions should be explored between channel leadership behavior and other organizational variables such as communication linkages, power relationships, level of channel commitment, autonomy and role perceptions.

References

Alderson, Wroe (1965), "Cooperation and Conflict in Marketing Channels," in *Dynamic Marketing Behavior,* Homewood, iL: Richard D. Irwin, Inc., 239–241, 244–258.

Alper, Mark I. and Robert A. Peterson (1972), "On the Interpretation of Canonical Analysis," *Journal of Marketing Research,* 9 (May), 1972.

Assael, Henry (1969), "Constructive Role in Inter-Organization Conflict," *Administrative Science Quarterly,* 14 (December), 573–575.

Baird, P. D., John Hay and J. M. Bailey (1977), "Government Regulation of Real Estate Franchising," *Real Property, Probate Trust Journal,* 12 (Fall), 580.

Bower, Joseph L. (1965), "The Role of Conflict in Economic Decision-Making Groups: Some Empirical Results," *Quarterly Journal of Economics,* 74 (May), 267–277.

Brown, James R. and Ralph L. Day (1981), "Measures of Manifest Conflict in Distribution Channels," *Journal of Marketing Research,* 18 (August), 263–274.

Bucklin, Louis P. (1968), "The Locus of Channel Control," in *Marketing and the New Science of Planning,* Robert L. King, ed., Chicago: American Marketing Association.

——— (1973), "A Theory of Channel Control," *Journal of Marketing,* 37 (January), 39–47.

Burke, P. J. (1966), "Authority Relations and Descriptive Behavior in Small Discussion Groups," *Sociometry,* 29 (September), 237–250.

Churchill, Gilbert, Jr. (1979), *Marketing Research: Methodical Foundations*, Hinsdale, IL: The Dryden Press, Inc.

Coleman, James S. (1957), *Community Conflict*. Glencoe, IL: Free Press.

Corwin, Ronald G. (1969), "Patterns of Organizational Conflict," *Administrative Science Quarterly*, 14 (March), 507–519.

Cronbach, L. J. (1960), *Essentials of Psychological Testing*, 2nd ed., New York: Harper and Row Company.

Davidson, William (1970), "Changes in Distribution Institutions," *Journal of Marketing*, 32 (January), 7.

Davis, Jerry (1979), "Big Business Enters Real Estate," *Real Estate Today*, 12 (August), 43–48.

Deutsch, Morten (1971), "Toward an Understanding of Conflict," *International Journal of Group Tensions*, 1 (January–March), 42–54.

El-Ansary, Adel I. and R. A. Robicheaux (1974), "A Theory of Channel Control: Revisited," *Journal of Marketing*, 38 (January), 2–7.

———— and Louis W. Stern (1972), "Power Measurement in the Distribution Channel," *Journal of Marketing Research*, 9 (February), 47–52.

Etgar, Michael (1976), "Channel Domination and Countervailing Power in Distribution Channels," *Journal of Marketing Research*, 13 (August), 154–162.

———— (1977), "Channel Environment and Channel Leadership." *Journal of Marketing Research*. 14 (May), 69–76.

———— (1978), "Selection of Effective Channel Control Mix," *Journal of Marketing*, 42 (July), 53–58.

———— (1979), "Sources and Types of Intrachannel Conflict," *Journal of Retailing*, 55 (Spring), 63–78.

Firat, Faud, Alice Tybout and Louis Stern (1974), "A Perspective on Conflict and Power in Distribution," in *Combined Proceedings of the AMA Fall and Spring Conference*, R. C. Curhan, ed., Chicago: American Marketing Association, 436–438.

Gamson, William A. (1966), "Rancorous Conflict in Community Politics," *American Sociologist Review*, 31 (April), 71–81.

Green, Paul E. (1978), *Analyzing Multivariate Data*, Hinsdale, IL: The Dryden Press.

Hellriegel, Don and John W. Slocum, Jr. (1974) "Organizational Climate: Measures, Research, and Contingencies," *Academy of Management Journal*, 17 (June), 255–280.

House, R. J., Alan C. Filley and Steven Kerr (1971), "Relation of Leader Consideration and Initiating Structure to R and D Subordinates' Satisfaction." *Administrative Science Quarterly*, 16 (March), 19–30.

———— and T. R. Mitchell (1974), "Path-Goal Theory of Leadership," *Journal of Contemporary Business*, 4 (Autumn), 81–97.

———— and J. R. Rizzo (1972), "Toward the Measurement of Organizational Practices: Scale Development and Validation," *Journal of Applied Psychology*, 56 (October), 388–396.

Hunt, S. D. and J. R. Nevin (1974), "Power in a Channel of Distribution: Sources and Consequences," *Journal of Marketing Research*, 11 (May), 186–93.

Ivancevich, John M., Andrew D. Szilagyi, Jr. and Marc J. Wallace, Jr. (1977), *Organizational Behavior and Performance*, Santa Monica, CA: Goodyear Publishing Company, Inc.

Katz, Daniel and Robert L. Kahn (1953), "Leadership Practices in Relation to Productivity and Morale," in *Group Dynamics: Research and Theory*, D. Cartwright and A. Zanders, eds., New York: Harper and Row, 612–628.

————, Nathan M. Maccoby and Nancy Morse (1950), *Productivity, Supervision and Morale in an Office Situation*, Ann Arbor: Survey Research Center, University of Michigan.

Kilborn, Peter (1979), "Corporate Giants Invade the Residential Market," *New York Times*, 28 (February 4), Section 3, 1F.

Lambert, Z. V. and R. M. Durand (1975), "Some Precautions in Using Canonical Analysis," *Journal of Marketing Research*, 12 (November), 468–475.

Little, Robert W. (1970), "The Marketing Channel: Who Should Lend This Extracorporate Organization?," *Journal of Marketing*, 34 (January), 31–38.

Lusch, R. F. (1976), "Sources of Power: Their Impact on Intrachannel Conflict," *Journal of Marketing Research*, 13 (November), 382–90.

Mallen, Bruce (1964), "Conflict and Cooperation in Marketing Channels," in *Reflections on Progress in Marketing*, L. G. Smith, ed., Chicago: American Marketing Association.

Nunnally, Jum C. (1967), *Psychometric Theory*, New York: McGraw-Hill Book Company, Inc.

Parsons, Talcott (1951), *The Social System*, Glencoe, IL: Free Press.

Pearson, M. and J. F. Monoky (1976), "The Role of Conflict and Cooperation in Channel Performance." in *Marketing: 1776–1976 and Beyond*, K. L. Bernhardt, ed., Chicago: American Marketing Association, 240–44.

Pondy, Louis (1967), "Organizational Conflict: Concepts and Models," *Administrative Science Quarterly*, 12 (September), 296–320.

Reve, Torger (1977), "Conflict and Performance in Distribution Channels," unpublished paper, Northwestern University.

Robicheaux, Robert A. and A. L. El-Ansary (1975), "A General Model for Understanding Channel Member Behavior." *Journal of Retailing*, 52 (Winter), 13–31.

Rosenberg, L. J. (1971), "Conflict Measurement in the Distribution Channel," *Journal of Marketing Research*, 8 (November), 437–42.

———— and L. W. Stern (1970), "Toward the Analysis of Conflict in Distribution Channels: A Descriptive Model," *Journal of Marketing*, 34 (October), 40–6.

———— and ———— (1971), "Conflict Measurement in the Distribution Channel," *Journal of Marketing Research*, 8 (November), 437–442.

Rosenbloom, Bert (1973), "Conflict and Channel Efficiency: Some Conceptual Models for the Decision Maker," *Journal of Marketing*, 37 (July), 26–30.

——— (1978), *Marketing Channels: A Management View*, Hinsdale, IL: Dryden Press.

Seashore, S. and D. Bowers (1970), "Durability of Organizational Change," *American Psychologist*, 25 (March), 227–233.

Speh, Thomas W. and E. H. Bonfield (1978), "The Control Process in Marketing Channels: An exploratory Investigation," *Journal of Retailing*, 54 (Spring), 13–26, 95–98.

Stern, Louis W. (1967), "The Concept of Channel Control," *Journal of Retailing*, 53 (Summer), 14–20.

——— and Adel I. El-Ansary (1977), *Marketing Channels*, Englewood, NJ: Prentice-Hall, Inc.

——— and R. H. Gorman (1969), "Conflict in Distribution Channels: An Exploration," in *Distribution Channels: Behavioral Dimensions*, Louis W. Stern, ed., Boston: Houghton Mifflin Company, 156–175.

——— and J. L. Heskett (1969), "Conflict Management in Interorganization Relations: A Conceptual Framework," in *Distribution Channels: Behavioral Dimensions*, Louis W. Stern, ed., Boston: Houghton Mifflin Company, 288–305.

———, Brian Sternthal and C. S. Craig (1973), "Managing Conflict in Distribution Channels: A Laboratory Study," *Journal of Marketing Research*, 10 (May), 169–179.

Stewart, D. K. and W. A. Love (1968), "A General Canonical Corrrelation Index," *Psychological Bulletin*, 70 (July), 160–163.

Stogdill, R. M. and Alvin Coons (1957), *Leader Behavior: Its Description and Measurement*, Columbus: Bureau of Education Research Monograph 88, Ohio State University.

Strauss, S. and E. Rosenstein (1970), "Worker's Participation: A Critical View," *Industrial Relations*, 4 (February), 197–214.

Sturdivant, F. D. and D. L. Granbois (1968), "Channel Interaction: An Institutional Behavioral View," *The Quarterly Review of Economics and Business*, 8 (Summer), 61–8.

Tannebaum, A. S. and F. H. Allport (1956), "Personality Structure and Group Structure: An Interpretive Study of Their Relationship Through an Event-Structure Hypothesis," *Journal of Abnormal and Social Psychology*, 53 (November), 272–280.

——— and R. A. Cooke (1974), "Control and Participation," *Journal of Contemporary Business*, 4 (Autumn), 45–46.

Vroom, V. H. (1959), "Some Personality Determinants of the Effects of Participation," *Journal of Abnormal and Social Psychology*, 59 (July), 322–327.

Walker, Bruce J. and N. H. Shooshtari (1979), "A Review of Channel Behavior Empirical Research in the 70's," *Proceedings*, Southern Marketing Association.

Walker, Orville C. (1970), "An Experimental Investigation of Conflict and Power in Marketing Channels," Ph.D. dissertation, University of Wisconsin, Madison.

Wilkinson, I. A. (1972), "Power and Influence Structures in Distribution Channels," *European Journal of Marketing*, 7 (Summer), 119–29.

CHAPTER **7**

Factor Analysis

LEARNING OBJECTIVES

Upon completing this chapter, you should be able to do the following:

- Differentiate factor analytic techniques from other multivariate techniques.

- State the major purposes of factor analytic techniques.

- Distinguish between exploratory and confirmatory uses of factor analytic techniques.

- Identify the differences between component analysis and common factor analysis models.

- Tell when component analysis and common factor analysis should be utilized.

- Identify the difference between R and Q factor analysis.

- Explain the concept of rotation of factors.

- Tell how to determine the number of factors to extract.

- Explain how to name a factor.

- Explain the purpose of factor scores and how to use them.

- Explain how to select surrogate variables for subsequent analysis.

- State the major limitations of factor analytic techniques.

CHAPTER PREVIEW

The multivariate statistical technique of factor analysis has found increased use during the past decade in all fields of business-related research. This chapter describes factor analysis, a technique particularly suitable for analyzing the patterns of complex, multidimensional relationships encountered by researchers and businesspeople. It defines and explains in broad, conceptual terms the fundamental aspects of factor analytic techniques. Factor analysis can be utilized to examine the underlying patterns or relationships for a large number of variables and to determine whether or not the information can be condensed or summarized in a smaller set of factors or components. Basic guidelines for presenting and interpreting the results of these techniques are also included to further clarify the methodological concepts.

KEY TERMS

Before starting the chapter, review the key terms to develop an understanding of the concepts and terminology used. Throughout the chapter the key terms appear in **boldface.** Other points of emphasis in the chapter are *italicized.* Also, cross-references within the key terms are in *italics.*

Anti-image correlation matrix Matrix of the partial correlations among variables after factor analysis, or the degree to which the factors "explain" each other in the results. The diagonal contains the *measures of sampling adequacy* for each variable, and the off-diagonal values are partial correlations among variables.

Bartlett test of sphericity Statistical test for the overall significance of all correlations within a correlation matrix.

Cluster analysis Multivariate technique with the objective of grouping respondents or cases with similar profiles on a defined set of characteristics. Similar to *Q factor analysis.*

Common factor analysis Factor model in which the factors are based on a reduced correlation matrix. That is, *communalities* are inserted in the diagonal of the *correlation matrix,* and the extracted factors are based only on the *common variance,* with *specific* and *error variance* excluded.

Common variance Variance shared with other variables in the factor analysis.

Communality Amount of variance an original variable shares with all other variables included in the analysis.

Component analysis Factor model in which the factors are based on the total variance. With component analysis, unities (1s) are used in the diagonal of the *correlation matrix;* this procedure computationally implies that all the variance is common or shared.

Correlation matrix Table showing the intercorrelations among all variables.

Eigenvalue Column sum of squared loadings for a factor; also referred to as the *latent root.* It represents the amount of variance accounted for by a factor.

Error variance Variance of a variable due to errors in data collection or measurement.

Factor Linear combination (variate) of the original variables. Factors also represent the underlying dimensions (constructs) that summarize or account for the original set of observed variables.

Factor indeterminacy Characteristic of *common factor analysis* such that several different factor scores can be calculated for a respondent, each fitting the estimated factor model. This means the factor scores are not unique for each individual.

Factor loadings   Correlation between the original variables and the factors, and the key to understanding the nature of a particular factor. Squared factor loadings indicate what percentage of the variance in an original variable is explained by a factor.

Factor matrix   Table displaying the *factor loadings* of all variables on each factor.

Factor rotation   Process of manipulating or adjusting the factor axes to achieve a simpler and pragmatically more meaningful factor solution.

Factor score   Composite measure created for each observation on each factor extracted in the factor analysis. The factor weights are used in conjunction with the original variable values to calculate each observation's score. The factor score then can be used to represent the factor(s) in subsequent analyses.

Latent root   See *eigenvalue.*

Measure of sampling adequacy   Measure calculated both for the entire correlation matrix and each individual variable evaluating the appropriateness of applying factor analysis. Values above .50 for either the entire matrix or an individual indicate appropriateness.

Oblique factor rotation   Factor rotation computed so that the extracted factors are correlated. Rather than arbitrarily constraining the factor solution so the factors are independent of one another, the analysis is conducted to express the relationship between the factors that may or may not be *orthogonal.*

Orthogonal   Mathematical independence of factor axes to each other (i.e., at right angles, or 90 degrees).

Orthogonal factor rotation   Factor rotation in which the factors are extracted so that their axes are maintained at 90 degrees. Each factor is independent of, or *orthogonal* from, all other factors. The correlation between the factors is determined to be zero.

Q factor analysis   Forms groups of respondents or cases based on their similarity on a set of characteristics.

R factor analysis   Analyzes relationships among variables to identify groups of variables forming latent dimensions (*factors*).

Specific variance   Variance of each variable unique to that variable and not explained or associated with other variables in the factor analysis.

Trace   Sum of the squares of the numbers on the diagonal of the correlation matrix used in the factor analysis. It represents the total amount of variance on which the factor solution is based. With component analysis, the trace is equal to the number of variables, based on the assumption that the variance in each variable is equal to 1. With common factor analysis, the trace is equal to the sum of the communalities on the diagonal of the reduced correlation matrix (also equal to the amount of common variance for the variables being analyzed).

VARIMAX   One of the most popular orthogonal factor rotation methods.

What Is Factor Analysis?

Factor analysis is a generic name given to a class of multivariate statistical methods whose primary purpose is to define the underlying structure in a data matrix. Broadly speaking, it addresses the problem of analyzing the structure of the interrelationships (correlations) among a large number of variables (e.g., test scores, test items, questionnaire responses) by defining a set of common underlying dimensions, known as **factors.** With factor analysis, the analyst can first identify the separate dimensions of the structure and then determine the extent to which each variable is explained by each dimension. Once these dimensions and the explanation of each variable are determined, the two primary uses for factor analysis—summarization and data reduction—can be achieved. In summarizing the data, factor analysis derives underlying dimensions that, when interpreted and understood, describe the data in a much smaller number of items than the original individual variables. Data reduction can be achieved by calculating scores for each underlying dimension and substituting them for the original variables.

Factor analysis is not like the dependence techniques discussed in earlier chapters (i.e., multiple regression, discriminant analysis, multivariate analysis of variance, or canonical correlation), where one or more variables are explicitly considered the criterion or dependent variables and all others are the predictor or independent variables. Factor analysis is an interdependence technique in which all variables are simultaneously considered, each related to all others. Although not a dependence technique, factor analysis still employs the concept of the variate, the linear composite of variables. In factor analysis, variates are formed in a manner similar to discriminant analysis or canonical correlation, but with a different purpose. In those dependence techniques, the variates were formed to maximize their predictive power. In factor analysis, the variates (factors) are formed to maximize their explanation of the entire variable set, not to predict a dependent variable(s). If we were to draw an analogy to dependence techniques, it would be that each of the observed (original) variables is a dependent variable that is a function of some underlying and latent set of factors (dimensions) that are themselves made up of all other variables. Thus, each variable is predicted by all others. Conversely, one can look at each factor (variate) as a dependent variable that is a function of the entire set of observed variables. Either analogy illustrates the differences in purpose between dependence (prediction) and interdependence (identification of structure) techniques.

Factor analytic techniques can achieve their purposes from either an exploratory or confirmatory perspective. There is continued debate concerning the appropriate role for factor analysis. Many researchers consider it only exploratory, useful in searching for structure among a set of variables or as a data reduction method. In this perspective, factor analytic techniques "take what the data give you" and do not set any a priori constraints on the estimation of components or the number of components to be extracted. For many, if not most, applications this use of factor analysis is appropriate. However, in other situations, the analyst has preconceived thoughts on the actual structure of the data, based on theoretical support or prior research. The analyst may wish to test hypotheses involving such issues as which variables should be grouped together on a factor or the precise number of factors. In these instances, the analyst requires that factor analysis take a confirmatory approach—that is, assess the degree to which the data meet the expected structure of the analyst. The methods we discuss in this chapter do not directly provide the necessary structure for formalized hypothesis testing. We explicitly address the confirmatory perspective of factor analysis in Chapter 11, where methods of performing confirmatory factor analysis are discussed. In this chapter, however, we view factor analytic techniques principally from an exploratory or nonconfirmatory viewpoint.

A Hypothetical Example of Factor Analysis

Assume that through qualitative research a retail firm has identified 80 different characteristics of retail stores and their service that consumers have mentioned as affecting their patronage choice among stores. The retailer wants to understand how consumers make decisions but feels that it cannot evaluate 80 separate characteristics or develop action plans for this many variables, as they are too specific. Instead, it would like to know if consumers think in more general evaluative dimensions rather than in just the specific items. To identify these dimensions, the retailer could commission a survey asking for consumer evaluations on each of these specific items. Factor analysis would then be used to identify the underlying evaluative dimensions. Specific items that correlate highly are assumed to be a ''member'' of that broader dimension. These dimensions become composites of specific variables, which in turn allow the dimensions to be interpreted and described. In our example, the factor analysis might identify such dimensions as product assortment, product quality, prices, store personnel, service, and store atmosphere as the evaluative dimensions used by the respondents. Each of these dimensions is comprised of specific items that are a facet of the broader evaluative dimension. From these findings, the retailer may then use the dimensions (factors) to define broad areas for planning and action.

Factor Analysis Decision Diagram

As we have done with previous multivariate techniques, we center the discussion of factor analysis on the six-stage model-building paradigm introduced in Chapter 1. Figure 7.1 shows the general steps followed in any application of factor analysis techniques. Factor analysis also adds an additional stage (stage 7) beyond the estimation, interpretation, and validation of the factor models that aids in selecting surrogate variables and computes factor scores for use in other multivariate techniques. A discussion of each stage follows.

Stage One: Objectives of Factor Analysis

The starting point in factor analysis, as with other statistical techniques, is the research problem. The general purpose of factor analytic techniques is to find a way of condensing (summarizing) the information contained in a number of original variables into a smaller set of new, composite dimensions or variates (factors) with a minimum loss of information—that is, to search for and define the fundamental constructs or dimensions assumed to underlie the original variables [10, 18]. More specifically, factor analysis techniques can meet any of three objectives:

1. Identify the structure of relationships among either variables or respondents. Factor analysis can examine either the correlations between the variables or the correlations between the respondents. For example, suppose you have data on

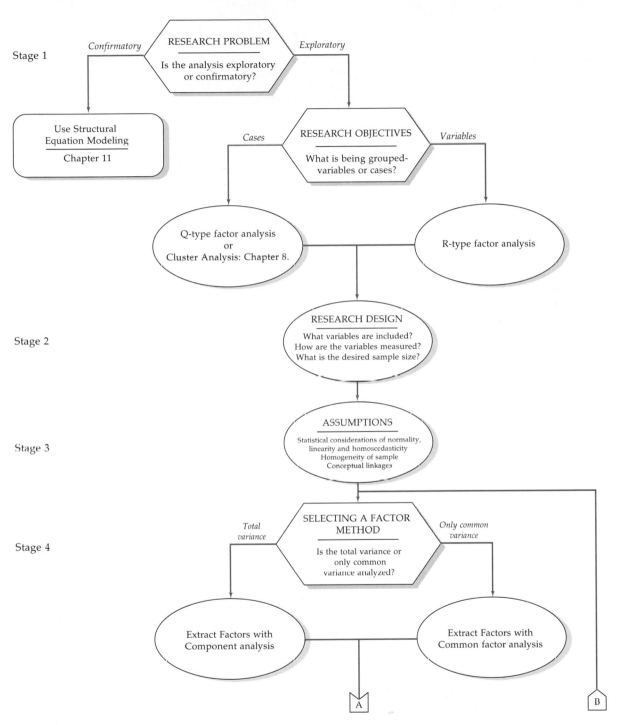

FIGURE 7.1 Factor analysis decision diagram.

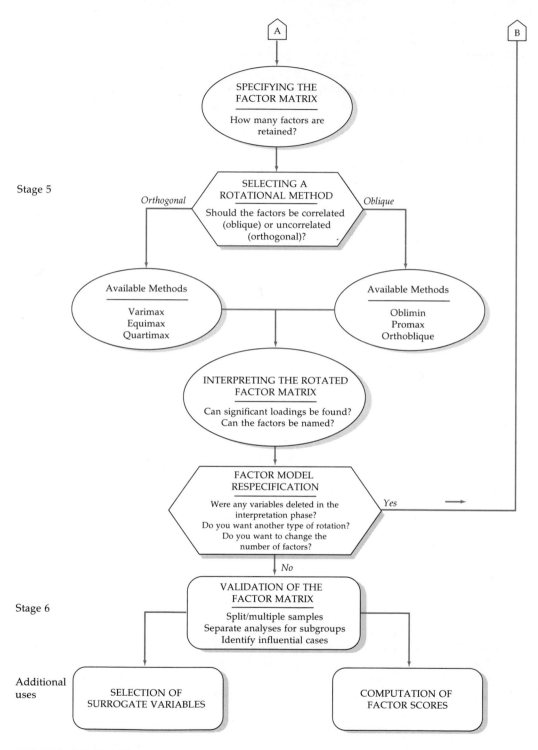

FIGURE 7.1 *Continued*

100 respondents in terms of 10 characteristics. If the objective of the research is to summarize the characteristics, the factor analysis would be applied to a correlation matrix of the variables. This most common type of factor analysis is referred to as R factor analysis. **R factor analysis** analyzes a set of *variables* to identify the dimensions that are latent (not easily observed). Factor analysis also may be applied to a correlation matrix of the individual respondents based on their characteristics. This is referred to as **Q factor analysis,** a method of combining or condensing large numbers of people into distinctly different groups within a larger population. The Q factor analysis approach is not utilized very frequently because of computational difficulties. Instead, most analysts utilize some type of **cluster analysis** to group individual respondents. Also see [23] for other possible combinations of groups and variable types.

2. Identify representative variables from a much larger set of variables for use in subsequent multivariate analyses (see below, Stage 7: Selecting Surrogate Variables for Subsequent Analysis).

3. Create an entirely new set of variables, much smaller in number, to partially or completely replace the original set of variables for inclusion in subsequent techniques, ranging from the dependence methods of regression, correlation, or discriminant analysis to cluster analysis, another interdependence technique (see below, "Stage 7: Using Factor Scores").

The first objective makes the identification of the underlying dimensions or factors ends in themselves; the estimates of the factors and the contributions of each variable to the factors (termed loadings) are all that is required for the analysis. The second objective relies on the factor loadings as well but uses them as the basis for identifying variables for subsequent analysis with other techniques. The third objective requires that estimates of the factors themselves (factor scores) be obtained; then the factor scores replace the original variables in uses such as independent variables in a regression, discriminant, or correlation analysis. The method of calculating and interpreting factor loadings is discussed later.

For data reduction and summarization factor analysis can be used either with preexisting sets of variables or with variables defined and selected by the researcher in a new research effort. When using an existing set of variables, the researcher should still consider the conceptual underpinnings of the variables and use judgment as to the appropriateness of the variables for factor analysis. These considerations can have substantial effects on the results. When used in a new research effort, factor analysis can also determine structure and/or create new composite scores for the original variables. For example, one of the first steps in constructing a summated scale (see Chapter 1) is to assess its dimensionality and the appropriateness of the selected variables through factor analysis. Thus, even though not truly confirmatory, exploratory factor analysis is used to evaluate the proposed dimensionality.

With the purpose of factor analysis specified, the researcher must then define the set of variables to be examined. In either R-type or Q-type factor analysis, the researcher specifies the potential dimensions that can be identified through the character and nature of the variables submitted to the factor analysis. In our hypothetical example, if no questions on store personnel were included, factor analysis would not be able to identify this dimension. The researcher must also remember that factor analysis will always produce factors. Thus, factor analysis is always a potential candidate for the "garbage in, garbage out" phenomenon. If the researcher indiscriminately includes a large number of variables and hopes that factor analysis will "figure it out," then the possibility of poor results is high. The "quality" and meaning of the derived factors reflects the conceptual underpinnings of the variables used in the analysis. The use of factor analysis as a data summarization technique does not exclude the need for a conceptual basis for any variables analyzed. Even if used solely for data reduction, factor analysis is most efficient when conceptually defined dimensions can be represented by the derived factors.

Stage Two: Designing a Factor Analysis

The design of a factor analysis involves three basic decisions: (1) calculation of the input data (a correlation matrix) to meet the specified objectives of grouping variables or respondents, (2) the design of the study in terms of number of variables, measurement properties of variables, and the types of allowable variables, and (3) the sample size necessary, both in absolute terms and as a function of the number of variables in the analysis.

Correlations Among Variables or Respondents

The first decision in the design of a factor analysis focuses on the approach used in calculating the correlation matrix to be used in either R-type or Q-type factor analysis. The analyst could derive the input data matrix from the computation of correlations between the variables. This would be an R-type factor analysis, and the resulting factor pattern would demonstrate the underlying relationships of the variables. The analyst could also elect to derive the correlation matrix from the correlations between the individual respondents. In a Q-type factor analysis, the results would be a **factor matrix** that would identify similar individuals. For example, if the individual respondents are identified by number, the resulting factor pattern might tell you that individuals 1, 5, 7, and 10 are similar. These respondents would be grouped together because they exhibited a high loading on the same factor. Similarly, respondents 2, 3, 4, and 8 would perhaps load together on another factor. We would label these individuals as similar. From the results of a Q factor analysis, we could identify groups or clusters of individuals demonstrating a similar pattern on the variables included in the analysis.

A logical question at this point would be, How does Q-type factor analysis differ from cluster analysis? The answer is that both approaches compare a series of responses to a number of variables and place the respondents in several groups. The difference is that the resulting groups for a Q-type factor analysis would be based on the intercorrelations between the means and standard deviations of the respondents. In a typical cluster analysis approach, groupings would be devised based on a distance measure between the respondents' scores on the variables being analyzed. To illustrate this difference, consider Table 7.1, which contains the scores of four respondents over three different variables. A Q-type factor analysis of these four respondents would yield two groups with similar variance structures. The two groups would consist of respondents A and C versus B and D. In contrast, the clustering approach would be sensitive to the distances among the respondents' scores and would lead to a grouping of the closest pairs. Thus, with a cluster analysis approach, respondents A and B would be placed in one group and C and D in the other group. If the researcher decides to employ

TABLE 7.1 Comparisons of Score Profiles for Q-type Factor Analysis and Hierarchical Cluster Analysis

Respondent	Variables		
	1	2	3
A	7	6	7
B	6	7	6
C	4	3	4
D	3	4	3

Q-type factor analysis, these distinct differences from traditional cluster analysis techniques should be noted. With the availability of other grouping techniques and the widespread use of factor analysis for data reduction and summarization, the remaining discussion in this chapter focuses on R-type factor analysis, the grouping of variables rather than respondents.

Variable Selection and Measurement Issues

The analyst also needs to answer two specific questions at this point: How are the variables measured? How many variables should be included? Variables for factor analysis are generally assumed to be of metric measurement. In some cases, dummy variables (coded 0–1), although considered nonmetric, can be used. If all variables are dummy variables, then specialized forms of factor analysis, such as Boolean factor analysis, are more appropriate [3]. The researcher should also attempt to minimize the number of variables included but still maintain a reasonable number of variables per factor. If a study is being designed to assess a proposed structure, be sure to include several variables (five or more) that may represent each proposed factor. The strength of factor analysis lies in finding patterns among groups of variables, and it is of little use in identifying factors composed of only a single variable. Finally, when designing a study to be factor analyzed, identify if possible several key variables (sometimes referred to as key indicants or marker variables) that reflect closely the hypothesized underlying factors. This will aid in interpreting the derived factors and assessing whether the results have practical interpretation value.

Sample Size > 100

Regarding the sample size question, the researcher generally would not factor analyze a sample of fewer than 50 observations, and preferably the sample size should be 100 or larger. As a general rule, the minimum is to have at least five times as many observations as there are variables to be analyzed, and the more acceptable range would be a ten-to-one ratio. Some researchers even propose a minimum of 20 cases for each variable. One must remember that with 30 variables, for example, there are 435 correlations in the factor analysis. At a .05 significance level, perhaps even 20 of those correlations would be deemed significant and appear in the factor analysis just by chance. The researcher should always try to obtain the highest cases-per-variable ratio to minimize the chances of "overfitting" the data, in this case deriving factors that are sample specific with little generalizability. The researcher may do this both by employing the most parsimonious set of variables, guided by conceptual and practical considerations, and by then obtaining an adequate sample size for the number of variables examined. When dealing with smaller sample sizes and/or a lower cases-to-variable ratio, the analyst should always interpret any findings cautiously. The issue of sample size will also be addressed in a later section on interpreting factor loadings.

Stage Three: Assumptions in Factor Analysis

The critical assumptions underlying factor analysis are more conceptual than statistical. From a statistical standpoint, the departures from normality, homoscedasticity, and linearity apply only to the extent that they diminish the observed correlations. Only normality is necessary if a statistical test is applied to the significance of the factors, but these tests are rarely used. In fact, some degree of multicollinearity is desirable, because the objective is to identify interrelated sets of variables.

In addition to the statistical bases for the correlations of the data matrix, the factor analyst must also ensure that the data matrix has sufficient correlations to justify the application of factor analysis. If visual inspection reveals no substantial number of correlations greater than .30, then factor analysis is probably inappropriate. The correlations among variables can also be analyzed by computing the partial correlations among variables, that is, the correlations between variables when the effects of other variables are accounted for. If "true" factors exist in the data, the partial correlation should be small, because the variable can be explained by the factors (variates with loadings for each variable). If the partial correlations are high, then there are no underlying "true" factors and factor analysis is inappropriate. SPSS [22] and SAS [19] provide the **anti-image correlation matrix,** which is just the negative value of the partial correlation, while BMDP [3] directly provides the partial correlation. In each case, larger partial or anti-image correlations are indicative of a data matrix perhaps not suited to factor analysis.

As another mode of determining the appropriateness of factor analysis, some measures examine the entire correlation matrix. The **Bartlett test of sphericity,** a statistical test for the presence of correlations among the variables, is one such measure. It provides the statistical probability that the correlation matrix has significant correlations among at least some of the variables. The factor analyst should note, however, that increasing the sample size causes the Bartlett test to become more sensitive to detecting correlations among the variables. Another measure to quantify the degree of intercorrelations among the variables and the appropriateness of factor analysis is the **measure of sampling adequacy (MSA).** The index ranges from zero to one, reaching one when each variable is perfectly predicted without error by the other variables. The measure can be interpreted with the following guidelines: .90 or above, marvelous; .80 or above, meritorious; .70 or above, middling; .60 or above, mediocre; .50 or above, miserable; and below .50, unacceptable [13, 14]. The MSA increases as (1) the sample size increases, (2) the average correlations increase, (3) the number of variables increases, or (4) the number of factors decreases [14]. The MSA and the same guidelines can be extended to individual variables as well. The factor analyst should first examine the MSA values for each variable and exclude those falling in the unacceptable range. Once the individual variables achieve an acceptable level, then the overall MSA can be evaluated and a decision made on continuance of the factor analysis.

The conceptual assumptions underlying factor analysis deal with the set of variables selected and the sample chosen. A basic assumption of factor analysis is that some underlying structure does exist in the set of selected variables. It is the responsibility of the factor analyst to ensure that the observed patterns are conceptually valid and appropriate for study with factor analysis, because the technique has no means to determine appropriateness other than the correlations among variables. As an example, mixing dependent and independent variables in a single factor analysis and then using the derived factors to support dependence relationships is inappropriate. The researcher must also ensure that the sample is homogeneous with respect to the underlying factor structure. Applying factor analysis to a sample of males and females for a set of items known to differ because of gender is inappropriate. When the two subsamples (males and females) are combined, the resulting correlations may be a poor representation of the unique structures of each group. Thus, whenever differing groups are expected in the sample, separate factor analyses should be performed, and the results should be compared to identify differences not reflected in the results of the combined sample.

Stage Four: Deriving Factors and Assessing Overall Fit

Once the variables are specified and the correlation matrix prepared, the researcher is ready to apply factor analysis in an attempt to identify the underlying structure of relationships. In doing so, decisions must be made concerning (1) the method of extracting the factors (common factor analysis versus components analysis) and (2) the number of factors selected to represent the underlying structure in the data. Selection of the extraction method depends upon the analyst's objective. Component analysis* is used when the objective is to summarize most of the original information (variance) in a minimum number of factors for prediction purposes. In contrast, common factor analysis is used primarily to identify underlying factors or dimensions reflecting what the variables share in common. Both of these factor models are discussed in more detail in the following sections.

Common Factor Analysis Versus Component Analysis

The analyst can utilize two basic methods to obtain factor solutions. They are known as **common factor analysis** and **component analysis.** To select the appropriate model, the analyst must understand something about the types of variance. For the purposes of factor analysis, total variance consists of three kinds: (1) **common,** (2) **specific** (also known as unique), and (3) **error.** These types of variance and their relationship to the factor model selection process are illustrated in Figure 7.2. Common variance is defined as that variance in a variable that is shared with all other variables in the analysis. Specific variance is that variance associated with only a specific variable. Error variance is the variance due to unreliability in the data-gathering process, measurement error, or a random component in the measured phenomenon. When using component analysis, one must consider the total variance and derive factors that contain small proportions of unique variance and, in some instances, error variance. However, the first few factors do not contain enough unique or error variance to distort the overall factor structure.

*Many texts refer to this approach as *principal components*. For our purposes, component analysis is the same as principal components analysis.

Principal components analysis

It is important in your research to state that you have used principal components analysis or common factor analysis, as opposed to simply stating that you have performed a "factor analysis", which is too imprecise.

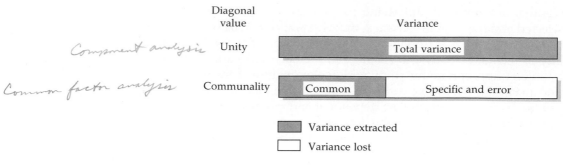

FIGURE 7.2 Types of variance carries into the factor matrix.

Specifically, with component analysis, unities are inserted in the diagonal of the correlation matrix, so that the full variance is brought into the factor matrix, as shown in Figure 7.2. Conversely, with common factor analysis, **communalities** are inserted in the diagonal. Communalities are estimates of the shared, or common, variance among the variables. Factors resulting from common factor analysis are based only on the common variance.

The common factor and component analysis models are both widely utilized. The selection of one model over the other is based on two criteria: (1) the objective of the researcher conducting the factor analysis and (2) the amount of prior knowledge about the variance in the variables. The component factor model is appropriate when the analyst is primarily concerned about prediction or the minimum number of factors needed to account for the maximum portion of the variance represented in the original set of variables, and when the factor analyst has prior knowledge suggesting that specific and error variance represent a relatively small proportion of the total variance. In contrast, when the primary objective is to identify the latent dimensions or constructs represented in the original variables and the researcher has little knowledge about the amount of unique error variance and therefore wishes to eliminate this variance, the common factor method is the appropriate model. With the more restrictive assumptions and its use of only the latent dimensions (shared variance), common factor analysis is often viewed as more theoretically based. However, although theoretically sound, common factor analysis has several problems. First, common factor analysis suffers from **factor indeterminacy,** which means that for any individual respondent, several different factor scores can be calculated from the factor model results [16]. There is no single unique solution as found in component analysis, but in most instances, the differences are not substantial. The second issue involves the calculation of the estimated communalities used to represent the shared variance. For larger-sized problems, the computations can take substantial computer time and resources. Also, the communalities are not always estimable or may be invalid (e.g., values greater than one or less than zero), requiring the deletion of the variable from the analysis (see the empirical example of common factor analysis later in this chapter).

The complications of common factor analysis have contributed to the widespread use of component analysis. While there remains considerable debate over which factor model is the more appropriate [2, 11, 15, 21], empirical research has demonstrated a similarity of the results in many instances [24]. In most applications, both component analysis and common factor analysis arrive at essentially identical results if the number of variables exceeds 30 [10] or the communalities exceed .60 for most variables. If the researcher is concerned with the assumptions of components analysis, then common factor analysis should also be applied to assess its representation of structure.

When a decision has been made on the factor model and the extraction method, the analyst is ready to extract the initial unrotated factors. By examining the unrotated factor matrix, the analyst can explore the data reduction possibilities for a set of variables and obtain a preliminary estimate of the number of factors to extract. Final determination of the number of factors must wait, however, until the factor matrix is rotated and the factors are interpreted.

Criteria for the Number of Factors to Be Extracted

How do we decide on the number of factors to extract? When a large set of variables is factored, the analysis first extracts the largest and best combinations of variables and then proceeds to smaller, less understandable combinations. In deciding when to stop factoring (that is, how many factors to extract), the analyst generally begins with some predetermined criterion, such as the a priori or the latent root criterion, to arrive at a specific number of factors to extract. (These two techniques are discussed in more detail later.) After the initial solution has been derived, the analyst makes several additional trial solutions—usually one less factor than the initial number and two or three more factors than were initially derived. Then, on the basis of information contained in the results of these several trial analyses, the factor matrices are examined, and the best representation of the data is used to assist in determining the number of factors to extract. By analogy, choosing the number of factors to be interpreted is something like focusing a microscope. Too high or too low an adjustment will obscure a structure that is obvious when the adjustment is just right. Therefore, by examining a number of different factor structures derived from several trial solutions, the analyst can compare and contrast to arrive at the best representation of the data. An exact quantitative basis for deciding the number of factors to extract has not been developed. However, the following stopping criteria for the number of factors to extract are currently being utilized.

Latent Root Criterion The most commonly used technique is the latent root criterion. This technique is simple to apply to either components analysis or common factor analysis. Only the factors having **latent roots** or <u>**eigenvalues**</u> greater <u>than 1</u> are considered significant; all factors with latent roots less than 1 are considered insignificant and are disregarded. The rationale for the latent root criterion is that any individual factor should account for the variance of at least a single variable if it is to be retained for interpretation. Using the eigenvalue for establishing a cutoff is probably most reliable when the number of variables is between 20 and 50. In instances where the number of variables is less than 20, there is a tendency for this method to extract a conservative number of factors (too few). When more than 50 variables are involved, however, it is not uncommon for too many factors to be extracted.

A Priori Criterion The a priori criterion is a simple yet reasonable criterion under certain circumstances. When applying it, the analyst already knows how many factors to extract before undertaking the factor analysis. The analyst simply instructs the computer to stop the analysis when the desired number of factors has been extracted. This approach is useful if the analyst is testing a theory or hypothesis about the number of factors to be extracted. It also can be justified in instances where the analyst is attempting to replicate another researcher's work and extract the same number of factors that was previously found.

A variant of the a priori criterion involves selecting enough factors to represent all the variables in the original data set. If theoretical or practical reasons necessitate each variable's communality being sufficient, then the researcher will include as many factors as necessary to adequately represent each of the original variables.

Percentage of Variance Criterion   The percentage of variance criterion is an approach in which the cumulative percentages of the variance extracted by successive factors are the criterion. The purpose is to ensure practical significance for the derived factors. No absolute threshold has been adopted for all applications. However, in the natural sciences the factoring procedure usually should not be stopped until the extracted factors account for at least 95 percent of the variance or until the last factor accounts for only a small portion (less than 5 percent). In contrast, in the social sciences, where information is often less precise, it is not uncommon for the analyst to consider a solution that accounts for 60 percent of the total variance (and in some instances even less) as a satisfactory solution.

Scree Test Criterion   Recall that with the component analysis factor model, the later factors extracted contain both common and unique variance. While all factors contain at least some unique variance, the proportion of unique variance is substantially higher in later than in earlier factors. The scree test is used to identify the optimum number of factors that can be extracted before the amount of unique variance begins to dominate the common variance structure [4]. The scree test is derived by plotting the latent roots against the number of factors in their order of extraction, and the shape of the resulting curve is used to evaluate the cutoff point. Figure 7.3 plots the first 18 factors extracted in a study by the authors. Starting with the first factor, the plot slopes steeply downward initially and then slowly becomes an approximately horizontal line. The point at which the curve first begins to straighten out is considered to indicate the maximum number of factors to extract. In the present case, the first 10 factors would qualify. Beyond 10, too large a proportion of unique variance would be included; thus these factors would not be acceptable. Note that in using the latent root criterion, only eight factors would have been considered. In contrast, using the scree test provides us with two more factors. As a general rule, the scree test results in at least one and sometimes two or three *more* factors being considered significant than does the latent root criterion [4].

Heterogeneity of the Respondents   Shared variance among variables is the basis for both common and component factor models. An underlying assumption is that shared variance extends across the entire sample. If the sample is heterogeneous with regard to at least one set of the variables, then the first factors will be those that are more homogeneous across the entire sample. Variables that are better discriminators between the subgroups of the sample will load on later factors, many times those not selected by the criteria discussed above [9]. When the objective is to identify factors that discriminate among the subgroups of a sample, the factor analyst should extract additional factors beyond those indicated by the methods above and examine the additional factors' ability to discriminate among the groups. If they prove less beneficial in discrimination, the solution can be run again and these later factors eliminated.

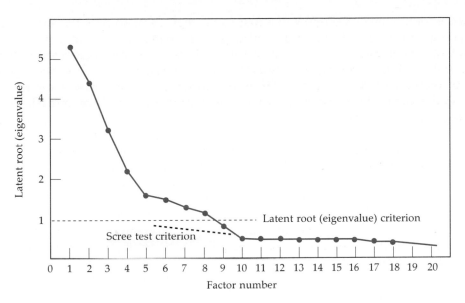

FIGURE 7.3 Eigenvalue plot for scree test criterion.

Summary of Factor Selection Criteria In practice, most factor analysts seldom use a single criterion in determining how many factors to extract. Instead, they initially use a criterion such as the latent root as a guideline for the first attempt at interpretation. After the factors have been interpreted, as discussed in the following sections, the practicality of the factors is assessed and factors included or excluded from another attempt at interpretation may be retained. Selecting the number of factors is interrelated with an assessment of structure, which is revealed in the interpretation phase. Thus, several factor solutions with differing numbers of factors are examined before the structure is well defined.

One word of caution in selecting the final set of factors. There are negative consequences for selecting either too many or too few factors to represent the data. If too few factors are used, then the correct structure is not revealed and important dimensions may be omitted. If too many factors are retained, then the interpretation becomes harder when the results are rotated (as discussed in the next section). While the factors are independent, you can just as easily have too many factors as having too few. As with other aspects of multivariate models, parsimony is important. The notable exception is when factor analysis is used strictly for data reduction and a set level of variance to be extracted is specified. The factor analyst should always strive to have the most representative *and* parsimonious set of factors possible.

Stage Five: Interpreting the Factors

Three steps are involved in the derivation of a final factor solution. First, the initial unrotated factor matrix is computed to assist in obtaining a preliminary indication of the number of factors to extract. In computing the unrotated factor matrix, the analyst is simply interested in the best linear combination of variables—best in the sense that the particular combination of original variables would account for more of the variance in the data as a whole than any other linear combination of variables. Therefore, the first factor may be viewed as the single best summary of linear relationships exhibited in the data. The second factor is defined as the second-best linear combination of the variables, subject to the constraint that it is orthogonal to the first factor. To be **orthogonal** to the first factor, the second one must be derived from the proportion of the variance remaining after the first factor has been extracted. Thus the second factor may be defined as the linear combination of variables that accounts for the most residual variance after the effect of the first factor has been removed from the data. Subsequent factors are defined similarly, until all the variance in the data is exhausted.

Unrotated factor solutions achieve the objective of data reduction, but the analyst must ask if the unrotated factor solution (while fulfilling desirable mathematical requirements) will provide information that offers the most adequate interpretation of the variables under examination. In most instances the answer to this question is no. The factor loading is the means of interpreting the role each variable plays in defining each factor. **Factor loadings** are the correlation of each variable and the factor. Loadings indicate the degree of correspondence between the variable and the factor, with higher loadings making the variable representative of the factor. The unrotated factor solution may or may not provide a meaningful patterning of variable loadings. If the unrotated factors are expected to be meaningful, the user may specify that no rotation be performed. Generally, rotation will be desirable because it simplifies the factor structure and it is usually difficult to determine whether unrotated factors will be meaningful or not. Therefore, the second step employs a rotational method to achieve simpler and theoretically more meaningful factor solutions. In most cases rotation of the factors improves the interpretation by reducing some of the ambiguities that often accompany initial unrotated factor solutions.

In the third step, the factor analyst assesses the need to respecify the factor model owing to (1) the deletion of a variable(s) from the analysis, (2) the desire to employ a different rotational method for interpretation, (3) the need to extract a different number of factors, or (4) the desire to change from one extraction method to another. Respecification of a factor model is accomplished by returning to the extraction stage, extracting factors, and interpreting them again.

Rotation of Factors

An important tool in interpreting factors is the **rotation of factors.** The term *rotation* means exactly what it implies. Specifically, the reference axes of the factors are turned about the origin until some other position has been reached. As indicated earlier, unrotated factor solutions extract factors in the order of their importance. The first factor tends to be a general factor with almost every variable loading significantly, and it accounts for the largest amount of variance. The second and subsequent factors are then based on the residual amount of variance. Each accounts for successively smaller portions of variance. The ultimate effect of rotating the factor matrix is to redistribute the variance from earlier factors to later ones to achieve a simpler, theoretically more meaningful factor pattern.

The simplest case of rotation is an orthogonal rotation in which the axes are maintained at 90 degrees. It is also possible to rotate the axes and not retain the 90-degree angle between the reference axes. When not constrained to being orthogonal, the rotational procedure is called an oblique rotation. Orthogonal and oblique factor rotations are demonstrated by Figures 7.4 and 7.5.

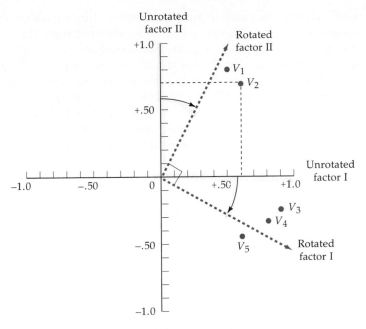

FIGURE 7.4 Orthogonal factor rotation.

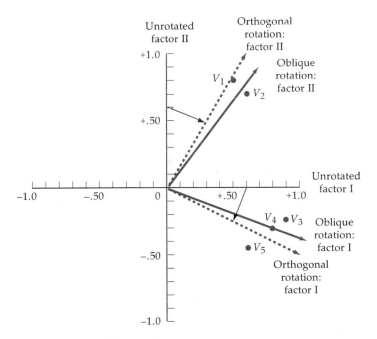

FIGURE 7.5 Oblique factor rotation.

An Illustration of Factor Rotation   Figure 7.4, in which five variables are depicted in a two-dimensional factor diagram, illustrates factor rotation. The vertical axis represents the unrotated factor II, and the horizontal axis represents unrotated factor I. The axes are labeled with a 0 at the origin and extend outward up to a +1.0 or a −1.0. The numbers on the axes represent the factor loadings. The five variables are labeled V^1, V^2, V^3, V^4, and V^5. The factor loading for variable 2 (V^2) on the unrotated factor II is determined by drawing a dashed line horizontally from the data point to the vertical axis for factor II. Similarly, a vertical line is drawn from variable 2 to the horizontal axis of the unrotated factor I to determine the loading of variable 2 on factor I. A similar procedure is followed for the remaining variables until all the loadings are determined for all the factor variables. The factor loadings for the unrotated and rotated solutions are displayed in Table 7.2 for comparison purposes. On the unrotated first factor, all the variables load fairly high. On the unrotated second factor, variables 1 and 2 are very high in the positive direction. Variable 5 is moderately high in the negative direction, while variables 3 and 4 have considerably lower loadings in the negative direction.

From visual inspection of Figure 7.4, it is obvious that there are two clusters of variables. Variables 1 and 2 go together, as do variables 3, 4, and 5. However, such patterning of variables is not so obvious from the unrotated factor loadings. By rotating the original axes clockwise, as indicated in Figure 7.4, we obtain a completely different factor loading pattern. Note that in rotating the factors, the axes are maintained at 90 degrees. This procedure signifies that the factors are mathematically independent and that the rotation has been **orthogonal**. After rotating the factor axes, variables 3, 4, and 5 load very high on factor I, and variables 1 and 2 load very high on factor II. Thus the clustering or patterning of these variables into two groups is more obvious after the rotation than before, even though the relative position or configuration of the variables remains unchanged.

The same general principles pertain to **oblique rotations** as they do to **orthogonal rotations.** The oblique rotation method is more flexible because the factor axes need not be orthogonal. It is also more realistic because the theoretically important underlying dimensions are not assumed to be uncorrelated with each other. In Figure 7.5 the two rotational methods are compared. Note that the oblique factor rotation represents the clustering of variables more accurately. This accuracy is a result of the fact that each rotated factor axis is now closer to the respec-

TABLE 7.2 Comparison Between Rotated and Unrotated Factor Loadings

Variables	Unrotated Factor Loadings		Rotated Factor Loadings	
	I	II	I	II
V_1	.50	.80	.03	.94
V_2	.60	.70	.16	.90
V_3	.90	−.25	.95	.24
V_4	.80	−.30	.84	.15
V_5	.60	−.50	.76	−.13

tive group of variables. Also, the oblique solution provides information about the extent to which the factors are actually correlated with each other.

Most factor analysts agree that many direct unrotated solutions are not sufficient; that is, in most cases rotation will improve the interpretation by reducing some of the ambiguities that often accompany the preliminary analysis. The major option available to the analyst in rotation is to choose an orthogonal or an oblique method. The ultimate goal of any rotation is to obtain some theoretically meaningful factors and, if possible, the simplest factor structure. Orthogonal rotational approaches are more widely used because all computer packages with factor analysis contain orthogonal rotation options, while the oblique methods are not as widespread. Orthogonal rotations are also utilized more frequently because the analytical procedures for performing oblique rotations are not as well developed and are still subject to considerable controversy. Several different approaches are available for performing either orthogonal or oblique rotations. However, only a limited number of oblique rotational procedures are available in most statistical packages; thus, the analyst will probably be forced to accept the one that is provided.

Orthogonal Rotation Methods   In practice, the objective of all methods of rotation is to simplify the rows and columns of the factor matrix to facilitate interpretation. In a factor matrix, columns represent factors, with each row corresponding to a variable's loading across the factors. By simplifying the rows, we mean making as many values in each row as close to zero as possible (i.e., maximizing a variable's loading on a single factor). By simplifying the columns, we mean making as many values in each column as close to zero as possible (i.e., making the number of "high" loadings as few possible). Three major orthogonal approaches have been developed: QUARTIMAX, VARIMAX, and EQUIMAX.

QUARTIMAX. The ultimate goal of a QUARTIMAX rotation is to simplify the rows of a factor matrix; that is, it focuses on rotating the initial factor so that a variable loads high on one factor and as low as possible on all other factors. In QUARTIMAX rotations, many variables can load high or near on the same factor because the technique centers on simplifying the rows. The QUARTIMAX method has not proved very successful in producing simpler structures. Its difficulty is that it tends to produce a general factor as the first factor on which most, if not all, of the variables have high loadings. Regardless of one's concept of a "simpler" structure, inevitably it involves dealing with clusters of variables; a method that tends to create a large general factor (i.e., QUARTIMAX) is not in line with the goals of rotation.

VARIMAX. In contrast to QUARTIMAX, the VARIMAX criterion centers on simplifying the columns of the factor matrix. With the VARIMAX rotational approach, the maximum possible simplification is reached if there are only 1s and 0s in a single column. That is, the VARIMAX method maximizes the sum of variances of required loadings of the factor matrix. Recall that in QUARTIMAX approaches, many variables can load high or near high on the same factor because the technique centers on simplifying the rows. With the VARIMAX rotational approach, there tend to be some high loadings (i.e., close to −1 or +1) and some loadings near 0 in each column of the matrix. The logic is that interpretation is easiest when the variable-factor correlations are close to either (a) +1 or −1, thus indicating a clear positive or negative association between the variable and the factor, or (b) close to 0, indicating a clear lack of association. This structure is fundamentally simple. Although the QUARTIMAX solution is analytically simpler than the VARIMAX solution, VARIMAX seems to give a clearer separation of the factors. In general, Kaiser's experiment [13, 14] indicates that the factor pattern obtained by VARIMAX rotation tends to be more invariant than that obtained by the QUARTIMAX method when different subsets of variables are analyzed. The VARIMAX method has proved very successful as an analytic approach to obtaining an orthogonal rotation of factors.

EQUIMAX. The EQUIMAX approach is a compromise between the QUARTIMAX and VARIMAX approaches. Rather than concentrating either on simplification of the rows or on simplification of the columns, it tries to accomplish some of each. EQUIMAX has not gained widespread acceptance and is used infrequently.

Oblique Rotation Methods   The oblique rotation is similar to orthogonal rotations, except that oblique rotations allow correlated factors instead of maintaining independence between the rotated factors. But where there were several choices among orthogonal approaches, there are typically only limited choices in most statistical packages for oblique rotations. For example, SPSS provides OBLIMIN; SAS has PROMAX and ORTHOBLIQUE; and BMDP provides DQUART, DOBLIMIN, and ORTHOBLIQUE. The objectives of simplification are comparable to the orthogonal methods, with the added feature of correlated factors. With the possibility of correlated factors, the factor analyst must take additional care to validate obliquely rotated factors, as they have an additional way (nonorthogonality) of becoming specific to the sample and not generalizable, particularly with small samples or a low cases-to-variable ratio.

Selecting Among Rotational Methods   No specific rules have been developed to guide the analyst in selecting a particular orthogonal or oblique rotational technique. In most instances, the analyst simply utilizes the rotational technique provided by the computer program. Most programs have the default rotation of VARIMAX, but all the major rotational methods are widely available [3, 19, 22]. However, there is no compelling analytical reason to favor one rotational method over another. The choice of an orthogonal or oblique rotation should be made on the basis of the particular needs of a given research problem. If the goal of the research is to reduce the number of original variables, regardless of how meaningful the resulting factors may be, the appropriate solution would be an orthogonal one. Also, if the researcher wants to reduce a larger number of variables to a smaller set of uncorrelated variables for subsequent use in a regression or other prediction technique, an orthogonal solution is the best. However, if the ultimate goal of the factor analysis is to obtain several theoretically meaningful factors or constructs, an oblique solution is appropriate. This conclusion is reached because, realistically, very few variables are uncorrelated, as in an orthogonal rotation.

Criteria for the Significance of Factor Loadings

In interpreting factors, a decision must be made regarding which factor loadings are worth considering. The following discussion presents three suggestions to aid in the interpretation of factor loadings:

1. The first suggestion is not based on any mathematical proposition but relates more to practical significance. It is a rule of thumb that has been used frequently by factor analysts as a means of making a preliminary examination of the factor matrix. In short, factor loadings greater than ±.30 are considered to meet the minimal level; loadings of ±.40 are considered more important; and if the loadings are ±.50 or greater, they are considered practically significant. Thus the larger the absolute size of the factor loading, the more important the loading in interpreting the factor matrix. Because the factor loading is the correlation of the variable and the factor, the squared loading is the amount of the variable's total variance accounted for by the factor. Thus, a loading of .30 translates to approximately 10 percent explanation, and a .50 loading denotes that 25 percent of the variance is accounted for by the factor. The factor analyst should realize that extremely high loadings (.80 and above) are not typical and that practical significance of the loadings is an important criterion. These guidelines are applicable when the sample size is 100 or larger. The emphasis in this approach is practical, not statistical, significance.

2. As pointed out previously, a factor loading represents the correlation between an original variable and its factor. In determining a significance level for the interpretation of loadings, an approach similar to determining the statistical significance of correlation coefficients could be used. However, research [7] has demonstrated that factor loadings have substantially larger standard errors than typical correlations; thus, factor loadings should be evaluated at considerably stricter levels. The factor analyst can employ the concept of statistical power discussed in Chapter 1 to specify factor loadings considered significant for differing sample sizes. With the stated objective of obtaining a power level of 80 percent, the use of a .05 significance level, and the proposed inflation of the standard errors of factor loadings, Table 7.3 contains the sample sizes necessary for each factor loading value to be considered significant. For example, in a sample of 100 respondents, factor loadings of .55 and above are significant. However, in a sample of 50, a factor loading of .75 is required for significance. In comparison with method 1, which denoted all loadings of .30 as having practical significance, this approach would consider loadings of .30 significant

TABLE 7.3 Guidelines for Identifying Significant Factor Loadings Based on Sample Size

Factor Loading	Sample Size Needed for Significance[a]
.30	350
.35	250
.40	200
.45	150
.50	120
.55	100
.60	85
.65	70
.70	60
.75	50

[a]Significance is based on a .05 significance level (α), a power level of 80 percent, and standard errors assumed to be twice those of conventional correlation coefficients.

Source: Computations made with *SOLO Power Analysis*, BMDP Statistical Software, Inc., 1993.

only for sample sizes of 350 or greater. These are quite conservative guidelines when compared with the guidelines of the previous section or even the statistical levels associated with conventional correlation coefficients. Thus, these guidelines should be used as a starting point in factor loading interpretation, with lower loadings considered significant and added to the interpretation based on other considerations. The next section details the interpretation process and the role that other considerations can play.

3. A disadvantage of methods 1 and 2 is that the number of variables being analyzed and the specific factor being examined are not considered. It has been shown that as the analyst moves from the first factor to later factors, the acceptable level for a loading to be judged significant should increase. The fact that unique variance and error variance begin to appear in later factors means that some upward adjustment in the level of significance should be included [13]. The number of variables being analyzed is also important in deciding which loadings are significant. As the number of variables being analyzed increases, the acceptable level for considering a loading significant decreases. Adjustment for the number of variables is increasingly important as one moves from the first factor extracted to later factors.

To summarize the criteria for the significance of factor loadings, the following guidelines can be stated: (1) the larger the sample size, the smaller the loading to be considered significant; (2) the larger the number of variables being analyzed, the smaller the loading to be considered significant; (3) the larger the number of factors, the larger the size of the loading on later factors to be considered significant for interpretation.

Interpreting a Factor Matrix

Interpreting the complex interrelationships represented in a factor matrix is no simple matter. By following the procedure outlined in the following paragraphs, however, one can simplify considerably the factor interpretation procedure.

1. Examine the factor matrix of loadings. If an oblique rotation has been used, two factor matrices of loadings are provided. The first is the factor pattern matrix that has loadings that represent the unique contribution of each variable to the factor. The second matrix is the factor structure matrix, which has simple correlations between variables and factors, but these loadings contain both the unique variance between variables and factors and the correlation among factors. As the correlation among factors becomes greater, it is harder to distinguish which variables load uniquely on each factor in the factor structure matrix. Most factor analysts report the results of the pattern matrix.

   Each column of numbers represents a separate factor. The columns of numbers are the factor loadings for each variable on each factor. For identification purposes, the computer printout usually identifies the factors from left to right by the numbers 1, 2, 3, 4, and so forth. It also identifies the variables by number from top to bottom. To further facilitate interpretation, the analyst should write the name of each variable in the left margin beside the variable numbers.

2. To begin the interpretation, the analyst should start with the first variable on the first factor and move horizontally from left to right, looking for the highest loading for that variable on any factor. When the highest loading (largest absolute factor loading) is identified, the analyst should underline it if it is signifi-

cant. The analyst should then go to the second variable and, again moving from left to right horizontally, look for the highest loading for that variable on any factor and underline it. This procedure should be continued for each variable until all variables have been underlined once for their highest loading on a factor. Recall that for sample sizes of less than 100, the lowest factor loading to be considered significant would in most instances be ±.30.

The process of underlining only the single highest loading as significant for each variable is an ideal that the analyst should strive for but can seldom achieve. When each variable has only one loading on one factor that is considered significant, the interpretation of the meaning of each factor is simplified considerably. In practice, however, many variables may have several moderate-size loadings, all of which are significant, and the job of interpreting the factors is much more difficult. The difficulty arises because a variable with several significant loadings must be considered in interpreting (labeling) all the factors on which it has a significant loading. Because most factor solutions do not result in a simple structure solution (a single high loading for each variable on only one factor), the analyst will, after underlining the highest loading for a variable, continue to evaluate the factor matrix by underlining all significant loadings for a variable on all the factors. Ultimately, the analyst tries to minimize the number of significant loadings on each row of the factor matrix (that is, make each variable associate with one factor).

MULTIPLE LOADINGS

3. Once all the variables have been underlined on their respective factors, the analyst should examine the factor matrix to identify variables that have not been underlined and therefore do not load on any factor. Also, the communalities for each variable are provided, representing the amount of variance accounted for by the factor solution for each variable. The factor analyst should view each variable's communality to assess whether it meets acceptable levels of explanation. For example, a researcher may specify that at least one-half of the variance of each variable must be accounted for. Using this guideline, the researcher would identify all variables with communalities less than .50 as not having sufficient explanation.

If there are variables that do not load on any factor or whose communalities are deemed too low, the analyst has two options: (1) interpret the solution as it is and simply ignore those variables, or (2) evaluate each of those variables for possible deletion. Ignoring the variables may be appropriate if the objective is solely data reduction, but the factor analyst must still note that the variables in question are poorly represented in the factor solution. Consideration for deletion should depend on the variable's overall contribution to the research as well as its communality index. If the variable is of minor importance to the study's objective or has an unacceptable communality value, the analyst may decide to eliminate the variable(s) and respecify the factor model by deriving a new factor solution with the nonloading variables eliminated.

4. When a factor solution has been obtained in which all variables have a significant loading on a factor, the analyst attempts to assign some meaning to the pattern of factor loadings. Variables with higher loadings are considered more important and have greater influence on the name or label selected to represent a factor. Thus the analyst will examine all the underlined variables for a particular factor and, placing greater emphasis on those variables with higher loadings, will attempt to assign a name or label to a factor that accurately reflects the variables loading on that factor. The signs are interpreted just as with any

other correlation coefficients. On each factor, like signs mean the variables are positively related, and opposite signs mean the variables are negatively related. In orthogonal solutions the factors are independent of one another. Therefore, the signs for a factor loading relate only to the factor that they appear on, not to other factors in the solution.

This label is not derived or assigned by the factor analysis computer program; rather, the label is intuitively developed by the factor analyst based on its appropriateness for representing the underlying dimensions of a particular factor. This procedure is followed for each extracted factor. The final result will be a name or label that represents each of the derived factors as accurately as possible.

In some instances, it is not possible to assign a name to each of the factors. When such a situation is encountered, the analyst may wish to label a particular factor or factors derived by that solution as "undefined". In such cases, the analyst interprets only those factors that are meaningful and disregards undefined or less meaningful ones. In describing the factor solution, however, the analyst indicates that these factors were derived but were undefinable and that only those factors representing meaningful relationships were interpreted.

As discussed earlier, the selection of a specific number of factors and the rotation method are interrelated. Several additional trial rotations may be undertaken, and by considering the initial criterion and comparing the factor interpretations for several different trial rotations, the analyst can select the number of factors to extract. In short, the ability to assign some meaning to the factors, or to interpret the nature of the variables, becomes an extremely important consideration in determining the number of factors to extract.

Stage Six: Validation of Factor Analysis

The sixth stage involves assessing the degree of generalizability of the results to the population and the potential influence of individual cases/respondents on the overall results. The issue of generalizability is critical for each of the multivariate methods, but it is especially relevant for the interdependence methods, because they describe a data structure that should be representative of the population as well. The most direct method of validating the results is to move to a confirmatory perspective and assess the replicability of the results, either with a split sample in the original data set or with a separate sample. The comparison of two or more factor model results has always been problematic. However, several options exist for the analyst wishing to make an objective comparison. The emergence of confirmatory factor analysis (CFA) through structural equation modeling has provided one option, but it is generally more complicated and requires additional software packages, such as LISREL or EQS [1, 12]. Chapter 11 discusses confirmatory factor analysis in greater detail. Apart from CFA, several other methods have been proposed, ranging from a simple matching index [5] to programs (FMATCH) designed specifically to assess the correspondence between factor matrices [20]. These methods have had sporadic use, owing in part to (1) their perceived lack of sophistication and (2) the unavailability of software or analytical programs to automate the comparisons. Thus, when CFA is not appropriate, these methods provide some objective basis for comparison.

Another aspect of generalizability is the stability of the factor model results. Factor stability is primarily dependent on the sample size and on the number of cases per variable. The researcher is always encouraged to obtain the largest sample possible and develop parsimonious models to increase the cases-to-variables ratio.

In addition to generalizability, another issue of importance to the validation of factor analysis is the detection of influential observations. Discussions in Chapter 2 on the indentification of outliers and in Chapter 3 on the influential observations in regression both have applicability in factor analysis. The analyst is encouraged to estimate the model with and without observations identified as outliers to assess their impact on the results. If omission of the outliers is justified, the results should have greater generalizability. Also, as discussed in Chapter 3, several measures of influence that reflect one observation's position relative to all others (e.g., covariance ratio) are applicable to factor analysis as well. Finally, recent research has proposed methods for identifying influential observations specific to factor analysis [6], but complexity has limited application of these methods.

Stage Seven: Additional Uses of the Factor Analysis Results

Depending upon the reason for applying factor analysis techniques, the researcher may stop with factor interpretation or proceed to other uses for factor analysis, including the computation of factor scores or the selection of surrogate variables for subsequent analysis with other statistical techniques. If the objective is simply to identify logical combinations of variables or respondents (objective 1), the analyst will stop with the factor interpretation. If the objective is to identify appropriate variables for subsequent application to other statistical techniques (objective 2), the analyst will examine the factor matrix and select the variable with the highest factor loading as a surrogate representative for a particular factor dimension. If the objective is to create an entirely new, smaller set of variables to replace the original set of variables for inclusion in a subsequent type of statistical analysis (objective 3), composite factor scores would be computed to represent each of the factors. The factor scores could then be used, for example, as the independent variables in a regression or discriminant analysis or as dependent variables in multivariate analysis of variance.

Selecting Surrogate Variables for Subsequent Analysis

If the researcher's objective is to identify appropriate variables for subsequent application with other statistical techniques (objective 2), the researcher could examine the factor matrix and select the variable with the highest factor loading on each factor as a surrogate representative for that particular factor. If there is only one factor loading for a variable that is substantially higher than all other factor loadings, the variable with the obviously higher loading would be selected for subsequent analysis to represent that factor. In many instances, however, the selection process is more difficult because two or more variables have loadings that are significant and fairly close to each other. In such cases, the analyst would have to examine critically the several factor loadings that are of approximately the same size and select only one as a representative of a particular dimension. This decision would be based on the researcher's a priori knowledge of theory that may suggest a particular variable would be more logically representative of the dimension. Also, the analyst may have knowledge suggesting that the raw data for a variable that is loading slightly lower are in fact more reliable than the raw data for the highest-loading variable. In such cases, the analyst may choose the variable that is loading slightly lower as the variable to represent a particular factor.

In instances where several high loadings complicate the selection of a single variable, factor analysis may be the basis for calculating a summed scale for use as the surrogate variable. In these instances, all the variables loading highly on a factor would be totaled. The total, or its average, could then be the surrogate variable. The objective, just as in the case of selecting a single variable, is to best represent the basic nature of the factor or component.

Using Factor Scores

When the analyst is interested in creating an entirely new and smaller set of composite variables to replace the original set, either in part or in whole, factor scores can be computed (objective 3). **Factor scores** are composite measures for each factor representing each subject. The original raw data measurements and the factor analytic results are utilized to compute factor scores for each individual. Conceptually speaking, the factor score represents the degree to which each individual scores high on the group of items that load high on a factor. Thus, an individual who scores high on the several variables that have heavy loadings for a factor surely will obtain a high factor score on that factor. The factor score, therefore, shows that an individual possesses a particular characteristic represented by the factor to a high degree. Most factor analysis computer programs compute scores for each respondent on each factor to be utilized in subsequent analysis. The analyst merely selects the factor score option, and these scores are saved for further use in subsequent analyses.

Creating factor scores to represent factor structures is not difficult for the analyst. Thus the question may arise whether factor scores or surrogate variables/summated scales should be used. Factor scores and summated scales have both advantages and disadvantages, and no clear-cut answer is available for all situations. Factor scores have the advantage of representing a composite of all variables loading on the factor, whereas surrogate variables represent only a single variable. However, a disadvantage of factor scores is that they are based on correlations with all the variables in the factor. Because these correlations are likely to be much less than 1.0, the scores are only approximations of the factors and, as such, are error-prone indicators of the underlying factors. The single surrogate variable is quite interpretable but will not represent all the ''facets'' of the factor and is also prone to measurement error. Finally, the summated scale is a compromise. It includes the variables loading highly on the factor and excludes those having little impact on the factor. It is easily replicated on subsequent samples, whereas exactly comparable factor scores are much harder to compute for other samples. This results because factor analysis of another sample will most probably not have the same weights used in calculating the factor scores and the factor analyst must calculate them manually if exact comparability is needed. But for summated scales, the variables are just averaged and are comparable between samples. Also, like surrogate variables, the summated scales are not necessarily orthogonal, whereas orthogonally rotated factors are othogonal or uncorrelated. The decision rule, therefore, would be that if data are used only in the original sample or orthogonality must be maintained, factor scores are quite suitable. If transferability is desired, then scales or surrogate variables are more appropriate. If the scale is a well-constructed, valid, and reliable instrument, the summated scale is probably the best alternative. But if the scale is untested and exploratory, with little or no evidence of reliability or validity, surrogate variables should probably be used.

An Illustrative Example

In the preceding sections, the major questions concerning the application of factor analysis have been discussed within the model-building framework introduced in Chapter 1. To clarify these topics further, we use an illustrative example of the application of factor analysis based on data from the database presented in Chapter 1. Our discussion of the empirical example also follows the six-stage model-building process. The first three stages, common to either component or common factor analysis, is discussed first. Then, stages four through six for component analysis will be discussed, along with examples of the additional uses of factor results. We conclude with an examination of the differences for common factor analysis in stages four and five.

Stage One: Objectives of Factor Analysis

Factor analysis can identify the structure of a set of variables as well as provide a process for data reduction. In our example, the perceptions of HATCO on seven attributes (X_1 to X_7) will be examined to (1) understand if these perceptions can be "grouped" and (2) reduce the seven variables to a smaller number. Even the relatively small number of perceptions examined here presents a complex picture of 21 separate correlations. By grouping the perceptions, HATCO will be able see the "big picture" in terms of understanding their customers and what the customers think about HATCO. If the seven variables can be represented in a smaller number of composite variables, then the other multivariate techniques can be made more parsimonious. Of course, this approach assumes that a certain degree of underlying order exists in the data being analyzed.

Stage Two: Designing a Factor Analysis

Understanding the structure of the perceptions of variables requires R-type factor analysis and a correlation matrix between variables, not respondents. All the variables are metric and constitute a homogeneous set of perceptions appropriate for factor analysis. Regarding the adequacy of the sample size, in this example there is a 14-to-1 ratio of observations to variables, which falls within acceptable limits. Also, the sample size of 100 provides an adequate basis for the calculation of the correlations between variables.

Stage Three: Assumptions in Factor Analysis

The underlying statistical assumptions impact factor analysis to the extent that they affect the derived correlations. Departures from normality, homoscedasticity, and linearity can diminish correlations between variables. These assumptions are examined in Chapter 1, where the reader is encouraged to review the findings.

The factor analyst must also assess the factorability of the correlation matrix. A first step is a visual examination of the correlations, identifying those that are statistically significant. Table 7.4 shows the correlation matrix for the seven perceptions of HATCO. Inspection of the correlation matrix reveals that 12 of the 21 correlations are significant at the .01 level. Thus over one-half of the correlations are significant. This provides an adequate basis for proceeding to the next level of examination of adequacy for factor analysis on both an overall basis and for each variable.

TABLE 7.4 Assessing the Appropriateness of Factor Analysis: Correlations, Measures of Sampling Adequacy, and Partial Correlations Among Variables

	Correlations Among Variables						
Variable	X_1	X_2	X_3	X_4	X_5	X_6	X_7
X_1 Delivery speed	1.00	−.34*	.51*	.05	.61*	.07	−.48*
X_2 Price level		1.00	−.48*	.27*	.51*	.18	.46*
X_3 Price flexibility			1.00	−.11	.06	−.03	−.44*
X_4 Manufacturer's image				1.00	.29*	.78*	.19
X_5 Service					1.00	.24*	−.05
X_6 Sales force's image						1.00	.17
X_7 Product quality							1.00

*Indicates correlations significant at the .01 level.

Overall Measure of Sampling Adequacy: .446
Bartlett Test of Sphericity: 567.5   significance .0000

	Measures of Sampling Adequacy and Partial Correlations*						
Variable	X_1	X_2	X_3	X_4	X_5	X_6	X_7
X_1 Delivery speed	.344						
X_2 Price level	.957	.330					
X_3 Price flexibility	.018	.155	.913				
X_4 Manufacturer's image	.149	.134	.095	.558			
X_5 Service	−.977	−.975	−.091	−.173	.288		
X_6 Sales force's image	−.060	−.045	−.085	−.766	.051	.552	
X_7 Product quality	−.016	−.141	.139	−.039	.088	−.092	.927

*Diagonal values are measures of sampling adequacy for individual variables; off-diagonal values are anti-image correlations (negative partial correlations).

The next step is to assess the overall significance of the correlation matrix with the Bartlett test. In this example, the correlations, when taken overall, are significant at the .0001 significance level (see Table 7.4). But this tests only for the presence of nonzero correlations, not the pattern of these correlations. The other overall test is the measure of sampling adequacy, which in this case falls in the unacceptable range (under .50) with a value of .446. Examination of the values for each variable identifies three variables (X_1, X_2, and X_5) that also have values under .50. Because X_5 has the lowest value, it will be omitted in the attempt to obtain a set of variables that can exceed the minimum levels on this measure.

Table 7.5 contains the correlation matrix for the revised set of variables (X_1, X_2, X_3, X_4, X_6, and X_7) along with the measures of sampling adequacy and Bartlett test value. In the reduced correlation matrix, 8 of the 15 correlations are statistically significant. As with the full set of variables, the Bartlett test shows that nonzero correlations exist at the significance level of .0001. The reduced set of variables collectively meets the necessary threshold of sampling adequacy. Each of the variables also exceeds the threshold value, indicating that the reduced set of vari-

TABLE 7.5 Assessing the Appropriateness of Factor Analysis for the Revised Set of Variables: Correlations, Measures of Sampling Adequacy, and Partial Correlations Among Variables

	Correlations Among Variables					
Variable	X_1	X_2	X_3	X_4	X_6	X_7
X_1 Delivery speed	1.00	−.34*	.51*	.05	.07	−.48*
X_2 Price level		1.00	.48*	.27*	.18	.46*
X_3 Price flexibility			1.00	−.11	−.03	−.44*
X_4 Manufacturer's image				1.00	.78*	.19
X_6 Sales force's image					1.00	.17
X_7 Product quality						1.00

*Indicates correlations significant at the .01 level.

Overall Measure of Sampling Adequacy: .665
Bartlett Test of Sphericity: 205.9   significance .0000

	Measures of Sampling Adequacy and Partial Correlations*					
Variable	X_1	X_2	X_3	X_4	X_6	X_7
X_1 Delivery speed	.721					
X_2 Price level	.074	.787				
X_3 Price flexibility	−.338	.301	.748			
X_4 Manufacturer's image	−.098	−.159	.081	.542		
X_6 Sales force's image	−.045	.025	−.081	−.769	.532	
X_7 Product quality	.331	−.253	.1149	−.024	−.097	.779

*Diagonal values are measures of sampling adequacy for individual variables, off-diagonal values are anti-image correlations (negative partial correlations).

ables meets the fundamental requirements for factor analysis. Finally, with the exception of one partial correlation, they are all fairly low, another indicator of the interrelationships among variables in the reduced set. These measures all indicate that the reduced set of variables is appropriate for factor analysis, and the analysis can proceed to the next stages.

Stage Four: Deriving Factors and Assessing Overall Fit

As noted earlier, factor analysis procedures are based on the initial computation of a complete table of intercorrelations among the variables (correlation matrix). This correlation matrix is then transformed through estimation of a factor model to obtain a factor matrix. The loadings of each variable on the factors are then interpreted to identify the underlying structure of the variables, in this case perceptions of HATCO. These steps of factor analysis, contained in stages four, five, and six, will be examined first for component analysis. Then, a common factor analysis will be performed and comparisons made between the two factor models.

The first step is to select the number of components to be retained for further analysis. Table 7.6 contains the information regarding the seven possible factors and their relative explanatory power as expressed by their eigenvalues. In addition to assessing the importance of each component, we can also use the eigenvalues to assist in selecting the number of factors. If we apply the latent root criterion, two components will be retained. The scree test (Figure 7.6), however, indicates that three factors may be appropriate. In viewing the eigenvalue for the third factor, it was determined that its low value (.597) relative to the latent root criterion value of 1.0 precluded its inclusion. If its eigenvalue was quite close to one, then the analyst might consider inclusion of the third factor as well. These results illustrate the need for multiple decision criteria in deciding the number of components to be retained. Still, the two factors retained represent 71 percent of the variance of the six variables.

Stage Five: Interpreting the Factors

The result of stage four is shown in Table 7.7, the unrotated component analysis factor matrix. To begin the analysis, let's explain the numbers included in the table. Three columns of numbers are shown. The first two are the results for the

TABLE 7.6 Results for the Extraction of Component Factors

Factor	Eigenvalue	Percent of Variance	Cumulative Percent of Variance
1	2.51349	41.9	41.9
2	1.73952	29.0	70.9
3	.59749	10.0	80.8
4	.52956	8.8	89.7
5	.41573	6.9	96.6
6	.20422	3.4	100.0

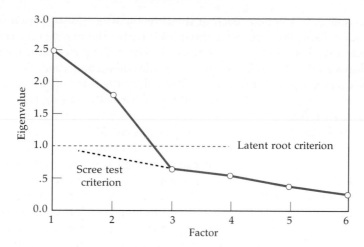

FIGURE 7.6 Scree test for component analysis.

two factors that are extracted (i.e., factor loadings of each variable on each of the factors). The third column provides summary statistics detailing how well each variable is "explained" by the two components and will be discussed in the next section. The first row of numbers at the bottom of each column is the column sum of squared factor loadings (eigenvalues) and indicates the relative importance of each factor in accounting for the variance associated with the set of variables being analyzed. Note that the sums of squares for the two factors are 2.51 and 1.74, respectively. As expected, the unrotated factor solution has extracted the factors in the order of their importance, with factor 1 accounting for the most variance and factor 2 slightly less. At the far right-hand side of the row of sums of squares is the number 4.25, which represents the total explained sum of squares. The total sum of squared factor loadings is obtained by adding the individual sums of squares for each of the factors. It represents the total amount of variance extracted by the factor solution.

TABLE 7.7 Unrotated Component Analysis Factor Matrix

		Factors		
Variables		*1*	*2*	*Communality*
X_1	Delivery speed	−.627	.514	.66
X_2	Price level	.759	−.068	.58
X_3	Price flexibility	−.730	.337	.65
X_4	Manufacturer's image	.494	.798	.88
X_6	Sales force's image	.425	.832	.87
X_7	Product quality	.767	−.168	.62
				Total
Sum of squares (eigenvalue)		2.51	1.74	4.25
Percentage of trace*		41.8	29.0	70.9

*Trace = 6.0 (sum of eigenvalues).

The percentages of **trace** for each of the three factors is also shown in the last row of values of Table 7.7. The percentages of trace for the two factors are 41.8 and 29.0, respectively. The percentage of trace is obtained by dividing each factor's sum of squares by the trace for the set of variables being analyzed. For example, if the sum of squares of 2.51 for factor 1 is divided by the trace of 6.0, the result will be the percentage of trace, or 41.8 percent for factor 1. By adding the percentages of trace for each of the two factors, we obtain the total percentage of trace extracted for the factor solution. The total percentage of trace can be used as an index to determine how well a particular factor solution accounts for what all the variables together represent. If the variables are all very different from one another, this index will be low. If the variables fall into one or more highly redundant or related groups, and if the extracted factors account for all the groups, the index will approach 100 percent. The index for the present solution shows that 70.9 percent of the total variance is represented by the information contained in the factor matrix. Therefore, the index for this solution is high, and the variables are in fact highly related to one another.

The row sum of squared factor loadings is shown at the far right side of the table. These figures, referred to in the table as communalities, show the amount of variance in a variable that is accounted for by the two factors taken together. The size of the communality is a useful index for assessing how much variance in a particular variable is accounted for by the factor solution. Large communalities indicate that a large amount of the variance in a variable has been extracted by the factor solution. Small communalities show that a substantial portion of the variance in a variable is unaccounted for by the factors. For instance, the communality figure of .65 for variable X_3 indicates that it has less in common with the other variables included in the analysis than does variable X_4, which has a communality of .88.

Having defined the various elements of the unrotated factor matrix, let's examine the factor loading patterns. As anticipated, the first factor accounts for the largest amount of variance and is a general factor, with every variable having a high loading. The loadings on the second factor show three variables (X_1, X_4, and X_6) also having high loadings. Based on this factor loading pattern, interpretation would be extremely difficult and theoretically less meaningful. Therefore, the analyst should proceed to rotate the factor matrix to redistribute the variance from the earlier factors to the later factors. Rotation should result in a simpler and theoretically more meaningful factor pattern.

Applying an Orthogonal (VARIMAX) Rotation

The VARIMAX rotated component analysis factor matrix is shown in Table 7.8. Note that the total amount of variance extracted is the same in the rotated solution as it was in the unrotated one, 70.9 percent. Two major differences are obvious, however. First, the variance has been redistributed so that the factor loading pattern is different, and the percentage of variance for each of the factors is different also. Specifically, in the VARIMAX rotated factor solution, the first factor accounts for 39.7 percent of the variance, but the second factor accounts for 31.2 percent. Thus the explanatory power has been distributed more evenly because of the rotation. Second, the interpretation of the factor matrix has been simplified. Recall that in the unrotated factor solution all variables loaded significantly on the first factor. In the rotated factor solution, however, variables X_4 and X_6 load

TABLE 7.8 VARIMAX Rotated Component Analysis Factor Matrix

		VARIMAX-rotated Loadings		
	Variables	Factor 1	Factor 2	Communality
X_1	Delivery speed	−.787	.194	.66
X_2	Price level	.724	.266	.58
X_3	Price flexibility	−.804	−.011	.65
X_4	Manufacturer's image	.102	.933	.88
X_6	Sales force's image	.025	.934	.87
X_7	Product quality	.764	.179	.62
				Total
Sum of squares (eigenvalue)		2.38	1.87	4.25
Percentage of trace*		39.7	31.2	70.9

*Trace = 6.0 (sum of eigenvalues).

significantly on factor 2, variables X_1, X_2, X_3, and X_7 load significantly on factor 1. No variable loads significantly on more than one factor. It should be apparent that factor interpretation has been simplified considerably by rotating the factor matrix.

Naming the Factors

When a satisfactory factor solution has been derived, the analyst usually attempts to assign some meaning to it. The process involves substantive interpretation of the pattern of factor loadings for the variables, including their signs, in an effort to name each of the factors. Before interpretation, a minimum acceptable level of significance for a factor loading must be selected. All significant factor loadings typically are used in the interpretation process. But variables with higher loadings influence to a greater extent the name or label selected to represent a factor.

Let's look at the results in Table 7.8 to illustrate this procedure. Our factor solution was derived from component analysis with a VARIMAX rotation of the six supplier perceptions of HATCO. Our cutoff point for interpretation purposes in this example is all loadings ±.57 or above (see Table 7.3). This is a conservatively high cutoff and may be adjusted if needed. But in our example, all the loadings fall substantially above or below this threshold, making interpretation quite straightforward.

Substantive interpretation is based on the significant higher loadings. Factor 1 has four significant loadings and factor 2 has two. For factor 1, we see two groups of variables. The first are price level (X_2) and product quality (X_7), both of which have positive signs. The two other variables, delivery speed (X_1) and price flexibility (X_3), both have negative signs. Thus product quality and price level vary together, as do delivery speed and price flexibility. However, the two groups move in opposite directions to each other. In our example, this result would indicate that as price quality and level increase, for example, delivery speed and price flexibility decrease, or vice versa. Turning next to factor 2, we note that variables X_4 (manufacturer's image) and X_6 (sales force's image) both relate to image components. Both variables are of the same sign, suggesting that these perceptions are quite similar among respondents.

We should note that quality (X_5) was not included in the factor analysis. When the factor-loading interpretations are presented, it must always be noted that quality was not included. If the results are used in other multivariate analyses, X_5 could be included as a separate variable, although it would not be assured to be orthogonal to the factor scores.

The process of naming factors has been demonstrated. You will note that it is not very scientific and is based on the subjective opinion of the analyst. Different analysts in many instances will no doubt assign different names to the same results because of the difference in their background and training. For this reason, the process of labeling factors is subject to considerable criticism. But if a logical name can be assigned that represents the underlying nature of the factors, it usually facilitates the presentation and understanding of the factor solution and therefore is a justifiable procedure.

Applying an Oblique Rotation

The VARIMAX rotation is orthogonal, meaning that the factors remain uncorrelated thoughout the rotation process. But in many situations, the factors need not be uncorrelated and may even be conceptually linked, which requires correlation between the factors. In our example, it is quite reasonable to expect that perceptual dimensions would be correlated; thus the application of an oblique rotation is justified. Table 7.9 contains the pattern and structure matrices with the factor loadings for each variable on each factor. As discussed earlier, the pattern matrix is typically for interpretation purposes, especially if the factors have a substantial correlation between them. In our instance, the correlation between the factors is only .12, so that the pattern and structure matrices have quite comparable loadings. Examining the variables loading highly on each factor, we note that the interpretation is exactly the same as we found with the VARIMAX rotation.

Stage Six: Validation of Factor Analysis

Validation of any factor analysis results is essential, particularly when attempting to define underlying structure among the variables. While it would be nice to always follow our use of factor analysis with some form of confirmatory factor analysis, such as structural equation modeling (see Chapter 11), this is often not feasible. We must look to other means such as split sample analysis or application to entirely new samples.

In this example, we split the sample into two equal samples of 50 respondents and reestimate the factor models to test for comparability. Table 7.10 contains the VARIMAX rotations for the two factor models, along with the communalities. As you can see, they are quite comparable in terms of both loadings and communalities for all six perceptions. One notable occurrence is the reversal of signs on factor 1 in split sample one versus split sample two. The interpretations of the relationships among the variables (e.g., as delivery speed gets higher, price level perceptions decrease) do not change because they are relative among the loadings in each factor.

TABLE 7.9 Oblique Rotation of Component Analysis Factor Matrix

| | | Oblique-rotation Loadings | | |
Variables		Factor 1	Factor 2	Communality[a]
Pattern Matrix				
X_1	Delivery speed	−.802	.248	.66
X_2	Price level	.704	.219	.58
X_3	Price flexibility	−.808	.043	.65
X_4	Manufacturer's image	.051	.931	.88
X_6	Sales force's image	−.025	.937	.87
X_7	Product quality	.759	.129	.62
Structure Matrix				
X_1	Delivery speed	−.773	.151	
X_2	Price level	.730	.304	
X_3	Price flexibility	−.802	−.054	
X_4	Manufacturer's image	.164	.938	
X_6	Sales force's image	.087	.934	
X_7	Product quality	.774	.220	

Factor Correlation Matrix

	Factor 1	Factor 2
Factor 1	1.00	
Factor 2	.121	1.00

[a]Communality values are not equal to the sum of the squared loadings owing to the correlation of the factors.

TABLE 7.10 Validation of Components Factor Analysis by Split-Sample Estimation with VARIMAX Rotation

| | | VARIMAX-rotated Loadings | | |
Variables		Factor 1	Factor 2	Communality
Split-Sample 1				
X_1	Delivery speed	.784	.138	.63
X_2	Price level	−.686	.260	.54
X_3	Price flexibility	.797	.145	.66
X_4	Manufacturer's image	−.097	.921	.86
X_6	Sales force's image	.049	.917	.84
X_7	Product quality	−.801	.176	.67
Split-Sample 2				
X_1	Delivery speed	−.798	.266	.71
X_2	Price level	.749	.227	.62
X_3	Price flexibility	−.798	−.163	.66
X_4	Manufacturer's image	.115	.939	.90
X_6	Sales force's image	.115	.942	.90
X_7	Product quality	.727	.201	.57

With these results we can be more assured that the results are stable within our sample. If possible we would always like to perform additional work through gathering additional respondents to ensure that the results generalize across the population.

Stage Seven: Additional Uses of the Factor Analysis Results

Selecting Surrogate Variables for Subsequent Analysis

Let's examine the data in Table 7.8 to clarify the procedure for selecting surrogate variables. First, recall that surrogate variables would be selected only when the rotation is orthogonal, because when the analyst is interested in using surrogate variables in subsequent analyses, he or she wants to observe to the extent possible the assumption that the independent variables should be uncorrelated with each other. Thus an orthogonal solution would be selected instead of an oblique one.

If we assume the factor analyst desired to select only a single variable for further use, rather than constructing a summated scale (see discussion below) or use factor scores (see next section), we would examine the magnitude of the factor loadings. Focusing on the factor loadings for factor 2, we see that the loading for variable X_4 is .933 and for variable X_6, .934. The selection of a surrogate is difficult in cases like this because the sizes of the loadings are essentially identical. However, if the analyst has no a priori evidence to suggest that the reliability or validity of the raw data for one of the variables is better than for the other, and if neither would be theoretically more meaningful for the factor interpretation, the analyst would select variable X_6 as the surrogate variable, knowing that it represents both image elements to a high degree. Given the high loadings for both variables, selection of only one would be sufficient because of the high degree of intercorrelation between them (shown by their extremely high loadings on the same factor/component). In contrast, the loadings for factor 1 are .724 for variable X_2 and .764 for X_7, while having comparable negative for values for X_1 ($-.787$) and X_3 ($-.804$). No single variable "represents" the component best; thus a summated scale would be appropriate.

A summated scale develops a composite value for a set of variables by such simple procedures as taking the average of the variables in the scale. This is much like the variates in other multivariate techniques, except that the weights for each variable are assumed to be equal in the averaging procedure. The usefulness of factor analysis is that it identifies which variables should be placed together in summated scales to ensure greater reliability of the scale. The reader is encouraged to review the section on summated scales in Chapter 1 and read the sections in Chapter 11 on scale construction.

Use of Factor Scores

Instead of calculating summated scales, the factor analyst could calculate factor scores for each of the two factors in our component analysis. In this way, each respondent would have two new variables (factor scores for factors 1 and 2) that could be substituted for the original six variables in other multivariate techniques. Table 7.11 illustrates the use of factor scores or summated scales as replacements for the original variables. We selected the example of identifying differences between respondents from small versus large firms (X_8). In the test of mean differ-

TABLE 7.11 Evaluating the Replacement of the Original Variables by Factor Scores or Summated Scales

Statistical Test	Mean Difference Between Groups of Respondents Based on X_8, Firm Size			
	Mean Scores		*F-Test*	
Measure	*Group 1:* *Small firms*	*Group 2:* *Large firms*	*F ratio*	*Signif.*
Original Variables				
X_1 Delivery speed[a]	4.19	2.50	64.7	.000
X_2 Price level	1.95	2.98	21.9	.000
X_3 Price flexibility[a]	8.62	6.80	70.2	.000
X_4 Mfg. Image	5.21	5.30	0.1	.709
X_6 Salesforce image	2.69	2.62	0.2	.674
X_7 Product quality	6.09	8.29	86.2	.000
Factor Scores				
Factor Score 1	−.640	.959	159.8	.000
Factor Score 2	.052	−.078	0.41	.525
Summated Scales				
Scale 1	3.81	5.49	156.7	.000
Scale 2	3.95	3.96	0.02	.957

Correlations Between Factor Scores and Summated Scales				
	Factor Scores		*Summated Scales[b]*	
	1	*2*	*1*	*2*
Factor Score 1	1.000	.000	.9953	.0749
Factor Score 2	.000	1.00	.0848	.9851
Summated Scale 1[b]	.9953	.0848	1.000	.1545
Summated Scale 2[b]	.0749	.9851	.1545	1.000

[a] Have negative factor loadings.

[b] Summated scales calculated as average score across items. For example, scale 1 is average of X_1, X_2, X_3, and X_7. Note that X_1 and X_3 are reverse-scaled, owing to their negative factor loadings.

ences between the two groups of respondents, we see that all the variables loading highly on factor 1 (X_1, X_2, X_3, and X_7) are significantly different between the respondents from small and large firms, while the variables loading highly on factor 2 (X_4 and X_6) do not have significant differences. The factor scores and summated scales should show similar patterns if they are truly representative of the variables. As seen in Table 7.11, the factor scores do differ in accordance with this pattern. Factor score 1 shows significant differences, while factor score 2 does not. Similar differences between the two groups are seen for the summated scales. Also, the summated scales correlate very highly with the factor scores. Thus, in this instance, both the factor scores and the summated scales accurately portray the variables they represent.

If the original variables are to be replaced by factor scores or summated scales, the factor analyst must decide on which to use. This decision is based on the need for replication in other studies (which favors use of summated scales) versus the desire of orthogonality of the measures (which favors factor scores). Table 7.11 also contains the correlation matrix of factor scores and summated scales. Because we employed an orthogonal rotation, the correlation between factor scores is .000. But the summated scales can be correlated and in this case the correlation is .1545. The researcher must ascertain the need for orthogonality versus replicability in selecting factor scores versus summated scales.

Common Factor Analysis: Stages 4 and 5

Common factor analysis is the second major factor analytic model that will be discussed. The primary distinction between component analysis and common factor analysis is that the latter considers only the common variance associated with a set of variables. This aim is accomplished by factoring a "reduced" correlation matrix with estimated initial communalities in the diagonal instead of unities. The differences between component analysis and common factor analysis occur only at the factor estimation and interpretation stages (stages four and five). Once the communalities are substitued on the diagonal, the common factor model extracts factors in a manner similar to component analysis. The factor analyst uses the same criteria for factor selection and interpretation. To illustrate the differences that can occur between common factor and component analysis, the following sections detail the extraction and interpretation of a common factor analysis of the six HATCO perceptions used in the component analysis.*

Stage Four: Deriving Factors and Assessing Overall Fit

The first step is to determine the number of factors to retain for examination and possible rotation. Table 7.12 shows the extraction statistics. If we were to employ the latent root criterion with a cutoff value of 1.0 for the eigenvalue, two factors would be retained. However, the scree analysis indicates that three factors be retained (see Figure 7.7). In combining these two criteria, we will retain two fac-

*Note that variable 5 was omitted from the analysis earlier in stage three. However, if it had been included, the communality could not have been estimated by the computer program in the initial extraction of factors. This is an example of one of the problems associated with communality estimation in common factor analysis. Thus, the common factor analysis still would have to have been performed with the six variables.

TABLE 7.12 Results for the Extraction of Common Factors

Factor	Eigenvalue	Percent of Variance	Cumulative Percent of Variance
1	2.51349	41.9	41.9
2	1.73952	29.0	70.9
3	.59749	10.0	80.8
4	.52956	8.8	89.7
5	.41573	6.9	96.6
6	.20422	3.4	100.0

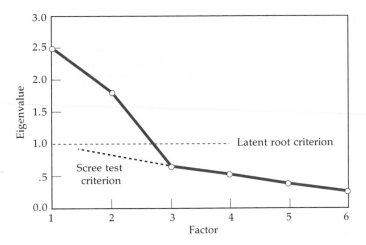

FIGURE 7.7 Scree test for common factor solution.

tors for further analysis because of the very low eigenvalue for the third factor and to maintain comparability with the component analysis. Again, as with the component analysis examined earlier, the analyst should employ a combination of criteria in determining the number of factors to retain.

The unrotated factor matrix (Table 7.13) shows that the common factor solution accounted for 58.6 percent of the total variance. The final common factor model sometimes differs from the initial extraction estimates (see Table 7.12), so be sure to evaluate the extraction statistics for the final common factor model. If the analyst was dissatisfied with the total variance explained, a common factor model extracting three factors could also be estimated. The reader will note that the communalities of each variable are lower than found in component analysis. This is due primarily to the lower overall variance explained, not the performance of any one variable. Again, exploration of a three-factor model could be made in an attempt to increase the communalities, as well as the overall variance explained. For our purposes, we will still interpret the two-factor solution.

TABLE 7.13 Unrotated Common Factor Matrix

		Factors		
	Variables	*1*	*2*	*Communality*
X_1	Delivery speed	−.484	.512	.50
X_2	Price level	.628	−.188	.43
X_3	Price flexibility	−.601	.401	.52
X_4	Manufacturer's image	.629	.687	.87
X_6	Sales force's image	.524	.665	.72
X_7	Product quality	.640	−.270	.48
				Total
	Sum of squares (eigenvalue)	2.07	1.45	3.52
	Percentage of trace*	34.5	24.1	58.6

*Trace = 6.0 (sum of eigenvalues).

TABLE 7.14 VARIMAX Rotated Common Factor Matrix

		VARIMAX-rotated Loadings		
Variables		Factor 1	Factor 2	Communality
X_1	Delivery speed	−.692	.133	.50
X_2	Price level	.620	.215	.43
X_3	Price flexibility	−.722	−.026	.52
X_4	Manufacturer's image	.109	.925	.87
X_6	Sales force's image	.037	.846	.72
X_7	Product quality	.677	.155	.48
				Total
Sum of squares (eigenvalue)		1.85	1.66	3.51
Percentage of trace*		30.8	27.7	58.5

*Trace = 6.0 (sum of eigenvalues).

Stage Five: Interpreting the Factors

Examining the unrotated loadings, we note the need for a factor matrix rotation. Turning then to the VARIMAX rotated common factor analysis factor matrix (Table 7.14), let's examine how it compares with the component analysis rotated factor matrix. The information provided in the common factor solution is similar to that provided in the component analysis solution. Sums of squares, percentage of variance, communalities, total sums of squares, and total variance extracted are all provided, just as with the component analysis solution.

Comparison of the information provided in the rotated common factor analysis factor matrix and the rotated component analysis factor matrix shows remarkable similarity. The primary differences between the component analysis and common factor analysis are the generally lower loadings in the common factor analysis, owing primarily to the lower communalities of the variables used in common factor analysis. Another comparison that may be useful to the analyst is the percentage of total variance explained by each factor. In component analysis the two rotated factors differed by 8 percent (39.7 percent versus 31.2 percent, respectively). In the common factor results, the rotation "spreads" the variance so that the two factors are almost equal in variance explained (30.8 percent for factor 1 and 27.7 percent for factor 2).*

Summary

The multivariate statistical technique of factor analysis has been presented in broad conceptual terms. Basic guidelines for interpreting the results were included to clarify further the methodological concepts. An example of the application of factor analysis was presented based on the database in Chapter 1.

*The interested reader is encouraged to explore the implications of adding a third factor to both the component analysis and the common factor analysis to assess the difference in interpretation and incremental variance explained.

Factor analysis can be a highly useful and powerful multivariate statistical technique for effectively extracting information from large databases. Factor analysis helps the investigator make sense of large bodies of interrelated data. When it works well, it points to interesting relationships that might not have been obvious from examination of the raw data alone or even a correlation matrix. Potential applications of factor analytic techniques to problem solving and decision making in business research are numerous. The use of these techniques will continue to grow as increased familiarity with the procedures is gained by academicians and practitioners.

Factor analysis is a much more complex and involved subject than might be indicated by this brief exposition. Three of the most frequently cited limitations are as follows: first, there are many techniques for performing factor analyses. Controversy exists over which technique is the best. Second, the subjective aspects of factor analysis (deciding how many factors to extract, which technique should be used to rotate the factor axes, which factor loadings are significant) are all subject to many differences in opinion. Third, the problem of reliability is real. Like any other statistical procedure, a factor analysis starts with a set of imperfect data. When the data change because of changes in the sample, the data-gathering process, or the numerous kinds of measurement errors, the results of the analysis also change. The results of any single analysis are therefore less than perfectly dependable. This problem is especially critical because the results of a single-factor analytic solution frequently look plausible. It is important to emphasize that plausibility is no guarantee of validity or even stability.

Questions

1. What are three problem situations in which factor analysis is the appropriate multivariate statistical technique to apply?
2. What is the difference between an orthogonal and an oblique factor rotation? When would the application of each approach be more appropriate?
3. What guidelines can you use to determine the number of factors to extract? Explain each briefly.
4. How do you use the factor-loading matrix to interpret the meaning of factors?
5. How and when should you use factor scores in conjunction with other multivariate statistical techniques?
6. What is the difference between Q-type factor analysis and cluster analysis?
7. When would the analyst use oblique factor analysis instead of an orthogonal factor analysis?
8. How can factor analysis be used in conjunction with other "dependence" multivariate techniques discussed in earlier chapters?

References

1. Bentler, Peter M. *EQS Structural Equations Program Manual.* Los Angeles: BMDP Statistical Software, 1992.
2. Borgatta, E. F., K. Kercher, and D. E. Stull. "A Cautionary Note on the Use of Principal Components Analysis." *Sociological Methods and Research* 15 (1986): 160–68.

3. BMDP Statistical Software, Inc. *BMDP Statistical Software Manual, Release 7,* vols. 1 and 2, Los Angeles: 1992.

4. Cattell, R. B. "The Scree Test for the Number of Factors." *Multivariate Behavioral Research* 1 (April 1966): 245–76.

5. Cattell, R. B., K. R. Balcar, J. L. Horn, and J. R. Nesselroade. "Factor Matching Procedures: An Improvement of the *s* index; with tables." *Educational and Psychological Measurement,* 29 (1969): 781–92.

6. Chatterjee, S., L. Jamieson, and F. Wiseman. "Identifying Most Influential Observations in Factor Analysis." *Marketing Science* 10 (Spring 1991): 145–60.

7. Cliff, N., and C. D. Hamburger. "The Study of Sampling Errors in Factor Analysis by Means of Artificial Experiments." *Psychological Bulletin* 68 (1967): 430–45.

8. Dillon, W. R. and M. Goldstein. *Multivariate Analysis: Methods and Applications.* New York: Wiley, 1984.

9. Dillon, W. R., N. Mulani, and D. G. Frederick. "On the Use of Component Scores in the Presence of Group Structure." *Journal of Consumer Research* 16 (1989): 106–12.

10. Gorsuch, R. L. *Factor Analysis.* Hillsdale, N.J.: Erlbaum, 1983.

11. Gorsuch, R. L. " Common Factor Analysis Versus Component Analysis: Some Well and Little Known Facts." *Multivariate Behavioral Research* 25 (1990): 33–39.

12. Joreskog, K. G., and D. Sorbom. *LISREL 8: Structural Equation Modeling with the SIMPLIS Command Language.* Mooresville, Ind.: Scientific Software International, 1993.

13. Kaiser, H. F. "A Second-Generation Little Jiffy." *Psychometrika* 35 (1970): 401–15.

14. Kaiser, H. F. "Little Jiffy, Mark IV." *Educational and Psychology Measurement* 34 (1974): 111–17.

15. Mulaik, S. A. "Blurring the Distinction Between Component Analysis and Common Factor Analysis." *Multivariate Behavioral Research* 25 (1990): 53–59.

16. Mulaik, S. A., and R. P. McDonald. "The Effect of Additional Variables on Factor Indeterminacy in Models with a Single Common Factor." *Psychometrika* 43 (1978): 177–92.

17. Nunnally, J. L. *Psychometric Theory,* 2d ed. New York: McGraw-Hill, 1978.

18. Rummel, R. J. *Applied Factor Analysis.* Evanston, Ill.: Northwestern University Press, 1970.

19. SAS Institute, Inc. *SAS User's Guide: Statistics, Version 6.* Cary, N.C.: 1990.

20. Smith, Scott M. *PC-MDS: A Multidimensional Statistics Package.* Provo, Utah: Brigham Young University Press, 1989.

21. Snook, S. C., and R. L. Gorsuch. "Principal Component Analysis Versus Common Factor Analysis: A Monte Carlo Study." *Psychological Bulletin* 106 (1989): 148–54.

22. SPSS, Inc. *SPSS Advanced Statistics Guide,* 4th ed. Chicago: 1990.

23. Stewart, D. W. "The Application and Misapplication of Factor Analysis in Marketing Research." *Journal of Marketing Research* 18 (February 1981): 51–62.

24. Velicer, W. F., and D. N. Jackson. "Component Analysis Versus Common Factor Analysis: Some Issues in Selecting an Appropriate Procedure." *Multivariate Behavioral Research* 25 (1990): 1–28.

The Organizational Context of Market Research Use

Rohit Deshpande

Largely as a function of developments in its environment, marketing is asking introspective questions about its own efficiency. At the beginning of the 1980s we have seen the rapid growth of the marketing function over the past two decades slowed under the impacts of inflation, raw material shortages, unemployment and recession. These economic changes necessitate a reassessment of strategies that had earlier proved successful. The drive now is to become leaner, more efficient in the use of available resources and more oriented toward the future (Wind 1980).

If we are to believe that the U.S. and other post-industrial economies are moving from an "Age of Product Technology" to a "Knowledgebased Society" (Bell 1976), we should be increasingly concerned with our ability to manage our corporate knowledge systems. The growth and even survival of today's business entities will depend on their strategies for handling and processing information. The more current this information, the greater the ability of managers to make policy decisions based upon it. In turn, the effectiveness of those decisions will be measured in terms of market information.

The marketing function is somewhat unique in that the information gathering and analysis processes in firms have been institutionalized as marketing research departments or divisions. Although these specialized information processing units have existed for some time, very little examination has been given to the effectiveness of research in providing information at the right place for the right decision. Additionally, it is only very recently that any attention has been paid to the factors that affect the usefulness of marketing research.

From Rohit Deshpande, "The Organizational Context of Market Research Use," *Journal of Marketing*, Vol. 46 (Fall 1982), pp. 91–101. Reprinted with permission.

The issue of examining marketing's R&D has not gone unnoticed. The critical costs of inadequate utilization of marketing tools and techniques have been mentioned recently by a special AMA/Marketing Science Institute joint commission (Myers, Massy and Greyser 1980). The commission's members were surprised at the relatively low rate of adoption at the line manager level of new marketing knowledge generated over a period encompassing the past 25 years. Their major recommendation was to develop better ways "to bridge the gaps between knowledge-generation and knowledge-utilization" (Myers, Greyser and Massy 1979, p. 27). Both marketing practitioners and academics support these observations and agree that much problem-oriented research is not used (Dyer and Shimp 1977, Ernst 1976, Kover 1976, Kunstler 1975). However, little formal research has been conducted in this area (Greenberg, Goldstucker and Bellenger 1977; Krum 1978; Luck and Krum 1981). Most observations about the factors affecting use of marketing research have been limited to introspective, albeit careful, analyses of personal experiences (Hardin 1973, Kunstler 1975, Newman 1962).

The issue of inadequate utilization of available research information is not unique to marketing. Underuse occurs in all areas of applied research activity. Most recently it has received much empirical attention in the policy sciences and has led to the creation of the area of inquiry called Knowledge Utilization (Caplan, Morrison and Stambugh 1975; Rich 1975; Weiss 1977; Weiss and Bucuvalas 1980). Developments in this area indicate that an understanding of the research use phenomenon lies in examining the organizational contexts in which policy decisions are made. The design of the decision-making structures of organizations sometimes provides clues as to why some of them are more efficient at using research than others.

As Day and Wind (1980) have commented, senior management has come to believe that focusing only on a customer-oriented search for competitive advantage may be shortsighted. There is a need to widen the scope of empirical attention in marketing by looking at relationships beyond those of the company and its customers. One set of these relationships deals with managers *within* an organization. Unless the structure of work relationships in a firm has been designed to optimize managerial effectiveness, the company-customer transactions will suffer and, in turn, negatively impact on the firm's long-term profitability. Yet the influence of organizational structure on the marketing function has seldom been studied systematically (Bonoma, Zaltman and Johnston 1977; Silk and Kalwani 1980; Spekman and Stern 1979). This issue is particularly important in the knowledge utilization area since parallel findings in the policy sciences, as mentioned earlier, indicate the importance of organizational design in influencing research use. In the pursuit of marketing effectiveness it may be useful to examine what forms of marketing organization appear best suited to manage the marketing research process efficiently (Wind 1980). This paper looks at the issue by surveying marketing managers in major U.S. business firms.

This paper does not intend to develop or extend a paradigm in organizational theory but attempts to look at why some consumer product companies make more use of marketing research than others. In the process, it is necessary to acquaint the reader with several organizational studies conducted in the past. Although these studies did not explicitly study marketing departments or market research–based decisions, it is possible to transfer the knowledge gained in those studies to the marketing area. This is the task of the following section.

Information Use in an Organizational Setting

Harold Wilensky, in his famous treatise on organizational intelligence, writes of barriers to the use of information in organizations: "Intelligence failures are rooted in structural problems that cannot be fully solved; they express universal dilemmas of organizational life that can, however, be resolved in various ways at varying costs. In all complex systems, hierarchy, specialization, and centralization are major sources of distortion and blockage of intelligence" (1967, p. 42). Why is this so? According to Wilensky, an organization that has a long hierarchial structure and emphasizes rank is likely to have much distortion occurring as information flows upward from junior through senior managerial levels. Due to the differential selective perception of information by different individuals, new knowledge takes on different shades of meaning as it passes from one person to the next. This distortion is further accentuated by the tendency of lower level managers to show themselves in the most favorable light to their superiors. In the case of the marketing organization, therefore, although senior marketing managers may wish to exert more effective control by centralizing information (and thereby its use), the knowledge with which they are provided may be a far cry from what was initially gathered by junior members of the marketing department.

Yet the attempt at resolving this problem by greatly decentralizing information collection and decision-making activities may not serve the need either. What Wilensky refers to as the dilemma of centralization may occur ". . . if intelligence is lodged at the top, too few officials and experts with too little accurate and relevant information are too far out of touch and too overloaded to function effectively; on the other hand, if intelligence is scattered throughout many subordinate units, too many officials and experts with too much specialized information may . . . delay decisions while they warily consult each other. . . . More simply, plans are manageable only if we delegate; plans are coordinated in relation to organizational goals only if we centralize" (1967, p. 58).

It appears therefore that the structure of a marketing organization may impact on the use of research information by its managers in one of two opposing ways. It is not entirely clear whether few rules and procedures and extensive decentralization of decison-making authority will help or hinder the organization's use of information.

It is helpful at this point to look at some past studies in the sociology of organizations to see whether an empirical resolution can be found to

Wilensky's dilemma. Two sociologists, Michael Moch and Edward Morse, recently studied the impact of organizational size and centralization on the adoption of innovations in 1,000 U.S. hospitals (Moch and Morse 1977). Their study found that in larger organizations a great deal of task specialization and role differentiation occurred. "Large organizations are in a better position to employ specialists and formally to differentiate responsibilities assigned to personnel in order to accommodate variation in input material. By employing specialists, the organization gains access to knowledge of new ideas, practices, and technical skills . . . [Also] formally differentiating task responsibilities to organizational personnel focuses their interest within specialized areas" (Moch and Morse 1977, p. 717). Additionally, these organizations tended to adopt innovations far more readily than those organizations that had less specialization of tasks and more centralization. This finding is supported by several other observers of social change in organizations (Hage and Aiken 1970; Pondy 1970; Zaltman, Duncan and Holbek 1973).

We can think of market research information that is new to a firm's marketing manager as an innovation for that organization. This is in keeping with the Rogers and Shoemaker (1971) conception of innovation as any set of ideas, practices or material artifacts perceived to be new by the relevant unit of adoption. According to Zaltman and Duncan (1977), people change their behavior when they define the situation as being different and requiring new or different behavior. If we then translate the above findings from organization theory into implications for marketing departments of firms, we can hypothesize that departments that are more structured will be less likely to adopt (or use) new research information than departments that are less structured.

However, a problem may exist. Looking once again at new research information as an innovation for the organization, awareness of the information is a function of the extent of work experience of the manager (Zaltman, Duncan and Holbek 1973). Presumably a marketing manager with several years of corporate experience behind him/her will be better able to judge the fit between new research information and its applicability to a specific decision-making situation.

Hence it is necessary to supplement any inquiry into the impact of structural arrangements on research use with an assessment of managers' work experiences.

These observations bring us to a statement of several formal hypotheses tested in this study. Before these can be stated, a description of the variables and their operationalization is provided.

Variables

Organizational Structure

Although several different concepts have been discussed in the sociology of organizations in measuring organizational structure, this study employs the two dimensions of *formalization* and *centralization*. These dimensions have been studied extensively in the context of organizational adoption of innovations (which is relevant to our case, as discussed earlier), and measures have been developed for thise dimensions that have been carefully validated and replicated across a wide variety of different organizations (public and private, large and small, both in the U.S. and in Europe).

Formalization, as defined in the work of Hall, Haas and Johnson (1967), is the degree to which rules define roles, authority relations, communications, norms and sanctions, and procedures. This dimension of organizational structure is an attempt to measure the flexibility that a manager enjoys when handling a particular task (such as the implementation of research recommendations). Centralization, as defined by Aiken and Hage (1968), looks at the delegation of decision-making authority throughout an organization and the extent of participation by managers in decision making.

The above concepts have been studied using two methods. The first, espoused in the work of Blau and Schoenherr (1971), Hinings and Lee (1971), Child (1972), and others focuses on "institutional" measures that look at the span of control, worker/supervisor ratios, distribution of employees across functional areas, and other indicants of an organization chart (Payne and Pugh 1976, Pugh et al. 1968). The second method uses questionnaires, with respondents indicating the extent of their agreement or disagreement with a series of statements dealing with issues such as the flexibility allowed in the handling of organizational tasks, the requirement for conformity with rules and guidelines, the amount of decentalization of authority, and so on (Aiken and Hage 1968, Hall 1972). This method has been empirically validated across a series of studies on different firms in both the U.S. and Europe.

However, the two methods do not produce identical results. Using a multitrait-multimethod matrix, Pennings (1973) found a low degree of convergence between institutional and questionnaire measures. He also discovered that although questionnaire measures of formalization and centralization were positively associated, the institutional measures produced negative correlations between the dimensions. These anomalies were criticized by several researchers on the grounds of instrument unreliability (Dewar, Whetten and Boje 1980; Pennings 1973; Seidler 1974). It is conceivable that some instrument bias can occur with self-report measures such as those used in the questionnaire method and also in the use of informants for the institutional method.

An alternative explanation has been suggested by Sathe (1978), who replicated Pennings' (1973) study with some modifications to increase internal validity and instrument reliability and suggested that the two methods were measuring different concepts. Since the institutional measure examines the structure of an organization in terms of the organization chart, the results reflect the structure as it was designed to operate. However, the questionnaire method, since it asks respondent managers to indicate their perceptions of participation in decisions, job flexibility, etc., taps the organizational structure as managers see it operating. "The questionnaire measures tend to reflect the degree of structure experienced by organizational members in work-related activities on a day-to-day basis and to the extent that such information is not biased, describe the *emergent* structure" (Sathe 1978, p. 234). This latter method is more pertinent to this study, since we are less interested in how a marketing organization was designed to function than in how an individual manager perceives the organization influencing his or her job. Therefore the questionnaire measures of formalization and centralization were used in this study.

The two structural dimensions are themselves conceptual aggregates of certain independent constructs. Formalization, for instance, is composed of measures tapping the extent to which jobs are codified (Job Codification), the degree to which rules are observed (Rule Observation), and the extent to which the specifics of tasks are stated (Job Specificity). Centralization is composed of the subdimensions of Participation in Decision Making and Hierarchy of Authority (the extent to which authority to make decisions affecting the firm is confined to higher levels of the hierarchy). Each of these dimensions is represented by a series of questions measured on a four or five point scale. The questions are displayed in Table 1.

Utilization of Research Information

Most of the work in defining and measuring what constitutes research use has been in nonmarketing areas, primarily political scient and public administration. Robert Rich, in his study of federal policymaking (1977), defines use as specific information coming to the desk of a decision maker, being read, and influencing the discussion of particular policies. In this sense the use of information is analogous to the use of a marketing research report being examined by a manager. Nathan Caplan and his co-workers have also looked largely at this instrumental type of information use in their study of 204 government officials (Caplan, Morrison and Stambaugh 1975). They define use in terms of familiarity of the officials with pertinent research and a consideration of an attempt to apply the research to some relevant policy areas. However, there is still much discussion as to how best to define research information use and the optimal way to measure it (Deshpande 1979, Larsen 1980, Weiss 1980). In this study, use of research information was defined and organizationalized in terms of whether a decision could have been made without it or whether the decision, when made without research, would have been very different from the decision for which research information was considered. Two questions were asked to determine

TABLE 1 Perceptions of Organizational Structure

Formalization Questions	*Centralization*

Response Categories: 1 Definitely true
2 More true than false
3 More false than true
4 Definitely false
5 Not applicable

Job Codification

(1) First, I felt that I was my own boss in most matters relating to the project.
(2) I could make my own decisions regarding the project without checking with anybody else.
(3) How things were done around here was left pretty much up to me.
(4) I was allowed to do almost as I pleased.
(5) I made up my own rules on this job.

Rule Observation

(6) I was constantly being checked on for rule violations.
(7) I felt as though I was constantly being watched to see that I obeyed all the rules.
(8) There was no specific rules manual relating to this project.
(9) There is a complete written job description for going about this task.

Job Specificity

(10) Whatever situation arose, we had procedures to follow in dealing with it.
(11) Everyone had a specific job to do.
(12) Going through the proper channels in getting this job done was constantly stressed.
(13) The organization kept a written record of everyone's performance.
(14) We had to follow strict operating procedures at all times.
(15) Whenever we had a problem we were supposed to go the same person for an answer.

Response Categories: 1 Never
2 Seldom
3 Often
4 Always

Participation in Decision Making

(1) How frequently did you usually participate in decisions on the adoption of new products.
(2) How frequently did you usually participate in decisions on the modification of existing products.
(3) How frequently did you usually participate in decisions to delete existent products.

Response Categories: 1 Definitely true
2 More true than false
3 More false than true
4 Definitely false
5 Not applicable

Hierarchy of Authority

(1) There could be little action taken on this project until a superior approved a decision.
(2) If I wished to make my own decisions, I would be quickly discouraged.
(3) Even small matters on this job had to be referred to someone higher up for a final answer.
(4) I had to ask my boss before I did almost anything.
(5) Any decision I made had to have my boss's approval.

research use. The first asked respondents to agree or disagree (on a five point Likert scale) with the statement. "Without this research information, the decisions made would have been very different." And the second, using the same response format, stated, "No decision would have been made without this research information."

Admittedly, several alternative methods of operationalizing research use do exist. Some of the literature cited above indeed defines use in different ways. However, in this study we are most concerned with the *so what?* or *impact* dimension of market research. Has there been any change caused by the presence of new information? Has the research affected managers' decision making in any way? What would have happened to the decisions if the research did not exist? These are the types of issues that this definition of research use attempts to get at. Additionally, the questions on research use are relatively more indirect than operationalizations that ask, "Did you actually use the market research?" As will be seen when the means and standard deviations of variables are described, the tendency toward positive bias is limited by utilizing more inferential methods of measuring use.

Interrelationship Between Concepts

Now that the major concepts have been defined and their operationalizations described, we can proceed to show how they are interrelated. Research in organization behavior indicates that firms that are more decentralized and less formalized are likely to adopt innovations quicker than those that are more structured (Hage and Aiken 1970; Moch and Morse 1977; Zaltman, Duncan and Holbek 1973). Additionally, as mentioned earlier, new research information can be thought of as an innovation that a manager may or may not decide to use (Deshpande and Zaltman 1981). The following propositions, therefore, flow from these considerations:

1. The greater the Job Codification[1] perceived by managers, the lower the utilization of market research information.
2. The greater the Rule Observation perceived by managers, the lower the utilization of market research information.
3. The greater the Job Specificity perceived by managers, the lower the utilization of market research information.
4. The lower the Participation in Decision Making, the lower the utilization of market research.
5. The greater the Hierarchy of Authority, the lower the utilization of market research.

Additionally, looking once again at new research information as an innovation to the orga-

nization, awareness of the information is a function of the work experience of the manager (Radnor, Rubinstein and Tansik 1970; Zaltman, Duncan and Holbek 1973). If we consider the years of experience in a firm or an industry as surrogates for work experience, then two further propositions are:

6. The greater the number of years of work experience in the firm, the greater the managers' perceived utilization of research.
7. The greater the number of years of work experience in the industry to which the firm belongs, the greater the managers' perceived utilization of research.

The questionnaire described below asked direct questions concerning number of years of work experience in the firm and the industry. The seven propositions were tested on the sample.

Sample and Research Methodology

Data used here come from a larger study of a sample of 92 managers who were questioned about marketing research projects in their companies (Deshpande and Zaltman 1982). The first stage of the study involved personal interviews conducted with 16 individuals in 7 firms (10 managers and 6 research suppliers). All 16 persons were selected on a convenience basis from large firms (all in the Fortune 500 sample) and from leading advertising and research agencies. The questionnaire used for the personal interviews was modified with structured queries for use, after a pilot test, in the mail survey that constituted the second stage of the data collection.

The sampling frame for managers was an *Advertising Age* listing of the 100 largest U.S. advertisers. Five hundred primarily product/brand and marketing managers in marketing divisions and firms (the universe of such managers in the frame) were selected. Firms dealing with industrial products or services were deleted, and out of 249 eligible respondents, 92 (37%) managers responded after one follow-up mailing. This rate of response, though good when compared with those in studies of similar organizations, required a detailed nonresponse analysis. A randomly se-

---

[1] Each of the organizational structure measures is briefly described at the end of the earlier section on Variables.

lected subsample of 50 nonrespondents was contacted directly by telephone to ascertain their reasons for not returning questionnaires. The major reasons for nonresponse concerned the lack of time to fill out the rather lengthy questionnaire. No subject matter related reasons for nonresponse were stated by any of the individuals contacted. Thus the actual replies received can be assumed to represent the valid responses of the total original sample (since the randomly selected nonrespondent's sample is assumed to be representative of all nonrespondents). Eligible respondents did not differ from nonrespondents in terms of organizational demographics or the salience of issues being studied.

Rather than asking managers questions about their general experiences with market research, it was felt necessary to get specific details on one such critical research experience per manager. In this manner it is possible to focus more precisely on the factors contributing to research use for that research project.

Accordingly, after preliminary questions regarding job title and work experience, the questionnaire asked respondents to focus on the most recently *completed* marketing research project with which they had been associated and for which a research report had already been presented.[2] The research incident preferred was to have contributed toward a consumer product (or service) strategy decision, i.e., the addition, modification or deletion of a product from the firm's line of offerings. In addition, the research for the scenario was to have been conducted by a research agency external to the marketing firm.[3] The questions regarding the use of the research were posed toward the end of the questionnaire to limit contamination of earlier responses.

[2] Clearly the one research project described by each manager may not be entirely representative of that firm's research experience. However, by having each manager describe the most recently completed research project, it is hoped that the sum of all such project experiences across the total sample would be representative for the sample of firms considered in this study.

[3] Since the study described here represents an initial exploration into the area of market research use by private firms, the investigation was not designed to include questions concerning internal marketing research departments or divisions.

Analysis

In order to improve the face validity of the 23 statements dealing with perceptions of organizational structure, the questions were altered slightly to make them market research project specific. For example, the original statement, "How frequently do you usually participate in decisions on the adoption of new programs?" was modified to read, ". . . in decisions on the adoption of new products?" These modifications are in keeping with recent suggestions to improve instrument validity (Dewar, Whetten and Boje 1980). However, such alterations limit the comparability of these questions to nonmarketing uses of the organizational structure measures. In order to ascertain whether the measures still retained construct validity (i.e., measure what they are supposed to), a factor analysis was conducted. This resulted in five factors explaining 70% of the overall variance. Table 2 shows the variables loading on each of the five factors.

It may be seen that the analysis produces a clean factor structure with items loading on the appropriate factors. With only a few items being deleted because of low or incorrect loading, the measures of the three formalization constructs (Job Codification, Rule Observation, Job Specificity) and the two centralization constructs (Participation in Decision Making, Hierarchy of Authority) show excellent validity. Additionally, internal reliability tests showed strong Cronbach alphas ranging from 0.73 through 0.92.

The next step, following the treatment suggested by developers of the measures, was to form cumulative, equally weighted indices for each of the five measures so as to develop scores for each case. As a further validity check, the sample was split randomly and Cronbach alphas were recalculated for the indices on each subsample. Alphas continued to be excellent with a range of 0.62 to 0.96.

In order to measure Research Utilization, a simple additive index was formed of the responses to the two questions on research use. (The mean response on each of the questions was 2.85 and 3.53 with standard deviations of 1.01 and 1.14, respectively. This indicates first, that manag-

TABLE 2 Factor Analysis of Perceptions of Organizational Structure

	Job Codification	Rule Observation	Job Specificity	Participation in Decision Making	Hierarchy of Authority
(1) My own boss	.67	.10	.24	.10	.18
(2) Make my own decisions	.76	.19	.08	.06	.17
(3) Doing things left up to me	.91	.05	.19	.01	.20
(4) Do almost as I pleased	.89	.08	.15	−.02	.15
(5) Made my own rules	.70	.18	.16	.03	.23
(6) Checked for rule violations	.24	.82	.25	−.06	.21
(7) Constantly being watched	.27	.91	.24	−.11	.14
(8) No rules manual*	.35	.11	.42	−.03	.07
(9) Complete job description*	.18	.31	.76	−.09	−.02
(10) Procedures for dealing with situations	.22	.10	.85	−.01	−.03
(11) Specific job to do	.14	−.09	.71	−.25	.07
(12) Going thru proper channels	.05	.08	.54	−.03	.30
(13) Written performance record	.15	.17	.69	.12	.18
(14) Strict operating procedures	.06	.07	.80	.06	.37
(15) Same person for problem referral	.15	.14	.56	.08	.33
(16) Adoption of new products	.11	.04	−.06	.61	.03
(17) Modification of existent products	.02	−.08	−.01	.83	.05
(18) Deletion of existent products	−.04	−.09	−.00	.81	.04
(19) Superior approves decision*	.16	.14	.08	.10	.25
(20) Discourage my own decision	.20	.07	.00	.04	.68
(21) Refer to superior for small matters	.19	.08	.36	.06	.63
(22) Had to ask boss	.22	.07	.25	−.03	.82
(23) Boss' approval required	.30	.19	.20	.12	.76

Cumulative variance explained by five factors: 70%

*Items deleted due to low item-factor correlations.

ers generally agreed that decisions, if made, would have been very different without the research information, and second, that the tendency for positive bias on these questions was not a major problem. This additive index of Research Utilization was then utilized as the dependent variable in an ordinary least squares regression. The five dimensions of organizational structure (Job Codification, Rule Observation, Job Specificity, Participation and Hierarchy of Authority) and the measures of industry and firm work experience were the predictor variables. Results of the regression analysis are displayed in Table 3. As this table shows, the overall regression equation explains 67.2% of the total variance, a result that is statistically significant at the 0.001 level. In order to test for internal validity of this result, the sample was split randomly into halves and the regression recomputed. Both standardized betas and the estimate of explained variance remained stable, suggesting the results reported here are not due to chance.

Now going beyond the summary R^2 statistic, a perusal of the contributions of individual independent variables produces interesting findings. First, it appears that the length of work experience of managers (in either the firm or the industry) is not a major determinant of the use of market research information. The standardized beta weights are small and statistically insignificant. This finding substantively rejects Propositions (6) and (7).

TABLE 3 Multiple Regression Analysis of Research Utilization with Perceptions of Organizational Structure and Work Experience

Dependent Variable: Utilization of Research Information

Independent Variables	Standardized Beta	F	Significance
Job Codification	−.38	8.013	.006
Rule Observation	−.07	0.304	.583
Job Specificity	−.38	6.000	.017
Participation in Decision Making	.48	21.650	.001
Hierarchy of Authority	−.38	6.857	.011
Number of years in firm	−.02	0.169	.897
Number of years in industry	.04	0.685	.794

Overall $F = 21.45$ significance = .001   adjusted $R^2 = .672$
Sample: Managers ($N = 92$)

Among the formalization variables, Job Codification and Specificity have significant impacts on the utilization of research (betas of −0.38 and −0.38, significant at 0.006 and 0.017 levels, respectively). In addition, the signs on these two coefficients are in the hypothesized direction, thus both substantively and statistically confirming Propositions (1) and (3). However, Rule Observation makes a nonsignificant contribution to explaining the variance in the dependent variable. Although the sign on its coefficient is as hypothesized (thus providing some reason for acceptance on substantive grounds), the low beta value of the Rule Observation coefficient leads to a rejection of Proposition (2).

Additionally, among the centralization variables both Participation in Decision Making and Hierarchy of Authority prove to have significant effects on the utilization of market research information. The coefficients of both of these variables are substantial, in the proposed direction, and statistically significant (beta of 0.48, significant at 0.001 and beta −0.38, significant at 0.011, respectively). This leads us to accept Propositions (4) and (5) as hypothesized.

Discussion

Briefly summarizing the results of the above analysis, it appears that managers who see themselves as operating in firms that are relatively decentralized and have few formalized procedures for carrying out marketing tasks are likely to make extensive use of market research information. This finding is true for this sample regardless of the extent of work experience of the managers. That is, both junior and senior managers in the marketing divisions of firms utilize research information more when they perceive a considerable amount of flexibility in how they should go about their tasks.

It is interesting to note that the extent to which a manager is perceived as observing organizational/departmental rules does not have great importance in determining whether market research information gets used. Managers appear to be less concerned (in a research utilization context) with whether they were constantly being watched to see they obeyed all the rules than with whether or not they were their own boss, whether or not they could make decisions on their own, or whether or not they were required to "go through channels." Looking at the questions constituting job codification and job specificity in Table 1 it seems that marketing managers tend to utilize research more when they feel that they, rather than their colleagues or bosses, are in control. These managers felt that they had substantial latitude in defining their roles ("I made up my own rules on this job." "I was allowed to do almost as I pleased"). They felt they were working in organizations or departments where strict operating procedures did not exist or, if they did exist, did not have to be followed with great attention to specific detail.

Although the question was not asked in this study, it is possible that when managers feel they have greater flexibility, they may also believe they have more freedom in doing their jobs. In the marketing research project situation this freedom can translate itself into a manager being more committed to research activity, being more involved with the entire research process, and as a result being more likely to use the findings from the research.

The feeling of freedom while handling marketing tasks is reinforced when managers participate frequently in product (or service) adoption, modification or deletion decisions. Together with the decentralization of decision-making authority comes added responsibility, which puts a burden on the managers' shoulders, the burden of eventual accountability for the decisions they make. A decision to launch a million dollar product is not one taken easily. Consequently, it is conceivable that managers will want to get as much corroborative and supportive evidence as possible before making the decision. This evidence is readily available in the form of market research data. Hence, the more decentralized the marketing operation of a company, the more likely it is that its managers will seek out research support.

The logic of the above argument may seem intuitively appealing and also provide a rationale for why managers in decentralized and less formalized marketing firms tend to be greater users of research. But this result is not trivial, since it is also possible to argue the opposite. Indeed this is what Wilensky (1967) has done, as mentioned earlier, imposing his "dilemma of centralization." Firms that are highly centralized and also have established more formally structured bureaucratic rules and procedures for handling marketing tasks are also likely to have centralized information systems. One would expect that in such a centralized system, research information goes directly to an individual at a senior level in the management hierarchy and, as such, is highly likely to be used in making strategic or policy decisions. However, when we consider the enormous amount of information that is likely to cross a decision maker's desk in such a centralized

management system, we can see immediately the difficulties that will result. As Rich (1975) found in the public policy setting, to avoid information overload problems senior policy makers ask their aides to distill information so that only the most critical and urgent issues reach their offices. In the distillation process a great deal of (perhaps pertinent) information is lost. Additionally, advocacy positions of junior aides are reflected in the reports reaching their superiors. The end product report that arrives on the policy maker's desk is frequently a far cry from the research that was originally executed, and the recommendations bear only a mild resemblance to those originally proposed by researchers.

The above scenario can be contrasted effectively with one where individual line managers are given the flexibility to take products from the R&D laboratory to test market shelves. Along with this flexibility comes, as mentioned earlier, accountability in terms of maintaining profit expectations. The line manager, who also oversees and interacts with the firm's research function, authorizes the collection of data concerning the potential viability of the product in the market place. The resulting market research information will then be carefully examined before a decision (largely based upon the information) is pronounced.

Conclusion

Very little empirical attention in marketing has been given to designing organizations or task groups to enhance managerial efficiency. Yet it is clear that even before considering company market transactions, it is important to ensure that the internal marketing operation functions effectively. One area of much significance is that involving the use of marketing research information. Although annual research transactions of larger business firms in the U.S. comprise expenditures of millions of dollars, the management of the research function to increase research productivity has received little scrutiny.

The study reported here indicates that marketing managers making consumer product strategy decisions are more likely to use research information when they see themselves working in a decentralized organization with few formal rules or procedures that must be observed. With increasing sophistication of research operations and growing uncertainties of the economic environment, this study has clear implications for senior marketing management. In order to enhance the efficient use of market research, line marketing managers should be allowed to operate in reasonably flexible task environments. This flexibility would allow managers a generous amount of freedom to participate extensively in product strategy decisions, coupled with accountability for demonstrating desired returns on product investment. The responsibilities with which the managers are entrusted would include overseeing the collection and analysis of marketing research information on the product in their charge. This would permit line marketing managers to be strongly involved in the research process, ensuring that the research information produced would be highly relevant to decisions that need to be made. The final result of the managers' commitment to the marketing research activity would be a more effective utilization of research.

The interesting results this study has provided reflect the perceptions of marketing managers. To complement these perceptions, further investigation in this area might examine marketing researchers working on the same research projects as managers. It would then be enlightening to compare manager and researcher perspectives on factors affecting the utilization of research on the same project. However, the insights provided here are no less important or valid. In fact, from a marketing firm's viewpoint it can be argued that the crucial element in considering the design of its organization is the task reality as seen by its own managers. This reality for marketing research projects has been reflected in the responses of marketing managers in this study. Their insights should help both managers in other companies and scientists and observers of the management of market research activity in an organization.

Future research in this area should replicate this study with an extension to industrial and services marketing firms. Additionally, since the domain of inquiry here concerned product and marketing managers, future work could be directed at a sample of marketing vice presidents or marketing directors. It is not yet known whether the findings from such samples would be similar to those reported here.

References

Aiken, M. and J. Hage (1968). "Organizational Independence and Intra-Organizational Structure," *American Sociological Review*, 33 (December), 912–930.

Bell D. (1976). *The Coming of Post-Industrial Society: A Venture in Social Forecasting*, New York: Basic Books.

Blau, P. M. and R. A. Schoenherr (1971). *The Structure of Organizations*, New York: Basic Books.

Bonoma, T. V., G. Zaltman and W. J. Johnston (1977), *Industrial Buying Behavior*, Cambridge, MA: Marketing Science Institute.

Caplan, N., A. Morrison and R. J. Stambaugh (1975), *The Use of Social Science Knowledge in Policy Decisions at the National Level*, Ann Arbor, MI: Institute for Social Research.

Child, J. (1972). Organization Structure and Strategies of Control: A Replication of the Aston Study." *Administrative Science Quarterly*, 17 (June), 163–177.

Day, G. S. and Y. Wind (1980), "Strategic Planning and Marketing: Time for a Constructive Partnership," *Journal of Marketing*, 44 (Spring), 7–8.

Deshpande, R. (1979). "The Use, Nonuse, and Abuse of Social Science Knowledge: A Review Essay," *Knowledge: Creation, Diffusion, Utilization*, 1 (September), 164–176.

——— and G. Zaltman (1981), "The Characteristics of Knowledge: Corporate and Public Policy Insights," in *Government Marketing: Theory and Practice*, M. P. Mokwa and S. E. Permut, eds., New York: Praeger, 270–278.

——— and ——— (1982), "Factors Affecting the Use of Market Research Information: A Path Analysis," *Journal of Marketing Research*, 19 (February), 14–31.

Dewar, R. D., D. A. Whetten and D. Boje (1980), "An Examination of the Reliability and Validity of the Aiken and Hage Scales of Centralization, Formalization, and Task Routineness," *Administrative Science Quarterly*, 25 (March), 120–128.

Dyer, R. F. and T. A. Shimp (1977), "Enhancing the Role of Marketing Research in Public Policy Decision Making," *Journal of Marketing*, 41 (January), 63–67.

Ernst, L. A. (1976), "703 Reasons Why Creative People Don't Trust Research," *Advertising Age*, 47 (February), 35–36.

Greenberg, B. A., J. L. Goldstucker and D. N. Bellenger (1977), "What Techniques Are Used by Marketing Researchers in Business," *Journal of Marketing*, 41 (April) 62–68.

Hage, J. and M. Aiken (1970), *Social Changes in Complex Organizations*, New York: Random House.

Hall, R. (1972), *Organizations, Structure and Process*, Englewood Cliffs, NJ: Prentice-Hall.

———, E. F. Haas and N. F. Johnson (1967), "Organizational Size, Complexity, and Formalization," *American Sociological Review*, 32 (December), 903–911.

Hardin, D. K. (1973), "Marketing Research and Productivity," in *Proceeding of the AMA Educators' Conference*, T. V. Greer, ed., Chicago: American Marketing Association, 169–171.

Hinings, C. R. and G. Lee (1971), ''Dimensions of Organization Structure and Their Context: A Replication,'' *Sociology*, 5 (February), 83–93.

Kover, A. J. (1976), ''Careers and Noncommunication: The Case of Academic and Applied Marketing Research,'' *Journal of Marketing Research*, 13 (November), 339–344.

Krum, J. R., (1978), ''B For Marketing Research Departments,'' *Journal of Marketing*, 42 (October), 8–12.

Kunstler, D. A. (1975), ''An Outline of AMA Research Divisions's Responsibilities, *Marketing News*, 8 (March 14), 12.

Larsen, J. K. (1980), ''Knowledge Utilization: What Is It?,'' *Knowledge: Creation, Diffusion, Utilization*, 1 (March), 421–442.

Luck, D. J. and J. R. Krum (1981), ''Conditions Conducive to the Effective Use of Marketing Research in the Corporation,'' *Report No. 81–100*. Cambridge, MA: Marketing Science Institute (May).

Moch, M. K. and E. V. Morse (1977), ''Size, Centralization, and Organizational Adoption of Innovations,'' *American Sociological Review*, 42 (October), 716–725.

Myers, J. G., S. A. Greyser and W. F. Massy (1979), ''The Effectiveness of Marketing's 'R&D' for Marketing Management: An Assessment,'' *Journal of Marketing*, 43 (January), 17–29.

————, W. F. Massy and S. A. Greyser (1980), *Marketing Research and Knowledge Development*, Englewood Cliffs, NJ: Prentice-Hall.

Newman, J. A. (1962), ''Put Research into Marketing Decisions,'' *Harvard Business Review*, 40 (March-April), 105–112.

Payne, R. and D. S. Pugh (1976), ''Organizational Structure and Climate,'' in *Handbook of Industrial and Organizational Psychology*, M. D. Dunnette, ed., Chicago: Rand-McNally, 1125–1173.

Pennings, J. (1973), ''Measures of Organizational Structure: A Methodological Note,'' *American Journal of Sociology*, 79 (November), 686–704.

Pondy, L. (1970), ''Toward a Theory of Internal Resource Allocation,'' in *Power in Organizations*, M. Zald, ed., Nashville: Vanderbilt University Press, 270–311.

Pugh, D. S., D. J. Hickson, C. R. Hinings and C. Turner (1968), ''Dimensions of Organization Structure,'' *Administrative Science Quarterly*, 13 (June), 65–105.

Radnor, M., A. Rubenstein and D. Tansik (1970), ''Implementation in Operations Research and R&D in Government and Business Organizations,'' *Operations Research*, 18 (November-December), 967–991.

Rich, R. F. (1975), "An Investigation of Information Gathering and Handling in Seven Federal Bureaucracies: A Case Study of the Continuous National Survey," Ph.D. dissertation, University of Chicago.

——— (1977), "Uses of Social Science Information by Federal Bureaucrats: Knowledge for Action Versus Knowledge for Understanding," in *Using Social Research in Public Policy Making*, C. H. Weiss, ed., Lexington, MA: D. C. Heath, 199–211.

Rogers, E. M. and F. F. Shoemaker (1971), *Communication of Innovations*, New York: The Free Press.

Sathe, V. (1978), "Institutional Versus Questionnaire Measures of Organizational Structure," *Academy of Management Journal*, 21 (June), 227–238.

Seidler, J. (1974), "On Using Informants: A Technique for Collecting Quantitative Data and Controlling Measurement Error in Organization Analysis," *American Sociological Review*, 39 (December), 816–831.

Silk, A. J. and M. U. Kalwani (1980), "Measuring Influence in Organizational Purchase Decisions," Unpublished working paper No. 1077–79. Sloan School of Management, Massachusetts Institute of Technology.

Spekman, R. E. and L. W. Stern (1979), "Environmental Uncertainty and Buying Group Structure: An Empirical Investigation," *Journal of Marketing*, 43 (Spring), 54–64.

Weiss, C. H., ed. (1977), *Using Social Research in Public Policy Making*, Lexington, MA: D. C. Heath.

——— (1980), "Knowledge Creep and Decision Accretion," *Knowledge: Creation, Diffusion, Utilization*, 1 (March), 381–404.

——— and M. J. Bucuvalas (1980), "Truth Tests and Utility Tests," *American Sociological Review*, 45 (April), 302–312.

Wilensky, H. (1967), *Organizational Intelligence: Knowledge and Policy in Governmental and Industry*, New York: Basic Books, Inc.

Wind, Y. (1980), "From the Editor: Marketing in the Eighties," *Journal of Marketing*, 44 (Winter), 7–9.

Zaltman, G. and R. Duncan (1977), *Strategies for Planned Change*, New York: John Wiley.

———, ——— and J. Holbek (1973). *Innovations and Organizations*. New York: John Wiley.

CHAPTER 8

Cluster Analysis

LEARNING OBJECTIVES

Upon completing this chapter, you should be able to do the following:

- Understand how interobject similarity is measured.

- Distinguish between the various distance measures.

- Differentiate between the clustering algorithms and their appropriate applications.

- Understand the differences between hierarchical and nonhierarchical clustering techniques.

- Understand how to select the number of clusters to be formed.

- Follow the guidelines for cluster validation.

- Construct profiles for the derived clusters and assess managerial significance.

- State the limitations of cluster analysis.

Academicians and market researchers often encounter situations best resolved by defining groups of homogeneous objects, whether they be individuals, firms, products, or even behaviors. Strategy options based on identifying groups within the population such as segmentation and target marketing would not be possible without an objective methodology. This same need is encountered in other areas, ranging from the physical sciences (e.g., classification of various animal groups—insects or mammals) to the social sciences (e.g., analysis of various psychiatric profiles). In all instances, the analyst is searching for a "natural" structure among the observations based on a multivariate profile.

The most commonly used technique for this purpose is cluster analysis. Cluster analysis is a technique for grouping individuals or objects into clusters so that objects in the same cluster are more like one another than they are like objects in other clusters. This chapter explains the nature and purpose of cluster analysis and guides the analyst in the selection and use of various cluster analysis approaches.

KEY TERMS

Before starting the chapter, review the key terms to develop an understanding of the concepts and terminology used. Throughout the chapter the key terms will appear in **boldface.** Other points of emphasis in the chapter will be *italicized.* Also, cross-references within the Key Terms will be in *italics.*

Agglomerative methods   *Hierarchical procedure* that begins with each object or observation in a separate cluster. In subsequent steps, object clusters that are closest together are combined to build a new aggregate cluster.

Algorithm   Set of rules or procedures; similar to an equation.

Average linkage   Agglomerative algorithm using the average distance from all objects (or individuals) in one cluster to all objects in another. At each stage, the two clusters with the smallest average distance are combined. This approach tends to combine clusters with small variances.

Centroid   The average or mean value of the objects contained in the cluster on each of the variables used in the cluster variate.

Centroid method   Agglomerative algorithm in which the distance between two clusters is the distance (typically Euclidean) between their *cluster centroid.* The two clusters with the smallest distance are grouped and a new centroid is computed. Thus cluster centroids migrate, or move, as cluster mergers take place.

City-block approach   Method of calculating distances based on the sum of the absolute differences of the coordinates for the objects. This method assumes the variables are uncorrelated and unit scales are compatible.

Cluster centroid   Average value of the objects contained in the cluster on all the variables in the *cluster variate.*

Cluster seeds   Initial centers or starting points for clusters. These individual values are selected to initiate nonhierarchical clustering procedures. Clusters are built around these preselected seeds.

Cluster variate   Set of variables or characteristics representing the objects to be clustered and used to calculate the similarity between objects.

Complete linkage   Agglomerative algorithm in which the clustering criterion is based on the maximum distance between objects in two clusters. At each stage of the agglomeration, the two clusters with the smallest maximum distance (or minimum similarity) are combined.

Criterion validity   Ability of clusters to show the expected differences on a variable not used to form the clusters. For example, if clusters are formed on performance ratings, clusters with higher performance ratings should also have higher satisfaction scores. If so, then criterion validity is supported.

Dendrogram   Graphical representation (tree graph) of the results of a clustering procedure in which the vertical axis consists of the objects or individuals and the horizontal axis consists of the number of clusters formed at each step of the procedure.

Divisive method   Clustering procedure, the opposite of *agglomerative method*, that begins with all objects in a single large cluster, that is divided into separate clusters based on the most dissimilar objects.

Entropy group   Group of objects or individuals independent of any cluster (they do not fit into any cluster).

Euclidean distance   Most commonly used measure of the similarity between two objects. Essentially, it is a measure of the length of a straight line drawn between two objects.

Hierarchical procedures   Stepwise clustering procedures involving a combination (or division) of the objects (clusters). The result is the construction of a hierarchy or treelike structure composed of separate clusters.

Interobject similarity   How similar two objects are based on their ratings on variables of interest. Similarity can be measured using "proximity" or "closeness" between each pair of objects. Likewise, "distance" or "difference" can be used.

Mahalanobis distance (D^2)   Standardized form of *Euclidean distance*. Data are standardized by scaling responses in terms of standard deviations, and adjustments are made for intercorrelations between the variables.

Nonhierarchical procedures   Instead of using a treelike construction process found in the *hierarchical procedures, cluster seeds* are used to group objects within a prespecified distance of the seeds.

Normalized distance function   Process that converts each raw data score to a standardized variate with zero mean and unit standard deviation, to remove the bias introduced by differences in scales of several variables.

Optimizing procedure   Nonhierarchical clustering procedure that allows for the reassignment of objects to another cluster from the original one on the basis of some overall optimizing criterion.

Parallel threshold method   Nonhierarchical clustering procedure that selects several cluster seeds simultaneously in the beginning. Objects within the threshold distances are assigned to the nearest seed. Threshold distances can be adjusted to include fewer or more objects in the clusters.

Predictive validity   See *criterion validity*.

Profile diagram   Graphical representation of data that aids in screening for outliers. Typically, the variables are listed along the horizontal axis and the value scale on the vertical axis. Object scores (original or standardized) are then plotted in the graphic plane.

Response-style effect   Series of systematic responses by a respondent that reflect a "bias" or consistent pattern. Examples include always responding that something is important or that something is poor performing across all attributes.

Row-centering standardization   See *within-case standardization.*

Sequential threshold method   Nonhierarchical procedure that begins by selecting one cluster seed. All objects within a prespecified distance are then included in that cluster. Subsequent cluster seeds are selected until all objects are grouped in a cluster.

Single-linkage method   Hierarchical clustering procedure based on the minimum distance between objects in one cluster and objects in another. The procedure finds two clusters with the shortest distance and combines them in a cluster until all objects are in one cluster.

Vertical icicle diagram   Graphic representation of clusters. The numbers of the objects are shown horizontally across the top and the number of clusters is shown vertically down the left side. This pilot aids in determining the appropriate number of clusters in the solution.

Ward's method   Hierarchical clustering procedure where the similarity used to join clusters is calculated as the sum of squares between the two clusters summed over all variables. Clusters with the greatest similarity are combined at each stage.

Within-case standardization   Standardization method in which a respondent's responses are not compared to the overall sample but only to his or her responses. Each respondent's average response is used to standardize his or her own responses.

What Is Cluster Analysis?

Cluster analysis is the name for a group of multivariate techniques whose primary purpose is to group objects based on the characteristics they possess. Cluster analysis classifies objects (e.g., respondents, products, or other entities) so that each object is very similar to others in the cluster with respect to some predetermined selection criterion. The resulting clusters of objects should then exhibit high internal (within-cluster) homogeneity and high external (between-cluster) heterogeneity. Thus, if the classification is successful, the objects within clusters will be close together when plotted geometrically, and different clusters will be far apart.

In cluster analysis, the concept of the variate is again a central issue, but in a quite different way from other multivariate techniques. The **cluster variate** is the set of variables representing the characteristics used to compare objects in the cluster analysis. Because the cluster variate includes only the variables used to compare objects, it determines the "character" of the objects. Cluster analysis is the only multivariate technique that does not estimate the variate empirically but instead uses the variate as specified by the researcher. The focus of cluster analysis is on the comparison of objects based on the variate, not estimation of the variate itself. This makes the researcher's definition of the variate a critical step in cluster analysis.

Cluster analysis has been referred to as Q analysis, typology, classification analysis, and numerical taxonomy. This variety of names is due in part to the usage of clustering methods in such diverse disciplines as psychology, biology, sociology, economics, engineering, and business. Although the names differ across disciplines, the methods all have a common dimension: classification according to natural relationships [1, 3, 5, 10, 16]. This common dimension represents the essence of all clustering approaches. As such, the primary value of cluster analysis lies in the classification of data, as suggested by "natural" groupings of the data themselves. Cluster analysis is comparable to factor analysis (see Chapter 7) in its objective of assessing structure. But cluster analysis differs from factor analysis in that cluster analysis groups objects, while factor analysis is primarily concerned with grouping variables.

Cluster analysis is a useful tool for data analysis in many different situations. For example, a researcher who has collected data by means of a questionnaire may be faced with a large number of observations that are meaningless unless classified into manageable groups. Cluster analysis can perform this data reduction procedure objectively by reducing the information from an entire population or sample to information about specific, smaller subgroups. For example, if we can understand the attitudes of a population by identifying the major groups within the population, then we have reduced the data for the entire population into profiles of a number of groups. In this fashion the researcher has a more concise, understandable description of the observations, with minimal loss of information.

Cluster analysis is also useful when a researcher wishes to develop hypotheses concerning the nature of the data or to examine previously stated hypotheses. For example, a researcher may believe that attitudes toward the consumption of diet versus regular soft drinks could be used to separate soft drink consumers into logical segments or groups. Cluster analysis can classify soft drink consumers by their attitudes about diet versus regular soft drinks, and the resulting clusters, if any, can be profiled for demographic similarities and differences.

These examples are just a small fraction of the types of applications of cluster analysis. Ranging from the derivation of taxonomies in biology for grouping all living organisms, to the psychological classifications based on personality and other personal traits, to the segmentation analyses of marketers, cluster analysis has always had a strong tradition of grouping individuals. This tradition has been extended to classifying objects, including the market structure, analyses of the similarities and differences among new products, and performance evaluations of firms to identify groupings based on the firms' strategies or strategic orientations. The result has been an explosion of applications in almost every area of inquiry, creating not only a wealth of knowledge on the use of cluster analysis but also the need for a better understanding of the technique to minimize its misuse.

How Does Cluster Analysis Work?

The nature of cluster analysis can be illustrated by a graphic presentation of a bivariate example. Suppose a marketing researcher has to determine relevant market segments in a small community. Assume further that a random sample of the population is selected and information for all objects is tabulated on two criteria:

1. Level of education
2. Brand loyalty

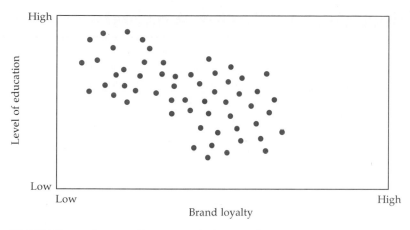

FIGURE 8.1 Scatter diagram of cluster observations.

All the respondents are then plotted on the scatter diagram in Figure 8.1.

Figure 8.1 shows that a definite relationship exists. In fact, the researcher could simply draw a line dividing the two groups and designate each as a cluster. In the terminology of cluster analysis, the researcher has identified two distinct clusters. Furthermore, if the researcher considers a profile representation of each cluster, it may be determined that the two clusters are uncorrelated or are even negatively correlated; that is, they may be quite dissimilar. Finally, the handful of respondents who are independent of either cluster may be designated, in the terminology of cluster analysis, the **entropy group.**

Cluster analysis is rather simple in the bivariate case, because the data are two-dimensional. In most marketing research studies, however, more than two variables are measured on each object, and the situation is much more complex.

Cluster Analysis Decision Process

Cluster analysis, like the other multivariate techniques discussed earlier, can be viewed from the six-stage model-building approach introduced in Chapter 1 (see Figure 8.2). Starting with research objectives that can be either exploratory or confirmatory, the design of a cluster analysis deals with the partitioning of the data set to form clusters, interpretation of the clusters, and validation of the results. The partitioning process determines whether and how clusters may be developed. The interpretation process involves understanding the characteristics of each cluster and developing a name or label that appropriately defines its nature. The third process involves assessing the validity of the cluster solution (i.e., determining its stability and generalizability), along with describing the characteristics of each cluster to explain how they may differ on relevant dimensions such as demographics. The following sections detail all these issues through the six stages of the model-building process.

Stage One: Objectives of Cluster Analysis

The primary goal of cluster analysis is to partition a set of objects into two or more groups based on the similarity of the objects for a set of specified characteristics (cluster variate). The most traditional use of cluster analysis has been for exploratory purposes. In an exploratory mode, cluster analysis is most often used to

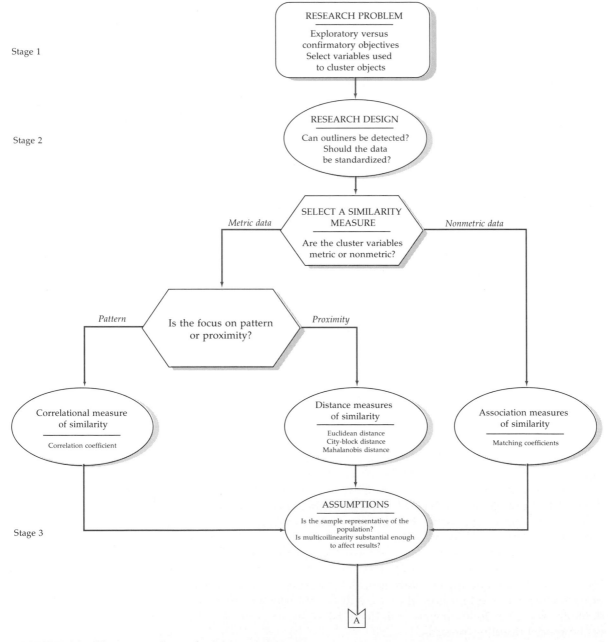

FIGURE 8.2 Cluster analysis decision process.

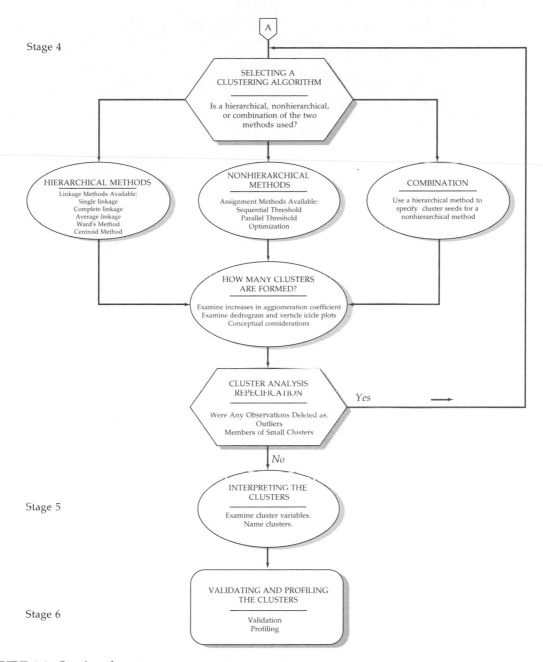

FIGURE 8.2 *Continued*

develop an objective classification of objects. But cluster analysis's partitioning ability can also generate hypotheses related to the structure of the objects. Yet while viewed principally as an exploratory technique, cluster analysis can also be used for confirmatory purposes. If a proposed structure can be defined for a set of objects, cluster analysis can be applied and the proposed structure can be compared to that derived from the cluster analysis.

In any application, the objectives of cluster analysis cannot be separated from the selection of variables used to characterize the objects to be clustered. Whether the objective is exploratory or confirmatory, the researcher effectively bounds the possible results by the variables selected for use. The derived clusters can only reflect the inherent structure of the data as defined by the variables.

Selecting the variables to be included in the cluster variate must be done with regard to both theoretical/conceptual and practical considerations. Any application of cluster analysis must have some rationale upon which variables are selected. Whether the rationale is based on an explicit theory, past research, or supposition, the researcher must realize the importance of including only those variables that (1) characterize the objects being clustered, and (2) relate specifically to the objectives of the cluster analysis. The cluster analysis technique has no means of differentiating the relevant from irrelevant variables. It only derives the most consistent, yet distinct, groups of objects across *all* variables. The inclusion of an irrelevant variable increases the chance that outliers will be created on these variables, which can have a substantive effect on the results. Thus, one should never include variables indiscriminately but instead choose the variables with the research objective as the criterion for selection.

In a practical vein, cluster analysis can be dramatically affected by the inclusion of only one or two inappropriate or undifferentiated variables [7]. The researcher is always encouraged to examine the results and to eliminate the variables that are not distinctive (i.e., that do not differ significantly) across the derived clusters. This procedure allows the cluster techniques to maximally define clusters based only on those variables exhibiting differences across the objects.

Stage Two: Research Design in Cluster Analysis

With the objectives defined and variables selected, the researcher must address three questions before starting the partitioning process: (1) Can outliers be detected and, if so, should they be deleted? (2) How should object similarity be measured? (3) Should the data be standardized? Many different approaches can be used to answer these questions. However, none of them has been evaluated sufficiently to provide a definitive answer to any of these questions, and unfortunately, many of the approaches provide different results for the same data set. Thus cluster analysis, along with factor analysis, is much more of an art than a science. For this reason, our discussion reviews these issues in a very general way by providing examples of the most commonly used approaches and an assessment of the practical limitations where possible.

The importance of these issues and the decisions made in later stages becomes apparent when we realize that while cluster analysis is seeking structure in the data, it must actually impose a structure through a selected methodology. Cluster analysis cannot evaluate all the possible partitions, because for even the relatively small problem of partitioning 25 objects into 5 nonoverlapping clusters there are 2.4×10^{15} possible partitions [1]. Instead, based on the decisions of the researcher, the technique identifies one of the possible solutions as "correct." From this viewpoint, the research design issues and the choice of methodologies made by the researcher have greater impact than perhaps with any other multivariate technique.

Detecting Outliers

In its search for structure, cluster analysis is very sensitive to the inclusion of irrelevant variables. But cluster analysis is also sensitive to outliers (objects that are very different from all others). Outliers can represent either (1) truly "aberrant" observations that are not representative of the general population, or (2) an undersampling of actual group(s) in the population that causes an underrepresentation of the group(s) in the sample. In both cases, the outliers distort the true structure and make the derived clusters unrepresentative of the true population structure. For this reason, a preliminary screening for outliers is always necessary. Probably the easiest way to conduct this screening is to prepare a graphic profile diagram such as that shown in Figure 8.3. The profile diagram lists the variables along the horizontal axis and the variable values along the vertical axis. Each point on the graph represents the value of the corresponding variable, and the points are connected to facilitate visual interpretation. Profiles for all objects are then plotted on the graph, a line for each object. Outliers are those objects with very different profiles, most often characterized by extreme values on one or more variables. Obviously, such a procedure becomes cumbersome with large numbers of objects (observations) or variables. In these instances, the procedures for identifying outliers discussed in Chapter 2 can be applied. By whatever means used, observations identified as outliers must be assessed for their representativeness of the population and deleted from the analysis if deemed unrepresentative.

Similarity Measures

The concept of similarity is fundamental to cluster analysis. **Interobject similarity** is a measure of correspondence or resemblance between objects to be clustered. If you remember in factor analysis, we created a correlation matrix between variables that was then used to group variables into factors. A comparable process occurs in cluster analysis. Here, the characteristics defining similarity are first

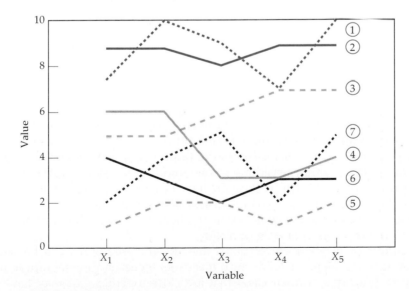

FIGURE 8.3 Profile diagram.

specified. Then, the characteristics are combined into a similarity measure calculated for all pairs of objects, just as we used correlations in factor analysis. In this way, any object can be compared to any other object through the similarity measure. The cluster analysis procedure then proceeds to group similar objects together into clusters.

Interobject similarity can be measured in a variety of ways, but three methods dominate the applications of cluster analysis: correlational measures, distance measures, and association measures. Each of the methods represents a particular perspective on similarity, dependent on both its objectives and type of data. Both the correlational and distance measures require metric data, while the association measures are for nonmetric data.

Correlational Measures

The interobject measure of similarity that probably comes to mind first is the correlation coefficient between a pair of objects measured on several variables. In effect, instead of correlating two sets of variables, we invert the objects' X variables matrix so that the columns represent the objects and the rows represent the variables. Thus, the correlation coefficient between the two columns of numbers is the correlation (or similarity) between the profiles of the two objects. High correlations indicate similarity and low correlations denote a lack of it. This procedure is followed in the application of Q-type factor analysis (see Chapter 7).

Correlational measures represent similarity by the correspondence of *patterns* across the characteristics (X variables). This is illustrated by examining the example of seven observations shown in Figure 8.3. A correlational measure of similarity does not look at the magnitude of the values but instead the patterns of values. In Table 8.1, which contains the correlations among observations, we can see two distinct groups. First, cases 1, 5, and 7 all have similar patterns and corresponding high intercorrelations. Likewise, cases 2, 4, and 6 also have high correlations among themselves but low or negative correlations with the other observations. Case 3 has low or negative correlations with all other cases, thereby perhaps forming a group by itself. Thus, correlations represent patterns across the variables much more than the magnitudes themselves. But correlational measures are rarely used because emphasis in most applications of cluster analysis is on the magnitudes of the objects, not the patterns of values.

Distance Measures

While correlational measures have an intuitive appeal and are used in many other multivariate techniques, they are not the most commonly used measure of similarity in cluster analysis. Distance measures of similarity, which represent similarity as the *proximity* of observations to one another across the variables in the cluster variate, are the similarity measure most often used. A simple illustration of this was shown in Figure 8.1, where clusters of observations were defined based on the proximity of observations to one another when each observation's scores on two variables were plotted graphically.

The difference between correlational and distance measures can be seen by referring again to Figure 8.3. Distance measures focus on the magnitude of the values and portray as similar cases that are close together, but may have very

TABLE 8.1 Calculating Correlational and Distance Measures of Similarity

Original Data

			Variables		
Case	X_1	X_2	X_3	X_4	X_5
1	7	10	9	7	10
2	9	9	8	9	9
3	5	5	6	7	7
4	6	6	3	3	4
5	1	2	2	1	2
6	4	3	2	3	3
7	2	4	5	2	5

Similarity Measure: Correlation

				Case			
Case	*1*	*2*	*3*	*4*	*5*	*6*	*7*
1	1.00						
2	−.147	1.00					
3	.000	.000	1.00				
4	.087	.516	−.824	1.00			
5	.963	−.408	.000	−.060	1.00		
6	−.466	.791	−.354	.699	−.645	1.00	
7	.891	−.516	.165	−.239	.963	−.699	1.00

Similarity Measure: Euclidean Distance

				Case			
Case	*1*	*2*	*3*	*4*	*5*	*6*	*7*
1	nc						
2	3.32	nc					
3	6.86	6.63	nc				
4	10.24	10.20	6.00	nc			
5	15.78	16.19	10.10	7.07	nc		
6	13.11	13.00	7.28	3.87	3.87	nc	
7	11.27	12.16	6.32	5.10	4.90	4.36	nc

nc = distances not calculated.

different patterns across the variables. Table 8.1 also contains distance measures of similarity for the seven cases, and we see a very different clustering of cases emerging than that found when using the correlational measures. With smaller distances representing greater similarity, we see that cases 1, 2, and 3 form one group, while cases 4, 5, 6, and 7 make up another group. These groups represent those with higher versus lower values. The choice of a correlational measure rather than the more traditional distance measure requires a quite different interpretation of the results by the researcher. Clusters based on correlational mea-

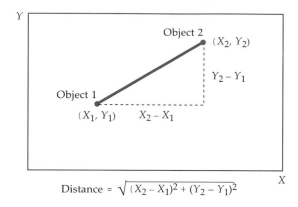

$$\text{Distance} = \sqrt{(X_2 - X_1)^2 + (Y_2 - Y_1)^2}$$

FIGURE 8.4 An example of Euclidean distance between two objects measured on two variables, X and Y.

sures may not have similar values but instead have similar patterns. Distance-based clusters have more similar values across the set of variables, but the patterns can be quite different.

Several distance measures are available. The most commonly used is **Euclidean distance**. An example of how Euclidean distance is obtained is shown geometrically in Figure 8.4. Suppose that two points in two dimensions have coordinates (X_1, Y_1) and (X_2, Y_2), respectively. The Euclidean distance between the points is the length of the hypotenuse of a right triangle, as calculated by the formula under the figure. This concept is easily generalized to additional variables. The Euclidean distance is used to calculate several specific measures, one being the simple Euclidean distance (calculated as described above) and the other is the squared, or absolute, Euclidean distance, where the distance value is the sum of the squared differences without taking the square root. The squared Euclidean distance has the advantage of not taking the square root, which speeds computations markedly, and is the recommended distance measure for the centroid and Ward's methods of clustering.

Several options not based on the Euclidean distance are also available. One of the most widely used alternative measures involves replacing the squared differences by the sum of the absolute differences of the variables. This procedure is called the absolute, or city-block, distance function. The **city-block approach** to calculating distances may be appropriate under certain circumstances [15], but it causes several problems. One is the assumption that the variables are not correlated with one another; if they are correlated, the clusters are not valid. Other measures that employ variations of the absolute differences or the powers applied to the differences (other than just squaring the differences) are also available in most cluster programs (SPSS, SAS, BMDP) [2, 13, 18].

A problem faced by all the distance measures that use unstandardized data is the inconsistencies between cluster solutions when the scale of the variables is changed. For example, suppose three objects, A, B, and C, are measured on two variables, probability of purchasing brand X (in percentages), and amount of time spent viewing commercials for brand X (in minutes or seconds), with the following results:

		Commercial Viewing Time	
Object	Purchase Probability (%)	(minutes)	(seconds)
A	60	3.0	180
B	65	3.5	210
C	63	4.0	240

From this information, distance measures can be calculated. In our example, we will calculate the following distance measures for each object pair: simple Euclidean distance, the absolute or squared Euclidean distance, and the city-block distance. The distance values, with smaller values indicating greater proximity and similarity, are shown below.

Distance based on minutes of viewing time

Object pair	Simple Euclidean Distance	Squared or Absolute Euclidean Distance	City-block Distance
A-B	5.025	25.25	5.5
A-C	3.162	10.00	4.0
B-C	2.062	4.25	2.5

As we can see, the most similar objects (with the smallest distance) are B and C, followed by A and C, with A and B the least similar (or least proximal). This ordering holds for all three distance measures, but the relative similarity or dispersion between objects is the most pronounced in the squared Euclidean distance measure.

The ordering of similarities can change markedly with only a change in the scaling of one of the variables. If we measured the viewing time in seconds instead of minutes, then differences would emerge. The new distance measures, based on viewing time in seconds, are shown below:

Distance based on seconds of viewing time

Object pair	Simple Euclidean Distance	Squared or Absolute Euclidean Distance	City-block Distance
A-B	30.41	925	35
A-C	60.070	3,609	63
B-C	30.06	904	32

We can now see that the similarity orderings have changed substantially. While B and C are still the most similar, the pair AB is now next most similar and is almost identical to the similarity of BC. Yet when we used minutes of viewing time, pair AB was the least similar by a substantial margin. What has occurred is that the scale of the viewing time variable has dominated the calculations, making purchase probability less significant in the calculations. The reverse was true, however, when we measured viewing time in minutes as purchase probability was dominant in the calculations. The researcher should thus note the tremendous impact that variable scaling can have on the final solution. Standardization of the variables, whenever possible conceptually, should be employed to avoid such instances as found in our example. The issue of standardization is discussed in the following section.

A commonly used measure of Euclidean distance that does directly incorporate a standardization procedure is the **Mahalanobis distance**. The Mahalanobis approach not only performs a standardization process on the data by scaling in terms of the standard deviations but also sums the pooled within-group variance-covariance, which adjusts for intercorrelations among the variables. Highly intercorrelated sets of variables in cluster analysis can implicitly overweight one set of variables in the clustering procedures, as is discussed in the next section. In short, the Mahalanobis generalized distance procedure computes a distance measure between objects comparable to the R^2 in regression analysis. While many situations are appropriate for use of the Mahalanobis distance, many programs do not include it as a measure of similarity. In such cases, the analyst usually selects the squared Euclidean distance.

In attempting to select a particular distance measure, the analyst should remember the following caveats. In most situations, different distance measures may lead to different cluster solutions. Thus, it is advisable to use several measures and compare the results with theoretical or known patterns. Also, when the variables are intercorrelated (either positively or negatively), the Mahalanobis distance measure is likely to be the most appropriate because it adjusts for intercorrelations and weights all variables equally. Of course, if the analyst wishes to weight the variables unequally, other procedures are available [8, 9].

Association Measures

Association measures of similarity are used to compare objects whose characteristics are measured only in nonmetric terms (nominal or ordinal measurement). As an example, respondents could answer yes or no on a number of statements. An association measure could assess the degree of agreement or matching between each pair of respondents. The simplest form of association measure would be the percentage of times there was agreement (both respondents said yes or both said no to a question) across the set of questions. Extensions of this simple matching coefficient have been developed to accommodate multicategory nominal variables and even ordinal measures. Many computer programs, however, have limited support for association measures, and the researcher is many times forced to first calculate the similarity measures and then input the similarity matrix into the cluster program. Reviews of the various types of association measures can be found in several sources [3, 16].

Standardizing the Data

With the similarity measure selected, the researcher must address only one more question: Should the data be standardized before similarities are calculated? In answering this question, the researcher must address several issues. First, most distance measures are quite sensitive to differing scales or magnitude among the variables. We saw this impact earlier when we changed from minutes to seconds on one of our variables. In general, variables with larger dispersion (i.e., larger standard deviations) have more impact on the final similarity value. Let us consider another example to illustrate this point. Assume that we want to cluster individuals on three variables—an attitude toward a product, age, and income. Now assume that we measured attitude on a seven-point scale of liking-disliking, while age was measured in years and income in dollars. If we plotted this on a three-dimensional graph, the distance between points (and their similarity) would be almost totally based on the income differences. The possible differences on attitude range from one to seven, while income may have a range perhaps a thousand times greater. Thus, graphically we would not be able to see any difference on the dimension associated with attitude. For this reason the researcher must be aware of the implicit weighting of variables based on their relative dispersion that occurs with distance measures.

The most common form of standardization is the conversion of each variable to standard scores (also known as Z scores) by subtracting the mean and dividing by the standard deviation for each variable. (Computer programs for this procedure are readily available.) This is the general form of a **normalized distance function,** which utilizes a Euclidean distance measure amenable to a normalizing transformation of the raw data. This process converts each raw data score into a standardized value with a zero mean and a unit standard deviation. This transformation, in turn, eliminates the bias introduced by the differences in the scales of the several attributes or variables used in the analysis.

Up to now we have discussed standardizing only variables. What about "standardizing" respondents or cases? Why would we ever do this? Let's take a simple example. Suppose we had collected a number of ratings on a ten-point scale from respondents on the importance of several attributes in their purchase decision for a product. We could apply cluster analysis and obtain clusters, but one very distinct possibility is that what we would get are clusters of people who said everything was important, some who said everything had little importance, and perhaps some clusters in between. What we are seeing are **response-style effects** in the clusters. Response-style effects are the systematic patterns of responding to a set of questions, such as yea-sayers (answer very favorably to all questions) or nay-sayers (answer unfavorably to all questions).

If we want to identify groups according to their response style, then standardization is not appropriate. But in most instances what is desired is the *relative* importance of one variable to another. In other words, is attribute 1 more or less important than the other attributes, and can clusters of respondents be found with similar patterns of importance? In this instance, standardizing by respondent would standardize each question not to the sample's average but instead to that respondent's average score. This **within-case** or **row-centering standardization** can be quite effective in removing response effects and is especially suited to many forms of attitudinal data [14]. We should note that this is similar to a correlational measure in highlighting the pattern across variables, but the proximity of cases still determines the similarity value.

Stage Three: Assumptions in Cluster Analysis

Cluster analysis, like factor analysis, is not a statistical inference technique where parameters from a sample are assessed as possibly being representative of a population. Instead, cluster analysis is an objective methodology for quantifying the structural characteristics of a set of observations. As such, it has strong mathematical properties but not statistical foundations. The requirements of normality, linearity, and homoscedasticity that were so important in other techniques really have little bearing on cluster analysis. The researcher must focus, however, on two other critical issues: representativeness of the sample and multicollinearity.

In very few instances does the researcher have a census of the population to use in the cluster analysis. Instead, a sample of cases is obtained and the clusters derived in the hope that they represent the structure of the population. The researcher must therefore be confident that the obtained sample is truly representative of the population. As mentioned earlier, outliers may really be only an undersampling of divergent groups that, when discarded, introduce bias in the estimation of structure. The researcher must realize that cluster analysis is only as good as the representativeness of the sample. Therefore, all efforts should be taken to ensure that the sample is representative and the results are generalizable to the population of interest.

Multicollinearity was an issue in other multivariate techniques because it made it difficult to discern the "true" impact of multicollinear variables. But in cluster analysis the effect is different because those variables that are multicollinear are implicitly weighted more heavily. Let's start with an example that illustrates its effect. Suppose that respondents are being clustered on ten variables, all attitudinal statements concerning a service. When multicollinearity is examined, we see that there are really two sets of variables, the first made up of eight statements and the second consisting of the remaining two statements. If our intent is to really cluster the respondents on the dimensions of the product (in this case represented by the two groups of variables), then use of the original ten variables will be quite misleading. Because each variable is weighted equally in cluster analysis, the first dimension will have four times as many chances (eight items to two items) to affect the similarity measure as does the second dimension. Thus, multicollinearity acts as a weighting process not apparent to the observer but affecting the analysis nonetheless. For this reason, the researcher is encouraged to examine the variables used in cluster analysis for substantial multicollinearity and if found, either reduce the variables to equal numbers in each set or use one of the distance measures, such as Mahalanobis distance, that compensates for this correlation. There is debate over the use of factor scores in cluster analysis, as some research has shown that the variables that truly discriminate among the underlying groups are not well represented in most factor solutions. Thus, when factor scores are used, it is quite possible that a poor representation of the true structure of the data will be obtained [11]. The researcher must deal with both multicollinearity and discriminability of the variables to arrive at the best representation of structure.

Stage Four: Deriving Clusters and Assessing Overall Fit

With the variables selected and the similarity matrix calculated, the partitioning process begins. The researcher must first select the clustering algorithm used for forming clusters and then make the decision on the number of clusters to be formed. Both decisions have substantial implications not only on the results that will be obtained but also on the interpretation that can be derived from the results. Each of these issues is discussed in the following sections.

Clustering Algorithms

The first major question to answer in the partitioning phase is, What procedure should be used to place similar objects into groups or clusters? That is, What clustering algorithm or set of rules is the most appropriate? This is not a simple question because hundreds of computer programs using different algorithms are available, and more are being developed. The essential criterion of all the algorithms, however, is that they attempt to maximize the differences between clusters relative to the variation within the clusters, as shown in Figure 8.5. The ratio of the between-cluster variation to the average within-cluster variation is then comparable to (but not identical to) the F ratio in analysis of variance.

Most commonly used clustering algorithms can be classified into two general categories: (1) hierarchical and (2) nonhierarchical. We discuss the hierarchical techniques first.

Hierarchical Cluster Procedures

Hierarchical procedures involve the construction of a hierarchy of a treelike structure. There are basically two types of hierarchical clustering procedures—agglomerative and divisive. In the **agglomerative methods**, each object or observation starts out as its own cluster. In subsequent steps, the two closest clusters (or individuals) are combined into a new aggregate cluster, thus reducing the number of clusters by one in each step. In some cases, a third individual joins the first two in a cluster. In others, two groups of individuals formed at an earlier stage may join together in a new cluster. Eventually, all individuals are grouped into one large cluster; for this reason, agglomerative procedures are sometimes referred to as buildup methods. An important characteristic of hierarchical proce-

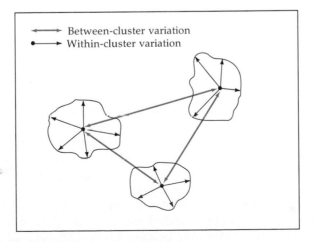

FIGURE 8.5 Cluster diagram showing between- and within-cluster variation.

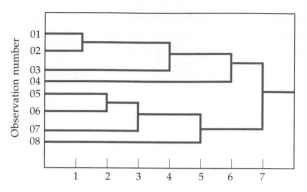

FIGURE 8.6 Dendogram illustrating hierarchical clustering.

dures is that the results at an earlier stage are always nested within the results in a later stage, causing its similarity to a tree. For example, a six-cluster solution is obtained by joining two of the clusters found at the seven-cluster stage. Because clusters are formed only by joining existing clusters, any member of a cluster can trace its membership in an unbroken path to its beginning as a single observation. This process is shown in Figure 8.6; the representation is referred to as a **dendrogram** or tree graph.

When the clustering process proceeds in the opposite direction to agglomerative methods, it is referred to as a **divisive method.** In divisive methods, we begin with one large cluster containing all observations (objects). In succeeding steps, the observations that are most dissimilar are split off and made into smaller clusters. This process continues until each observation is a cluster in itself. In Figure 8.6, agglomerative methods would move from left to right, and divisive methods would move from right to left. Because most commonly used computer packages use agglomerative methods and divisive methods act almost as agglomerative methods in reverse, we will focus on the agglomerative methods in our subsequent discussions.

Five popular agglomerative procedures used to develop clusters are (1) single linkage, (2) complete linkage, (3) average linkage, (4) Ward's method, and (5) centroid. These rules differ in how the distance between clusters is computed.

Single Linkage   The **single-linkage procedure** is based on minimum distance. It finds the two individuals (objects) separated by the shortest distance and places them in the first cluster. Then the next-shortest distance is found, and either a third individual joins the first two to form a cluster or a new two-individual cluster is formed. The process continues until all individuals are in one cluster. This procedure has also been called the nearest-neighbor approach.

The distance between any two clusters is the shortest distance from any point in one cluster to any point in the other. Two clusters are merged at any stage by the single shortest or strongest link between them. Problems occur, however, when clusters are poorly delineated. In such cases, single linkage procedures form long, snakelike chains, and eventually all individuals are placed in one chain. Individuals at opposite ends of a chain may be very dissimilar. An example of this arrangement is shown in Figure 8.7.

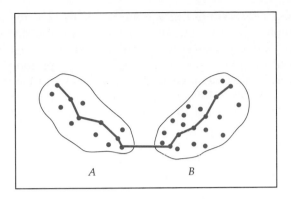

FIGURE 8.7 Example of single linkage joining dissimilar points A and B.

Complete Linkage   The **complete linkage** procedure is similar to single linkage except that the cluster criterion is based on maximum distance. For this reason, it is sometimes referred to as the furthest-neighbor approach or as a diameter method. The maximum distance between individuals in each cluster represents the smallest (minimum-diameter) sphere that can enclose all objects in clusters. This method is called complete linkage because all objects in a cluster are linked to each other at some maximum distance or by minimum similarity. We can say that within-group similarity equals group diameter. This technique eliminates the snaking problem identified with single linkage.

Figure 8.8 shows how the shortest (single linkage) and longest (complete linkage) distances represent similarity between groups. Both measures reflect only one aspect of the data. The use of the shortest distance reflects only a single pair of objects (the closest), while the complete linkage reflects again a single pair, this time the two most extreme. It is thus useful to visualize the measures as reflecting the similarity of most similar pair or least similar pair of objects.

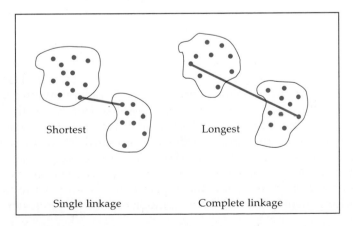

FIGURE 8.8 Comparison of distance measures for single linkage and complete linkage.

Average Linkage   The **average linkage** method starts out the same as a single linkage and complete linkage, but the cluster criterion is the average distance from all individuals in one cluster to all individuals in another. Such techniques do not depend on extreme values, as do single linkage or complete linkage, and partitioning is based on all members of the clusters rather than on a single pair of extreme members. Average linkage approaches tend to combine clusters with small variances. They also tend to be biased toward the production of clusters with approximately the same variance.

Ward's Method   In **Ward's method** the distance between two clusters is the sum of squares between the two clusters summed over all variables. At each stage in the clustering procedure, the within-cluster sum of squares is minimized over all partitions (the complete set of disjoint or separate clusters) obtainable by combining two clusters from the previous stage. This procedure tends to combine clusters with a small number of observations. It is also biased toward the production of clusters with approximately the same number of observations.

Centroid Method   In the **centroid method** the distance between two clusters is the distance (typically squared Euclidean or simple Euclidean) between their centroids. Cluster centroids are the mean values of the observations on the variables in the cluster variate. In this method, every time individuals are grouped, a new centroid is computed. Cluster centroids migrate as cluster mergers take place. In other words, there is a change in a cluster centroid every time a new individual or group of individuals is added to an existing cluster. These methods are the most popular with biologists but may produce messy and often confusing results. The confusion occurs because of reversals, that is, instances when the distance between the centroids of one pair may be less than the distance between the centroids of another pair merged at an earlier combination. The advantage of this method is that it is less affected by outliers than are other hierarchical methods.

Nonhierarchical Clustering Procedures

In contrast to hierarchical methods, **nonhierarchical procedures** do not involve the treelike construction process. Instead, they assign objects into clusters once the number of clusters to be formed is specified. Thus, the six-cluster solution is not just a combination of two clusters from the seven-cluster solution, but is based only on finding the best six-cluster solution. In a simple example, the process works this way. The first step is to select a **cluster seed** as the initial cluster center, and all objects (individuals) within a prespecified threshold distance are included in the resulting cluster. Then, another cluster seed is chosen and the assignment continues until all objects are assigned. Then, objects may be reassigned if they are closer to another cluster than the one originally assigned. There are several different approaches for selecting cluster seeds and assigning objects that we will discuss in the next section. Nonhierarchical clustering procedures are frequently referred to as *K*-means clustering and typically use one of the following three approaches for assigning individual observations to one of the clusters [4, p. 428].

Sequential Threshold   The **sequential threshold method** starts by selecting one cluster seed and includes all objects within a prespecified distance. When all objects within the distance are included, a second cluster seed is selected, and all objects within the prespecified distance are included. Then a third seed is selected, and the process continues as before. When an object is clustered with a seed, it is no longer considered for subsequent seeds.

Parallel Threshold   In contrast, the **parallel threshold method** selects several cluster seeds simultaneously in the beginning and assigns objects within the threshold distance to the nearest seed. As the process evolves, threshold distances can be adjusted to include fewer or more objects in the clusters. Also, in some methods, objects remain unclustered if they are outside the prespecified threshold distance from any cluster seed.

Optimization   The third method, referred to as **optimizing procedure,** is similar to the other two except that it allows for reassignment of objects. If, in the course of assigning objects, an object becomes closer to another cluster that is not the cluster it was originally assigned, then an optimizing procedure will switch the object to the more similar (closer) cluster.

Nonhierarchical procedures are available in a number of computer programs, including all the major statistical packages (e.g., SAS [12, 13], SPSS [17, 18], and BMDP [2]). The sequential threshold procedure (e.g., FASTCLUS program in SAS) is an example of a nonhierarchical clustering program designed for large data sets. After the researcher specifies the maximum number of clusters allowed, the procedure begins by selecting cluster seeds, which are used as initial guesses of the means of the clusters. The first seed is the first observation in the data set with no missing values. The second seed is the next complete observation (no missing data) that is separated from the first seed by a specified minimum distance. The default option is a zero minimum distance. After all seeds have been selected, the program assigns each observation to the cluster with the nearest seed. The analyst can specify that the cluster seeds are revised (updated) by calculating seed cluster means each time an observation is assigned. In contrast, the parallel threshold methods (e.g., QUICK CLUSTER in SPSS [17, 18]) establish the seed points as user-supplied points or select them randomly from all observations.

The major problem faced by all nonhierarchical clustering procedures is how to select the cluster seeds. For example, with a parallel threshold option, the initial and probably the final cluster results depend on the order of the observations in the data set, and shuffling the order of the data is likely to affect the results. Specifying the initial cluster seeds as in the parallel threshold procedure can reduce this problem. But even selecting the cluster seeds randomly will produce different results for each set of random seed points. Thus, the researcher must be aware of the impact of the cluster seed selection process on the final results.

Should Hierarchical or Nonhierarchical Methods Be Used?

A definitive answer to this question cannot be given for two reasons. First, the research problem at hand typically may suggest one method or the other. Second, both methods are evolving rapidly, and what we learn with future applications may suggest one over the other.

In the past, hierarchical clustering techniques were more popular, with Ward's method and average linkage being probably the best available [7]. Hierarchical procedures do have the advantage of being fast and therefore taking less computer time. But they can be misleading because undesirable early combinations may persist throughout the analysis and lead to artificial results. Of specific concern is the substantial impact of outliers on hierarchical methods, particularly with the complete linkage method. To reduce this possibility, the analyst may wish to cluster analyze the data several times, each time deleting problem observations or outliers. The deletion of cases, even those not found to be outliers, can many times distort the solution. Thus, the researcher must employ extreme care in the deletion of observations for any reason.

Also, while computations of the clustering process are relatively fast, hierarchical methods are not amenable to analyzing very large samples. As sample size increases, the data storage requirements increase dramatically. For example, a sample of 400 cases requires storage of approximately 80,000 similarities, and this increases to almost 125,000 for a sample of 500. Even given today's technological advances, problems of this size exceed the capacity of most personal computers, thus limiting the application in many instances. The researcher may take a random sample of the original observations to reduce its size but must now question the representativeness of the sample taken from the original sample.

Nonhierarchical methods have gained increased acceptability and are applied increasingly. Their use, however, depends on the ability of the researcher to select the seed points according to some practical, objective, or theoretical basis. In these instances, nonhierarchical methods have several advantages over hierarchical techniques. The results are less susceptible to the outliers in the data, the distance measure used, and the inclusion of irrelevant or inappropriate variables. These benefits are realized, however, only with the use of nonrandom (i.e., specified) seed points; thus, the use of nonhierarchical techniques with random seed points is markedly inferior to the hierarchical techniques. Even a nonrandom starting solution does not guarantee an optimal clustering of observations. In fact, in many instances the researcher will get a different final solution for each set of specified seed points. How is the researcher to select the "correct" answer? Only by analysis and validation can the researcher then select what is considered the "best" representation of structure, realizing there are many alternatives that may be as acceptable.

Another approach is to use *both* methods (hierarchical and nonhierarchical) to gain the benefits of each [7]. First, a hierarchical technique can establish the number of clusters, profile the cluster centers, and identify any obvious outliers. After outliers are eliminated, the remaining observations are then clustered by a nonhierarchical method, with the cluster centers from the hierarchical results as the initial seed points. In this way, the advantages of the hierarchical methods are complemented by the ability of the nonhierarchical methods to "fine-tune" the results by allowing the switching of cluster membership.

How Many Clusters Should Be Formed?

A major issue with all clustering techniques is how to select the number of clusters. There are many criteria and guidelines for approaching the problem. Unfortunately, no standard, objective selection procedure exists. The distances between clusters at successive steps may serve as a useful guideline, and the analyst may choose to stop when this distance exceeds a specified value or when the successive distances between steps makes a sudden jump. These distances are sometimes called error variability measures. Also, some intuitive conceptualization of theoretical relationship may suggest a natural number of clusters. In the final analysis, however, it is probably best to compute a number of different cluster solutions (e.g., two, three, four) then decide among the alternative solutions by using a priori criteria, practical judgment, common sense, or theoretical foundations. Also, one might start this process by specifying some criteria based on practical considerations such as saying, "My findings will be more manageable and easier to communicate if I have three to six clusters," and then solving for this number of clusters and selecting the best alternative after evaluating all of them. The cluster solutions will be improved by restricting the solution according to conceptual aspects of the problem.

Should the Cluster Analysis Be Respecified

When an acceptable cluster analysis solution is identified, the researcher should examine the fundamental structure represented in the defined clusters. Of particular note are widely disparate cluster sizes or clusters of only one or two observations. Researchers must examine widely varying cluster sizes from a conceptual perspective, comparing the actual results with the expectations formed in the research objectives. More troublesome are single-member clusters, which may be outliers not detected in earlier analyses. If a single-member cluster (or one of very small size compared with other clusters) appears, the analyst must decide if it represents a valid structural component in the sample or if it should be deleted as unrepresentative. If any observations are deleted, especially when hierarchical solutions are employed, the researcher should rerun the cluster analysis and start the process of defining clusters anew.

Stage Five: Interpretation of the Clusters

The interpretation stage involves examining the cluster variate to name or assign a label that accurately describes the nature of the clusters. To clarify this process, let's refer to the example of diet versus regular soft drinks. Assume that an attitude scale was developed that consisted of statements regarding consumption of soft drinks. Individuals were asked to evaluate these statements on a seven-point scale. Examples of statements are "Diet soft drinks taste harsher," "Regular soft drinks have a fuller taste," "Diet drinks are healthier," and so forth. Assume further that demographic and soft drink consumption data were also collected.

When starting the interpretation process, one measure frequently used is the cluster's centroid. If the clustering procedure was preformed on the raw data, this would be a logical description. If the data were standardized or if the cluster analysis was performed using factor analysis (component factors), the analyst would have to go back to the raw scores for the original variables and compute average profiles using these data. The use of modal profiles is also a possibility, as is assessment of the variability within the clusters.

Continuing with our soft drink example, in this stage we examine the average score profiles on the attitude statements for each group and assign a descriptive label to each. Average score profiles could be developed with discriminant analysis. For example, two of the groups (clusters) may have favorable attitudes about diet soft drinks and the third cluster negative attitudes. Moreover, of the two favorable clusters, one may exhibit favorable attitudes toward only diet soft drinks, while the other may display favorable attitudes toward both diet and regular soft drinks. From this analytical procedure, one would evaluate each cluster's attitudes and develop substantive interpretations to facilitate labeling each. For example, one cluster might be labeled "Health- and Calorie-Conscious," whereas another might be labeled "Get a Sugar Rush."

Stage Six: Validation and Profiling of the Clusters

Validation includes attempts by the analyst to assure that the cluster solution is representative of the general population, and thus is generalizable to other objects and stable over time. The most direct approach in this regard is to cluster analyze separate samples, comparing the cluster solutions and assessing the correspondence of the results. This approach, however, is often impractical because of time or cost constraints or the unavailability of objects (particularly consumers) for multiple cluster analyses. In these instances, a common approach is to split the sample into two groups. Each is cluster analyzed separately, and the results are then compared. A modified form is to obtain cluster centers from one group and employ them with the other groups to define clusters, then compare results between the two groups [6].

The profiling stage involves describing the characteristics of each cluster to explain how they may differ on relevant dimensions. Just as in the interpretation stage, this typically involves the use of discriminant analysis or some other appropriate statistic. The procedure begins with the clusters identified (labeled). The analyst utilizes data not previously included in the cluster procedure to profile the characteristics of each cluster. These data typically are demographic characteristics, psychographic profiles, consumption patterns, and so forth. Using discriminant analysis, the analyst compares average score profiles for the clusters. The categorical dependent variable is the previously identified clusters, and the independent variables are the demographics, psychographics, and so on. From this analysis, assuming statistical significance, the analyst could conclude, for example, that the "Health- and Calorie-Conscious" cluster from our previous example consists of better-educated, higher-income professionals who are moderate consumers of soft drinks. In short, the profile analysis focuses on describing not what directly determines the clusters but the characteristics of the clusters after they have been identified. Moreover, the emphasis is on the characteristics that differ significantly across the clusters and those that could predict membership in a particular attitude cluster.

The researcher may also attempt to establish some form of **criterion** or **predictive validity.** To do so, the researcher selects a variable(s) not used to form the clusters but known to vary across the clusters. In our example, we may know from past research that attitudes toward diet soft drinks vary by age. Thus, we can statistically test for the differences in age between those clusters favorable to diet soft drinks and those that are not.

An Illustrative Example

To illustrate the application of cluster analysis techniques, let's turn to the HATCO database. The seven perceptions of HATCO provide a basis for illustrating one of the most common uses of cluster analysis—the formation of customer segments. In our example, we will follow the stages of the model-building process, starting with the setting of objectives, addressing research design issues, and finally the actual partitioning of respondents into clusters and the interpretation and validation of the results. The following sections detail these procedures through each of the stages.

Stage One: Objectives of the Cluster Analysis

We begin by cluster analyzing the ratings by HATCO customers as to the performance of HATCO on the seven attributes (X_1 to X_7). Our objective is to segment customers into groups with similar perceptions of HATCO. Once identified, HATCO can then formulate strategies with different appeals for the separate groups. A primary concern is that the seven attributes used to form the clusters be adequate in scope and detail. From the examples in other chapters with the various multivariate techniques, we have found that these variables have sufficient predictive power to justify their use as the basis for segmentation.

Stage Two: Research Design of the Cluster Analysis

The first step is to identify any outliers in the sample before partitioning begins. In our example, the sample has been examined for outliers and found to have no strong candidates for deletion. (See Chapter 2 and the discussion on outlier detection). The next issues involve the choice of similarity measure. Given that the set of seven variables is metric, squared Euclidean distances were chosen as the similarity measure, because Mahalanobis distance (D^2) was not available. Correlational measures were not employed because the derivation of segments should consider the magnitude of the perceptions (favorable versus unfavorable) as well as the pattern. This is best accomplished with a distance measure of similarity. Finally, no form of standardization was used. The standardization of variables was not undertaken because all variables were on the same scale, and within-case standardization was not appropriate because the magnitude of the perceptions was an important element of the segmentation objectives.

Stage Three: Assumptions in Cluster Analysis

For purposes of illustration, the sample is considered a representative sample of HATCO customers. Still left to resolve is the impact of multicollinearity on the implicit weighting of the results. The analysis of multicollinearity detailed in Chapter 3 identified only minimal levels that should not impact the cluster analysis in any substantial manner.

Stage Four: Deriving Clusters and Assessing Overall Fit

In our example, we follow the approach of employing both hierarchical and non-hierarchical methods in combination. The first step in the partitioning stage first uses the hierarchical procedure to identify the appropriate number of clusters.

Then, in step 2 we use nonhierarchical procedures to fine-tune the results even further by utilizing the hierarchical results as a basis for generating the seed points.*

Step 1: Hierarchical Cluster Analysis

The first question to ask is, Which clustering algorithm should we use? Ward's method was chosen to minimize the within-cluster differences and to avoid problems with "chaining" of the observations found in linkage methods. Table 8.2 contains the results of the cluster analysis, including the cases being combined at

*The CLUSTER (hierarchical) and QUICK CLUSTER (nonhierarchical) procedures of SPSS are used in this example. The necessary control cards are contained in Appendix A.

TABLE 8.2 Agglomeration Schedule of Hierarchical Cluster Analysis Using Ward Method

	Cluster Combined			Stage Cluster First Appears		
Stage	Cluster 1	Cluster 2	Coefficient	Cluster 1	Cluster 2	Next Stage
1	15	20	.000000	0	0	60
2	5	42	.005000	0	0	94
3	24	27	.010000	0	0	74
4	47	61	.020000	0	0	78
5	19	28	.040000	0	0	60
6	67	90	.070000	0	0	39
7	18	92	.105000	0	0	65
8	51	77	.140000	0	0	72
9	33	62	.175000	0	0	63
10	36	41	.210000	0	0	45
11	85	87	.260000	0	0	69
12	65	79	.310000	0	0	68
13	43	46	.360000	0	0	76
14	25	44	.410000	0	0	63
15	38	63	.475000	0	0	54
16	69	81	.555000	0	0	52
17	94	98	.650000	0	0	73
18	56	91	.745000	0	0	66
19	50	72	.840000	0	0	52
20	75	99	.950000	0	0	62
21	1	95	1.060000	0	0	72
22	16	73	1.170000	0	0	61
23	37	48	1.280000	0	0	58
24	11	100	1.405000	0	0	69
25	4	89	1.545000	0	0	62
26	84	88	1.685000	0	0	45
27	2	83	1.825000	0	0	82
28	29	78	1.965000	0	0	61
29	3	71	2.105000	0	0	75
30	23	32	2.245000	0	0	66
31	17	64	2.435000	0	0	83
32	12	76	2.650000	0	0	67

TABLE 8.2 *Continued*

	Cluster Combined			Stage Cluster First Appears		
Stage	Cluster 1	Cluster 2	Coefficient	Cluster 1	Cluster 2	Next Stage
33	8	68	2.865000	0	0	70
34	9	74	3.130000	0	0	55
35	52	60	3.420000	0	0	57
36	10	34	3.755000	0	0	43
37	26	59	4.105000	0	0	64
38	49	97	4.525000	0	0	81
39	7	67	4.995000	0	6	77
40	13	21	5.515000	0	0	51
41	82	93	6.040000	0	0	91
42	40	54	6.565000	0	0	53
43	10	30	7.096667	36	0	50
44	66	80	7.631667	0	0	59
45	36	84	8.189167	10	26	70
46	22	55	8.749167	0	0	71
47	6	70	9.409167	0	0	57
48	45	86	10.239167	0	0	53
49	39	96	11.079167	0	0	68
50	10	53	11.965001	43	0	56
51	13	35	13.025002	40	0	71
52	50	69	14.467502	19	16	65
53	40	45	15.970001	42	48	73
54	14	38	17.558334	0	15	59
55	9	58	19.213335	34	0	67
56	10	31	21.260836	50	0	58
57	6	52	23.515835	47	35	88
58	10	37	25.868692	56	23	75
59	14	66	28.244358	54	44	80
60	15	19	30.704357	1	5	77
61	16	29	33.179356	22	28	78
62	4	75	35.714355	25	20	74
63	25	33	38.536854	14	9	64
64	25	26	41.567688	63	37	84
65	18	50	44.878521	7	52	76
66	23	56	48.546021	30	18	87
67	9	12	52.279022	55	32	80
68	39	65	56.214024	49	12	89
69	11	85	60.251522	24	11	87
70	8	36	64.364021	33	45	83
71	13	22	68.580025	51	46	90
72	1	51	73.082527	21	8	84
73	40	94	77.886696	53	17	85
74	4	24	82.785027	62	3	82
75	3	10	88.133278	29	58	79
76	18	43	93.522446	65	13	92
77	7	15	98.976730	39	60	86
78	16	47	104.835060	61	4	90
79	3	57	111.624947	75	0	91

TABLE 8.2 *Continued*

Stage	Cluster Combined		Coefficient	Stage Cluster First Appears		Next Stage
	Cluster 1	Cluster 2		Cluster 1	Cluster 2	
80	9	14	118.529945	67	59	81
81	9	49	126.007111	80	38	86
82	2	4	134.772522	27	74	85
83	8	17	143.875015	70	31	88
84	1	25	156.719177	72	64	92
85	2	40	170.259476	82	73	89
86	7	9	185.589966	77	81	94
87	11	23	201.109970	69	66	93
88	6	8	218.440796	57	83	93
89	2	39	236.111191	85	68	96
90	13	16	258.730957	71	78	95
91	3	82	281.428284	79	41	97
92	1	18	305.026886	84	76	95
93	6	11	333.080566	88	87	96
94	5	7	364.897583	2	86	98
95	1	13	398.081848	92	90	98
96	2	6	446.282684	89	93	97
97	2	3	522.980835	96	91	99
98	1	5	614.953796	95	94	99
99	1	2	994.751709	98	97	0

each stage of the process and the clustering coefficient. The coefficient (fourth column) is the within-cluster sum of squares. For the other linkage methods, the agglomeration coefficient is the squared Euclidean distance between the two cases of clusters being combined. In either case, small coefficients indicate that fairly homogenous clusters are being merged. The joining of two very different clusters results in a large coefficient. The analyst looks for large increases in the value, similar to the scree test in factor analysis.

How many clusters should we have? Because the data involve profiles of HATCO customers and our interest is in identifying types or profiles of these customers that may form the bases for differing strategies, a manageable number of clusters would be in the range of two to five. The analyst must now select the final cluster solution from these possible cluster solutions.

The clustering (agglomeration) coefficient shows rather large increases in going from four to three clusters (522.9 − 446.3 = 76.6), three to two clusters (614.9 − 522.9 = 92.0), and two to one cluster (994.7 − 614.9 = 379.8). To help identify large relative increases in the cluster homogeneity, we calculated the percentage change in the clustering coefficient for ten to two clusters (see Table 8.3). Because the largest increases were observed in going from two to one cluster, the two-cluster solution was selected.

A **vertical icicle diagram** can also be used to evaluate a cluster solution (see Figure 8.9). This diagram shows the number of objects horizontally across the top and the number of clusters vertically down the left side. The 1 at the top left

TABLE 8.3 Analysis of Agglomeration Coefficient for Hierarchical Cluster Analysis

Number of Clusters	Agglomeration Coefficient	Percentage Change in Coefficient to Next Level
10	258.7	8.9
9	281.4	8.5
8	305.0	9.2
7	333.1	9.3
6	364.9	9.3
5	398.1	12.1
4	446.3	17.0
3	522.9	17.6
2	614.9	61.9
1	994.7	—

occurs when all objects are grouped to form one cluster; the 2 shows the two-cluster solution. To see which objects are found in each group of the two-cluster solution, look for the blank spots in the row of Xs beside the 2. To do so for the three-cluster solution, look for the blank spots in the row of Xs beside the 3 on the left side of the figure. The number of objects is obtained by counting the number of objects between the blanks. Looking at two clusters in the vertical icicle plot in Figure 8.9, we note that the size of the clusters is equal, whereas when we move on to three clusters the sizes are very unequal (50, 19, and 31). Thus the data again suggest that two clusters is the best solution. Finally, Figure 8.10 shows the dendrogram for this solution. It provides a quick visual overview of the clustering process and shows which observations are found in each cluster.

The vertical icicle plot (Figure 8.9), dendrogram (Figure 8.10), and agglomeration schedule (Table 8.2) all provide a means of identifying outliers in the sample. The vertical icicle plot and dendrogram permit a visual inspection for outliers. In the vertical icicle plot, an outlier would be a "narrow" icicle extending almost to the top. In the dendrogram, an outlier would be a "long" branch that did not join until very late. In the agglomeration schedule, the researcher can ascertain the presence of single-member clusters quite easily with many of the computer programs. The example in Table 8.2 from SPSS [18] shows on the left side of the agglomeration coefficient the clusters being combined. In the right-hand columns, the steps at which each cluster was formed are noted. An observation that has never been joined into a cluster has a stage of zero. So we see in the first 42 stages that single observations are being joined together. Only at stage 43 does the cluster analysis first join a cluster formed at another stage. We can use this information also to identify single observations that are joined very late in the clustering process—potential outliers. Looking backward from stage 99 in Table 8.2, we see that at stage 94 (six clusters) a cluster formed at stage 2 was joined. This means that if we selected a seven-cluster solution, one of the clusters would have only two observations. We can also see that the last single-member cluster to be joined occurred at stage 79. Thus, if analysis is confined to a smaller number of clusters (say, ten or fewer), then the research has only one potential problem (the two-member cluster) to deal with. In this case, the selection of two clusters eliminated the need for any further respecification of the cluster analysis.

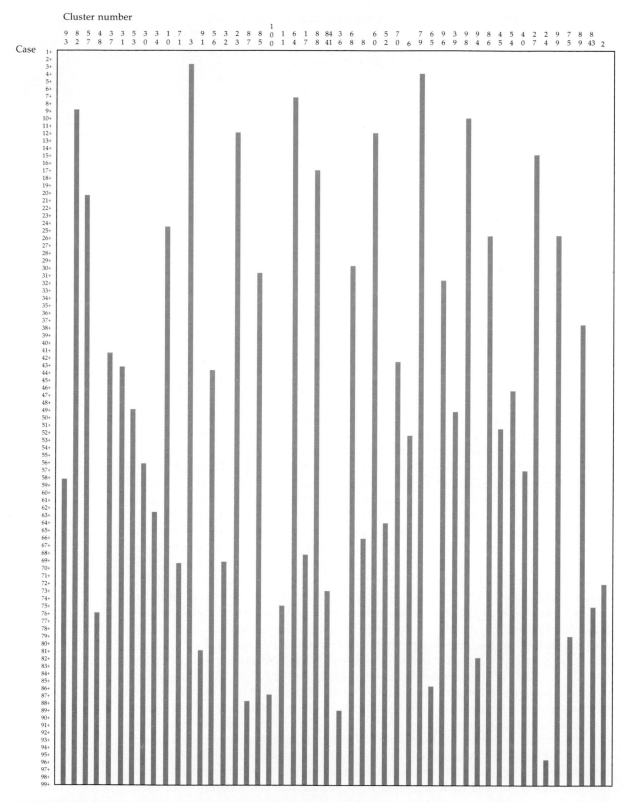

FIGURE 8.9 Vertical icicle plot for hierarchical cluster analysis using Ward's method.

Cluster number

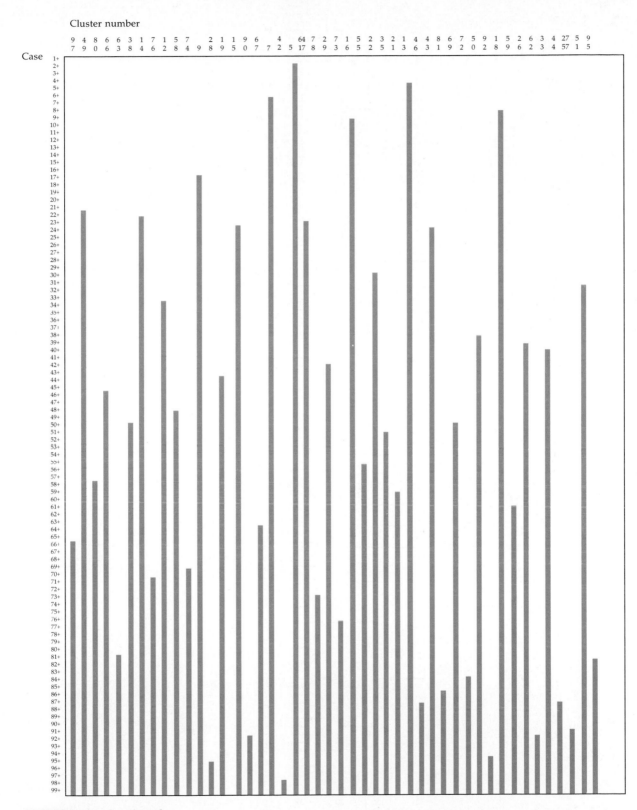

FIGURE 8.9 *Continued*

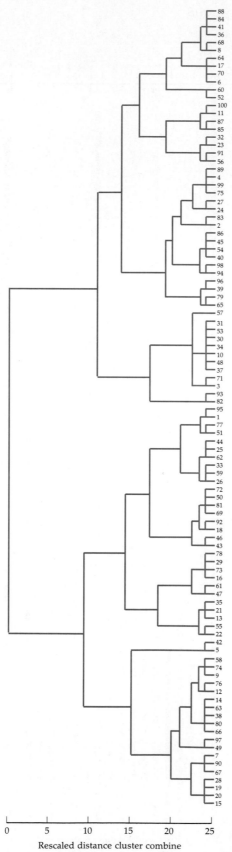

FIGURE 8.10 Dendrogram for hierarchical cluster and analysis using Ward's method.

TABLE 8.4 Results of Nonhierarchical Cluster Analysis with Initial Seed Points from Hierarchical Results

| | Mean Values | | | | | | | |
Cluster	X_1	X_2	X_3	X_4	X_5	X_6	X_7	Cluster Size
Classification cluster centers								
1	4.40	1.39	8.70	5.09	2.94	2.65	5.91	
2	2.43	3.22	6.74	5.69	2.87	2.87	8.10	
Final cluster centers								
1	4.38	1.58	8.90	4.92	2.96	2.52	5.90	52
2	2.57	3.21	6.80	5.60	2.87	2.82	8.13	48

Significance Testing of Differences Between Cluster Centers

Variable	Cluster Mean Square	Degrees of Freedom	Error Mean Square	Degrees of Freedom	F Value	Probability
X_1	81.5631	1	.9298	98	87.72	.000
X_2	66.4571	1	.7661	98	86.76	.000
X_3	109.6372	1	.8233	98	133.17	.000
X_4	11.3023	1	1.1778	98	9.56	.003
X_5	.1883	1	.5682	98	.33	.566
X_6	2.1233	1	.5786	98	3.67	.058
X_7	123.3719	1	1.2797	98	96.40	.000

Step 2: Nonhierarchical Cluster Analysis

The second step uses nonhierarchical techniques to adjust or fine-tune the results from the hierarchical procedures. In performing the cluster analysis, the researcher should take the initial seed points from the results in step 1; in this case, these are the cluster centroids on the seven-cluster variables (X_1 to X_7). Using these values as seed points, the procedure defined two groups, with the centroids shown in Table 8.4. Just as found with the hierarchical methods, the groups were of equal size (50 customers per cluster). Again, only X_5 showed no differences between the clusters. The similarity of the results from the two methods confirms the hierarchial results.

Stage Five: Interpretation of the Clusters

Information essential to the interpretation and profiling stages is provided in Table 8.5. For each cluster, the mean value (centroid) on each of the seven rating variables is provided. Beside the means are listed the univariate F ratios and levels of significance comparing the differences between the group means. For the interpretation stage, we consider only the seven perceptual variables because they were used in the cluster solution. Looking at the levels of significance for these variables, we note that six of the seven exhibit significantly different patterns. Only X_5, service, was not different between the two groups. Thus, in interpreting and ultimately labeling the cluster, we focus on the six significant variables and their respective group means. Interpretation of the means of the clusters for these

TABLE 8.5 Group Means and Significance Levels for Two-group Nonhierarchical Cluster Solution

		Interpretation of the Clusters			
		Cluster			
	Variable	*1*	*2*	*F Ratio*	*Significance*
X_1	Delivery speed	4.460	2.570	105.00	.0000
X_2	Price level	1.576	3.152	76.61	.0000
X_3	Price flexibility	8.900	6.888	111.30	.0000
X_4	Image	4.926	5.570	8.73	.0039
X_5	Service	2.992	2.840	1.02	.3141
X_6	Sales force image	2.510	2.820	4.17	.0438
X_7	Product quality	5.904	8.038	82.68	.0000

		Profiling the Clusters			
		Cluster			
Other variables of interest		*1*	*2*	*F Ratio*	*Significance*
X_9	Usage level	49.88	42.32	21.312	.0000
X_{10}	Satisfaction level	5.16	4.38	26.545	.0000

variables shows that cluster 1 focuses their attention, relative to cluster 2, on delivery speed (X_1) and price flexibility (X_3). Cluster 2 is the opposite, focusing on the factors of price level, image, sales force's image, and product quality.

Stage Six: Validation and Profiling of the Clusters

The process of validation and profiling is accomplished in several steps. First, validity is assessed by applying alternative cluster methods and comparing the solutions. Then the clusters are profiled on two additional measures (X^9, usage level; and X^{10}, satisfaction) that are indicative of the potential for differentiated strategies between the clusters.

As a validity check on the cluster solution, a second nonhierarchical analysis was performed, this time allowing the procedure to randomly select the initial seeds for the two-cluster solution. As discussed earlier, the analyst should perform several analyses to obtain a stable solution. The results in Table 8.6 also confirm the consistency of the results, as the cluster sizes are almost exact (48 and 52), and the cluster centroids are very similar. Thus, management should feel confident that "true" differences do exist among customers in terms of their needs in a supplier.

For the profiling stage, we focus on variables not included in the cluster solution (refer again to Table 8.5). In the present case, we will consider variables X_9 (usage level) and X_{10} (satisfaction level). Note that the univariate F ratios show that the group means for both variables are significantly different. The profiling process here shows that customers in cluster 1, which rated HATCO higher on delivery speed and price flexibility, had higher levels of usage and satisfaction with HATCO. The implication of the findings here, from a HATCO managerial

TABLE 8.6 Results of Nonhierarchical Cluster Analysis with Randomly Selected Initial Seed Points

Cluster	Mean Values							Cluster Size
	X_1	X_2	X_3	X_4	X_5	X_6	X_7	
Classification Cluster Centers								
1	4.95	1.14	9.03	6.55	3.21	3.79	5.09	
2	1.76	2.70	6.87	5.50	1.97	2.70	8.45	
Final Cluster Centers								
1	4.47	1.57	8.93	4.99	2.99	2.57	5.78	48
2	2.63	3.10	6.94	5.49	2.84	2.75	8.07	52

Significance Testing of Differences Between Cluster Centers

Variables	Cluster Mean Square	Degrees of Freedom	Error Mean Square	Degrees of Freedom	F Value	Probability
X_1	84.3339	1	.9016	98	93.5415	.000
X_2	58.6837	1	.8454	98	69.4175	.000
X_3	98.5164	1	.9367	98	105.1700	.000
X_4	6.2640	1	1.2292	98	5.0958	.026
X_5	.5883	1	.5641	98	1.0428	.310
X_6	.7477	1	.5927	98	1.2616	.264
X_7	131.1200	1	1.2007	98	109.2055	.000

perspective, is that an effort should be made to improve performance with respect to the factors considered poorer by cluster 2. Such improvements will, it is hoped, raise both the usage and the satisfaction levels to those found among customers in cluster 2.

Summary

Cluster analysis can be a very useful data reduction technique. But because its application is more of an art than a science, it can easily be abused (misapplied) by the analyst. Different interobject measures and different algorithms can and do affect the results. The analyst needs to consider these problems and, if possible, replicate the analysis under varying conditions. If the analyst proceeds cautiously, cluster analysis can be very helpful in identifying latent patterns suggesting useful groupings (clusters) of objects.

Questions

1. What are the three stages in the application of cluster analysis?
2. What is the purpose of cluster analysis, and when should it be used instead of factor analysis?

3. What should the analyst remember when selecting a distance measure to use in cluster analysis?
4. How does the analyst know whether to use hierarchical or nonhierarchical cluster techniques? Under which conditions would each approach be used?
5. How can you decide the number of clusters to have in your solution?
6. What is the difference between the interpretation stage and the profiling stage?
7. How do analysts use a vertical icicle?

References

1. Anderberg, M. *Cluster Analysis for Applications*. New York: Academic Press, 1973.
2. BMDP Statistical Software, Inc. *BMDP Statistical Software Manual, Release 7*, vols. 1 and 2, Los Angeles, 1992.
3. Everitt, B. *Cluster Analysis*, 2d ed. New York: Halsted Press, 1980.
4. Green, P. E. *Analyzing Multivariate Data*. Hinsdale, Ill.: Holt, Rinehart & Winston, 1978.
5. Green, P. E., and J. Douglas Carroll. *Mathematical Tools for Applied Multivariate Analysis*. New York: Academic Press, 1978.
6. McIntyre, R. M. and R. K. Blashfield. "A Nearest-Centroid Technique for Evaluating the Minimum-Variance Clustering Procedure." *Multivariate Behavioral Research* 15 (1980): 225–38.
7. Milligan, G. "An Examination of the Effect of Six Types of Error Perturbation on Fifteen Clustering Algorithms." *Psychometrica* 45 (September 1980): 325–42.
8. Morrison, D. "Measurement Problems in Cluster Analysis." *Management Science* 13, no. 12 (August 1967): 775–80.
9. Overall, J. "Note on Multivariate Methods for Profile Analysis" *Psychological Bulletin* 61, no. 3 (1964): 195–98.
10. Punj, G., and D. Stewart. "Cluster Analysis in Marketing Research: Review and Suggestions for Application." *Journal of Marketing Research* 20 (May 1983): 134–48.
11. Rohlf, F. J. , "Adaptive Hierarchical Clustering Schemes." *Systematic Zoology* 19 (1970): 58–82.
12. SAS Institute, Inc. *SAS User's Guide: Basics, Version 6*. Cary, N.C.: 1990.
13. SAS Institute, Inc. *SAS User's Guide: Statistics, Version 6*. Cary, N.C.: 1990.
14. Schaninger, C. M., and W. C. Bass. "Removing Response-Style Effects in Attribute-Determinance Ratings to Identify Market Segments." *Journal of Business Research* 14 (1986): 237–52.
15. Shephard, R. "Metric Structures in Ordinal Data." *Journal of Mathematical Psychology* 3 (1966): 287–315.
16. Sneath, P. H. A. and R. R. Sokal. *Numerical Taxonomy*. San Francisco: Freeman Press, 1973.
17. SPSS, Inc. *SPSS User's Guide*, 4th ed. Chicago: 1990.
18. SPSS, Inc. *SPSS Advanced Statistics Guide*, 4th ed. Chicago: 1990.

A Typology of Consumer Dissatisfaction Response Styles

Jagdip Singh

Researchers and practitioners have begun to recognize that the study of consumers' responses to marketplace dissatisfaction has significant implications for such key phenomena as brand loyalty and repurchase intentions (Diener and Greyser 1978; LaBarbera and Mazursky 1983; TARP 1986; Day 1984), market feedback mechanisms and new product development (*Business Week* 1984; Fornell and Wernerfelt 1987; Etzel and Silverman 1981), and consumer welfare (Andreasen 1984). For retailers this is of particular interest because a retail establishment is often the first avenue for consumer complaints. Not surprisingly, therefore, the *Journal of Retailing* devoted a full issue (Fall 1981) to highlight the contribution of the consumer satisfaction/dissatisfaction research and spur further advances in this area. As a result, researchers have begun to systematically examine why and how consumers respond to perceived dissatisfaction. Within this literature, the study of dissatisfaction response styles appears to be one of the central areas of research. By "response styles" we imply a unique set of responses that one or more consumers utilize to deal with a particular dissatisfying situation.[1] Early efforts in this area date back at least 15 years and, since then, this area has received a disproportionate share of research attention. Specifically, questions along the line of, What distinct styles do consumers use to communicate their dissatisfaction with products and services? and Do different people use different styles? have received significant research attention (e.g., Mason and Himes 1973; Warland, Herrmann, and Willits 1975; Wall, Dickey, and Talarzyk 1977; Barnes and Kelloway 1980; Bearden and Mason 1984; Moyer 1984).

Despite the large number of studies, a clear understanding of these questions remains elusive (Strahle and Day 1985; Day 1980; Grønhaug and Zaltman 1981). This state of affairs is attributable to at least four reasons. First, most researchers have tended to utilize simplistic response styles, such as (a) to take some action: the "complainers," and (b) to take no action: the "non-complainers" (e.g., Warland, Herrmann, and Willits 1975), although recently finer distinctions have been offered (e.g., Day and Landon's (1977) notions of public and private actions). Second, previous researchers have generally tended to utilize some deductivist criteria in proposing response styles. Relatively less attention has been directed, however, to understand what distinct styles consumers actually utilize to deal with their dissatisfactions. Because consumers often engage in multiple responses, it is likely that styles differ due to the specific combination of responses utilized (e.g., complaint to retailer and negative word-of-mouth versus complaint to retailer only). Third, in profiling consumers using different styles, a preponderance of past studies have utilized demographic and personality variables only. As a result, these studies have often been described as providing inconsistent findings and modest descriptive and explanatory power (Grønhaug and Zaltman 1981). Fourth, efforts toward developing a clear and explanatory rich profile of consumers is hampered by the inconsistent definition of the response styles. For instance, Morganosky and Buckley (1986) define two styles, complainers and non-complainers,

---

[1]Note, we use the term "complaint responses" rather than "complaint actions." The use of actions is less satisfactory since it connotes some directed behavior (e.g., complain to seller). Unlike actions, complaint responses implies all plausible reactions to dissatisfaction, including no-action, negative word-of-mouth communication to friends and relatives, filing a suit (among others). We maintain this distinction throughout the paper.

From Jagdip Singh, "A Typology of Consumer Dissatisfaction Response Styles," *Journal of Retailing*, vol. 66, no. 1 (Spring 1990), pp. 57–99. Reprinted by permission.

based on agreement or disagreement with the statement, "If I buy clothes I am not satisfied with, I take them back to the store and complain." In contrast, Grønhaug and Zaltman (1981) define the styles, activists and passivists, as those who either complained to some formal party (e.g., store, consumer agency) or took no action. Such inconsistencies undermine the goals of systematic and cumulative research into the focal questions.

In terms of retailing practice, the lack of a valid typology and the inconsistent findings (which may be because of the former) in previous research are especially troublesome. This is because while on one hand researchers have carefully documented the hidden agenda of consumer complaints (cf. Day et al. 1981), and the adverse effects it has on retailers (cf. Etzel and Silverman 1981), on the other hand little guidance is offered to retailers on how to deal effectively with this phenomenon. This is a disturbing void in the retailing literature. Consequently, in outlining the directions for future research in this area, Etzel and Silverman (1981) suggest that much more needs to be understood about the way the consumers complain so that retail managers do not become debilitated by such complaints. Instead, they should be able to use consumer complaints strategically (e.g., feedback mechanism), and in turn possibly enhance profitability.

The aim of this research is to address some of the preceding gaps. More specifically, this study attempts to achieve three main objectives. First, unlike much of past research, this study utilizes empirical data to isolate distinct clusters of response styles based on consumers' responses to a specific dissatisfying experience. That is, this research does not offer yet another typology on the basis of some logical, deductivist criteria. Instead, it investigates questions along the lines of What typology do the data suggest for consumer response styles? In addition, the episodic (i.e., specific dissatisfying experience) focus of this study responds to the growing recognition among CCB scholars that such a focus will yield better explanation of CCB (Day 1980; Day 1984). Second, this study explicitly investigates the internal and external validity of the proposed typology. This issue is of critical importance because while the empirical classification techniques (e.g., cluster analysis) hold considerable promise, they are at

an embryonic stage of development and require converging evidence regarding the stability and applied importance of the identified sub-types (Lorr 1983; Owens and Schoenfeldt 1979; Punj and Steward 1983). Finally, the third objective is to investigate the characteristics that differentiate among consumers who utilize different response styles. Based on a review of the literature, several demographic, personality/attitudinal and episode-specific characteristics are utilized. We begin with a brief review of literature concerning the focal issues.

Literature Review

Response Styles: Conceptual and Empirical Issues

Because consumers can, and often do engage in multiple responses, it is important to distinguish between a taxonomy of responses and a typology of response styles. Recently, several researchers have offered taxonomies for responses (Day and Landon 1977; Singh 1988). These attempts categorize the various dissatisfaction responses into two or more groups so that responses within a group are similar to each other, but different from responses categorized in other groups (cf. Singh's notions of voice, private and third party). In this sense, a taxonomy is the classification of variables, not people.

By contrast, a typology of response styles is a partitioning of people. That is, the aim in developing a typology is to determine two or more groups of consumers such that within each group consumers tends to engage in similar response styles when faced with a dissatisfying situation. Across groups, however, these styles are expected to differ significantly. In general, there is no necessary condition that a taxonomy of responses will be isomorphic with a typology of response styles. Rather, the fact that consumers engage in multiple responses argues against such correspondence. Thus, it is critical that the typology of response styles be developed independent of a taxonomy of responses.

Mason and Himes (1973) and Warland et al. (1975) are the early studies that proposed a typology of response styles. Most subsequent studies have tended to utilize the typology developed by these researchers (see Warland 1977 for an exception). Based on data for consumer dissatisfaction, both of these studies proposed an almost identical typology. In the terminology proposed by Warland et al., these styles are: (1) Upset-No action; and (2) Upset-Action.[2] The first group represents dissatisfied consumers who did not take any action. Consumers who took some action(s) fall in the second group. Subsequently researchers have used different labels for these styles, such as, complainers and non-complainers (Shuptrine and Wenglorz 1980; Bearden 1983) and activists and non-activists (Pfaff and Blivice 1977). However, the criterion for grouping has remained intact.

Although Warland et al.'s typology appears to have received some acceptance, several concerns need mention. First, the typology aggregates the wide variety of plausible responses into a mono-category. This is especially troublesome because the different responses (e.g., complain to friend, Better Business Bureau, etc.) are known to differ in terms of the effort involved and individual tendencies (cf. Richins 1983a; Singh 1988). Such real-life differences are obfuscated by Warland et al.'s typology. Second, note that in Warland et al.'s typology the partitioning of consumers is not based on rigorously analyzing response styles across respondents. Instead, it is based on a categorization criterion logically deduced by the researchers. Thus, what is open to question is not whether no-action and action are two valid response styles. Rather, the question is if the procedure adopted by Warland et al. might have ignored other distinct styles.

How, then, should one develop a typology for dissatisfaction response styles? Based on studies in other areas (e.g., Furse, Punj, and Stewart 1984; Owens and Schoenfeldt 1979), a two-step approach is suggested. In the first step, a parsimonious structure for the multiple responses should be ascertained. This will involve analyzing (e.g., factor analysis) the responses to examine if the data can be represented in fewer dimensions. In the second step, a partitioning algorithm (e.g.,

cluster analysis) should be used to group consumers so that within each cluster the consumers are homogeneous with respect to the response dimensions but not so across clusters.

The preceding is not meant to disparage the important work done by Warland et al. more than a decade ago. Rather, the aim is to identify gaps in our current knowledge and attempt to fill them. Development of a valid typology of response styles that faithfully represents how consumers actually resolve their dissatisfactions appears to be the critical next step.

Profile Issues: Discriminating Consumer Characteristics

DEMOGRAPHIC CHARACTERISTICS In general, consistent results have been obtained with respect to income, education, occupation, and age. Complainers tend to earn higher incomes, have more education, have professional jobs, and are younger (Zaichowsky and Liefeld 1977; Moyer 1984). Bearden, Teel, and Crockett (1980) have advanced resource and perceived risk arguments to support the preceding findings. Upscale consumers tend to have more resources, in terms of information and self-confidence, to deal with marketplace problems. In addition, such consumers tend to perceive less risk or embarrassment in complaining. Thus, these consumers are hypothesized to be more frequent complainers.

However, the discriminatory power of these variables is modest. For instance, in Warland et al.'s study all the demographic variables taken together could not explain more than about 8 percent of the variance. For other demographic variables such as sex, race, and marital status, the results thus far have been mixed. In addition, Grønhaug and Zaltman have questioned if demographic variables possess unique discriminating power over and above other variables (e.g., personality).

PERSONALITY/ATTITUDINAL CHARACTERISTICS The role of these variables is explained by learning and personality theories (cf. Howard 1977; Landon 1977). The learning model posits that complaining behavior is a function of prior learning (e.g., past behavior, knowledge of unfair practices, information about consumer rights and complaint

---

[2] Warland, Hermann, and Willits (1975) also proposed a third group of consumers; those who are not upset. Because the aim of this study is to examine the response styles of dissatisfied consumers, this third grouping is not utilized.

channels), which helps in the formation of attitudes, which in turn determine behaviors. Thus, variables such as prior complaining experience, attitude toward businesses, consumerism issues, and attitude toward complaining have been investigated.

The personality model hypothesizes that a consumer's predispositional nature influences his/her complaining behaviors (Landon 1977; Richins 1983b). Two types of studies can be identified that utilize the personality model. First, some studies have tended to use personality measures that are not specific to marketplace problems. Instead, these measures reflect general predispositions. Examples are Cattell's personality measure (Zaichowsky and Liefeld 1977), self-confidence (Grønhaug and Zaltman 1981), and assertiveness (Bearden and Mason 1984). The second type of personality measures are more specific; they pertain to consumer's interactions with the marketplace. Examples of such measures are consumer alienation (Bearden and Mason 1984) and assertiveness/aggressiveness (Richins 1983b).

Though not impressive, results using these variables have been more encouraging than demographic variables. For instance, using a varied set of personality/attitudinal variables, such as personal competence and consumer alienation, Bearden (1983) explained 13 percent of the variance. In general, complainers tend to have more (compared to non-complainers) prior experience of complaining, have more positive attitude toward complaining, are more self-confident, and more assertive. In regard to alienation and consumerism issues, the results are mixed, however.

EPISODE-SPECIFIC CHARACTERISTICS Landon (1977) and Day (1980) among others have long argued that episode-specific measures hold the key to an understanding of why consumers complain. For instance, Day (1980) observes:

> . . . The great majority of consumer complaints and redress seeking actions appear to be motivated by specific aspects of experiences with particular products or services. Research at this level seems more promising at this time than research on generalized feelings about the business system.

Such episode-specific variables include, cost/benefit evaluations, attributions of blame, prob-

ability of successful redress, and the type of product or service involved in the dissatisfaction.

The most frequently studied episode-specific variable appears to be the nature of product or service involved (Shuptrine and Wenglorz 1980; Bearden and Mason 1984). As expected, significant variation across products/services was found. Although other episode-specific variables have recently been examined to study complaint behaviors (e.g., Richins 1983a), their adoption in understanding different response styles has been slow. For instance, Bearden's (1983) study is one of the few studies that examined such variables (e.g., cost disconfirmation) and could explain about 12 percent of the variance.

Based on the preceding review, two directions for future research can be suggested. First, it is critical that the distinct styles be identified using a partitioning methodology that uncovers similarities among consumers with respect to complaint responses. Second, in profiling these consumer groupings, it appears essential to utilize more episode-specific variables as suggested by Day (1980). Also, the discriminating power and the unique contribution for each of the preceding characteristics should be mapped out in future research. The following research was designed as an initial study to address these issues.

The Study

Research Design

In order to address the preceding issues, several additional factors were considered. First, it was important to ensure that: (a) complaint responses as well as intentions were measured: (b) the domain of possible complaint options was exhaustively tapped; and (c) multiple response options were allowed. The measurement of intentions is necessary to obtain the dimensions that underlie the various complaint responses. As noted by Singh (1988) in the CCB literature, the delineation of the unique dimensions that underlie complaint responses is more appropriate when based on the relative intensity with which consumers evaluate different CCB options. Note, the behavior data (i.e., did/didn't do) do not reveal this relative intensity. Consider, for instance, a dissatisfied consumer who strongly intended to complain about poor service, but could not actually do so because of uncontrollable situational considerations (e.g., unplanned trip). In such cases, consumers' behaviors provide little information about what complaint response(s) was strongly (or weakly) preferred by the consumer. This renders the behavior data unsuitable for dimension analysis.

Second, the use of episode-specific variables required that a dissatisfaction episode be the contextual setting for the questionnaire. To accomplish this, the respondents were asked to recall a dissatisfying experience that they could remember most clearly. Respondents were then asked to report the complaint responses they utilized in this experience. Landon (1980) has suggested that using a recall-based design results in several undesirable biases (e.g., due to memory lapses, consistency factors, etc). Recognizing these problems, we attempted to supplement the questionnaire with a modified approach. Specifically, the respondents were asked to imagine that a dissatisfying experience similar to the one they had described occurred again. By eliciting a future incident that is similar to the past experience, the design attempts to control for dissatisfying experiences that may have no relevance or are unfamiliar to the respondent. There is precedence for such design in satisfaction research (e.g., Scammon and Kennard 1983). Various episode-specific measures about what respondents feel/think/intend-to-do were then obtained with respect to this future incident. This approach ensures that respondents do not have to retrieve from their memory information about their psychological states (e.g., expectancy-value, intentions) prior to a past dissatisfying experience.

Third, because of the response variation across product/service categories, it was important to focus on some specific categories. Four service categories were selected for this study. The selection of services rather than products was guided by two factors: (a) several studies report that services entail greater dissatisfaction than products (Best and Andreasen 1977); and (b) few studies have investigated service dissatisfactions. The specific service categories selected were: (a) grocery shopping; (b) automotive repair; (c) medical care; and (d) banking and financial services. These service categories were purposely selected in order to obtain complaint data over a range of dissatisfactions. For instance, several studies re-

port that dissatisfaction is among the lowest for grocery shopping and among the highest for automotive repair (Day and Bodur 1977).

Data Collection

Four different questionnaires were developed, one for each of the four service categories. The population of interest was defined as households who had a dissatisfying experience with a specific service category (e.g., grocery shopping). Sampling frames for such populations are not easily available, however (Robinson 1979). In accord with an alternative procedure suggested by Robinson, and often used in CCB research, a random sample of households were asked to preselect themselves; that is, to recall a recent dissatisfying experience with a specific service category. For each service category, a random sample of 1,000 households in the Southwest were mailed a questionnaire packet. Reminder cards and telephone call-backs were used as follow-up techniques.

The number of responses received from the four surveys were as follows: grocery shopping = 176; automotive repair = 155; medical care = 166; and banking services = 172. True response rates cannot be estimated precisely since this involves calculating the proportion: (households who responded)/(household who had a dissatisfying experience with a specific service category). As noted above, the denominator of this term is an elusive number. However, the telephone callbacks provided some estimate for this term. In all 2,000 callbacks were made, 500 for each survey. Telephone numbers were obtained from the criss-cross directory. A contact rate of about 80 percent was achieved after up to three call-backs. Of those contacted, at least 70 percent stated that they could not recall a dissatisfying experience. In contrast, households who had experienced some dissatisfaction were eager to participate and "let someone know" about their problem. This implies a dissatisfaction rate of about 30 percent in a random sample. This rate is higher than comparable studies in the CCB literature (Best and Andreasen 1977). Using the rather liberal estimate of 30 percent, an estimate for the response rates in the four surveys is: grocery = 59 percent; automotive = 52 percent; medical = 56 percent; banking = 57 percent.

Nevertheless, measures were examined for possible non-response effect. Wave analysis was performed by classifying responses into four categories based on the postmark date. Responses to the various measures (see details below) were examined for significant differences in the four waves. The null hypotheses for no systematic differences could not be rejected (F values < 2; $p > 0.05$). The demographic profile of the four samples is in Table 1. Because the survey instructions requested that the questionnaire be completed by that person in the household who deals most frequently with a service category (e.g., grocery), Table 1 depicts variability in some demographic characteristics. This variability is consistent with the extant findings in CCB research (e.g., War-

land, Herrmann, and Willits 1975). Cases with missing values were deleted, leaving usable responses from 104 (banking) to 125 (medical care).

Measures of Study: Definitions and Measurement Characteristics

Well-tested measures for some constructs were not available in the CCB literature. Thus, directions were sought from the conceptual deductions provided by researchers (e.g., Day 1984; Richins 1983a). The selection of items was based on three criteria: (a) elementary measurement consistent with previous research (in terms of conceptual definitions); (b) similarity of concep-

TABLE 1 Demographic Characteristics of the Four Samples (All values are in percentages)

	Grocery	Auto Repair	Medical Care	Banking
1. Sex				
Males	27	67	30	47
Females	73	33	70	53
2. Age[a]				
≤ 25 yrs.	22	8	10	9
$> 25, \leq 35$ yrs.	30	18	22	30
$> 35, \leq 50$ yrs.	23	39	29	27
> 50 yrs.	25	35	39	34
3. Marital status				
Married	66	77	70	69
Other	34	23	30	31
(Separated/Divorced/Widow)				
4. Occupation[a]				
Professionals	14	28	25	32
White collar	27	27	21	25
Blue collar	32	26	15	15
Retired	27	19	39	38
5. Education				
High school	27	13	23	22
College	53	57	55	53
Graduate	20	30	22	25
6. Race				
White	91	93	88	92
Other	9	7	12	8
7. Income[a]				
$\leq 20,000$	34	17	29	31
$> 20,000, \leq 30,000$	19	13	23	22
$> 30,000, \leq 50,000$	32	39	29	31
$> 50,000$	15	31	19	16

[a]For the sake of clarity, some categories were combined for this characteristic.

tual content in items across the four service categories; and (c) multiple item measures for each construct. The initial drafts were pretested with faculty and staff, and the necessary modifications and revisions were made. A summary of the basis, conceptual definitions, item composition, and measurement characteristics of the various constructs is provided in Table 2. The specific items and the corresponding factor coefficients are given in Appendix I.

COMPLAINT INTENTIONS/BEHAVIORS The complaint intentions were measured by a ten-item, six-point

''very likely—very unlikely'' Likert scale. For the past dissatisfaction experience, behaviors were measured on a seven-item, dichotomous category (Yes/No) scale. Multiple responses were allowed. To mitigate the effect of response biases, the behavior and intentions items were separated by two pages containing other measures.

Initially the dimensional structure of the intentions data was ascertained. The complaint intentions were factor analyzed in one of the four data-sets, and the obtained structure subjected to confirmatory analysis in the other three data-sets (see Singh 1988). The automotive repair data was ran-

TABLE 2 Measurement Summary for the Constructs of Study

Construct/Dimension	Conceptual Definition/Approach	Based on	No. of Items	Coeff. Alpha
1. Complaint intentions	A six-point Likert scale indicating	Day (1984) Singh (1988)	10	0.83[a]
a. Voice[b]	Propensity for actions directed at the seller/manufacturer.		3	0.75
b. Private	Propensity for complaining to friends/relatives and/or exit.		3	0.77
c. Third party	Propensity for complaining to parties not involved in the exchange (Better Business Bureau: lawyer).		4	0.84
2. Expectancy-value (EV)	A six-point Likert scale indicating:	Day (1984) Bagozzi (1982)	9	0.70[a]
a. EV-Voice	EV judgments for consequences resulting from Voice responses.		3	0.89
b. EV-Private	EV judgments for consequences resulting from Private responses.		3	0.52
c. EV-Third party	EV judgments for consequences resulting from third-party responses.		3	0.75
3. Attitudes toward complaining	A six-point Likert scale indicating:	Richins (1982)	8	0.72[a]
a. Personal norms	individual norms regarding complaining to sellers/providers.		5	0.67
b. Societal benefits	individual beliefs about societal benefits resulting from complaining.		3	0.66
4. Consumer alienation	A six-point Likert scale measuring individual beliefs about the practices followed by the providers of a particular service (e.g., medical care).	Allison (1987)	7	0.80

TABLE 2 *Continued*

Construct/Dimension	Conceptual Definition/Approach	Based on	No. of Items	Coeff. Alpha
5. Prior experience of complaining	Behavioral measure		—	—
a. Voice	Number of times complained to seller/provider in the last one year.			
b. Private	Six-point frequency scale for talking to friends/relatives about problems with the focal service.			
c. Third party	Number of times complained to any third party, such as Better Business Bureau, newspaper, lawyer.			
6. Demographics	As appropriate		—	—
a. Sex	Male/Female			
b. Age	Nine-category scale			
c. Education	Three-category scale			
d. Income	Seven-category scale			
e. Occupation	Five-category scale			
f. Marital Status	Married, Single, or Divorced			

[a] Reliability for the linear composite of individual dimensions. Coefficient alpha for the individual dimensions is also reported.
[b] The item measuring intentions to engage in no-actions is reverse scored and belongs to this dimension.

domly selected for the initial exploratory analysis. For the correlation matrix of the ten auto repair CCB intentions, the breaks-in-scree-plot criterion suggested that a three-dimensional structure is plausible (eigenvalues: 3.76, 1.61, 1.33). The rotated factor pattern is provided in Appendix II. The dimensions were labeled as (see Appendix I): (a) *Voice,* that is, responses directed toward objects directly involved in the dissatisfying relationship (e.g., salesperson, retailer, provider); (b) *Private,* that is, negative word-of-mouth communication (e.g., to friends/relatives) and exit from exchange relationship; and (c) *Third party,* that is, complain to formal agencies not involved in the exchange relationship (e.g., Better Business Bureau, newspaper, lawyer).

Next, the confirmatory analysis (using LISREL VI), which constrains all cross-loadings to zero, was performed. The results are summarized in Table 3 (see Singh 1988 for more details). For the auto repair data, as expected, the proposed three-dimensional structure fits the data fairly well:

$\chi^2 = 49.6$, $df = 32$, $p = 0.024$, GFI = 0.89, Adjusted GFI = 0.82, and RMR = 0.065 (see first column in Table 3). In addition, each of the dimensions achieves a composite reliability of 0.69 or higher, and satisfies Fornell and Larcker's (1981) criterion for discriminant validity (i.e., variance extracted > variance shared). The results from the other three data appear to support the three-dimensional structure as well: bank: $\chi^2 = 51.07$; $df = 32$; $p > 0.01$; GFI = 0.88, Adjusted GFI = 0.80, RMR = 0.06; medical: $\chi^2 = 63.69$; $df = 32$; $p < 0.01$; GFI = 0.87, Adjusted GFI = 0.77, RMR = 0.06; and grocery shopping: $\chi^2 = 69.2$; $df = 32$; $p < 0.01$; GFI = 0.87, Adjusted GFI = 0.77, and RMR = 0.08. Although χ^2 statistic is significant (at $p = 0.01$) for grocery and medical care data, the other indicators of goodness-of-fit (e.g., GFI, RMR) are comparable across the three data. Table 3 provides additional support for the three-dimensional structure. In general, irrespective of the service category, the item loadings are high and significant, composite reliabilities exceed 0.80

TABLE 3 Estimated Maximum Likelihood Parameters for the Three-Dimensional Model of CCB Intentions[a]

Parameter[b]	Auto Repair	Grocery	Medical	Bank
λ_{11}[d]	.49[c]	.93[c]	.69[c]	.70[c]
λ_{21}	.72 (4.20)	.39 (3.03)	.78 (6.93)	.90 (6.54)
λ_{32}	.73[c]	.50[c]	.73[c]	.43[c]
λ_{41}	.76 (4.17)	.62 (3.55)	.79 (6.95)	.69 (6.31)
λ_{52}	.73 (6.76)	.56 (4.49)	.71 (7.15)	.53 (3.56)
λ_{62}	.81 (7.05)	.91 (5.02)	.86 (7.75)	.93 (3.10)
λ_{73}	.78[c]	.79[c]	.83[c]	.83[c]
λ_{83}	.53 (5.50)	.63 (7.22)	.74 (8.68)	.69 (7.17)
λ_{93}	.90 (8.78)	.93 (10.46)	.83 (9.97)	.82 (8.69)
$\lambda_{10.3}$	.61 (6.44)	.55 (6.10)	.68 (7.94)	.71 (7.42)
ϕ_{21}	.32 (2.26)	.14 (1.25)	.42 (3.27)	.09 (.78)
ϕ_{31}	.44 (2.87)	.21 (1.97)	.45 (3.60)	.37 (2.82)
ϕ_{32}	.54 (3.87)	.71 (3.83)	.54 (4.16)	.39 (2.30)
$\rho\xi_1$[e]	.69	.80	.80	.81
$\rho\xi_2$	.80	.53	.80	.68
$\rho\xi_3$	.80	.83	.85	.85
$\rho_{vc}\xi_1$[f]	.44	.47	.57	.59
$\rho_{vc}\xi_2$	.58	.28	.57	.44
$\rho_{vc}\xi_3$	.52	.545	.60	.59
GFI	.892	.869	.867	.883
Adj. GFI	.815	.775	.772	.799
RMR	.065	.079	.061	.066
χ^2	49.60	69.23	63.69	51.07
df	32	32	32	32

[a] T-values are in parentheses.

[b] Standardized parameter value.

[c] Corresponding λ set to 1.0 to fix scale of measurement.

[d] The first subscript refers to the item, and the second to the underlying factor. The item subscripts correspond to the items listed in the footnote of Appendix II.

[e] $\rho\xi_1$ denotes the reliability of the construct ξ_1 (i.e., "voice" in this case).

[f] $\rho_{vc}\xi_1$ is the variance extracted by ξ_1 in accord with Fornell and Larcker (1981).

(exceptions being the private dimension in grocery and bank data), and all dimensions achieve discriminant validity. More importantly, in a detailed analysis, Singh (1988) shows that the preceding dimensional structure is a better representation of data than any other competing taxonomy available in the CCB literature. Furthermore, the residuals among the ten CCB dimensions were generally small, although two fairly large values (i.e., standardized residual > 2) were present in the case of medical and grocery data. No systematic pattern in these residuals across data sets was evident. Plausible explanations for these residuals include idiosyncrasies of the question wording, method artifacts, or peculiarities due to the service category (e.g., see Singh 1988). Allowing the corresponding error terms to correlate produced nonsignificant χ^2 values (all with $df = 30$; medical = 43.52, $p = 0.053$, and grocery = 42.08, $p = 0.07$). However, this reestimation did not change any of the parameters significantly. For the preceding reasons, it appears safe to conclude that the three-dimensional structure for CCB intentions is reasonable basis for further analysis (also see Singh 1988 for more validation information).

DEMOGRAPHIC MEASURES The measures used in this study were sex, age, education, income, occupation, and marital status. Sex was coded as a 0/1 variable (0 = male). Age was measured on a nine-category scale, ranging from less than 25 years to over 60 years. Three categories (i.e., high school, college, and graduate school) were utilized to measure education of the respondents. An eight-category scale was used to assess income. However, the categories $90,001–110,000 and over $110,000 were combined because of small cell size in the latter category. Occupation was assessed by an open-ended item. The responses obtained were classified into a five-category scale representing: retired/unemployed, blue collar worker, housewife, teacher/white collar staff, and professionals/managers. Although marital status was measured by a five-point scale, the responses were grouped into two categories (coded 0/1) representing married and other (e.g., single, divorced). This was done so as to avoid small cell sizes.

PERSONALITY/ATTITUDINAL MEASURES Three constructs were included: consumer alienation, attitudes toward complaining, and prior expereinces. Each of these constructs has been suggested as a potential discriminator (Bearden and Mason 1984; Richins 1982; Moyer 1984). Below, we discuss each of these constructs and provide results of the analysis utilized to confirm that the measurement characteristics (e.g., reliability, factor structure) of the respective scales are consistent with previous research.

Consumer alienation was conceptualized as a consumer's interaction style was measured by seven items adapted from Allison (1978). The correlation matrix of the seven items was input to common factor analysis. The breaks-in-scree-plot criterion indicated that a single factor underlies the seven alienation items (eigenvalue = 3.25) and explains 47 percent of the total variance. The alpha reliability of this measure is 0.81.

Attitude toward complaining was measured by eight items adapted from Richins (1982). Upon factor analyzing the correlation matrix, a two-factor solution appeared tenable. The first two eigenvalues (2.74 and 1.26) together explain 50 percent of the total variance and the scree plot indicated a clear break after the second eigen-

value. Upon rotation using VARIMAX, a clear and interpretable factor structure with negligible cross-loadings was obtained. Five items loaded on the first factor and three on the second factor (see Appendix I). Like Richins (1982), the first and second factors can be interpreted as the pesonal norm and societal benefits dimensions. The alpha reliabilities for the personal norm and societal benefits dimensions are 0.67 and 0.66, respectively.

The prior experience was measured by four items and categorized (post-hoc) into voice, private, and third-party experience. Given the three-dimensional structure of CCB reported earlier, this option was more desirable from a substantive standpoint than either lumping all experiences together or ignoring the nature of interdependencies among them. Voice experience was the number of times a respondent had complained to any party directly involved in previous dissatisfying exchanges. A period of one year was specified so as to provide a common time frame as well as to limit the time span within the recall ability of most respondents. The private experience was measured by how often a respondent would talk to friends and relatives about a dissatisfying experience and was obtained on a six-point, often-never scale. Finally, the third-party experience was the number of times a respondent had complained to any third party (i.e., Better Business Bureau, consumer agency, lawyer) concerning problems with the focal service category.

EPISODE-SPECIFIC MEASURES Day (1984), and Richins (1983a) have suggested episode-specific constructs such as the probability of obtaining redress and payoff (i.e., costs/benefits) from complaint actions. Note redress is merely a consequence of voice actions. Other consequences from voice are plausible (e.g., better service in future, such as with banking). In addition, private and third-party actions may have consequences of their own. As such Day's notion of probability of redress can be generalized to the "probability of obtaining some (desirable) consequences"; in other words, the concept of "expectancy". Similarly, the notions of payoff (costs/benefits) correspond closely with the concept of "value" of consequences resulting from some actions. The use of expectancy and value concepts is desirable since it allows the use of a theoretical framework for the conceptualization and operationalization of these concepts (Fishbein and Ajzen 1975; Bagozzi 1982). In addition, this does not pose requirements of new terminology. Thus, consistent with the expectancy-value model, expectancies and values were measured for several consequences resulting from the various complaint responses and then combined multiplicatively to arrive at expectancy-value judgments.

Eighteen items were used to measure expectancy and value judgments for nine consequences from the various complaint responses. In particular, the focus groups were used to identify the set of consequences that might result from the various options of complaint responses. The consequences selected for the final questionnaire were based on two criteria: (a) saliency, that is, most frequent mention by individuals in the focus groups; and (b) parsimony, that is, conceptually relevant to dissatisfactions/complaints for each of the four service categories. This was essential to allow similar items (in content) to be used in the four surveys. These measures were operationalized and worded in a manner specifically suggested by Bagozzi (see Appendix 1).

Corresponding to each expectancy item, a value item was included. Usual measures of value utilize a good/bad scale. Initial pretests indicated that such a measure neither provided any significant variation in responses nor evoked cost/benefit trade-offs. Thus, a different approach suggested by Bagozzi was adopted. This particular measure essentially asks the respondent to evaluate the costs involved in voice actions (for instance) given that some particular benefit was sure to occur (e.g., refund). A desirable feature of this approach is that it allows for trade-offs inherent in complaint decisions. In fact, some consumers may choose not to voice (for instance) even if the benefits were sure to occur merely due to the prohibitive costs (e.g., time and effort) involved. The expectancy and value items were, however, physically separated by one page on the questionnaire to mitigate the effects of response-set bias.

Upon multiplying the expectancies and values, nine expectancy-value (EV) judgments were obtained. As in the case of personality and attitudinal constructs, preliminary analyses were conducted to assess the measurement properties of individual constructs. Upon subjecting to common factor analysis, a three-factor solution was supported. The first three eigenvalues (3.21, 1.80, and 1.19) together explain 69 percent of the total variance. In addition, the scree plot suggested a clear break after the third eigenvalue. Upon rotation using VARIMAX, a clear and interpretable factor structure resulted; three items each for EV-Voice, EV-Private, and EV-Third party (see Appendix 1 for specific items). The alpha reliabilities for the three dimensions were 0.89, 0.52, and 0.75 respectively. Clearly, the reliability for EV-Private is less than satisfactory, and results should be evaluated in light of this limitation. However, given that the episode-specific measures are in their early stages of development and the initial nature of this study, it appears fruitful to retain EV-Private for further analysis.

Method of Analysis

A schematic diagram for the method of analysis is displayed in Figure 1. Cluster analysis procedures were utilized to develop a typology of dissatisfaction response styles. However, these procedures are not based on probabilistic statistics. As a result, there is often no single best solution to a clustering problem. In addition, one does not have specified alpha levels to guide selection of a particular set of clusters from several alternative solutions. Consequently issues concerning the validity and stability of cluster solutions are critically important (Punj and Steward 1983). In the light of these issues, we devised a three-step data analysis approach.

In the first step, the aim was to ascertain the optimal number of clusters (n) based on iterative cluster analysis and internal validation of alternative solutions. For cluster analysis, the CCB intentions data were used. The intentions data were more appropriate (than behavior data) because they are obtained on a graded scale (i.e., ranging in order from high [e.g., very likely] to low [e.g., very unlikely], possibly with interval properties). Such graded data facilitate the use of a distance metric in cluster analysis, thus retaining information for the elevation, shape, and scatter in the data (Lorr 1983). Furthermore, consistent with the recommendation of Bailey (1974), we believe that

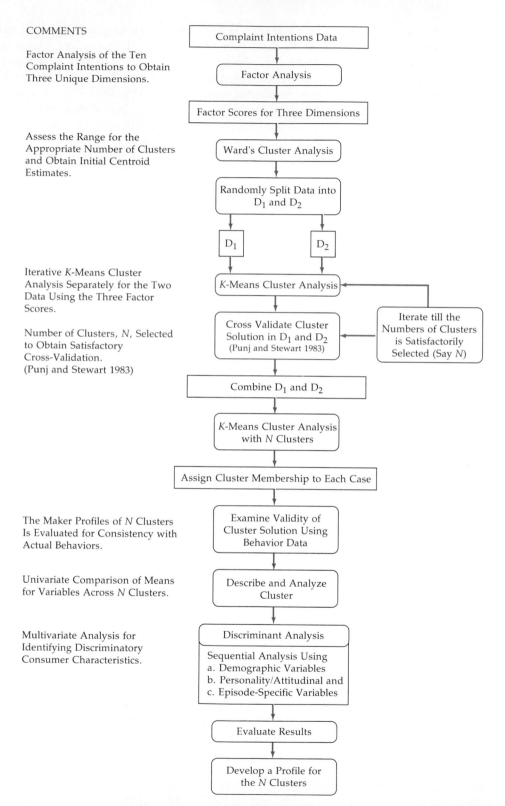

COMMENTS

Factor Analysis of the Ten
Complaint Intentions to Obtain
Three Unique Dimensions.

Assess the Range for the
Appropriate Number of Clusters
and Obtain Initial Centroid
Estimates.

Iterative *K*-Means Cluster
Analysis Separately for the Two
Data Using the Three Factor
Scores.

Number of Clusters, *N*, Selected
to Obtain Satisfactory
Cross-Validation.
(Punj and Stewart 1983)

The Maker Profiles of *N* Clusters
Is Evaluated for Consistency with
Actual Behaviors.

Univariate Comparison of Means
for Variables Across *N* Clusters.

Multivariate Analysis for
Identifying Discriminatory
Consumer Characteristics.

Complaint Intentions Data

Factor Analysis

Factor Scores for Three Dimensions

Ward's Cluster Analysis

Randomly Split Data into
D_1 and D_2

D_1 D_2

K-Means Cluster Analysis

Cross Validate Cluster
Solution in D_1 and D_2
(Punj and Stewart 1983)

Iterate till the
Numbers of Clusters
is Satisfactorily
Selected (Say *N*)

Combine D_1 and D_2

K-Means Cluster Analysis
with *N* Clusters

Assign Cluster Membership to Each Case

Examine Validity of
Cluster Solution Using
Behavior Data

Describe and Analyze
Cluster

Discriminant Analysis

Sequential Analysis Using
a. Demographic Variables
b. Personality/Attitudinal and
c. Episode-Specific Variables

Evaluate Results

Develop a Profile for
the *N* Clusters

FIGURE 1 The method of analysis utilized in the study.

the CCB dimensions rather than the raw variables should be used as input to cluster analysis. This is because raw variables contain interdependencies (i.e., which reflect the number of variables in each dimension and their intercorrelations) that are likely to bias the cluster analysis results. By contrast, the use of dimensions removes such interdependencies by representing the data by a relatively independent and parsimonious set of factors (Lorr 1983). Not surprisingly, therefore, applications of cluster analysis in marketing have often recommended and utilized dimensions from factor analysis (e.g., Furse, Punj, and Stewart 1984). Additionally, the use of CCB dimensions is also theoretically appealing because these dimensions are conceptually meaningful and interpretable, and represent an acceptable taxonomy of complaint responses (Singh 1988).

Consistent with the research design adopted for this study, a dissatisfaction episode was utilized as the unit of analysis. (Respondents were asked to recall any one dissatisfying episode with the focal service category which they could remember most clearly.) Because of this level of analysis, it makes sense to pool episodes across the four service categories for analysis. The implication of this approach is that it seeks to answer the question. "Do consistent and stable patterns of dissatisfaction response styles emerge across several diverse episodes in relatively different contextual settings?" This strategy is likely to yield a more generalizable typology than separate analysis for each of the four individual service categories. Nevertheless, to examine this issue, additonal analysis was conducted at the service category level, and points of differences were noted.

In accord with the procedure recommended by Punj and Stewart (1983), the 465 cases were randomly split into two data, D_1 and D_2, containing 241 and 224 cases respectively. D_1 was the test sample, and D_2 the internal validation sample. In essence, the test sample was utilized to generate the possible alternative cluster solutions to the classification problem. The internal validation sample was then used to select the most optimum solution among these alternatives based on its stability and reproducibility. Because the different styles are not expected to be nested in each other, a non-hierarchical cluster approach,

K-means, was selected. However, the use of K-means requires an a priori specification of the number of clusters to be extracted as well as their centroids. The K-means procedure is known to be sensitive to this a priori specification (Punj and Stewart 1983). To obtain some idea about the number of clusters, Ward's clustering method was initially utilized. Euclidean distances between consumers in the test sample were input for analysis. These distances were based on the factor scores for the three CCB dimensions (i.e., voice, private, and third party). Using the initial centroid estimates from the Ward's method, K-means cluster analysis was performed for several different cluster values suggested by the dendograms produced by the Ward's method. The optimal n was chosen based on the internal validation of the various cluster solutions (cf. Figure 1). This procedure is essentially a cross-validation of the D_2 sample utilizing a constrained and unconstrained solution for each alternative cluster value (i.e., n). For a given n, the constrained solution classifies all cases in D_2 based on the cluster analysis results from the test sample. By contrast, the unconstrained solution poses no restrictions. The chance corrected coefficient of agreement, $Kappa$, was computed for the two solutions of D_2 cases, for each n (Fleiss and Cohen 1973; Lorr 1983). The optimal n was chosen so as to maximize $Kappa$. Once the optimal n was determined, the data (D_1 and D_2) were $Kappa$. Once the optimal n was determined, the data (D_1 and D_2) were pooled and input into K-means cluster analysis with the number of clusters specified at the optimal value.

In the second step, the optimal typology uncovered in the first step was subjected to external validation. Note, this typology is based on CCB intentions. Thus, although it is obvious that the response styles for the various clusters would differ in terms of CCB intentions, it is not certain that similar differences would be found for actual behaviors. To the extent that the marker styles (i.e., distinct identifying characteristics or styles for each of the cluster groups) based on CCB dimensions faithfully reproduce differences in actual behavior, the typology would tend to be valid. Initially the seven Yes/No behavior items were grouped into three types of behaviors—voice, private, and third party. Next, for the n

cluster typology, specific hypotheses were developed for the variation in actual behaviors based on the marker response styles. These hypotheses were then tested using contingency tables and chi-square statistic.

Finally, discriminant analysis was used to determine the characteristics that differ across the n cluster groups. The variables were sequentially entered into discriminant analysis. The discriminating power of the total set as well as sub-set of variables (e.g., demographic) was evaluated by several criteria: (a) Wilks's lambda, (b) variance explained, I^2 and (c) percentage correct classified. The Wilks's lambda provides a multivariate test of the null hypothesis that, in the population, the means of all discriminant functions are equal in the obtained groups. The I_2 is analogous to R^2 in multiple regression and is a measure of the amount of the variance in the dependent variable accounted for by the predictor variables acting together as a set (Peterson and Mahajan 1976). The percentage of cases correctly classified into their respective groups by utilizing the discriminant functions affords another basis to evaluate the discriminant model. Also, more information was gained by evaluating the discrimination functions in terms of the associated canonical correlation and the eigenvalue. The significant discriminant functions were then interpreted on the basis of the standardized coefficients.

Findings

Cluster Analysis and Internal Validation

As noted above, factor scores estimated from a three-factor rotated solution for the ten intentions items were utilized. Because the cluster analysis is known to be sensitive to outliers, the data was first examined for outlying observations. Note, the factor scores are standardized variables. Consequently, values exceeding $+3.0$ and below -3.0 are potential outliers. Upon examination, it was determined that none of the observations could be classified as a potential outlier. Thus, it appeared safe to conduct cluster analysis with the entire data.

The initial cluster analysis (i.e., using Ward's method) of Euclidean distances between the cases in D_1 suggested between three and seven clusters. Consequently, using the initial centroid estimates from the Ward's method, K-means cluster analysis was performed for five different cluster values (i.e., $n = 3, 4, 5, 6$ and 7). Next, the coefficient of agreement between the constrained and unconstrained solution of D_2 cases was computed for each of the five alternatives. The three, four, five, and six cluster solutions produced $Kappa,$ the chance corrected coefficient of agreement, of 0.70, 0.89, 0.80 and 0.78, respectively. A high degree of confusion among the cluster assignments existed in the seven-cluster solution. Because the decision criterion is to maximize $Kappa$, the four-cluster solution appeared optimal. However, before accepting the four-cluster solution, a check for pooling the data from the four service categories was conducted. For this check, data from each of the individual services was also analyzed separately. The four-cluster solution produced the maximum $Kappa$ for the automotive repair, medical care, and banking services data. This solution was identical (with minor variations) to the one obtained by pooling the four data. For the case of grocery shopping, the five-cluster solution was marginally better ($Kappa = 0.86$) than the four-cluster solution ($Kappa = 0.82$). One of the five clusters in this case had relatively small membership (7). Because of this reason and the greater stability of the four-cluster solution across the individual services data, the four-cluster solution was selected as the most appropriate representation of data. A final four-cluster solution on the basis of all the cases (465) was then developed. Based on the cluster means for the derived factor scores and the cluster sizes (see Table 4), the clusters were labeled as follows:

1. **Passives,** classifying 14 percent of the sample represents dissatisfied consumers whose intentions to complain are below average on all three factors—especially for voicing complaints to sellers/providers of the dissatisfying service. In the face of dissatisfaction, this group is least likely to take any action. As such, this group appears to be consistent with the noncomplainer segment reported in past research.

2. **Voicers,** classifying 37 percent of the respondents, characterizes dissatisfied consumers who are below average on private and third-

TABLE 4 A Summary of Cluster Descriptors for the Proposed Typology[a]

	Cluster 1 (Passives)	Cluster 2 (Voicers)	Cluster 3 (Irates)	Cluster 4 (Activists)
Complaint factor[b]				
Factor 1—Voice	−1.54	0.39	−0.05	0.32
Factor 2—Private	−0.55	−0.56	1.17	0.15
Factor 3—Third party	−0.21	−0.58	−0.42	1.18
Number of cases	66	171	97	131
Percentage of respondents	14	37	21	28

[a]Note, that the cluster descriptors are based on factor scores that have a mean of zero and standard deviation of one. For instance, −0.05 (see third column, first row) indicates just about average activity on a particular factor.
[b]The factor scores were derived from intentions data pooled across the four service categories.

party actions. In particular, these consumers have little desire to engage in negative W-O-M or switch patronage or go to third parties. Instead, they are the highest on the voice factor, implying that these consumers actively complain to the service provider to obtain redress.

3. **Irates,** classifying 21 percent of the respondents, represents angry consumers who depcit *above average* private responses (i.e., negative word-of-mouth to friends/relatives, stop patronage of retailer), and just about average tendency to complain directly to sellers/providers. In other words, these consumers not only complain directly to the service provider but also switch patronage and/or engage in negative W-O-M. However, these consumers are less likely to take third-party actions.

4. **Activists,** classifying 28 percent of the respondents, represents dissatisfied consumers who are characterized by above average complaint activity on all three dimensions—especailly for complaining to third parties (Better Business Bureau, court actions, newspaper). This bears a close resemblance to the notion of the consumer activist—one who belongs to a consumerist agency and actively utilizes all channels of complaining, not only to seek individual redress but for social good (Grikscheit and Granzin 1985).

Although prior studies have identified an activist consumer segment, the activist cluster in this study is defined precisely. For instance, Grøhaug and Zaltman categorize consumers who take (any) one or more actions following a dissat-isfaction as activists. In contrast, we define the activist consumer as one who intends to engage in all, *not any*, dimensions of complaint responses (i.e., voice, private, and third party). In addition, the irate and the voicer segments have not been identified in prior studies.

Corresponding to each of the preceding clusters, a response style can be identified that characterizes how a set of consumers deal with their dissatisfactions. In all, four unique styles are posited: (a) no-action; (b) voice actions only; (c) voice and private actions; (d) voice, private, and third-party actions; corresponding to passive, voicer, irate, and activist segments.

External Validity Check

Initially hypotheses were developed for the variability of actual behaviors across the four clusters. Based on Table 4, it was expected that for the typology to be valid: (a) voice behaviors should be more frequent among voicers and activists, and the least evident in passives; (b) private behaviors should be more common among irates, and the least evident among voicers and passives; and (c) third-party behaviors should be more common among activists and the least frequent among irates and voicers.

Results from the analysis of the preceding hypotheses are in Table 5. Variation across the four clusters is significant for voice (chi-square = 43.7, $p < 0.001$), private (chi-square = 37.35, $p < 0.001$) and third-party behaviors (chi-square = 9.26, $p = 0.02$). In addition, the hypothesized pattern is strongly supported. Voice behaviors are exercised

TABLE 5 Validity Check for the Proposed Typology (Contingency Table for Response Style-behavior Consistency)

Actual Complaint Behaviors[a]	Passives	Voicers	Irates	Activists	Chi-Square Value[b]
Voice CCB					
Raw responses	27	143	62	94	43.7
					(0.00)
% of respondents	40.9	83.6	63.9	71.8	
Private CCB					
Raw responses	24	66	68	82	37.35
					(0.00)
% of respondents	36.4	38.6	70.1	62.6	
Third-party CCB					
Raw responses	2	4	6	13	9.26
					(0.02)
% of respondents	3.0	2.3	6.2	9.9	

[a] Because behaviors are dichotomous (Yes/No type) variables, we present data for yes responses only. Note that the no responses can be derived from this table.

[b] Tests the null hypotheses that the actual CCB behaviors (e.g., Voice) are independent of the response style. P-value is in parentheses.

by 83.6 percent and 71.8 percent of voicers and activists respectively. However, only 40.9 percent of the passives noted that they voiced. Likewise, fully 70.1 percent of the irates reported taking private actions. In contrast, only 38.6 percent and 36.4 percent of the voicers and passives indicated that they took such actions. For third-party behaviors also, 10 percent of the activists reported taking legal actions and/or complained to a consumer agency. In contrast, only 6.2 percent and 2.3 percent of the irates and voicers reported such actions, respectively.

Although a complete correspondence was not expected (because of situational considerations), the preceding results appear to unequivocally suggest that the response styles of the proposed typology indeed reflect differences in actual behaviors. This correspondence is especially noteworthy in view of the fact that the behaviors represent actions taken for a specific past episode and the CCB intentions are behavioral tendencies for a future but similar episode. Thus, the four response styles uncovered in the first step of analysis appear to be a reasonably valid typology.

Differentiating Consumer Characteristics

Except for the demographics, all measures vary significantly across the four groups (F-values range from 5.67 to 36.89). Age is the exceptional demographic variable which varies significantly (F-value = 3.62, $p < 0.05$). Furthermore, while the voicers have the highest EV-Voice (mean = 16.47, $F = 17.07$), the activists are the most alienated from the marketplace (mean = 3.08, $F = 11.44$), evaluate consequences from voice, and particularly private (mean = 18.33, $F = 39.95$), and third-party actions (mean = 13.40, $F = 22.65$) more positively than any other group, and possess the most positive attitude towards complaining (mean = 4.39 [$F = 15.18$] and 4.80 [$F = 5.92$]). In sharp contrast, passives have significantly less positive attitudes toward complaining (mean = 3.55 and 4.25; the lowest across clusters), evaluate the consequences from all forms of complaint responses the least positively, and are also the least alienated from the marketplace (mean = 2.69). The irate profile lies somewhere in between the passives and the activists. In order to obtain more precise evaluation, the discriminant analysis results were examined.

DISCRIMINANT ANALYSIS Results from sequentially entering the predictor variables are summarized in Table 6. In order to examine if the assumptions of multiple discriminant analysis were violated, initially the Box's M test statistic was evaluated to test the null hypotheses of the equality of var-covariance matrices across the four groups. For the full model, the following statistics were obtained: Box's $M = 834$; approx. $F = 1.49$; $df = 513,203123$; $p < 0.01$. However, this test is sensitive to sample size. In this context, note that the corresponding F-value is only 1.49 suggesting that departure from the null hypothesis, if any, is not large.

Overall, the final discrimination model produced the following statistics: Wilks's' lambda = 0.45, chi-square = 352, $df = 54$, $p < 0.01$, $I^2 = 55$ percent, and percentage correctly classified = 59 percent. In other words, the null hypothesis for the equivalence of discriminant functions across the four groups is rejected resoundingly. In addition, the functions explain 55 percent of the variance in the clusters. Also, compared to the expected percentage of correct assignments of 25 percent, the obtained classification accuracy (59 percent) represents over two-fold improvement ($\chi^2 = 282.6$, $df = 9$, $p < 0.001$). These results, therefore, suggest that reasonable discrimination has been achieved among the four clusters.

Also, the episode-specific variables—EV-judgments and the nature of service—contribute the

TABLE 6 Summary of Sequential Discriminant Analysis

Variables Entered[a]	Wilks's λ[b]	$\Delta\lambda$	Variance[c] Explained—I^2	ΔI^2	% Correct Classified	$\Delta\%$
Demographics	0.95	—	4.3%	—	32	
Sex						
Age						
Income						
Education						
Occupation						
Marital status						
Alienation	0.88	0.07	11.6%	7.3%	35	3
Prior experience	0.79	0.09	20%	8.4%	38	3
Voice						
Private						
Third party						
Attitudes	0.75	0.04	25%	5%	41	3
P. norms						
S. benefits						
Service category	0.63	0.12	36%	11%	50	9
Expectancy-value	0.45	0.18	55%	19%	59	9
Voice						
Private						
Third party						

[a] Variables were entered in a sequential manner in the order depicted.
[b] All values are statistically significant at $p = 0.05$.
[c] Based on Peterson and Mahajan (1976). Computed as follows (N is the number of observations, k is the number of groups, and λ_i is the ith eigenvalue):

$$I^2 = 1 - \frac{N}{(N-k)(1+\lambda_1)(1+\lambda_2)(1+\lambda_3)+1}$$

most to the discrimination among the four clusters. The EV-judgments affect a decrease in Wilks's lambda of 0.18 and add 19 percent explained variance to the discrimination function.

Likewise, the nature of service adds 11 percent to explained variance. Taken together, the episode-specific variables explain more than half of the total explained variance. In addition, these variables improve classification by 18 percent. This appears to support Day's contention that the episode-specific variables are more powerful predictors of complaint responses.

As a set, personality variables were the next most significant discriminators. Prior experience and consumer alienation are especially potent characteristics with contributions to explained variance of 8.4 percent and 7.3 percent, respectively. Attitude toward complaining increased I^2 by 5 percent. In terms of classification accuracy, each of the preceding variables contributes 3 percent. Taken together, the personality variables explain over one third of the total explained variance. By contrast, the demographics have modest discriminatory power. The explained variance is only of the order of 4 percent. This suggests that while demographics are not inconsequential, better explanations of response styles will stem from consumers' personalities, past behaviors, and episode-related characteristics.

For the overall model, the discriminant function coefficients and other statistics are displayed in Table 7. All three functions are significant at $p = 0.01$, chi-square values of 352 ($df = 54$), 147 ($df = 34$), and 56 ($df = 16$) with variance ex-

TABLE 7 Standardized Discriminant Function Coefficients[a] and Associated Statistics

Consumer Characteristic	Function 1	Function 2	Function 3
Expectancy-value			
EV-Voice	−0.02	**0.75**	−0.14
EV-Private	**0.57**	0.04	0.03
EV-Third party	0.17	−0.08	**0.84**
Service category			
Dummy 1[b]	**0.73**	−0.28	0.21
Dummy 2[c]	**0.58**	**−0.32**	0.14
Dummy 3[d]	**0.44**	−0.00	**0.36**
Attitudes toward complaining			
Personal norm	0.18	**0.37**	−0.06
Societal benefits	0.06	0.02	0.16
Prior experience			
Voice	0.09	**0.34**	−0.13
Private	0.22	0.14	0.24
Third party	−0.12	−0.10	0.28
Consumer alienation	0.21	0.00	**−0.32**
Demographics			
Age	−0.03	**0.32**	**−0.33**
Eigenvalue	0.59	0.23	0.14
Canonical correlation	0.61	0.43	0.35
Group centroids			
Passives	−1.10	−0.84	−0.15
Voicers	−0.52	0.50	−0.25
Irates	0.70	−0.39	−0.40
Activists	0.73	0.04	0.73

[a]Coefficients greater than 0.30 are in boldface.
[b]This dummy is a contrast between grocery and automotive repair categories.
[c]This dummy is a contrast between grocery and medical care categories.
[d]This dummy is a contrast between grocery and banking categories.

plained values of 62 percent, 24 percent, and 14 percent, respectively. Note, the first function is mainly dominated by EV-private (0.57) and differences due to service categories. In addition, the prior experience of private actions and consumer alienation load 0.22 and 0.21, respectively. Examination of group centroids suggests that this function appears to discriminate between irates/ activists and passives/voicers. This suggests that irates/activists are more alienated and prone to private actions than passives/voicers. As such, the passives' style of no-action *cannot* be wholly attributed to the feelings of alienation and/or powerlessness. Instead, it may stem from their attitudes that complaining is not good or from episode-specific EV jugements, or both factors may play a role. The second function has high loadings for EV-voice (0.75), prior experience of voice actions (0.34), and personal norms dimension of the attitudes measure (0.37). Because the personal norms measure represents predispositions to complain to sellers/providers, the second function appears to reflect different aspects of voicing directly to marketers. Consistent with this, this function appears to differentiate between the voicers and all other groups. The third function has high loadings for EV-third party (0.84), prior experience of third-party actions (0.28), age (−0.33), and consumer alienation (−0.32). The negative loadings for alienation and age signify greater prominence of nonalienated and younger respondents. Group centroids indicate that this function distinguishes activists from all other groups. This suggests that while activists are likely to engage in all forms of complaint responses, the third-party responses are a key marker (i.e., a distinct identifying characteristic) for this segment. Based on the preceding analysis, the thumbnail sketches for the four cluster groups are provided in Table 8.

TABLE 8 Thumbnail Sketches for the Four Cluster Groupings

Cluster 1 (Passives)

likely to be less alienated from the marketplace

tend to have less positive attitude toward complaining due to its social benefits

tend to feel less positive toward complaining because of personal norms

Less positive evaluation of consequences of third-party responses

Less positive evaluation of consequences of private responses

Less positive evaluation of consequences of voice responses

Somewhat likely to be younger

Cluster 2 (Voicers)

likely to be less alienated from the marketplace

tend to have positive attitude toward complaining due to its social benefits

tend to feel more positive toward complaining because of personal norms

Less positive evaluation of consequences of third-party responses

Less positive evaluation of consequences of private responses

Very positive evaluation of consequences of voice responses

Somewhat likely to be older

Cluster 3 (Irates)

likely to be more alienated from the marketplace

tend to have positive attitude toward complaining due to its social benefits

tend to feel more positive toward complaining because of personal norms

Less positive evaluation of consequences of third-party responses

Very positive evaluation of consequences of private responses

Somewhat positive evaluation of voice consequences

Somewhat likely to be older

Cluster 4 (Activists)

likely to be more alienated from the marketplace

tend to have very positive attitude toward complaining due to its social benefits

tend to feel very positive toward complaining because of personal norms

Very positive evaluation of consequences of third-party responses

Very positive evaluation of consequences of private responses

Very positive evaluation of voice consequences

Somewhat likely to be younger

Discussion

Classification occupies a central role in the systematic understanding and prediction of any phenomena (Hunt 1983). The CCB literature lacks a logically developed and empirically supported schema for the classification of dissatisfied consumers. For the retailer interested in dealing with customer dissatisfaction, most prior research has offered a two-group typology, consisting of complainers and noncomplainers (e.g., see Etzel and Silverman 1981). However, such broad classification is less likely to offer specific guidelines to retailers in improving the effectiveness of their complaint handling mechanisms. As such, Etzel and Silverman (1981) observe for the case of the noncomplainer segment, "It is not enough [for retailers] to get noncomplainers to complain more. Research should be done to understand this group more fully. Who are they? How large a group are they? Why are they hesitant to complain? Can they be satisfied the same way as complainers?" Similar gaps can be identified for the complainer segment.

The research reported here has attempted to address some of the preceding gaps in the previous literature. In particular, the research focused on: (a) identifying consumer groupings with unique response styles; (b) validating the obtained typology with behavior data; and (c) profiling the characteristics of the obtained groups. It would be foolhardy to claim that this study has addressed the preceding objectives in a definitive way. Rather, we consider this study as an initial step toward a programmatic and systematic research into the focal questions. Below, we first discuss limitations of our research. Next, we discuss the findings, provide implications for retail managers, and outline areas for future research.

Certain limitations of the study shuld be noted. The results are based on cross-sectional data collected from households in the southwest U.S. Though care was taken to randomly select households, conduct several follow-ups, and analyze data for nonresponse bias, it is not clear how the geographic restriction or the less than perfect response rates may have affected the results. In part, the purposive selection of the service categories and conducting four independent surveys allowed some check on the validity of the results.

As is true for most design decisions, the price for this selection is paid in the coin of uncertainty about the focal processes for other product/service categories. Future research across different product/service categories and geographic areas can address these limitations.

In addition, the research relies on respondents' recall of a past dissatisfying episode, and seeks to capture their intentions to complain in future episodes. Although this approach is shown to be internally consistent (i.e., due to cross-validation) and externally valid (i.e., because of validation by behavior data), problems in such recall-based designs are acknowledged. The projective nature of some of the measures also argues for future replication and validation of the proposed typology. Finally, the fact that the results could be contaminated by method variance is a legitimate concern. However, it should be noted that the external validation analysis reveals patterns of actual behaviors that are not uniformly high in the four cluster groups. Instead, these patterns vary systematically in a fashion consistent with CCB intentions. This appears to suggest that the method variation, while likely to be present, does not play a dominant role in the data.

This paper extends previous research on the typology of dissatisfaction response styles and lends additional empirical evidence to support the existence of such styles. In sharp contrast to the two broad styles reported in more previous research (i.e., to take no action and to take some action), the results reported here suggest the presence of four consumer clusters with distinct response styles; namely the passives, voicers, irates, and activists. Although the notion of a passive style is articulated in most studies done to date, our findings suggest the presence of three additional styles not revealed in previous research. Several points argue in favor of our findings. First, unlike many previous studies, we allow for multiple responses to dissatisfaction. In addition, these responses reproduce the three dimensions of complaint responses—voice, private, and third party—which have been shown to represent the domain of all complaint responses (Singh 1988). Second, we utilized a methodological approach (see Figure 1) that has been successfully employed in other areas of marketing research (e.g., Furse, Punj, and Stewart 1984) and has been suggested as the recommended method when the aim is to determine a parsimonious structure for similarities among people (Owens and Schoenfeldt 1979). Previous research has not tackled the focal issues with such an approach. Third, we subjected the proposed typology to internal and external validity checks. The internal validity focused on stability and reproducibility of response styles by utilizing a hold-out sample. The four sub-types in the proposed typology evidenced a high degree of internal validity (*Kappa* = 0.89). The external validity check examined the consistency between the identified response styles and actual behaviors of consumers in the four groups. Our data reveals a high degree of consistency (cf. Table 5). As such, these groupings cannot be easily attributed to random or idiosyncratic factors. Rather, they stem from systematic differences among consumers. Thus, we seem to have enough evidence at this point to indicate that the proposed four-group typology represents distinct and valid response styles, and is deserving of greater research attention.

This presence of distinct response styles argues for an alternate perspective on the understanding of the CCB pehnomena. Most previous studies attempt to build CCB models that are based largely on the frameworks of consumer decision-making (Landon 1977; Day 1984). In particular, these models tend to treat the dependent construct as a binary (yes/no) choice, such as, to complain or to not complain (cf. Day 1984). There are situations where prediction of such binary choice is appropriate (e.g., for managerial guidelines). However, our results appear to suggest that because consumers typically engage in multiple complaint responses, they in fact do not make a simple binary choice (e.g., to complain/not complain). Rather, it seems clear that complex choices are made, such as to engage in a sub-set of complaint behaviors (e.g., voice and private). We refer to these sub-sets as response styles. We believe that the explanation and prediction of consumers' response styles is a fruitful avenue of future research that heretofore has remained unexplored.

As an initial step in this direction of research, this study examined the differentiating characteristics of consumers who use different response styles. This is the second key contribution of this paper. The results indicate that distinct profiles emerge for the four consumer groups (cf. Table 8). Of the variables examined, the episode-specific variables (i.e., EV jugments and the nature of dissatisfying service) and personality variables (i.e., prior experiences, alienation, and attitudes toward complaining) were potent discriminators and contribute about 54 percent and 37 percent, respectively, to the total explained variance. This suggests that although episode-specific variables influence a consumer's choice of a response style, they are not sufficient. Instead, it appears that consumers' decisions stem from not only what they see in a dissatisfying situation (i.e., EV judgments) but also what they bring to the situation by way of enduring predispositions (i.e., toward the marketplace and complaining) and previous experiences. This supports the notion of enduring response styles; that is, a specific consumer may consistently use similar style in different dissatisfaction contexts (over time), at least within a particular product/service category. Consider for instance the voicer segment. Results from discriminant analysis appear to suggest that voicers are distinguishable from all other groups due in part to distinct personality characteristics. In particular, voicers have more positive attitudes toward complaining (i.e., personal norm dimension), have had greater prior experience of voice actions, and are not alienated from the marketplace. Likewise, other personality variables are implicated in isolating unique characteristics of passives (e.g., alienation), irates (e.g., prior private actions), and activists (e.g., prior third-party actions). However, although the notion of enduring response styles appears plausible, more research is needed to address this issue.

This notion of enduring response styles raises interesting implications for retail managers, and offers new avenues for improving the effectiveness of complaint handling. Thus far the prescriptions for retial managers have focused on increasing the frequency of voice actions (Etzel and Silverman 1981; Day et al. 1981). Such prescriptions have often ignored other CCB actions (e.g., negative word-of-mouth). The proposed typology offers a precise approach to segment-dissatisfied customers. Of the four distinct response styles identified only one—voicers—is the most favorable from a retailer's perspective. This is so because dissatisfied voicers neither engage in negative word-of-mouth, nor do they switch patronage. Instead, they only complain directly to the retailer. Retailers can utilize the results of this study in several ways. First, they could assess the distribution of the four styles among their customers. Lower percentage of voicers is a cause for alarm; just as is a higher percentage of irates and activists. In response to Etzel and Silverman's observation that retailers need to know how large the different groups are, the cluster sizes reported here (i.e., see Table 4) provide an initial benchmark for comparison. It may be useful in future research to develop such benchmarks within each category of retail establishments (e.g., department stores, grocery outlets), so that individual retailers can assess how well they are doing compared to their competitors.

Second, retail managers could institute programs so as to make the voicer style a more attractive option, as compared to other styles. Directions for such programs would stem from analyzing how voicers differ from other groups in terms of the various consumer characteristics (e.g., as shown in this study). In particular, this study reveals that voicers are likely to have positive attitudes toward complaining and not be alienated from the marketplace. Retailers can foster such attitudes and disposition in their customer base by acting individually as well as collectively. For instance, individual retailers can attempt to engender positive attitudes toward complaining by rewarding customers whose complaints result in identifying strategic and/or implementation gaps in a retailer's marketing plan (e.g., see *Business Week* 1984). More importantly, retailers need to seriously examine, and thereby address (if present), customer alienation. Such a disposition not only undermines the retailers' attempt to build loyalty, but also appears to be the key factor that differentiates between irates/activists and voicers. As such, retailers may find it useful to focus on customer alienation as a critical component of the complaint handling mechanism, and monitor such levels in their customer base at periodic intervals. Corrective strat-

egies should be set in motion as soon as alienation levels exceed preset acceptance limits.

Furthermore, it is apparent from this study that the episodic variables are the major factors in consumers' dissatisfaction response styles. This heightens the importance of customer interactions (e.g., with customer-service personnel) that occur in retailers' establishments. As such, the understanding and control of the customers' encounters during a retail visit appear to offer the most efficient and effective strategy for converting customer dissatisfaction into retailers' advantage (e.g., by increasing voicers). This implies that retailers must learn to manage the hundreds and thousands of customer encounters that occur everyday in their premises. Although this appears formidable, the emerging literature on service encounters (e.g., Solomon et al. 1985) is likely to offer promising guidelines for such management.

Surprisingly, the demographic characteristics in this study did not support previous results. In particular, three variables that have often been found to discriminate between complainers and noncomplainers—income, education, and occupation—failed to significantly distinguish among the four clusters obtained in this study. Instead, only age varied significantly across the four groups. This appears to support Grønhaug and Zaltman's contention that demographics hold little, if any, explanatory power. These findings call into question the usefulness of attempts to profile complainers or activists on the basis of demographic characteristics.

Nevertheless, using selected consumer characteristics, we were able to explain 55 percent of the total variance. This explanation level is significantly higher than most previous studies in the CCB literature. Future researchers may wish to further explore the robustness of the proposed typology. More work is also needed to refine measurements of some constructs (e.g., value). Furthermore, future research using other predictor variables (e.g., attributions of blame) and across different product/service contexts should be especially rewarding.

APPENDIX I Operational Measures[a] and Factor Coefficients[b] for the Constructs of Study

Coefficient	Item/Construct

Complaint intentions—Voice

How likely is it that you would:

0.67   forget the incident and do nothing? (reverse scored).

0.67   definitely complain to the store manager on your next trip?

0.74   go back or call the repair shop immediately and ask them to take care of the problem?

Complaint intentions—Private

How likely is it that you would:

0.59   decide not to use the repair shop again?

0.60   speak to your friends and relatives about your bad experience?

0.85   convince your friends and relatives not to use that repair shop?

Complaint intentions—Third party

How likely is it that you would:

0.76   complain to a consumer agency and ask them to make the repair shop take care of your problem?

0.63   write a letter to the local newspaper about your bad experience?

0.82   report to a consumer agency so that they can warn other consumers?

0.61   take some legal action against the repair shop/manufacturer?

Consumer alienation

Strongly agree—Strongly disagree statements:

0.72   Most companies care nothing at all about the consumer.

0.58   Shopping is usually an unpleasant experience.

0.49   Consumers are unable to determine what products will be sold in the stores.

0.62   In general, companies are plain dishonest in their dealings with the consumer.

0.61   Business firms stand behind their products and guarantees (reverse scored).

0.62   The consumer is usually the least important consideration to most companies.

0.65   As soon as they make a sale, most businesses forget about the buyer.

Attitudes toward complaining—Personal norms

Strongly agree—Strongly disagree statements:

0.50   It bothers me quite a bit if I do not complain about an unsatisfactory product.

0.51   It sometimes feels good to get my dissatisfaction and frustration with the product off my chest by complaining.

0.57   I often complain when I'm dissatisfied with business or products because I feel it is my duty to do so.

0.42   People are bound to end up with unsatisfactory products once in a while, so they should not complain (reverse scored).

0.51   I don't like people who complain to stores, because usually their complaints are unreasonable (reverse scored).

Attitudes toward complaining—Social benefits

0.70   By making complaints about unsatisfactory products, in the long run the quality of products will improve.

0.52   By complaining about defective products. I may prevent other consumers from experiencing the same problem.

0.59   People have a responsibility to tell stores when a product they purchase is defective.

APPENDIX I *Continued*

Coefficient	Item/Construct

Expectancy—Voice[c]

Assume you reported the incident to the repair shop; how likely is it that the repair shop would:

0.75	take appropriate action to take care of your problem (refund, etc.)?
0.96	solve your problem and give better service to you in the future?
0.83	be more careful in future and everyone would benefit:

Expectancy—Private[c]

Assume you mentioned the problem to your friends and relatives who use the *same repair shop;* how likely is it that they would:

0.51	be more careful when using that repair shop?
0.71	stop using that repair shop altogether?
0.38	help you solve your problem?

Expectancy—Third party[c]

Assume that you reported the incident to a consumer agency such as the Better Business Bureau; how likely is it that they would:

0.35	take no action? (reverse scored).
0.91	make the repair shop take care of your problem?
0.84	solve your problem and ensure that the repair shop is careful in future?

Value—Voice[c]

How likely is it that you would report the incident to the repair shop if you were pretty sure that the repair shop would:

0.75	take appropriate action to take care of your problem (refund, etc.)?
0.96	solve your problem and give better service to you in the future?
0.83	be more careful in future and everyone would benefit?

Value—Private[c]

How likely is it that you would mention the incident to your friends and relatives if you were pretty sure that they would:

0.51	be more careful when using that repair shop?
0.71	stop using that repair shop altogether?
0.38	help you solve your problem?

Value—Third party[c]

How likely is it that you would report the incident to a consumer agency, such as the Better Business Bureau, if you were pretty sure that they would:

0.35	take no action? (reverse scored).
0.91	make the repair shop take care of your problem?
0.84	solve your problem and ensure that the repair shop is careful in future?

[a] All items are presented for the automotive repair category. Items were modified slightly to be relevant to other service categories.

[b] Because all cross-loadings were small in comparison, only the highest loadings are shown. Full factor analysis results are available from the authors.

[c] The Expectancy and Value items were combined multiplicatively before factor analyzing. For this reason, though the Expectancy and Value items are shown separately, the loadings pertain to the corresponding EV judgment.

APPENDIX II Rotated Factor Pattern for the Ten CCB Intentions Items from Auto Repair Data

Item[a]	Factor 1 (Third Party)	Factor 2 (Private)	Factor 3 (Voice)
CCB1	−.04	.13	.519
CCB2	.17	.05	.708
CCB3	.22	.654	.22
CCB4	.18	.082	.721
CCB5	.17	.743	.02
CCB6	.25	.762	.13
CCB7	.673	.29	.24
CCB8	.559	.14	−.08
CCB9	.878	.14	.22
CCB10	.574	.18	.12
Eigenvalue	1.77	1.74	1.48

[a]The items were as follows: How likely is it that you would: (1) forget about the incident and do nothing? (reverse scored); (2) definitely complain to the store manager on your next trip? (3) decide not to use the repair shop again? (4) go back or call the repair shop immediately and ask them to take care of your problem? (5) speak to your friends and relatives about your bad experience? (6) convince your friends and relatives not to use that repair shop? (7) complain to a consumer agency and ask them to make the repair shop take care of your problem? (8) write a letter to the local newspaper about your bad experience? (9) report to the consumer agency so that they can warn other consumers? (10) take some legal action against the repair shop/manufacturer?

References

Allison, N. (1978), "A Psychometric Development of a Test for Consumer Alienation from the Marketplace," *Journal of Marketing Research*, **15** (November), 565–575.

Andreasen, A. (1984), "Consumer Satisfaction in Loose Monopolies: The Case of Medical Care," *Journal of Public Policy and Marketing*, **2**, 122–135.

Bagozzi, R. P. (1982), "A Field Investigation of Causal Relations Among Cognitions, Affect, Intentions and Behavior," *Journal of Marketing Research*, **24** (November), 562–584.

Bailey, K. D. (1974), "Cluster Analysis" in David R. Heise (ed.), *Sociological Methodology*, San Francisxo, CA: Jossey-Bass.

Barnes, J., and K. Kelloway (1980), "Consumerists: Complaining Behavior and Attitudes Toward Social and Consumer Issues," in Kent B. Monroe (ed.), *Advances in Consumer Research*, Vol. 8, 329–334.

Bearden, W. (1983), "Profiling Consumers Who Register Complaints Against Auto Repair Services," *Journal of Consumer Affairs*, **17** (Winter), 315–335.

Bearden, W., and J. Mason, eds. (1984), "An Investigation of Influences on Consumer Complaint Reports," in T. Kinnear (ed.), *Advances in Consumer Research*, Vol. 11, 490–495.

Bearden, W., M. Crockett, and J. Teel (1980), "A Past Model of Consumer Complaint Behavior," in *Marketing in the 80s: Changes and Challenges*, Proceedings of the American Marketing Association, pp 101–104.

Best, A., and A. Andreasen (1977), "Consumer Responses to Unsatisfactory Purchases: A Survey of Perceiving Defects, Voicing Complaints and Obtaining Redress," *Law and Society Review*, **11** (Spring), 701–742.

Business Week (1984), "Making Service a Potent Marketing Tool," (June 11), 164–170.

Day, R. (1984), "Modeling Choices Among Alternative Responses to Dissatisfaction," in T. Tinnear (ed.), *Advances in Consumer Research*, Vol. 11, 496–499.

——— (1980), "Research Perspectives on Consumer Complaining Behavior," in Lamb and Dunne (eds.), *Theoretical Developments in Marketing*, Chicago, IL: American Marketing Association, 211–215.

Day, R., K. Gabricke, T. Schaetzle, and F. Staubach (1981), "The Hidden Agenda of Consumer Complaining," *Journal of Retailing*, **57** (Fall), 86–106.

Day, R., and M. Bodur (1977), "A Comprehensive Study of Satisfaction with Consumer Services," in R. Day (ed.), *Consumer Satisfaction, Dissatisfaction and Complaining Behavior*, Bloomington, IN: Indiana University Press, 64–70.

Day, R., and E. Landon, Jr., (1977), "Toward a Theory of Consumer Complaining Behavior," in Woodside, Sheth, and Bennett (eds.), *Consumer and Industrial Buying Behavior*, Amsterdam: North Holland Publishing Co., 425–437.

Diener, B., and S. Greyser (1978), "Consumer Views of Redress Needs," *Journal of Marketing*, (October), 21–27.

Etzel, M., and B. Silverman (1981), "A Managerial Perspective on Directions for Retail Customer Dissatisfaction Research," *Journal of Retailing*, **57** (Fall), 124–136.

Fishbein, M., and I. Ajzen (1975) *Belief, Attitude, Intention, and Behavior: An Introduction to Theory and Research*, Reading, MA: Addison-Wesley.

Fleiss, J., and J. Cohen (1973), "The Equivalence of Weighted Kappa and the Intraclass Correlation Coefficients as Measures of Reliability," *Educational and Psychological Measurements*, **33**, 613–619.

Fornell, C., and B. Wernerfelt (1987), "Defensive Marketing Strategy by Customer Complaint Management," *Journal of Marketing Research*, **24** (November), 337–346.

Fornell, C., and D. Larcker (1981), "Evaluating Structural Equations Models with Unobservable Variables and Measurement Error," *Journal of Marketing Research*, **18** (February), 39–50.

Furse, D., G. Punj, and D. Stewart (1984), "A typology of Individual Search Strategies Among Purchasers of New Automobiles," *Journal of Consumer Research*, **10** (March), 417–431.

Grikscheit, G., and K. Granzin (1975), "Who are the Consumerists?," *Journal of Business Research*, **3** (January), 1–12.

Grønhaug, K., and G. Zaltman (1981), "Complainers and Non-complainers Revisited: Another Look at the Data," in Kent Monroe (ed.), *Advances in Consumer Research*, Vol. 8, 83–87.

Howard, J. A. (1977), *Consumer Behavior: Application of Theory*, New York, NY: McGraw-Hill.

Hunt, Shelby (1983), *Marketing Theory: The Philosophy of Marketing Science*, Homewood, IL: Irwin Publications.

LaBarbera, P., and D. Mazursky (1983), "A Longitudinal Assessment of Consumer Satisfaction/Dissatisfaction: The Dynamic Aspect of the Cognitive Process," *Journal of Marketing Research*, **20** (November), 393–404.

Landon, E. L., Jr. (1980), "The Direction of Consumer Complaint Research," in Jerry Olson (ed.), *Advances in Consumer Research*, Vol. 7, 335–338.

——— (1977). "A Model of Consumer Complaint Behavior," in R. Day (ed.), *Consumer Satisfaction, Dissatisfaction, and Complaining Behavior*, Bloomington, IN: Indiana University Press, 31–35.

Lorr, M. (1983), *Cluster Analysis for Social Scientists*, San Francisco, CA: Jossey-Bass Publishers.

Mason, J. B., and S. H. Himes, Jr. (1973), "An Exploratory Behavioral and Socio-economic Profile of Consumer Action About Dissatisfaction with Selected Household Appliances," *Journal of Consumer Affairs*, **7** (Winter), 121–127.

Morganosky, M., and M. Buckley (1986), "Complaint Behavior: Analysis by Demographics, Lifestyle and Consumer Values," in Wallendorf and Anderson (eds.), *Advances in Consumer Research*, Vol. 14, 223–226.

Moyer, M. (1984), "Characteristics of Consumer Complainants: Implications for Marketing and Public Policy," *Journal of Public Policy and Marketing*, Vol. 3, 67–84.

Owens, W., and L. Schoenfeldt (1979), "Toward a Classification of Persons," *Jouranl of Applied Psychology*, **65** (October) 569–607.

Peterson, R., and V. Mahajan (1976), "Practical Significance and Partitioning Variance in Discriminant Analysis," *Decision Sciences*, **7**, 649–658.

Pfaff, M., and S. Blivice (1977), "Socioeconomic Correlates of Consumer and Citizen Dissatisfaction and Activism," in R. Day (ed.), *Consumer Satisfaction, Dissatisfaction and Complaining Behavior*, Bloomington, IN: Indiana University Press, 115–123.

Punj, G., and D. Stewart (1983), "Cluster Analysis in Marketing Research: Review and Suggestions for Application," *Journal of Marketing Research*, **20** (May), 134–148.

Richins, M. (1983a), "Negative Word-of-Mouth by Dissatisfied Consumers: A Pilot Study," *Journal of Marketing*, **47** (Winter), 68–78.

——— (1983b), "An Analysis of Consumers Interaction Styles in the Marketplace," *Journal of Consumer Research*, **10** (June), 73–82.

——— (1982), "An Investigation of Consumers Attitudes Towards Complaining,'' in Andrew Mitchell (ed.), *Advances in Consumer Research*, Vol. 9, 502–506.

Robinson, L. (1979), ''Consumer Complaint Behavior: A Review with Implications for Future Research,'' in Hunt and Day (eds.), *New Dimensions of Consumer Satisfaction and Complaining Behavior*, Vol. 3, 41–50.

Scammon, D., and L. Kennard (1983), ''Improving Health Care Strategy Planning Through the Assessment of Perceptions of Consumers, Providers and Administrators,'' *Journal of Health Care Marketing*, **3** (Fall), 9–17.

Shuptrine, K., and G. Wenglorz (1980), ''Comprehensive Identification of Consumers' Marketplace Problems and What They Do About Them,'' in Kent B. Monroe, *Advances in Consumer Research*, Vol. 8, 687–692.

Singh, J. (1988), ''Consumer Complaint Intentions and Behaviors: Definitional and Taxonomical Issues,'' *Journal of Marketing*, (January), 93–107.

Solomon, M., C. Surprenant, J. Czepiel, and E. Gutman (1985), ''A Role Theory Perspective on Dyadic Interactions: The Service Encounter,'' *Journal of Marketing*, **49** (Winter), 99–111.

Strahle, W., and R. Day (1985), ''Sex Roles, Lifestyles, Store Types and Compalining Behaviors,'' in K. Hunt and R. Day (eds.), *Consumer Satisfaction, Dissatisfaction and Complaining Behaviors*, Bloomington, IN: Indiana University Press, 59–66.

TARP (1986), *Consumer Complaint Handling in America: An Update Study*, Washington, D.C.: White House Office of Consumer Affairs.

Wall, M., L. Kickey, and W. Talarzyk (1977), ''Predicting and Profiling Consumer Satisfaction and Propensity to Complain,'' in R. Day (ed.), *Consumer Satisfaction, Dissatisfaction and Complaining Behavior*, Bloomington, IN: Indiana University Press, 91–101.

Warland, Rex (1977), ''A Typology of Consumer Complaints,'' in R. Day (ed.), *Consumer Satisfaction, Dissatisfaction and Complaining Behavior*, Bloomington, IN: Indiana University Press, 144–146.

Warland, R., R. Herrmann, and J. Willits (1975), ''Dissatisfied Consumers: Who Gets Upset and Who Takes Action,'' *Journal of Consumer Affairs*, (Winter), 148–163.

Zaichowsky, J., and J. Liefeld (1977), ''Personality Profiles of Consumer Complaint Writers,'' in R. Day (ed.), *Consumer Satisfaction, Dissatisfaction and Complaining Behavior*, Bloomington, IN: Indiana University Press, 124–129.

CHAPTER 9

Multidimensional Scaling

LEARNING OBJECTIVES

Upon completing this chapter, you should be able to do the following:

- Appreciate how spatial representation of data can clarify underlying relationships.

- Determine the number of dimensions represented in the data.

- Interpret the spatial maps so the dimensions can be understood.

- Understand the differences between similarities data and preference data as used in multidimensional scaling.

- Distinguish between the various approaches to applying multidimensional scaling.

CHAPTER PREVIEW

CHAPTER PREVIEW

Multidimensional scaling (MDS) is a series of techniques that helps the analyst to identify key dimensions underlying respondents' evaluations of objects. For example, multidimensional scaling is often used in marketing to identify key dimensions underlying customer evaluations of products, services, or companies. Other common applications include the comparison of physical qualities (e.g., food tastes or various smells), perceptions of political candidates or issues, and even the assessment of cultural differences between distinct groups. Multidimensional scaling techniques can infer the underlying dimensions from a series of similarity or preference judgments provided by respondents about objects. Once the data are in hand, multidimensional scaling can help determine (1) what dimensions respondents use when evaluating objects, (2) how many dimensions they may use in a particular situation, (3) the relative importance of each dimension, and (4) how the objects are related perceptually.

KEY TERMS

Before starting the chapter, review the key terms to develop an understanding of the concepts and terminology used. Throughout the chapter the key terms will appear in **boldface.** Other points of emphasis in the chapter will be *italicized.* Also, cross-references within the key terms will be in *italics.*

Aggregate analysis   Approach to MDS in which a *perceptual map* is generated for a group of respondents' evaluations of stimuli. This composite perceptual map may be created by the analyst in finding a few "average" or representative subjects or by the computer program.

Compositional method   Alternative approach to the more traditional *decompositional methods* of perceptual mapping that derive overall similarity or preference evaluations from evaluations of separate attributes by each respondent. These separate attribute evaluations are combined (composed) for an overall evaluation. The most common examples of compositional methods are the techniques of factor analysis and discriminant analysis.

Confusion data   Procedure to obtain respondents' perceptions of similarities data. Respondents indicate the similarities between pairs of stimuli. The pairing (or "confusing") of one stimulus with another is taken to indicate similarity. Also known as subjective clustering.

Contingency table   Crosstabulation of two nonmetric or categorical variables where the entries are the frequencies of responses that fall into each "cell" of the matrix. For example, if three brands are rated on four attributes, the brand-by-attribute contingency table would be a three-row by four-column table. The entries would be the number of times a brand was rated as having an attribute.

Correspondence analysis   *Compositional approach* to perceptual mapping that relates categories of a *contingency table.* Most applications involve a set of objects and attributes, with the results portraying both objects and attributes in a common perceptual map.

Crosstabulation table   See *contingency table.*

Decompositional method   Perceptual mapping methods associated with MDS techniques. In these techniques, the respondent provides only an overall evaluation of similarity or preference between objects. This set of overall evaluations is then "decomposed" into a set of "dimensions" that best represent the objects' differences.

Degenerate solution   MDS solution that is invalid because of (1) inconsistencies in the data or (2) too few objects compared with the dimensionality of the solution. While the computer program may indicate a valid solution, the analyst should disregard the degenerate solution and examine the data for the cause.

Derived measures   Procedure to obtain respondents' perceptions of similarities data. Derived similarities are typically based on "scores" given to stimuli by respondents, which are then combined in some manner. The semantic differential scale is frequently used to elicit such "scores."

Dimensions   Features of an *object*. A particular object can be thought of as possessing both perceived/subjective dimensions (i.e., expensive, fragile, etc.) and objective dimensions (i.e., color, price, features).

Disaggregate analysis   Approach to MDS in which the researcher generates perceptual maps on a respondent-by-respondent basis. The results may be difficult to generalize across respondents. Therefore, the analyst may attempt to create fewer maps by some process of *aggregate analysis,* where the results of respondents are combined.

Disparities   Differences in the computer-generated distances representing similarity and the distances provided by the respondent.

Ideal point   Point on a perceptual map that represents the most preferred combination of perceived attributes (according to the respondents). A major assumption is that the position of the ideal point (relative to the other objects on the perceptual map) would define relative preference such that objects further from the ideal point should be preferred less.

Importance/performance grid   Two-dimensional approach for assisting the analyst in labeling dimensions. The respondents' perceptions of the importance (e.g., as measured on a scale of "extremely important" to "not at all important") and performance (e.g., as measured on a scale of "highly likely to perform" to "highly unlikely to perform") for each brand or product/service on various attributes are plotted with "importance" on one axis and "performance" on the opposing axis.

Index of fit   Squared correlation index (R^2) that may be interpreted as indicating the proportion of variance of the *disparities* (optimally scaled data) that can be accounted for by the MDS procedure. It measures how well the raw data fit the MDS model. This index is an alternative to the *stress* measure for determining the number of dimensions. Similar to measures of covariance in other multivariate techniques, measures of .60 or greater are considered acceptable.

Initial dimensionality   Before beginning an MDS procedure, the researcher must specify how many dimensions or features are represented in the data. The MDS procedure then uses this "initial dimensionality" as a starting point in selecting the best spatial configuration for the data.

Multiple correspondence analysis   Form of *correspondence analysis* that involves three or more categorical variables related in a common perceptual space.

Object   Can be any stimulus, including tangible objects (product or physical object), actions (e.g., service), sensory perceptions (smell, taste, sights), or even thoughts (ideas, slogans, etc.), that can be compared and evaluated by the respondent.

Objective dimensions   Physical or tangible characteristics of an object that have an objective basis of comparison. For example, a product has size, shape, color, weight, and so on.

Perceived dimensions   Respondents may "subjectively" attach features to an object that represent its intangible characteristics. Examples include "quality," "expensive," and "good-looking." These perceived dimensions are unique to the individual and may bear little correspondence to the objective dimensions of the object.

Perceptual map   Visual representation of a respondent's perceptions of objects on two or more dimensions. Usually this map has opposite levels of dimensions on the ends of the X and Y axes, such as "sweet" to "sour" on the ends of the X axis and "high-priced" to "low-priced" on the ends of the Y axis. Each object then has a spatial position in the perceptual map reflecting the relative similarity or preference to other objects with regard to the dimensions of the perceptual map.

Preference data   Preference implies that stimuli are judged by the respondent in terms of dominance relationships; that is, the stimuli are ordered in preference with respect to some property. Direct ranking, paired comparisons, and preference scales are frequently used to determine respondent preferences.

Projections   Points defined by perpendicular lines from an object to the vector. Projections are used in determining the preference order with vector representations.

Similarities data   When collecting this type of data, the researcher is trying to determine which of the objects are most similar to each other and which are the most dissimilar. Implicit in similarities measurement is the ability to compare all pairs of objects. Three procedures to obtain similarities data are paired comparison of objects, *confusion data*, and *derived measures*.

Similarity scale   Arbitrary scale, say, from -5 to $+5$ that allows the representation of an ordered relationship between objects from the most similar (closest) to the least similar (farthest apart). This type of scale is appropriate only for representing a single dimension.

Spatial map   See *perceptual map*.

Stress measure   Measure of the proportion of the variance of the *disparities* (optimally scaled data) that is not accounted for by the MDS model. This type of measurement varies according to the type of program and the data being analyzed. The stress measure helps to determine the appropriate number of dimensions to include in the model.

Subjective clustering   See *confusion data*.

Subjective dimension   See *perceived dimension*.

Subjective evaluation   Method of determining how many dimensions are represented in the MDS model. The analyst makes a "subjective inspection" of the spatial maps and asks, Does the configuration look reasonable? The objective is to obtain the best fit with the least number of dimensions.

Unfolding   Representation of an individual respondent's preferences within a common (aggregate) stimulus space derived for all respondents as a whole. The individual's preferences are "unfolded" and portrayed as the best possible representation within the aggregate analysis.

Vector   Method of portraying an ideal point or attribute in a perceptual map. Involves the use of *projections* to determine an object's order on the vector.

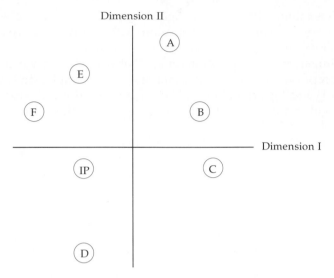

FIGURE 9.1 Illustration of a multidimensional "map" of perceptions of six industrial suppliers.

What Is Multidimensional Scaling?

Multidimensional scaling (MDS), also known as **perceptual mapping,** is a procedure that allows a researcher to determine the perceived relative image of a set of objects (firms, products, ideas, or other items associated with commonly held perceptions). The purpose of MDS is to transform consumer judgments of similarity or preference (e.g., preference for stores or brands) into distances represented in multidimensional space. If objects A and B are judged by respondents to be the most similar compared with all other possible pairs of objects, MDS techniques will position objects A and B so that the distance between them in multidimensional space is smaller than the distance between any other two pairs of objects. The resulting **perceptual maps** show the relative positioning of all objects, as shown in Figure 9.1.

Multidimensional scaling is based on the comparison of **objects.** Any object (product, service, image, aroma, etc.) can be thought of as having both perceived and objective **dimensions.** For example, HATCO's management may see their product (a lawn mower) as having two color options (red versus green), a 2-horsepower motor, and a 24-inch blade. These are the **objective dimensions.** On the other hand, customers may (or may not) see these attributes. Customers may also perceive the HATCO mower as expensive-looking or fragile. These are **perceived dimensions.** Two products may have the same physical characteristics (objective dimensions) but be viewed differently because the different brands are perceived to differ in quality (a perceived dimension) by many customers. Thus, two differences between objective and perceptual dimensions are very important:

1. The dimensions perceived by customers may not coincide with (or may not even include) the objective dimensions assumed by the researcher.
2. The evaluations of the dimensions (even if the perceived dimensions are the same as the objective dimensions) may not be independent and may not agree. For example, one soft drink may be judged sweeter than another because the first has a fruitier aroma, although both contain the same amount of sugar.

The challenge to the researcher is first to understand the perceived dimensions and then to relate them to objective dimensions, if possible. Additional analysis is needed to assess which attributes predict the position of each object in both perceptual and objective space.

A note of caution must be raised, however, concerning the interpretation of dimensions. Because this process is more of an art than a science, the analyst must resist the temptation to allow personal perception to affect the qualitative dimensionality of the perceived dimensions. Given the level of analyst input, caution must be taken to be as objective as possible in this critical, yet still rudimentary, area.

A Simplified Look at How Multidimensional Scaling Works

To make possible a better understanding of the basic procedures in multidimensional scaling, we first present a simple example to illustrate the basic concepts underlying it and the procedure by which MDS transforms similarity judgments into the corresponding spatial positions.

We begin by assuming that we have gathered the most basic type of data from respondents concerning the similarities or dissimilarities among a set of six products (candy bars). The data are typically gathered by having respondents give simple global responses such as these:

- Rate the similarity of products A and B on a 10-point scale.
- Is product A more similar to B than to C?
- I like product A better than product B.

From these simple responses, a picture may be drawn that reveals a pattern. The following example illustrates this process: a set of 15 index cards was prepared, each containing a pair of candy bars. These cards, one of which is shown here, represent all the pairs of the six products:

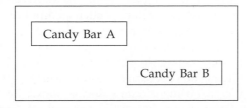

A respondent was then asked to rank the following 15 candy bar pairs, where a rank of 1 is assigned to the pair of candy bars that is most similar and a rank of 15 indicates the pair that is least alike. The results (rank orders) for all pairs of candy bars for one respondent are as follows:

Candy Bar	A	B	C	D	E	F
A	—	2	13	4	3	8
B		—	12	6	5	7
C			—	9	10	11
D				—	1	14
E					—	15
F						—

This respondent thought that candy bars D and E were the most similar and that E and F were the least similar. If one wanted to illustrate similarity, a first attempt would be to draw a single **similarity scale** and fit all the candy bars to it. This represents a one-dimensional portrayal of similarity. For example, placing A, B, and C on an arbitrary scale might yield the following:

$$\begin{array}{c} \qquad\qquad\qquad\qquad \text{A}\quad\text{B}\qquad\qquad\qquad\qquad \text{C} \\ \hline -5 \;\; -4 \;\; -3 \;\; -2 \;\; -1 \;\; 0 \;\; 1 \;\; 2 \;\; 3 \;\; 4 \;\; 5 \;\; 6 \;\; 7 \;\; 8 \;\; 9 \end{array}$$

This scale shows the distances $\overline{AB}$, $\overline{BC}$, and $\overline{AC}$ in an ordered relationship. (Note: the line over each pair of letters indicates that the expression refers to the distance [similarity] between the pair A and B.) From these values, we can create an ordered relationship from most similar ($\overline{AB}$), which are the closest, to least similar ($\overline{AC}$), which are the farthest apart. Given our placement of three products on the scale, $\overline{AB}$ is the closest pair, $\overline{BC}$ is the next closest, and $\overline{AC}$ is the farthest apart. This placement matches exactly the rank order of these three pairs based on the rankings given earlier.

Now let's add a fourth product, candy bar D. The set of ordered relationships for four products, derived from the responses of similarity above, show $\overline{AD} < \overline{BD} < \overline{CD} < \overline{BC} < \overline{AC}$. Now try to place D on the chart.

$$\begin{array}{c} \qquad\qquad\quad \text{D}\qquad\qquad \text{A}\quad\text{B}\qquad\qquad\qquad\qquad \text{C} \\ \hline -5 \;\; -4 \;\; -3 \;\; -2 \;\; -1 \;\; 0 \;\; 1 \;\; 2 \;\; 3 \;\; 4 \;\; 5 \;\; 6 \;\; 7 \;\; 8 \;\; 9 \end{array}$$

It can't be done. In one dimension, if $\overline{AB} < \overline{AD} < \overline{BD}$, then $\overline{CD}$ cannot be less than BC, no matter where you place D.

It is clear that if the person judging the similarity between the candy bars had been thinking of a simple rule of similarity, such as amount of chocolate, all the pairs could be placed on a single arbitrary scale that should reproduce the perceived single quality (e.g., chocolate) used to judge the pairs.

Because one-dimensional scaling does not fit the data well, a two-dimensional solution can be attempted. The procedure is too tedious to attempt by hand, so a computer produced the two-dimensional solution shown in Figure 9.2.

Examining this solution, we see that in the ordered relationships given by the respondent, from most similar to least similar ($\overline{DE}$, $\overline{AB}$, $\overline{AE}$, $\overline{AD}$, $\overline{BE}$, $\overline{BD}$, $\overline{BF}$, $\overline{AF}$,

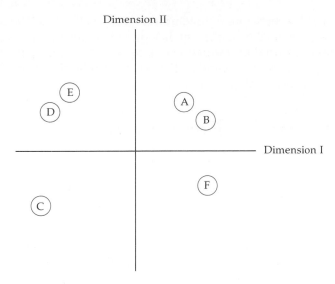

FIGURE 9.2 A two-dimensional perceptual map.

$\overline{CD}$, $\overline{CE}$, $\overline{CF}$, $\overline{BC}$, $\overline{AC}$, $\overline{DF}$, $\overline{EF}$) are all preserved. This result could lead to several conclusions. The most notable is that the respondent was probably thinking of at least two features of candy bars when making this evaluation. The conjecture that at least two features (dimensions) were considered is based on the inability to represent the respondent's perceptions in one dimension. However, we are still not aware of what attributes the respondent used in this evaluation.

Comparing MDS to Other Interdependence Techniques

Multidimensional scaling can be compared to the other interdependence techniques (factor and cluster analysis) based on its approach to defining structure. Factor analysis groups variables into variates that define underlying dimensions in the original set of variables. Variables that highly correlate are grouped together. Cluster analysis groups observations according to their profile on a set of variables (the cluster variate). Observations in close proximity to each other are grouped together. But multidimensional scaling differs in two key aspects:

1. Each respondent provides evaluations of all objects being considered, so that a solution can be obtained for each individual that is not possible in cluster analysis or factor analysis. As such, the focus is not on the objects themselves but instead on how the individual perceives the objects. The structure being defined is the perceptual dimensions of comparison for individual(s). Once the perceptual dimensions are defined, the relative comparisons among objects can also be made.

2. Multidimensional scaling, unlike the other multivariate techniques, does not use a variate. Instead, the "variables" that would make up the variate (i.e., the perceptual dimensions of comparison) are inferred from global measures of similarity among the objects. In a simple analogy, it is like providing the dependent variable (similarity among objects) and figuring out what the inde-

pendent variables (perceptual dimensions) must be. This has the advantage of reducing the influence of the researcher by not requiring the specification of the variables to be used in comparing objects, as was required in cluster analysis. But it also has the disadvantage that the researcher is not really sure what variables the respondent is using to make the comparisons.

A Decision Framework for Perceptual Mapping

Perceptual mapping encompasses a wide range of possible methods, including multidimensional scaling, but all these techniques can be viewed through the model-building process introduced in Chapter 1. These steps represent a decision framework, depicted in Figure 9.3, within which all perceptual mapping techniques can be applied and the results evaluated. The six stages are discussed in the following sections.

Stage One: Objectives of Multidimensional Scaling

Perceptual mapping, and multidimensional scaling in particular, is most appropriate for achieving two objectives:

1. as an exploratory technique to identify unrecognized dimensions affecting behavior
2. as a means of obtaining comparative evaluations of objects when the specific bases of comparison are unknown or undefinable

In multidimensional scaling, it is not necessary to specify the attributes of comparison for the respondent. All that is required is to specify the objects and make sure that the objects share a common basis of comparison. This flexibility makes multidimensional scaling particularly suited to image and positioning studies where the dimensions of evaluation may be too global or emotional/affective to be measured by conventional scales. Some multidimensional scaling methods combine the positioning of objects and subjects in a single overall map. In these techniques, the relative positions of objects and consumers make segmentation analysis much more direct.

A common characteristic of each objective is the lack of specificity in defining the standards of evaluation of the objects. The strength of perceptual mapping is its ability to "infer" dimensions without the need for defined attributes. The flexibility and inferential nature of multidimensional scaling places a greater responsibility on the researcher to "correctly" define the analysis. Conceptual as well as practical considerations are essential for multidimensional scaling to achieve its best results. To ensure this success, the researcher must define a multidimensional scaling analysis through three key decisions: selecting the objects that will be evaluated, deciding whether similarities or preference is to be analyzed, and choosing whether the analysis will be performed at the group or individual level.

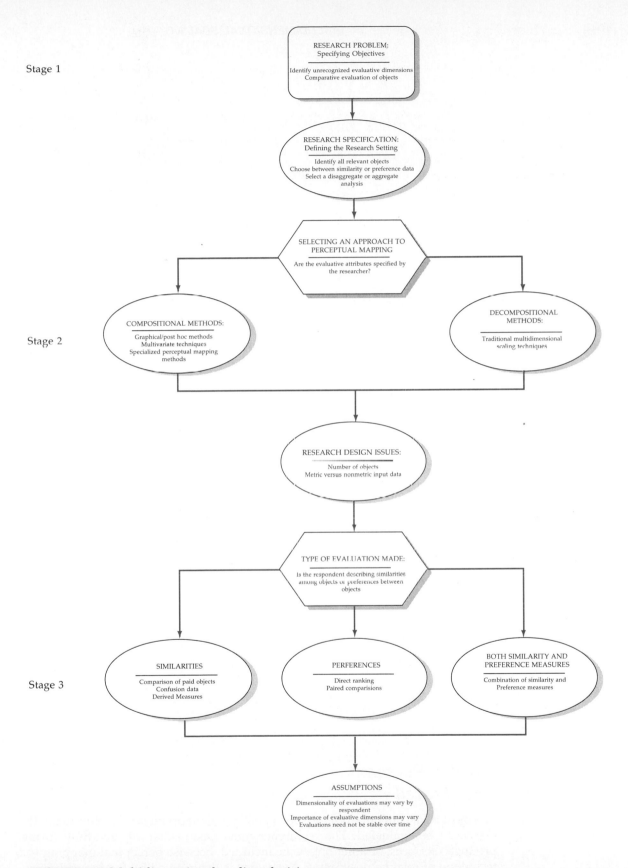

Stage 1

RESEARCH PROBLEM:
Specifying Objectives

Identify unrecognized evaluative dimensions
Comparative evaluation of objects

RESEARCH SPECIFICATION:
Defining the Research Setting

Identify all relevant objects
Choose between similarity or preference data
Select a disaggregate or aggregate
analysis

SELECTING AN APPROACH TO
PERCEPTUAL MAPPING

Are the evaluative attributes specified by
the researcher?

Stage 2

COMPOSITIONAL METHODS:

Graphical/post hoc methods
Multivariate techniques
Specialized perceptual mapping
methods

DECOMPOSITIONAL
METHODS:

Traditional multidimensional
scaling techniques

RESEARCH DESIGN ISSUES:

Number of objects
Metric versus nonmetric input data

TYPE OF EVALUATION MADE:

Is the respondent describing similarities
among objects or preferences between
objects

Stage 3

SIMILARITIES

Comparison of paid objects
Confusion data
Derived Measures

PERFERENCES

Direct ranking
Paired comparisions

BOTH SIMILARITY AND
PREFERENCE MEASURES

Combination of similarity and
Preference measures

ASSUMPTIONS

Dimensionality of evaluations may vary by
respondent
Importance of evaluative dimensions may vary
Evaluations need not be stable over time

FIGURE 9.3 Multidimensional scaling decision process.

493

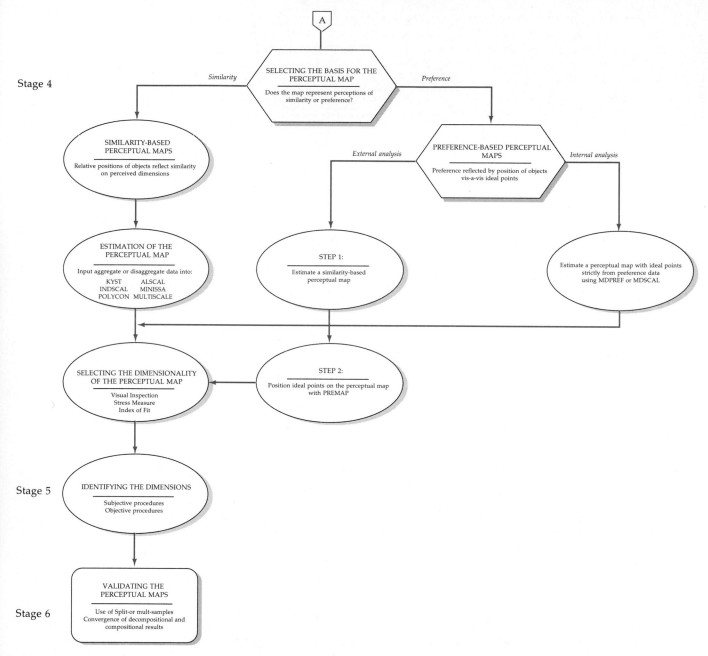

FIGURE 9.3 *Continued*

Identification of All Relevant Objects to Be Evaluated

The most basic, but important, issue in perceptual mapping is the definition of the objects to be evaluated. The researcher must ensure that all "relevant" firms, products/services, or other objects be included, because perceptual mapping is a technique of relative positioning. Relevancy is determined by the research questions to be addressed. The perceptual maps resulting from any of the methods can be greatly influenced by either the omission of objects or the inclusion of inappropriate ones [8, 21]. If "irrelevant" or noncomparable objects are included, the researcher is forcing the technique not only to infer the perceptual dimensions that distinguish among comparable objects but also to infer those dimensions that distinguish among noncomparable objects as well. This task is beyond the scope of multidimensional scaling and results in a solution that addresses neither question well.

Similarities Versus Preference Data

Having selected the objects for study, the researcher must next select the basis of evaluation: similarity versus preference. To this point, we have discussed perceptual mapping and MDS mainly in terms of similarity judgments. In providing **similarities data,** the respondent does not apply any "good-bad" aspects of evaluation in the comparison. The assessment "good-bad' is done, however, within **preference data,** where we assume that differing combinations of perceived attributes are valued more highly than others. Both bases of comparison can be used to develop perceptual maps, but with differing interpretations. Similarity-based perceptual maps represent attribute similarities and perceptual dimensions of comparison but do not reflect any direct insight into the determinants of choice. Preference-based perceptual maps do reflect preferred choices but may not correspond in any way to the similarity-based positions, because respondents may base their choices on entirely different dimensions or criteria from those on which they base comparisons. There is no optimal base for evaluation, but the decision between similarities and preference data must be made with the ultimate research question in mind, because they are fundamentally different in what they represent.

Aggregate Versus Disaggregate Analysis

In considering similarities or preference data, we are taking respondent's perceptions of stimuli and creating outputs of representations of stimulus proximity in t-dimensional space (where the number of dimensions t is less than the number of stimuli). The researcher can generate this output on a subject-by-subject basis (producing as many maps as subjects), known as a **disaggregate analysis.** One of the distinctive characteristics of multidimensional scaling techniques is their ability to estimate solutions for each respondent, which can then be represented separately for each respondent. The advantage is to represent the unique elements of each respondent's perceptions. The disadvantage is that the researcher must identify the common elements across respondents.

Multidimensional scaling techniques can also combine respondents and create fewer perceptual maps by some process of **aggregate analysis.** The aggregation may take place either before or after scaling the subjects' data. Before scaling, the simplest approach is for the researcher to find the "average" evaluations for all respondents and obtain a single solution for the group of respondents as a whole. To identify groups of similar respondents, the researcher may cluster analyze the subjects' responses to find a few average or representative subjects and then develop maps for the cluster's "average respondent." The researcher may also develop maps for each individual and cluster the maps according to the coordinates of the stimuli on the maps. It is recommended that "average" evaluations be used rather than clustering the individual perceptual maps because minor rotations of essentially the same map can cause problems in creating reasonable clusters by the second approach.

A specialized form of **disaggregate analysis** is available with INDSCAL (INdividual Differences SCALing) [5] and its variants, which have characteristics of both disaggregate and aggregate analyses. INDSCAL assumes that all individuals share a common or group space (an aggregate solution) but that the respondents individually weight the dimensions, including zero weights when totally ignoring a dimension. As a first step, INDSCAL derives the perceptual space shared by all individuals, just as do other aggregate solutions. However, individuals are also portrayed in a special group space map. Here the respondents' position is determined by their weights for each dimension. Respondents positioned closely together employ similar combinations of the dimensions from the common group space. Moreover, the distance of the individual from the origin is an approximate measure of the proportion of variance for that subject accounted for by the solution. Thus a position farther from the origin indicates better fit. Being at the origin means "no fit" because all weights are zero. If two or more subjects or groups of subjects are at the origin, separate group spaces need to be configured for each of them. In this analysis, the researcher is presented with not only an overall representation of the perceptual map but also the degree to which each respondent is represented by the overall perceptual map. These results for each respondent can then be used to group respondents and even identify different perceptual maps in subsequent analyses.

The choice of aggregate or disaggregate analysis is based on the study objectives. If the focus is on an understanding of the overall evaluations of objects and the dimensions employed in those evaluations, an aggregate analysis is the most suitable. But if the objective is to understand variation among individuals, then a disaggregate approach is the most helpful.

Stage Two: Research Design of Multidimensional Scaling

While MDS looks quite simple computationally, the results, as with other multivariate techniques, are heavily influenced by a number of key issues that must be resolved before the research can proceed. We cover four of the major issues, ranging from discussions of research design (selecting the approach and objects or stimuli for study) to specific methodological concerns (metric versus nonmetric methods and data collection methods).

Selection of Either a Decompositional (Attribute-free) or Compositional (Attribute-based) Approach

Perceptual mapping techniques can be classified by the nature of the responses obtained from the individual concerning the object. One type, the **decompositional method,** measures only the overall impression or evaluation of an object and then attempts to derive spatial positions in multidimensional space reflecting these perceptions. These techniques are typically associated with multidimensional scaling. The **compositional method** is an alternative method that employs several of the multivariate techniques we have already discussed that are used in forming an impression or evaluation based on a combination of specific attributes. Each approach has advantages and disadvantages that are addressed in the next paragraphs. Our discussion will center on the distinctions between the two approaches and then the focus will be primarily on the decompositional methods.

Decompositional or Attribute-free Approach

Commonly associated with the techniques of multidimensional scaling, decompositional methods rely on global or overall measures of similarity, from which the perceptual maps and relative positioning of objects are formed. They have two distinct advantages. First, they require only that respondents give their overall perceptions of objects; respondents do not detail the attributes used in this evaluation. Second, because each respondent gives a full assessment of similarities among all objects, perceptual maps can be developed for individual respondents or aggregated to form a composite map.

Decompositional methods have disadvantages as well. First, the researcher has no objective basis provided by the respondent on which to identify the basic "dimensions" of evaluation of the objects (i.e., the correspondence of perceptual and objective dimensions). In many instances, the usefulness to managers of attribute-free studies is restricted because the studies provide little guidance for specific action. For example, the inability to develop a direct link between actions by the firm (the objective dimension) and market positions of their products (the perceptual dimension) many times diminishes the value of perceptual mapping. Moreover, the researcher has little guidance, other than generalized guidelines or a priori beliefs, in determining both the dimensionality of the perceptual map and the representativeness of the solution. Although some overall measures of fit are available, they are nonstatistical, and thus decisions about the final solution involve substantial researcher judgment.

Characterized by the generalized category of multidimensional scaling techniques, a wide range of possible decompositional techniques is available. As will be discussed in the remainder of this chapter, selection of a specific method requires decisions regarding the nature of the respondent's input (rating versus ranking), whether similarities or preferences are obtained, and whether individual or composite perceptual maps are derived. Among the most common multidimensional scaling programs are KYST, MDSCAL, PREFMAP, MDPREF, INDSCAL, ALSCAL, MINISSA, POLYCON, and MULTISCALE. Detailed descriptions of the programs and sources for obtaining them are available in [24, 27].

Compositional or Attribute-based Approach

Compositional methods include some of the more traditional multivariate techniques (e.g., discriminant analysis or factor analysis), as well as methods specifically designed for perceptual mapping, such as correspondence analysis. A principle common to them all, however, is the assessment of similarity in which a defined set of attributes is considered in developing the similarity between objects.

As you might expect, one advantage of this approach is the explicit description of the dimensions of perceptual space. Because the respondent provides detailed evaluations across numerous attributes for each object, the evaluative criteria represented by the dimensions of the solution are much easier to ascertain. Second, these methods provide a direct method of representing both attributes and objects on a single map, with several methods providing the additional positioning of respondent groups. This information provides unique managerial insight into the competitive marketplace.

There are four primary disadvantages to compositional techniques. First, the similarity between objects is limited to only the attributes rated by the respondents. If salient attributes are omitted, there is no opportunity for the respondent to incorporate them, as there would be if a single overall measure were provided. Second, the researcher must *assume* some method of combining these attributes to represent overall similarity. The chosen method may or may not represent the respondent's thinking. Third, the data collection effort is substantial, especially as the number of choice objects increases. Finally, results are not typically available for the individual respondent.

The techniques of this type can be grouped into one of three basic groups:

1. *Graphical or post hoc approaches.* Included in this set are analyses such as semantic differential plots or **importance/performance grids,** which rely on researcher judgment and univariate or bivariate representations of the objects.
2. *Conventional multivariate statistical techniques.* These techniques, especially *factor analysis* and *discriminant analysis,* are particularly useful in developing a dimensional structure among numerous attributes and then representing objects on these dimensions.
3. *Specialized perceptual mapping methods.* Notable in this class is *correspondence analysis,* developed specifically to provide perceptual mapping with only qualitative or nominally scaled data as input.

Selecting Between Compositional and Decompositional Techniques

Perceptual mapping can be performed with both the compositional and decompositional techniques, but each technique has specific advantages and disadvantages that must be considered in view of the research objectives. If perceptual mapping is undertaken in the "spirit" of either of the two basic objectives discussed earlier, the decompositional or attribute-free approaches are the most appropriate. If, however, the research objectives shift to the portrayal among objects on a defined set of attributes, then the compositional techniques become the preferred alternative. Our discussion of the compositional methods in past chapters has illustrated their uses and application, as well as their strengths and weaknesses. The researcher must always remember the alternatives that are available in the event that the objectives of the research change. Thus, we will focus our remaining discussions on the decompositional approaches, with the exception of our discussion of correspondence analysis, an emerging compositional technique particularly suited to perceptual mapping. As such, we also consider as synonymous the terms *perceptual mapping* and *multidimensional scaling* unless necessary distinctions are made.

Objects: Their Number and Selection

Before beginning any perceptual mapping study, the researcher must address several questions about the objects being evaluated. First and foremost, are the objects really comparable? An implicit assumption in perceptual mapping is that there are common characteristics, either objective or perceived, that the respondent could use for evaluations. It is not possible for the researcher to "force" the respondent to make comparisons by creating pairs of noncomparable objects. Even if responses are given in such a forced situation, their usefulness is questionable.

A second question deals with the number of objects to be evaluated. In deciding how many objects to include, the researcher must balance two desires: a smaller number of objects to ease the effort on the part of the respondent versus the required number of objects to obtain a stable multidimensional solution. A suggested guideline for stable solutions is to have more than four times as many objects as dimensions desired [10]. Thus at least five objects are required for a one-dimensional perceptual map, nine objects for a two-dimensional solution, and so on. When using the method of evaluating pairs of objects for similarity (discussed in more detail in the next section), the respondent must make 36 comparisons of the nine objects—a substantial task. And a three-dimensional solution suggests at least 13 objects be evaluated, necessitating the evaluation of 78 pairs of objects. As you can see, a trade-off must be made between the dimensionality accommodated by the objects (and the implied number of underlying dimensions that can be identified) and the effort required on the part of the respondent.

The number of objects also affects the determination of an acceptable level of fit. Many times, having less than the suggested number of objects for a given dimensionality causes an inflated estimate of fit. Similar to the overfitting problem we found in regression, falling below the recommended guidelines of at least four objects per dimension greatly reduces the stress measure without respect to the input data. In an empirical study, research has demonstrated that quite acceptable levels of stress are obtained more than 50 percent of the time with *random similarities* if seven objects were fitted to three dimensions. If the seven objects were fitted to a four-dimensional solution, a stress of zero was found half the time [19]. Thus we must be aware of violating the guidelines for the number of objects per dimension and the impact this has on both the measures of fit and the validity of the resulting perceptual maps.

Nonmetric Versus Metric Methods

The original multidimensional scaling programs were truly nonmetric, meaning that they required only nonmetric input but they also provided only nonmetric (rank order) output. The nonmetric output, however, limited the interpretability of the perceptual map. Therefore, all MDS programs used today produce metric output. The metric multidimensional positions can be rotated about the origin, the origin can be changed by adding a constant, the axes can be flipped (reflection), or the entire solution can be uniformly stretched or compressed, all without changing the relative positions of the objects.

Because all programs today produce metric output, the distinction is based on the *input measures of similarity*. Nonmetric methods, distinguished by the nonmetric input typically generated by rank ordering pairs of objects, are more flexible in that they do not assume any specific type of relationship between the calculated distance and the similarity measure. However, they are more likely to result in degenerate or nonoptimal solutions. Metric methods assume that input as well as output are metric. This assumption allows us to strengthen the relationship between the final output dimensionality and the input data. Rather than assuming that only the ordered relationships are preserved in the input data, we can assume that the output preserves the interval and ratio qualities of the input data. Even though the assumptions underlying metric programs are harder to support conceptually in many cases, the results of nonmetric and metric procedures applied to the same data are often very similar. Thus selection of the input data type is mostly based on the preferred mode of data collection, not on substantive differences in the results.

Collection of Similarity or Preference Data

As noted above, the primary distinction among multidimensional scaling programs is the type of data (metric versus nonmetric) used to represent similarity and preferences. We address issues associated with making similarity-based and preference judgments. For many of the data collection methods, either metric (ratings) or nonmetric (rankings) data may be collected. In some instances however, the responses are limited to only one type of data.

Similarities Data

When collecting similarities data, the researcher is trying to determine which items are the most similar to each other and which are the most dissimilar. Implicit in similarities measurement is the ability to compare all pairs of objects.* If, for example, all pairs of objects of the set A, B, C (i.e., AB, AC, BC) are rank-ordered, then all pairs of objects can also be compared. Assume that the pairs were ranked AB = 1, AC = 2, and BC = 3 (where 1 is most similar). Clearly, the pair AB is more similar than the pair AC, pair AB is more similar than pair BC and pair AC is more similar than pair BC.

Several procedures are commonly used to obtain respondents' perceptions of the similarities among stimuli. Each procedure is based on the notion that the relative differences between any pair of stimuli must be measured so that the researcher can determine whether the pair is more or less similar to any other pair. We will discuss three procedures commonly used to obtain respondents' perceptions of similarities: comparison of paired objects, confusion data, and derived measures.

Comparison of Paired Objects   By far the most widely used method of obtaining similarity judgments is paired objects, where the respondent is asked simply to rank or rate the similarity of all pairs of objects. If we have stimuli A, B, C, D, and E, we could rank pairs AB, AC, AD, AE, BC, BD, BE, CD, CE, and DE from most similar to least similar. If, for example, pair AB is given the rank of 1, we would

---

*The terms *dissimilarities* and *similarities* often are used interchangeably to represent measurement differences between objects.

assume that the respondent sees that pair as containing the two stimuli that are the most similar, in contrast to all other pairs. This procedure would provide a nonmetric measure of similarity. Metric measures of similarity would involve a rating of similarity (e.g., from 1 "Not at All Similar" to 10 "Very Similar"). Either form (metric or nonmetric) can be used in most MDS programs.

Confusion Data   The pairing (or "confusing") of stimulus I with stimulus J is taken to indicate similarity. Also known as **subjective clustering,** the typical procedure for gathering these data is to place the objects whose similarity is to be measured (e.g., ten candy bars) on small cards, either descriptively or with pictures. The respondent is asked to sort the cards into stacks so that all the cards in a stack represent similar candy bars. Some researchers tell the respondents to sort into a fixed number of stacks; others say, "Sort into as many stacks as you like." In either situation, the data result in an aggregate similarities matrix similar to a cross-tabulation table. These data then indicate which products appeared together most often and are therefore considered the most similar. Collecting data in this manner allows only for the calculation of aggregate similarity, because the responses from all individuals are combined to obtain the similarities matrix.

Derived Measures   **Derived measures** of similarity are typically based on scores given to stimuli by respondents. For example, subjects are asked to evaluate 3 stimuli (cherry soda, strawberry soda, and lemon-lime drink) on 2 semantic differential scales, like the following:

The 2 × 3 matrix could be evaluated for each respondent (e.g., correlation, index of agreement, etc.) to create similarity measures. There are three important assumptions here:

1. The researcher has selected the appropriate dimensions to measure with the semantic differential.
2. The scales can be weighted (either equally or unequally) to achieve the similarities data for a subject or group of subjects.
3. Even if the weighing of scales can be determined, all individuals have the same weights.

Of the three procedures we have discussed, the derived measure is the least desirable in meeting the "spirit" of MDS—that the evaluation of objects be made with minimal influence by the researcher.

Collecting Preference Data

Preference implies that stimuli should be judged in terms of dominance relationships; that is, the stimuli are ordered in terms of the preference for some property. For example, brand A is preferred over brand C. The two most common procedures for obtaining preference data are direct ranking and paired comparisons.

Direct Ranking   Each respondent ranks the objects from most preferred to least preferred, as in the following example:

Rank from most preferred (1) to least preferred (5).
_____ Candy bar A
_____ Candy bar B
_____ Candy bar C
_____ Candy bar D
_____ Candy bar E

Paired Comparisons   A respondent is presented with all possible pairs and asked to indicate which member of each pair is preferred, as in this example:

Please circle the preferred candy bar in each pair:

A   B
A   C
A   D
A   E
B   C
B   D
B   E
C   D
C   E
D   E

Preference data allow the researcher to view the location of objects in a perceptual map where distance implies differences in preference. This procedure is useful, because an individual's perception of objects in a preference context may be different from that in a similarity context; that is, a particular dimension may be very useful in describing the differences between two objects but may not be consequential in determining preference. Therefore, two objects could be perceived as different in a similarity-based map but be similar in a preference-based spatial map; that is, two objects may be quite different (e.g., two different brands of candy bars) and placed far apart in a similarity-based map but may be preferred equally or about the same and therefore positioned close to each other on a preference map.

In summary, the collection procedures for both similarity and preference data have the common purpose of obtaining a series of unidimensional responses that represent the respondents' judgments. These judgments then serve as inputs to the many multidimensional scaling procedures that define the underlying multidimensional pattern leading to these judgments.

Stage Three: Assumptions of Multidimensional Scaling Analysis

Multidimensional scaling, while having no restraining assumptions on the methodology, type of data, or form of the relationships among the variables, does require that the researcher accept several tenets about perception, including the following:

1. Each respondent will not perceive a stimulus to have the same dimensionality (although it is thought that most people judge in terms of a limited number of characteristics or dimensions). For example, some persons might evaluate a car in terms of its horsepower and appearance, while others do not consider these factors at all but instead assess it in terms of cost and interior comfort.

2. Respondents need not attach the same level of importance to a dimension, even if all respondents perceive this dimension. For example, two persons may perceive a cola drink in terms of its level of carbonation, but one considers this dimension unimportant while the other considers it very important.

3. Judgments of a stimulus in terms of either dimensions or levels of importance need not remain stable over time. In other words, one may not expect persons to maintain the same perceptions for long periods of time.

In spite of these assumptions and the differences we can expect between individuals, MDS attempts to represent perceptions spatially so that any common underlying relationships can be examined. The purpose of employing a multidimensional scaling technique lies not only in understanding each separate individual but also in identifying the shared perceptions and evaluative dimensions within the sample of respondents.

Stage Four: Deriving the MDS Solution and Assessing Overall Fit

The variety of computer programs for multidimensional scaling is rapidly expanding, particularly for use on the personal computer. We employ a series of MDS applications to illustrate the types of input data, the differing types of spatial representations, and the interpretational alternatives. Our objective is to provide an overview of multidimensional scaling that will allow you to understand readily the differences between these programs. However, in no way can we provide a complete discussion of an area that has increased dramatically in both applications and knowledge of use in the past decade. Thus, we refer the user interested in specific program applications to other texts devoted solely to multidimensional scaling [10, 11, 17, 19, 24].

Determining an Object's Position in the Perceptual Map

The first task involves the positioning of objects to best reflect the similarity evaluations provided by the respondents. MDS programs follow a common process for determining the optimal positions, much as we did in the simple example in the preceding section. This process can be described in five steps:

1. Select an initial configuration of stimuli (S_k) at a desired **initial dimensionality** (*t*). A number of options for obtaining the initial configuration are available. The two most widely used are a configuration either (1) applied by the researcher based on previous data or (2) generated by selecting pseudorandom points from an approximately normal multivariate distribution.

2. Compute the distances between the stimuli points and compare the relationships (observed versus derived) with a measure of fit. Once a configuration is

found, the interpoint distances between stimuli (d_{ij}) in the starting configuration are compared with distance measures ($\hat{d}_{ij}$) derived from the similarity judgments (s_{ij}). The two distance measures are then compared by a measure of fit, typically a measure of stress. (Fit measures are discussed in a later section.)

3. If the measure of fit does not meet a predefined stopping value you have selected, find a new configuration where the measure of fit is further minimized. The program determines the directions in which the best improvement in fit can be obtained and then moves the points in the configuration in those directions in small increments.

4. The new configurations are evaluated and adjusted until satisfactory stress is achieved.

5. Once satisfactory stress has been achieved, the dimensionality is reduced by one, and the process is repeated until the lowest dimensionality with an acceptable measure of fit has been reached.

The need for a computer program versus hand calculations becomes apparent as the number of objects and dimensions increases. For example, assume that we have 10 products to be evaluated. Each respondent must now rank all 45 pairs of products from most similar (1) to least similar (45). We place the 10 points (representing the 10 products) randomly on a sheet of graph paper and measure the distances between every pair of points (45 distances). We then calculate the stress of the solution, a measure that shows the rank-order agreement between the Euclidean (straight-line) distances of the plotted objects and the original 45 ranks. If the straight-line distances do not agree with the original ranks, move the 10 points and try again. The process becomes intractable as the number of objects increases and the differences in perception and dimensions used in the evaluation increases. The computer is used only to replace the manual calculations and allow for a more accurate and detailed solution.

The primary criterion in all instances for finding the best representation of the data is preservation of the ordered relationship between the original rank data and the derived distances between points. The stress measure is simply a measure of how well (or poorly) the ranked distances on the map agree with the ranks given by the respondents.

In evaluating a perceptual map, the analyst should always be aware of **degenerate solutions,** which are derived perceptual maps that are not accurate representations of the similarity responses. Most often they are caused by inconsistencies in the data or an inability of the MDS program to reach a stable solution. They are characterized most often by either a circular pattern where all objects are shown to be equally similar or a clustered solution where the objects are grouped at two ends of a single dimension. In both cases, the computer program was unable to differentiate among the objects for some reason. The analyst should then reexamine the research design to see where the inconsistencies occurred.

Selecting the Dimensionality of the Perceptual Map

The objective of this step is the selection of a spatial configuration in a specified number of dimensions. The determination of how many dimensions are actually represented in the data is generally reached through one of three approaches: subjective evaluation, scree plots of the stress measures, or an overall index of fit.

The spatial map is a good starting point for the evaluation. The number of maps necessary for interpretation depends on the number of dimensions. A map is produced for each combination of dimensions. One objective of the analyst should be to obtain the best fit with the smallest possible number of dimensions. Interpretation of solutions derived in more than three dimensions is extremely difficult and usually is not worth the improvement in fit. The analyst typically makes a **subjective evaluation** of the spatial maps and determines whether the configuration looks reasonable. This question must be considered, because at a later stage the dimensions will need to be interpreted and explained.

A second approach is to use a **stress measure,** which indicates the proportion of the variance of the **disparities** not accounted for by the MDS model. This measurement varies according to the type of program and the data being analyzed. Kruskal's [18] stress is the most commonly used measure for determining the model's goodness of fit. It is defined by the following equation:

$$\text{Stress} = \sqrt{\frac{(d_{ij} - \hat{d}_{ij})^2}{(d_{ij} - \bar{d})2}}$$

where $\bar{d}$ is the average distance ($\Sigma\, d_{ij}/n$) on the map. The stress value becomes smaller as the estimated $\hat{d}_{ij}$ approaches the original d_{ij}. Stress is minimized when the objects are placed in a configuration so that the distances between the objects best matches the original distances.

A problem found in using stress, however, is analogous to that of R^2 in multiple regression, in that stress always improves with increased dimensions. (Remember that R^2 always increased with additional variables.) A trade-off must then be made between the fit of the solution and the number of dimensions. As was done for the extraction of factors in factor analysis, we can plot the stress value against the number of dimensions to determine the best number of dimensions to utilize in the analysis [19]. For example, in the scree plot in Figure 9.4, the elbow indicates that there is substantial improvement in the goodness of fit when the number of dimensions is increased from one to two. Therefore, the best fit is obtained with a relatively low number (two) of dimensions.

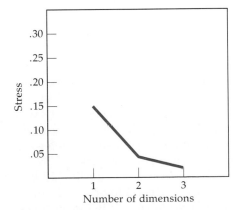

FIGURE 9.4 Use of the scree plot to determine the appropriate dimensionality.

An **index of fit** sometimes used is a squared correlation index that can be interpreted as indicating the proportion of variance of the disparities (optimally scaled data) accounted for by the multidimensional scaling procedure. In other words, it is a measure of how well the raw data fit the multidimensional scaling model. The R^2 measure on multidimensional scaling represents essentially the same measure of variance as it does in other multivariate techniques. Therefore, it is possible to use similar measurement criteria; that is, measures of .60 or better are considered acceptable. Of course, the higher the R^2, the better the fit.

Incorporating Preferences into Multidimensional Scaling

Up to this point, we have concentrated on developing perceptual maps based on similarity judgments. However, perceptual maps can also be derived from preferences. The objective is to determine the preferred mix of characteristics for a set of stimuli that predicts preference, given a set configuration of objects [9, 10]. In doing so, a joint space is developed portraying both the objects (stimuli) and the subjects (ideal points). A critical assumption is the homogeneity of perception across individuals for the set of objects. This allows all differences to be attributed to preferences, not perceptual differences.

Ideal Points

The term **ideal point** has been misunderstood or misleading at times. We can assume that if we locate (on the derived perceptual map) the point that represents the most preferred combination of perceived attributes, we have identified the position of an ideal object. Equally, we assume that the position of this ideal point (relative to the other products on the derived perceptual map) defines relative preference so that products farther from the ideal should be less preferred. An ideal point is positioned so that the distance from the ideal conveys changes in preference. Consider, for example, Figure 9.5. When preference data on the six candy bars were obtained from the person indicated by the dot (•), the point (•) was positioned so that increasing the distance from it indicated declining preference. One may assume that this person's preference order is C, F, D, E, A, B. To imply that the ideal candy bar is exactly at the point (•) or even beyond (in the direction shown by the dashed line from the origin) can be misleading. The ideal point simply defines the ordered preference relationship among the set of six candy bars for that respondent. Although ideal points individually may not offer much insight, clusters of them can be very useful in defining segments. Many respondents with ideal points in the same general area represent potential market segments of persons with similar preferences, as indicated in Figure 9.6.

Two approaches have generally been used to determine ideal points: explicit and implicit estimation. Explicit estimation proceeds from the direct responses of subjects. This procedure may involve asking the subject to rate a hypothetical ideal on the same attributes on which the other stimuli are rated. Alternatively, the respondent is asked to include among the stimuli used to gather similarities data a hypothetical ideal stimulus (brand, image, etc.).

When asking respondents to conceptualize an ideal of anything, we typically run into problems. Often the respondent conceptualizes the ideal at the extremes of the explicit ratings used or as being similar to the most preferred product from

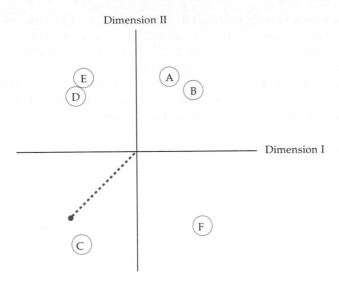

● Indicates respondent's ideal point

FIGURE 9.5 A respondent's ideal point within the
perceptual map.

among those with which the respondent has had experience. Also, the respondent must not think in terms of similarities but preferences, which is often difficult with relatively unknown objects. Often these perceptual problems lead the researcher to use implicit ideal point estimation.

There are several procedures for implicitly positioning ideal points (see the following section for a more detailed description). The basic assumption underlying most procedures is that derived measures of ideal points' spatial positions are

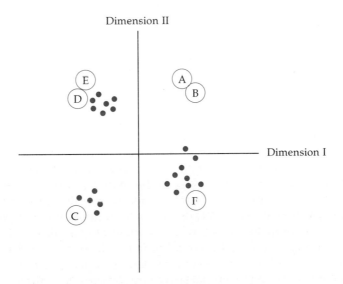

● Indicates a single respondent's ideal point

FIGURE 9.6 Incorporating multiple ideal points in
the perceptual map.

maximally consistent with individual respondents' preferences. Srinivasan and Shocker [25] assume that the ideal point for all pairs of stimuli is determined so that it violates with least harm the constraint that it be closer to the most preferred in each pair than it is to the least preferred.

In summary, there are many ways to approach ideal point estimation, and no one best method has been demonstrated. The choice depends on the researcher's skills and the multidimensional scaling procedure selected.

Internal Versus External Analysis of Preference Data

Implicit positioning of the ideal point from preference data can be accomplished through either an internal or an external analysis. *Internal analysis of preference data* refers to the development of a spatial map shared by both stimuli and subject points (or vectors) solely from the preference data. As an example of this approach, MDPREF [6] or MDSCAL [18], two of the more widely used programs of this type, allow the user to find configurations of stimuli and ideal points by assuming (1) no difference between subjects at all, (2) separate configurations for each subject, or (3) a single configuration with individual ideal points. By gathering preference data, the researcher can represent both stimuli and respondents on a single perceptual map.

Internal analyses must make certain assumptions, however, in deriving both the perceptual map of stimuli and ideal points. The objects' positions are calculated based on **unfolding** of preference data for each individual. The results reflect perceptual dimensions that are "stretched" and weighted to predict preference. One characteristic of internal estimation methods is that they typically employ a vector representation of the ideal point (see the following section for a discussion of vector versus point representations), whereas external models can estimate either vector or point representations.

External analysis of preference data refers to fitting ideal points (based on preference data) to stimulus space developed from similarities data obtained from the same subjects. For example, we might scale similarities data individually, examine the individual maps for commonality of perception, and then scale the preference data for any group identified in this fashion. If this procedure is followed, the researcher has to gather both preference and similarities data to achieve external analysis.

PREFMAP [7] was developed solely to perform external analysis of preference data. Because the similarity matrix defines the objects in the perceptual map, the researcher can now define both attribute descriptors (assuming that the perceptual space is the same as the evaluative dimensions) and ideal points for individuals. PREFMAP provides estimates for a number of different types of ideal points, each based on different assumptions as to the nature of preferences (i.e., vector versus point representations and equal versus differential dimension weights).

It is generally accepted [10, 11, 24] that external analysis is clearly preferable in most instances. This conclusion is based on computational difficulties with internal analysis procedures and on the confounding of differences in preference with differences in perception. In addition, the saliences of perceived dimensions may change as one moves from perceptual space (are the stimuli similar or dissimilar?) to evaluative space (which stimulus is preferred?).

We will illustrate the procedure of external estimation in our example of perceptual mapping with multidimensional scaling at the end of the chapter.

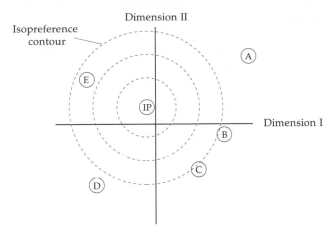

Preference order (highest to lowest): E > C > B > D > A

Ⓐ Object ⅠⓅ Ideal Point

FIGURE 9.7 Point representation of an ideal point.

Vector Versus Point Representations

The discussion of perceptual mapping of preference data has emphasized an ideal point that portrays the relationship of an individual's preference ordering for a set of stimuli. The most easily understood method of portraying the ideal point is to use a single point. Preference ordering is measured by the straight-line (Euclidean) distances from the ideal point (IP) to all the points representing the objects. We are assuming that the direction of distance from the IP is not critical, only the relative distance. An example is shown in Figure 9.7. Here, the ideal point as positioned indicates that the most preferred object is E, followed by object C, then B, D, and finally A.

The ideal point can also be shown as a **vector.** To calculate the preferences in this approach, perpendicular lines (also known as **projections**) are drawn from the objects to the vector. Preference increases in the direction the vector is pointing. The preferences can be read directly from the order of the projections. Figure 9.8 illustrates the vector approach for two subjects with the same set of stimuli positions. For subject 1, the vector has the direction of lower preference in the bottom left-hand corner to higher preference in the upper right-hand corner. When the projection for each object is made, the preference order (highest to lowest) is A, B, C, E, D. However, the same objects have a quite different preference order for subject 2, ranging from the most preferred, E, to the least preferred, C. In this manner, each subject can be represented by a separate vector. In the vector approach, there is no single ideal point, but it is assumed that the ideal point is at an infinite distance outward on the vector.

Although either the point or the vector representations can indicate what combinations of attributes are more preferred, these observations are often not borne out by further experimentation. For example, Raymond [23] cites an example in which the conclusion was drawn that people would prefer brownie pastries on the basis of degree of moistness and chocolate content. When the food technicians applied this result in the laboratory, they found that their brownies made to the

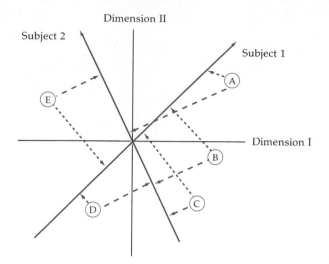

Preference order (highest to lowest): Subject 1: A > B > C > E > D
Subject 1: E > A > D > B > C

FIGURE 9.8 Vector representations of two ideal points:
subjects 1 and 2.

experimental specification became chocolate milk. One cannot always assume
that the relationships found are independent or linear, or that they hold over time,
as noted previously. However, multidimensional scaling is a beginning in understanding perceptions and choice that will expand considerably as applications
extend our knowledge of both the methodology and human perception.

Summary

Preference data are best examined using an external analysis as a means to better
understand both the perceptual differences between objects based on similarity-
judgments and the preference choices made within this perceptual map of objects.
In this manner, the researcher can distinguish between both types of perceptual
evaluations and more accurately understand the perceptions of individuals in the
"true spirit" of multidimensional scaling.

Stage Five: Interpreting the MDS Results

Once the perceptual map is obtained, the two approaches again diverge in their
interpretation of the results. For compositional methods, the perceptual map must
be validated against other measures of perception, because the positions are totally defined by the attributes specified by the researcher. For example, discriminant analysis results might be applied to a new set of objects or respondents,
assessing the ability to differentiate with these new observations.

For decompositional methods, the most important issue is the description of the
perceptual dimensions and their correspondence to attributes. A number of de-

scriptive techniques will be discussed later to "label" the dimensions, as well as to integrate preferences (for objects and attributes) with the similarity judgments. Again, in line with their objectives, the decompositional methods provide an initial look into perceptions from which more formalized perspectives may emerge.

Because other chapters in this text have dealt with many of the compositional techniques, the remainder of this chapter focuses on decompositional methods, primarily the various techniques used in multidimensional scaling. A notable exception is the discussion of a compositional approach, correspondence analysis, that, to a degree, bridges the gap between the two approaches in its flexibility and methods of interpretation.

Identifying the Dimensions

As we discussed in Chapter 7, identifying underlying dimensions is often a difficult task. Multidimensional scaling techniques have no built-in procedure for labeling the dimensions. The researcher, having developed the maps with a selected dimensionality, can adopt several procedures, either subjective or objective.

Subjective Procedures

Interpretation must always include some element of researcher or respondent judgment, and in many cases this proves adequate for the questions at hand. A quite simple, yet effective, method is labeling (by visual inspection) the dimensions of the perceptual map by the respondent. Respondents may be asked to interpret the dimensionality subjectively by inspecting the maps, or a set of "experts" may evaluate and identify the dimensions. Although there is no attempt to quantitatively link the dimensions to attributes, this approach may be the best available if the dimensions are believed to be highly intangible or affective/emotional in content, so that adequate descriptors could not be devised.

In a similar manner, the researcher may describe the dimensions in terms of known (objective) characteristics. In this way, the correspondence is made between objective and perceptual dimensions directly, although these relationships are not a result of respondent feedback but of the researcher's judgment.

Objective Procedures

As a complement to the subjective procedures, a number of more formalized methods have been developed. The most widely used method, PROFIT (PROperty FITting) [4], collects attribute ratings for each object and then finds the best correspondence of each attribute to the derived perceptual space. The attempt is to identify the determinant attributes in the similarity judgments made by individuals. Measures of fit are given for each attribute, as well as their correspondence with the dimensions. The analyst can then determine which attributes best describe the perceptual positions and are illustrative of the dimensions. The need for correspondence between the attributes and the defined dimensions diminishes with the use of metric output, as the dimensions can be rotated freely without any changes in interpretation.

Selecting Between Subjective and Objective Procedures

For either subjective or objective procedures, the researcher must remember that, although a dimension can represent a single attribute, it usually does not. More common is to collect data on several attributes, associate them either subjectively or empirically with the dimensions where applicable, and determine labels for each dimension using multiple attributes, similar to factor analysis. Many researchers suggest that using attribute data to help label the dimensions is the best alternative. The problem, however, is that the researcher may not include all the important attributes in the study. Thus, the researcher can never be totally assured that the labels represent all relevant attributes.

Both types of procedures illustrate the difficulty of labeling the axes. This task cannot be left until completion, as the dimensional labels are essential for further interpretation and use of the results. The researcher must select the type of procedure that best suits both the objectives of the research and the available information. Thus, the researcher must plan for deriving the dimensional labels as well as the estimation of the perceptual map.

Stage Six: Validating the MDS Results

Validation in multidimensional scaling is as important as in any other multivariate technique. Owing to the highly inferential nature of MDS, this effort should be directed toward ensuring the generalizability of the results both across objects and to the population. But validation efforts are problematic. The only output of MDS that can be used for comparative purposes is the relative positions of the objects. Thus, while the positions can be compared, the underlying dimensions have no basis for comparison. If the positions vary, the researcher cannot determine whether the objects are viewed differently, whether the perceptual dimensions vary, or both. Moreover, systematic methods of comparison have not been developed and integrated into the statistical programs. The researcher is left to improvise with procedures that may address general concerns but are not specific to multidimensional scaling results.

What options are available? The most direct approach is a split- or multisample comparison, where either the original sample is divided or a new sample is collected. In either instance, the researcher must then find a means of comparing the results. Most often the comparison between results is done visually or with a simple correlation of coordinates. While some matching programs are available, such as FMATCH [27], the researcher must still determine how many of the disparities are due to differences in object perceptions, differing dimensions, or both.

Another method is to obtain a convergence of MDS results by applying both decompositional and compositional methods to the same sample. The decompositional method(s) could be applied first, along with interpretation of the dimensions to identify key attributes. Then one or more compositional methods, particularly correspondence analysis, could be applied to confirm the results. The analyst must realize this is not true validation of the results as being generalizable but does confirm the interpretation of the dimension. From this point, validation efforts with other samples and other objects could be undertaken to demonstrate generalizability to other samples.

Correspondence Analysis

Up to this point we have discussed the traditional decompositional approaches to multidimensional scaling. In the past, the compositional approaches have relied on traditional multivariate techniques like discriminant and factor analysis. But recent developments have combined aspects of both methods to form potent tools for perceptual mapping.

Correspondence analysis is a recently developed interdependence technique that facilitates dimensional reduction and conducts perceptual mapping [2, 3, 12, 14, 20]. It can be classified as a compositional technique because it creates the perceptual map based on the association between objects and a set of descriptive characteristics or attributes. Thus the association is based on the attributes specified by the researcher. Among the compositional techniques, factor analysis is the most similar, but correspondence analysis has been extended past the applications of factor analysis. Its most direct application is portraying the "correspondence" of categories of variables, particularly those measured in nominal terms. This correspondence then becomes the basis for developing perceptual maps. The benefits of correspondence analysis lie in its unique abilities for representing rows and columns (e.g., brands *and* attributes) in a joint space.

Stage One: Objectives of Correspondence Analysis

Researchers are constantly faced with the need to "quantify the qualitative data" found in nominal variables. Correspondence analysis differs from other interdependence techniques in its ability to accommodate both nonmetric data and nonlinear relationships. In its most basic form, correspondence analysis employs a **contingency table,** which is the crosstabulation of two categorical variables. It then transforms the nonmetric data to a metric-level form and performs dimensional reduction similar to factor analysis. In addition, correspondence analysis performs a type of perceptual mapping similar to multidimensional scaling, where categories are represented in the multidimensional space. Proximity indicates the level of association among row or column categories. Its uses are not, however, just for relating objects to attributes. For example, the categories of a question (e.g., low, medium, and high) could be analyzed to determine the positions of the categories. Categories positioned closely together indicate high association and the possibility of combining the categories into a single category.

As an example of its perceptual mapping application, respondents' brand preferences can be crosstabulated on demographic variables (e.g., gender, income categories, occupation) by indicating how many people preferring each brand fall into each category of the demographic variables. Through correspondence analysis, the association, or "correspondence," of brands and the distinguishing characteristics of those preferring each brand are then shown in a two- or three-dimensional map of both brands and respondent characteristics. Brands perceived as similar are located in close proximity to each other. Likewise, the most distinguishing characteristics of respondents are also determined by the proximity of the demographic variable categories to each other. Correspondence analysis provides a multivariate representation of interdependence for nonmetric data not possible with other methods.

Stage Two: Research Design of Correspondence Analysis

Correspondence analysis requires only a rectangular data matrix (crosstabulation) of nonnegative entries. The rows and columns do not have predefined meanings (i.e., attributes do not always have to be rows, etc.) but instead represent the responses to one or more categorical variables. The categories for a row or column need not be a single variable but can represent any set of relationships. A prime example is the "pick any" method [15, 16], where respondents are given a set of objects and characteristics. The respondents then indicate which objects, if any, are described by the characteristics. Note that the respondent may choose any number of objects for each characteristic, rather than a prespecified number (i.e., choose only the object best described or the top two objects). In this situation, the crosstabulation table would be the total number of times each object was described by each characteristic.

The crosstabulation of more than two variables in a multiway matrix form is known as **multiple correspondence analysis.** In a procedure quite similar to two-way analysis, the additional variables are "fitted" so that all the categories are placed in the same multidimensional space.

Stage Three: Assumptions in Correspondence Analysis

Correspondence analysis shares a relative freedom from assumptions with the more traditional MDS techniques. The use of strictly nonmetric data in its simplest form (crosstabulated data) represents linear and nonlinear relationships equally well. The lack of assumptions, however, must not cause the researcher to neglect the efforts needed in ensuring the comparability of objects and, because this is a compositional technique, the completeness of the attributes used.

Stage Four: Deriving the Correspondence Analysis Results and Assessing Overall Fit

Once the crosstabulation table is presented to correspondence analysis, the frequencies for any row/column combination of categories are related to all other combinations based on the marginal frequencies. This procedure yields a conditional expectation very similar to an expected Chi-square value. Once obtained, these values are normalized, and then a process much like factor analysis defines lower-dimensional solutions. These "factors" simultaneously relate the rows and columns in a single joint plot. The result is a single representation of categories of both rows and columns (i.e., brands *and* attributes) in the same plot. A number of computer programs are available to perform correspondence analysis, both for mainframe and personal computer use. Among the more popular programs are ANACOR and HOMALS, available with SPSS [26], CA from BMDP [1], CORRAN and CORRESP from PC-MDS [27], and MAPWISE [22].

As in factor analysis, the analyst must first identify the appropriate number of dimensions and their importance. Eigenvalues are derived for each dimension, indicating the relative contribution of dimensions in explaining the variance in the categories. The analyst can select the number of dimensions representing the level of explained variance desired. As discussed with regard to perceptual mapping, interpretation is facilitated by a three-dimensional or lower representation.

Stage Five: Interpretation of Results

Once the dimensionality has been established, the researcher can then identify a category's association with other categories by proximity. At this time, there is debate on the appropriateness of comparing between row and column categories. In a strict sense, the proximities should be compared only within rows or within columns. Thus, in our brands by attributes example, comparisons of association should be made only among brands or attributes. The correspondence of a brand to an attribute category is not, however, represented by the proximity of the categories. Variations on the original technique are proposed to eliminate this restriction and to make all categories comparable among each other [3, 22], but there is still disagreement as to their success [13]. Although direct comparisons are not possible, the general correspondence does hold, and distinct patterns can be distinguished.

Stage Six: Validation of the Results

The compositional nature of correspondence analysis provides more specificity for the researcher to validate the results. As with all MDS techniques, an emphasis must be made to ensure generalizability through split- or multisample analyses. However, as with other perceptual mapping techniques, the generalizability of the objects (individually and as a set) must also be established. The sensitivity of the results to the addition or deletion of an object can be evaluated, as well as the addition or deletion of an attribute. The goal is to assess whether the analysis is dependent on only a few objects and/or attributes. In either instance, the analyst must understand the "true" meaning of the results in terms of the objects and attributes.

Overview of Correspondence Analysis

Correspondence analysis presents the analyst with a number of advantages. First, the simple crosstabulation of multiple categorical variables, such as product attributes versus brands, can be represented in a perceptual space. This approach allows the researcher either to analyze existing responses or to gather responses at the least restrictive measurement type, the nominal or categorical level. For example, the respondent need rate only yes/no for a set of objects on a number of attributes. These responses can then be aggregated in a crosstabulation table and analyzed. Other techniques such as factor analysis require interval ratings of each attribute for each object.

Second, CA portrays not only the relationships between the rows and columns but also the relationships between the categories of either the rows or the columns. For example, if the columns are attributes, multiple attributes in close proximity would all have similar profiles across products. This forms a group of attributes quite similar to a factor from principal components analysis.

Finally, and most important, CA provides a joint display of row and column categories in the same dimensionality. Certain program modifications allow for interpoint comparisons where relative proximity is directly related to higher association among separate points [2, 22]. When these comparisons are possible, they allow row and column categories to be examined simultaneously. An analysis of this type would enable the analyst to identify groups of products characterized by attributes in close proximity.

With the advantages of correspondence analysis come a number of disadvantages or limitations. First, the technique is descriptive and not at all appropriate for hypothesis testing. If the quantitative relationship of categories is desired, methods such as loglinear models are suggested. Correspondence analysis is best suited for exploratory data analysis. Second, correspondence analysis, as with many dimensionality-reducing methods, has no method for conclusively determining the appropriate number of dimensions. As with other similar methods, the researcher must balance interpretability versus parsimony of the data representation. Finally, the technique is quite sensitive to outliers, either in terms of rows or columns (e.g., attributes or brands). Also for purposes of generalizability, the problem of omitted objects or attributes is critical.

Illustration of Multidimensional Scaling and Correspondence Analysis

To demonstrate the use of multidimensional scaling techniques, we will examine data gathered in a series of interviews with company representatives from a cross section of potential customers. In the course of the perceptual mapping analysis, we will apply both decompositional and compositional methods. The discussion will first examine the initial three stages of the model-building process that are common to both methods. The discussion then focuses on the next two stages for decompositional methods. After the discussion of the decompositional methods is complete, it is followed by a discussion of a compositional method, correspondence analysis, applied to the same sample of respondents. A final section details the final stage by discussing the validation of the analysis through comparison of the results from both types of methods.

Stage One: Objectives of Perceptual Mapping

The purpose of the research was to explore HATCO's image and competitiveness. This exploration included addressing the perceptions in the market of HATCO and nine major competitors, as well as an investigation of preferences among potential customers. The data were analyzed in a two-phase plan: (1) identification of the position of HATCO in a perceptual map of major competitors in the market with an understanding of the dimensions comparison used by major customers, and (2) assessment of the preferences toward HATCO relative to major competitors. Before proceeding with a discussion of the results, we briefly describe the data collection process.

Stage Two: Research Design
of the Perceptual Mapping Study

The HATCO image study comprised depth interviews with 18 midlevel management personnel from different firms selected as representative of the potential customer base existing in the market. The nine competitors, plus HATCO, represented all the major firms in this industry and collectively had more than 85 percent of total sales. In the course of the interviews, three types of data were collected: similarity judgments, attribute ratings of firms, and preferences for each firm in different buying situations.

Similarity Data

The starting point for data collection was in obtaining the perceptions of the respondents concerning the similarity/dissimilarity of HATCO and nine competing firms in the market. Similarity judgments were made with the comparison-of-paired-objects approach. The 45 pairs of firms [(10 × 9)/2] were presented to the respondents, who indicated how similar each was on a nine-point scale, with one being "Not at All Similar" and nine being "Very Similar." Note that the values have to be transformed because increasing values for the similarity ratings indicate greater similarity, the opposite of a distance measure of similarity.

Attribute Ratings

In addition to the similarity judgments, ratings of each firm for eight attributes were obtained by two methods. The attributes included product quality, management orientation, service quality, delivery speed, price level, sales force image, price flexibility, and manufacturing image. In the first method, each firm was rated on a six-point scale for each attribute. In the second method, each respondent was asked to pick the firms best characterized by each attribute. As with the "pick any" method [15, 16], the respondent could pick any number of firms for each attribute.

Preference Evaluations

The final data assessed the preferences of each respondent for the ten firms in three different buying situations: a straight-rebuy, a modified-rebuy, and a new-buy situation. In each situation, the respondents ranked the firms in order of preference for that particular type of purchase. For example, in the straight-rebuy situation, the respondent indicated the most preferred firm for the simple reordering of products (rank order = 1), the next most preferred (rank order = 2), and so on. Similar preferences were gathered for the remaining two buying situations.

Stage Three: Assumptions in Perceptual Mapping

The assumptions of multidimensional scaling and correspondence analysis deal primarily with the comparability and representativeness of the objects being evaluated and the respondents. With regard to the sample, the sampling plan emphasized obtaining a representative sample of HATCO customers. Moreover, care was taken to obtain respondents of comparable position and market knowledge. Because HATCO and the other firms serve a fairly distinct market, all the firms evaluated in the perceptual mapping should be known, ensuring that positioning discrepancies can be attributed to perceptual differences among respondents.

Multidimensional Scaling: Stages Four and Five

Having specified the ten firms to be included in the image study, HATCO's management specified that both decompositional (MDS) and compositional (CA) approaches be employed in constructing the perceptual maps. We first discuss a series of decompositional techniques, then examine a compositional approach to perceptual mapping.

Stage Four: Deriving the Multidimensional Scaling Results and Assessing Overall Fit

INDSCAL [5] was used to develop both a composite or aggregate perceptual map and the measures of the differences between respondents in their perception.* The 45 similarity judgments from the 18 respondents were input into the program. The results were analyzed to assess the appropriate dimensionality and portray the results in a perceptual map.

To determine the most appropriate dimensionality, the researchers considered both the index of fit at each dimensionality and the ability of the researcher to interpret the solution. Table 9.1 shows the index of fit for solutions of one to four dimensions. As we can see, there is substantial improvement in moving from one to two dimensions, after which the improvement diminishes somewhat and remains consistent as we increase in the number of dimensions. Balancing this improvement in fit against the increasing difficulty of interpretation, we note that the two- or three-dimensional solutions seemed the most appropriate. For purposes of illustration, the two-dimensional solution was selected for further analyses, but the methods we will discuss could just as easily be applied to the three-dimensional solution. The researcher is encouraged to explore other solutions to assess whether any substantively different conclusions would be reached based on the dimensionality selected.

The two-dimensional aggregate perceptual map is shown in Figure 9.9. HATCO is most closely associated with firm A, with respondents considering them almost identical. Firms B and I were considered the next most similar based on their proximity. But this similarity is based primarily on Dimension II, because there are substantial differences between HATCO and firms B and I on Dimension I. These differences are reflected in their relative positions in the perceptual map.

*The data input and program control cards for each multidimensional scaling program (and correspondence analysis) are contained in Appendix A. For a more detailed explanation of the specific details of program execution, the reader is referred to [10, 24].

TABLE 9.1 Determining the Appropriate Dimensionality: Measures of Fit
  Versus Dimensionality

| Dimensionality | Measure of Fit | |
of the Solution	Value	Increase
1	.34663	—
2	.51598	.16935
3	.62743	.11145
4	.72147	.09404

Similar comparisons can be made among all sets of firms. To understand the
sources of these differences, however, the analyst must interpret the dimensions.

In addition to developing the composite perceptual map, INDSCAL also provides the means for assessing one of the assumptions of MDS, the homogeneity of
respondents' perceptions. Weights are calculated for each respondent indicating
the correspondence of their own perceptual space and the aggregate perceptual
map. These weights provide a measure of comparison because respondents with
similar weights have similar individual perceptual maps. INDSCAL also provides a measure of fit for each subject by correlating the computed scores and the
respondent's original similarity ratings. Table 9.2 contains the weights and measures of fit for each respondent. As we see from examining the data, the respondents are quite homogeneous in their perceptions, because the weights show few
substantive differences on either dimension, and no distinctive "clusters" of individuals emerge. Moreover, all the respondents are well represented by the composite perceptual map, with the lowest measure of fit being .66.

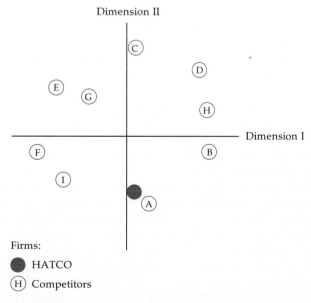

Firms:

⬤ HATCO

Ⓗ Competitors

FIGURE 9.9 Perceptual map of HATCO and
  major competitors.

TABLE 9.2 Measures of the Individual Differences in Perceptual Mapping: Dimension Weights and Respondent-Specific Measures of Fit

| Subject | Subject Weights | | Measure of Fit* |
	Dimension I	Dimension II	
1	.521	.413	.672
2	.382	.571	.694
3	.387	.573	.698
4	.457	.631	.787
5	.540	.437	.702
6	.432	.602	.749
7	.467	.609	.776
8	.517	.429	.679
9	.363	.573	.686
10	.515	.554	.764
11	.399	.548	.685
12	.573	.503	.770
13	.396	.603	.729
14	.369	.546	.665
15	.401	.532	.674
16	.362	.591	.699
17	.440	.577	.733
18	.505	.542	.749

*Measure of fit: correlation of computed scores and original data for each subject.

Stage Five: Interpretation of the Results

Once the perceptual map has been established, we can begin the process of interpretation. HATCO gathered ratings of the firms on eight attributes descriptive of typical strategies followed in this industry. The ratings for each firm were then averaged across the respondents for a single overall rating. To provide an objective means of interpretation, PROFIT [4], a vector model, was used to match the ratings to the firm positions in the perceptual map. The results of applying the ratings data to the composite perceptual map are shown in Figure 9.10. This shows that there are three distinct "groups" or dimensions of attributes. The first involves attributes 1 (delivery speed), 2 (price level), 3 (price flexibility), which are all pointed in the same direction, and 5 (service), which is in the direction opposite to that of the three price-oriented variables. This directional difference indicates a negative correspondence of service versus the three other variables. The second set of variables reflects more global evaluations, consisting of the two image variables (sales force and manufacturer), along with the strategic orientation variable. Finally, product quality runs almost perpendicular to the price/service dimension, indicating a separate and distinct evaluative dimension.

To interpret the dimensions, the analyst looks for attributes closely aligned with the axis. In this instance, the groups of attributes are slightly angled from the original axes. However, because the perceptual map is a point representation, the axes can be rotated without any impact on the relative positions. If we rotate the axes slightly counterclockwise (much as is done in factor analysis), we would

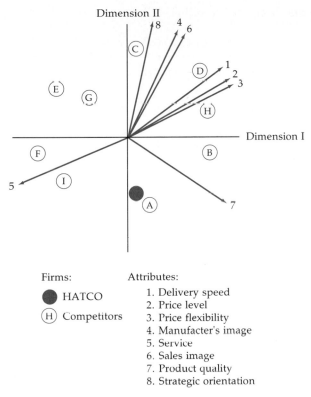

Firms:

● HATCO

Ⓗ Competitors

Attributes:

1. Delivery speed
2. Price level
3. Price flexibility
4. Manufacter's image
5. Service
6. Sales image
7. Product quality
8. Strategic orientation

FIGURE 9.10 Perceptual map with vector representations of attributes.

now have a dimension of price/service versus a second dimension of product quality. Although it is not necessary to actually perform the rotation because firms can be compared directly on the attribute vectors, many times rotation can aid in gaining a more fundamental understanding of the perceived dimension. Rotation is especially helpful in solutions involving more than two dimensions.

To determine the values for any firm on an attribute vector, we need to calculate the projections from the firm to the vector. To assist in interpretation, the program provides the projection values for each attribute, which are listed as the second row of values for each variable in Table 9.3. Also included are the original ratings (values in the first row) to see how well the vector represents the respondent's actual perceptions. For example, the order of the firm's ratings on product quality (from highest to lowest) were H, B, A, HATCO, F, E, D, C, I, and G. Using the vector projections, we see that the order of firms is B, A, HATCO, H, D, I, G, C, F, and E. There is a fairly close correspondence between the original and calculated values, particularly among the top six firms. A statistical measure of fit for each attribute is the correlation between the original ratings and the vector projections. In the case of product quality, the correlation is .654. The analyst should not expect a perfect fit for several reasons. First, the perceptual map is based on the overall evaluation, which may not be directly comparable to the ratings. Second, the ratings are averaged across respondents, making their values determined by differences between individuals as well as differences between firms. Given these factors, the level of fit for the attributes individually and collectively is acceptable.

TABLE 9.3 Interpreting the Perceptual Map with PROFIT

Original Attribute Ratings/Projections on Fitted Vectors

Variables	HATCO	Firm A	Firm B	Firm C	Firm D	Firm E	Firm F	Firm G	Firm H	Firm I	Fit*
Product	6.94	7.17	7.67	3.22	4.78	5.11	6.56	1.61	8.78	3.17	.654
quality	.3065	.3955	.3994	−.2593	.0204	−.4214	−.3022	−.2557	.1914	−.0747	
Strategic	4.00	1.83	6.33	7.67	6.00	5.78	5.50	6.11	7.50	4.17	.816
orientation	−.4078	−.4244	−.1026	−.4477	.4316	.2129	−.1677	.1845	.2096	−.3837	
Overall service	6.94	5.67	3.39	3.67	3.67	6.94	6.44	7.22	4.94	6.11	.753
	.1006	.0231	−.3182	−.1927	−.4739	.2261	.4978	.0793	−.4236	.4816	
Delivery speed	4.00	3.39	7.33	6.11	7.50	4.22	7.17	4.33	8.22	5.56	.505
	−.2663	−.2217	.1664	.3483	.5315	−.0502	−.4196	.0357	.3895	−.5135	
Price level	5.16	3.47	6.41	5.88	6.06	4.94	5.29	4.82	8.35	4.65	.666
	−.2193	−.1633	.2184	.3062	.5246	−.1073	−.4536	.0001	.4088	−.5144	
Sales force's	5.11	1.22	5.78	7.89	6.56	3.28	4.28	6.94	8.67	4.72	.715
image	−.3571	−.3430	.0307	.4209	.5068	.0887	−.3070	.1178	.3140	−.4718	
Price flexibility	5.33	3.72	6.33	5.56	6.39	4.72	5.28	5.22	7.33	5.11	.668
	−.2014	−.1414	.2361	.2897	.5197	−.1274	−.4638	−.0131	.4139	−.5124	
Manufacturer's	4.17	1.56	6.06	8.22	7.72	4.28	3.89	6.33	7.72	5.06	.825
image	−.3631	−.3517	.0188	.4250	.5022	.1002	−.2957	.1242	.3059	−.4658	

*Fit is measured as the correlation between the original attribute ratings and the vector projections.

Incorporating Preferences in the Perceptual Map

Up to this point we have dealt only with judgments of firms based on similarity, but many times we may wish to extend the analysis to the decision-making process and to understand the respondent's preferences for the objects (in this case, firms). To do so, we can employ several MDS techniques that allow for the estimation of ideal points, from which preferences for the objects can be determined. In this example, we employ an external method of preference formation (PREFMAP [7]) that utilizes the aggregate perceptual maps derived in the prior section.

Preferences were measured by asking respondents to detail their preferences for firms in three purchasing situations. For our purposes, we will examine the preferences for firms in the new-buy situation. The program inputs include the coordinates of firms in the aggregate perceptual map and the preferences of six respondents.

The program can estimate both point and vector-based ideal points. In this situation, HATCO management decided on the point representations, which resulted in the derivation of ideal points for the six respondents, plus an ideal point for the "average" subject. The results are shown in Figure 9.11.

There are three groups of respondents (also known as subjects) when preferences are considered. The first is a single-member group (subject 1) whose location is directly adjacent to firm A and HATCO. Given its proximity, these are its two most preferred firms. Preference is directly related to distance; thus preference for other firms is markedly lower. The second group contains subjects 2, 4, and 6, plus the "average" subject. This group is situated among firms E, G, and C. The final group comprises subjects 3 and 5, both of which are located quite far away from the other subjects and firms. Still, preference is represented by the distance from the respondents to the firms.

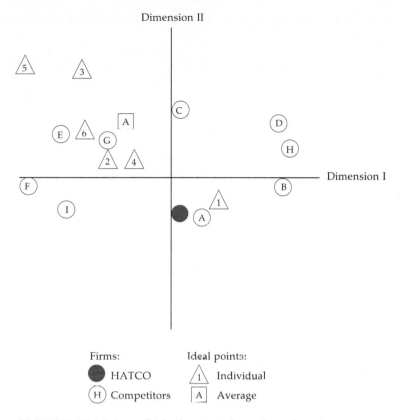

FIGURE 9.11 Map of ideal points for selected and average respondents: new-buy purchasing situation.

The ideal points for a single respondent are of little use, but collectively they indicate possible segments existing among the respondents. In our situation, there seems to be a large segment with higher preferences for firms C, E, and G. HATCO has a shared advantage in preference (with firm A) for only one subject. This example examined only six respondents, and inclusion of all the respondents would present a more complete analysis of the possible market segments.

Overview of the Decompositional Results

The decompositional methods employed in this image study illustrate both the advantages and disadvantages of this approach. The use of overall similarity judgments provides a perceptual map based on only the relevant criteria, chosen by each respondent. However, the attribute-free techniques also demonstrate the notable difficulty of interpreting the perceptual map in terms of specific attributes. The researcher is required to infer the bases for comparison among objects without direct confirmation from the respondent.

Correspondence Analysis: Stages Four and Five

An alternative to attribute-free perceptual mapping is correspondence analysis, a compositional method based on eight binary firm ratings provided for each firm (i.e., the yes/no ratings of each firm on each attribute). In this attribute-based

TABLE 9.4 Cross-Tabulated Frequency Data of Attribute Descriptors for HATCO and Competing Firms

	Variables	HATCO	Firm A	Firm B	Firm C	Firm D	Firm E	Firm F	Firm G	Firm H	Firm I
X_7	Product quality	4	3	1	13	9	6	3	18	2	10
X_8	Strategic orientation	15	16	15	11	11	14	16	12	14	14
X_5	Overall service	15	14	6	4	4	15	14	13	7	13
X_1	Delivery speed	16	13	8	13	9	17	15	16	6	12
X_2	Price level	14	14	10	11	11	14	12	13	10	14
X_6	Sales force's image	7	18	13	4	9	16	14	5	4	16
X_3	Price flexibility	6	6	14	10	11	8	7	4	14	4
X_4	Manufacturer's image	15	18	9	2	3	15	16	7	8	8

method, the perceptual map is a joint space, showing both attributes and firms in a single representation. However, the positions of firms are relative not only to the other firms included in the analysis but also to the attributes selected as well.

Stage Four: Deriving the Correspondence Analysis

Preparing the data for analysis involves creating a crosstabulation matrix relating the attributes (represented as rows) to the ratings of firms (the columns). The individual entries in the matrix are the number of times a firm is rated as possessing a specific attribute. Thus simple frequencies are provided for each firm across the entire set of attributes (see Table 9.4).

Correspondence analysis is based on a transformation of the Chi-square value into a metric measure of distance. The Chi-square value is calculated as the actual frequency of occurrence minus the expected frequency. Thus a negative value indicates, in this case, that a firm was rated less often than would be expected. The expected value for a cell (any firm/attribute combination in the crosstabulation table) is based on how often the firm was rated on other attributes and how often other firms were rated on this attribute. (In statistical terms, the expected value is based on the row [attribute] and column [firm] marginal probabilities.) Table 9.5 contains the transformed (metric) Chi-square values for each cell of the crosstabulation table. High positive values indicate a strong degree of "correspondence" between the attribute and firm, and negative values have an opposite

TABLE 9.5 Measures of Similarity in Correspondence Analysis: Chi-Square Distances

	Variables	HATCO	Firm A	Firm B	Firm C	Firm D	Firm E	Firm F	Firm G	Firm H	Firm I
X_7	Product quality	−1.27	−1.83	−2.08	3.19	1.53	−.86	−1.73	4.07	−1.42	.97
X_8	Strategic orientation	.02	−.13	.76	−.01	.04	−.73	.07	−.60	1.07	−.20
X_5	Overall service	1.08	.40	−1.10	−1.52	−1.48	.57	.59	.65	−.36	.53
X_1	Delivery speed	.68	−.51	−.95	.95	−.27	.40	.20	.86	−1.15	−.37
X_2	Price level	.19	−.19	−.30	.37	.42	−.30	−.54	.08	.20	.23
X_6	Sales force's image	−1.32	−1.49	1.15	−1.54	.23	.81	.55	−1.80	−1.44	1.39
X_3	Price flexibility	−1.02	−1.28	2.37	1.27	1.71	−.73	−.83	−1.59	2.99	−1.66
X_4	Manufacturer's image	1.24	1.69	−.01	−2.14	−1.76	.72	1.32	−1.07	.10	−.85

TABLE 9.6 Determining the Appropriate Dimensionality: Correspondence Analysis

Dimension	Eigenvalue	Percent of Total	Cumulative Total
1	.0765	53.13	53.13
2	.0478	33.19	86.32
3	.0153	10.61	96.93
4	.0027	1.84	98.77
5	.0008	.56	99.33
6	.0006	.40	99.73
7	.0004	.27	100.00

interpretation. For example, the high values for firms A and F with the product quality attribute indicate that they should be located close together on the perceptual map if at all possible. Likewise, the high negative values for firm C on the same variable would indicate that its position should be far from the attribute's location.

Correspondence analysis satisfies all these relationships simultaneously by performing an analysis (similar to factor analysis) to produce dimensions representing the Chi-square distances. To determine the dimensionality of the solution, the researcher examines the cumulative percent of variation explained, much as in factor analysis, and determines the appropriate dimensionality. Table 9.6 contains the eigenvalues and cumulative percent of variation explained for each dimension up to the maximum of seven. Again, the researcher balances the desire for increased explanation versus interpretability. A two-dimensional solution in this situation explains 86 percent of the variation, whereas increasing to a three-dimensional solution adds only an additional 10 percent. Thus a two-dimensional solution was deemed adequate for further analysis.

Stage Five: Interpreting the Correspondence Analysis Results

The attribute-based perceptual map shows the relative proximities of both firms and attributes (see Figure 9.12). If we focus on the firms first, we see that the pattern of firm groups is similar to that found in the MDS results. Firms A, I, and HATCO, and firms C, D, H, and B form similar groups. However, the relative proximities of the members in each group differ somewhat from the MDS solution. Also, firm G is more isolated and distinct, and firms F and E are seen as more similar to HATCO.

These and other comparisons highlight the differences between the two methods and their results. The correspondence analysis results provide a means for directly comparing the similarity or dissimilarity of firms and the associated attributes, whereas MDS allows only for the comparison of firms. But the correspondence analysis solution is conditioned on the set of attributes included. It assumes that all attributes are appropriate for all firms and that the same dimensionality applies to each firm. Thus the resulting perceptual map should always be viewed only in the context of both the firms and attributes included in the analysis.

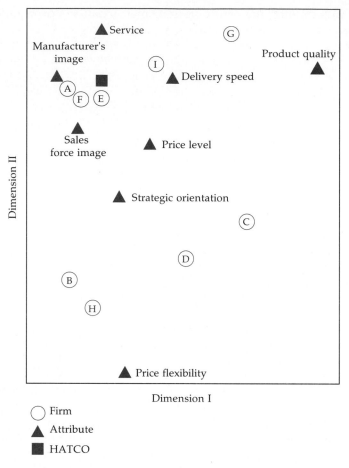

FIGURE 9.12 Perceptual mapping with compositional
methods: correspondence analysis.

Correspondence analysis is a quite flexible technique applicable to a wide range
of issues and situations. The advantages of the joint plot of attributes and objects
must always be weighed against the inherent interdependencies that exist and the
potentially biasing effects of a single inappropriate attribute or firm, or perhaps
more important, the omitted attribute of a firm. Yet, the method still provides a
powerful tool for gaining managerial insight into the relative position of firms
and the attributes associated with those positions.

Stage Six: Validation of the Results

Perhaps the strongest internal validation of this analysis is to assess the convergence between the results from the separate decompositional and compositional
techniques. Each technique employs different types of consumer responses, but
the resulting perceptual maps are representations of the same perceptual space
and should correspond. If the correspondence is high, the researcher can be assured that results reflect the problem as depicted. The researcher should note that

this type of convergence does not address the generalizability of the results to other objects or samples of the population.

Comparison of the two approaches, shown in Figures 9.10 and 9.12, can take two perspectives: interpreting the axes and examining the relative positioning of objects and interpreting the axes. Let's start by examining the positioning of firms. When Figures 9.9 and 9.11 are rotated to obtain the same perspective, we see quite similar patterns of firms, reflecting two groups: firms B, H, D and C versus firms E, F, G and I. While the relative distances among firms do vary between the two perceptual maps, we still see HATCO associated strongly with Firm A in each perceptual map. Correspondence analysis produces more distinction between the firms, but its objective is to define firm positions as a result of differences; thus it will generate more distinctiveness in its perceptual maps.

The interpretation of axes and distinguishing characteristics also shows similar patterns in the two perceptual maps. For the decompositional method shown in Figure 9.10, we noted in the earlier discussion that by rotating the axes we would obtain a clearer interpretation. If we rotate counterclockwise, then dimension I becomes associated with delivery speed, price level, and price flexibility, while dimension II reflects product quality. The remaining attributes are not associated strongly with either axis. In the correspondence analysis (Figure 9.12), we see that there are really three attribute groups representing evaluative dimensions: (1) product quality, (2) price flexibility and (3) the remaining attributes in a third group. This compares quite favorably with the decompositional results except in the case of price flexibility, which was not distinguished as a separate dimension.

Overall, while some differences do exist, owing to the characteristics of each approach, the perceptual maps do provide some internal validity to the perceptual maps. Perceptual differences may exist for a few attributes, but the overall patterns of firm positions and evaluative dimensions are supported by both approaches.

The disparity of the price flexibility attribute illustrates the differences in the two approaches. The decompositional method determines position based on overall judgments, with attributes applied only as an attempt to explain the positions. The compositional method positions firms according to the selected set of attributes, thus creating positions based on the attributes. Moreover, each attribute is weighted equally, thus potentially distorting the map with irrelevant attributes. These differences do not make either approach better or optimal but instead must be understood by the analyst to ensure selection of the method most suited to the research objectives.

Summary

Multidimensional scaling is a set of procedures that may be used to display the relationships tapped by data representing similarity or preference. It has been used successfully (1) to illustrate market segments based on preference judgments, (2) to determine which products are more competitive with each other (similar), and (3) to deduce what criteria people use when judging objects (products, companies, advertisements, etc.). Multidimensional scaling can reveal relationships that appear to be obscured when one examines only the numbers resulting from a study. A visual perceptual map does much to emphasize the relationships between the stimuli under study. However, great care must be taken when attempting to use this technique. Misuse is common. The researcher should become very familiar with the technique before using it and should view the output as only the first step in the determination of perceptual information.

Questions

1. How does MDS differ from cluster analysis? Conjoint analysis?
2. What is the difference between preference data and similarities data, and what impact does it have on the results of MDS procedures?
3. How are ideal points used in MDS procedures?
4. How do metric and nonmetric MDS procedures differ?
5. How can the analyst determine when the "best" MDS solution has been obtained?
6. How does the analyst identify the dimensions in MDS? Compare this procedure with the procedure for factor analysis.
7. Compare and contrast correspondence analysis with the MDS techniques.

References

1. BMDP Statistical Software, Inc. *BMDP Statistical Software Manual, Release 7*, vols. 1 and 2. Los Angeles: 1992.
2. Carroll, J. Douglas, Paul E. Green, and Catherine M. Schaffer. "Interpoint Distance Comparisons in Correspondence Analysis." *Journal of Marketing Research* 23 (August 1986): 271–80.
3. ———. "Comparing Interpoint Distances in Correspondence Analysis: A Clarification." *Journal of Marketing Research* 24 (November 1987): 445–50.
4. Chang, J. J., and J. Douglas Carroll. "How to Use PROFIT, a Computer Program for Property Fitting by Optimizing Nonlinear and Linear Correlation," unpublished paper. Murray Hill, N.J.: Bell Laboratories, 1968.
5. ———. "How to use INDSCAL, a Computer Program for Canonical Decomposition of *n*-way Tables and Individual Differences in Multidimensional Scaling," unpublished paper. Murray Hill, N.J.: Bell Laboratories, 1969.
6. ———. "How to Use MDPREF, a Computer Program for Multidimensional Analysis of Preference Data," unpublished paper. Murray Hill, N.J.: Bell Laboratories, 1969.
7. ———. "How to Use PREFMAP and PREFMAP2—Programs Which Relate Preference Data to Multidimensional Scaling Solution," unpublished paper. Murray Hill, N.J.: Bell Laboratories, 1972.
8. Green, P. E. "On the Robustness of Multidimensional Scaling Techniques." *Journal of Marketing Research* 12 (February 1975): 73–81.
9. Green, P. E., and F. Carmone. "Multidimensional Scaling: An Introduction and Comparison of Nonmetric Unfolding Techniques." *Journal of Marketing Research* 7 (August 1969), 33–41.

10. Green, Paul E., Frank J. Carmone, and Scott M. Smith. *Multidimensional Scaling: Concept and Applications*. Boston: Allyn & Bacon, 1989.

11. Green, P. E., and Vithala Rao. *Applied Multidimensional Scaling*. New York: Holt, Rinehart and Winston, 1972.

12. Greenacre, Michael J. *Theory and Applications of Correpondence Analyses*. London: Academic Press, 1984.

13. ———. "The Carroll-Grenn-Schaffer Scaling in Correspondence Analysis: A Theoretical and Empirical Appraisal." *Journal of Marketing Research* 26 (August 1989): 358–65.

14. Hoffman, Donna L., and George R. Franke. "Correspondence Analysis: Graphical Representation of Categorical Data in Marketing Research." *Journal of Marketing Research* 23 (August 1986): 213–27.

15. Holbrook, Morris B., William L. Moore, and Russell S. Winer. "Constructing Joint Spaces from Pick-Any Data: A New Tool for Consumer Analysis." *Journal of Consumer Research* 9 (June 1982): 99–105.

16. Levine, Joel H. "Joint-Space Analysis of 'Pick-Any' Data: Analysis of Choices from an Unconstrained Set of Alternatives." *Psychometrika* 44 (March 1979): 85–92.

17. Lingoes, James C. *Geometric Representations of Relational Data*. Ann Arbor, Mich.: Mathesis Press, 1972.

18. Kruskal, Joseph B., and Frank J. Carmone. "How to Use M-D Scal. Version 5-M, and Other Useful Information," unpublished paper. Murray Hill, N.J.: Bell Laboratories, 1967.

19. Kruskal, Joseph B., and Myron Wish. *Multidimensional Scaling*. Sage University Paper Series on Quantitative Applications in the Social Sciences, 07–011. Beverly Hills, Calif.: Sage, 1978.

20. Lebart, Ludovic, Alain Morineau, and Kenneth M. Warwick. *Multivariate Descriptive Statistical Analysis: Correspondence Analysis and Related Techniques for Large Matrices*. New York: Wiley, 1984.

21. Maholtra, Naresh. "Validity and Structural Reliability of Multidimensional Scaling." *Journal of Marketing Research* 24 (May 1987): 164–73.

22. Market ACTION Research Software, Inc. *MAPWISE: Perceptual Mapping Software*. Peoria, Ill.: Business Technology Center, Bradley University, 1989.

23. Raymond, Charles. *The Art of Using Science in Marketing*. New York: Harper & Row, 1974.

24. Schiffman, Susan S., M. Lance Reynolds, and Forrest W. Young. *Introduction to Multidimensional Scaling*. New York: Academic Press, 1981.

25. Srinivasan, V., and A. D. Schocker. "Linear Programming Techniques for Multidimensional Analysis of Preferences." *Psychometrica* 38 (September 1973): 337–69.

26. SPSS, Inc. *SPSS User's Guide*, 4th ed. Chicago: 1990.

27. Smith, Scott M. *PC—MDS: A Multidimensional Statistics Package*. Provo, Utah: Brigham Young University, 1989.

Product Positioning: An Application of Multidimensional Scaling

Yoram Wind
Patrick J. Robinson

Product positioning, a construct frequently referred to by marketing and advertising practitioners, is rarely if ever mentioned in the professional marketing literature. This has occurred in a situation where there seems to be a marginal acceptance of the relevance (and sometimes even importance) of product positioning as a diagnostic device which provides operational guidelines for new product development efforts, redesign of existing products and design of advertising and distribution strategies, while little attention is given to its measurements.

This paper is concerned with just this latter issue—that quality of one research approach—multidimensional scaling and related techniques—to the measurement of product positioning. While many papers [8, 9, 13] emphasize the methodology of multidimensional scaling, and offer excellent espository discussion of these techniques, this paper offers a discussion of these techniques and emphasizes the position of this methodology to one set of marketing terms—product positioning. This will be done by highlighting a number of studies in which various multidimensional scaling techniques have been used to establish product positioning. This is preceded by a brief discussion on the future and measurement of product positioning. The paper concludes with some comments on the measurement of product positioning and the relevance of positioning studies as guidelines for marketing strategies.

On the Nature and Measurement of Product Positioning

The term product (brand) positioning refers to the place a product occupies in a given market. Conceptually, the origin of the positioning concept can be related to the economist's work on market structure, competitive position of the firm and the concepts of substitution and competition among products. Marketing also has been concerned with such phenomena as product differentiation [14] and market position analysis, an interest which ranges from simple market share statistics to various approaches (such as Markov processes) for forecasting changes in a firm's market position [1].

More recently, increasing attention has been given to product image. This suggests a new perspective on product positioning, one which focuses on *consumers' perceptions* concerning the place a product occupies in a given market. In this context, the word positioning encompasses most of the common meanings of the word position—position as a place (what place does the specific product occupy in its relevant market?), a rank (how does the given product fare against its competitors on various evaluative dimensions?), and a mental attitude (customer attitudes—the cognitive, affective and action tendencies) toward the given product.

Given this view, the product (brand) positioning should be assessed by measuring consumers' or organizational buyers' *perception* and *preference* for the given product in relation to its realistic competitors (both branded and generic). The necessity to determine product positioning, not only on the basis of its perception (perceived similarity to other products) but also on consumers'

From Yoram Wind and Patrick J. Robinson, "Product Positioning: An Application of Multidimensional Scaling," (1972) pp. 155–175. Reprinted by permission.

preferences for it (overall preference as well as preference under various conditions—scenarios) is a key premise of the "ideal" approach to product positioning,[1] this is based on the premise that customer behavior is a function of *both* their perceptions and preferences and the recognition that buyers may differ with respect to both their perception and the preference for a product.

A somewhat different approach has been taken by Stefflre and his associates in their "Market Structure" analysis [2, 3, 15]. In this new product development procedure, the first few research steps are concerned directly with establishing the market structure, i.e., "determine which items (products and brands) consumers see as constituting a market and the 'position' of each item in the market vis-à-vis the other items."[2] This analysis positions the various brands based on consumers' *perceptions* (similarity) of the various brands and utilizes certain multidimensional scaling procedures. This "positioning" analysis dos not utilize preference data in *conjunction with* similarities data but occasionally uses data on patterns of brand-to-brand substitution obtained from large scale purchase panel data, when these are available.

Whether one uses perceptions, preferences, or both as basis for product positioning, it is essential to start with the appropriate set of competing products. This set, which in many cases includes products outside the immediate product class of the product in question, can be generated by marketing experts based on their experience and analysis of existing information or from unstructured depth interviews with consumers. Identifying a broad set of competing items is quite crucial since it constitutes the stimulus set for the positioning study. In designing the stimulus set it is sometimes desirable to include two types of products and brands—brands in the same product class and products and brands outside the product class—which may be used by consumers as substitutes for the product in question. For example, in a study of soups, one could include a set of different soup brands, forms, types, and fla-

vors as the primary set as well as a set of soup substitutes—sandwich, salad, coffee, various snacks, etc.

Given the stimulus set of brands and products, the next step is to determine consumers' perceived product positioning. This can be done by eliciting (a) consumers' perceptions *via* a variety of available procedures for (overall) similarity measurement, or (b) consumers' preferences—overall and under a variety of usage and purchase conditions—or (c) both consumers' perceptions and preferences.

The data collection procedure employed in each of these cases depends on the number of items in the stimulus set, the total length of the interview, and the researcher's preferences. If one deals with a set of 20 to 40 brands, sorting or rating will be a much more appropriate task than strict ranking which is feasible when a smaller competitive set is included in the study.

Whatever the data collection procedure is, the data should provide in its original or transformed form a $P \times P$ matrix of product similarities (or dissimilarity) at the desired level of analysis—individual, segment, or total market. This matrix serves as the input data for the appropriate set of multidimensional scaling and clustering programs. The basic idea underlying this analysis is that a market (consumers' perceptions of the various brands) can be conceived as a multidimensional space in which individual brands are positioned. A product's positioning is determined from its position on the relevant dimensions of the similarity space, its position on the various product attribute vectors (if a joint space analysis is undertaken), and its position with respect to other brands, as may be obtained *via* cluster analysis.

A prototypical product positioning procedure is summarized in Figure 1. In designing such a study there are a number of research decisions which have to be made concerning the following topics.

Stimulus Definition

The brands and products selected for inclusion in the stimulus set can be defined in a number of ways and may be presented in terms of the physi-

---

[1] In certain empirical studies, because of cost considerations, one may elect to use only preference data as the basis for product positioning.

[2] Barnett [2].

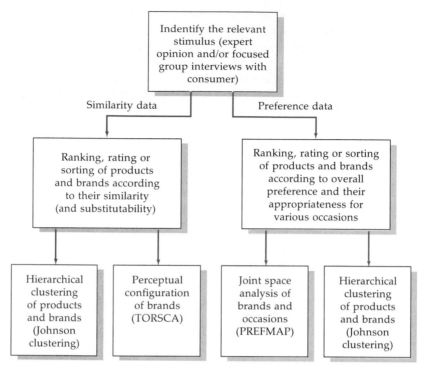

FIGURE 1 Determining product positioning.

cal objects or services themselves, names of the items, verbalized profile descriptions of the items, or a mixture of the above. As might be imagined, respondent evaluations may easily differ, depending upon the manner in which the stimuli are described and semantically encoded.

Task Definition

Defining the respondent's task requires an explicit decision. The task could consist of having the respondent react to the stimuli in terms of: overall evaluation of the objects in terms of preference or in terms of their relative similarity; judgments of objects' similarity or preference (or other types of orderings) according to a set of prespecified scenarios (problem solving conditions) or according to a set of prespecified attributes (other than overall similarity of preference). Alternatively, one can collect objective data on the characteristics of the stimuli, in which case an objective performance space rather than a subjective brand space will result.

Response Definition

Assuming that the stimulus set and task have been defined, the researcher must still contend with specifying the nature of the subject's response. Verbal and non-verbal responses may involve ratings judgments, ranking, including strict orderings or weak orderings (those including the possibility of ties) and assignments to prespecified classes. Even given the restriction of responses to those that represent subjective, verbalized judgments, there are numerous ways for eliciting ratings, rankings, and category assignments.

Some Illustrative Studies

To illustrate the applicability of multidimensional scaling to the study of product positioning we will review briefly the results of a number of commercial studies. All of these studies were originally designed for some other purpose such as evaluating a set of new concepts, evaluating new product designs, or promotional programs.[3] They did include, however, as an integral part of the study an examination of product positioning by each of the relevant market segments.

The studies to be reported cover a wide range of products including calculators, diet products, medical journals, financial services, and retail stores. They also utilize a variety of analytical techniques, and develop maps of product positioning based on perceptions, preferences, or a combination of the two.

Positioning a Calculator

As with any product, calculators can be positioned in an objective space or a subjective space. In a business and scientific calculating machine

---

[3]Common to all these studies is the concern with *evaluating* alternative stimuli, and not *generating* them. Occasionally, as a by-product of these studies one can get some ideas for new products. Yet, there are more efficient procedures for generating new product ideas which do not require a positioning analysis.

study, objective data were obtained for 40 calculator models on each of 20 performance characteristics (e.g., control features, maximum number of index registers, minimum and maximum storage size, memory cycle time, output display speed, etc.). From these performance data similarities were computed between each pair of calculator models across the full set of 20 objective variables. The result of this analysis is presented in Figure 2 in a form of a "tree diagram" which resulted from a hierarchical clustering program [11]. The tree diagram portrays the clusters of models that emerged from this analysis.

This tree diagram may be thought of as being similar to a family tree in the sense that close relations and "branches" of the family that have a closely related pedigree, are shown as being lined in close proximity. That is, the sooner two calculators join, from left to right, the more similar the calculators are in terms of objective performance.

Recognizing that objective positioning does not necessarily correspond to the perceived position of the various calculators, a study was conducted

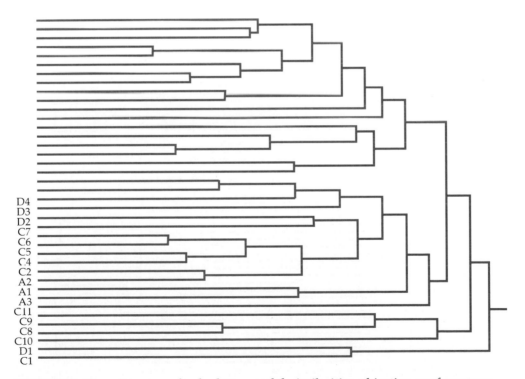

FIGURE 2 Tree diagram of calculator model similarities objective performance evaluation.

in which various calculator users (representing professional and managerial positions in a variety of using industries) were asked, among other things to group the 40 models according to their impressions of the various calculators regarding their similarities and differences from a performance standpoint. These data were analyzed *via* various multidimensional scaling [16] and hierarchical clustering procedures. The results of the clustering program are portrayed as a tree diagram in Figure 3. A comparison of this figure with Figure 2 (the objective space) indicates considerable difference between the subjective configuration and the objective one. The major difference was that some calculator users tended to view all models of a given manufacturer as a group with exception of one model of calculator that was viewed as dissimilar to the others of the same make. Calculators of a given manufacturer or of certain capabilities were grouped together depending on which respondent segment was concerned. When clustered on *objective* performance data the various models did *not* maintain their manufacturer separation nor the same capability configurations. This disparity in the perceived and objective positioning suggests the importance of not restricting positioning analysis

to objective performance data base. It may satisfy engineering specifications, but buyers and users may see things very differently. Furthermore the perceived product positioning differed considrably among the various user groups, i.e., Toshiba customers, for example, viewed the calculator market differently from non-Toshiba customers, suggesting the need for coupling the product positioning analysis with market segmentation.

Among the policy implications stemming from the positioning aspects of this study were distinct sales promotion appeals (emphasizing or deemphasizing the objective product attributes) for certain models aimed at different market segments.

Positioning New Diet Products

The calculator study illustrated a positioning based on contrasting objective and subjective performance evaluations of existing industrial products (calculators). The same approach, viz., gaining better understanding of a product positioning *via* a comparison of its objective and subjective evaluations, was undertaken in a study of diet food items. In this case one of the objectives of the

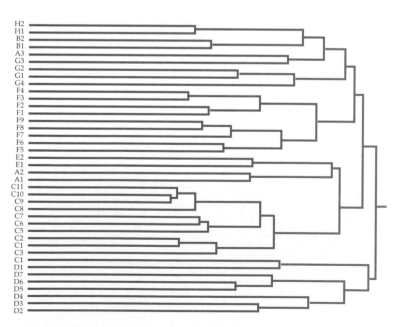

FIGURE 3 Tree diagram of calculator model similarities subjective performance evaluation (segment 1).

study was to assess the positioning of some new diet products. The respondents were women who were on a diet and were asked to group food items (some of which were diet products such as Metrecal and some were not, such as pudding, potato chips, milk shakes, etc.) and 13 concepts of new diet products according to their similarity.

In addition they were asked to evaluate the products and concepts according to their overall preference and preference for serving and eating in various eating occasions (scenarios) such as: for lunch, short crash diet to lose weight quickly, to improve appearance, at dinner, when I am by myself, etc. Following this they also rated the products and concepts on a set of 12 attributes including calories, nutrition, taste, and convenience of preparation, vitamins, cholesterol and fillingness. This resulted in three sets of subjective data:

1. Product similarity data
2. Product preference rankings (overall and by scenario)
3. Product rankings on various product attributes

These data were subjected to a variety of multidimensional scaling programs. The 40 × 40 product similarity data across subjects were submitted to the TORSCA multidimensional scaling program and to the Johnson clustering program. The product evaluative data (both preference for overall usage and usage under certain occasions, and attributes) were submitted to Carroll and Chang's joint space (PREFMAP) program [4].

In addition to these data, a group of food technicians evaluated the various products according to their actual "objective" attributes. These data were also submitted to an appropriate set of multi-dimensional scaling programs enabling a comparison of the subjective-objective maps to be made.

Figure 4 presents the two-dimensional configuration of the 40 items as derived from the products similarity data. The product clusters were determined by the Johnson hierarchical clustering program and incorporated into Figure 4. An examination of this figure suggests a number of clusters:

A cluster of high calorie snack/dessert products such as cookies, candies, milk shakes, etc.

A cluster of "non-natural" diet concepts and products including the new diet concepts and Metrecal.
A cluster of natural diet foods such as yogurt, fruit salad, celery sticks, etc.
A cluster of solid, meal-like items.

Examination of any given product or concept reveals its position with respect of other products and concepts as well as its position on the two dimensions. If one is interested in the positioning of new diet concepts, Figure 4 suggests the following conclusions:

1. The new concepts will compete with other "artificial" diet products such as Metrecal and Weight Watchers Complete Meals.
2. The concepts seem to be positioned opposite their "natural" counterparts. Hence if dimension 2 is reviewed as "fattening-healthy" it may suggest that the "fattening" attribute of the natural products may "rub off" on the diet concepts, leading to overestimation of the fattening attributes of concepts such as a "diet cookie."

Further insights into the position of each concept and product was gained by an examination of the joint space configuration of the products and concepts and their perceived attributes. Figure 5 presents the results of this analysis. Looking at the solid line vectors (subjective evaluations) and the relation of the various products to them suggests:

"Natural" diet products such as yogurt are perceived as being both nutritious and healthy.
"Natural" meal items such as steak are perceived as filling and rich in proteins.
High-calorie snacks and desserts, such as potato chips and candies, are perceived as being "fatty" (fat, cholesterol, and carbohydrates), sugary, and high in calories. They are, however, convenient for preparation.
The new diet concepts are perceived as expensive, not tasty, less nutritious than natural diet products and poor on health, protein and fillingness.

A comparison of the "subjective" evaluation configuration with the objective data was done in two stages. First a two- and three-dimensions

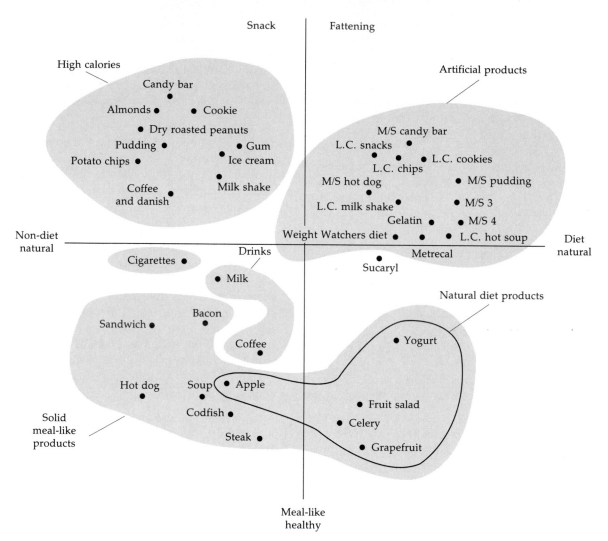

FIGURE 4 Two-dimensional perceptual configuration of twenty-seven food products and thirteen new diet concepts.

"objective" configuration of the 40 products was derived and compared to the subjective configuration. This suggested that dieters do not perceive diet and non-diet products according to their objective attributes.

A more direct comparison was undertaken *via* a joint space analysis, the results of which are presented in Figure 5. A comparison of the discrepancies between the objective (broken line) and subjective (solid line) vectors suggest that the greatest discrepancy (50% or greater) exists with regard to fillingness, carbohydrates, proteins, and vitamins. Fairly high congruence (less than a 20% discrepancy) is found with respect to calories, sugar, fat, cholesterol, and convenience of preparation. This suggests that dieters are more conscious of and interested—hence knowledgeable with respect to—this latter set of attributes.

Positioning of a Journal

The concept of positioning can apply not only to industrial and consumer products but also to professional journals. A publisher of a medical journal was concerned with the positioning of his journal. A study was undertaken among a sample

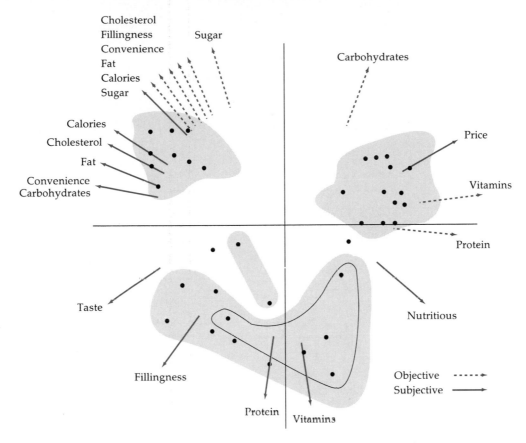

FIGURE 5 Joint space configuration: two-dimensional perceptual
configuration of forty products and their "perceived" and objective
attributes.

of physicians who were presented with a set of 10
medical journals and asked to rank them according to 6 scenarios such as "general overall preference," "general reading preference," etc. These
data provided the input for a journal dissimilarity
matrix, across scenarios.

These data were then submitted to the PREF-
MAP joint space multidimensional scaling program which finds vector directions in the perceptual map (obtained in an earlier phase *via*
INDSCAL multidimensional program) [5]. The
relationships of the vectors to the journals can be
interpreted in terms of preference evaluations.
Figure 6 presents the results of this analysis. Interpretation of the dimensionality of this figure
shows that the primary attribute by which medical journals are judged are technical versus non-
technical (horizontal axis) and the specialized

versus general (vertical axis). We see, for example, that *JAMA* is perceived as most technical
whereas *Human Sexuality* is seen as the least technical journal.

Observing the six preference vectors illustrates
the popularity of *JAMA*, the *New England Journal
of Medicine,* and *Modern Medicine.* These journals
carrying high general preference are those that
are viewed as informative as well. The scenarios,
"best when in a hurry" and "greatest breadth of
appeal" show directions favorable to the news
journals such as as *Modern World News* and *Medical Tribune.*

The specific position of any of the journals can
easily be determined by examining its relations to
other journals and its relative position on the two
dimensions and the various preference vectors.

In this study one of the objectives was to help

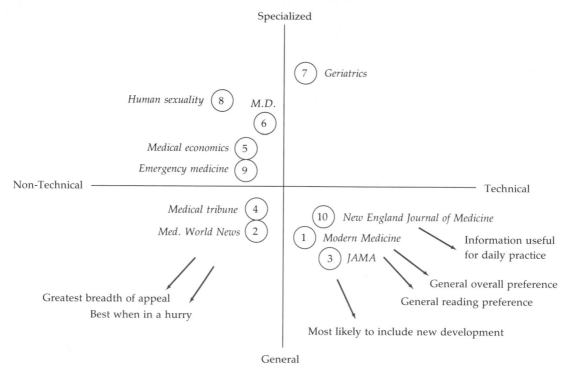

FIGURE 6 Joint space of journals and (vector directions of) evaluative scenarios—From PREFMAP Computer Program.

the journal in question present itself more effectively to its prospective space-buyers insofar as its relative position and strengths versus other vehicles. There was substantial "surprise value" and yet intuitively satisfying insights for management in the positioning and relative strengths and weaknesses of the various publications. Solicitation policy and various editiroal, format, and content changes were clearly identified for improvement and exploitation.

Positioning of Financial Services

The three studies described so far illustrated a number of direct attempts at positioning a product based on consumers' perceptions and preferences as well as pointing out the discrepancy between "objective" and "subjective" product evaluation. Product positioning can also be achieved, however, by other more indirect methods. One such procedure is described in the next study. This study was concerned with the evalua-

tion and positioning of a number of new financial services. In this case, in addition to the customary evaluation of new and existing services on a set of evaluative scales, the respondents (male heads of households) were also asked to pick up from a set of 12 occupations—the five occupations whose holders would be most likely to use the given service. Upon selection of the five occupations, respondents were further asked to rank them from most to least likely to use the given service. These data and the similarity configuration of the various financial services (obtained in an earlier phase, *via* the TORSCA program) were submitted to a joint space program [6]. The results of this analysis are presented in Figure 7.

An examination of this map suggests a clear division of the set of financial services based on their perceived prestige into high prestige services such as "investment fund" or "special telephone advice" as opposed to "financial programming" and "monetary counseling" services which are viewed as low prestige services. In

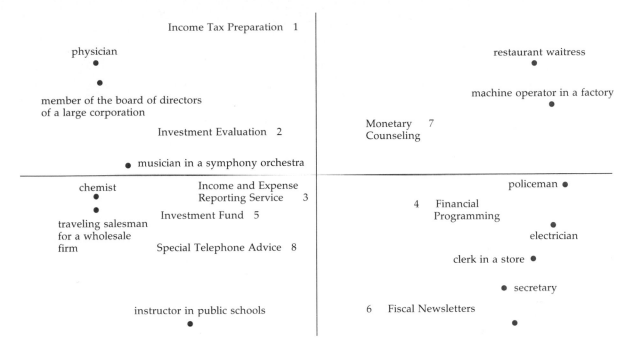

FIGURE 7

addition a number of services have either a very wide appeal that cuts across all occupations (as might be the case with income expense reporting services).

This indirect service positioning was consistent with the results obtained from the more direct position based on perceived similarity of the concepts and existing services. It provided further guidelines for promotional strategy than the more common approach, by suggesting the most appropriate type of people (occupations) to be portrayed in promoting the service. For example, the use of testimonials of blue-collar workers for one service and testimonials of professionals such as physician and chemist for another.

Positioning of a Retail Store

Our last illustrative study presents a new dimension in positioning studies. Whereas all of the previous studies focus on a product/service/journal positioning at a point in time, the current study is concerned with the changes in a retail store positioning over time. The study was based on housewives' evaluation of various retail stores

in a given metropolitan area over a period of two years. The data for the first three months and last three months were grouped separately and analyzed for the overall market and various *a priori* segments.

Figure 8 presents the results of matching (*via* the Cliff Match procedure [7] separate TORSCA maps of 9 stores for each of the time periods. The stores in this map have been disguised, but the configuration is the actual configuration that was derived from the actual data. The dimensions of this map could be interpreted as high prestige/relatively expensive vs. low prestige/discount (the horizontal axis) and width of assortment (vertical axis). The change in the perceived position of each store is traced by the dotted line which links the store position in the two time periods. Examination of the magnitude and direction of changes in the stores' positions can provide management with considerable insight into their market position and changes in it as they occurred over the relevant study period.

As in the other positioning studies, the time series analysis can be extended to cover changes not only in the stores but also in the stores rela-

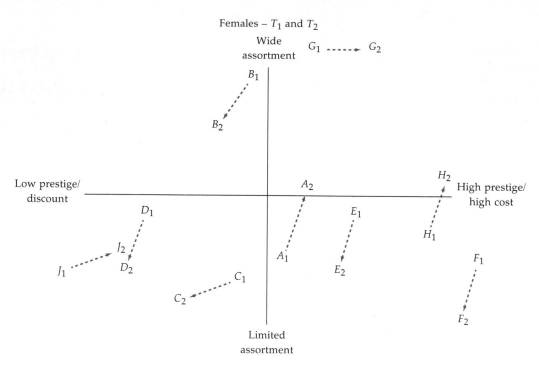

FIGURE 8 Two-dimensional configuration of nine stores.

tive to various evaluative scales (such as good service, easy credit, best value, etc.) and/or various products for which the store may be appropriate (e.g., appliances, clothes, etc.). The analytical procedure is quite similar and the only difference being that instead of developing a "super space" (*via* the Cliff Match program) based on the results of the TORSCA program, the input data to the matching program are the results of the joint space analyses of stores and evaluative scales or stores and products.

These analyses were of diagnostic significance in revealing relative competing store positions and the extent and direction of movement over a known time interval. As with any time series analysis, certain implications were revealed and of relevance to those concerned with relative standings and trends. More explicit and detailed insights stem from examining store policy image positioning and shifts over time. Clearly, significant differences in one's own position may result not only from one's own but one's competitor's

policy shifts and positioning and tracing such moves may suggest a number of strategy implications.

Conclusions

The five studies which were described briefly in this paper illustrate the wide applicability of the concept of positioning to a variety of products (calculators, diet foods, and medical journals), services (financial), and retail stores. They further demonstrate the usefulness of utilizing multidimensional scaling and clustering techniques in these types of studies.

Multidimensional scaling techniques include a set of computer algorithms that permit the researcher to develop perceptual and evaluative "maps" (i.e., geometric configurations) that summarize how people perceive various stimulus objects (products, services, stores, and the like) as being similar or different, and how much stimulus objects are evaluated on a variety of evaluative scales. As such, this set of techniques is especially suited for the portraying of the perceived position of a product.

Moreover, nonmetric multidimensional scaling and clustering techniques utilize data that need only be rank ordered (relative similarity or preference for a set of stimuli) which facilitates the respondent's task and places the burden of analysis on the researcher.

In view of the heterogeneous nature of every market the real value of product positioning is revealed, however, only when the positioning is coupled with an appropriate market segmentation analysis. In the studies reported here and in a variety of other studies which included positioning analyses, a segmentation analysis was always included, enabling one to conduct both an overall and segmented positioning analysis. The scope of this paper did not enable us to elaborate on the differences in product positioning by segment. Yet, in all of the studies reported above the findings were quite conclusive in suggesting that such differences do exist and hence justify the value of conducting separate positioning analysis for each market segment.

Product positioning when coupled with market segmentation can provide useful guidelines for the design and coordination of the firm's marketing strategy.

As with segmentation studies the value of a separate positioning study is quite limited. Applying the concept of positioning requires that each marketing research project which is concerned with evaluating a given marketing strategy (e.g., concept evaluation, advertising evaluation, and the like) should include a section on the perceived positioning of the given product or service. This should provide a useful addition to almost any marketing study.

Finally, it is hoped that employing multidimensional scaling and related techniques in positioning studies with further the utilization of these and other recently developed techniques in other areas of marketing and image research, hence contributing to improved new directions in the measurement of attitudes and market behavior.

References

Alderson, W. and Green, P. E., *Planning and Problem Solving in Marketing* (Homeward, Ill.: Richard D. Irwin, 1964), especially pp. 170–192.

Barnett, N. L., "Developing Effective Advertising for New Products." *Journal of Advertising Research.* 8 (December 1968), pp. 13–20.

Barnett, N. L., "Beyond Market Segmentation." *Harvard Business Review* 27, (January–February 1969), pp. 152–166.

Carroll, J. D., and Chang, J. J., "Relating Preference Data to Multidimensional Scaling Solutions via a Generalization of Coombs' Unfolding Model," mimeographed. Bell Telephone Laboratories. Murray Hill, N.J., 1967.

Carroll, J. D., and Chang, J. J., "A New Method for Dealing with Individual Differences in Multidimensional Scaling," mimeographed. Bell Telephone Laboratories, Murray Hill, N.J., 1969.

Chang, J. J., Carroll, J. D., "How to Use MDPREF, A Computer Program for Multidimensional Analysis of Preference Data," mimeographed, Bell Telephone Laboratories, Murray Hill, N.J., 1969.

Cliff, N., "Orthogonal Rotation to Congruence," *Psychometrika*, 31 (1966), pp. 33–42.

Green, P. E. and Carmone, F. J., *Multidimensional Scaling and Related Techniques in Marketing Analysis* (Boston: Allyn and Bacon, 1970).

Green, P. E. and Roa, V., *Applied Multidimensional Scaling: A Comparison of Approaches and Algorithms* (New York: Holt Rinehart Inc., 1972).

Howard, N. and Harris, B., "A Hierarchical Grouping Routine, IBM 360-65 FORTRAN IV Program." University of Pennsylvania Computer Center, Philadelphia, Pa., 1966.

Johnson, S. C., "Hierarchical Clustering Schemes," *Psychometrika.* 32 (1967), pp. 241–54.

Kuehn, A. A. and Day, R. L., "Strategy of Product Quality," *Harvard Business Review,* 40 (November–December 1962), pp. 100–110.

Silk, A. J., "Preference and Perception Measures in New Product Development: An Exposition and Review," *Industrial Management Review,* 11 (Fall 1969).

Smith, W. R., "Product Differentiation and Market Segmentation as Alternative Marketing Strategies." *Journal of Marketing,* 21 (July 1956).

Stefflre, V., "Marketing Structure Studies: New Products for Old Markets and New Markets (Foreign) for Old Products," in Bass, King, and Pessemier (editors) *Applications of the Sciences in Marketing* (New York: John Wiley and Sons, 1968), pp. 251–268.

Young, F. W. and Torgerson, W. S., "TORSCA—FORTRAN IV Program for Shepart-Kruskal Multidimensional Scaling Analysis," *Behavioral Science,* 12 (1967), p. 498.

A Reduced-Space Approach to the Clustering of Categorical Data in Market Segmentation

Paul E. Green
Catherine M. Schaffer
Karen M. Patterson

Introduction

The value of clustering techniques in marketing research and market segmentation has been widely recognized, if we are to judge by the volume of articles that have appeared on the topic. A recently prepared bibliography (Dickinson 1986), lists some 17 pages of citations dealing with cluster analysis. Punj & Stewart (1983) discuss a number of clustering studies in marketing (as well as some of the major problems that still beset their application).

One of these problems concerns the clustering of profile data (typically, individuals by characteristics) when the characteristics are all qualitative, multistate attributes. As is generally known, if all characteristics are measured, continuous variables (e.g., ratio or interval-scaled), it is easy to develop a dissimilarity coefficient for each pair of individuals, based on some type of distance measure. Even if the variables all consist of ranked variables, it is not too difficult to find suitable monotonic transformations to convert them to "measured" variables (Lingoes 1977), prior to computing a distance-based proximity measure.

If the characteristics are *dichotomous* attributes, a wide variety of dissimilarity coefficients are available. For example, the Proximities program in the SPSSX statistical package lists some 27 different matching coefficients, based on functions of frequencies in a 2×2 contingency table for each pair of attributes (SPSSX User's Guide 1986).

But, in the case of multistate attributes (where the states, in general, are not ordered along some continuum), the clustering literature has relatively little to say. Some authors have advocated the use of dummy variable coding of the states, followed by application of a matching coefficient based on dichotomous data (Sneath & Sokal 1973). However, few applications appear to have been made of profiles characterized by several attributes, some (or all) of which may be multistate (Romesburg 1984).

In marketing research it is not unusual to obtain categorical data in which each individual is characterized as falling into some category on each of several attributes. For example, a buyer of deodorant may be simultaneously classified by: (a) name of brand most recently purchased; (b) favorite product form (e.g., stick, roll-on, spray); (c) type of occupation; (d) age group; (e) sex and (f) marital status. The researcher is often interested in clustering buyers in terms of their profile commonalities across all of the attributes.

Correspondence Analysis

Recently, simple (or two-way) correspondence analysis (Hoffman & Franke 1986) has been described as a multidimensional scaling method for spatially portraying categorical data, originally expressed as cross tabulations. Row and column labels (e.g., brand names by product forms) are represented as points in a joint space in which (squared) interpoint distances can be related to a

From "A Reduced-Space Approach to the Clustering of Categorical Data in Market Segmentation." Paul E. Green, Catherine M. Schaffer, and Karen M. Patterson, *Journal of the Market Research Society*, vol. 30, no. 3 (1988), pp. 267–88. Reprinted by permission.

chi square metric. With appropriate scaling, Carroll, Green & Schaffer (1986, 1987) show that *both* within-set and between-set distances are comparable.

Carroll & Green (1987) describe ways in which simple correspondence analysis can be extended to multiple correspondence analysis, where individuals are classified on three or more qualitative attributes. Again, the attribute levels can be portrayed as points in a Euclidean space. Individuals can also be represented in the space as centroids of the coordinates representing the attribute categories in which they fall.

Hence, in both simple and multiple correspondence analysis, attribute categories and/or individuals are represented as points in a reduced Euclidean space (usually of only two or three dimensions, for clarity of presentation).

Purpose of the Paper

The purpose of the current paper is, first, to utilize the reduced space approach to find spatial coordinate positions of attribute categories (or individuals). Following this, a standard proximity measure can be calculated and used to cluster the attribute categories and/or individuals on the reduced space (which is not limited to only two or three domensions) as a *complementary* way to portray the mapping-based relationships in the original data.

There is ample support for the combined use of MDS and cluster analysis in the analysis of a common set of profile data (Green & Rao 1972; Green, Wind & Jain 1973, Arabie *et al.* 1981; Srivastava, Leone & Shocker 1981). However, to data we have seen little discussion of the usefulness of correspondence analysis as a prelude to developing product/market segments via cluster analysis.[1]

Three types of real-world applications of the approach are described in this paper. The first application involves physicians' preferences for various types of promotional activities (e.g., symposia, clinical reports, detailing, etc.) by which information about a newly introduced antihypertensive product could be disseminated to the medical community. Simple correspondence is first used to develop a joint-space representation of physician-type segments and the alternative promotional activities being evaluated. The coordinates of the joint space are then used to find Euclidean distances for each pair of points which, in turn, are submitted to a hierarchical grouping program. The clusters obtained from the grouping program provide a complementary way to promotional activities, and physician-activity pairs.

The second application involves a different study of physicians' descriptions of the last patient seen for the treatment of hypertension. In this study each respondent picked one category from each of nine attributes that best described his/her most recently treated hypertensive patients. Multiple correspondence analysis was used to develop a joint space of patient descriptors and physicians (the respondents). Physicians were then clustered via a *K*-means program, using their coordinate values from the reduced space as input data.

The third application entails commercial farmers' evaluations of 18 benefits that a new herbicide (designed for soybean crops) might provide. In this case the data were of the 'pick k/n' variety; that is, multiple selections were permitted. Respondents were again clustered in the reduced space via *K*-means clustering. The clusters were then submitted as a new categorical variable to obtain a second reduced space for showing the segment positions *vis-à-vis* the 18 herbicide benefit positions.

In all three applications, technical details are suppressed; these can be found in the correspondence analysis papers, cited earlier. In addition, textbooks are available that describe both simple and multiple correspondence methods (Greenacre 1984; Lebart, Morineau & Warwick 1984; Heiser 1981; Meulman 1982; Gifi 1981; Nishisato 1980). Our primary audience for the paper is industry practitioners, interested in the practical utility of the approach to market segmentation problems.

---

[1] However, Lebart, Morineau & Warwick (1984) describe an approach in which correspondence analysis and clustering are applied sequentially to survey data. Aside from this, we have seen little mention of the tandem technique described here.

TABLE 1 Input Data for Example 1

	News-letter	Clinical reports	Use trials	Ex-hibits	Detail-ing	Dinner mtgs	Direct mail	Journal cards	Samp-ling	Sym-posia	Audio/visual	Total
Cardiologist –Innov CRD-1	14	43	43	27	25	26	11	8	17	41	16	271
Cardiologist –Non-innov –CRD-2	9	30	29	17	31	17	5	3	6	34	22	203
General practitioner –Innov –GP-1	13	23	23	18	52	34	7	15	15	19	25	244
–Non-innov GP-2	13	26	25	21	56	29	8	14	17	16	29	254
Nephrologist –Innov NEP-1	2	15	12	10	9	10	3	7	6	14	8	96
–Non-innov NEP-2	1	5	9	5	6	3	1	3	6	6	4	49
Internist –Innov IM-1	2	8	8	6	3	3	0	2	4	9	3	48
–Non-innov IM-2	6	11	11	7	6	3	2	5	3	7	1	62
Family physician –Innov FP-1	2	5	9	3	11	9	2	4	0	3	3	51
–Non-innov Fp-2	8	10	10	4	21	11	6	3	5	8	11	97
Total	70	176	179	118	220	145	45	64	79	157	122	1375

Attitudinal statements

1 I like to try new and different things.
2 I often prescribe new medications before my colleagues do.
3 I go to more medical conventions than most of my colleagues.
4 I spend more of my time keeping up with new medical developments than most doctors.
5 I spend more time reading medical journals than most doctors.

Example 1: Two-Way Correspondence Analysis and Hierarchical Grouping

The first application is the most straightforward of the three. The input data consisted of a 10 × 11 frequencies matrix (Table 1) obtained from a sample of 404 physicians. [Each respondent was] asked to select those promotional activities: Newsletter, Clinical Reports, etc, that he/she would most prefer a drug firm to implement in introducing a new anti-hypertensive drug. A total of 1,375 responses (an average of 3.4 selec-tions per respondent) were obtained. Physicians were also classified by specialty (cardiologist, general practitioner, nephrologist, internist, and family physician) and by proneness to innovate (high versus low), based on their first principal component score developed from responses to five psychographic statements (see Table 1).[2]

[2] The original responses represented degree of agreement on a seven-point equal-interval scale. These data were factor analyzed; the first principal component accounted for 78.4% of the data variability. Within speciality, physicians were then arrayed by their first principal component score and divided into two groups (high versus low innovativeness), based on a cut at the median of the within-speciality array.

Table 1 represents a conventional cross tabulation of frequencies which can be analyzed by simple correspondence analysis. Solutions were obtained for two, three and four dimensions, with accounted-for variation of 58.9%, 73.7% and 84.4%, respectively. Illustratively, Figure 1 shows the two-dimensional joint space plot of physician segments and promotional activities. As noted, along the horizontal axis the specialty physicians (nephrologists, cardiologists and internists) are separated from the generalists (general practitioners and family physicians). We also observe that the effect of a high versus low "innovativeness" is seen to vary by physician specialty.

Cluster Analysis of Joint-Space Coordinates

While the two-dimensional map of Figure 1 provides some help in interpreting the space, one wonders whether a cluster analysis of higher-dimensional coordinates might provide enhanced interpretive value. To that end, the coordinates of the four-dimensional solution (accounting for 84.4% of the variation) were clustered via a hierarchical, average-linkage program. Figure 2 shows the resulting dendrogram (Johnson 1967).

As observed from Figure 2, the higher-dimensional solution provides useful ancillary information, beyond that provided by the map. For example, we observe that innovative cardiologists and nephrologist cluster together, along with the more "scientific" information sources: clinical reports, clinical trials and exhibits. This cluster is later joined by innovative internists and symposia.

Both innovative and non-innovative general practitioners cluster tightly, along with the less "prestigious" information sources of detailing and audio-visual. This cluster is then joined by noninnovative family physicians and direct mail promotion. Rather isolated clusters are noted in the case of: (a) innovative family physicians and dinners; (b) non-innovative nephrologists and

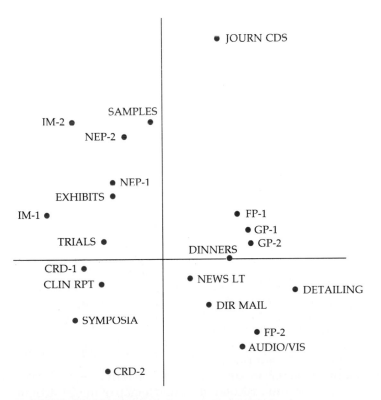

FIGURE 1 Joint space obtained from simple correspondence analysis (example 1).

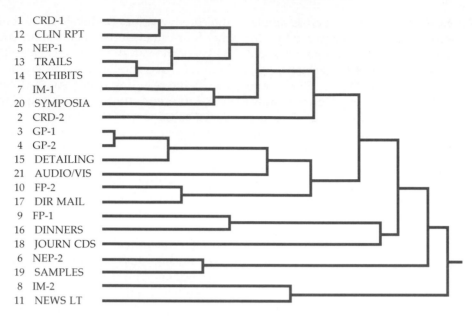

FIGURE 2 Average linkage dendrogram (example 1).

samples; and (c) non-innovative internists and newsletters.

In general, the associated cluster analysis brings out relationships between specialty/ innovativeness and preferred promotional activities. At the high end of the prestige scale are clinical reports, clinical trials, exhibits, and symposia; these are typically associated with specialist physicians. General practitioners favor detailing, and family physicians appear to favor direct mail and dinner meetings. Since the cluster solution entailed four-dimensional coordinates (rather than the two-dimensional coordinates utilized in Figure 1), more of the original information was preserved in the cluster solution of Figure 2.

Example 2: Multiple Correspondence Analysis and K-Means Clustering

The preceding illustration showed how hierarchical grouping can be used to develop a dendrogram of nested clusters of *a priori* selected segments (specialty/innovativeness) versus promotional activities. In contrast, our second illustration starts out with a respondent by attribute category matrix, as illustrated in Table 2.

The respondents in this application consisted of 300 doctors (cardiologists, internists and general practitioners/family physicians). Each doctor described the last patient he/she treated for hypertension, according to the multiple-choice attributes, shown in Table 2. For example, respondent 1 described his last patient as:

- Middle aged
- Female
- Non-Caucasian
- Normal weight
- Moderate/high income
- Sedentary
- Mild hypertension
- Previous diuretic therapy
- No previous beta blocker therapy

The basic data thus consisted of a 300 × 9 matrix of physicians by the attribute categories which they picked as best describing their last hypertensive patient.

Multiple Correspondence Analysis

The objective of multiple correspondence is to find a map of the 20 attribute categories, shown in Table 2. Through appropriate scaling (Carroll & Green 1987), all squared distances are compara-

TABLE 2 Characteristics of Last Patient Treated for Hypertension (Example 2)

A Age of patient
  1 Young
  2 Middle
  3 Elderly
B Sex
  1 Male
  2 Female
C Race
  1 Non-Caucasian
  2 Caucasian
D Weight
  1 Nomal
  2 Obese
E Economic situation
  1 No fee (third party pays)
  2 Low income
  3 Moderate income (or higher)

F Patient's activity level
  1 Active
  2 Sedentary

G Severity of hypertension
  1 Mild
  2 Moderate
H Previous diuretic therapy?
  1 D—Yes
  2 D No
I Previous beta blocker therapy?
  1 B—Yes
  2 B—No

Sample Input Data

Respondent #	A	B	C	D	E	F	G	H	I
001	2	2	1	1	3	2	1	1	2
002	3	1	1	2	2	2	2	1	1
003	3	1	2	2	1	2	1	2	2
⋮	⋮	⋮	⋮	⋮	⋮	⋮	⋮	⋮	⋮
300	1	2	1	1	3	1	2	1	2

ble. Respondent (i.e., physician) coordinates can also be found; if so, each respondent is plotted at the centroid of the nine categories that he/she picked. Given the fact that each respondent picks one and only one category from each of the nine attributes, the data could be described as a nine-way contingency table. However, the same information is obtained in the 300 × 9 input matrix of qualitative profile data, illustrated in Table 2.

Multiple correspondence was applied to this input matrix. Accounted-for variation was 62.2%, 75.7%, and 85.3% in 2, 3, and 4 dimensions, respectively.[3] For ease of presentation, only the two-dimensional configuration of 20 patient categories is plotted in Figure 3. (However, the program also computed coordinates for each of the 300 respondents). Along the horizontal dimen-

sion we note a separation of elderly, no fee or low income, sedentary, non-Caucasian, females who are obese and have no previous diuretic or beta blocker treatment history (but suffer from moderate hypertension) from the rest of the categories. It may be viewed as a socioeconomic dimension. The vertical dimension shows those respondents with previous drug therapy separated from those without.

K-Means Cluster Analysis

For illustrative purposes we elected to cluster the 300 respondents based on the four-dimensional solution coordinates. First, the 300 × 4 matrix of respondent coordinates (not shown in Figure 3) were submitted to a hierarchical grouping program. The hierarchical tree was then cut at 3, 4, 5, and 6 levels to obtain an initial partition of the 300 points, from which cluster centroids were computed as initial seed points. Following this, Quick

[3] Accounted-for variation for this application was computed by a measure proposed by Benzécri, described in Greenacre (1984, p. 145).

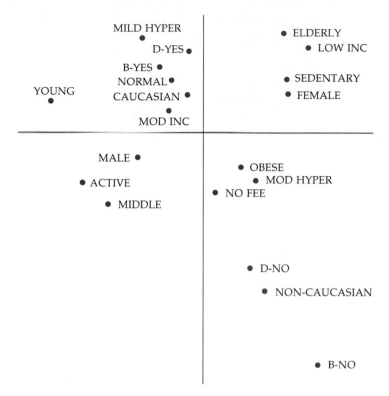

FIGURE 3 Configuration obtained from multiple correspondence analysis (example 2).

Cluster (an SPSSX *K*-means program) was used to reassign points, according to the standard *K*-means algorithm.

For illustrative purposes, only the results of the four-cluster solution are presented. Cluster sizes were 56, 47, 126, and 71, respectively, for clusters 1 to 4. We note from Table 3 that clusters 1 and 4 are male dominated, middle ages, actives, with moderate incomes and no previous diuretic history. However, they differ on normal versus obese, mild versus moderate hypertension, and previous beta blocker history.

Clusters 2 and 3 are female dominated, elderly, sedentary types who suffer from moderate hypertension. However, they differ by normal versus obese, income level, and previous diuretic, beta blocker histories. (We note that in this particular cluster solution, Caucasian versus non-Caucasian does not discriminate across clusters.) In sum, example 2's multiple correspondence analysis is used as a preliminary procedure to portray individual responses as point coordinates (centroids) in a reduced space. These spatial coordinates are then clustered via *K*-means to obtain a segmentation of the respondents by commonality of patient type; segment characteristics are summarized by the profile descriptions in Table 3.

Example 3: Multiple Correspondence Analysis and Clustering Applied to "Pick *k/n*" Data

In marketing research one of the more typical kinds of data collection is to show respondents a list of benefits, brands, etc. and ask them to pick the ones that are most important, most preferred, most descriptive, etc. Such data are called "pick *k/n*" data (where $k < n$). Multiple correspondence analysis can be used to scale this class of data, as well. Here, however, it is the *person-trial* that is

TABLE 3 Cluster Profiles for the Four-Group Solution (Example 2)

	Cluster			
Characteristic	1 N = 56	2 N = 47	3 N = 126	4 N = 71
Young	0%	0%	0%	5%
Middle	89*	26	7	69
Elderly	11	74	93	
Male	94	40	46	95
Female	6	60	54	5
Non-Caucasian	30	28	6	1
Caucasian	70	72	94	99
Normal	36	23	53	61
Obese	64	77	47	39
No fee	43	49	23	29
Low income	11	28	31	4
Moderate income	46	23	46	67
Active	66	5	4	90
Sedentary	34	95	96	10
Mild	26	2	41	65
Moderate	74	98	59	35
D—Yes	32	4	76	44
D—No	68	96	24	56
B—Yes	54	42	65	29
B—No	46	58	35	71

*Boxed items denote highest percentage within attribute.

the unit of association, so that at any given trial one and only one item is picked.[4]

In example 3 the sample consisted of 202 commercial growers of soybeans. The study's sponsor, a large producer of agricultural chemicals, wished to introduce a new herbicide that displayed several features/benefits of potential interest to commercial growers. Table 4 shows the

[4]Selections are not independent across trials, within person. As such, we use correspondence analysis strictly as a data reduction, exploratory technique.

list of 18 benefits that constituted the set of interest. The sponsor wished to see which of the proposed product's benefits (a subset of those in Table 4) would be most attractive from a market positioning viewpoint.

Each respondent was shown the set of 18 benefits and asked to choose eight benefits that he/she would most desire in a soybean herbicide. The response data for a given respondent thus consisted of a vector of eight 1's and ten 0's, where 1 denotes picked and 0, otherwise, as illustrated in Table 4.

TABLE 4 List of 18 Herbicide Benefits (Example 3)

A   Saves fuel (SV FUEL)
B   Consistent control everytime (CONTROL)
C   Won't leach out of soil (NO LEACH)
D   Controls black nightshade (NIGHTSHADE)
E   Easy to apply (EASY APP)
F   Season long control (SEAS CONT)
G   Only required 1 pass (1 PASS)
H   Costs less than current herbicides (COSTS LESS)
I   Guaranteed by manufacturer (GUARANT)
J   Increases yield over current herbicides (HI YIELD)
K   Can surface apply (SURFACE AP)
L   No carryover into next growing season (NO CARRYO)
M   Fewer tillage passes than current herbicides (LITT TILL)
N   Saves time in application (SAVES TIME)
O   Superior control of other broadleaves (BROADLEAVES)
P   Won't stunt early crop growth (NO STUNT)
Q   Requires little rain to work (NO RAIN)
R   Superior grass control (GRASSES)

Sample input data

Respondent #	A	B	C	D	E	F	G	H	I	J	K	L	M	N	O	P	Q	R
001	1	0	1	0	1	1	0	1	0	1	0	1	1	0	0	0	0	0
002	0	1	1	0	0	0	1	0	1	0	1	0	0	1	0	0	1	1
003	1	0	0	1	0	0	1	1	0	0	0	0	1	1	1	0	0	1
⋮	⋮	⋮	⋮	⋮	⋮	⋮	⋮	⋮	⋮	⋮	⋮	⋮	⋮	⋮	⋮	⋮	⋮	⋮
202	0	0	1	1	1	0	0	0	1	0	1	0	0	1	0	1	1	0

The 202 × 18 matrix of respondents by benefits was then submitted to a multiple correspondence program, and the 18 benefits were scaled, along with the respondent points, positioned as centroids of the benefit coordinates that were picked. Accounted-for variability was 38.4%, 55.1%, and 70.8% in 2, 3 and 4 dimensions, respectively (considerably lower than that of example 2). The two-dimensional configuration of the 18 benefits appears in Figure 4.

In contrast to the configuration of Figure 3, the present example shows a high concentration of points near the origin of Figure 4's configuration. This suggests relatively *high agreement* across respondents for several of the benefits (e.g., season long control, superior grass control, easy to apply, no carryover into next growing season, etc.).

Along the horizontal dimension convenience benefits (saves fuel, fewer tillage passes, saves time) are separated from "guaranteed by manufacturer"; however, the interpretation of this configuration is not as clear as that of example 2 (Figure 3).

K-Means Cluster Analysis

In accordance with the procedure followed earlier, the four-dimensional respondent coordinates were then clustered via the *K*-means procedure. A 3-group clustering, consisting of 95, 46, and 61 respondents, respectively, was selected for illustrative purposes. As Table 5 indicates, cluster 1 indicates highest incidences on seven of the 18 benefits; cluster 2 on five, and cluster 3 on six of the 18 benefits.

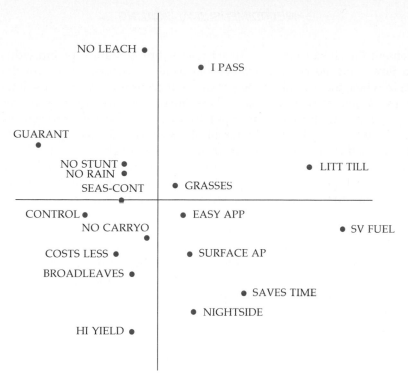

FIGURE 4 Configuration obtained from multiple correspondence analysis of pick k/n data (example 3).

TABLE 5 Cluster Profiles for the Three-Group Solution (Example 3). See Table 4 for complete descriptive cluster

Characteristic (PLOTTING SYMBOL)	1 N = 95	2 N = 46	3 N = 61
SV FUEL	0%	65%	2%
CONTROL	79	50	69
NO LEACH	0	13	59
NIGHTSHADE	40	39	23
EASY APP	53	43	46
SEAS CONT	67	41	72
1 PASS	7	30	56
COSTS LESS	63	30	46
GUARANT	18	2	26
HI YIELD	50	28	15
SURFACE AP	39	59	34
NO CARRYO	74	61	59
LITT TILL	5	76	15
SAVES TIME	23	43	13
BROADLEAVES	66	30	43
NO STUNT	40	24	46
NO RAIN	71	35	82
GRASSES	79	83	77

*Boxed items denote highest percentage within row.

It is also useful to supplement the cluster profile data of Table 5 with a new multiple correspondence analysis solution in which the original set of 18 benefits of Table 4 is augmented by a nineteenth attribute, namely the cluster number that each respondent was assigned to, via the K-means cluster analysis. All that needs to be done is to append an additional column to the input data illustrated in Table 4; the last column simply indicates each respondent's cluster number: 1, 2, or 3.

The second multiple correspondence analysis was then carried out, in which the 18 benefit positions were augmented with three additional points, each designating a cluster number. Results appear in Figure 5 (for the two-dimensional solution). We first note that the addition of a nineteenth attribute has very little impact on changing the positions of the original 18 attribute points.[5]

Cluster 1's position is closest to the mass of central points near the origin (e.g., high yield, superior broadleaf control, etc.), reflecting cluster 1's relatively large size. Cluster 2 is positioned quite far from the origin and close to "saves fuel" and "fewer tillage passes"). Cluster 3 is also relatively far from the origin and is nearest to "won't leach out of soil" and "works as well in 1 pass as in 2." These results are in reasonable accord with the profile summaries of Table 5, given that only two dimensions are plotted. The cluster positions in Figure 5 thus provide a graphical supplement to the numerical summary of Table 5. In particular, they support the notion that cluster 1 is the "mainstream" cluster while clusters 2 and 3 reflect less typical preference patterns.

Concluding Comments

The purpose of the paper has been to illustrate (by three real-world examples) how correspondence analysis can provide a useful data reduction tool for developing product positioning maps as well as providing a preliminary spatial representation of multistate categorical data. The reduced space representation of individuals (as well as categories) provides a basis for developing proximity measures between pairs of individuals. These proximity measures can then be used as input to various clustering procedures (e.g., hierarchical and/or K-means) for finding product/market segments.

Example 1 illustrated how hierarchical grouping techniques can enhance the joint space map obtained from two-way correspondence analysis. Example 2 showed how multiple correspondence and K-means clustering can be used in tandem to provide a category map and a clustering of individuals who exhibit similar patterns in their choices of categories. Example 3 showed how multiple correspondence analysis can be adapted to pick k/n data. This example also illustrated how the cluster solution can be mapped back into the original category configuration for visual backup to the more usual profile summaries of Tables 3 and 5.

Fortunately, hierarchical grouping and K-means packages are widely available, both as single programs and as parts of general statistical packages (e.g., SPSSX, BMDP, and SAS). Simple and multiple correspondence analysis programs (several designed for the personal computer) are also available (Lebart & Morineau 1982; Nishisato & Nishisato 1983; Smith 1986; Greenacre 1986; Market Action 1987).

In sum, the illustrations presented here can be readily implemented via more or less standard statistical packages. While not a panacea for coping with the problem of clustering individuals, characterized by qualitative, multistate attributes, the reduced space approach provides a workable solution. Other methods for obtaining reduces spaces (e.g., DeSarbo & Hoffman 1987; van der Heijden & de Leeuw 1985; Takane 1983; Levine 1979) can, and should be, explored. Greenacre (1986) also describes an interesting two-way, correspondence analysis procedure for separately clustering rows and columns of a contingency matrix. The large quantities of categorical response data that are routinely collected in marketing research would seem to justify continued research in finding ways to use this information more effectively. It is hoped that applications researchers will find these tools of practical utility in market segmentation and product positioning.

[5] Other approaches to positioning the cluster-based categories use supplementary point fitting (see Greenacre 1984, pp. 70–76).

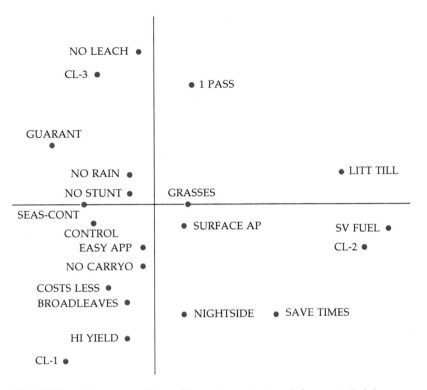

FIGURE 5 Augmented configuration obtained from pick k/n data *(example 3)*.

References

Arabie, P.; Carroll, J. D.; DeSarbo, W. S. & Wind, Y. (1981). Overlapping clustering: a new method for product positioning. *Journal of Marketing Research* **18**, 3, August 310–317.

Carroll, J. D. & Green, P. E. (1987). A new approach to the multiple correspondence analysis of categorical data. Working Paper, Wharton School, University of Pennsylvania, July.

Carroll, J. D.; Green, P. E. & Schaffer, C. M. (1986). Interpoint distance comparisons in correspondence analysis. *Journal of Marketing Research* **23**, August, 271–280.

Carroll, J. D.; Green, P. E. & Schaffer, C. M. (1987). Comparing interpoint distances in correspondence analysis: a clarification. *Journal of Marketing Research* **24**, 4, November, 445–450.

DeSarbo, W. S. & Hoffman, D. L. (1987). Constructing MDS joint spaces from binary choice data: a multidimensional unfolding threshold model for marketing research. *Journal of Marketing Research* **24**, 1, February, 40–54.

Dickinson, J. R. (1986). *The bibliography of marketing research methods*. Lexington, MA: Lexington Books, 580–597.

Gifi, A. (1981). *Non-linear multivariate analysis*. Leiden, The Netherlands: Department of Data Theory, University of Leiden.

Green, P. E. & Rao, V. R. (1972). *Applied multidimensional scaling*. New York: Holt, Rinehart and Winston, pp. 37–39.

Green, P. E.; Wind, Y. & Jain, A. K. (1973). Analyzing free-response data in marketing research. *Journal of Marketing Research* **10**, 1, February, 45–52.

Greenacre, M. J. (1984). *Theory and application of correspondence analysis*. London: Academic Press, Inc.

Greenacre, M. J. (1986). Clustering the rows and columns of a contingency table. Report 86/3. August, Department of Statistics, University of South America.

Greenacre, M. J. (1986). Correspondence analysis: program for the IBM PC and compatibles. Version 1.1, June, University of South Africa.

Heiser, W. J. (1981). *Unfolding analysis of proximity data*. Leiden, The Netherlands: Department of Data Theory. University of Leiden.

Hoffman, D. L. & Franke, G. R. (1986). Correspondence analysis: graphical representation of categorical data in marketing research. *Journal of Consumer Research* **9**, June, 99–105.

Johnson, S. C. (1967). Hierarchical clustering schemes. *Psychometrika* **32**, September, 241–254.

Lebart, L. & Morineau, A. (1982). SPAD: a system of FORTRAN programs for correspondence analysis. *Journal of Marketing Research* **19**, 4, November, 608–609.

Lebart, L.; Morineau, A. & Warwick, K. M. (1984). *Multivariate descriptive statistical analysis: correspondence analysis and related techniques for large matrices*. New York: John Wiley & Sons, Inc.

Levine, J. H. (1979). Joint-space analysis of "pick-any" data analysis of choices from an unconstrained set of alternatives. *Psychometrika* **44,** March, 85–92.

Lingoes, J. C. (1977). A general survey of the Guttman–Lingoes non-metric program series. In *Geometric representations of relational data*, J. C. Lingoes, E. E. Roskam & J. Borg (Eds.).

Market Action, Inc. (1987). *STRATMAP Manual*. Peoria Ill: Bradley University, Lovelace Technology Center.

Meulman, J. (1982). *Homogeneity analysis of incomplete data*. Leiden, The Netherlands: DSWO Press.

Nishisato, S. (1980). *Analysis of categorical data: dual scaling and its applications*. Toronto: University of Toronto Press.

Nishisato, S. & Nishisato, I. (1983). *An introduction to dual scaling*. Islington, Ontario: MicroStats.

Punj, G. & Stewart, D. W. (1983). Cluster analysis in marketing research: review and suggestions for applications. *Journal of Marketing* **20,** May, 134–148.

Romesburg, H. C. (1984). Cluster analyses for researcher. Belmont, CA: Lifetime Learning Publications, 158 159.

Smith, S. (1986). PC programs for MDS and clustering: programs and user's manual. School of Business, Brigham Young University.

Sneath, P. H. A. & Sokal, R. R. (1973). *Numerical taxonomy*. San Francisco: W. H. Freeman, pp 150–152.

SPSSX User's Guide, 2nd ed (1986). McGraw-Hill Book Company, pp. 737–742.

Srivastava, R. K.; Leone, R. P. & Shocker, A. D. (1981). Market structure analysis: hierarchical clustering of products based on substitution-in-use. *Journal of Marketing* **45**, Summer, 38–48.

Takane, Y. (1983). Choice model analysis of the "pick any/n" type of binary data. European psychometric and classification meetings, July, Jouy-en-Josas, France.

van der Heijden, P. G. M. & de Leeuw, J. (1985). Correspondence analysis used complementary to loglinear analysis. *Psychometrika* **50**, December, 429–448.

CHAPTER 10

Conjoint Analysis

LEARNING OBJECTIVES

Upon completing this chapter, you should be able to do the following:

- Explain the many managerial uses of conjoint analysis.

- Understand the guidelines for selecting the variables to be examined by conjoint analysis, as well as their values.

- Formulate the experimental plan for simple conjoint analysis, including how to design factorial designs and the impact of choosing from rank choice versus ratings as the measure of preference.

- Assess the relative importance of the predictor variables and each of their levels in affecting consumer judgments.

- Apply a choice simulator to conjoint results for the prediction of consumer judgments of new attribute combinations.

- Examine the implications for selecting a main effects model versus a model involving interaction terms and demonstrate approaches for establishing the validity of one model versus the other.

Since the mid-1970s, conjoint analysis has attracted considerable attention as a method that portrays consumers' decisions realistically as trade-offs among multiattribute products or services. It has gained widespread acceptance and use in many industries with usage rates increasing up to tenfold in the 1980s [31]. This acceleration has coincided with the widespread introduction of computer programs that integrate the entire process, from generating the combinations of predictor variable values to be evaluated to creating choice simulators for predicting consumer choices across a wide number of alternative product/service formulations. Access to these programs has increased even further with the introduction of personal computer–based programs. Today several widely employed packages can be accessed by any researcher with a personal computer [4, 5, 6, 25, 26, 27, 28, 29].

Conjoint analysis is closely related to traditional experimentation. For example, a chemist in a soap manufacturing plant may want to know the effect of the temperature and pressure in the soap-making vats on the density of the resulting bar of soap. The chemist could conduct a laboratory experiment to measure these relationships. Once the experiments were conducted, they could be analyzed with ANOVA (analysis of variance) procedures like those discussed in Chapter 5. In situations involving human behavior, we often also need to conduct "experiments" with the factors we control. (For example, should the bar of soap be slightly or highly fragranced? Should it be promoted as a cosmetic or a cleaner/deodorizer? Which of three prices should be charged?) Conjoint analysis developed from the need to analyze the effects of predictor variables (the factors we control) that are often qualitatively specified or weakly measured [12, 14]. It is actually a family of techniques and methods, all theoretically based on the models of information integration and functional measurement [20].

In terms of the basic dependence model discussed in Chapter 1, conjoint analysis can be expressed as

$$Y_1 = X_1 + X_2 + X_3 + \cdots + X_N$$

$$\text{\textit{(nonmetric, metric)}} \quad \text{\textit{(nonmetric)}}$$

Conjoint analysis is best suited for understanding consumers' reactions to and evaluations of predetermined attribute combinations that represent potential products or services. While maintaining a high degree of realism, it provides the analyst with insight into the composition of consumer preferences. The flexibility and uniqueness of conjoint analysis arise primarily from (1) its ability to accommodate either a metric or a nonmetric dependent variable, (2) the use of categorical predictor variables, and (3) the quite general assumptions about the relationships of independent variables with the dependent variable.

KEY TERMS

Before starting the chapter, review the key terms to develop an understanding of the concepts and terminology used. Throughout the chaper the key terms will appear in **boldface**. Other points of emphasis in the chapter will be *italicized*. Also, cross-references within the key terms will be in *italics*.

Additive model Also known as a *main effects* model, it is based on the additive composition rule, which assumes individuals just "add up" the *part-worths* to calculate an overall or "total worth" score indicating preference. It is the simplest conjoint model in terms of the number of evaluations and the estimation procedure required.

Choice-based approach   Alternative form of collecting responses and estimating the conjoint model. The primary difference is that respondents select a *full profile* from a set of profiles (known as a *choice set*) instead of just rating or ranking each profile separately.

Choice set   Set of stimuli from the *full profile method* constructed through experimental design principles and used in the *choice-based approach*.

Choice simulator   Once the conjoint *part-worths* have been estimated for each respondent, the choice simulator analyzes a set (or sets) of stimuli and predicts both individual and aggregate choices for each stimulus in the set. This procedure allows the researcher to assess many "what-if" scenarios, including possible product/service configurations and attribute combinations.

Compositional model   Class of multivariate models that base the dependence relationship on observations from the respondent regarding both the dependent and the independent variables. Such models calculate or "compose" the dependent variable from the respondent-supplied values for the independent variables. Principal among such methods are regression analysis and discriminant analysis. These models are in direct contrast to *decompositional models*.

Composition rule   Rule used in combining attributes to produce a judgment of relative value or utility for a product/service. For illustration, a person is asked to evaluate four objects. The person is assumed to evaluate the attributes of the four objects and to create some overall relative value for each. The rule may be as simple as creating a mental weight for each perceived attribute and adding the weights for an overall score (additive rule), or it may be a more complex procedure involving interaction terms (interactive rule).

Conjoint variate   Combination of variables (known as *factors*) specified by the analyst that constitute the total worth or utility of the stimuli. The analyst also specifies all the possible values for each variable. These values are known as *levels*.

Decompositional model   Class of multivariate models that "decompose" the respondent's preference. This class of methods presents the respondent with a set of independent variables, usually in the form of a hypothetical or actual product/service, and then asks for an overall evaluation or preference of the product/service. Once given, the preference is "decomposed" by relating the known attributes of the product (which become the independent variables) to the evaluation (dependent variable).

Factor   Variable the researcher manipulates to measure any effect on another variable. In conjoint analysis, the predictor variables (factors) are nonmetric. Factors must be represented by two or more values (also known as *levels*).

Factorial design   Method of designing stimuli for evaluation by generating *all* possible combinations of levels. For example, a three-factor experiment with three levels per factor ($3 \times 3 \times 3$) would result in 27 stimuli.

Fractional factorial design   As an alternative to a factorial design, this approach uses only a subset of the possible stimuli needed to estimate the results based on the assumed composition rule. Its primary task is to reduce the number of evaluations collected while still maintaining orthogonality among the *part-worth* estimates. The simplest design is an *additive model* in which only main effects are estimated. If selected interactions are included, then additional stimuli are created. The design can be created either by referring to published sources or by using computer programs that accompany most conjoint analysis packages.

Full-profile method   Method of presenting stimuli to the respondent for evaluation that consists of a complete description of the stimuli across all attributes.

Interaction effects or terms   In assessing value, a person may assign a unique value to specific combinations of features. For example, a person is evaluating mouthwash products. This person has an average preference for both the color red and a brand X. If he or she evaluated this combination (red and brand X) like all other combinations, then the red brand X product would have an expected overall preference rating somewhere in the middle of all possible stimuli. Assume, however, that the red brand X mouthwash is the most preferred, even above other combinations of attributes (color and brand) that had higher evaluations of the individual features. This unique evaluation of a combination that is greater (or could be less) than expected based on the separate judgments indicates a two-way interaction. If flavor (e.g., mint) is now added to the combinations, a three-way interaction occurs.

Interattribute correlation   Correlation among attributes, also known as environmental correlation, that makes combinations of attributes unbelievable or redundant. A negative correlation depicts the situation in which two attributes are naturally assumed to operate in different directions, such as horsepower and gas mileage. As one increases, the other is naturally assumed to decrease. Thus, because of this correlation, all combinations of these two attributes (e.g., high gas mileage and high horsepower) are not believable. The same effects can be seen for positive correlations, where perhaps price and quality are assumed to be positively correlated. It may not be believable to find a high-price/low-quality product in such a situation. The presence of strong interattribute correlations requires that the researcher closely examine the stimuli presented to respondents and avoid unbelievable combinations that are not useful in estimating the part-worths.

Level   Specific value of a factor variable. Each factor variable must be represented by two or more levels, but the number of levels typically never exceeds 4 or 5. If the factor variable is metric, it must be reduced to a smaller number of specific value levels. For example, the many possible values of size and price must be represented by a small number of levels:

Size: 10, 12, or 16 ounces
Price: $1.19, $1.39, or $1.99

If the variable is nonmetric, the original values can be used as in these examples:

Color: red or blue
Brand: brand X, brand Y, or brand Z
Fabric softener additive: present or absent

Main effects   Direct effect of each factor (predictor) variable on the dependent variable.

Orthogonality   Mathematical constraint requiring that the *part-worth* estimates be independent of each other. In conjoint analysis orthogonality refers to the ability to measure the effect of changing each attribute level and to separate it from the effects of changing other attribute levels and from experimental error.

Pairwise comparison method   Method of presenting a pair of *stimuli* to a respondent for evaluation, with the respondent selecting one stimuli as the preferred.

Part-worth   Estimate from conjoint analysis of the overall preference or utility associated with each value of each attribute used to define the product/service.

Preference structure   Representation of both the relative importance or worth of each attribute and the impact of individual levels within each attribute in affecting preference.

Stimulus   See *treatment.*

Trade-off method   Method of presenting stimuli to respondents where attributes are depicted two at a time and respondents rank all combinations of the levels in terms of preference.

Treatment   Specific set of levels (one per factor variable) evaluated by respondents (also known as a *stimulus*). One method of defining treatments is achieved by taking all combinations of all levels. For example, three factors with two levels each would create eight ($2 \times 2 \times 2$) treatments. However, in many conjoint analyses, the total number of combinations is too large for a respondent to evaluate them all. In these instances, some subset of treatments/stimuli are created according to a systematic plan, most often a *fractional factorial design.*

What Is Conjoint Analysis?

Conjoint analysis is a multivariate technique used specifically to understand how respondents develop preferences for products or services. It is based on the simple premise that consumers evaluate the value or utility of a product/service/idea (real or hypothetical) by combining the separate amounts of utility provided by each attribute. Conjoint analysis is unique among multivariate methods in that the researcher first constructs a set of real or hypothetical products or services by combining the selected levels of each attribute. These hypothetical products are then presented to respondents who provide only their overall evaluations. Thus the researcher is asking the respondent to perform a very realistic task—choosing among a set of products. Respondents need not tell the researcher anything else, such as how important an attribute is to them or how well the product performs on a number of attributes. Because the researcher constructed the hypothetical products/services in a specific manner, the importance of each attribute and each value of each attribute can be determined from the respondents' overall ratings.

    To be successful, the analyst must be able to describe the product or service in terms of both its attributes and all relevant values for each attribute. We will use the term **factor** when describing a specific attribute or other characteristic of the product or service. The possible values for each factor are called **levels.** In conjoint terms, we describe a product or service in terms of *its level on the set of factors* characterizing it. When the analyst selects the factors and the levels to describe a product/service according to a specific plan, the combination is known as a **treatment** or **stimulus.**

e.g. orange juice with different characteristics. (factors)

A Hypothetical Example of Conjoint Analysis

As an example, assume that HATCO is trying to decide which attributes affect choice in the canned dog food market. The variables selected as affecting the purchase decision are as follows:

Factor	*Level*
Brand name	Arf versus Mr. Dog
Ingredients	All meat versus meat and fiber supplements
Can size	6 ounce versus 12 ounce

If three attributes (factors) with two values (levels) each are selected, then eight ($2 \times 2 \times 2$) combinations can be formed. Three examples of the eight possible combinations (stimuli) are:

- Mr. Dog all-meat dog food in a 6-ounce can
- Arf meat-and-fiber dog food in a 12-ounce can
- Arf all-meat dog food in a 6-ounce can

Respondents would be asked either to rank-order the eight stimuli from highest to lowest in terms of preference or to rate each combination on a preference scale (perhaps a 1-to-10 scale). We can see why conjoint analysis is also called "trade-off analysis," because in making a judgment on a hypothetical product, respondents must consider both the "good" and "bad" characteristics of the product to form a preference. Thus, respondents must weigh all attributes simultaneously in making their judgments.

By constructing specific combinations (treatments or stimuli), the analyst is attempting to understand a respondent's **preference structure**. The preference structure "explains" not only how important each factor is in the overall decision but also how the differing levels within a factor influence the formation of an overall preference. In our example, conjoint analysis would assess the relative impact of each brand name (Arf versus Mr. Dog), each can size (6 ounces versus 12 ounces), and the different ingredient options (all meat versus meat and fiber) in determining a person's overall preference. This overall preference, which represents the total "worth" of an object, can be thought of as based on the **part-worths** for each level. The general form of a conjoint model can be shown as

$$\text{Total Worth for product}_{ij..n} = \text{Part-worth of level}_i \text{ for factor}_1 +$$
$$\text{part-worth of level}_j \text{ for factor}_2 + \cdots +$$
$$\text{part-worth of level}_n \text{ for factor}_m$$

where the product/service has m attributes, each having two or more levels. The product consists of level$_i$ of factor$_1$, level$_j$ of factor$_2$, . . . up to level $_n$ for factor$_m$.

In our example, a simple additive model would predict preference of any dog food combination based on the formula

$$\text{Preference} = \text{Brand effect} + \text{Ingredient effect} + \text{Size effect}$$

The preference for a specific dog food product can be directly calculated from the part-worth values. For example, the preference for Mr. Dog all-meat dog food in a 12-ounce can is

$$
\begin{aligned}
\text{Preference} = \text{ } & \text{Part-worth of Mr. Dog brand +} \\
& \text{Part-worth of all-meat ingredients +} \\
& \text{Part-worth of 12-ounce can}
\end{aligned}
$$

The Managerial Uses of Conjoint Analysis

Before discussing the statistical basis of conjoint analysis, we should understand the technique in terms of its role in decision making and strategy development. The flexibility of conjoint analysis gives rise to its application in almost any area in which decisions are studied. Conjoint analysis assumes that any set of objects (e.g., brands, companies) or concepts (e.g., positioning, benefits, images) is evaluated as a bundle of attributes. Having determined the contribution of each attribute to the consumer's overall evaluation, the marketing researcher could then

1. Define the object or concept with the optimum combination of features.
2. Show the relative contributions of each attribute and each level to the overall evaluation of the object.
3. Use estimates of purchaser or customer judgments to predict market shares among objects with differing sets of features (other things held constant).
4. Isolate groups of potential customers who place differing importance on the features to define high and low potential segments.
5. Identify marketing opportunities by exploring the market potential for feature combinations not currently available.

Comparing Conjoint Analysis with Other Multivariate Methods

Conjoint analysis differs from other multivariate techniques in three distinct areas: (1) its decompositional nature, (2) the fact that estimates can be made at the individual level, and (3) the flexibility in terms of relationships between dependent and independent variables.

Compositional Versus Decompositional Techniques

Conjoint analysis is termed a **decompositional model** because the analyst needs to know only a respondent's overall preference for an object and the characteristics of the object. In conjoint analysis, the analyst specifies the ratings on each attribute and then asks the respondent only for the overall preference. In this way conjoint analysis can *decompose* the preference to determine the value of each attribute.

Conjoint analysis differs from **compositional models** like discriminant analysis and many regression applications, where the analyst collects ratings from the respondent on many product characteristics (e.g., favorability toward color, style, specific features, etc.) and then relates these ratings to some overall preference rating to develop a predictive model. With regression and discriminant analysis the analyst is calculating or "composing" the overall preference from the respondent's evaluations of the product on *each* attribute.

Specifying the Conjoint Variate

Conjoint analysis employs a variate quite similar in form to what we have seen in other multivariate techniques. The **conjoint variate** is a linear combination of effects of the independent variables (factors) on a dependent variable. The important difference is that in the conjoint variate, the researcher specifies both the variables *and* their values. The only value provided by the respondent is the dependent measure. The variable values specified by the researcher are then used by conjoint analysis to decompose the respondent's response into effects for each variable, much as is done in regression analysis. This feature illustrates the common characteristics shared by conjoint analysis and experimentation, whereby the design of the project is critical to its success. For example, if a variable or effect was not anticipated in the research design, then it will not be available for analysis. For this reason, a researcher may be tempted to include a number of variables that *might* be relevant. However, conjoint analysis is limited in the number of variables it can include, so the researcher cannot just include additional questions to compensate for a lack of clear conceptualization of the problem.

Separate Models for Each Individual

Conjoint analysis differs from almost all other multivariate methods in that it can be carried out at the individual level, meaning that the analyst generates a separate "model" for predicting preference for *each* respondent. Most other multivariate methods take a single measure of preference (observation) from each respondent and then perform the analysis using all respondents simultaneously. In fact, many methods *require* that a respondent provide only a single observation (the assumption of independence) and then develop a common model for all respondents, fitting each respondent with varying degrees of accuracy (represented by the errors of prediction for each observation, such as residuals in regression). In conjoint analysis, however, estimates can be made for either the individual (disaggregate) or groups of individuals (aggregate). At the disaggregate level, each respondent rates enough attribute combinations for the analysis to be performed separately for each person. Predictive accuracy is calculated for each person, rather than only for the total sample. The individual results can then be aggregated to portray an overall model as well. Many times, however, the researcher selects an aggregate analysis method that performs the estimation of part-worths for the group of respondents as a whole. Aggregate analysis can provide a means for reducing the data collection task through more complex designs (discussed in later sections) or provide for greater statistical efficiency by using more observations in the estimation. In selecting between the two approaches, the researcher must balance the benefits gained by aggregate methods with the insights provided by the separate models obtained through disaggregate conjoint methods.

Types of Relationships

Conjoint analysis is not limited at all in the types of relationships required between the dependent and independent variables. As we have discussed in earlier chapters, most dependence methods assume that a linear relationship exists, meaning that the dependent variable increases (or decreases) in equal amounts for each unit change in the independent variable. Conjoint analysis, however, can make separate predictions for the effects of each level of the independent variable and does not assume they are related at all. Thus conjoint analysis can easily handle nonlinear relationships, even the complex curvilinear relationship in which one value is positive, the next negative, and the third positive again. As we discuss later, however, the simplicity and flexibility of conjoint analysis compared with the other multivariate methods is based on a number of assumptions made by the analyst.

Designing a Conjoint Analysis Experiment

Although conjoint analysis places the fewest demands on the respondent in terms of both the number and types of responses needed, the analyst must make a number of key decisions in designing the experiment and analyzing the results. Figure 10.1 shows the general steps followed in the design and execution of a conjoint analysis experiment. The discussion follows the model-building paradigm introduced in Chapter 1. The decision process is initiated with a specification of the objectives of conjoint analysis. Because conjoint analysis is very similar to an experiment, the conceptualization of the research is critical to its success. After the objectives have been defined, the issues related to the actual research design are addressed and the assumptions are evaluated. The decision process then considers the actual estimation of the conjoint results, the interpretation of the results, and the methods used to validate the results. The discussion ends with an examination of the use of conjoint analysis results in further analyses such as market segmentation and choice simulators. Each of these decisions stems from the research question and the use of conjoint analysis as a tool in understanding the respondent's preferences and judgment process.

Stage One: The Objectives of Conjoint Analysis

As with any statistical analysis, the starting point is the research question. In conjoint analysis, experimental design in the analysis of consumer decisions has two objectives:

1. To determine the contributions of predictor variables and their respective values to the determination of consumer preferences. For example, how much does fragrance contribute to the willingness to buy soap? Which fragrance

level is the best? How much change in willingness to buy soap can be accounted for by differences between the levels of fragrance?

2. To establish a valid model of consumer judgments useful in predicting the consumer acceptance of any combination of attributes, even those not originally evaluated by consumers. In doing so, the issues addressed include these: Do the respondent's choices indicate a simple linear relationship between the predictor variables and choices? Is a simple model of "adding up" the value of each attribute sufficient, or do we need to add more complex evaluations of preference to mirror the judgment process adequately?

The respondent reacts only to what the analyst provides in terms of attribute combinations. Are these the actual attributes used in making a decision? Are other attributes, particularly attributes of a more qualitative nature such as emotional reactions, important as well? These and other questions require the research question to be framed around two major issues: (1) Can you describe all the attributes that give utility or value to the product or service being studied? (2) What are the key decision criteria involved in the choice process for this type of product or service? These questions need to be resolved before proceeding into the design phase of a conjoint analysis because they will provide critical guidance for key decisions in each stage.

Defining the Total Worth of the Object

The analyst must first be sure to define the total worth of the object. To represent the respondent's judgment process accurately, all attributes that potentially *create or detract* from the overall worth of the product/service should be included. It is essential that both positive and negative factors be considered, either because (1) focusing on only positive factors will seriously distort the respondents' judgments or (2) respondents can subconsciously employ the negative factors, even though not provided, and thereby render the experiment invalid. For example, if exploratory focus groups are employed to assess the types of characteristics considered when evaluating the object, the analyst must be sure to address what makes the object unattractive, as well as attractive. Fortunately, the omission of a single factor has only a small impact on the estimates for other factors when using an additive model [24].

Specifying the Determinant Factors

In addition, the analyst must be sure to include all determinant factors (drawn from the concept of determinant attributes [3]). The goal is to include the factors that best *differentiate* between the objects. Many attributes may be considered important but also may not differentiate in making choices because they do not vary substantially between objects. For example, safety in automobiles is a very important attribute, but it would not be determinant in most cases because all cars meet strict government standards and thus are considered safe, at least at an acceptable level. However, other features, such as gas mileage, are both important *and* much more likely to be used to decide among different car models. Thus the analyst should always strive to identify the key determinant variables, because they are pivotal in the actual judgment decision.

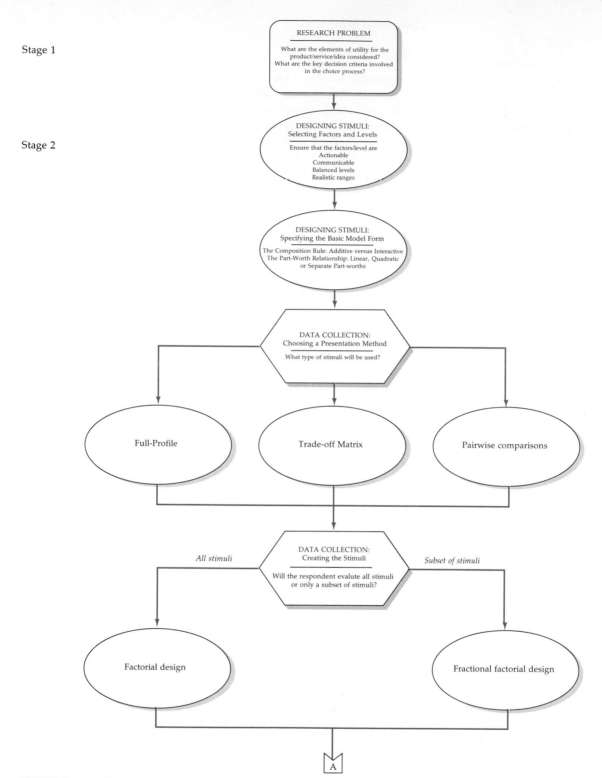

Stage 1

Stage 2

RESEARCH PROBLEM

What are the elements of utility for the
product/service/idea considered?
What are the key decision criteria involved
in the choice process?

DESIGNING STIMULI:
Selecting Factors and Levels

Ensure that the factors/level are
Actionable
Communicable
Balanced levels
Realistic ranges

DESIGNING STIMULI:
Specifying the Basic Model Form

The Composition Rule: Additive versus Interactive
The Part-Worth Relationship: Linear, Quadratic
or Separate Part-worths

DATA COLLECTION:
Choosing a Presentation Method

What type of stimuli will be used?

Full-Profile

Trade-off Matrix

Pairwise comparisons

DATA COLLECTION:
Creating the Stimuli

Will the respondent evalute all stimuli
or only a subset of stimuli?

All stimuli

Subset of stimuli

Factorial design

Fractional factorial design

A

FIGURE 10.1 Conjoint analysis decision process.

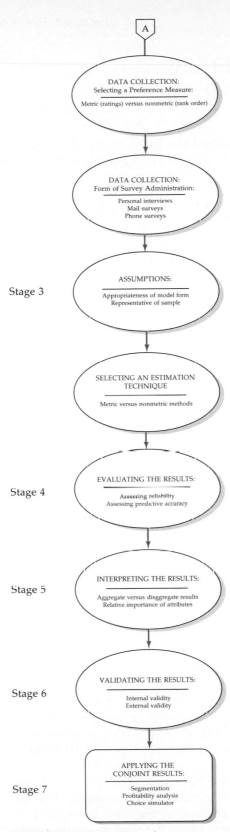

FIGURE 10.1 *Continued*

Stage Two: The Design of a Conjoint Analysis

Having resolved the issues stemming from the objectives of the research, the analyst shifts attention to the particular issues involved in designing and executing the conjoint analysis experiment. For example, how does one decide which specific combinations of attribute levels to present to the respondent for evaluation? In addition to specifying the combinations (treatments), the analyst must also decide on such issues as which attributes to include, how many levels of each, how to measure preference and collect data, and what estimation procedure to use. We examine all these and other necessary issues in designing a conjoint study in the following sections.

Designing Stimuli

The experimental foundations of conjoint analysis place great importance on the design of the stimuli evaluated by respondents. The design involves specifying the conjoint variate by selecting the factors and levels to be included in constructing the stimuli. The analyst must then specify the assumed model of preference to allow for the composition of the stimuli. These design issues are important because they affect both the accuracy of the results and their managerial relevance.

Determining the Factors to Be Used and Selecting the Levels of Each Factor

The first decisions an analyst must make involve the attributes used to describe the product or service and the possible values for each attribute. Having determined the general types of attributes from the nature of the research problem, the researcher must now create specific factors and levels suitable for use in a conjoint analysis experiment. The analyst must consider a number of issues relating to the type and character of the variables and levels selected, including the following:

Actionable Measures   The factors and levels must be capable of being put into practice, meaning the attributes must be distinct and represent a single concept. They must not be "fuzzy" attributes like "overall quality" or "convenience." These concepts are imprecise because of perceptual differences among individuals as to what they actually mean (as compared with actual differences as to how they feel about them). If these fuzzy factors cannot be defined more precisely, the analyst may use a two-stage process. A preliminary conjoint study defines what determines judgments of quality or convenience. Then the factors identified as important in the preliminary study are included in the larger study.

Communicable Measures   The factors and levels must be easily communicated for a realistic evaluation. For example, it is difficult to describe the actual fragrance of a perfume or the "feel" of a hand lotion. Written descriptions have a difficult time in capturing sensory effects, unless the respondent sees the product firsthand, smells the fragrance, or uses the lotion.

Number of Attributes The number of attributes included in the analysis directly affects the statistical efficiency and reliability of the results. As factors and levels are added, the increased number of parameters to be estimated requires either a larger number of stimuli or a reduction in the reliability of parameters. This problem is similar to those we encountered in regression when the number of observations was insufficient to estimate valid coefficients. It is especially important in conjoint analysis because each respondent generates the required number of observations, and therefore the problem cannot be "solved" by adding more respondents.

Balanced Number of Levels Analysts should attempt, as best they can, to balance the number of levels across factors. It has been found that the estimated relative importance or worth of a variable (defined as range of estimated parameters) increases as the number of levels increases, even if end points stay the same [32, 33]. It is conjectured that the refined categorization calls attention to the attribute and causes consumers to focus on that factor more than on others.

Range of the Attribute Levels Set the range (low to high) of the levels somewhat outside existing values but not at an unbelievable level. Setting levels outside existing levels has a tendency to reduce **interattribute correlation,** but it also can reduce believability. Completely unacceptable levels can also cause substantial problems. Before excluding a level, however, the analyst must ensure that it is truly unacceptable, because many times people still pick products that have what they term unacceptable levels. If an unacceptable level is found after the experiment has been administered, the recommended solutions are either to eliminate all stimuli that have unacceptable levels or to reduce part-worth estimates of the offending level to such a low level that any objects containing that level will not be chosen.

The analyst must also apply the criteria of practical relevance and feasibility in defining the levels. Levels that are impractical or would never be used in realistic situations can artificially affect the results. An example illustrates this point. Assume that in the normal course of business activity, the range of prices varies about 10 percent around the average market price. If a price level 20 percent lower is included that would not realistically be offered, its inclusion will markedly distort the results. Respondents would logically be most favorable to such a price level. When the part-worth estimates are made and the importance of price is calculated, this will artificially make price more important than the role it would actually play in day-to-day decisions. The analyst must apply the criteria of feasibility and practical relevance to all attribute levels to ensure that stimuli are not created that will be favorably viewed by the respondent but never have a realistic chance of occurring.

Attribute Multicollinearity Multicollinearity among the factors is a problem that must be remedied. The correlation among factors (known as interattribute or environmental correlation) denotes a lack of **orthogonality** among the attributes. In such cases, the parameter estimates are affected just as in regression (see Chapter 3 for a discussion of multicollinearity and its impact). Moreover, attribute multicollinearity usually results in unbelievable combinations of two or more factors. For example, horsepower and gas mileage are generally thought to be negatively correlated. As a result, how believable is an automobile with the highest levels of both horsepower and gas mileage? The problem lies not in the levels themselves but in the fact that they cannot realistically be combined in all combinations, a required capability for parameter estimation.

If multicollinearity creates unrealistic stimuli, the researcher is encouraged to create "superattributes" that combine the aspects of correlated attributes. In our example of horsepower and gas mileage, perhaps a factor of "performance" could be substituted. As an example of positively correlated attributes, factors of store layout, lighting, and decor may be better addressed by a single concept, such as "store atmosphere." In all cases, when these superattributes are added, they should be made as actionable and specific as possible. If it is not possible to define the broader factors, then eliminate one of them. Refined experimental designs and estimation techniques are also available if these simple remedies are impractical or ineffective [30].

Specifying the Basic Model Form

For conjoint analysis to explain a respondent's preference structure only from overall evaluations of a set of stimuli, the analyst must make two key decisions regarding the underlying conjoint model. These decisions affect both the design of the stimuli and the analysis of respondent evaluations.

The Composition Rule: Selecting an Additive Versus an Interactive Model   The most wide-ranging decision by the analyst involves the specification of the respondent's **composition rule.** The composition rule describes how the respondent combines the part-worths of the factors to obtain overall worth.

The most common, basic composition rule is an **additive model,** with which the respondent simply "adds up" the values for each attribute (the part-worths) to get the total value for a combination of attributes (product or service). For example, assume that a product has two factors and the part-worths are 3 and 4. Then the total worth would simply be 7. The additive model accounts for the majority (up to 80 or 90 percent) of the variation in preference in almost all cases and suffices for most applications.

The composition rule using **interaction effects** is similar to the additive form in that it assumes the consumer sums the part-worths to get an overall total across the set of attributes. It differs in that it allows for certain combinations of levels to be more or less than just their sum. Using our previous example, we see that an interactive model allows for the sum of the two levels to be either more or less than seven, the result of the additive model. In our dog food example, a respondent may really like a certain brand (Arf), but only with a certain type of ingredients (all meat). In this case, the brand has a low part-worth except when combined with another specific level (all meat) of dog food ingredients. We say that brand and ingredients are interacting as well as using the additive effects for each factor. The interactive form corresponds to the statement "The whole is greater (or less) than the sum of its parts."

Many times, adding interaction terms to models decreases predictive power because the reduction in statistical efficiency (more part-worth estimates) is not offset by increases in predictive power gained from the interactions. The interactions predict substantially less variance than the additive effects, most often not exceeding a 5 to 10 percent increase in explained variance. Interaction terms are most likely to be substantial where attributes are less tangible, particularly where aesthetic or emotional reactions play a large role. The increased importance of interaction terms comes from the inability to depict the actual differences between certain attributes, with the "unexplained" portions associated with only certain levels of an attribute.

The choice of a composition rule determines the types and number of treatments or stimuli that the respondent must evaluate, along with the form of estimation method used. An additive form requires fewer evaluations from the respondent, and it is easier to obtain estimates for the part-worths. However, the interactive form may be a more accurate representation of how respondents actually value a product/service.

The analyst does not know with certainty which form is the best but must instead understand the implications of either choice on both the design of the study and the results obtained. We will examine the need for making this choice and the trade-offs associated with choosing either form at various points in our discussion.

Selecting the Part-worth Relationship: Linear, Quadratic, or Separate Part-worths
The flexibility of conjoint analysis in handling different types of variables comes from the assumptions the analyst makes regarding the relationships of the part-worths within a factor. In making decisions about the composition rule, the analyst is deciding how factors relate to one another in the respondent's decision process. In defining the type of relationship, the analyst is focusing on how the *levels of a factor* are related.

Conjoint analysis gives the analyst three alternatives, ranging from the most restrictive (a linear relationship) to the least restrictive (separate part-worths), with the ideal point, or quadratic model, falling in between. Figure 10.2 illustrates the differences between the three types of relationships. The linear model is the simplest yet most restricted form, because we estimate only a single part-worth (similar to a regression coefficient) that is multiplied by the level's value. In the quadratic form, also known as the ideal model, the assumption of strict linearity is relaxed, so that we have a simple curvilinear relationship. The curve can turn either upward or downward. Finally, the part-worth form is the most general, allowing for each level to have its own part-worth estimate. When using the separate part-worths, the number of estimated values increases quickly as we add factors and levels because each new level has a separate part-worth estimate.

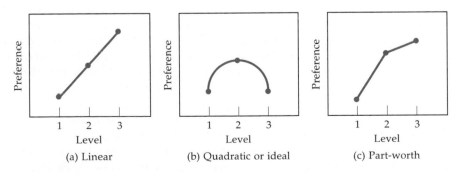

FIGURE 10.2 The three basic types of relationships between factor levels in conjoint analysis.

The type of relationship can be specified for each factor separately, thus allowing for a mixture if needed. This choice does not affect how the treatments or stimuli are created, but it does impact how and what types of part-worths are estimated by conjoint analysis. If we can reduce the number of part-worths estimated for any given set of stimuli, the calculations will be more efficient and reliable. But we must consider the trade-off between these gains and the accurate representation of how the consumer actually forms overall preference.

The analyst has several approaches to deciding on the type of relationship for each factor. The conjoint model can be estimated first as a part-worth model, and the different part-worth estimates can be examined visually to detect whether a linear or a quadratic form is appropriate. In many instances, the general form is apparent, and the model can be reestimated with relationships specified for each variable as justified. Alternatively, the analyst can assess the changes in predictive ability under different combinations of relationships for one or more variables. This approach is not recommended without at least some theoretical or empirical evidence for the possible type of relationship considered (e.g., prior estimates of part-worths). Without such support, the results may have high predictive ability but little use in decision making. In all instances, the analyst must balance predictive ability with the intended use of the study and the degree of managerial relevance and interpretation needed.

Data Collection

Having specified the factors and levels, plus the basic model form, the analyst next decides on the type of presentation of the stimuli (trade-off versus full profile), the type of response variable, and the method of data collection. The objective is to convey to the respondent the attribute combinations (stimuli) in the most realistic and efficient manner possible. Most often the stimuli are presented in written descriptions, although physical or pictorial models can be quite useful for aesthetic or sensory attributes.

Choosing a Presentation Method

The **trade off, full-profile,** and **pairwise comparison methods** are the three methods of stimulus presentation most widely associated with conjoint analysis. The trade-off method compares attributes two at a time by ranking all combinations of levels (see Figure 10.3). It has the advantages of being simple for the respondent and easy to administer, and it avoids information overload by presenting only two attributes at a time. However, usage of this method has decreased dramatically in recent years owing to several limitations: (1) a sacrifice in realism by using only two factors at a time, (2) the large number of judgments necessary for even a small number of levels, (3) a tendency for respondents to get confused or follow a routinized response pattern because of fatigue, (4) the inability to employ pictorial or other nonwritten stimuli, (5) the sole use of nonmetric responses, and (6) its inability to use fractional factorial stimuli designs to reduce the number of comparisons made. Recent studies have shown that the third approach, pairwise comparisons, has now displaced trade-off methods for second place in commercial applications [31].

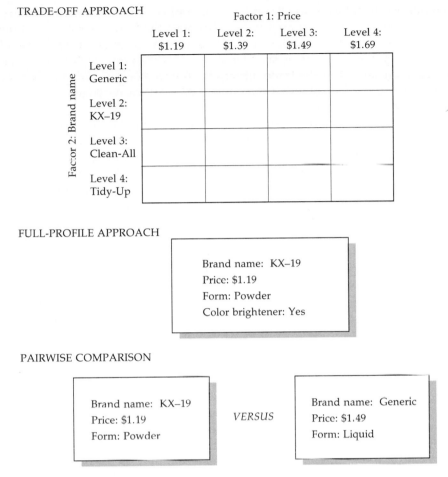

TRADE-OFF APPROACH

FULL-PROFILE APPROACH

PAIRWISE COMPARISON

FIGURE 10.3 Examples of the trade-off and full-profile methods of presenting stimuli.

The most popular method is the full-profile presentation, principally because of its ability to reduce the number of comparisons through the use of fractional factorial designs. In this approach, each stimulus is described separately, most often on a profile card (see Figure 10.3 for an example). This approach elicits fewer judgments, but each is more complex and the judgments can be either ranked or rated. Among its advantages are a more realistic description achieved by defining levels of each factor in a stimulus, a more explicit portrayal of the trade-offs among *all* factors and the existing environmental correlations among the attributes, and possible use of more types of preference judgments, such as intentions to buy, likelihood of trial, and chances of switching—all difficult to answer with the trade-off method.

The full-profile method is not flawless and faces two major limitations. First, as the number of factors increases, so does the possibility of information overload. The respondent is tempted to simplify the process by focusing on only a few factors, when in an actual situation he or she would consider all factors. Second, the order in which factors are listed on the stimulus card may have an impact on the evaluation. Thus the researcher needs to rotate the factors across respondents when possible to minimize order effects. The full-profile method is recommended when the number of factors is six or fewer. When the number of factors ranges from seven to ten, then the trade-off approach becomes a possible compromise along with the profile method. If the number of factors exceeds ten, then alternative methods are suggested [13].

The third method, the pairwise combination, combines the two other presentation methods. The pairwise combination is a comparison of two profiles (see Figure 10.3), with the respondent most often using a rating scale to indicate strength of preference for one profile versus the other [18]. The distinguishing characteristic of the pairwise comparison is that the profile does not contain all the attributes, as does the full profile method, but instead only a few attributes at a time are selected in constructing profiles. It is similar to the trade-off method in that pairs are evaluated, but in the case of the trade-off method the pairs being evaluated are attributes, while in the pairwise comparison method the pairs are profiles with multiple attributes. The pairwise comparison method is also instrumental in many specialized conjoint designs, such as adaptive conjoint analysis (ACA) [25], which is used in conjunction with a large number of attributes (see below for a more detailed discussion of dealing with a large number of attributes).

Creating the Stimuli

Once the factors and levels have been selected and the presentation method chosen, the analyst turns to the task of creating the treatments or stimuli for evaluation by the respondents. In the case of the trade-off method, all possible combinations of attributes are used. If five attributes were involved, ten trade-off matrices representing all possible pairwise attribute combinations would be evaluated. This can quickly lead to heavy burdens on the respondent as the number of attributes increases.

The other two methods—full-profile and pairwise comparison—involve the evaluation of one stimulus at a time (full-profile) or pairs of stimuli (pairwise comparison). In simple experiments with a small number of factors and levels (like those discussed earlier where three factors with two levels each resulted in eight combinations), the respondent may evaluate each stimulus in the full-profile method or all possible pairs. This approach is known as a **factorial design,** when all combinations are used. But as the number of factors and levels increases, this design becomes impractical. If the researcher is interested in assessing the impact of 4 variables with 4 levels for each variable, 256 stimuli (4 levels × 4 levels × 4 levels × 4 levels) would be created in a full factorial design for the full-profile method. This is obviously too many for one respondent to evaluate and still give consistent, meaningful answers. An even greater number of pairs of stimuli would be created for the pairwise combinations of profiles with differing numbers of attributes.

In deciding on the subset of stimuli to be used in the full-profile method to evaluate in the pairwise combination method, the researcher can use a **fractional factorial design.** In the full-profile method, the fractional factorial design selects a sample of possible stimuli, with the number of stimuli depending on the type of composition rule assumed to be used by respondents. The simplest and most

TABLE 10.1 Alternative Fractional Factorial Designs for an Additive Model (Main Effects) with Four Factors at Four Levels Each

Stimulus	Levels for*			
	Factor 1	Factor 2	Factor 3	Factor 4
Design 1				
1	3	2	3	1
2	3	1	2	4
3	2	2	1	2
4	4	2	2	3
5	1	1	1	1
6	4	3	4	1
7	1	3	2	2
8	2	1	4	3
9	2	4	2	1
10	3	3	1	3
11	1	4	3	3
12	3	4	4	2
13	1	2	4	4
14	2	3	3	4
15	4	4	1	4
16	4	1	3	2
Design 2				
1	2	3	1	4
2	4	1	2	4
3	3	3	2	1
4	2	2	4	1
5	1	1	1	1
6	1	4	4	4
7	4	2	1	3
8	2	4	2	3
9	3	2	3	4
10	3	4	1	2
11	4	3	4	2
12	1	3	3	3
13	2	1	3	2
14	3	1	4	3
15	1	2	2	2
16	4	4	3	1

*The numbers in the columns under Factor 1 through Factor 4 are the levels for each factor. For example, the first stimulant in design 1 consists of level 3 for Factor 1 and Factor 3, a value of 2 on Factor 2, and 1 for Factor 4.

popular composition rule is the additive model, which assumes only main effects for each factor with no interactions. In our example of 4 factors at 4 levels, only 16 stimuli are needed to estimate the main effects for the 4 factors if the full-profile method is used. The 16 stimuli must be carefully constructed for **orthogonality** to ensure the correct estimation of the main effects. Table 10.1 shows two possible sets of 16 orthogonal stimuli. Note that any number of combinations of stimuli can be used, so the researcher is encouraged to examine the set of stimuli and see whether any are unbelievable or unsatisfactory. If undesirable stimuli are found, another design can be created and checked for suitability.

The remaining 240 possible stimuli in our example are needed if all 11 interaction terms are to be estimated. As discussed earlier, the researcher may decide that selected interactions are important and should be included in the model estimation. In this case, the fractional factorial design must include additional stimuli to accommodate the interactions. Published guides and computer programs are available to aid in creating the stimuli [1, 7, 15, 22].

If a stimulus is unacceptable, perhaps unbelievable, and a better alternative design cannot be found, then the unacceptable stimulus can be deleted. Although the design will not be totally orthogonal (i.e., it will be somewhat correlated), it will not violate any assumptions of conjoint analysis. It will, however, add to the problems of multicollinearity similar to those in regression (i.e., instability of the estimates when levels are slightly changed and a lessened ability to assess the unique impact of each attribute). In practical terms, interattribute correlations should be minimized but do not need to be zero if small correlations (.20 or less) will add to realism. Most problems are found in the case of negative correlations, as between gas mileage and horsepower. The average interattribute correlation can be reduced by adding uncorrelated factors, so that with a realistic number of factors (e.g., 6 factors), the average intercorrelation would be close to .20, which has inconsequential effects.

The published guides for fractional factorial designs and the conjoint program components that will design the stimuli to maintain orthogonality make the generation of full-profile stimuli quite easy. When using pairwise comparisons, the number may be quite large and complex, so that most often interactive computer programs are used that select the optimal sets of pairs as the questioning proceeds.

Selecting a Measure of Consumer Preference

The researcher must also select the measure of preference: rank-ordering versus rating (i.e., a 1-to-10 scale). Although the trade-off method employs only ranking data, the pairwise comparison method can evaluate the preference either by obtaining a rating of preference of one stimulus over the other or just a binary measure of which is preferred.

The full-profile method can accommodate both ranking and rating methods, and each preference measure has certain advantages and limitations. Obtaining a rank-order preference measure (i.e., rank-ordering the stimuli from most to least preferred) has two major advantages: (1) it is likely to be more reliable, because ranking is easier than rating with a reasonably small number (20 or fewer) stimuli, and (2) it provides more flexibility in estimating different types of composition rules. It has, however, one major drawback: it is extremely difficult to administer, because the ranking process is most commonly performed by sorting stimulus cards into the preference order, and this sorting can be done only in a personal interview setting.

The alternative is to obtain a rating of preference on a metric scale. Metric measures are easily analyzed and administered, even by mail, and allow conjoint estimation to be performed by multivariate regression. However, respondents can be less discriminating in their judgments than when they are rank-ordering. Also, given the large number of stimuli evaluated, it is useful to expand the number of response categories over that found in most consumer surveys. A rule of thumb is to have 11 categories (i.e., rating from 0 to 10 or 0 to 100 in increments of ten) for 16 or fewer stimuli and expand to 21 categories for more than 16 stimuli [20].

Survey Administration

In the past, the complexity of the conjoint analysis task has led most often to the use of personal interviews to obtain the conjoint responses. Personal interviews enable the interviewer to explain the sometimes more difficult tasks associated with conjoint analysis. Recent developments in interviewing methods, however, have now made conducting conjoint analyses feasible both through the mail (with both pencil-and-paper questionnaires and computer-based surveys) and by phone. If the survey is designed to ensure that the respondent can assimilate and process the stimuli properly, then all of the interviewing methods produce relatively equal predictive accuracy [2]. The use of disk-by-mail computerized interviewing has greatly simplified the conjoint task demands on the respondent and made such developments as adaptive conjoint analysis [25] widely available.

Stage Three: Assumptions of Conjoint Analysis

Conjoint analysis has the least restrictive set of assumptions involving the estimation of the conjoint model. The structured experimental design and the generalized nature of the model make most of the tests performed in other dependence methods unnecessary. Yet while the statistical assumptions may be less, the conceptual assumptions are perhaps greater than with other methods. As mentioned earlier, the analyst must specify the general form of the model (main effects versus interactive model) before the research is designed. This "builds in" this decision and makes it impossible to test alternative models once the research is designed and the data collected. Conjoint analysis is not like regression, for example, where additional effects (interaction or nonlinear terms) can be easily evaluated. The analyst must make this decision concerning model form and must design the research accordingly. Thus, conjoint analysis, while having few statistical assumptions, is very theory-driven in its design, estimation, and interpretation.

Stage Four: Estimating the Conjoint Model and Assessing Overall Fit

The options available to the researcher in terms of estimation techniques have increased dramatically in recent years. The development of techniques in conjunction with specialized methods of stimulus presentation (e.g., the adaptive conjoint techniques [25]) is just one improvement of this type. The analyst, however, in obtaining the results of a conjoint analysis study, must address the issues of selecting the estimation method and evaluating the results. We address these issues briefly in the following sections.

Selecting an Estimation Technique

Rank-order evaluations require a modified form of analysis of variance specifically designed for ordinal data. Among the most popular and best-known computer programs are MONANOVA (MONotonic ANalysis Of VAriance) and LINMAP. These programs give estimates of attribute part-worths, so that the rank order of their sum (total worth) for each treatment is correlated as closely as possible to the observed rank order. If a metric measure of preference is obtained (e.g., ratings rather than rankings), then many methods, even multiple regression, can estimate the part-worths for each level. Most computer programs available today can accommodate either type of evaluation, as well as estimate any of the three types of relationships (linear, ideal point, and part-worth).

A simple example of the statistical relationships and calculations underlying conjoint analysis and the impact of interactions in calculating preference is provided in Appendix 10A.

Evaluating the Results

Conjoint analysis results must be examined to assess the accuracy of the estimated models at both the individual and aggregate levels. The ability of the conjoint model to predict consumer preferences accurately can be assessed for both metric and nonmetric responses. The objective in assessing reliability is to ascertain how consistently the model predicts across the set of preference evaluations given by each person. For the rank-order data, correlations based on the actual and the predicted ranks (e.g., Spearman's rho or Kendall's tau) are used. If a metric rating is obtained, then a simple Pearson correlation, just like that used in regression, is appropriate. Again, the actual and predicted preferences are correlated for each person. These values can then be tested for statistical significance.

In most conjoint experiments, however, the number of stimuli does not substantially exceed the number of parameters, and there is always the potential for "overfitting" the data. Researchers are thus strongly encouraged to measure model accuracy not only on the original stimuli but also with a validation or holdout set of stimuli. In a procedure similar to a holdout sample in discriminant analysis, the researcher prepares more stimulus cards than needed, and the respondent rates all of them at the same time. Parameters from the estimated conjoint model are then used to predict preference for the new set of stimuli, which is compared with actual responses to assess model reliability.

If an aggregate estimation technique is used, then researchers can also use a holdout sample of respondents in each group to assess predictive accuracy. This method is not feasible for disaggregate results because there is no "generalized" model to apply to the holdout sample, as each respondent in the estimation sample has individualized model estimates.

Stage Five: Interpreting the Results

Aggregate Versus Disaggregate Analysis

The customary approach to interpreting conjoint analysis is disaggregate; that is, each respondent is modeled separately, and the fit of the model is examined for each respondent. Then the researcher appraises the behavior of each respondent relative to the assumptions of the model. This approach also allows for the exclusion of respondents who show such poor preference structure that they did not perform the task expected of them.

Interpretation can also take place with aggregate results. Whether the model estimation is made at the individual level and then aggregated or the estimates are made on a set of respondents, the analysis fits one model to the aggregate of the responses. As one might expect, this process generally yields poor results when one is trying to estimate what any single respondent would do and poor results when trying to interpret the values that each of the attributes has for any single respondent. Unless the researcher is definitely dealing with a population exhibiting homogeneous behavior with respect to the factors, aggregate analysis should not be used. However, many times aggregate analysis more accurately predicts aggregate behavior, such as market share. Thus the researcher must identify the primary purpose of the study and employ the appropriate level of analysis.

Assessing the Relative Importance of Attributes

In addition to portraying the impact of each level with the part-worth estimates, conjoint analysis can assess the relative importance of each factor. Because the part-worth estimates are typically scaled to appear on a common scale, the greatest contribution to overall utility of preference, and hence the most important factor, is the factor with the highest range (low to high) of part-worths. To provide a consistent basis of comparison across respondents, the range values are standardized by dividing each one by the sum of all range values. The result is an importance value for each factor that totals 100 percent for each individual across all factors.

The analyst must always remember the impact on the importance values of an extreme or practically infeasible level. If such a level is found, it should be deleted from the analysis or the importance values should be reduced to reflect only the range of feasible levels.

Stage Six: Validation of the Conjoint Results

The validation of conjoint analysis occurs both internally and externally. Internal validation involves confirmation that the selected composition rule (i.e., additive versus interactive) is appropriate. In most instances the analyst is limited in empirically assessing the validity of the two-model forms in a full study, owing to the data collection demands. Thus, the analyst should compare the model forms in a pretest study to confirm which model is appropriate. External validation involves the issue of sample representativeness. While there is no evaluation of sampling error in the individual-level models, the analyst must always be assured that the sample is representative of the population of study. This becomes especially important when the conjoint results are used for segmentation of choice simulation purposes (see the next section for a more detailed discussion of these uses of conjoint results).

Stage Seven: Applying Conjoint Analysis Results

Typically, the conjoint models are estimated at the individual level (one per individual) and used in one or more of the following areas of decision support. Their purpose is to use the conjoint results to represent the decision processes of individuals. With individual-level results, conjoint analysis can provide a model of preference for each individual. Aggregate results can represent groups of individuals and a means of predicting their decisions for any number of situations.

Segmentation

One of the most common uses of individual-level conjoint analysis results is to group respondents with similar part-worths or importance values to identify segments. The estimated conjoint part-worth utilities can be used solely or in combination with other variables (e.g., demographics) to derive respondent groupings that are most similar in their preferences [10]. In the dog food example, we might find one group for which brand is the most important feature, while another group might value ingredients most highly. The researcher is interested in knowing the presence of such groups and their relative magnitude.

Profitability Analysis

A complement to the product design decision is a marginal profitability analysis of the proposed product design. If the cost of each feature is known, the cost of each ''product'' can be combined with the expected market share and sales volume to predict its viability. This process might point to a combination of attributes with a smaller share as the most profitable because of an increased profit margin resulting from the low cost of particular components. Both individual and aggregate results can be used in this analysis.

Conjoint Simulators

At this point, the researcher still understands only the relative importance of the attributes and the impact of specific levels. But how does conjoint analysis achieve its other primary objective—using what-if analysis to predict the market share that a stimulus (real or hypothetical) is likely to capture in various competitive scenarios of interest to management? This is the role played by **choice simulators,** which follow a three-step process:

1. Estimate and validate conjoint models for each respondent (or group).
2. Select the sets of stimuli to test according to possible competitive scenarios.
3. Simulate the choices of all respondents (or groups) for the specified sets of stimuli and predict market share for each stimulus by aggregating their choices.

After the conjoint model has been estimated, the researcher can specify any number of sets of stimuli for simulation of consumer choices. Among the possible uses are assessing (1) the impact of adding a product to an existing market; (2) the increased potential from a multiproduct or multibrand strategy, including estimates of cannibalism; or (3) the impact of deleting a product/brand from the market. In each case, the researcher provides the set of stimuli representing the market, and the choices of respondents are then simulated.

Choice simulators typically use two types of rules in predicting which stimulus is selected [9]. The first is the maximum utility model, which assumes that the respondent chooses the stimulus with the highest predicted utility score. This is most appropriate in cases of markets with individuals of widely different preferences and in situations involving sporadic, nonroutine purchases. The alternative choice rule is a purchase probability measure, where predictions of choice probability sum to 100 percent over the set of stimuli tested. This approach is best suited to repetitive purchasing situations where purchases may be more tied to usage situations over time. The two most common methods of making these predictions are the BTL (Bradford-Terry-Luce) and logit models, which make quite similar predictions in almost all situations [16].

Special Topics in Conjoint Analysis

Up to this point we have dealt with relatively simple conjoint analysis applications involving the traditional conjoint methods of data collection and a reasonably small number of attributes. But real-world applications many times involve 20 to 30 attributes or require a more realistic choice task than used in our earlier discussions. Recent developments have been made to overcome these problems encountered in many conjoint studies, and two specific issues—a large number of attributes and more realistic choice tasks—will be discussed next.

Conjoint Analysis with a Large Number of Factors

The full-profile and trade-off methods start to become unmanageable with between 6 and 8 attributes, yet many commercial studies need to incorporate up to 25 or 30 attributes. In these cases, some reduced form of conjoint analysis must be used to simplify the data collection effort and still represent a realistic choice decision. The two basic options are self-explicated models and hybrid/adaptive models.

In the self-explicated model, the respondent provides a rating of the desirability of each level of an attribute and then rates the relative importance of the attribute overall. Part-worths are then calculated by a simple multiplication of the two values. This approach is compositional in that ratings are made on the components of utility, rather than just overall preference. However, it is not in the spirit of conjoint analysis and is closer to traditional multiattribute models. Also, inter-attribute correlations may cause substantial biases because of "double counting" of correlated factors. A primary concern is whether respondents can assess the relative importance of attributes accurately, because research shows that they underestimate certain factors, perhaps because they want to give socially desirable answers. Finally, the researcher does not really set the stage for evaluation. Instead, consumers use their experience as a basis for judgment, because they never perform a choice task (rating the set of hypothetical combinations of attributes). This lack of realism is a critical limitation in new-product applications. However, if the number of factors cannot be reduced to a manageable level within traditional conjoint methods, then a self-explicated model may be an alternative method.

A second approach is the hybrid model, so termed because it combines the self-explicated and part-worth conjoint models [13]. It combines self-explicated values with evaluations of a small subset (from three to nine) of stimuli selected from a larger fractional factorial design. The sets of stimuli differ among respondents such that, although each respondent evaluates only a small number, all of the stimuli are evaluated by a portion of the respondents. One variant of this approach is Adaptive Conjoint Analysis (ACA), developed by Sawtooth Software [25]. It employs self-explicated ratings to reduce the factorial design size and make the process more manageable. Results as to its relative predictive ability are mixed at this time, but it is a promising alternative when the number of attributes is large [11, 19].

When we are faced with a number of factors that cannot be accommodated in the conjoint methods discussed to this point, the self-explicated and hybrid models preserve at least a portion of the underlying principles of conjoint analysis. In comparing these two extensions, we have found the self-explicated method to have slightly lower reliability. When the hybrid models and self-explicated methods are compared with full-profile methods, the results are mixed, with slightly better performance achieved by the hybrid/adaptive method [17]. Although more research is needed to confirm the comparisons across methods, the empirical studies indicate that each of the methods reaches acceptable results, and the analyst should choose the method that best meets the study objectives and the research design.

Choice-based Conjoint

In recent years, researchers in the area of conjoint analysis have focused their efforts in a new direction. As discussed earlier in the chapter, one conducts a conjoint exercise to understand the respondent's decision-making process and to predict behavior in the marketplace. Traditional conjoint analysis assumes that the judgment task, based on ranking or rating, captures the choices the respondent will make in the marketplace. In other words, a direct link is assumed between the judgements exhibited in the conjoint exercise and actual choice. Some researchers have argued that this may not be a realistic way of depicting a respondent's actual decision process. Although of practical value, current approaches for using the results of conjoint studies to forecast respondent choices are inadequate. This inadequacy is primarily due to a lack of formal theory linking these measured judgements to choices [21]. An alternative conjoint task with inherent face validity is to allow the respondent to choose a profile/scenario from a set of alternative stimuli known as a **choice set.** Such a method of obtaining responses is known as a **choice-based approach.**

Advantages and Disadvantages of Choice-based Conjoint

The growing acceptance of choice-based conjoint analysis among marketing research practitioners is primarily due to the belief that obtaining preferences by having respondents choose a single preferred stimuli from among a set of stimuli is more realistic, and thus a better method of approximating actual decision processes. Moreover, choice-based conjoint provides an option of *not* choosing any of the presented profiles/scenarios by including a no-choice option among the profiles. Traditional conjoint assumes respondents' preferences will always be allocated among a set of alternatives, while the choice-based approach allows for market contraction if all the alternatives in a choice set are unattractive.

Choice-based models, which are estimated at the aggregate level, have several other desirable features. The values and statistical significance of all estimates, including interactions, are easily reported; market share predictions for new stimuli are easily produced, and the researcher has the added assurances that "choices" among stimuli were used to calibrate the model. However, aggregate choice models hinder segmentation of the market. Choice-based conjoint is not capable of estimating a separate conjoint model for each respondent; thus it is not possible to group respondents according to their conjoint model results as discussed earlier. In contrast, the ratings-based models described earlier are well suited to segmentation studies but face the problem of cumbersome tests for the statistical significance of the part-worth estimates. Thus, the results may be difficult to summarize and the simulation of choice shares can be problematic.

Choice-based conjoint has other disadvantages as well. Each choice set contains several stimuli and each stimulus contains several factors at different levels. The stimuli are similar to the full-profile presentation forms discussed earlier but do not have to include all factors. Therefore, the respondent must process a considerable amount of information before making a choice in each choice set. Sawtooth Software, developers of the Choice-Based Conjoint system (CBC), believe those choices involving more than six attributes are likely to confuse and overwhelm the respondent [26]. Although the choice-based method does mimic more closely actual decisions, the inclusion of too many attributes creates a formidable task that ends up with less information than would have been gained through the rating of each stimulus individually.

In practice, ratings-based, rankings-based, and choice-based approaches allow for similar types of analyses, simulations, and reporting, even though the models that estimate the results are different. Though models estimated from choice data have not yet been subjected to thorough empirical tests, some researchers believe they presumably have an advantage in predicting choice behavior.

However, preliminary evidence suggests that the superiority of choice-based conjoint over ratings/ranking-based conjoint is not clearly supported. Empirical tests indicate little difference between individual-level ratings-based models adjusted to take nonchoice into account and the generalized multinomial logit choice-based models [23]. The conclusion was that both ratings-based and choice-based models predicted equally well. Other research compared the two approaches to conjoint (ratings-based or choice-based) in terms of the ability to predict shares in a holdout sample [8]. Both approaches predict holdout sample choices well, with neither approach dominant and the results mixed in different situations. Ultimately, the decision to use one method over the other is dictated by the objectives and scope of the study, the analyst's familiarity with each method, and the available software to properly analyze the data.

A Simple Illustration of Full-profile and Choice-based Conjoint

A cellular phone company wishes to estimate the market potential for three service options:

ICA	Itemized call accounting	$2.75/month service charge
CW	Call waiting	$3.50/month service charge
TWC	Three-way calling	$3.50/month service charge

The base service is $14.95 and $0.50 per minute of calling time.

The conjoint analysis could be presented with full profiles. The resulting profiles would be as follows:

	Levels of Factors (1 = present, 0 = absent)		
Profile	*ICA*	*CW*	*TWC*
Profile 1	0	0	0
Profile 2	1	0	0
Profile 3	0	1	0
Profile 4	1	1	0
Profile 5	0	0	1
Profile 6	1	0	1
Profile 7	0	1	1
Profile 8	1	1	1

That is, Profile 1 would represent the base service with no options; Profile 2 is the base service plus itemized call accounting; . . .; Profile 8 is the base service plus itemized call accounting, plus call waiting, plus three-way calling. In a full-profile conjoint, the respondent would be asked to rate or rank these eight profiles. In a choice-based approach, the respondent may be shown many choice sets, each having many alternatives. The respondent is then asked to choose only one of the profiles in the choice set ("most preferred" or "most liked") or select the "no choice" option. The preparation of profiles and choice sets is based on experimental design principles [20].

One possible configuration of choice sets and the profiles in each is given below:

Choice Set	Profiles in Choice Set
1	Profiles 1, 2, 4, 5, 6, and No Choice
2	Profiles 2, 3, 5, 6, 7, and No Choice
3	Profiles 1, 3, 4, 6, 7, 8, and No Choice
4	Profiles 2, 4, 5, 7, 8, and No Choice
5	Profiles 3, 5, 6, 8, and No Choice
6	Profiles 4, 6, 7, and No Choice
7	Profiles 1, 5, 7, 8, and No Choice
8	Profiles 1, 2, 6, 8, and No Choice
9	Profiles 1, 2, 3, 7, and No Choice
10	Profiles 2, 3, 4, 8, and No Choice
11	Profiles 1, 3, 4, 5, and No Choice

Notice that the number of profiles varies across choice sets. Also, the number of choices made (one choice for each of 11 choice sets) is actually more in this case than just rating or ranking each of the profiles. The advantage of the choice-based approach is the additional realism and the ability to estimate a number of interaction terms not possible with this number of full profiles. After each respondent has chosen an alternative for each choice set, the data are aggregated across respondents (or segments or some other grouping of respondents) to obtain the conjoint part-worths for each level. From these results, the contributions of each attribute can be assessed and the likely market shares of competing profiles can be estimated.

An Illustration of Conjoint Analysis

In the following sections we examine the steps in an application of conjoint analysis. The discussion follows the model-building process introduced in Chapter 1 and focuses on (1) design of the stimuli, (2) estimation and interpretation of the conjoint part-worths, and (3) application of a conjoint simulator to predict market shares for a new product formulation.*

Stage One: Objectives of the Conjoint Analysis

In developing a new industrial cleaner, HATCO decided on a conjoint analysis experiment to assist in understanding the needs of its industrial customers. In an adjunct study to the one described in Chapter 1, HATCO also commissioned a conjoint analysis experiment among 100 industrial customers. Marketing research and consultation with the product development group identified five factors as the determinant attributes in the targeted segment of the industrial cleaner market. The five attributes are shown in Table 10.2.

---

*The CATEGORIES option of SPSS-PC is used in the design, analysis, and choice simulator phases of this example. The necessary control cards are shown in Appendix A.

TABLE 10.2 Attributes and Levels for the HATCO Conjoint Analysis
Experiment

Attribute	Description	Level
MIXTURE	Form of the product	1. Premixed liquid 2. Concentrated liquid 3. Powder
NUMAPP	Number of applications per container	1. 50 2. 100 3. 200
GERMFREE	Addition of disinfectant to cleaner	1. Yes 2. No
BIOPROT	Cleaner in biodegradable formulation	1. No 2. Yes
PRICE	Price per typical application	1. 35 cents 2. 49 cents 3. 79 cents

Stage Two: Design of the Conjoint Analysis

The decisions at this phase are (1) the levels of each attribute to be included in the stimuli, (2) the composition rule for respondents, and (3) the method of data collection. In addressing the first decision, focus group research established specific levels for each attribute (see Table 10.2). The levels were deemed actionable and communicable through a small-scale pretest and evaluation study. The price levels were taken from the range found in existing products to ensure they were not unrealistic. The product type did not suggest intangible factors that would contribute to interattribute correlation, and the attributes were defined to minimize interattribute correlation.

To ensure realism and allow for the use of ratings rather than rankings, HATCO decided to use the full-profile method of obtaining respondent evaluations. After careful consideration, HATCO researchers felt confident in assuming that an additive composition rule was appropriate. In choosing the additive rule, they were also able to use a fractional factorial design to avoid the evaluation of all 108 possible combinations ($3 \times 3 \times 2 \times 2 \times 3$). The stimulus design component of the computer program generated a set of 418 full-profile descriptions (see Table 10.3), allowing for the estimation of the orthogonal main effects for each factor.

The conjoint analysis experiment was administered during a personal interview. After collecting some preliminary data, the respondents were handed a set of 22 cards, each containing one of the full-profile stimulus descriptions. They were also presented with a foldout form that had seven response categories, ranging from ''Not at All Likely to Buy'' to ''Certain to Buy.'' Respondents were instructed to place each card in the response category best describing their purchase intentions. After initially placing the cards, they were asked to review their placements and rearrange any cards if necessary. The validation stimuli were rated at the same time as the other stimuli but withheld from the analysis at the estimation stage. Upon completion, the interviewer recorded the category for each card and proceeded with the interview.

TABLE 10.3 Set of 18 Full-profile Stimuli Used in the HATCO Conjoint Analysis Experiment

Card Number	Product Form	Number of Applications	Disinfectant Quality	Biodegradability	Price per Application
Stimuli Used in Estimation of Part-Worths					
1	Concentrate	200	Yes	No	35 cents
2	Powder	200	Yes	No	35 cents
3	Premixed	100	Yes	Yes	49 cents
4	Powder	200	Yes	Yes	49 cents
5	Powder	50	Yes	No	79 cents
6	Concentrate	200	No	Yes	79 cents
7	Premixed	100	Yes	No	79 cents
8	Premixed	200	Yes	No	49 cents
9	Powder	100	No	No	49 cents
10	Concentrate	50	Yes	No	49 cents
11	Powder	100	No	No	35 cents
12	Concentrate	100	Yes	No	79 cents
13	Premixed	200	No	No	79 cents
14	Premixed	50	Yes	No	35 cents
15	Concentrate	100	Yes	Yes	35 cents
16	Premixed	50	No	Yes	35 cents
17	Concentrate	50	No	No	49 cents
18	Powder	50	Yes	Yes	79 cents
Holdout/Validation Stimuli					
19	Concentrate	100	Yes	No	49 cents
20	Powder	100	No	Yes	35 cents
21	Powder	200	Yes	Yes	79 cents
22	Concentrate	50	No	Yes	35 cents

Stage Three: Assumptions in Conjoint Analysis

The relevant assumption in conjoint analysis is the specification of the composition rule and thus the model form used to estimate the conjoint results. In this situation, the nature of the product, the tangibility of the attributes, and the lack of intangible or emotional appeals justifies the use of an additive model. HATCO felt confident in using an additive model for this industrial decision-making situation. Moreover, it simplified the design of the stimuli and facilitated the data collection efforts.

Stage Four: Estimating the Conjoint Model and Assessing Overall Model Fit

The estimation of part-worths and the relative importance of each attribute was first performed for each respondent separately, and the results were then aggregated to obtain an overall result. Separate part-worth estimates were made for all levels initially, with examination of the individual estimates undertaken later to examine the possibility of placing constraints on a factor relationship form (i.e., employ a linear or quadratic relationship form). Table 10.4 shows the results for

TABLE 10.4 Conjoint Analysis Experiment Results for the Overall Sample and Selected Respondents

Part-Worth Estimates

	Product Form			Number of Applications			Disinfectant		Biodegradable		Price per application			Relative Importance of Factors					Predictive Accuracy	
	Premixed	Concentrate	Powder	50	100	200	Yes	No	No	Yes	$.35	$.49	$.79	1	2	3	4	5	Estimation	Holdout
Overall Sample																				
	-.331	.192	.139	-.416	.021	.396	.486	-.486	-.077	.077	.962	.069	1.031	11.75	18.23	21.82	3.44	44.76	.988	.957
Selected Respondents																				
1.	.056	.611	-.556	.444	.611	-1.056	.208	-.208	.542	-.542	1.444	.944	-2.389	14.29	20.41	5.10	13.27	46.94	.929	.986
2.	-.444	.389	.056	-.944	-.278	1.222	-1.083	1.083	-.417	.417	.389	.889	-1.278	10.20	26.53	26.53	10.20	26.53	.949	.616
3.	-1.111	1.222	-.111	-.611	.556	.056	.083	-.083	-.417	.417	.556	1.056	-1.611	32.56	16.28	2.33	11.63	37.21	.776	.660
4.	-1.722	.944	.778	-1.056	-.056	1.111	.792	-.792	-.083	.083	1.944	-.222	-1.722	26.02	21.14	15.45	1.63	35.77	.907	.893
5.	-.611	.389	.222	-.444	.222	.222	-.417	.417	-.542	.542	2.556	.056	-2.611	11.43	7.62	9.52	12.38	59.05	.851	.759

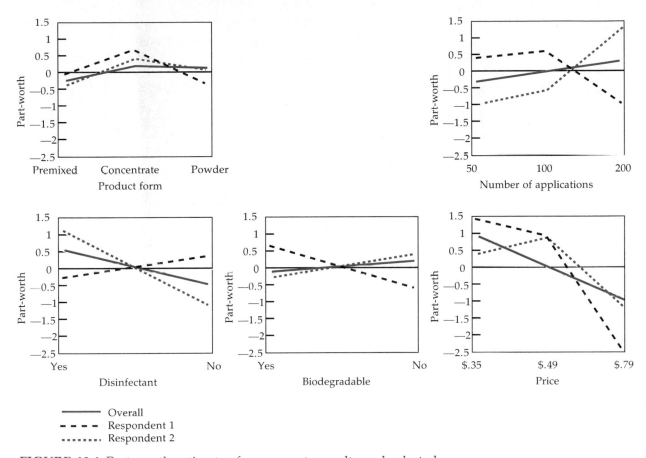

FIGURE 10.4 Part-worth estimates for aggregate results and selected respondents.

the overall sample, as well as for five respondents. Examination of the overall results suggests that perhaps a linear relationship can be estimated for the price variable. However, review of the individual results shows that only two of the six respondents had part-worth estimates for the price factors that were of a generally linear pattern. Thus application of a linear form for the price factor would severely distort the relationship among levels. Therefore, all levels were estimated as separate part-worth values.

Stage Five: Interpreting the Results

Figure 10.4 shows a diversity of part-worth estimates for every factor. For example, the overall results show little impact for the biodegradable feature. However, when the individual results are shown, we see that respondent 1 values the feature, while respondent 2 does not. Contrasting results are also seen for the disinfectant feature and the number of applications per container. These differences illustrate the value of conjoint analysis performed for each respondent instead of relying solely on aggregate results.

Figure 10.5 compares the derived importance values of each factor for both the aggregate results and the disaggregate results of two respondents. Although we see

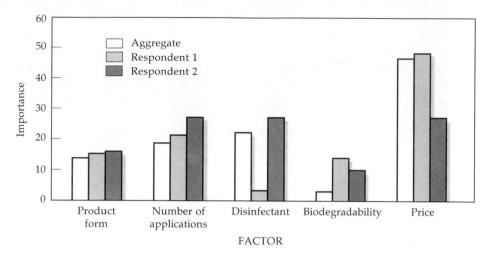

FIGURE 10.5 Factor importance for aggregate results and selected respondents.

a general consistency in the results, each respondent has unique aspects differing from each other and from the aggregate results. The greatest differences are seen for the features of biodegradability and disinfectant quality. Both respondents place more importance on biodegradability than is reflected in the aggregate results. But for the disinfectant feature, respondent 1 has little interest, and respondent 2 is slightly above average. Finally, respondent 2 is much less price sensitive than either the aggregate results or respondent 1. In summary, the results show two distinct and unique individuals, pointedly illustrating the benefits of conjoint results estimated for each individual instead of for the aggregate. One extension of conjoint analysis is to define groups of respondents with similar part-worth estimates or importance values of the factors using cluster analysis. These segments may then be profiled and assessed for their unique preference structures and market potential.

Stage Six: Validation of the Results

The high levels of reliability across respondents confirmed the additive composition rule for this set of respondents. The major validity issue is the representativeness of the sample. In this situation, HATCO would most likely proceed to a larger-scale project with greater coverage of its customer bases to ensure representativeness. Another consideration is the inclusion of noncustomers, especially if the goal is to understand the entire market, not just HATCO customers.

Stage Seven: Application of a Choice Simulator

In addition to understanding the aggregate and individual preference structures of the respondents, HATCO also used the conjoint results to simulate choices among three possible products. The three products tested were as follows:

Product 1. A premixed cleaner in a handy-to-use size (50 applications per container) that was environmentally safe (biodegradable) while still meeting all sanitary standards (disinfectant) at only 79 cents per application.

Product 2. An industrial version of product 1 with the environmental and sanitary features, but in a concentrate form in large containers (200 applications) at the low price of 49 cents per application.

Product 3. A real cleanser value in powder form in economical sizes (200 applications per container) for the unbelievable price of 35 cents per application.

The choice simulator then calculated the preference estimates for the products for each respondent. Predictions of the expected market shares were made with two choice models: the maximum utility model and a probabilistic model. The maximum utility model counts the number of times each of the three products had the highest utility across the set of respondents. As seen in Table 10.5, product 1 was preferred (it had the highest predicted preference value) for only 5 percent of the respondents. Product 2 was next, preferred by 25.5 percent, and the most preferred was product 3, with 69.5 percent.*

A second approach to predicting market shares is a probability model, either the BTL or logit model. Both models assess the relative preference of each product and estimate the proportion of times a respondent or the set of respondents will purchase a product. In the HATCO analysis, the aggregated predicted preference values for the products were 2.4, 4.8, and 5.7 for products 1, 2, and 3, respectively. The predicted market shares using the BTL models are then calculated by

$$\text{Market share}_{\text{PRODUCT 1}} = 2.4/(2.4 + 4.8 + 5.7) = 18.6\%$$
$$\text{Market share}_{\text{PRODUCT 2}} = 4.8/(2.4 + 4.8 + 5.7) = 37.2\%$$
$$\text{Market share}_{\text{PRODUCT 3}} = 5.7/(2.4 + 4.8 + 5.7) = 44.2\%$$

Similar results are obtained using the logit probabilistic model and are shown in Table 10.5 as well. Using the model recommended in situations involving repetitive choices (probability models), as is the case with an industrial cleaner, HATCO

*The fractional percentages are due to tied predictions among products 2 and 3.

TABLE 10.5 Choice Simulator Results for the Three-product Formulations

| | | *Market Share Predictions* | | |
| | | | *Probabilistic Models* | |
Product Formulation	*Aggregate Predicted Preference Scores*	*Maximum Utility Model (%)*	*BTL (%)*	*Logit (%)*
1	2.4	5.0	18.6	9.5
2	4.8	25.5	37.2	32.6
3	5.7	69.5	44.2	57.9

has market share estimates indicating an ordering of product 3, product 2, and finally product 1. It should be remembered that these results are aggregated across the entire sample, and the market shares may differ within specific segments of the respondents.

Summary

Conjoint analysis places more emphasis on the ability of the researcher/manager to theorize about the behavior of choice than it does on analytical technique. The appropriateness of the experimental design and the assumptions concerning the model form and types of relationships among variables are more critical than the choice of estimation technique. As such, it should be viewed primarily as exploratory, because many of its results are directly attributable to basic assumptions made during the course of the design and the execution of the study. Moreover, although conjoint analysis has strong theoretical foundations, its applications have increased dramatically without the corresponding theoretical development. The critical interplay between the assumed conceptual model of decision making and the appropriate elements of the conjoint analysis makes this a unique multivariate method. The researcher must accurately assess many facets of the decision-making process to employ conjoint analysis. Our focus has been on providing a better understanding of the principles of conjoint analysis experiments and how they represent the consumer's choice process. It is hoped that this understanding will enable researchers to avoid misapplication of this relatively new and powerful technique whenever faced with the need to understand choice judgments and preference structures.

An Illustration of Conjoint Analysis

Preview

To provide the interested reader with the principles and calculations underlying conjoint analysis and its estimation of part-worths, we offer a simple numerical example. In this discussion, we examine the simplest compositional rule (additive) and then briefly outline more complex assumptions (allowing interactions) about how people make choices. To illustrate the additive model, assume that the dog food experiment discussed previously was conducted with respondents who owned dogs and fed them canned food. Each respondent was given eight dummy cans of dog food in a simulated display and asked to rank his or her choices in order of preference for purchase. The eight cans (stimuli) are described in Table 10A.1, along with the ranks given by two respondents.

TABLE 10A.1 Product Descriptions and Ranking for Dog Food Experiment

	Product Description		Ranks Given by Respondents	
Size	Ingredients	Brand	Respondent 1	Respondent 2
6 ounce	All meat	Arf	1	1
6 ounce	All meat	Mr. Dog	2	2
6 ounce	Meat and fiber	Arf	5	6
6 ounce	Meat and fiber	Mr. Dog	6	5
12 ounce	All meat	Arf	3	4
12 ounce	All meat	Mr. Dog	4	3
12 ounce	Meat and fiber	Arf	7	7
12 ounce	Meat and fiber	Mr. Dog	8	8

If we assume the additive model is appropriate, the overall preference for a combination of attributes is represented as

$$\text{Preference}_{SIB} = PW_S + PW_I + PW_B$$

where the overall preference for a combination of size, ingredients, and brand (Preference$_{SIB}$) is composed of PW_S, PW_I, and PW_B, the estimated part-worths for the levels of size, ingredients, and brand, respectively.

To calculate the coefficients, we must specify the composition rule to be employed. If the additive model strictly holds, a simple difference from the mean (ANOVA) should apply. Because the average rank (of the eight combinations) is 4.5, we can calculate the impact of each factor, termed its **main effect,** as differences from this average. For example, the average ranks for the two levels of size are

$$6 \text{ ounce} = (1 + 2 + 5 + 6) \div 4 = 3.5$$
$$12 \text{ ounce} = (3 + 4 + 7 + 8) \div 4 = 5.5$$

The average ranks and deviations for each factor from the overall average rank (4.5) for respondent 1 are given in Table 10A.2.

Most often we use a small number to indicate higher rank and a more preferred stimulus. If that indication is used, we reverse all signs so that positive part-worths will now indicate higher preference. The coefficients are then calculated by the following:

1. Squaring the deviations and finding their sum (in this case the total is 10.5).
2. Multiplying each squared deviation by a standardizing value calculated as the number of levels divided by the sum of squared deviations (in this case $6 \div 10.5$ or .571).
3. Taking the square root of the standardized squared deviation to get the actual coefficient.

For example, for the first level of meat (all meat), the deviation of 2 (remember we reverse signs) is squared and then multiplied by .571 to get 2.284 ($2^2 \times .571 =$

TABLE 10A.2 Average Ranks and Deviations for Respondent 1

Factor Level	Average Rank	Deviation
6 ounce	3.5	−1
12 ounce	5.5	+1
All meat	2.5	−2
Meat and fiber	6.5	+2
Arf	4.0	−.5
Mr. Dog	5.0	+.5

TABLE 10A.3 Average Ranks, Deviations, and Coefficients for the Factor Levels

Factor Level	Average Rank	Deviation	Coefficient
6 ounce	3.5	−1	.77
12 ounce	5.5	+1	−.77
All meat	2.5	−2	1.55
Meat and fiber	6.5	+2	−1.55
Arf	0.0	0.0	0.0
Mr. Dog	0.0	0.0	0.0

2.284). To calculate the coefficient for this level, we then take the square root of 2.284 for a coefficient of 1.511. This process yields the following coefficients for each level:

Size		Ingredients		Brand	
6 Ounce	12 Ounce	Meat	Meat and Fiber	Arf	Mr. Dog
.756	−.756	1.511	−1.511	.378	−.378

Because the part-worth estimates are in a common scale, we can compute the relative importance of each factor directly. The importance of a factor is represented by the range of its levels (i.e., the difference between the highest and lowest values). The ranges are then "standardized" by dividing each range by the sum of the ranges. For example, for respondent 1, the ranges are 1.512 [.756 − (−.756)], 3.022 [1.511 − (−1.511)] and .756 [.378 − (.378)]. The sum total of ranges is 5.294. The relative importance for size, ingredients, and brand is calculated as 1.512 ÷ 5.294, 3.022 ÷ 5.294, and .756 ÷ 5.294, or 29 percent, 57 percent, and 14 percent, respectively.

We can follow the same procedure for the second respondent and calculate the average ranks, deviations, and coefficients for each level, as shown in Table 10A.3.

To examine the ability of this model to predict the actual choices of the respondents, we predict preference order by summing the coefficients for the different combinations of factor levels and rank-ordering the resulting scores. The calculations for respondent 1 for each stimulus are shown in Table 10A.4.

We can also compare the predicted preference order to the respondent's actual preference order for a measure of predictive accuracy. Note that the total part-worth values have no real meaning except as a means of developing the preference order and, as such, are not compared across respondents. The predicted preference orders for both respondents are given in Table 10A.5.

The estimated part-worths predict the preference order perfectly for respondent 1. However, because the weights of zero were calculated for the brand name for the second respondent, the compositional rule is incapable of predicting a difference between brands within the ingredient and can-size combinations. For

TABLE 10A.4 Predicted Part-worth and Rank for the Combinations of Factor Levels

Product Description					Predicted Value	
Size		Ingredients		Brand	Part-worth Total	Predicted Rank
6 ounce (.756)	+	All meat (1.51)	+	Arf (.378)	= 2.646	1
6 ounce (.756)	+	All meat (1.51)	+	Mr. Dog (−.378)	= 1.886	2
6 ounce (.756)	+	Meat and fiber (−1.51)	+	Arf (.378)	= −.374	5
6 ounce (.756)	+	Meat and fiber (−1.51)	+	Mr. Dog (−.378)	= −1.134	6
12 ounce (−.756)	+	All meat (1.51)	+	Arf (.378)	= 1.134	3
12 ounce (−.756)	+	All meat (1.51)	+	Mr. Dog (−.378)	= .374	4
12 ounce (−.756)	+	Meat and fiber (−1.51)	+	Arf (.378)	= −1.886	7
12 ounce (−.756)	+	Meat and fiber (−1.51)	+	Mr. Dog (−.378)	= −2.646	8

example, both combinations of all meat in 6-ounce cans are predicted as equal because the only difference between them, brand name, has a part-worth of zero. It appears that brand name may be a random choice, given size and ingredients. When two or more stimuli have equal total-worth predictions, then the rank orders are averaged. In the example for respondent 2, the first two stimuli were both the highest value; thus their average rank (ranks 1 and 2) would be 1.5.

TABLE 10A.5 Predicted Rank Orders for the Two Respondents

	Predicted Rank Order	
Original Rank	Respondent 1	Respondent 2
1	1	1.5
2	2	1.5
3	3	3.5
4	4	3.5
5	5	5.5
6	6	5.5
7	7	7.5
8	8	7.5

When Is an Interactive Composition Rule Appropriate?

Up to this point, we have considered only a situation in which the additive model is the appropriate compositional rule. We can, however, posit a situation where the respondent makes choices in which interactions appear to influence the choices. Assume a third respondent made the following preference ordering:

		Size	
Brand	Ingredients	6 Ounce	12 Ounce
Arf	Meat	1	2
	Meat and fiber	3	4
Mr. Dog	Meat	7	8
	Meat and fiber	5	6

These ranks were formed assuming that this respondent prefers Arf and normally prefers all meat over meat and fiber. However, a bad experience with Mr. Dog made the respondent select meat and fiber over all meat only if it was Mr. Dog. This choice is termed an **interaction effect** between the factors of brand and ingredients. If we consider only an additive model, we obtain the following coefficients:

Size		Ingredients		Brand	
6 Ounce	12 Ounce	Meat	Meat and Fiber	Arf	Mr. Dog
.42	−.42	0.0	0.0	1.68	−1.68

Using these coefficients to calculate preference orders for the combinations yields the following:

Actual rank	1	2	3	4	5	6	7	8
Predicted rank	1.5	3.5	1.5	3.5	5.5	7.5	5.5	7.5

The predictions are obviously less accurate, given that we know interactions exist. Also, the coefficients are misleading because the main effects of brand and ingredients are confounded by the interactions. If we were to proceed with only an additive model, we would be violating one of the principal assumptions and making potentially quite inaccurate predictions.

Checking for Interactions

To examine for first-order interactions is a reasonably simple task. For the previous preference data, we can form three two-way matrices of preference order. For example, in the first box, the two preference orders (one for each brand) for each combination of meat and ingredients are listed and then summed. To check for interactions, the diagonal values are then added and the difference calculated. If the total is zero, then no interaction exists. As seen in the example, the only interaction found is between brand and ingredients. It is unusual not to see some slight differences indicating some interactions. As the difference gets greater, the impact of the interaction increases, and it is up to the researcher to decide when the interactions pose enough problems in prediction to warrant the increased complexity of estimating coefficients for the interaction terms.

Summary

Conjoint analysis is quite sensitive to the analyst's decisions regarding both the type of composition rule to be used and the nature of the relationship among factor levels. In our example, we have illustrated the basic computational procedures underlying conjoint analysis, whether calculated by hand or through a statistical program. In gaining a better understanding of the computations involved, the reader should be better able to assess the implications of the decisions needed in the execution of a conjoint analysis experiment.

Questions

1. Ask three of your classmates to evaluate choice combinations based on these variables and levels relative to the choice of a textbook for a class, and specify the compositional rule you think they will use. Collect information with both the trade-off and full-profile methods.

Factor	*Level*
Depth	a. Goes into great depth on each subject
	b. Introduces each subject in a general overview
Illustrations	a. Each chapter includes humorous pictures
	b. Illustrative topics are presented
	c. Each chapter includes graphics to illustrate the numeric issues
References	a. General references are included at the end of the textbook
	b. Each chapter includes specific references for the topics covered

How difficult was it for respondents to handle the wordy and slightly abstract concepts they were asked to evaluate? Which presentation method was easier for the respondents?

2. Using either the simple numerical procedure discussed in Appendix 9A or a computer program, analyze the data from the experiment in question 1.

3. Design a conjoint analysis experiment with at least four variables and two levels of each variable that is appropriate to a marketing decision. Define the compositional rule you will use, the experimental design for creating stimuli, and the analysis method. Use at least five respondents to support your logic.

4. What are the practical limits of conjoint analysis in terms of variables or types of values for each variable? What type of choice problems are best suited to analysis with conjoint analysis? Which are least well served by conjoint analysis?

References

1. Addelman, S. "Orthogonal Main-Effects Plans for Asymmetrical Factorial Experiments." *Technometrics* 4 (1962): 21–46.
2. Akaah, I. "Predictive Performance of Self-Explicated, Traditional Conjoint, and Hybrid Conjoint Models under Alternative Data Collection Modes." *Journal of the Academy of Marketing Science* 19 (1991): 309–14.
3. Alpert, M. "Definition of Determinant Attributes: A Comparison of Methods." *Journal of Marketing Research* 8, no. 2 (1971): 184–91.
4. Bretton-Clark. *Conjoint Analyzer*. New York: Bretton-Clark, 1988.
5. Bretton-Clark. *Conjoint Designer*. New York: Bretton-Clark, 1988.
6. Bretton-Clark. *Simgraf*. New York: Bretton-Clark, 1988.
7. Conner, W. S., and M. Zelen. *Fractional Factorial Experimental Designs for Factors at Three Levels*, Applied Math Series S4. Washington, D.C.: National Bureau of Standards, 1959.
8. Elrod, T., J. J. Louviere, and K. S. Davey. "An Empirical Comparison of Ratings-Based and Choice-Based Conjoint Models."*Journal of Marketing Research* 29 (1992): 368–77.
9. Green, P. E., and A. M. Kreiger. "Choice Rules and Sensitivity Analysis in Conjoint Simulators." *Journal of the Academy of Marketing Science* 16 (Spring 1988): 114–27.
10. Green, P. E., and A. M. Kreiger. "Segmenting Markets with Conjoint Analysis." *Journal of Marketing* 55 (October 1991): 20–31.
11. Green, P. E., A. M. Kreiger, and M. K. Agarwal. "Adaptive Conjoint Analysis: Some Caveats and Suggestions." *Journal of Marketing Research* 28 (May 1991): 215–22.
12. Green, P. E., and V. Srinivasan. "Conjoint Analysis in Consumer Research: Issues and Outlook." *Journal of Consumer Research* 5 (September 1978): 103–23.
13. Green, P. E., and V. Srinivasan. "Conjoint Analysis in Marketing: New Developments with Implications for Research and Practice." *Journal of Marketing* 54, no. 4 (1990): 3–19.
14. Green, P. E., and Y. Wind. "New Way to Measure Consumers' Judgments." *Harvard Business Review* 53 (July-August 1975): 107–17.
15. Hahn, G. J., and S. S. Shapiro. *A Catalog and Computer Program for the Design and Analysis of Orthogonal Symmetric and Asymmetric Fractional Factorial Experiments*, Report No. 66-C-165. Schenectady, N.Y.: General Electric Research and Development Center, 1966.
16. Huber, J., and W. Moore. "A Comparison of Alternative Ways to Aggregate Individual Conjoint Analyses," in *Proceedings of the AMA Educator's Conference*, ed. L. Landon, pp. 64–68. Chicago: American Marketing Association, 1979.
17. Huber, J., D. R. Wittink, J. A. Fielder, and R. L. Miller. "The Effectiveness of Alternative Preference Elicitation Procedures in Predicting Choice." *Journal of Marketing Research* 30 (February, 1993): 105–14.

18. Johnson, R. M. "A Simple Method for Pairwise Monotone Regression." *Psychometrika* 40 (June 1975): 163–68.

19. Johnson, R. M. "Comment on Adaptive Conjoint Analysis: Some Caveats and Suggestions." *Journal of Marketing Research* 28 (May 1991): 223–25.

20. Louviere, J. J. *Analyzing Decision Making: Metric Conjoint Analysis*. Sage University Paper series on Quantitative Applications in the Social Sciences, vol. 67. Beverly Hills, Calif.: Sage, 1988.

21. Louviere, J. J., and G. Woodworth. "Design and Analysis of Simulated Consumer Choice or Allocation Experiments: An Approach Based on Aggregate Data." *Journal of Marketing Research* 20 (1983): 350–67.

22. McLean, R., and V. Anderson. *Applied Factorial and Fractional Designs*. New York: Marcel Dekker, 1984.

23. Oliphant, K., T. C. Eagle, J. J. Louviere, and D. Anderson. "Cross-Task Comparison of Ratings-Based and Choice-Based Conjoint." *Sawtooth Software Conference Proceedings* (1992): 383–404.

24. Reibstein, D., J. E. G. Bateson, and W. Boulding. "Conjoint Analysis Reliability: Empirical Findings." *Marketing Science* 7 (Summer 1988): 271–86.

25. Sawtooth Software. *Adaptive Conjoint Analysis*. Evanston, Ill.: Sawtooth Software, 1993.

26. Sawtooth Software. *Choice-Based Conjoint*. Evanston, Ill.: Sawtooth Software, 1993.

27. Sawtooth Software. *Conjoint Value Analysis*. Evanston, Ill.: Sawtooth Software, 1993.

28. Smith, Scott M. *PC-MDS: A Multidimensional Statistics Package*. Provo, Utah: Brigham Young University Press, 1989.

29. SPSS, Inc. *SPSS Categories*. Chicago: 1990.

30. Steckel, J., W. S. DeSarbo, and V. Mahajan. "On the Creation of Acceptable Conjoint Analysis Experimental Design." *Decision Sciences* 22, no. 2 (1991): 435–42.

31. Wittink, D. R., and P. Cattin. "Commercial Use of Conjoint Analysis: An Update." *Journal of Marketing* 53 (July 1989): 91–96.

32. Wittink, D. R., L. Krishnamurthi, and J. B. Nutter. "Comparing Derived Importance Weights Across Attributes." *Journal of Consumer Research* 8 (March 1982): 471–74.

33. Wittink, D. R., L. Krishnamurthi, and D. J. Reibstein. "The Effect of Differences in the Number of Attribute Levels on Conjoint Results." *Marketing Letters* 1, no. 2 (1990): 113–29.

Strategic Marketing Applications of Conjoint Analysis: An HMO Perspective

Michael D. Rosko
Michael DeVita
William F. McKenna
Lawrence R. Walker

Introduction

In response to rising expenditures for Medicare beneficiaries. PL 98–21 was enacted on April 20, 1983. This law provides for prospective payment of inpatient hospital services rendered to Medicare beneficiaries. Although prospective payment has been shown to slow the growth of hospital expenditures, this system of reimbursement provides a number of perverse incentives to hospitals (Rosko, 1984a). Furthermore, PL 98–21 creates direct incentives for efficiency only for hospitals and ignores other participants, i.e., physicians and patients, who are also responsible for rising health care costs. Accordingly, the Medicare prospective payment system has been called an interim stopgap measure by those who argue for a more comprehensive cost-control program (Durenburg, 1983). For example, proponents of a market-oriented system have called for the formation of health maintenance organizations (HMOs) as a key element of their strategy for cost-containment (Enthoven, 1980; Ellwood and McClure, 1976). However, there is some concern that this is not an acceptable delivery system for all population groups, especially the poor and the elderly (Galdblum and Trieger, 1982).

Less than five percent of the Medicare population has enrolled in HMOs due to a variety of legislative and personal barriers. Nevertheless, Galdblum and Trieger (1982) argue that Medicare beneficiaries will enroll in HMOs if benefit packages are attractive. This optimistic prognosis is based upon the high enrollment rate at three demonstration sites where over 20,000 Medicare beneficiaries were enrolled. The authors suggest that if legislative barriers are dropped, HMOs would market their plans to this group more intensely and, consequently, the percentage of Medicare beneficiaries enrolled would increase dramatically.

The assessment by Galdblum and Trieger (1982) leads to the question, "How can HMOs market their plans more effectively to the elderly?" Little information is currently available to help answer this question. Titus (1982) reports that most of the investigations of HMO enrollment choice studied the under-65 population. Furthermore, most of the HMO studies were not designed with HMO marketing as their ultimate goal and, consequently, their results have limited marketing applicability (Acito, 1978).

Obviously, if HMOs offered a very comprehensive benefit package while charging a low premium, they should be able to improve their market penetration. However, if they wish to maintain their financial solvency, more generous benefit/service packages may have to be matched by higher premiums. Thus, it is likely that HMOs will have to trade-off attractive benefits or attributes in order to maximize enrollments. The purpose of this study is to demonstrate how conjoint measurement can be used to quantify this type of trade-off in order to facilitate strategic planning decisions.

Conjoint measurement is a multivariate technique which can be used to decompose a set of overall responses to multiattribute alternatives so that the utility of each attribute level can be in-

"Strategic Marketing Applications of Conjoint Analysis: An HMO Perspective," Michael D. Rosko, Michael DeVita, William F. McKenna, and Lawrence R. Walker. *Journal of Health Care Marketing*, vol. 5. no. 4 (Fall 1985), pp. 27–38. Reprinted by permission.

ferred from the respondents' global evaluations of the alternatives. The solution technique involves a type of analysis of variance in which the respondents' overall preferences serve as a dependent variable and the predictor variables are represented by the various attribute levels making up each alternative (Green and Tull, 1973). A more detailed description of conjoint analysis is provided by Malhotra and Jain (1982) and Rosko and McKenna (1983).

Recent articles have demonstrated the utility of conjoint analysis for health care marketing (Malhotra and Jain, 1982; Rosko and McKenna, 1983; and Akaah and Becherer, 1983). The emphasis of the previous studies has been to show that conjoint analysis can be used to measure consumer preferences for attributes of health care services at either the individual or segment level of aggregation. Although the authors of the previous studies have concluded their articles with discussions of how conjoint analysis can be used to facilitate strategic marketing decisions, they have typically done so in a very brief manner with a lack of integration of the marketing mix.

The purpose of this article is to demonstrate how data from a conjoint analysis study can be used to help determine the most appropriate marketing mix for an operational HMO which is entering a new market, i.e., the geriatric population. This article includes two features which are absent in the previous health care applications of conjoint analysis: (1) external validation of results; and (2) a demonstration of how conjoint analysis can be used to simulate market responses to changes in the provider's marketing mix.

Research Methods

Study Site

In 1980, GHP (a disguised name for an actual staff-model HMO with four office locations) was awarded a contract by Health Care Financing Administration to market a comprehensive supplementary health benefits plan to Medicare beneficiaries. The plan, which was called "GHP/Medicare Plus," provided benefits which exceeded the number and level of services typically provided by the combination of Medicare Part A

and B plans, plus supplementary insurance. A variety of physician, hospital and other services were covered under GHP/Medicare Plus and were provided by GHP without limits, copayments or deductibles. The premium for GHP/Medicare Plus was only $0.60 more per month than Blue Cross 65 Special, a Medicare supplementary insurance plan which dominated the market in the area.

In January, 1981, GHP initiated a direct mail campaign in which promotional material about GHP was sent to over 500,000 Medicare beneficiaries in the area. Six months later, 116 families had enrolled in GHP/Medicare Plus. At the time of this writing, total enrollment had increased to over 600.

Application of Conjoint Measurement

An additive main-effects compensatory model was used to retrospectively predict HMO enrollment choice. This model uses the assumption that the individuals who are given an opportunity to enroll in an HMO will select that option if it yields more utility than their current health care arrangements. This type of model follows the tradition of Lancaster's (1966) theory of consumer behavior which assumes that an individual derives utility from the properties or characteristics of a good and not from the good itself. Additional support for the use of this type of model is provided by a number of studies which found that compensatory models can approximate noncompensatory decision-making processes quite well (Green and DeVita, 1975; Dawes and Corrigan, 1974; and Olshavsky and Acito, 1980).

The health system attributes were generated from a literature review of studies of the determinants of HMO enrollment choice, as well as previous conjoint measurement studies. Acito (1978) provided a very comprehensive review of the earlier articles. Information from focus group interviews of Medicare beneficiaries that were conducted by a marketing research firm for GHP also guided attributed generation. The attributes and their levels are shown in Table 1. The attribute levels were constructed to conform with actual levels encountered by the study population. Two criteria guided the selection of attributes. First, the attribute should play an important part in the

TABLE 1 Attributes and Levels Used in Conjoint Ranking Tasks

Attribute	Level	
Length of time needed to make nonemergency appointment	(A) 1 day	(B) 5 days
	(C) 10 days	(D) 15 days
Monthly premiums	(A) $12	(B) $20
	(C) $28	(D) $36
Waiting time in physician's office	(A) 10 min.	(B) 20 min.
	(C) 30 min.	(D) 60 min.
Travel time to physician's office	(A) 5 min.	(B) 15 min.
	(C) 30 min.	(D) 60 min.
Type of physician practice	(A) Group	(B) Solo
Physician's office hours	(A) 8:30 A.M.–5:30 P.M. Mon., Wed., Fri.; 8:30 A.M.–9:00 P.M. Tues., Thurs.; 8:30–11:30 A.M., Sat.	
	(B) 8:30 A.M.–5:30 P.M. Mon., Wed., Fri.; 8:30 A.M.–9:00 P.M. Tues., Thurs.;	
	(C) 8:30 A.M.–5:30 P.M. Mon.–Fri.	
	(D) 9:00 A.M.–3:00 P.M. Mon.–Fri.	
Routine office visits	(A) Complete coverage	
	(B) Partial/no coverage	
Routine eye and hearing examinations	(A) Complete coverage	
	(B) No coverage	
Prescription drugs	(A) Complete coverage	
	(B) No coverage	
Skilled nursing home care	(A) Complete coverage	
	(B) Partial coverage	

consumer's decision to enroll in an HMO. Second, it should be actionable by the producer. For example, convenience has been found to be a determinant of enrollment in HMOs (Juba, Lave and Shaddy, 1980; Ashcraft et al., 1978). However, expressing this attribute as more convenient or very convenient is not actionable. In contrast, defining convenience in terms of hours of operation or average waiting time is actionable.

A convenience sample of 97 Medicare beneficiaries were asked to supply data by using the full-profile approach. Fifty-eight of the respondents (hereafter called enrollees) were enrolled in GHP/Medicare Plus; the remaining 39 (hereafter called non-enrollees) were not enrolled in an HMO. None of the enrollees had been a member of GHP for more than six months when this study was conducted. All of the non-enrollees had received and read promotional material (which had been sent to enrollees and non-enrollees approximately seven months before this study) regarding

GHP/Medicare Plus, and had expressed some interest in obtaining further information about GHP prior to the initiation of this study. Accordingly, the respondents should be viewed as being more receptive of HMOs than the general population of the elderly. Therefore, it is likely that their preferences of HMO attributes (i.e., type of physician practice and health benefits coverage) may be greater than those of the typical Medicare beneficiary.

A previous study of consumer preferences for ambulatory care found that the full-profile technique resulted in higher internal validity scores than the trade-off method (Rosko and McKenna, 1983). In order to reduce the complexity of the respondents' evaluation task, two sets of profile cards were used. First, the respondents were asked to consider 16 profile cards pertaining to types of services covered (i.e., office visits, vision and hearing care, prescription medication, and skilled nursing home care), the type of physician

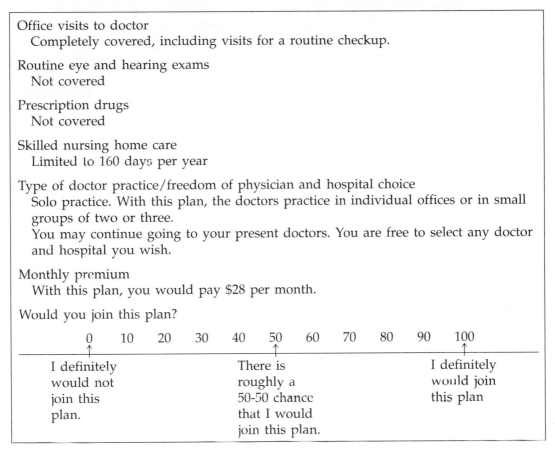

Office visits to doctor
    Completely covered, including visits for a routine checkup.

Routine eye and hearing exams
    Not covered

Prescription drugs
    Not covered

Skilled nursing home care
    Limited to 160 days per year

Type of doctor practice/freedom of physician and hospital choice
    Solo practice. With this plan, the doctors practice in individual offices or in small groups of two or three.
    You may continue going to your present doctors. You are free to select any doctor and hospital you wish.

Monthly premium
    With this plan, you would pay $28 per month.

Would you join this plan?

    0   10   20   30   40   50   60   70   80   90   100

    I definitely   There is   I definitely
    would not   roughly a   would join
    join this   50-50 chance   this plan
    plan.   that I would
                              join this plan.

FIGURE 1 Illustrative cost/coverage profile card for plan #8.

practice (i.e., HMO or current arrangements) and monthly premium. An illustrative "coverage/cost card" is shown in Figure 1. The respondents were asked to evaluate each plan separately and rate each on a scale of 0 (definitely would not join this plan) to 100 (definitely would join this plan). The rating scale was in increments of ten.

After the respondent completed the evaluation of the first 16 profile cards s/he was asked to rate 25 profile cards pertaining to convenience. An illustrative "convenience card" is shown in Figure 2. Type of physician practice and monthly premium appeared in both sets of cards, so that utilities of all of the attributes could be expressed on a common scale. The attribute levels on both sets of cards were determined by an orthogonal factorial design. This type of fractional factorial design permits an estimation of all main-effects while minimizing the number of profile cards to

be evaluated by the respondents (Adelman, 1962).

Utilities, or parts-worth values, were estimated for each individual by ordinary least squares regression. Each profile card served as an observation. The dependent variable was the rating of the card. Binary variables, set equal to 1 if the attribute level was present on the profile card and set equal to 0 otherwise, were entered for $n - 1$ levels of each attribute. These binary coded variables served as the independent variable set. The beta coefficients of the independent variables are the estimated parts-worth utility scores.

Results

Presented in Table 2 are average estimated utilities for each level of each attribute for the enrollees and the non-enrollees. Table 3 provides the

Type of doctor practice/freedom of physician and hospital choice
    Group Practice. With this plan the doctors practice in a one-stop health center
    where you receive all types of medical care. You must select a family doctor from
    among those associated with the plan. If you require hospitalization you must use
    the hospital associated with the plan in your area.

Length of time needed to make appointment
    One day

Office hours
    9:00 A.M.–3:00 P.M., Monday through Friday

Travel time to doctor's office
    Five minutes

Waiting time to see doctor
    60 minutes

Monthly premium
    With this plan you would pay $28 per month.

Would you join this plan?

| 0 | 10 | 20 | 30 | 40 | 50 | 60 | 70 | 80 | 90 | 100 |

I definitely
would not
join this
plan.

There is
roughly a
50-50 chance
that I would
join this plan.

I definitely
would join
this plan

FIGURE 2 Illustrative cost/convenience profile card for plan #5.

TABLE 2 Estimated Average Utility Values for Health System Attributes

Attribute	Estimated Utility	
	Enrollees	Non-Enrollees
Length of time needed to make non-emergency appointment		
(A) 1 day	3.67	3.46
(B) 5 days	1.94	2.23
(C) 10 days	0.92	0.95
(D) 15 days	0.00	0.00
Monthly premiums		
(A) $12	3.01	2.33
(B) $20	2.04	1.62
(C) $28	0.66	0.54
(D) $36	0.00	0.00
Waiting time in physician's office		
(A) 10 min.	1.81	1.67
(B) 20 min.	1.52	1.50
(C) 30 min.	0.70	0.27
(D) 60 min.	0.00	0.00

TABLE 2 *Continued*

Attribute	Estimated Utility	
	Enrollees	Non-Enrollees
Travel time to physician's office		
(A) 5 min.	1.61	1.25
(B) 15 min.	0.90	0.50
(C) 30 min.	0.58	0.33
(D) 45 min.	0.00	0.00
Type of physician practice		
(A) Solo	1.20	2.56
(B) HMO	0.00	0.00
Physician's office hours		
(A) 8:30 A.M.–5:30 P.M., Mon., Wed., Fri. 8:30 A.M.–9:00 P.M. Tue., Thurs. 8:30 A.M.–11:30 A.M. Sat.	0.71	1.01
(B) 8:30 A.M.–5:30 P.M., Mon., Wed., Fri. 8:30 A.M.–9:00 P.M. Tue., Thurs.	0.52	0.61
(C) 8:30 A.M.–5:30 P.M. Mon.–Fri.	0.29	0.26
(D) 9:00 A.M.–3:00 P.M. Mon.–Fri.	0.00	0.00
Routine office visits		
(A) Complete coverage	0.62	0.50
(B) Partial/no coverage	0.00	0.00
Routine eye and hearing examinations		
(A) Complete coverage	0.46	0.38
(B) No coverage	0.00	0.00
Prescription drugs		
(A) Complete coverage	0.40	0.38
(B) No coverage	0.00	0.00
Skilled nursing home care		
(A) Complete coverage	0.42	0.18
(B) Partial	0.00	0.00

TABLE 3 Estimated Importance of Weight of Health System Attributes

Attribute	Status			Benefit Segment		
	Total	Enrolled	Not Enrolled	1	2	3
Length of time needed to make non-emergency appointment	23.2	23.0	23.5	12.4	23.2	38.3
Monthly premiums	17.5	18.3	16.3	29.8	6.8	12.9
Physicians' office hours	13.8	14.4	12.8	15.3	13.6	11.8
Type of physician practice	13.0	10.6	16.5	8.8	21.2	9.0
Waiting time in physician's office	11.3	11.2	11.5	10.6	15.4	7.6
Travel time to physician's office	9.3	10.0	8.3	9.5	10.3	7.9
Routine office visit coverage	4.0	4.0	4.0	4.8	2.7	4.4
Routine eye and hearing examination coverage	2.9	2.9	3.0	3.1	2.5	3.1
Prescription drug coverage	2.7	2.7	2.7	3.1	2.3	2.7
Skilled nursing facility coverage	2.1	2.7	1.3	2.5	1.7	2.2

average importance weight for each of the attributes. The formula for calculating the importance weight (IW_1) of the ith attribute is

$$(1) \quad IW_1 = \frac{\text{range}_{Ai}}{\sum\limits_{i=1}^{m} \text{range}_{Ai}}$$

where range$_{Ai}$ equals the utility score of the most preferred level of the ith attribute minus the utility score of the least preferred level of the ith attribute.[1]

The importance weights for health system attributes were very similar for enrollees and non-enrollees. There was no intergroup difference in the rankings of the cost and convenience attributes. However, the non-enrollees viewed *type of practice* as the second most important attribute, while the enrollees on the average viewed this as the fifth most important attribute. These results are not surprising. The estimated utilities for this attribute indicate that both groups prefer solo practice to HMOs. However, the enrollees indicated only a relatively slight preference for solo practice; consequently, the importance weight of this group for this attribute is lower than that of the non-enrollees who reveal a marked preference for solo practice.

Titus (1982) reported a number of barriers to HMO enrollment facing the Medicare population over 65 years of age. These include: (1) knowledge deficits; (2) lack of exposure to information about HMOs; (3) general resistance to change; and (4) difficulties of integrating the "new" with old insurance patterns. Thus, it was not surprising that even the enrollees did not view a closed-panel HMO as their most preferred form of physician practice arrangement.

If the results from this demonstration represent predecision making values, one could expect that the enrollees opted for supplementary insurance coverage by an HMO because they were relatively indifferent about type of physician practice. Thus, they could be compensated more easily for limited physician choice by other attributes of the HMO. Apparently, most of the non-enrollees had

such strong feelings for solo practice as indicated by the estimated utilities for this attribute level that GHP could not induce them to enroll in a closed-panel HMO. However, the study's results suggest that the non-enrollees would be more predisposed to enroll in a well-designed IPA-type HMO, especially if their regular doctor participated in the IPA.

The relative magnitudes of the importance weights of the other attributes were not surprising. *Length of time needed to make an appointment* was considered to be the most important health systems attribute by both groups of respondents. Its importance weight was 23.0 among the enrollees, and 23.5 among the non-enrollees. This result is reasonable because the attribute reflects access to the provider. *Monthly premium* had the second highest importance weight among enrollees ($IW = 18.3$), and the third highest importance weight among non-enrollees ($IW = 16.3$). Cost has been found in several studies to be an important determinant of HMO enrollment (Juba, Lave and Shaddy, 1980; Tessler and Mechanic, 1975). The importance weights of the other convenience attributes, i.e., *waiting time, travel time* and *office hours*, were lower for both groups. The elderly retired population, in general, should have a lower value of time (i.e., opportunity cost) than those in the workforce. Thus, it was expected that waiting time and travel time would be viewed by the respondents as relatively unimportant in comparison with the other attributes. Similarly, office hours for routine, nonemergency visits should not be very important for those who are not constrained by the demands of employment. It is likely that these convenience attributes would have higher importance weights in a study of an employed population.

The coverage attributes had relatively low importance weights among both enrollees and non-enrollees. However, the former group attached greater importance to these attributes. Low importance weights for individual coverage attributes are not surprising for a population which already has extensive coverage (i.e., Medicare Parts A and B). If the importance weights of the four coverage attributes are summed to develop a new attribute, "comprehensiveness of coverage" which has two levels—high and low—importance weights of 12.3 and 11.0 are found for en-

---

[1] The Howard-Harris clustering algorithm was used with importance weight data to form three market segments for which average importance are also displayed in Table 3.

rollees and non-enrollees respectively. These values would make comprehensiveness of coverage the fourth most important attribute among the enrollees and the sixth most important attribute among the non-enrollees.

Validity

Validity of the results was assessed by examining the estimated utilities and by prediction of enrollment choice. In addition, the reasonableness of the importance weights (as discussed in the previous section) supports the validity of the results. An inspection of Table 2 indicates that the mean values of attribute levels conformed to the implicit assumption of a monotonically increasing relationship between the attribute levels. Obviously a priori specification could not be made for an attribute such as type of practice because attitudes toward this attribute may be formed on the basis of previous experiences and medical history of individual respondents.

The strongest test of a model's validity is its ability to predict behavior. None of the published conjoint measurement studies of consumer preferences for HMOs have assessed predictive validity. This study tested predictive validity retrospectively. Obviously, a prospective test would be preferred; however, logistical problems precluded this. However, since the preferences for health systems attributes were measured indirectly, the enrollees' experience with the HMO may not have seriously affected their evaluations of the profile cards. The fact that even enrollees revealed a preference for solo practice provides some support for this contention. Nevertheless, the reader should be cautioned that dissonance reduction may have affected some of the responses.

Both sets of respondents were asked to supply information about their current health systems arrangements in terms of the attribute levels shown in Table 1. In addition, the enrollees were asked to supply information about their health care systems arrangements prior to their enrollment in GHP. The non-enrollees were asked to record their perceptions of the attribute levels of GHP. From this information, it was possible to determine the attribute levels (real or perceived) of each of the alternative delivery systems facing

each respondent. Next, the utilities of the two delivery systems available to each respondent were calculated.[2]

It was predicted that the respondents would enroll in the delivery system which yielded them the highest level of utility. If the two alternatives had the same total utility, it was assumed that the respondent would not enroll in the HMO. The behavior of over 78% of the respondents was correctly predicted. The model's predictive power was similar for both groups of respondents. Enrollment choice was correctly predicted for 79.3% of the enrollees, and for 76.8% of the non-enrollees.

Discussion

The growth of HMO enrollment has disappointed many observers of the health care field. This study has evaluated a technique which may assist HMO managers in marketing their plans. The relatively high value for this study's predictive validity measure provides assurance that conjoint measurement can be a valuable technique to use in marketing and planning health services. Accordingly, this article concludes with a demonstration as to how data from a conjoint analysis study could be used in strategic marketing decisions pertaining to product, price, place and promotion. For illustrative purposes, the following discussion will assume the existence of data gained from a probability-based sample which could be generalized to a larger population.

One of the areas in which conjoint measurement is particularly well-suited is the development of new products and services (Green and Srinivasan, 1978). Data from a conjoint measurement study can be used to simulate market responses to changes in the levels of attributes or to

---

[2] Total utility is given by the expression:

$$(2) \quad U(x_i) = \sum_{i=1}^{m} U_{ij}(x_{ij})$$

where

$U(x_i)$ is the total utility for the jth respondent.

x_{ij} is a vector representing the attribute level of the ith attribute facing the jth respondent.

U_{ij} is the utility of the attribute level facing the jth respondent.

the addition of new attributes. For example, the elderly, who tend to have more chronic problems than younger population groups, may view coverage for prescription drugs as a desirable benefit. However, the addition of this benefit is expensive to the HMO and, thus, many HMOs (including the one studied in this project) do not provide this benefit.

The appropriate economic criterion for adding this benefit is that its contribution to total revenue must exceed its contribution to total expense. Information about the incremental cost of drug coverage is readily available from actuarial records. However, since prescription drug coverage has not been offered to this group, information about the incremental revenue (i.e., changes in enrollment) associated with offering this new benefit is not available from existing data. However, conjoint measurement data can be used to stimulate the market response to the addition of new benefits, provided these benefits were measured in the original experiment.

The first step in a simulation of market response to the addition of prescription drug coverage to the GHP/Medicare Plus benefits package was to establish base-line enrollment for GHP. This was done in the manner described in the section on predictive validity. Once again the criterion for enrollment choice is the maximization of total utility function is given by equation 2. The results, which are shown in Table 4, indicate that 55 of the 97 respondents would be expected to enroll in GHP. Next, market response was simulated to the addition of prescription drug coverage. As Table 2 shows, the addition of this benefit is expected to increase the utility of GHP to the average respondent by about 0.39. The simulation indicates that this would increase projected HMO enrollment from 55 to 62, a 12.7% increase from base-line enrollment. This is a much greater increase in enrollment than that expected from the addition of an attribute whose group mean importance weight is only 2.7. However, this apparent contradiction is explained by the fact that a number of individuals valued prescription drug coverage very highly. This information would not be available from traditional aggregate analysis methods such as multiple regression or discriminant analysis. However, it highlights one of the unique advantages of conjoint analysis which is able to measure preference at the individual level of aggregation. This, of course, also permits market segmentation studies to be performed.

Although the addition of prescription drug coverage by GHP is expected to increase enrollment in GHP/Medicare Plus by 12.7%, it is not clear that it should add this benefit. Ideally, the criteria that marginal revenue should exceed or equal marginal costs should be used. However, cost data were not given by GHP. Accordingly, this problem was analyzed by assuming that the HMO would pass the cost of this new benefit directly to all enrollees. If more consumers are predicted to enroll in a plan which costs more but has more benefits (i.e., the addition of prescription drugs), it may be advisable for the HMO to

TABLE 4 Predicted Changes in HMO Enrollment Associated with Addition of Prescription Drug Benefits and Premium Increase

Item of Change		Total Enrollment	Change in Enrollment from Base-line	Percentage Change from Base-line
Prescription drugs added as a benefit		62	7	12.7
Prescription drugs added as a benefit, and monthly premium increased by:	$1	60	5	9.1
	$2	58	3	5.5
	$3	56	1	1.8
	$4	54	−1	−1.8
Base-line configuration		55	—	—

expand its coverage. Presented in Table 4 are the expected changes in GHP/Medicare Plus enrollment when prescription drugs are included as a benefit and monthly premiums are increased by $1, $2, $3 and $4, respectively. The results show that increases in enrollment are expected only if the HMO increases premium charges by less than $4. The next step would be for the HMO to contact an actuary to determine if prescription drug benefits for this population will amount to less than $4 per month. Other marketing mix applications of conjoint analysis are provided in the following sections.

Product

Besides market share simulation, data from a conjoint analysis study can be used to: (1) determine the relative importance of attributes; (2) form market segments on the basis of benefits sought; (3) target attractive segments; and (4) determine which segments are most likely to be attracted to the HMO in its current form or as it varies its marketing mix.

The importance weights of HMO attributes, which are derived from estimated part-worth values, allow the HMO manager to assess the relative importance of each attribute used in the study. Since these importance weights are estimated for each individual, it is possible to form segments according to benefits (i.e., attributes) sought. This was done by the use of the Howard-Harris clustering algorithm, and three segments (clusters) emerged. Segment 1 was composed of 38 respondents (24 of whom were enrollees) who tended to be more sensitive to premium levels than other respondents. Segment 2 was composed of 32 respondents (14 of whom were enrollees) who were most concerned with type of physician practice. The third segment was composed of 27 respondents (20 of whom were enrollees) who were concerned with days needed to make a routine appointment. Discriminant analysis verified the existence of three distinct, and relatively homogeneous, segments and that the attributes described above had statistically significant ($p < 0.05$) discriminant power.

The criteria which the HMO should follow in targeting market segments are substantiality and accessibility. The first criterion suggests that the segment to be targeted must be large or profitable enough to be worth considering for separate market cultivation (Kotler, 1982). The second criterion implies that the HMO must be able to effectively focus the marketing efforts on chosen segments. As Table 3 indicates, Segment 2 is composed of individuals who value freedom to choose physicians very highly. Accordingly, it is unlikely that a staff-model HMO, such as GHP, will enroll many individuals from this segment irrespective of how attractive the other features of the HMO might be. The other segments are more promising. Segment 1 is price conscious and Segment 3 is concerned with access (i.e., days needed to make a routine appointment). The HMO can influence the level of either attribute. However, the strategies for attracting these two segments may be conflicting. For example, one way to reduce the number of days' wait for an appointment is to increase the ratio of primary care physicians to enrollees. This, of course, would raise operating expenses which would, in turn, lead to higher premiums. This would make the HMO less attractive to Segment 1.

The data collected in this study suggest two reasons why the HMO should give more emphasis to attracting members from Segment 1. First, it is larger than Segment 3. Second, it may be more profitable. The HMO is required to charge all GHP/Medicare Plus enrollees the same premiums for the same benefits package. Thus, profits are a function of services provided to each enrollee. Given the constraint upon premiums, it may be necessary for the HMO to avoid adverse risk selection (Luft, 1982). Another analysis (Rosko, 1984b) suggests the emphasis upon access by the members of Segment 3 may be due to higher perception of need for health services, as well as a higher incidence of previous health problems than for the members of Segment 1. Although it is regrettable that fiscal constraints may preclude some HMOs from serving the very sick, Luft (1980) reports that HMOs may have altered their attributes to avoid adverse risk selection. Accordingly, the results suggest that it may be preferable for this HMO to emphasize low premium and to de-emphasize access in its marketing strategy for the elderly.

Price

It is well recognized that management should have information about the price elasticity of demand. Demand curves can easily be estimated for existing products by multiple regression analysis. However, this technique is not feasible for new products because, by definition, there is no data available to correlate quantities demanded at various market prices. However, data from a conjoint analysis study can be used to simulate changes in enrollment associated with a change in monthly premium. As Table 5 indicates, an $8 increase (or 40%) in premium from the base price of $20 would cause predicted enrollment to drop from 55 to 41 (a 25% decrease). Since the percentage price increase is greater than the percentage change in enrollment, an inelastic demand relationship is suggested at the current premium of $20. Microeconomic theory indicates that if the HMO wishes to maximize revenue it should increase its premium. However, it should be noted that the HMO may have another objective such as market penetration which is not consistent with revenue maximization. In that case, the demand curve estimated by conjoint analysis would still provide useful information to management.

Place

Distribution strategy can have a significant effect upon enrollment in an HMO. It is well recognized that the demand for ambulatory health services is inversely related to explicit as well as implicit costs (Acton, 1975). The former category includes out-of-pocket charges, or direct payments. Implicit costs include transportation costs, as well as opportunity costs, such as the patient's value of time. Accordingly, it is clear that an HMO can influence its enrollment by locating its physician offices closer to its target market. A closed-panel HMO can operationalize this strategy by moving its physician offices to a more centralized location or by opening satellite clinics. An IPA-model HMO may enroll doctors who practice in areas which are not currently served by its participating physicians.

Regarding the second type of implicit cost—opportunity cost of time—the HMO may reduce patient waiting time by increasing the ratio of physicians to enrollees or by establishing a more effective scheduling system.

The effect of changes in distribution policy on the market share of GHP was simulated by comparing predicted levels of enrollment under two

TABLE 5 Predicted Changes in HMO Enrollment Associated with Changes in Health Plan Attribute Levels

Item of Change		Total Enrollment	Change in Enrollment	Percentage
HMO changes monthly premium by:	$1	51	−4	−7.3
	$2	50	−5	−9.1
	$8	41	−14	−25.4
	−$8	62	7	12.7
HMO provides full coverage for doctor visits, nursing homes and R_x		69	14	25.5
Access attributes set at:				
Most favorable levels		77	22	40.0
Least favorable levels		19	−36	−65.5
HMO changes structure to IPA with solo practitioners		66	11	20.0
Base-line configuration		55	—	—

scenarios: (1) access attributes offered at most favorable levels; and (2) access attributes offered at least favorable levels. The access attributes used included: length of time needed to make a nonemergency appointment, waiting time in physician's office, travel time to physician office, and physician's office hours. In Table 2, Level A and Level D correspond to the most favorable and least favorable levels of these attributes, respectively. The results of the simulation in Table 5 reveal that, when access levels are set at their most favorable levels, enrollment may increase as much as 40% above base-line enrollment. Conversely, if access attributes are set at the least favorable levels, enrollment may be reduced by as much as 65.5% of its projected base-line amount. Although space limitations preclude further analysis, it should be clear that changes in each of the access-related attributes could be simulated in order to estimate their individual effect on enrollment.

Promotion

The estimated importance weights for HMO attributes can be used to help develop a promotional strategy. The HMO should emphasize factors which were revealed to be important (i.e., length of time needed to make an appointment, monthly premium, physician office hours), and which give the HMO an advantage over other health care providers.

The results suggest that group practice has an unfavorable image relative to solo practice. This suggests a need to develop a promotional campaign to improve the image of group practice. Another interesting finding was that the non-enrollees perceived less coverage by the HMO than was actually provided. Indeed, as Table 5 shows, the simulations suggested that potential HMO market share would increase by 20% if everyone in the market was fully aware of the extent of coverage for routine office visits, eye/hearing exams and nursing home care. Obviously, the HMO needs to communicate the availability of these benefits better.

Finally, the HMO can use information from a conjoint analysis study in developing either a diffused or segmented promotional campaign. As discussed earlier, since conjoint results are estimated for each individual, it is possible to determine importance weights for each market segment.

Summary and Conclusion

This study has demonstrated that conjoint measurement can be used to validly predict the decisions of Medicare beneficiaries when confronted with the opportunity to enroll in an HMO offering coverage to supplement their Medicare plan. An estimation procedure (i.e., ordinary least squares) was used which is much more accessible to health care planners than the techniques used in most of the previous health care applications of conjoint analysis (i.e., MONANOVA or LIN-MAP).

Finally, a data collection technique was used (i.e., interval rating scales of profile cards) which is easier for the respondent to complete than ranking tests which were used in previous studies. The most serious limitations of this demonstration study were its small sample size and the reliance on a sample which was not generated from probability-based methods.

References

Acito, Franklin (1978), "Consumer Preferences For Health Care Services: An Exploratory Investigation," Unpublished doctoral dissertation. State University of New York at Buffalo.

——— (1982), "Consumer Decision Making and Health Maintenance Organizations: A Review," *Medical Care* 16 (January), 1–13.

Acton, Jan (1975), "Non-monetary Factors in the Demand for Medical Services: Some Empirical Evidence," *The Journal of Political Economy* 83 (June), 595–614.

Adelman, Sidney (1962), "Orthogonal Main-Effect Plans for Assymetrical Factorial Experiments," *Technometric* 4 (February), 21–46.

Akaah, Ishmael and Richard Becherer (1983). "Integrating a Consumer Orientation into the Planning of HMO Programs: An Application of Conjoint Segmentation" *Journal of Health Care Marketing* 3 (Spring), 9–18.

Ashcraft, Marie, Roy Penchansky, Sylvester Berki, Robert Fortus and John Gray (1978), "Expectations and Experience Of HMO Enrollees After One Year: An Analysis of Satisfaction. Utilization, and Costs." *Medical Care* (January), 14–32.

Dawes, Robin and Bernard Corrigan (1974), ''Linear Models in Decision Making,'' *Psychological Bulletin* 81 (February), 95–106.

Durenburg, David (1983), Address to National Conference on Hospital-Medical Public Policy Issues, Washington, DC, April 16.

Ellwood, Paul and Walter McClure (1976) *Health Delivery Reform,* Excelsior, MN: Interstudy.

Enthoven, Alain (1980). *Health Plan,* Reading, MA: Addison-Wesley Publishing Company.

Galblum, Trudi and Sidney Trieger (1982), ''Demonstrations of Alternative Delivery Systems Under Medicare and Medicaid,'' *Health Care Financing Review* 3 (March), 1–11.

Green, Paul and Michael DeVita (1975), ''The Robustness of Linear Models Under Correlated Attribute Conditions,'' *Proceedings of the 1975 Educators' Conference,* Chicago: American Marketing Association, 108.

——— and V. Srinivasan (1978), ''Conjoint Analysis in Consumer Research: Issues and Outlook,'' *Journal of Consumer Research* 5 (September), 103–123.

——— and D. Tull (1973). *Research for Marketing Decisions,* 4th Edition, Englewood Cliffs, NJ: Prentice-Hall.

Juba, David, J. Lave and J. Shaddy (1980), ''An Analysis of the Choice of Health Benefits Plans,'' *Inquiry* 17 (Spring), 62–71.

Kotler, Phillip (1982). *Marketing Management,* Englewood, NJ: Prentice-Hall, Inc.

Lancaster, Kelvin (1966), ''A New Approach to Consumer Theory,'' *Journal of Political Economics* 74 (April), 132–157.

Luft, Harold (1982), ''Health Maintenance Organizations and the Rationing of Care,'' *Health and Society* 60 (Spring), 268–306.

Malhotra, Naresh and Arun Jain (1982), ''A Conjoint Analysis Approach to Health Care Marketing and Planning,'' *Journal of Health Care Marketing* 2 (Spring), 35–44.

McClain, John and Vithala Rao (1974), "Trade-offs and Conflicts in Evaluation of Health System Alternatives: Methodology for Analysis," *Health Services Research* 9 (Spring), 35–52.

McClure, Walter (1981), "Structure and Incentive Problems in Economic Regulation of Medical Care," *Milbank Memorial Fund Quarterly* 59 (Spring), 107–144.

Olshavsky, R. and Franklin Acito (1980), "An Information Processing Probe Into Conjoining Analysis," *Decision Sciences* 11 (July), 451–470.

Parker, Barnett and V. Srinivansan (1976), "A Consumer Preference Approach to the Planning of Rural Primary Health Care Facilities," *Operations Research* 24 (September/October), 991–1025.

Rosko, Michael (1984a), "The Impact of Prospective Payment: A Multi-Dimensional Analysis of New Jersey's SHARE Program," *Journal of Health Politics, Policy, and Law* 9 (Spring), 81–101.

——— (1984b), "Correlates of HMO Benefits Sought by the Elderly," Working paper, Widener University, Chester, PA.

——— and William McKenna (1983), "Modeling Consumer Choices of Health Plans: A Comparison of Two Techniques," *Social Science and Medicine* 17 (July), 421–429.

Tessler, Richard and David Mechanic (1975), "Factors Affecting the Choice Between Prepaid Group Practice and Alternative Insurance Programs," *Milbank Memorial Fund Quarterly* 53 (Spring), 149–171.

Titus, Sandra (1982), "Barriers to the Health Maintenance Organization for the Over 65's," *Social Science and Medicine* 16 (October), 1767–1774.

11

Structural Equation Modeling

LEARNING OBJECTIVES

Upon completing this chapter, you should be able to do the following:

- Understand the role of causal relationships in statistical analysis.

- Represent a series of causal relationships in a path diagram.

- Translate a path diagram into a set of equations for estimation.

- Appreciate the role and influence of a variable's measurement properties on the results of a statistical analysis.

- Differentiate between observable and unobservable variables and their part in structural equation modeling.

- Evaluate the results of a structural equation modeling analysis for its support of the proposed relationships and the possible areas of improving the results.

- Apply structural equation modeling techniques to such problems as confirmatory factor analysis, path analysis, and simultaneous equation estimation.

One of the primary objectives of multivariate techniques is to expand the researcher's explanatory ability and statistical efficiency. Multiple regression, factor analysis, multivariate analysis of variance, discriminant analysis, and the other methods discussed in previous chapters all provide the researcher with powerful tools for addressing a wide range of managerial and theoretical questions. But they all share one common limitation: each technique can examine only a single relationship at a time. Even the techniques allowing for multiple dependent variables, such as multivariate analysis of variance and canonical analysis, still represent only a single relationship between the dependent and independent variables.

All too often, however, the researcher is faced with a set of interrelated questions. For example, what variables determine a store's image? How does that image combine with other variables to affect purchase decisions and satisfaction at the store? And finally, how does satisfaction with the store result in long-term loyalty to it? This series of issues has both managerial and theoretical importance. Yet none of the multivariate techniques we have examined allow us to address all these questions with a single comprehensive method. For this reason, we are now going to examine the technique of **structural equation modeling (SEM),** an extension of several multivariate techniques we have already studied, most notably multiple regression and factor analysis.

As we briefly discussed in Chapter 1, structural equation modeling examines a series of dependence relationships simultaneously. It is particularly useful when one dependent variable becomes an independent variable in subsequent dependence relationships. This set of relationships, each with dependent and independent variables, is the basis of structural equation modeling. The basic formulation of structural equation modeling in equation form is

$$
\begin{aligned}
Y_1 &= X_{11} + X_{12} + X_{13} + \ldots + X_{1n} \\
Y_2 &= X_{21} + X_{22} + X_{23} + \ldots + X_{2n} \\
Y_m &= X_m1 + X_m2 + X_m3 + \ldots + X_{mn}
\end{aligned}
$$

(metric)   (metric, nonmetric)

Structural equation modeling has been used in almost every conceivable field of study, including education, marketing, psychology, sociology, management, testing and measurement, health, demography, organizational behavior, biology, and even genetics. The reasons for its attractiveness to such diverse areas is twofold: (1) it provides a straightforward method of dealing with multiple relationships simultaneously while provoding statistical efficiency, and (2) its ability to assess the relationships comprehensively has provided a transition from **exploratory** to **confirmatory analysis.** This transition corresponds to greater efforts in all fields of study toward developing a more systematic and holistic view of problems. Such efforts require the ability to test a series of relationships constituting a large-scale model, a set of fundamental principles, or even an entire theory. These are a task for which structural equation modeling is well suited.

Before starting the chapter, review the key terms to develop an understanding of the concepts and terminology used. Throughout the chapter the key terms appear in **boldface.** Other points of emphasis in the chapter are *italicized*. Also, cross-references within the key terms are in *italics*.

Biserial correlation   Correlation measure used to replace the product-moment correlation when a metrically measured variable is associated with a binary measure. Also see *polyserial correlation*.

Causal relationship   Dependence relationship between two or more variables in which the researcher clearly specifies that one or more variables "cause" or create an outcome represented by at least one other variable. Must meet the requirements for *causation*.

Causation   Principle by which cause and effect is established between two variables. It requires that there be sufficient degree of association (correlation) between the two variables, that one variable occur before the other (i.e., one variable is clearly the outcome of the other), and that there be no other reasonable causes for the outcome. Although in its strictest terms causation is rarely found, in practice strong theoretical support can make empirical estimation of causation possible.

Competing models strategy   Compares the proposed model with a number of alternative models in an attempt to demonstrate that no better-fitting model exists. This is particularly relevant in structural equation modeling because a model can be shown only to have acceptable fit, but acceptable fit alone does not guarantee that another model will not fit better.

Confirmatory analysis   Use of multivariate technique to test (confirm) a prespecified relationship. For example, suppose we hypothesize that only two variables should be predictors of a dependent variable. If we empirically test for the significance of these two predictors and the nonsignificance of all others, this test is a confirmatory analysis. It is the opposite of *exploratory analysis*.

Confirmatory modeling strategy   Strategy in which a single model is assessed statistically for its fit to the observed data. This approach is actually less rigorous than the *competing models strategy* because it does not consider alternative models which might fit better than the proposed model.

Construct   Concept that the researcher can define in conceptual terms but cannot be directly measured (e.g., the respondent cannot articulate a single response that will totally and perfectly provide a measure of the concept) or measured without error (see *measurement error*). Constructs are the basis for forming causal relationships as they are the "purest" possible representation of a concept. A construct can be defined in varying degrees of specificity, ranging from quite narrow concepts (e.g., total household income) to more complex or abstract concepts (intelligence or emotions). Yet no matter what its level of specificity, a construct cannot be measured directly and perfectly but must be approximately measured by *indicators*.

Cronbach alpha   Commonly used measure of *reliability* for a set of two or more construct indicators. Values range between 0 and 1.0, with higher values indicating higher reliability among the indicators.

Degrees of freedom (df)   Defined in formal terms as the number of bits of information available to estimate the sampling distribution of the data after all model parameters have been estimated. In practical terms, the degrees of freedom are the number of nonredundant correlations/covariances in the input matrix minus the number of estimated coefficients. The analyst attempts to maximize the degrees of freedom available while still obtaining the best-fitting model. Each estimated coefficient "uses up" a degree of freedom. A model can never estimate more coefficients than the number of nonredundant correlations/covariances, meaning that zero is the lower bound for the degrees of freedom for any model.

Endogenous construct   Construct or variable that is the dependent or outcome variable in at least one causal relationship. In terms of a path diagram, there are one or more arrows *leading into* the endogenous construct or variable.

Exogenous construct   Construct or variable that acts only as a predictor or "cause" for other constructs or variables in the model. In path diagrams, the exogenous variables have only causal arrows leading out of them and are not predicted by any other variables in the model.

Exploratory analysis   The opposite of *confirmatory analysis,* exploratory analysis defines possible relationships in only the most general form and then allows the multivariate technique to estimate a relationship(s). The researcher is not looking to "confirm" any relationships specified prior to the analysis, but instead lets the method and the data define the nature of the relationships. An example is stepwise multiple regression, in which the method adds predictor variables until some criterion is met.

Goodness-of-fit   Degree to which the actual/observed input matrix (covariances or correlations) is predicted by the estimated model. Goodness-of-fit measures are computed only for the total input matrix, making no distinction between exogenous and endogenous constructs or indicators.

Heywood cases   One of the most common types of *offending estimates,* a Heywood case occurs when the estimated error term for an indicator becomes negative, which is a nonsensical value. The problem is remedied either by deleting the indicator or by constraining the measurement error value to be a small positive value.

Identification   Degree to which there is a sufficient number of equations to "solve for" each of the coefficients (unknowns) to be estimated. Models can be underidentified (can't be solved), just identified (number of equations equals number of estimated coefficients with no degrees of freedom), or overidentified (more equations than estimated coefficients and the degrees of freedom greater than zero). The analyst desires to have an overidentified model for the most rigorous test of the proposed model. Also see *degrees of freedom.*

Independence model   See *null model.*

Indicator   Observed value *(manifest variable)* used as a measure of a concept or *latent construct* that cannot be measured directly. The researcher must specify which indicators are associated with each construct.

Latent construct or variable   Operationalization of a *construct* in structural equation modeling, a latent variable cannot be measured directly but can be represented or measured by one or more variables *(indicators)*. For example, a person's attitude toward a product can never be measured so precisely that there is no uncertainty. But by asking various questions we can assess the many aspects of the person's attitude. In combination, the answers to these questions give a reasonably accurate measure of the latent construct (attitude) for an individual.

Manifest variable   Observed value for a specific item or question, obtained either from respondents in response to questions (as in a questionnaire) or from observation by the researcher. Manifest variables are used as the *indicators* of latent constructs or variables.

Measurement error   Degree to which the variables we can measure (the *manifest* variables) do not perfectly describe the *latent construct(s)* of interest. Sources of measurement error can range from simple data entry errors to definition of constructs (e.g., abstract concepts such as patriotism or loyalty that mean many things to different people) that are not perfectly defined by any set of manifest variables. For all practical purposes, all constructs have some measurement error even with the best indicator variables. However, the researcher's objective is to minimize the amount of measurement error. Structural equation modeling can take measurement error into account in order to provide more accurate estimates of the causal relationships.

Measurement model   Submodel in structural equation modeling that (1) specifies the indicators for each construct, and (2) assesses the reliability of each construct for estimating the causal relationships. The measurement model is similar in form to factor analysis; the major difference lies in the degree of control provided the researcher. In factor analysis, the researcher can specify only the number of factors, but all variables have loadings (i.e., act as indicators) for *each* factor. In the measurement model, the researcher specifies which variables are indicators of each construct, with variables having no loadings other than those on its specified construct.

Model   Specified set of dependence relationships that can be tested empirically—an operationalization of a *theory*. The purpose of a model is to concisely provide a comprehensive representation of the relationships to be examined. The model can be formalized in a path diagram or a set of structural equations.

Model development strategy   Strategy for structural modeling that anticipates model respecification as a theoretically driven method of improving a tentatively specified model. This allows exploration of alternative model formulations that may be supported by theory. It does not correspond to an exploratory approach in which model respecifications are made atheoretically.

Modification indices   Calculated for each unestimated relationship possible in a specified model. The modification index values for a specific unestimated relationship indicate the improvement in overall model fit (the reduction in the Chi-square statistic) that is possible if a coefficient is calculated for that untested relationship. The researcher should use the modification indices only as a guideline for model improvements of those relationships that can theoretically be justified as possible modifications.

Nested models   Models that have the same constructs but differ in terms of the number or types of causal relationships represented. The most common form of nested model occurs when a single relationship is added or deleted from another model. Thus the model with fewer estimated relationships is "nested" within the more general model.

Null model   Baseline or comparison standard used in incremental fit indices. The null model is hypothesized to be the simplest model that can be theoretically justified. The most common example is a single construct model related to all indicators with no measurement error.

Offending estimates   Any value that exceeds its theoretical limits. The most common occurrences are negative error variances (the minimum value should be zero—no measurement error) or a very large standard error. The researcher must correct the offending estimate with one of a number of remedies before the results can be interpreted for overall model fit and the individual coefficients can be examined for statistical significance.

Parsimony   Degree to which a model achieves fit for each estimated coefficient. The objective is not to minimize the number of coefficients or to maximize the fit but to maximize the *amount of fit per estimated coefficient* and avoid "overfitting" the model with additional coefficients that achieve only small gains in model fit.

Path analysis   Employing simple bivariate correlations to estimate the relationships in a system of structural equations. The method is based on specifying the relationships in a series of regressionlike equations (portrayed graphically in a *path diagram*) that can then be estimated by determining the amount of correlation attributable to each effect in each equation simultaneously. When employed with multiple relationships among latent constructs and a measurement model, it is then termed *structural equation modeling*.

Path diagram   Graphical portrayal of the complete set of relationships among the model's constructs. Causal relationships are depicted by straight arrows, with the arrow emanating from the predictor variable and the arrowhead "pointing" to the dependent variable. Curved arrows represent correlations between constructs or indicators, but no causation is implied.

Polychoric correlation   Measure of association employed as a replacement for the product-moment correlation when both variables are ordinal measures with three or more categories.

Polyserial correlation   Correlation used as a substitute for the product-moment correlation when one variable is measured on an ordinal scale and the other variable is metrically measured.

Reliability   A set of latent construct indicators consistent in their measurements. In more formal terms, reliability is the degree to which a set of two or more indicators "share" in their measurement of a construct. The indicators of highly reliable constructs are highly intercorrelated, indicating that they all are measuring the same latent construct. As reliability decreases, the indicators become less consistent and thus are poorer indicators of the latent construct. Reliability can be computed as 1.0 minus the measurement error. Also see *validity*.

Specification error   Lack of model fit resulting from the omission of a relevant variable from the proposed model. Tests for specification error are quite complicated and involve numerous trials between alternative models. The researcher can avoid specification error to a high degree by using only theoretical bases for constructing the proposed model. In this manner, the researcher is less likely to "overlook" a relevant construct for the model.

Structural equation modeling   Multivariate technique combining aspects of multiple regression (examining dependence relationships) and factor analysis (representing unmeasured concepts—factors—with multiple variables) to estimate a series of interrelated dependence relationships simultaneously.

Structural model   Set of one or more dependence relationships linking the model *constructs*. The structural model is most useful in representing the interrelationships of variables *between* dependence relationships.

Tetrachoric correlation   Correlation used when two binary measures are being measured for association. Also see *polychoric correlation*.

Theory   Strictly defined as a systematic set of relationships providing a consistent and comprehensive explanation of a phenomenon. In practice, a theory is a researcher's attempt to specify the entire set of dependence relationships explaining a particular set of outcomes. A theory may be based on ideas generated from one or more of three principal sources: (1) prior empirical research, (2) past experiences and observations of actual behavior, attitudes, or other phenomena, and (3) other theories that provide a perspective for analysis. Thus, theory building is not the exclusive domain of academic researchers but has an explicit role for practitioners as well. For any researcher, theory provides a means to address the "big picture" and assess the relative importance of various concepts in a series of relationships.

Unidimensionality   Similar to the concept of *reliability*, a unidimensional construct is one in which the set of indicators has only one underlying trait or concept in common. From the match between the chosen indicators and the theoretical definition of the construct, the researcher must establish both conceptually and empirically that the indicators are reliable and valid measures of only the specified construct before establishing unidimensionality.

Unobserved concept or variable   See *latent construct or variable.*

Validity   Ability of a construct's indicators to measure accurately the concept under study. For example, how accurate is our measure of household income or intelligence? Validity is determined to a great extent by the researcher, because the original definition of the construct or concept is proposed by the researcher and must be matched to the selected indicators or measures. Validity does not guarantee reliability, and vice versa. A measure may be accurate (valid) but not consistent (reliable). Also, it may be quite consistent but not accurate. Thus validity and reliability are two separate but interrelated conditions.

Variance extracted measure   Amount of "shared" or common variance among the *indicators* or *manifest variables* for a construct. Higher values represent a greater degree of shared representation of the indicators with the construct.

What Is Structural Equation Modeling?

Structural equation modeling (SEM) encompasses an entire family of models known by many names, among them covariance structure analysis, latent variable analysis, confirmatory factor analysis, and often simply LISREL analysis (the name of one of the more popular software packages). Resulting from an evolution of multiequation modeling developed principally in econometrics and merged with the principles of measurement from psychology and sociology, structural equation modeling has emerged as an integral tool in both managerial and academic research [7, 16, 20, 28, 29, 37, 39, 42, 44, 51, 56]. As might be expected for a technique with such widespread use and so many variations in applications, many researchers are uncertain about what constitutes **structural equation modeling.** Yet all structural equation modeling techniques are distinguished by two characteristics: (1) estimation of multiple and interrelated dependence relationships, and (2) the ability to represent **unobserved concepts** in these relationships and account for measurement error in the estimation process.

Accommodating Multiple Interrelated Dependence Relationships

The most obvious difference between structural equation modeling and other multivariate techniques is the use of separate relationships for each of a set of dependent variables. In simple terms, SEM estimates a series of separate, but interdependent, multiple regression equations simultaneously by specifying the **structural model** used by the statistical program. First, the researcher draws upon theory, prior experience, and the research objectives to distinguish which independent variables predict each dependent variable. In our earlier example, we first wanted to predict store image. We then wanted to use store image to predict satisfaction, both of which in turn were used to predict store loyalty. Thus, some dependent variables become independent variables in subsequent relationships, giving rise to the interdependent nature of the structural model. Moreover, many of the same variables will affect each of the dependent variables, but with differing effects. The structural model expresses these relationships among independent and dependent variables, even when a dependent variable becomes an independent variable in other relationships.

The proposed relationships are then translated into a series of structural equations (similar to regression equations) for each dependent variable. This feature sets structural equation modeling apart from techniques discussed previously that accommodate multiple dependent variables—multivariate analysis of variance and canonical correlation—in that they allow only a *single* relationship between dependent and independent variables.

Incorporating Variables That We Do Not Measure Directly

The estimation of multiple interrelated dependence relationships is not the only unique element of structural equation modeling. SEM also has the ability to incorporate **latent variables** into the analysis. A latent variable is a hypothesized and unobserved concept that can only be approximated by observable or measured variables. The observed variables, which we gather from respondents through various data collection methods (surveys, tests, observation, etc.), are known as **manifest variables.** Yet why would we want to use a latent variable that we did not measure instead of the exact data (manifest variables) the respondents provided? Although this may sound like a nonsensical or "black box" approach, it has both practical and theoretical justification.

Improving Statistical Estimation

Statistical theory tells us that a regression coefficient is actually composed of two elements: the "true" or structural coefficient between the dependent and independent variable and the **reliability** of the predictor variable. Reliability is the degree to which the independent variable is "error-free" [15]. In all the multivariate techniques to this point, we have assumed we had no error in our variables. But we know from both practical and theoretical perspectives that we cannot perfectly measure a concept and that there is always some degree of **measurement error.** For example, when asking about something as straightforward as household income, we know some people will answer incorrectly, either overstating or understating the amount or not knowing it precisely. The answers provided have some measurement error and thus affect the estimation of the "true" structural coefficient.

The impact of measurement error (and the corresponding lowered reliability) can be shown from an expression of the regression coefficient as

$$\beta_{y \cdot x} = \beta_s \times \rho_x$$

where $\beta_{y \cdot x}$ is the regression coefficient, β_s is the structural coefficient, and ρ_x is the reliability of the predictor variable. Because all dependence relationships are based on the correlation (and resulting regression coefficient) between variables, we would hope to "strengthen" the correlations used in the dependence models and make them more accurate estimates of the structural coefficients by first accounting for the correlation attributable to any number of measurement problems.

Representing Theoretical Concepts

Measurement error is not just caused by inaccurate responses but occurs when we use more abstract or theoretical concepts, such as attitude toward a product or motivations for behavior. With concepts such as these, the researcher tries to design the best questions to measure the concept. The respondents also may be somewhat unsure about how to respond or may interpret the questions in a way that is different from what the researcher intended. Both situations can give rise to measurement error. But if we know the magnitude of the problem, we can incorporate the reliability into the statistical estimation and improve our dependence model.

Specifying Measurement Error

How do we account for measurement error? Structural equation modeling provides the **measurement model,** which specifies the rules of correspondence between manifest and latent variables. The measurement model allows the researcher to use one or more variables for a single independent or dependent concept and then estimate (or specify) the reliability. For example, the dependent variable might be a concept represented by a set of questions (also known as a scale). In the measurement model the researcher can assess the contribution of each scale item as well as incorporate how well the scale measures the concept (its reliability) into the estimation of the relationships between dependent and independent variables. This procedure is similar to performing a factor analysis of the scale items and using the factor scores in the regression. We discuss these similarities and specific details in a later section.

The Role of Theory in Structural Equation Modeling

Throughout the discussion of structural equation modeling, we refer to the need for theoretical justification for specifications of the dependence relationships, modifications to the proposed relationships, and many other aspects of estimating a **model.** "Theory" provides the rationale for almost all aspects of structural equation modeling. For purposes of this chapter, **theory** can be defined as a systematic set of relationships providing a consistent and comprehensive explanation of a phenomenon. From this definition, we see that theory is not the exclusive domain of academia but can be rooted in experience and practice obtained by observation of real-world behavior. While theory is often a primary objective of academic research, practitioners may develop or propose a set of relationships that are as complex and interrelated as any academically based theory. Thus researchers from both academia and industry can benefit from the unique analytical tools presented in structural equation modeling.

From a practical perspective, a theory-based approach to structural equation modeling is a necessity because the technique must be almost completely specified by the researcher. Whereas with other multivariate techniques the researcher may have been able to specify a basic model and allow default values in the statistical programs to "fill in" the remaining estimation issues, structural equation modeling has none of these features. Though the seven-step process we discuss makes these decisions straightforward, each component of the structural and measurement models must be explicitly defined. Moreover, any model modifications must be made through specific actions by the researcher. The need for a "theoretical" model to guide the estimation process becomes especially critical when model modifications are made. Because of the flexibility of structural equation modeling, the chances for "overfitting" the model or developing a model with little generalizability are quite high. Thus, when we stress the need for theoretical justification, our intent is to gain a recognition by the researcher that structural equation modeling is a confirmatory method, guided more by theory than empirical results.

Developing a Modeling Strategy

One of the most important concepts a researcher must learn regarding multivariate techniques is that there is no single "correct" way to apply them. Instead, the researcher must formulate the objectives of the analysis and apply the appropriate technique in the most appropriate manner to achieve the desired objectives. In some instances, the relationships are strictly specified and the objective is a confirmation of the relationship. At other times, the relationships are loosely recognized and the objective is the discovery of relationships. In each extreme instance and points in between, the researcher must formulate the use of the technique in accordance with the research objectives.

The application of structural equation modeling follows this same tenet. Its flexibility provides the analyst with a powerful analytical tool appropriate for many research objectives. But the researcher must define these objectives as guidelines in a modeling strategy. The use of the term *strategy* is designed to denote a plan of action toward a specific outcome. In the case of structural equation modeling, the ultimate outcome is always the assessment of a series of relationships. However, this can be achieved through many avenues. For our purposes, we will define three distinct strategies in the application of structural equation modeling. The most direct application is a **confirmatory modeling strategy,** wherein the analyst specifies a single model and structural equation modeling is used to assess its significance. But while this may seem to be the most rigorous application, it actually is not the most stringent test of a proposed model. If the proposed model has acceptable fit by whatever criteria are applied, the researcher has not "proved" the proposed model but only confirmed that it is one of several possible acceptable models. Several different models might have equally acceptable model fit. Thus, the more rigorous test is achieved by comparing alternative models.

Obtaining an acceptable level of fit for both the overall model and the measurement and structural models does not assure the researcher that the "best" model has been found. Numerous alternative models may provide an even better fit. As a means of evaluating the estimated model with alternative models, overall model comparisons can be performed in a **competing models strategy.** The strongest test of a proposed model is to identify and test competing models that represent truly different hypothetical structural relationships. When comparing these models, the researcher comes much closer to a test of competing "theories," which is a much stronger test than just a slight modification of a single "theory."

The **model development strategy** differs from the prior two strategies in that while a model is proposed, the purpose of the modeling effort is to improve the model through modifications of the structural and/or measurement models. In many applications, theory can provide only a starting point for development of a theoretically justified model that can be empirically supported. Thus, the analyst must employ structural equation modeling not just to empirically test the model but also to provide insights into its respecification. One note of caution must be made. The analyst must be careful not to employ this strategy to the extent that the final model has acceptable fit but cannot be generalized to other samples or populations. The respecification of a model must always be made with theoretical support rather than just empirical justification.

Steps in Structural Equation Modeling

The true value of structural equation modeling comes from the benefits of using the structural and measurement models simultaneously, each playing distinct roles in the overall analysis. To ensure that both models are correctly specified and the results are valid, a seven-step process will be discussed (see Figure 11.1). The reader will note that this process differs from the six-stage model-building approach introduced in Chapter 1 and used in our discussions of the other multivariate methods. The introduction of this separate process for structural equation modeling does not invalidate the model-building approach for other multivariate techniques but just accentuates the uniqueness of structural equation modeling.

The seven steps in structural equation modeling are (1) developing a theoretically based model, (2) constructing a path diagram of causal relationships, (3) converting the path diagram into a set of structural equations and measurement equations, (4) choosing the input matrix type and estimating the proposed model, (5) assessing the identification of the model equations, (6) evaluating the results for goodness-of-fit, and (7) making the indicated modifications to the model if theoretically justified.

Step One: Developing a Theoretically Based Model

Structural equation modeling is based on **causal relationships,** in which the change in one variable is assumed to result in a change in another variable [39]. We encountered this type of statement when we defined a dependence relationship, such as is found in regression analysis. Causal relationships can take many forms and meanings, from the strict causation found in physical processes, such as a chemical reaction, to the less well-defined relationships encountered in behavioral research, such as the "causes" of educational achievement or the "reasons" why we purchase one product rather than another. The strength and conviction with which the researcher can assume **causation** between two variables lies not in the analytical methods chosen but in the theoretical justification provided to support the analyses. Although in many instances not all of the established criteria for making causal assertions (i.e., sufficient association between the two variables, temporal antecedence of the cause versus the effect, lack of alternative causal variables, and a theoretical basis for the relationship) are strictly met, strong causal assertions can possibly be made if the relationships are based on a theoretical rationale.* Thus we caution any researcher against assuming that the techniques discussed in this chapter will by themselves provide a means of "proving" causation without having some guiding theoretical perspective. Using these techniques in an "exploratory" manner is invalid and misleads the researcher more often than it provides appropriate results.

The most critical error in developing theoretically based models is the omission of one or more key predictive variables, a problem known as **specification error.** The implication of omitting a significant variable is to bias the assessment of the importance of other variables [52]. For example, assume that two variables (a and b) both were predictors of c. If we included both a and b in our analysis, we would get the correct assessment of their relative importance as shown by their estimated coefficients. But if we left variable b out of the analysis, the coefficient for variable a would be different. This difference is the result of the coefficient for variable a reflecting not only its effect on c but the effect it shared with b as well. This shared effect, however, is controlled for when both variables are included in the analysis.†

The desire to include all variables, however, must be balanced against the practical limitations of structural equation modeling. Though no theoretical limit on the number of variables in the models exists, practical concerns occur even before the limits of most computer programs are met. Most often, interpretation of the results, particularly statistical significance, becomes quite difficult as the number of concepts becomes large (exceeding 20 concepts). We would never encourage the researcher to omit a concept solely because the number of variables is becoming large, but we should stress the benefits of parsimonious and concise theoretical models.

Step Two: Constructing a Path Diagram of Causal Relationships

Up to this point, we have expressed causal relationships only in terms of equations. But there is another method of portraying these relationships called **path diagrams,** which are especially helpful in depicting a series of causal relationships. Before examining path diagrams, however, we must first introduce the concept of **construct.** A construct is a theoretically based concept that acts as a "building block" used to define relationships. A construct can represent a concept as simple as age, income, or gender or as complex as socioeconomic status,

---

*For a more detailed discussion of the various perspectives on the requirements for causation, see [5, 41].

†We encourage readers who still do not fully understand the impact of omitted variables to review the material in Chapter 3 concerning specification error and its effects.

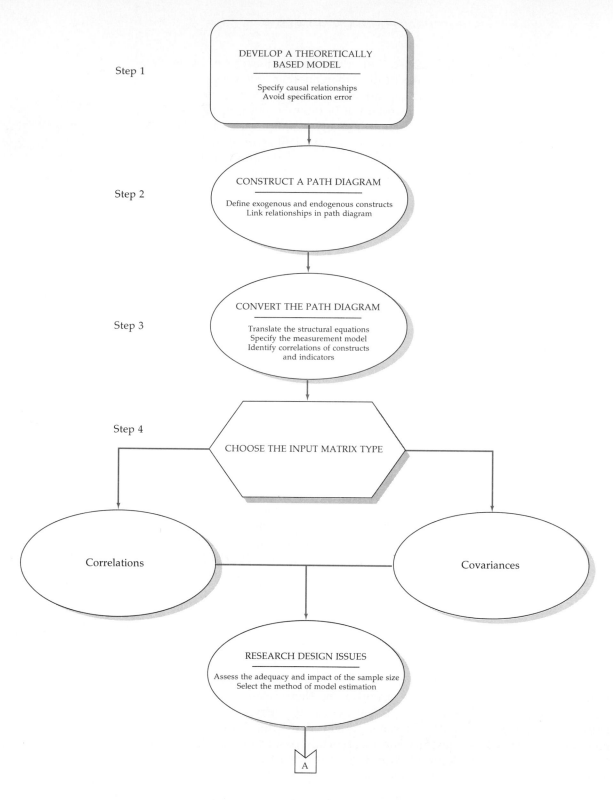

Step 1

DEVELOP A THEORETICALLY
BASED MODEL

Specify causal relationships
Avoid specification error

Step 2

CONSTRUCT A PATH DIAGRAM

Define exogenous and endogenous constructs
Link relationships in path diagram

Step 3

CONVERT THE PATH DIAGRAM

Translate the structural equations
Specify the measurement model
Identify correlations of constructs
and indicators

Step 4

CHOOSE THE INPUT MATRIX TYPE

Correlations

Covariances

RESEARCH DESIGN ISSUES

Assess the adequacy and impact of the sample size
Select the method of model estimation

A

FIGURE 11.1 A seven-step process for structural equation modeling.

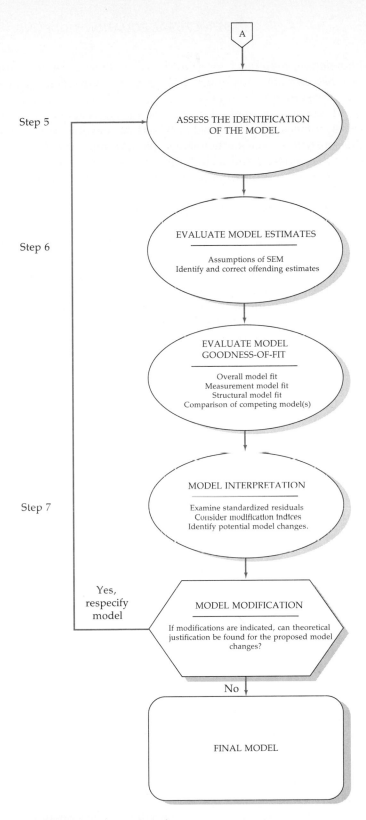

FIGURE 11.1 *Continued*

knowledge, preference, or attitude. A researcher defines path diagrams in terms of constructs and then finds variables to measure each construct. For example, we may ask someone what his or her age is and use this as a measure of the construct age. Likewise, we may ask a series of questions about a person's opinions and use this as a measure of attitude. Both sets of questions provide numerical values for the constructs. We can then assess the questions for the amount of measurement error they possess and include this in the estimation process. *But from now on, we will use the term* construct *to represent a particular concept, no matter how it is measured.*

In constructing a path diagram, we represent the relationships between constructs with arrows. A straight arrow indicates a direct causal relationship from one construct to another. A curved line between constructs indicates just a correlation between constructs. Figure 11.2 shows three examples of relationships depicted by path diagrams, with the corresponding equations also given. Figure 11.2a shows a simple two-construct model. Both X_1 and X_2 are predictor constructs for Y_1, with the curved arrow between X_1 and X_2 showing the effects of intercorrelation (multicollinearity) on the prediction. We can show this relationship with a single equation, much as we saw in our discussion of multiple regression.

In Figure 11.2b, we add a second dependent construct, Y_2. Now, in addition to the model and equation shown in Figure 11.2a, we add a second equation showing the relationship between X_2 and Y_1 with Y_2. Here we can first see the unique role played by structural equation modeling when more than one relationship "shares" constructs. We want to know the effects of X_1 on Y_1, the effects of X_2 on Y_1, and simultaneously the effects of X_2 and Y_1 on Y_2. If we did not estimate them in a consistent manner, we would not be assured of representing their "true and

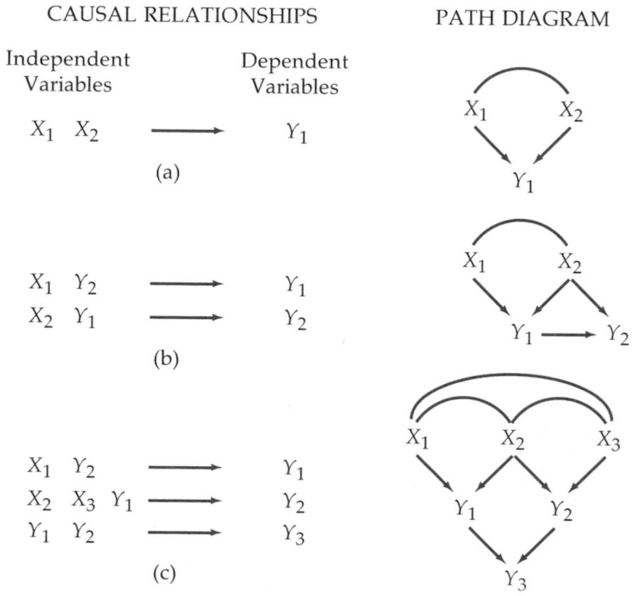

FIGURE 11.2 Representing causal relationships through path diagrams.

separate" effects. For example, such a technique is needed to show the effects of X_2 on both Y_1 and Y_2.

The relationships become even more intertwined in Figure 11.2c, where we have three dependent constructs, each related to the others and to the independent constructs. It is not possible to express all the relationships in either Figure 11.2b or 11.2c in a single equation. Separate equations are required for each dependent construct. The need for a method that can estimate all the equations simultaneously is met with structural equation modeling.

Two assumptions underlie path diagrams. First, all causal relationships are indicated. Thus theory is the basis for inclusion *or* omission of any relationship. It is just as important to justify why a causal relationship *does not exist* between two constructs as it is to justify the existence of another relationship. But the objective is to model the relationships among constructs with the smallest number of causal paths or correlations among constructs that can be theoretically justified. The second assumption relates to the nature of the causal relationships that are assumed to be linear. Just as was encountered with other multivariate techniques, nonlinear relationships cannot be directly estimated in structural equation modeling, but modified structural models can approximate nonlinear relationships [38, 50].

Before we proceed, let us introduce some terminology that will help in distinguishing among the constructs in our path diagrams. All constructs in a path diagram can be placed into one of two classes of constructs: exogenous or endogenous. **Exogenous constructs,** also known as source variables or independent variables, are not "caused" or predicted by any other variables in the model; that is, there are no arrows pointing to these constructs. In our examples in Figure 11.2, X_1, X_2, and X_3 were the exogenous constructs. In Figures 11.2a, 11.2b, and 11.2c, the Y constructs are **endogenous,** meaning that they are predicted by one or more other constructs. Endogenous constructs can predict other endogenous constructs (this is where we see the interrelationships that point to the need for structural models), but an exogenous construct can be causally related only to endogenous constructs. As you can see, the distinction between endogenous and exogenous is determined solely by the researcher, just as was done in deciding between what were the independent and dependent variables in regression.

Step Three: Converting the Path Diagram into a Set of Structural Equations and Specifying the Measurement Model

After developing the theoretical model and portraying it in a path diagram, the analyst is ready to specify the model in more formal terms through a series of equations that define (1) the structural equations linking constructs, (2) the measurement model specifying which variables measure which constructs, and (3) a set of matrices indicating any hypothesized correlations among constructs or variables. The objective is to link operational definitions of the constructs to theory for the appropriate empirical test. Each type of equation or matrix will be described. In the discussions, all equations will be specified in general terminology. Readers interested in specific mathematical terminology, such as the notation used in LISREL, are referred to Appendix 11A.

Structural Model

Translating a path diagram into a series of structural equations is a straightforward procedure. First, each endogenous construct (any construct with one or more straight arrows leading into it) is the dependent variable in a separate equation. Then the predictor variables are all constructs at the end, or "tails," of the straight arrows leading into the endogenous variable. It is that simple.

Figure 11.3 illustrates this translation process for each of the path diagrams in Figure 11.2. As we see, each endogenous variable (the Y_is) can be predicted either by exogenous variables (X_js) or by other endogenous variables. For each hypothesized effect, we will estimate a structural coefficient (b_{jm}). Also, because we know that we will have prediction error, just as in multiple regression, we include an error term (ϵ_i) for each equation as well. The error term represents the sum of the effects due to specification error and random measurement error. It is not possible to separate these two sources of error except in special situations.

Measurement Model

We have referred to the measurement model in general terms up to this point, but we now define it in specific terms. But before doing so, let's review the foundations of factor analysis (Chapter 7), which are quite analagous to the measurement model.

Remember that in factor analysis, each individual variable was "explained" by its loading on each factor. The objective was to best represent all the variables in a small number of factors. The factors related to "underlying dimensions" in the data, which we then had to interpret and label. We can represent these relationships mathematically as follows for an example of five variables and three factors:

	Factor Loadings		
Variable	Factor 1	Factor 2	Factor 3
V_1	L_{11}	L_{12}	L_{13}
V_2	L_{21}	L_{22}	L_{23}
V_3	L_{31}	L_{32}	L_{33}
V_4	L_{41}	L_{42}	L_{43}
V_5	L_{51}	L_{52}	L_{53}

The value for each factor (factor score) is calculated by the loadings on each variable (e.g., for Factor 1 = $L_{11}V_1 + L_{21}V_2 + L_{31}V_3 + L_{41}V_4 + L_{51}V_5$, where V_1 to V_5 are the actual data values for each variable). Also, the predicted value for each variable is calculated by the loadings of the variable on each factor. However, every variable has a factor loading on each factor; thus each factor is always a composite of all variables, although their loadings vary in magnitude.

As you can perhaps now see, a factor is actually a latent variable, defined by the loadings of all the variables. To specify the measurement model, we make the transition from factor analysis where the analyst had no control over which variables describe each factor to a confirmatory mode where the researcher specifies which variables define each construct (factor).

The manifest variables we collect from the respondents are termed **indicators** in the measurement model, because we use them to measure, or "indicate," the

	ENDOGENOUS VARIABLE	=	EXOGENOUS VARIABLES	+	ENDOGENOUS VARIABLES	+	ERROR
	Y_1		X_1 X_2 X_3		Y_1 Y_2 Y_3		ε_i

Path Diagram

Figure 11.2A	Y_1	$=$	$b_1X_1 + b_2X_2$			$+$	ε_1

Figure 11.2B	Y_1	$=$	$b_1X_1 + b_2X_2$			$+$	ε_1
	Y_2	$=$	b_3X_2	$+ b_4Y_1$		$+$	ε_2

Figure 11.2C	Y_1	$=$	$b_1X_1 + b_2X_2$			$+$	ε_1
	Y_2	$=$	$b_3X_2 + b_4X_3 + b_5Y_1$			$+$	ε_2
	Y_3		$b_6Y_1 + b_7Y_2$			$+$	ε_3

FIGURE 11.3 Translation of path diagrams into structural equations.

latent constructs (factors). Assume in our example that V_1 and V_2 are now hypothesized to be indicators of construct A, V_3 and V_4 are indicators of construct B, and V_5 is a single indicator of construct C. The measurement model would be expressed as follows:

Indicator Loadings on Constructs

Variable	Construct A	Construct B	Construct C
V_1	L_1		
V_2	L_2		
V_3		L_3	
V_4		L_4	
V_5			L_5

How and why does this configuration differ from the factor analysis loadings we discussed before? The most obvious difference is the much smaller number of loadings. In the exploratory mode of principal-components factor analysis, the researcher cannot control the loadings. In the measurement model, however, the researcher has complete control over which variables describe each construct. In the example, each variable was an indicator of only one construct; thus there was a lower number of loadings. Although a variable may be an indicator for more than one construct, this method is not recommended except in specific situations with strong theoretical rationale. The analyst specifies a measurement model for both the exogenous constructs and the endogenous constructs in exactly this manner.

Once the measurement model has been specified, the analyst must then provide for the reliability of the indicators. Reliability can be incorporated in one of two ways: (1) empirically estimated or (2) specified by the researcher (also termed *fixing* the value). For empirical estimation, the researcher specifies the loading matrix as described, along with an error term for each indicator variable (because we don't expect to predict each indicator perfectly). When the structural and measurement models are estimated, the loading coefficients will provide estimates of the reliabilities of the indicators and the overall construct. We illustrate this procedure in step 6.

In some instances it is appropriate for the researcher to specify, or "fix," the reliabilities. To "fix" the reliability of an indicator in a correlation matrix, the researcher specifies the loading value as the square root of the desired or estimated reliability, or specifies the error term of that variable as 1.0 minus the desired reliability value.* This procedure can be simply done in statistical programs through a single statement for each variable. The three most common instances of fixing the reliability occur when using (1) single-item measures, (2) previously established scales with known reliabilities, or (3) a two-stage analysis estimating first the measurement model and then the structural model. Exhibit 11.1 examines these three instances in more detail.

---

*In specifying the reliabilities, the analyst may specify either the loading value, the error term, or both. Because specifying either the loading or error terms automatically determines the other value, we recommend that both be set for greatest model parsimony and that a coefficient not be used for estimating a value that could be specified.

EXHIBIT 11.1

Situations Justifying the Specification of Reliabilities for the Measurement Model

The specification of reliabilities for the indicator(s) of any latent construct may seem to be counter to the objectives of structural equation modeling; however, in at least three situations it just justified and strongly recommended. In one instance, empirical estimation of the reliability is not possible, yet the analyst may know that measurement error still exists. In another, the indicators may have been used extensively; therefore, the reliabilities are known before use. And finally, a two-stage process in which the reliabilities are first assessed and then specified in the estimation process explicitly separates the two empirical processes and provides insight into each separately. A brief discussion of each situation follows.

Single-item Measures

With single-item measures, it is not possible to empirically estimate the reliability; thus the researcher is faced with two possibilities. First, set ("fix") the reliability at 1.0, indicating that there is no measurement error in the indicator. Yet as discussed before, we know this is erroneous in almost all instances, because if for no other reason, reliability is affected by the quality of data collection. For example, gender may be "perfect" or very close (99%), with error due only to coding errors. However, income may have a higher level of error (e.g., 10%) owing to reporting bias and the level of measurement. Most often, therefore, the researcher should make some estimate of the reliability and specify the value for single-item indicators. A number of recommended approaches are provided in [38].

Use of Validated Scales or Measures

Many times a researcher employs a scale or measure that has been extensively tested in previous research. If the objective in its use is replication of the effects found in prior studies, then the reliability of the scale or measure should be fixed at previously established levels. This is an example of the researcher specifying reliabilities to maintain "control" over the meaning of the constructs. By fixing the reliability, the researcher "forces" an indicator to have the amount of variance appropriate for the construct and maintains a specific meaning for the construct.

Two-stage Analysis

Many researchers are now proposing a two-stage process of structural equation modeling in which the measurement model is first estimated, much like factor analysis, and then the measurement model is "fixed" in the second stage when the structural model is estimated [3, 49, 64]. The rationale of this approach is that accurate representation of the reliability of the indicators is best accomplished in two stages to avoid the interaction of measurement and structural models. While we cannot truly evaluate the measurement and structural models in isolation, we must consider the po-

tential for within-construct versus between-construct effects in estimation, which can be substantial and result in what Burt [21] terms "interpretational confounding."

A single-stage analysis with the simultaneous estimation of both structural and measurement models is the best approach when the model possesses both strong theoretical rationale and highly reliable measures, resulting in more accurate relationships and decreasing the possibility for the structure or measurement interaction. However, when faced with measures that are less reliable or theory that is only tentative, the researcher should consider a staged approach to maximize the interpretability of both measurement and structural models. Considerable debate has emerged on the appropriateness of this approach and those instances in which it is justified, both on conceptual and empirical grounds [4, 32, 33].

Correlations Among Constructs and Indicators

In addition to the structural and measurement models, the researcher can also specify correlations between the exogenous constructs or between the endogenous constructs. Many times exogenous constructs are correlated, representing a "shared" influence on the endogenous variables. Correlations among the endogenous constructs, however, have fewer appropriate applications and are not recommended for typical use, as they represent correlations among the structural equations that confound their interpretation. Finally, the indicators in the measurement model can also be correlated separately from the construct correlations. This method is to be avoided except in specific situations, such as a study in which there are known effects from the measurement or data collection process on two or more indicators, or a longitudinal study where the same indicator is collected in two periods of time [5]. For a more complete discussion of the distorting effects of correlated indicators, see [35].

Step Four: Choosing the Input Matrix Type and Estimating the Proposed Model

Structural equation modeling also differs from other multivariate techniques in that it uses only the variance/covariance or correlation matrix as its input data. Individual observations can be input into the programs, but they are converted into one of the two types of matrices before estimation. The focus of structural equation modeling is not on individual observations but on the pattern of relationships across respondents. Analysis for outliers is possible only before the covariance or correlation matrices are calculated, and testing of the data for meeting the assumptions is done separately.

Input for the program is a correlation or variance/covariance matrix of all indicators used in the model. The measurement model then specifies which indicators correspond to each construct, and the latent construct scores are then employed in the structural model. To provide some insight into how the correlations between constructs are used to estimate structural coefficients, Appendix 11B illustrates the fundamental principles of path analysis—the estimation process underlying the structural model.

Covariances Versus Correlations

An important issue in interpreting the results is the use of the variance/covariance matrix versus the correlation matrix. Structural equation modeling was initially formulated for use with the variance/covariance matrix (hence its common designation as covariance structure analysis). The covariance matrix has the advantage of providing valid comparisons between different populations or samples, something not possible when models are estimated with a correlation matrix. Interpretation of the results, however, is somewhat more difficult when using covariances because the coefficients must be interpreted in terms of the units of measure for the constructs. Correlation matrices have a common range that makes possible direct comparisons of the coefficients within a model.

The correlation matrix, however, has gained widespread use in many applications. The correlation matrix in structural equation modeling is simply a "standardized" variance/covariance matrix in which the scale of measurement of each variable is removed by dividing the variances or covariances by the product of the standard deviations. Use of correlations is appropriate when the objective of the research is only to understand the pattern of relationships between constructs, but not to explain the total variance of a construct. Another appropriate use is to make comparisons across different variables, because the covariances are affected by the scale of measurement. Coefficients obtained from the correlation matrix are always in standardized units, similar to beta weights in regression, and range between -1.0 and $+1.0$. Moreover, research has shown that the correlation matrix provides more conservative estimates of the significance of coefficients and is not upwardly biased as previously thought [26].

In summary, the researcher should employ the variance/covariance matrix any time a true "test of theory" is being performed, as the variance/covariances satisfy the assumptions of the methodology and are the appropriate form of the data for validating causal relationships. However, often the researcher is concerned only with patterns of relationships, not with total explanation as needed in theory testing, and the correlation matrix is acceptable. Any time the correlation matrix is used, the analyst should cautiously interpret the results and generalizability to different situations.

Types of Correlations or Covariances Used

The most widely used means of computing the correlations or covariances between manifest variables is the Pearson product-moment correlation. This is also the most common form of correlation used in multivariate analysis, making it quite easy for the analyst to compute the correlation or covariance matrices. An assumption of the product-moment correlation, however, is that both variables are metrically measured. This makes the product-moment correlation inappropriate for use with nonmetric (ordinal or binary) measures. To allow for incorporation of the nonmetric measures into structural equation models, the analyst must employ different types of correlation measures. If both variables are ordinal with three or more categories (polychotomous), then the **polychoric correlation** is appropriate. If both nonmetric measures are binary, then the **tetrachoric correlation** can be used. For instances in which a metric measure is related to a polychotomous ordinal measure, the **polyserial correlation** represents the relationship. If a binary measure is related to a metric measure, the **biserial correlation** is used.

Sample Size

Even though individual observations are not needed as with all other multivariate methods, the sample size plays an important role in the estimation and interpretation of SEM results. Sample size, as in any other statistical method, provides a basis for the estimation of sampling error. The critical question in structural equation modeling is, How large a sample is needed? Maximum likelihood estimation (MLE) has been found to provide valid results with sample sizes as small as 50, but a sample this small is not recommended. It is generally accepted that the minimum sample size to ensure appropriate use of MLE is 100. As we increase the sample size above this value, the MLE method increases in its sensitivity to detect differences among the data. As the sample size becomes large (exceeding 400 to 500), the method becomes "too sensitive" and almost any difference is detected, making all **goodness-of-fit** measures indicate poor fit [22, 53, 60]. While there is no correct sample size, recommendations are for a size ranging between 100 to 200. One approach is always to test a model with a sample size of 200, no matter what the original sample size was, because 200 is proposed as being the "critical sample size" [40].

The sample size should also be large enough compared with the number of estimated parameters but with an absolute minimum of 50 respondents. A minimum recommended level is five observations for each estimated parameter. Note that this approach differs from the concept of degrees of freedom discussed later and concerns the number of original respondents used to calculate the covariance or correlation matrix.

Model Estimation

Once the structural and measurement model are specified and the input data type is selected, the analyst must choose the computer program for estimation. The most widely used program is LISREL (LInear Structural RELations) [45, 46, 48], a truly flexible model for a number of research situations (cross-sectional, experimental, quasi-experimental, and longitudinal studies). LISREL has found applications across all fields of study [8] and has become almost synonymous with structural equation modeling. However, a number of alternative programs exist, among them EQS [9, 10], COSAM [34], and PLS [65]. EQS places less stringent assumptions on the multivariate normality of the data, and PLS is better suited to prediction yet limited for interpretation purposes. LISREL, EQS, and PLS are all available in versions that can be run on personal computers [10, 14, 46, 48].

Step Five: Assessing the Identification
of the Structural Model

During the estimation process, the most likely cause of the computer program "blowing up" or producing meaningless or illogical results is a problem in the identification of the structural model. An **identification** problem, in simple terms, is the inability of the proposed model to generate unique estimates. It is based on the principle that we must have a separate and unique equation to estimate each coefficient. It is reflected in the dictum "You must have more equations than unknowns" that we all learned in algebra when defining a series of equations. However, as structural models become more complex, there is no guaranteed approach for ensuring that the model is identified [19].

One approach (if results are actually obtained) is to look for the possible symptoms of an identification problem. These include (1) very large standard errors for one or more coefficients, (2) the inability of the program to invert the information matrix, (3) wildly unreasonable estimates or impossible estimates such as negative error variances, or (4) high correlations (±.90 or greater) among the estimated coefficients. The next section (step 6) discusses interpreting these results in more detail.

Though LISREL will perform a simple test for identification during the estimation process, the analyst can also perform tests when the equation is identified to see whether the results are unstable because of the level of identification. First, the model can be reestimated several times, each time with different starting values.* If the results do not converge at the same point each time, the identification should be examined more thoroughly. The second test to assess the identification's effect on a single coefficient is first to estimate the model and obtain the coefficient estimate. Then "fix" the coefficient to its estimated value and reestimate the equation (a discussion on fixing coefficients will be seen in a later section). If the overall fit of the model varies markedly, identification problems are indicated.

If an identification problem is indicated, the researcher should first look to three common sources: (1) a large number of estimated coefficients relative to the number of covariances or correlations, indicated by a small number of **degrees of freedom**—similar to the problem of overfitting the data found in many other multivariate techniques; (2) the use of reciprocal effects (two-way causal arrows between two constructs); or (3) failure to fix the scale of a construct. (We discuss this procedure in our analysis of the example data.)

The only solution for an identification problem is to define more constraints on the model—that is, eliminate some of the estimated coefficients. The researcher should follow a structured process, gradually adding more constraints (deleting paths from the path diagram) until the problem is remedied. In doing so, the researcher is attempting to achieve what is termed an overidentified model (which has more equations than unknowns). Overidentified models have some degrees of freedom with which to assess, if possible, the amount of sampling and measurement error and thus provide better estimates of the "true" causal rela-

---

*The computer programs provide the researcher a way to specify an initial value for any coefficient, a "starting point" for the estimation process. If a starting value is not specified by the analyst, the programs automatically compute them by one of several methods.

tionships. To achieve this end, a four-step process is recommended [38]: (1) Build a theoretical model with the minimum number of coefficients (unknowns) that can be justified. If identification problems are encountered, proceed with remedies in this order: (2) Fix the measurement error variances of constructs if possible, (3) fix any structural coefficients that are reliably known, and (4) eliminate troublesome variables. If identification problems still exist, the researcher must reformulate the theoretical model to provide more constructs relative to the number of causal relationships examined.

Step Six: Evaluating Goodness-of-Fit Criteria

The first step in evaluating the results is to assess the degree to which the data and proposed models meet the assumptions of structural equation modeling. An initial inspection is made for "offending estimates"—estimated coefficients that violate accepted ranges or indicate problems in other areas of the model. Once established as meeting the necessary assumptions and providing acceptable estimates, the goodness-of-fit must then be assessed at several levels: first for the overall model and then for the measurement and structural models separately.

Assumptions

Structural equation modeling shares three assumptions with the other multivariate methods we have studied: independent observations, random sampling of respondents, and the linearity of all relationships. In addition, structural equation modeling is more sensitive to the distributional characteristics of the data, particularly the departure from multivariate normality (critical in the use of LISREL) or a strong kurtosis (skewness) in the data. Some programs, such as EQS, are less sensitive to nonnormal data, but the data should be evaluated no matter which program is used. Generalized least squares, an alternative estimation method, can adjust for these violations, but this method quickly becomes impractical as the model size and complexity increase; thus its use is limited. A lack of multivariate normality is particularly troublesome because it substantially inflates the Chisquare statistic and creates upward bias in critical values for determining coefficient significance. Although the structural equation programs do not have built-in diagnostic procedures for testing these assumptions, they can be tested with conventional methods (see a more detailed discussion in Chapter 2) or through programs such as PRELIS [47].

Offending Estimates

Once the assumptions have been satisfied at acceptable levels, the results are first examined for **offending estimates.** These are estimated coefficients in either the structural or measurement models that exceed acceptable limits. The most common examples of offending estimates are (1) negative error variances or nonsignificant error variances for any construct, (2) standardized coefficients exceeding or very close to 1.0, or (3) very large standard errors associated with any estimated coefficient. If offending estimates are encountered, the researcher must first resolve each occurrence before evaluating any specific results of the model, as changes in one portion of the model can have significant effects on other results.

Several approaches for resolution have already been examined in the discussion of identification issues. If identification problems are corrected and problems still exist, several other remedies are available. In the case of negative error variances (also known as **Heywood cases,** one possibility is to fix the offending error variances to a very small positive value (.005) [13, 26]. While meeting the practical requirements of the estimation process, the remedy only masks the underlying problem and must be considered when interpreting the results. If correlations in the standardized solution exceed 1.0, or two estimates are correlated highly, then the researcher should consider elimination of one of the constructs or should ensure that true discriminant validity has been established among the constructs. In many instances, such situations are the result of atheoretical models, established without sufficient theoretical justification or modified solely on the basis of empirical considerations.

Overall Model Fit

Once the researcher has established that the data meet the assumptions and that there are no offending estimates, the next step is to assess the overall model fit with one or more goodness-of-fit measures. Goodness-of-fit is a measure of the correspondence of the actual or observed input (covariance or correlation) matrix with that predicted from the proposed model.

In developing any statistical model, the researcher must guard against "overfitting" the model to the data. As you remember in regression, the researcher maintained a certain ratio (perhaps five to one) between the number of estimated coefficients and the number of respondents. This ratio should be maintained in structural equation modeling as well. However, the researcher has additional concerns in SEM with the size of the covariance or correlation matrix relative to the number of estimated coefficients. The difference between the number of correlations/covariances and the actual number of coefficients in the proposed model is termed **degrees of freedom.** The researcher should strive for a larger number of degrees of freedom, all other things being equal. In doing so, the model achieves **parsimony**—the achievement of better or greater model fit for each estimated coefficient.

The number of degrees of freedom for a proposed model is calculated as

$$\text{df} = \frac{1}{2}[(p + q)(p + q + 1)] - t$$

where p = the number of endogenous indicators, q = the number of exogenous indicators, and t = the number of estimated coefficients in the proposed model. The first portion of the equation calculates the nonredundant size of the correlation/covariance matrix (i.e., the lower or upper half of the matrix plus diagonal). Then each estimated coefficient "uses up" a degree of freedom. Thus the better fit we can achieve with fewer coefficients, the better test of the model and the more confidence we can have that the results are not a result of overfitting the data.

The goodness-of-fit measures fall into three types: (1) absolute fit measures, (2) incremental fit measures, or (3) parsimonious fit measures. The absolute fit measures assess only the overall model fit (both structural and measurement models collectively), with no adjustment for the degree of "overfitting" that might occur. The incremental fit measures compare the proposed model to a comparison model specified by the researcher. Finally, the parsimonious fit measures "adjust" the measures of fit to provide a comparison between models with differing numbers of estimated coefficients, the purpose being to determine the amount of fit achieved by each estimated coefficient.

The researcher is faced with the question of which measures to choose. No single measure or set of measures have been agreed upon as the only measures needed. As structural equation modeling has evolved in recent years, goodness-of-fit measures have been continually developed, and additional measures (not discussed here) have been proposed [55]. The researcher is encouraged to employ one or more measures from each class. Assessing the goodness-of-fit of a model is more a relative process than one with absolute criteria. The application of multiple measures will enable the researcher to gain a consensus across types of measures as to the acceptability of the proposed model. Appendix 11C describes several measures within each class of goodness-of-fit measures in more detail, including how to calculate measures not provided by the computer program. An excellent review of the various goodness-of-fit measures and their application in a number of situations is contained in various sources [18, 48].

Measurement Model Fit

Once the overall model fit has been evaluated, the measurement of each construct can then be assessed for unidimensionality and reliability. **Unidimensionality** is an assumption underlying the calculation of reliability and is demonstrated when the indicators of a construct have acceptable fit on a single-factor (one-dimensional) model. This topic is beyond the scope of this discussion, but the interested reader is referred to [2, 3, 4, 32, 33]. The use of reliability measures, such as **Cronbach's alpha** [24], does not ensure unidimensionality but instead assumes it exists. The researcher is encouraged to perform unidimensionality tests on all multiple-indicator constructs before assessing their reliability.

The next step is to examine the estimated loadings and to assess the statistical significance of each one. If statistical significance is not achieved, the researcher may wish to eliminate the indicator or attempt to transform it for better fit with the construct.

Beyond examination of the loadings for each indicator, the principal approach used in assessing the measurement model is the **composite reliability** and **variance extracted** measures for each construct. **Reliability** is a measure of the internal consistency of the construct indicators, depicting the degree to which they "indicate" the common latent (unobserved) construct. More reliable measures provide the researcher with greater confidence that the individual indicators are all consistent in their measurements. A commonly used threshold value for acceptable reliability is .70, although this is not an absolute standard, and values below .70 have been deemed acceptable if the research is exploratory in nature.

We should note, however, that reliability does not ensure **validity.** Validity is the extent to which the indicators "accurately" measure what they are supposed to measure. For example, several measures of how and why consumers purchase products may be quite reliable, but the researcher may mistakenly assume they measure brand loyalty when in fact they are indicators of purchase intentions. In this instance, the indicators are a reliable set of measures but an invalid measure of brand loyalty. The issue of validity rests on the researcher's specification of indicators for a latent construct. The means for assessing validity in its many forms are reviewed in [18].

Another measure of reliability is the **variance extracted measure.** This measure reflects the overall amount of variance in the indicators accounted for by the latent construct. Higher variance extracted values occur when the indicators are truly representative of the latent construct. The variance extracted measure is a complementary measure to the construct reliability value. Recommendations typically suggest that the variance extracted value for a construct should exceed .50.

Exhibit 11.2 details the formulas used in calculating these measures for each construct. Actual examples of the calculations are provided in the confirmatory factor analysis example later in this chapter.

Structural Model Fit

The most obvious examination of the structural model involves the significance of estimated coefficients. Structural equation modeling methods provide not only estimated coefficients but also standard errors and calculated t values for each coefficient. If we can specify the significance level we deem appropriate (e.g., .05), then each estimated coefficient can be tested for statistical significance (i.e., that it is different from zero) for the hypothesized causal relationship. However, given

EXHIBIT 11.2

Calculating the Reliability and Variance Extracted for a Latent Construct

The reliability and variance extracted for a latent construct must be computed separately for each multiple indicator construct in the model. Although LISREL and other programs do not compute them directly, all the necessary information is readily provided. This exhibit provides the formulas used in calculation of each of the measures. An actual example of the calculations can be found in the confirmatory factor analysis example in a later section of this chapter.

Construct Reliability

The composite reliability of a construct is calculated as

$$\text{Construct reliability} = \frac{(\Sigma \text{ std. loading})^2}{(\Sigma \text{ std. loading})^2 + \Sigma \, \epsilon_j}$$

where the standardized loadings are obtained directly from the program output, and ϵ_j is the mea-

surement error for each indicator [31]. The measurement error is 1.0 minus the reliability of the indicator, which is the square of the indicator's standardized loading. The indicator reliabilities should exceed .50, which roughly corresponds to a standardized loading of .7.

Variance Extracted

The variance extracted measure is calculated as

$$\text{Variance extracted} = \frac{\Sigma \text{ std. loading}^2}{\Sigma \text{ std. loading}^2 + \Sigma \epsilon_j}$$

This measure is quite similar to the reliability measure but differs in that the standardized loadings are squared before summing them [31]. Guidelines suggest that the variance extracted value should exceed .50 for a construct.

the statistical properties of MLE and its characteristics at smaller sample sizes, the researcher is encouraged to be conservative in specifying a significance level, choosing smaller levels (.025 or .01) instead of the traditional .05 level.

The selection of a critical value also depends on the theoretical justification for the proposed relationships. If a positive or negative relationship is hypothesized, then a one-tailed test of significance can be employed. However, if the researcher cannot prespecify the direction of the relationship, then a two-tailed significance test must be used. The difference is in the critical t values used to assess significance. For example, for the .05 significance level, the critical value for a one-tailed test is 1.645, but it increases to 1.96 for a two-tailed test. Thus the researcher can more accurately detect differences if stronger theory can be utilized in model specification.

As another means of evaluation, the researcher should examine the standardized solution where the estimated coefficients all have equal variances and a maximum value of 1.0. Thus the coefficients closely approximate effect sizes shown by beta weights in regression. Coefficients near zero have little, if any, substantive effect, whereas an increase in values corresponds to increased importance in the causal relationships.

As a measure of the entire structural equation, an overall coefficient of determination (R^2) is calculated, similar to that found in multiple regression. Although no test of statistical significance can be performed, it provides a relative measure of fit for each structural equation.

The results of structural equation modeling can be affected by multicollinearity just as was found in regression. Here the researcher must be aware of the correlations among construct estimates in the structural equation modeling results. The computer programs provide a correlation matrix of the estimated values for the latent constructs. If large values appear, then corrective action should be taken. This action may include the deletion of one construct or the reformulation of causal relationships. While no limit has been set that defines what are high correlations, values exceeding .90 should always be examined, and many times correlations exceeding .80 can be indicative of problems.

Comparison of Competing or Nested Models

The more common modeling strategies—a competing models or model development strategy—involve the comparison of model results to determine the best-fitting model from a set of models. In a competing models strategy, the researcher postulates a number of alternative models. The objective is to fit the "best" from among the set of models. In a model development strategy, the researcher starts with an initial model and engages in a series of model respecifications, each time hoping to improve the model fit while maintaining accordance with the underlying theory.

To assist in comparing models, a large number of measures have been developed to assess model fit. One class of measures assesses the overall model fit in absolute terms, providing specific measures of the fit. One drawback to these measures is that they do not account for the number of relationships used in obtaining the model fit. To measure model parsimony, a series of parsimonious fit measures have been proposed. Their objective is to determine the "fit per coefficient," because the absolute fit will always improve as estimated coefficients are added.

A comprehensive procedure for this purpose was proposed by Anderson and Gerbing [3], in which a series of competing models are specified. Differences between models can be shown to be simply the difference in the χ^2 values for the different models. This χ^2 difference can then be tested for statistical significance with the appropriate degrees of freedom being the difference in the number of estimated coefficients for the two models. The only requirement is that the number of constructs and indicators remains the same, so that the null model is the same for both models (i.e., they are **nested models**). The effect of adding or deleting one or more causal relationships can also be tested in this manner by making comparisons between models with and without the relationships. If the models become nonnested (have a different number of indicators or constructs), the researcher must rely on the parsimonious fit measures described earlier, as the χ^2 difference test is not appropriate in this instance.

Step Seven: Interpreting and Modifying the Model

Once the model is deemed acceptable, the researcher may wish to examine possible model modifications to improve the theoretical explanation or the goodness-of-fit. Before addressing some approaches for identifying model modification, we caution the researcher to make such modifications with care and only after obtaining theoretical justification for what empirically is deemed significant. Modifications to the original model should be made only after deliberate consideration. If modifications are made, the model should be cross-validated (i.e., estimated on a separate set of data) before the modified model can be accepted.

Where can the researcher look for model improvements? The first indication comes from examination of the residuals of the predicted covariance or correlation matrix. The standardized residuals (also called normalized residuals) provided by the programs represent the differences between the observed correlation/covariance and the estimated correlation/covariance matrix. With later versions of LISREL (version 7 and higher), the calculation of residuals was improved and the standard for assessing "significant" residuals has changed (the prior threshold was ±2.0). Residual values greater than ±2.58 are now to be considered statistically significant at a .05 level. Significant residuals indicate a substantial prediction error for a pair of indicators (i.e., one of the correlations or covariances in the original input data). A standardized residual indicates only that a difference exists but lends no insight into how it may be reduced. The researcher must identify a remedy by the addition or modification of the causal relationships [23, 36].

Another aid in assessing the fit of a specified model is the **modification indices,** which are calculated for each nonestimated relationship. The modification index values correspond approximately to the reduction in Chi-square that would occur if the coefficient was estimated. A value of 3.84 or greater suggests that a statistically significant reduction in the Chi-square is obtained when the coefficient is estimated. Though modification indices can be useful in assessing the impact of theoretically based model modifications, the researcher should never make model changes based solely on the modification indices. This atheoretical approach is totally contrary to the "spirit" of the technique and should be avoided in all instances. Model modification must have a theoretical justification before being considered, and even then the researcher should be quite skeptical about the changes [52].

As model modifications are made, the researcher must return to step 4 of the seven-step process and reevaluate the modified models. If extensive model modifications are anticipated, the data should be divided into two samples, one providing the basis for model estimation and modification, the other sample providing for validation of the final model.

A Recap of the Seven-Step Process

Structural equation modeling gives the researcher more flexibility than any of the other multivariate methods we have discussed. But along with this flexibility comes the potential for inappropriate use of the method. One overriding concern in any application of structural equation modeling should be a steadfast reliance on a theoretically based foundation for the proposed model and any modifications. Structural equation modeling, when applied correctly, provides a strong confirmatory test to a series of causal relationships. However, when the method is applied in an "exploratory" manner, the researcher is faced with the rather high probability of falling prey to "data snooping" or "fishing" and identifying relationships that have little generalizability by simply capitalizing on the relationships specific to the database being studied.

Two Illustrations of Structural Equation Modeling

Structural equation modeling can address a wide variety of causal relationships. Among the two most common types of analyses performed are confirmatory factor analysis and the estimation of a series of structural equations. We illustrate both these analyses in this section by following the seven-step process for structural equation modeling. First, a confirmatory factor analysis will examine a two-factor solution developed from the factor analysis performed in Chapter 7. Then, a series of structural relationships will be proposed from a new database dealing with an expanded study of supplier characteristics that customers deem important.

A Confirmatory Factor Analysis

Factor analysis, as discussed in Chapter 7, is concerned with exploring the patterns of relationships among a number of variables. These patterns are represented by what are termed principal components or, more commonly, factors. As variables load highly on a factor, they become descriptors of the underlying dimension. Only upon examination of the loadings of the variables on the factors, however, does the researcher identify the character of the underlying dimension.

At this point, the reader may now see the similarity between the objectives of factor analysis and the measurement model in structural equation modeling. The factors are, in measurement model terms, the latent variables. Each variable acts as an indicator of each factor (because every variable has a loading for each factor). Used in this manner, factor analysis is primarily an exploratory technique because of the researcher's limited control over which variables are indicators of which latent construct (i.e., which variables "load" on each factor). Structural equation modeling, however, can play a confirmatory role because the analyst has complete control over the specification of indicators for each construct. Moreover, structural equation modeling allows for a statistical test of the goodness-of-fit for the proposed confirmatory factor solution, which is not possible with principal components/factor analysis. Confirmatory factor analysis is particularly useful in the validation of scales for the measurement of specific constructs [58].

Step One: Developing a Theoretically Based Model

The confirmatory use of structural equation modeling can be illustrated by a synthesis of the principal components and common factor analysis examples in Chapter 7. In review, six variables measured the respondents' perceptions of HATCO on the supplier characteristics of X_1, delivery speed; X_2, price level; X_3, price flexibility; X_4, manufacturer's image; X_6, sales force's image; and X_7, product quality. Variable X_5, service, was omitted from the factor analyses and will be deleted in this example as well to maintain comparability.

The factor analyses in Chapter 7 both indicated the existence of two dimensions (factors). First, the four objective measures (delivery speed, price level, price flexibility, and product quality) were the principal descriptors of one dimension. The second dimension was consistently characterized by the two image variables. These two dimensions of supplier perceptions are characterized as specific strategy actions versus more global or affective evaluations. Thus the hypothesized model will posit two factors (*strategy* and *image*), with each set of variables now acting as indicators of the separate constructs. There is also no reason to expect uncorrelated perceptions; thus the factors will be allowed to correlate as well.

Step Two: Constructing a Path Diagram of Causal Relationships

The next step is to portray the relationships in a path diagram. In this case, the two hypothesized factors are considered exogenous constructs. The path diagram, including the variables measuring each construct, is shown in Figure 11.4. The correlation between perceptions is represented by the curved line connecting the two constructs.

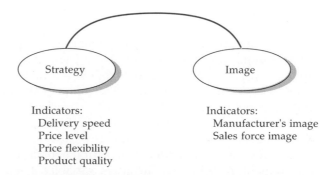

FIGURE 11.4 Path diagram for confirmatory factor analysis.

TABLE 11.1 A Two-Construct Measurement Model

	Variables	Indicator Loadings on Constructs	
		Strategy	*Image*
X_1	Delivery speed	L_1	
X_2	Price level	L_2	
X_3	Price flexibility	L_3	
X_4	Manufacturer's image		L_4
X_6	Sales force's image		L_5
X_7	Product quality	L_6	

Step Three: Converting the Path Diagram into a Set of Structural Equations and Specifying the Measurement Model

Because all the constructs in the path diagram are exogenous, we need only consider the measurement model and the associated correlation matrices for exogenous constructs and indicators. With no structural model, the measurement model constitutes the entire structural equation modeling effort (hence we refer to *confirmatory factor analysis*).

The measurement model can be represented simply by a two-construct model (strategy and image) as shown in Table 11.1. In addition, the two constructs are hypothesized to be correlated, and no within-construct correlated measurements are proposed.

For the interested reader, the appropriate LISREL notation (as discussed in Appendix A) is as shown in Table 11.2.

TABLE 11.2 LISREL Notation for the Measurement Model

Exogenous Indicator		*Exogenous Constructs*		*Error*
X_1	=	$\lambda_{11}^x \xi_1$	+	δ_1
X_2	=	$\lambda_{21}^x \xi_1$	+	δ_2
X_3	=	$\lambda_{31}^x \xi_1$	+	δ_3
X_4	=	$\lambda_{42}^x \xi_2$	+	δ_4
X_5	=	$\lambda_{52}^x \xi_2$	+	δ_5
X_6	=	$\lambda_{61}^x \xi_1$	+	δ_6

Correlation Among Exogenous Constructs (ϕ)

	ξ_1	ξ_2
ξ_1	—	
ξ_2	ϕ_{21}	—

TABLE 11.3 Correlation Matrix for Confirmatory Factor Analysis

Variables	Delivery Speed (X_1)	Price Level (X_2)	Price Flexibility (X_3)	Manufacturer's Image (X_4)	Sales Force's Image (X_6)	Product Quality (X_7)
X_1 Delivery speed	1.000					
X_2 Price level	−.349	1.000				
X_3 Price flexibility	.509	−.487	1.000			
X_4 Manufacturer's image	.050	.272	−.116	1.000		
X_6 Sales force's image	.077	.186	−.034	.788	1.000	
X_7 Product quality	−.483	.470	−.448	.200	.177	1.000

Step Four: Choosing Input Matrix Type and Obtaining Model Estimates

Structural equation modeling will accommodate either a covariance or a correlation matrix. For purposes of confirmatory factor analysis, either type of input matrix can be employed. However, because the objective is an exploration of the pattern of interrelationships, correlations are the preferred input data type. The correlation matrix of the six variables is shown in Table 11.3.

LISREL [45, 48] is used for estimation of the measurement model and the construct correlations. One note on estimation of the measurement model for constructs with more than one variable: Because of the estimation procedure, the construct must be made "scale invariant," meaning that the indicators of a construct must be "standardized" in a way to make constructs comparable [46, 51]. There are two common approaches to this procedure. First, one of the loadings in each construct can be set to the fixed value of 1.0. The second approach is to estimate the construct variance directly. Either approach results in exactly the same estimates, but for purposes of theory testing, the second approach (estimating construct variance) is recommended [3]. In this example, the second approach will be employed. The complete set of control cards used for the LISREL analysis is contained in Appendix A.

Step Five: Assessing the Identification of the Structural Model

Identification is a relatively simple matter in confirmatory factor analysis, and the diagnostic procedures of the program are sufficient to detect identification problems. The most common problem would occur if multiple variables were hypothesized to be indicators for two or more constructs. It is possible for identification problems to arise in such instances, but they are not encountered in most confirmatory factor analyses, and their chances of occurrence are minimized by the use of strong theoretical foundations for specification of the measurement model.

Step Six: Evaluating Goodness-of-Fit Criteria
Assumptions

The basic assumptions of structural modeling, similar to other multivariate methods, were examined in Chapter 2, and readers who wish may review those results. For purposes of illustration, the variables are deemed as meeting the assumptions, and we move to examining the results for offending estimates.

Offending Estimates

Table 11.4 contains the LISREL estimates for the measurement model and the construct correlations. As we see in examining the measurement model results, we have the occurrence of a loading for manufacturer's image (X_4) greater than 1.0 (known as a Heywood case). The reader should also note a corresponding negative error measurement value for the same variable. Such estimates are theoretically inappropriate and must be corrected before the model can be interpreted and the goodness-of-fit assessed.

TABLE 11.4 Confirmatory Factor Analysis Results: Initial Model

Construct Loadings

	Variables	Exogenous Construct	
		Strategy	Image
X_1	Delivery speed	.644	.000
X_2	Price level	−.654	.000
X_3	Price flexibility	.722	.000
X_4	Manufacturer's image	.000	**1.151**
X_6	Sales force's image	.000	.685
X_7	Product quality	−.689	.000

Correlations Among Latent Constructs (Factors)

	Strategy	Image
Strategy	1.000	
Image	−.175	1.000

Measurement Error for Indicators

	Variables	Delivery Speed (X_1)	Price Level (X_2)	Price Flexibility (X_3)	Manufacturer's Image (X_4)	Sales Force's Image (X_6)	Product Quality (X_7)
X_1	Delivery speed	.585					
X_2	Price level	.000	.573				
X_3	Price flexibility	.000	.000	.479			
X_4	Manufacturer's image	.000	.000	.000	−.325		
X_6	Sales force's image	.000	.000	.000	.000	.531	
X_7	Product quality	.000	.000	.000	.000	.000	.526

TABLE 11.5 Confirmatory Factor Analysis Results: Revised Model

Construct Loadings (t values in parentheses)

Variables	Exogenous Construct	
	Strategy	Image
X_1 Delivery speed	.643 (6.263)	.000
X_2 Price level	−.654 (−6.381)	.000
X_3 Price flexibility	.718 (7.118)	.000
X_4 Manufacturer's image	.000	.997 (14.001)
X_6 Sales force's image	.000	.790 (9.467)
X_7 Product quality	−.692 (−6.821)	.000

Handwritten notes:

in Lisrel 8:
LAMDA X
0.82 = loading
(0.03) = significance level
28.22 = 2-tailed t

Correlations Among Latent Constructs (t value in parentheses)

	Strategy	Image
Strategy	1.000	
Image	−.202 (−1.826)	1.000

Measurement error for indicators

Variables	X_1	X_2	X_3	X_4	X_6	X_7
X_1 Delivery speed	.586					
X_2 Price level	.000	.573				
X_3 Price flexibility	.000	.000	.484			
X_4 Manufacturer's image	.000	.000	.000	.005		
X_6 Sales force's image	.000	.000	.000	.000	.376	
X_7 Product quality	.000	.000	.000	.000	.000	.521

Model Respecification

Several remedies are possible for correcting the offending estimate, including dropping the offending variable. In this situation, the variable will be retained and the corresponding error variance (−.325 on the diagonal of the measurement error matrix) will be set to a small value (.005) to ensure that the loading will now be less than 1.0. The model is then reestimated.

The results of the respecified model are shown in Table 11.5. In examining the results, no offending estimates are present; thus we can proceed to assessing the goodness-of-fit of the confirmatory factor analysis.

Overall Model Fit

The first assessment of model fit must be done for the overall model. In confirmatory factor analysis, overall model fit portrays the degree to which the specified indicators represent the hypothesized constructs. The latest version of LISREL (version 8) now provides a full range of goodness-of-fit measures as described in Appendix 11C. For purposes of the confirmatory factor analysis, we focus on only a limited number of measures from each type. A more detailed discussion of all the available measures will occur in the second example of estimating a full structural model and evaluating competing models. The following measures represent the three types of overall model fit measures useful in confirmatory factor analysis.

Absolute Fit Measures   LISREL provides absolute goodness-of-fit measures, and these are shown in Table 11.6. The first measure is the likelihood ratio Chi-square statistic. The value ($\chi^2 = 15.75$, nine degrees of freedom) has a statistical significance level of .072, above the minimum level of .05, but not above the more conservative levels of .10 or .20. This statistic shows some support for believing that the differences of the predicted and actual matrices are nonsignificant, indicative of acceptable fit. Moreover, the sample size of 100 is within the acceptable range for application of this measure. The researcher must note, however, that the potential exists for the χ^2 with sample sizes of 100 or less to denote no differences even when the model has no significant relationships. Thus additional measures of fit must be employed. The GFI has a value of .948, which is quite high, but it is not adjusted for model parsimony. Finally, the root mean square residual (RMSR) indicates that the average residual correlation is .075, deemed acceptable given the rather strong correlations in the original correlation matrix. While all

TABLE 11.6 LISREL Goodness-of-Fit Measures for Confirmatory Factor Analysis: Revised and Null Models

Revised Model

Chi-square (χ^2)	15.75
Degrees of freedom	9
Significance level	.072
Goodness-of-fit index (GFI)	.949
Adjusted Goodness-of-fit index (AGFI)	.881
Root mean square residual (RMSR)	.075

Null Model

Chi-square (χ^2)	206.86
Degrees of freedom	15
Significance level	.000

The expected number of degrees of freedom for the revised model is 8 [$\frac{1}{2}$ (6)(7) − 13]. But since the measurement error of X_4 was set at .005 and not estimated, an additional degree of freedom is available.

the measures fall within acceptable levels, the incremental fit and parsimonious fit indices are needed to ensure acceptability of the model from other perspectives.

Incremental Fit Measures   The next goodness-of-fit measures assess the incremental fit of the model compared to a null model. In this case, the null model is hypothesized as a single-factor model with no measurement error. The estimation results for the null model are also shown in Table 11.6. (The control cards for null model estimation are shown in Appendix A.)

The null model has a χ^2 value of 206.86 with 15 degrees of freedom. With this information, we can now calculate the two incremental fit measures. The calculations are as follows:

Tucker-Lewis Measure (TL)

$$\text{Tucker Lewis} = \frac{(X^2_{\text{NULL}}/df_{\text{NULL}}) - (X^2_{\text{PROPOSED}}/df_{\text{PROPOSED}})}{(X^2_{\text{NULL}}/df_{\text{NULL}}) - 1}$$

$$\text{Tucker Lewis} = \frac{(206.86/15) - (15.75/9)}{(206.86/15) - 1} = .9414$$

Normed Fix Index (NFI)

$$\text{Normed Fit Index (NFI)} = \frac{(X^2_{\text{NULL}} - X^2_{\text{PROPOSED}})}{X^2_{\text{NULL}}}$$

$$\text{Normed Fit Index (NFI)} = \frac{(206.86 - 15.75)}{206.86} = .9239$$

Both incremental fit measures exceed the recommended level of .90, further supporting acceptance of the proposed model.

Parsimonious Fit Measures   The final measures of the overall model assess the parsimony of the proposed model by evaluating the fit of the model versus the number of estimated coefficients (or conversely, the degrees of freedom) needed to achieve that level of fit. Two measures appropriate for direct model evaluation are the AGFI and the normed Chi-square. The AGFI is provided by the LISREL program. The AGFI value of .881 is close to the recommended level of .90., and thus marginal acceptance can be given on this measure. The second measure is the normed Chi-square (χ^2/df), which has a value of 1.75 (15.75/9). This falls well within the recommended levels of 1.0 to 2.0. Combined with the AGFI, this result allows conditional support to be given for model parsimony.

In summary, the various measures of overall model goodness-of-fit lend sufficient support to deeming the results an acceptable representation of the hypothesized constructs.

Measurement Model Fit

Now that the overall model has been accepted, each of the constructs can be evaluated separately by (1) examining the indicator loadings for statistical significance and (2) assessing the construct's reliability and variance extracted. First, our examination of the *t* values associated with each of the loadings indicates that for each variable they exceed the critical values for the .05 significance level (critical value = 1.96) and the .01 significance level as well (critical value = 2.576). Thus all variables are significantly related to their specified constructs, verifying the posited relationships among indicators and constructs.

Estimates of the reliability and variance extracted measures for each construct are now needed to assess whether the specified indicators are sufficient in their representation of the constructs. Computations for each measure are shown in Table 11.7. Both constructs (.772 and .893) exceed the recommended level of .70, although we must remember that the reliability of construct 2 (image) is inflated as a result of respecification of the measurement error for variable X_4 to almost zero to eliminate the negative error variance value.

For the variance extracted measures, construct 1 (strategy) has a value of .459, falling somewhat short of the recommended 50 percent. Construct 2 (image), with

TABLE 11.7 Reliability and Variance Extracted Estimates for Constructs in the Confirmatory Factor Analysis

Reliability

$$\text{Construct Reliability} = \frac{(\text{Sum of standardized loadings})^2}{(\text{Sum of standardized loadings})^2 + \text{sum of indicator measurement error}}$$

Sum of Standardized Loadings
Strategy = .643 + .654[a] + .718 + .692[a] = 2.707
Image = .997 + .790 = 1.787

Sum of Measurement Error[b]
Strategy = .585 + .573 + .484 + .521 = 2.164
Image = .005 + .376 = .381

Reliability Computation
$$\text{Strategy} = \frac{(2.707)^2}{(2.707)^2 + 2.164} = .772$$
$$\text{Image} = \frac{(1.787)^2}{(1.787)^2 + .381} = .893$$

Variance Extracted

$$\text{Variance Extracted} = \frac{\text{Sum of squared standardized loadings}}{\text{Sum of squared standardized loadings} + \text{sum of indicator measurement error}}$$

Sum of Squared Standardized Loadings
Strategy = $.643^2 + .654^2 + .718^2 + .692^2 = 1.836$
Image = $.997^2 + .790^2 = 1.618$

Variance Extracted Computation
$$\text{Strategy} = \frac{1.836}{1.836 + 2.164} = .459$$
$$\text{Image} = \frac{1.618}{1.618 + .381} = .809$$

---

[a] For purposes of computing the reliability of a construct, the signs of a loading can be ignored. However, if the researcher wishes to compute a summed measure (such as a scale total) for the indicators, the variables with negative loadings must be reversed in magnitude. This ensures that the negative indicators do not offset the positive indicators.

[b] Indicator measurement error can be calculated as $1 - (\text{Standardized loading})^2$ or the diagonal of the measurement error correlation matrix (theta delta matrix) in the LISREL output.

a value of .809, again exceeds the recommended level substantially. The lower level of variance extracted for construct 1 indicates that more than half of the variance for the specified indicators is not accounted for by the construct. This finding may lead the researcher to explore additional loadings for these indicators on the other construct if theoretically justified.

Another coefficient estimated in the measurement model is the correlation between the two constructs. The t value for the obtained correlation of $-.202$ is 1.826, which falls below the critical value for the .05 significance level, although only slightly. This indicates that there may be weak support at best (only at the .10 significance level) for believing that the constructs are correlated.

Summary

The overall model goodness-of-fit results and the measurement model assessments lend substantial support for confirmation of the proposed two-factor model. Although acceptable results were achieved with this model, the next step explores possible modifications that may improve on the model results if theoretically justified.

Step Seven: Interpreting and Modifying the Model

Possible modifications to the proposed model may be indicated through examination of the normalized residuals and the modification indices. However, in any instance the proposed modifications must first have theoretical justification before a respecified model can be tested. In this example, the only possible modifications would be the possibility of a variable acting as an indicator for both constructs.

The normalized residuals are shown in Table 11.8, and examination reveals only one value exceeding 2.58. Thus only one correlation from the original input matrix has a statistically significant residual. This falls within the acceptable range of one in 20 residuals exceeding 2.58 strictly by chance. Table 11.8 also contains the modification indices for the measurement model. Two of the indicators (delivery speed and price level) have modification indicators above the suggested level (3.84) for possible model respecification. In this instance, the modification indices indicate that these variables might be indicators on the second construct, as well as the first (i.e., multiple loadings).

Theoretical support for such a structure cannot be found; thus the model is not respecified. If model respecification is based *only* on the values of the modification indices, the researcher is capitalizing on the uniqueness of these particular data, and the result will most probably be an atheoretical, but statistically significant, model that has little generalizability and limited use in testing causal relationships.

Summary

The confirmatory factor analysis provides adequate support for the proposed model based on the common factor analyses performed in Chapter 7. However, because the model has a data inconsistency (the Heywood case for X_4, manufacturer's image), the researcher should gather additional data and test the model on a new correlation matrix to ensure generalizability across multiple samples.

TABLE 11.8 Normalized Residuals and Modification Indices for Confirmatory Factor Analysis

Normalized Residuals

Variables		X_1	X_2	X_3	X_4	X_6	X_7
X_1	Delivery speed	.000					
X_2	Price level	1.826	.000				
X_3	Price flexibility	1.535	−.595	.000			
X_4	Manufacturer's image	2.707	2.157	.508	.000		
X_6	Sales force's image	2.234	1.028	1.067	.000	.000	
X_7	Product quality	−1.091	.523	1.959	1.011	.868	.000

Modification Indices

		Construct Loadings	
		Strategy	Image
X_1	Delivery speed	.000	7.359
X_2	Price level	.000	4.634
X_3	Price flexibility	.000	.269
X_4	Manufacturer's image	.321	.000
X_6	Sales force's image	.321	.000
X_7	Product quality	.000	1.029

		Indicator Measurement Error					
		X_1	X_2	X_3	X_4	X_6	X_7
X_1	Delivery speed	.000					
X_2	Price level	3.333	.000				
X_3	Price flexibility	2.357	.355	.000			
X_4	Manufacturer's image	1.712	2.182	.216	.321		
X_6	Sales force's image	.182	.046	.927	.321	.000	
X_7	Product quality	1.191	.274	3.837	.038	1.009	.000

Estimating a Path Model with Structural Equation Modeling

The next example of structural equation modeling examines a set of two causal relationships, relating customers' perceptions of HATCO to their usage level of HATCO products and their satisfaction with HATCO. The structural model allows for an understanding not only of the relative importance of the various perceptions of HATCO in determining purchases but also of possible sources of customers' satisfaction with HATCO.

To avoid duplication of the confirmatory factor analysis just examined, an additional survey was conducted among a new set of HATCO customers. One hundred thirty-six customers were asked to rate HATCO on a number of possible purchase determinants. This survey is similar to the original database introduced in Chapter 1 in that respondents were asked to rate HATCO on a 0 to 10 scale (0 = very poor and 10 = excellent), indicating how well HATCO performs on that attribute. The causal relationships thus relate to determining what elements of HATCO's performance determine the level of customers' purchases and their satisfaction. It differs in the attributes and the types of relationships examined.

As in the prior example, the discussion follows the seven-step process, illustrating the issues and interpretation at each stage. The reader is encouraged to refer to our earlier discussions for specific details on issues raised in the example.

Step One: Developing a Theoretically Based Model

The purpose of this analysis is to better understand how customers' perceptions of HATCO affect their behavior and attitudes. As already described, 136 customers were asked to provide evaluations on 13 attributes across the entire spectrum of ways in which they might interact with and evaluate HATCO. The attributes were derived by a two-stage process. First, all regional managers were participants in a series of focus groups in which these issues, among others, were discussed. From these discussions, a series of 27 attributes was identified as possible evaluative items. This set of 27 items was then presented to the district managers, who rated each item in terms of their perceptions of its importance in customer purchase decisions and satisfaction. Concurrently, a sample of customers of other firms in the industry were also asked for their perceptions as to the importance of each attribute. Analysis of the set of 27 items for both samples resulted in the final set of 13 items composing three general areas of evaluation. The items and evaluation dimensions are shown in Table 11.9.

When the dimensions of evaluation that will act as the independent variables in this analysis have been established, the final step is to determine their relationships to purchase decisions and satisfaction. Because satisfaction is based on a customer's evaluation of past experiences with HATCO, the appropriate causal

TABLE 11.9 Evaluative Dimensions and Firm
Attributes for the New HATCO Survey

Evaluative Dimension	Firm Attribute
Firm and product factors	Product quality
	Invoice accuracy
	Technical support
	Introduction of new products
	Reliable delivery
	Customer service
Price-based factors	Product value
	Low-price supplier
	Negotiation position (policies)
Buying relationship	Mutuality of interests
	Integrity/honesty
	Flexibility
	Problem resolution

relationship is for current satisfaction levels to be predicted by level of purchases from HATCO. Although a causal relationship can be posed depicting purchases arising *from* satisfaction, this method would require a greater degree of control over longitudinal aspects of data collection necessary to establish the temporal order. However, we know that purchase level reflects past purchase behavior, so that it can be a valid predictor of current satisfaction. The researcher must always examine each proposed relationship from a theoretical perspective to ensure that the results are conceptually valid. In this case, the purchase-to-satisfaction link could be easily reversed and estimates of the relationship made, because the correlation works equally well for both relationships. But the results could be potentially misleading because there is a conceptual flaw in the reversed relationship if estimated with these data.

Step Two: Constructing a Path Diagram of the Causal Relationships

Having developed a set of causal relationships, we next portray the relationships in a path diagram (see Figure 11.5). As we see, the three evaluative dimensions act as the exogenous variables, each related to the product usage level, which is one of the two endogenous variables. Product usage is then posited to be the sole predictor of satisfaction. The evaluative dimensions could also be sources of satisfaction; however, these causal relationships are not proposed initially, but instead will be explored through the testing of alternative model specifications (i.e., competing models). The path diagram also indicates that the three exogenous dimensions are all proposed to be intercorrelated. Although the evaluative dimensions are proposed to be distinct, it is recognized that some perceptions are shared, and thus there are correlations among the constructs.

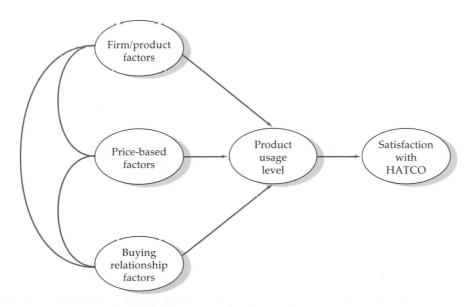

FIGURE 11.5 Path diagram for structural equations model.

Step Three: Converting the Path Diagram into Structural Equations and Specifying the Measurement Model

The path diagram provides the basis for specification of the structural equations and the proposed correlations (1) between exogenous constructs and (2) between structural equations. The specification of a structural equation for each endogenous construct (usage level and satisfaction) must specify the relationships to both the exogenous constructs and other endogenous constructs as well. The four coefficients to be estimated in the structural equations can be expressed as follows:

Constructs

Endogenous Variable	Exogenous			Endogenous	
	Firm/Product Factors	Price-Based Factors	Buying Relationship	Usage Level	Satisfaction
Usage level	b_1	b_2	b_3		
Satisfaction				b_4	

Note that the error terms for each equation were omitted, but they must be included when the model is estimated. In addition to the structural equations, the correlations between each pair of exogenous constructs must also be estimated, although no correlations are proposed between the structural equations.

The measurement model specifies the correspondence of indicators to constructs. For the two endogenous variables, only single indicators are available. Therefore, the researcher has to specify their reliability. Though estimates of the level of measurement error could be made, for illustration purposes the measurement error will be set at zero (indicating perfect measurement of both constructs). The relaxation of this assumption then provides the basis for an alternative model. (This alternative model will not be tested in our discussion, but the reader is encouraged to undertake the analysis and examine the effects of varying levels of measurements error.)

The exogenous constructs are measured by 13 indicators reflecting the dimensions described previously. We will not make a formal specification of the measurement model for the exogenous constructs, but the reader can refer to the prior example for more specific detail on this process if needed. Also, there are no instances for which indicators should be correlated; thus no measurement error correlations are allowed, either in the initial model or as a basis for model modifications.

Step Four: Choosing Input Matrix Type and Estimating Model

With the model completely specified, the next step is to select the type of input matrix (covariances or correlations) to be used for model estimation. When testing a series of causal relationships, covariances are the preferred input matrix type. In this example, correlations are used for both practical and theoretical reasons. From a practical perspective, correlations are more easily interpreted, and diagnosis of the results is more direct. From a theoretical perspective, the primary purpose of the analysis is to examine the *pattern* of relationships among the exogenous and endogenous constructs. For this purpose the correlation matrix is an acceptable input matrix. Moreover, the model most likely suffers from specification error, which is the omission of relevant variables from the model. Both theory and experience would suggest additional predictors of a customer's satisfaction with HATCO. Thus, given this limitation, the researcher should only draw conclusions about the patterns of relationships rather than about the predictive ability of the constructs [18]. For the purposes stated, the example will employ the correlation matrix (see Table 11.10). The sample size of 136 customers falls within the acceptable limits for use of structural equation modeling.

As in the prior example, the structural equation model was estimated with the LISREL program [48]. As noted in our discussion of confirmatory factor analysis, each scale must be made scale invariant through one of two methods. In this analysis we employ the method not used earlier (i.e., we set the loading of one variable per construct to a value of 1.0) to illustrate the procedure. The structural and measurement model coefficients are unaffected by the selected approach. Thus, the variables with fixed loadings of 1.0 are X_1, X_7, and X_{10}.

Step Five: Assessing the Identification of the Structural Model

Before examining the results, the researcher must be assured that the model is identified. The LISREL program assesses the identification of the model and highlights almost all problems. For this model, no identification problems were indicated. As an example of what might cause an identification problem, the specification of a reciprocal (two-way) relationship between usage level and satisfaction would in most instances result in an identification problem with at least one of the structural equations. While reciprocal relationships can be estimated, they require model constraints in other aspects of the structural equations. Interested readers may refer to [18, 38].

Step Six: Evaluating Goodness-of-Fit Criteria

Being assured that the model is correctly specified and the estimation process is not constrained by identification problems, the researcher proceeds to evaluate the specific results for the proposed model. If the assumptions underlying structural equation modeling are met, the estimated coefficients are evaluated along with the overall model fit. Each of these steps is discussed in the following sections.

Assumptions

Though the design of the study assured independent and random responses, the 15 variables still must be assessed for their distributional characteristics, particularly normality and kurtosis. The procedures described in Chapter 2 for assessing these assumptions were employed. No variable was found to have significant departure from normality or pronounced kurtosis.

TABLE 11.10 Correlation Matrix for the Structural Equation Model

	Variables	Y_1	Y_2	X_1	X_2	X_3	X_4	X_5	X_6	X_7	X_8	X_9	X_{10}	X_{11}	X_{12}	X_{13}
Y_1	Usage	1.000														
Y_2	Satisfaction	0.411	1.000													
X_1	Product quality	0.288	0.168	1.000												
X_2	Invoice accuracy	0.359	0.159	0.785	1.000											
X_3	Technical support	0.268	0.141	0.676	0.637	1.000										
X_4	Introduction of new products	0.212	0.081	0.581	0.622	0.627	1.000									
X_5	Reliable delivery	0.250	0.060	0.632	0.644	0.538	0.699	1.000								
X_6	Customer service	0.305	0.127	0.690	0.667	0.551	0.625	0.692	1.000							
X_7	Product value	0.328	0.133	0.293	0.263	0.336	0.290	0.207	0.174	1.000						
X_8	Low-price supplier	0.268	0.046	0.184	0.124	0.230	0.260	0.110	0.254	0.301	1.000					
X_9	Negotiation position	0.142	0.104	0.289	0.249	0.299	0.219	0.155	0.299	0.307	0.676	1.000				
X_{10}	Mutuality of interests	0.329	0.063	0.327	0.292	0.112	0.292	0.326	0.253	0.179	0.114	0.173	1.000			
X_{11}	Integrity/honesty	0.519	0.077	0.413	0.385	0.265	0.346	0.259	0.262	0.347	0.225	0.205	0.411	1.000		
X_{12}	Flexibility	0.510	0.090	0.330	0.272	0.192	0.151	0.210	0.225	0.348	0.259	0.162	0.555	0.532	1.000	
X_{13}	Problem resolution	0.341	0.173	0.154	0.173	0.182	0.153	0.107	0.102	0.272	0.052	0.093	0.202	0.411	0.425	1.000

Offending Estimates

Having met the assumptions of structural equation modeling, the results are first examined for nonsensical or theoretically inconsistent estimates. The three most common offending estimates are negative error variances, standardized coefficients exceeding or very close to 1.0, or very large standard error. Examination of the standardized results in Table 11.11 reveals no instances of any of these problems.

Overall Model Fit

Before evaluating the structural or measurement models, the researcher must assess the overall fit of the model to ensure that it is an adequate representation of the entire set of causal relationships. Each of the three types of goodness-of-fit measures will be used.

Absolute Fit Measures Three measures of the most basic measures of absolute fit are the likelihood-ratio Chi-square (χ^2), the goodness-of-fit index (GFI), and the root mean square residual (RMSR) (see Table 11.12). The Chi-square value of 178.714 with 85 degrees of freedom is statistically significant at the .000 significance level. Because the sensitivity of this measure is not overly affected by the sample size of 136, the researcher may conclude that significant differences exist.

TABLE 11.11 Structural Equation Model Results: Standardized Parameter Estimates

Structural Model

Structural Equation Coefficients (t Values in Parentheses)

Endogenous Constructs		Endogenous Constructs		Exogenous Constructs			
		Usage Level	Satisfaction	Firm/ Product Factors	Price-Based Factors	Buying Relationship	Structural Equation Fit (R^2)
Usage Level	=	.000	.000	.056 (.621)	.038 (.428)	.615 (4.987)	.433
Satisfaction	=	.411 (5.241)	.000	.000	.000	.000	.169

Correlations Among the Exogenous Constructs (t Values in Parentheses)

Exogenous Constructs	Exogenous Constructs		
	Firm/Product Factors	Price-Based Factors	Buying Relationship
Firm/Product Factors	1.000		
Price-Based Factors	.355 (2.724)	1.000	
Buying Relationship	.465 (3.746)	.353 (2.489)	1.000

TABLE 11.11 *Continued*

Measurement Model
Construct Loadings (t Values in Parentheses)

		Exogenous Constructs		
	Indicators	*Firm/Product Factors*	*Price-Based Factors*	*Buying Relationship Factors*
X_1	Product quality	.863 (.000)[a]	.000	.000
X_2	Invoice accuracy	.857 (12.855)	.000	.000
X_3	Technical support	.747 (10.297)	.000	.000
X_4	Introduction of new products	.759 (10.550)	.000	.000
X_5	Reliable delivery	.781 (11.020)	.000	.000
X_6	Customer service	.803 (11.524)	.000	.000
X_7	Product value	.000	.404 (.000)[a]	.000
X_8	Low-price supplier	.000	.811 (4.181)	.000
X_9	Negotiation position	.000	.821 (4.169)	.000
X_{10}	Mutuality	.000	.000	.603 (.000)[a]
X_{11}	Integrity	.000	.000	.729 (6.216)
X_{12}	Flexibility	.000	.000	.786 (6.447)
X_{13}	Problem resolution	.000	.000	.506 (4.786)

[a] Values were not calculated, because loading was set to 1.0 to fix construct variance.

However, we must also note that the χ^2 test becomes more sensitive as the number of indicators rises. With this in mind, we will examine a number of other measures. The GFI value of .865 is at a marginal acceptance level, as is the RMSR value is 0.76. The RMSR must be evaluated in light of the input correlation matrix, and in this context the residual value is relatively high.

As a complement to these basic measures, the analyst can also examine other absolute fit measures. The root mean square error of approximation has a value of .0904, which falls just outside the acceptable range of .08 or less. Three other measures—the noncentrality index, the scaled noncentrality index, and the expected cross-validation index—are all used in comparisons among alternative models. There is no established range of acceptable values for these measures. They are examined in the next section when competing models are analyzed.

TABLE 11.12 Goodness-of-Fit Measures for the Structural Equation Model

LISREL-Provided Measures

Absolute Fit Measures

Chi-square (χ^2) of estimated model:	178.71[a]
Degrees of freedom:	85
Significance level:	.000[a]
Noncentrality parameter (NCP)	93.714
Goodness-of-fit index (GFI):	.865[a]
Root mean square residual (RMSR):	.076[a]
Root mean square error of approximation (RMSEA):	.0904
P-Value of close fit (RMSEA < .05)	.000
Expected cross-validation index (ECVI)	
EVCI for estimated model	1.842
EVCI for saturated model	1.778
EVCI for independence model	7.927

Incremental Fit Measures

Chi-square (χ^2) of null or independence model:	1040.194
Degrees of freedom:	105
Adjusted goodness-of-fit index (AGFI):	.810[a]
Tucker-Lewis (TLI) or Non Normed fit index (NNFI):	.876
Normed fit index (NFI):	.828

Parsimonious Fit Measures

Parsimony normed fit index (PNFI):	.670
Parsimony goodness-of-fit index:	.613
Aikaike information criterion (AIC):	
Estimated model	248.7
Saturated model	240.0
Independence model	1070.2
Comparative fit index (CFI):	.900
Incremental fit index (IFI):	.902
Relative fit index (RFI):	.788
Critical N (CN):	90
Sample size: 136 respondents	

Calculated Measures of Overall Model Goodness-of-Fit

Normed Chi-square	$\dfrac{178.71}{85} = 2.10$
Scaled Noncentrality parameter (SNCP)	$\dfrac{93.714}{136} = .689$

---

[a] Measures provided directly by all versions of LISREL. Other measures provided only by version 8 of LISREL and must be calculated separately for earlier versions.

All of the absolute fit measures indicate that the model is marginally acceptable at best. This should not halt further examination of the results, however, unless the overall model fit was deemed so poor that both the structural and measurement models would be invalid. Moreover, the other types of fit measures will provide different perspectives on the acceptability of the model fit.

Incremental Fit Measures   In addition to the overall measures of fit, a model can be evaluated in comparison to a baseline or null model. In this instance, the null model is a single-factor model with no measurement error. The null model has a

χ^2 value of 1,040.19 with 105 degrees of freedom. Although we gain a substantial reduction in the Chi-square value owing to the estimated coefficients in the model, all incremental fit measures provide only marginal support. The AGFI, Tucker-Lewis index, and normed fix index all fall slightly below the desired threshold of .90. Although the .90 threshold has no statistical basis, practical experience and research have demonstrated its usefulness in distinguishing between acceptable and unacceptable models. However, all the incremental fit measures exceeded .80 and the true test comes with the comparison of the proposed model against alternative or competing models.

Parsimonious Fit Indices   This final type of measure provides a basis for comparison between models of differing complexity and objectives. One applicable measure for evaluating a single model is the normed Chi-square measure. With a computed value of 2.10, it falls within some threshold limits for this measure but exceeds other limits. Thus, again, only marginal support is provided. Three other parsimonious fit measures are available (the parsimonious normed fit index, the parsimonious goodness-of-fit index, and the Aikaike information criteria), but they are suited only for intermodel comparisons and will be utilized in a later section.

A review of the three types of overall measures of fit reveals a consistent pattern of marginal support for the overall model as proposed. As noted earlier, the truer test for the overall model is a comparison to a series of alternative models, which is carried out in step 7.

Measurement Model Fit

Though only marginal support was found for the overall model, it was deemed sufficient to proceed and assess the measurement model fit as well. The first step is an examination of the loadings, particularly focusing on any nonsignificant loading. Referring to Table 11.11, one sees that all the indicators were statistically significant for the proposed constructs. Because no indicators had loadings so low that they should be deleted and the model reestimated, the reliability and variance extracted measures need to be computed.

Table 11.13 contains the computations for both the reliability and the variance extracted measures. In terms of reliability, all three exogenous constructs exceed the suggested level of .70. In terms of variance extracted, the firm/product-factor construct exceeds the threshold value of .50. The price-based construct only slightly misses the .50 guideline, and the buying-relationship construct has a somewhat lower value (.442). Thus, for all three constructs, the indicators are sufficient in terms of how the measurement model is now specified.

Structural Model Fit

Having assessed the overall model and aspects of the measurement model, the analyst is now prepared to examine the estimated coefficients themselves for both practical and theoretical implications. Review of Table 11.11 reveals that both the structural equations contain statistically significant coefficients. For the causal relationship linking the three evaluative dimensions with usage level, we find that only one dimension, the buying relationship, is statistically significant. Therefore, the more personal aspects of the transaction, such a mutuality, integrity, flexibility in dealing, and problem resolution, have the only distinct and

TABLE 11.13 Reliability and Variance Extracted Estimates for Exogenous Constructs in the Structural Equation Model

Reliability

$$\text{Construct Reliability} = \frac{(\text{Sum of standardized loadings})^2}{(\text{Sum of standardized loadings})^2 + \text{Sum of indicator measurement error}}$$

Sum of Standardized Loadings
Firm/products factors = .863 + .857 + .747 + .759 + .781 + .803 = 4.81
Price-based factors = .404 + .811 + .821 = 2.036
Buying relationship = .603 + .729 + .786 + .506 = 2.624

Sum of Measurement Error[a]
Firm/product factors = .256 + .265 + .442 + .424 + .390 + .355 = 2.132
Price-based factors = .837 + .342 + .327 = 1.506
Buying relationship = .636 + .469 + .382 + .744 = 2.231

Reliability Computation

$$\text{Firm/product factors} = \frac{(4.81)^2}{(4.81)^2 + 2.132} = .915$$

$$\text{Price-based factors} = \frac{(2.036)^2}{(2.036)^2 + 1.506} = .735$$

$$\text{Buying relationship} = \frac{(2.624)^2}{(2.624)^2 + 2.231} = .755$$

Variance Extracted

$$\text{Variance Extracted} = \frac{\text{Sum of squared standardized loadings}}{\text{Sum of squared standardized loadings} + \text{Sum of indicator measurement error}}$$

Sum of Squared Standardized Loadings
Firm/product factors = $.863^2 + .857^2 + .747^2 + .759^2 + .781^2 + .803^2 = 3.868$
Price-based factors = $.404^2 + .811^2 + .821^2 = 1.495$
Buying relationship = $.603^2 + .729^2 + .786^2 + .506^2 = 1.769$

Variance Extracted Computation

$$\text{Firm/product factors} = \frac{3.868}{3.868 + 2.132} = .644$$

$$\text{Price-based factors} = \frac{1.495}{1.495 + 1.506} = .498$$

$$\text{Buying relationship} = \frac{1.769}{1.769 + 2.231} = .442$$

---

[a]Indicator measurement error can be calculated as $1 - (\text{standardized loading})^2$ or the diagonal of the measurement error correlation matrix (theta delta matrix) in the LISREL output.

substantive impact on increasing product usage levels. Thus, though HATCO must not ignore the other aspects of the business, emphasis should be placed on the maintenance of existing buying relationships and development of new ones focusing on these features. Moreover, the combined effect of these three factors achieves an R^2 value of .433, or 43.3 percent of the variance in usage levels. For the structural equation predicting satisfaction, the other endogenous construct (usage level) was statistically significant. Thus a significant causal relationship has been identified and may now be used as a basis for the formulation of alternative or competing models. The low R^2 value (.169) is to be expected because it is based only on the correlation between usage level and satisfaction.

Also of interest are the correlations between the evaluative dimensions (exogenous constructs). In this example, each construct is significantly correlated with the other exogenous constructs. This correlation reveals that, whereas the buying relationship is fundamental to increases in usage levels, all three constructs are interwoven, and the firm must not focus exclusively on any single dimension.

As a final means of examining the results, the correlations between estimated constructs (endogenous variables) should be reviewed and high values should be noted as an indication of an unacceptable level of intercorrelated constructs. For the endogenous constructs of this study, the correlation of their estimated values is only .411, too low to indicate troublesome intercorrelation.

Competing Models

The final approach to model assessment is to compare the proposed model with a series of competing models, which act as alternative explanations to the proposed model. In this way, the researcher can determine whether the proposed model, regardless of overall fit (within reasonable limits), is acceptable, because no other similarly formulated model can achieve a higher level of fit. This step is particularly important when the Chi-square statistic indicates no significant differences in overall model fit, because there may always be a better-fitting model, even in the case of nonsignificant differences. For a systematic approach to specification of competing models, see [3].

For the purpose of this example, two alternative models are proposed (see Figure 11.6 for these models expressed as path diagrams). The first model (COMPMOD1) adds the three exogenous constructs as predictors of satisfaction. In the second model (COMPMOD2), only buying relationship is added as a predictor of satisfaction because it was already a significant predictor of usage levels. As a means of comparison, a set of goodness-of-fit measures will be calculated for each model and then compared to determine which of the three is the most parsimonious.

Table 11.14 compares the three models on all three types of fit measures. For the absolute fit measures, COMPMOD1 has the lowest χ^2 value, but this model also has the largest number of estimated parameters (and thus lowest degrees of freedom). COMPMOD1 also is lowest on the GFI and RMSR measures. COMPMOD2 excels on the measures that attempt to minimize sample size and sample-specific estimates of model fit. The NCP, SNCP, RMSEA, and ECVI all have their lowest values with COMPMOD2. The estimated model, while not achieving the best fit on any of these measures, is still very close on all measures, making it a viable alternative for acceptance along with the other two competing models.

Two of the incremental fit measures favor COMPMOD2 (AGFI and NNFI), while COMPMOD1 has the best performance on the NFI measure. Again, the estimated model is quite close with no substantive differences. The parsimonious fit measures are the final set to be considered. Here, the estimated model has the best fit as measured by the PNFI and PGFI, while COMPMOD2 excels on the normal Chi-square and Akaike information criterion.

The results across all three types of measures show mixed results, sometimes favoring the estimated model or one of the competing models. If the focus is limited to the parsimonious fit indices, which account for the model parsimony, the results are still split between the original estimated model and the second

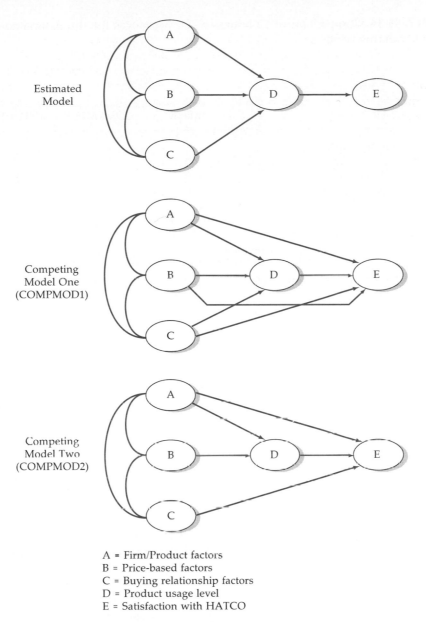

A = Firm/Product factors
B = Price-based factors
C = Buying relationship factors
D = Product usage level
E = Satisfaction with HATCO

FIGURE 11.6 Path diagrams of estimated model and competing models (COMPMOD1 and COMPMOD2).

competing model (COMPMOD2). Thus, although the fit of the proposed model did not exceed the recommended guidelines in many instances, we accept the proposed model with reservations until additional constructs can be added, measures refined, or causal relationships respecified. The selection of one of the competing models must be made on both theoretical and empirical bases, just as was done in formulating the original theoretical model that was the basis for the estimated model. This example illustrates the value of testing competing models for all structural equation models to assure the researcher that the proposed or revised model is truly the "best" model available.

TABLE 11.14 Comparison of Goodness-of-Fit Measures for the Estimated and Competing Models

Goodness-of-Fit Measure	Estimated Model	Competing Models	
		COMPMOD1	COMPMOD2
Absolute Fit Measures			
Likelihood-ration Chi-square (χ^2)	178.714	174.45	175.397
degrees of freedom	85	82	84
Noncentrality parameter (NCP)	93.714	92.450	91.397
Scaled Noncentrality parameter (SNCP)	.689	.679	.672
Goodness-of-fit index (GFI)	.865	.867	.866
Root mean square residual (RMSR)	.0759	.0741	.0752
Root mean square error of approximation (RMSEA)	.0904	.0914	.0898
Expected Crocc-validation index (ECVI)	1.842	1.855	1.833
Incremental Fit Measures			
Adjusted goodness-of-fit index (AGFI)	.801	.805	.809
Tucker-Lewis index (TLI) or (NNFI)	.876	.873	.878
Normed fit index (NFI)	.828	.832	.831
Parsimonious Fit Measures			
Parsimonious normed fit index (PNFI)	.670	.650	.665
Parsimonious goodness-of-fit index (PGFI)	.613	.592	.606
Normed Chi-square	2.103	2.127	2.088
Akaike information criteria	248.714	250.450	247.397

Step Seven: Interpreting and Modifying the Model

The final step examines some diagnostic elements (residuals and modifiction indices) that may indicate potentially significant model modifications. As noted earlier, we are assuming that all model modifications will have theoretical support before being implemented.

Table 11.15 contains the normalized residuals from the proposed model when estimated by LISREL. Identifying residuals exceeding the threshold value of ±2.58 finds 11 potentially significant residuals. This exceeds the guideline of having only 5 percent of the normalized residuals exceeding the threshold value, which in this case is six residuals (5% * 120). We can note only one consistent pattern among the residuals. The variable X_7 is connected with a majority of the residuals exceeding 2.58. This observation indicates that perhaps the elimination of only this single indicator will create a substantial improvement in fit. If this modification is made, the χ^2 value decreases to 146.94 with 71 degrees of freedom. Using the PFI index, we note that the computed value for this reduced indicator model is .557, which is not greater than the original proposed model. Thus, although model fit will improve, model parsimony will not. If we exclude X_7, the normalized residuals marginally meet the guideline. The analyst's decision regarding the exclusion of X_7 should be based on theoretical grounds.

TABLE 11.15 Normalized Residuals for the Structural Equation Model

	Y_1	Y_2	X_1	X_2	X_3	X_4	X_5	X_6	X_7	X_8	X_9	X_{10}	X_{11}	X_{12}	X_{13}
Y_1 Usage	.000														
Y_2 Satisfaction	.000	.000													
X_1 Product quality	-.542	.546	.000												
X_2 Invoice accuracy	1.526	.436	3.138	.000											
X_3 Technical support	.052	.409	1.430	-.154	.000										
X_4 Introduction of new products	-1.196	-.384	-3.423	-1.292	1.839	.000									
X_5 Reliable delivery	-.600	-.693	-2.039	-1.233	-1.495	3.569	.000								
X_6 Customer service	.437	.124	-.012	-1.117	-1.724	.559	2.499	.000							
X_7 Product value	2.972	1.040	2.293	1.887	2.985	2.363	1.243	.780	.000						
X_8 Low price supplier	1.402	-.593	-1.454	-2.736	.255	.757	-2.176	.438	-1.103	.000					
X_9 Negotiation position	-2.788	.147	.370	-.013	1.469	-.034	-1.383	1.287	-1.058	3.810	.000				
X_{10} Mutuality of interests	-1.582	-1.347	1.396	.851	-1.468	1.199	1.641	.427	1.174	-.941	-.035	.000			
X_{11} Integrity/Honesty	1.385	-1.704	2.357	1.830	.188	1.492	-.096	-.181	3.186	.308	-.128	-.847	.000		
X_{12} Flexibility	-.189	-1.794	.310	-.930	-1.450	-2.287	-1.416	-1.323	3.150	.735	-1.478	2.956	-2.310	.000	
X_{13} Problem resolution	.214	.485	-.742	-.436	.085	-.370	-1.105	-1.271	2.467	-1.343	-.775	-1.977	1.063	.836	.000

Note: Underlined values are residuals exceeding the suggested guideline of ± 2.58.

Another indication of possible model respecifications is the modification index. Table 11.16 contains the modification indices obtained during estimation of the proposed model. Looking for values exceeding 3.84 reveals that only in the measurement model are modifications suggested. We should note that the two highest modification indices are for X_7, the variable that the residual analysis suggested dropping. Besides X_7, double loadings are suggested for X_8 and X_9, although the anticipated reduction in Chi-square is minimal and not appropriate for model respecification.

Review of the Structural Equation Modeling Process

The seven-step process has empirically investigated a series of causal relationships with interrelated dependent (endogenous) constructs. The estimated model, while not achieving the recommended levels of fit, may represent the best available model until further research identifies improvements in theoretical relationships or measurement of the constructs. Although several model modifications improve model fit, they do not increase model parsimony, and little theoretical support can be found for these respecifications of theory. Thus, in situations such as these, the researcher should be very cautious about the type and the extent of model respecifications undertaken.

Summary

Structural equation modeling is a technique combining elements of both multiple regression and factor analysis that enables the researcher not only to assess quite complex interrelated dependence relationships but also to incorporate the effects of measurement error on the structural coefficients at the same time. Upon completion of this chapter, both the academic researcher and the industry analyst should be able to recognize the benefits afforded by structural equation modeling. Although SEM is useful in many instances, it should be used only in a confirmatory mode, leaving exploratory analyses to other multivariate techniques.

Questions

1. What are the similarities between structural equation modeling and the multivariate techniques discussed in earlier chapters?
2. Why should a researcher assess measurement error and incorporate it into the analysis?
3. Using the definition of a theory presented in this text, describe "theories" that might be of interest to academic and industry analysts.
4. Specify a set of causal relationships and then represent them in a path diagram.
5. What are the criteria by which the researcher should decide whether a model has achieved an acceptable level of fit?
6. Using the path diagram for question 4 (or one prepared for this question), specify at least two alternative models with the supporting theoretical rationale.

TABLE 11.16 Modification Indices for the Structural Equation Model

Measurement Model
Exogenous Construct Indicators

		Exogenous Constructs		
	Variables	*Firm/ Product Factors*	*Price- Based factors*	*Buying Relationship*
X_1	Product quality	.000	.040	2.224
X_2	Invoice accuracy	.000	1.574	.690
X_3	Technical support	.000	2.009	.872
X_4	Introduction of new products	.000	.543	.276
X_5	Reliable delivery	.000	3.565	.570
X_6	Customer service	.000	1.294	.589
X_7	Product value	6.242	.000	15.289
X_8	Low-price supplier	3.907	.000	.000
X_9	Negotiation position	.489	.000	3.916
X_{10}	Mutuality	1.094	.107	.000
X_{11}	Integrity	2.959	.574	.000
X_{12}	Flexibility	3.484	.003	.000
X_{13}	Problem resolution	.778	.766	.000

Structural Model
Structural Equations

	Endogenous Constructs		Exogenous Constructs		
Endogenous Construct	*Usage Level*	*Satisfaction*	*Firm/ Product Factors*	*Price- Based Factors*	*Buying Relationship*
Usage	.000	2.729	.000	.000	.000
Satisfaction	.000	.000	.016	.026	3.249

Correlations Between Exogenous Constructs

	Exogenous Constructs		
	Firm/ Product Factors	*Price- Based Factors*	*Buying Relationship*
Firm/product factors	.000		
Price-based factors	.000	.000	
Buying relationship	.000	.000	.000

Correlations Between Endogenous Constructs (Structural Equations)

	Usage Level	Satisfaction
Usage level	.000	
Satisfaction	2.729	.000

A Mathematical Representation in LISREL Notation

The discussion of structural equation modeling to this point has relied on terminology similar to that used in our discussion of multiple regression and other dependence relationships. However, a researcher attempting either (1) to read the program manuals, articles, and texts on this topic or (2) to publish in academic journals using this technique will need a more formal introduction to the mathematical notation used in structural equation modeling. Because of its widespread application, LISREL [45, 46, 48] has become the standard for notation. This appendix first presents the complete formulation of the set of causal relationships and measurement relationships in LISREL notation. Then, an example will progress from a path diagram through the structural equations into the appropriate LISREL notation for model specification.

LISREL Notation

The entire LISREL model can be expressed in terms of eight matrices, two defining the structural equations, two defining the correspondence of indicators and constructs, one for the correlation of exogenous constructs, and finally three detailing the error for structural equations and measurement of exogenous and endogenous variables. Table 11A.1 lists each of the matrices, a brief description, and notation for the matrix and its elements.

The subscripts in Table 11A.1 indicate the order (size) of each matrix:

m = Number of exogenous constructs
n = Number of endogenous constructs
p = Number of exogenous construct indicators
q = Number of endogenous construct indicators

TABLE 11A.1 Matrices of the LISREL Model

			Notation	
Matrix	Description		Matrix	Element
Structural Model:				
Beta	Relationships of endogenous to endogenous constructs		B	β_{nn}
Gamma	Relationships of exogenous to endogenous constructs		Γ	γ_{nm}
Phi	Correlation among exogenous constructs		Φ	ϕ_{mm}
Psi	Correlation of structural equations or endogenous constructs		Ψ	ψ_n
Measurement Model:				
Lambda-X	Correspondence of exogenous indicators		Λ_X	λ^x_{pm}
Lambda-Y	Correspondence of endogenous indicators		Λ_Y	λ^y_{qn}
Theta-delta	Matrix of prediction errors for indicators of exogenous constructs		Θ_δ	δ_{pp}
Theta-epsilon	Matrix of prediction errors for indicators of endogenous constructs		Θ_ϵ	ϵ_{qq}

Finally, the constructs and indicators also have specific notation:

$$\xi = \text{Exogenous constructs}$$
$$\eta = \text{Endogenous construct}$$
$$X = \text{Indicator of exogenous construct}$$
$$Y = \text{Indicator of endogenous construct}$$

These matrices are used to form the basic equations for both structural and measurement models. The equation for the structural model is

$$\eta = \Gamma\xi + B\eta + \zeta$$

And the equations for the measurement model are as follows:

Exogenous Constructs

$$X = \Lambda_X\xi + \delta$$

Endogenous Constructs

$$Y = \Lambda_Y\eta + \epsilon$$

A more straightforward presentation, however, is through the actual equations (structural and measurement). For purposes of illustration, there are two endogenous constructs ($n = 2$) and three exogenous constructs ($m = 3$), with four indicators of both endogenous and exogenous constructs ($Y = 4$, $X = 4$). *No* causal relationships are expressed; rather, the complete equations are shown. An example of specifying a path model in LISREL terms is addressed in the example that follows.

Structural Model

Endogenous Construct		Exogenous Construct		Endogenous Construct		Error
η_1	=	$\gamma_{11}\xi_1 + \gamma_{12}\xi_2 + \gamma_{13}\xi_3$	+	$\beta_{12}\eta_2$	+	ζ_1
η_2	=	$\gamma_{21}\xi_1 + \gamma_{22}\xi_2 + \gamma_{23}\xi_3$	+	$\beta_{21}\eta_1$	+	ζ_2

Measurement Models

Exogenous Indicator		Exogenous Construct		Error
X_1	=	$\lambda_{11}^x\xi_1 + \lambda_{12}^x\xi_2 + \lambda_{13}^x\xi_3$	+	δ_1
X_2	=	$\lambda_{21}^x\xi_1 + \lambda_{22}^x\xi_2 + \lambda_{23}^x\xi_3$	+	δ_2
X_3	=	$\lambda_{31}^x\xi_1 + \lambda_{32}^x\xi_2 + \lambda_{33}^x\xi_3$	+	δ_3
X_4	=	$\lambda_{41}^x\xi_1 + \lambda_{42}^x\xi_2 + \lambda_{43}^x\xi_3$	+	δ_4

Endogenous Indicator		Endogenous Constructs		Error
Y_1	=	$\lambda_{11}^y\eta_1 + \lambda_{12}^y\eta_2$	+	ϵ_1
Y_2	=	$\lambda_{21}^y\eta_1 + \lambda_{22}^y\eta_2$	+	ϵ_2
Y_3	=	$\lambda_{31}^y\eta_1 + \lambda_{32}^y\eta_2$	+	ϵ_3
Y_4	=	$\lambda_{41}^y\eta_1 + \lambda_{42}^y\eta_2$	+	ϵ_4

Structural Equations Correlations

Among Exogenous Constructs (Phi ϕ)

	ξ_1	ξ_2	ξ_3
ξ_1	—		
ξ_2	ϕ_{21}	—	
ξ_3	ϕ_{31}	ϕ_{32}	—

Among Endogenous Constructs (Psi ψ)

	η_1	η_2
η_1	—	
η_2	ψ_{21}	—

Measurement Model (Indicator) Correlations

Among Exogenous Indicators (Theta-delta Θ_δ)

	X_1	X_2	X_3	X_4
X_1	—			
X_2	$\Theta_{\delta 21}$	—		
X_3	$\Theta_{\delta 31}$	$\Theta_{\delta 32}$	—	
X_4	$\Theta_{\delta 41}$	$\Theta_{\delta 42}$	$\Theta_{\delta 43}$	—

Among Endogenous Indicators (Theta-epsilon Θ_ϵ)

	Y_1	Y_2	Y_3	Y_4
Y_1	—			
Y_2	$\Theta_{\epsilon 21}$	—		
Y_3	$\Theta_{\epsilon 31}$	$\Theta_{\epsilon 32}$	—	
Y_4	$\Theta_{\epsilon 41}$	$\Theta_{\epsilon 42}$	$\Theta_{\epsilon 43}$	—

When the structural and measurement equations are actually estimated, the error terms become part of the between-construct and between-indicator matrices described. For example, the δ_p and ϵ_p values are actually estimated as the diagonal values of the theta-delta and theta-epsilon matrices, respectively. Likewise, the structural equation errors (ζ_n) are the diagonals of the Psi matrix.

From a Path Diagram to LISREL Notation

We now illustrate the complete process of moving from the path diagram to the complete LISREL notation, ready for input into any of the structural equation modeling programs. The path diagram in Figure 11A.1 describes the set of causal relationships to be examined in this example. There are three endogenous (Y, or

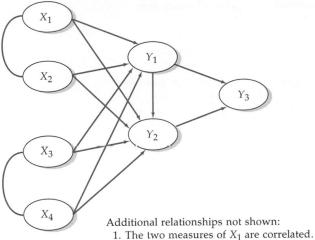

Additional relationships not shown:
1. The two measures of X_1 are correlated.
2. The equations for Y_1 and Y_3 are correlated.
3. The two measures of Y_2 are correlated.
4. There are two measures per construct.

FIGURE 11A.1 Path diagram of causal relationships.

dependent) variables related to four exogenous (X, or independent) variables. Each construct is measured by two variables (V_1 to V_{14}). Also, the measures of X_1 are correlated, as are the measures of Y_2. Moreover, correlations exist between two independent variable pairs (X_1-X_2 and X_3-X_4) and two dependent variables (Y_1-Y_3).*

Constructing Structural Equations from the Path Diagram

The first step is to translate the path diagram into a series of structural equations for each endogenous variable. The equations are as follows:

Endogenous variable	=	Exogenous variables	+	Endogenous variables	+ Error
		X_1 $\quad$ X_2 $\quad$ X_3 $\quad$ X_4		Y_1 $\quad$ Y_2 $\quad$ Y_3	
Y_1	=	$b_1X_1 + b_2X_2 + b_3X_3 + b_4X_4$	+		+ ϵ_1
Y_2	=	$b_5X_1 + b_6X_2 + b_7X_3 + b_8X_4$	+	b_9Y_1	+ ϵ_2
Y_3	=			$b_{10}Y_1 + b_{11}Y_2$	+ ϵ_3

*Do not confuse the use of the X and Y variables in the path diagram with the designation of exogenous and endogenous indicators as X and Y, respectively, in the LISREL notation. Indicators in the path diagram, which are not shown in Figure 11A.1, are the variables V_1 to V_{14}.

TABLE 11A.2 Constructs and Indicators for
the LISREL Notation

Exogenous		Endogenous	
Construct	Indicators	Construct	Indicators
X_1	V_1, V_2	Y_1	V_9, V_{10}
X_2	V_3, V_4	Y_2	V_{11}, V_{12}
X_3	V_5, V_8	Y_3	V_{13}, V_{14}
X_4	V_7, V_8		

Denoting the Correspondence of Indicators and Constructs

When the structural equations have been specified, the measurement of each construct must be defined. In this example, each construct will have two indicators. The correspondence of indicators and constructs is as shown in Table 11A.2.

Specifying the LISREL Structural and Measurement Model Equations

We are now ready to specify the set of equations for both the structural and measurement models. Only those coefficients to be estimated are included. First, we define the structural and measurement equations (see Tables 11A.3 and 11A.4).

The reader should note that the original variables (V_1 to V_{14}) must be identified as either exogenous or endogenous indicators. Variables V_1 to V_8 correspond to X_1 to X_8, respectively (the exogenous construct indicators), and V_9 to V_{14} correspond to Y_1 to Y_6 (the endogenous construct indicators). Using this notation, we present the measurement model equations in Table 11A.4.

TABLE 11A.3 Structural Model Equations

Endogenous Construct		Exogenous Constructs				Endogenous Constructs			Error
		ξ_1	ξ_2	ξ_3	ξ_4	η_1	η_2	η_3	
η_1	=	$\gamma_{11}\xi_1 +$	$\gamma_{12}\xi_2 +$	$\gamma_{13}\xi_3 +$	$\gamma_{14}\xi_4 \quad +$			$+$	ζ_1
η_2	=	$\gamma_{21}\xi_1 +$	$\gamma_{22}\xi_2 +$	$\gamma_{23}\xi_3 +$	$\gamma_{24}\xi_4 \quad +$	$\beta_{21}\eta_1$		$+$	ζ_2
η_3	=					$\beta_{31}\eta_1 +$	$\beta_{32}\eta_2$	$+$	ζ_3

TABLE 11A.4 Measurement Model Equations

Exogenous Indicator		*Exogenous Constructs*					Error
		ξ_1	ξ_2	ξ_3	ξ_4		Error
X_1	=	$\lambda_{11}^x\xi_1$				+	δ_1
X_2	=	$\lambda_{21}^x\xi_1$				+	δ_2
X_3	=		$\lambda_{32}^x\xi_2$			+	δ_3
X_4	=		$\lambda_{42}^x\xi_2$			+	δ_4
X_5	=			$\lambda_{53}^x\xi_3$		+	δ_5
X_6	=			$\lambda_{63}^x\xi_3$		+	δ_6
X_7	=				$\lambda_{74}^x\xi_4$	+	δ_7
X_8	=				$\lambda_{84}^x\xi_4$	+	δ_8

Exogenous Indicator		*Exogenous Constructs*				Error
		η_1	η_2	η_3		Error
Y_1	=	$\lambda_{11}^y\eta_1$			+	ϵ_1
Y_2	=	$\lambda_{21}^y\eta_1$			+	ϵ_2
Y_3	=		$\lambda_{32}^y\eta_2$		+	ϵ_3
Y_4	=		$\lambda_{42}^y\eta_2$		+	ϵ_4
Y_5	=			$\lambda_{53}^y\eta_3$	+	ϵ_5
Y_6	=			$\lambda_{63}^y\eta_3$	+	ϵ_6

Specifying the Structural Equation Correlations

Next, there are two correlation matrices pertaining to the structural equations. The first is the Phi matrix, which denotes the correlations among the exogenous constructs. From the path diagram, we note that there are two correlations of this type between constructs 1-2 and 3-4. These are noted in the matrix in Table 11A.5.

There is also a relationship among the structural equations for two endogenous constructs (η_1 and η_3). This is represented in the Psi matrix (Table 11A.6).

TABLE 11A.5 Exogenous Constructs

	ξ_1	ξ_2	ξ_3	ξ_4
ξ_1	—			
ξ_2	ϕ_{21}	—		
ξ_3			—	
ξ_4		ϕ_{43}		—

TABLE 11A.6 Endogenous Constructs

	η_1	η_2	η_3
η_1	—		
η_2		—	
η_3	ψ_{31}		—

TABLE 11A.7 Measurement Error Correlation for Exogenous Indicators (Theta-Delta)

	X_1	X_2	X_3	X_4	X_5	X_6	X_7	X_8
X_1	—							
X_2	$\theta_{\delta21}$	—						
X_3			—					
X_4				—				
X_5					—			
X_6						—		
X_7							—	
X_8								—

Measurement Model (Indicator) Correlations

The final two matrices portray any within-construct measurement errors among the indicators of the exogenous and endogenous constructs. Note that while the correlations can theoretically be between any exogenous or endogenous indicators, they cannot occur between indicators of different types.

For purposes of illustration, two measurement error correlations are hypothesized. The first is between two exogenous indicators (X_1 and X_2), as shown in the theta-delta matrix (Table 11A.7). The second correlation is between two endogenous indicators (Y_3 and Y_4), which is represented in the theta-epsilon matrix (Table 11A.8).

We have now completed the specification of the LISREL matrices needed for estimation of the proposed model. Figure 11A.2 illustrates the complete model, with all relationships portrayed in a causal model format. In the graphical portrayal of the structural and measurement models, some specific conventions are followed. First, as with path diagrams, straight arrows indicate causal relationships, and curved arrows denote correlations. Second, indicators are represented by rectangles; constructs are represented by ellipses.

TABLE 11A.8 Measurement Error Correlation for Endogenous Indicators (Theta-Epsilon)

	Y_1	Y_2	Y_3	Y_4	Y_5	Y_6
Y_1	—					
Y_2		—				
Y_3			—			
Y_4			$\theta_{\epsilon43}$	—		
Y_5					—	
Y_6						—

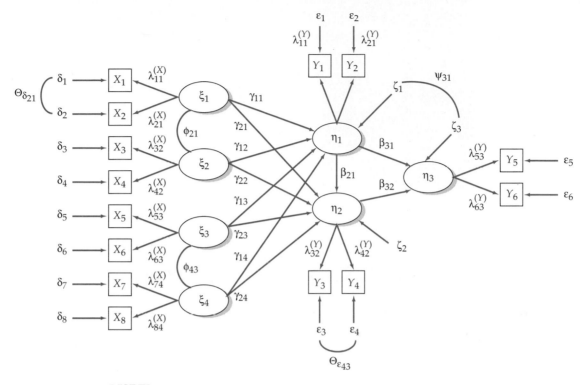

FIGURE 11A.2 LISREL notation for structural equation modeling.

Summary

Specification of any causal relationship can be incorporated directly into one of the eight matrices in the LISREL notation. Although it is not necessary to specify the models in this notation initially, after this review of the notational form used in structural equation modeling, researchers should feel equipped to portray their relationships in either format. Moreover, as the analyst becomes more acquainted with the notation, more advanced relationships can easily be incorporated as well.

Path Analysis: A Method of Computing Structural Coefficients

What is the purpose of developing path diagrams? Path models are the basis for path analysis, which is the basis for the empirical estimation of the strength of each causal relationship depicted in the path model. **Path analysis** is based on calculating the strength of the causal relationships from the correlations or covariances among the constructs.

The simple (bivariate) correlation between any two constructs can be represented as the sum of the compound paths of causal relationships connecting these points. A compound path is a path along the arrows that follows three rules:

1. After going forward on an arrow, the path cannot go backward again. But the path can go backward as many times as necessary before going forward.
2. The path cannot go through the same construct more than once.
3. The path can include only one curved arrow (correlated construct pair).

These may seem to be quite complicated rules, but they are really quite simple. For example, refer to the path diagram in Figure 11B.1. With two exogenous constructs (X_1 and X_2) that are correlated and one endogenous variable (Y), the single causal relationship can be stated as

$$Y = b_1 X_1 + b_2 X_2$$

The path analysis rules allow us to use the simple correlations between constructs to estimate the causal relationships represented by the coefficients b_1 and b_2. For ease in referring to the paths, the causal paths are labeled A, B, and C. Causal path A is X_1 correlated with X_2, path B is the effect of X_1 predicting Y, and path C shows the effect of X_2 predicting Y.

Path analysis uses the simple correlations also shown in Figure 11B.1 to estimate the causal paths using the three rules given earlier. For example, the correla-

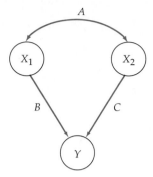

Bivariate correlations:

	X_1	X_2	Y
X_1	1.0		
X_2	.50	1.0	
Y	.60	.70	1.0

Correlations as compound paths:

$\mathrm{Cor}_{X_1 X_2} = A$

$\mathrm{Cor}_{X_1 X} = B + AC$

$\mathrm{Cor}_{X_2 X} = C + AB$

Solving for the structural coefficients:

$.50 = A$

$.60 = B + AC$

$.70 = C + AB$

Substitution of $A = .50$

$.60 = B + .50C$

$.70 = C + .50B$

Solving for B and C

$B = .33$

$C = .53$

FIGURE 11B.1 Calculating structural coefficients with path analysis.

tion of X_1 and Y ($r_{x1y} = .50$) can be represented as two causal paths: B and $A \times C$. The symbol B represents the direct path from X_1 to Y, and the other path (a compound path) follows the curved arrow from X_1 to X_2 and then to Y. Likewise, the correlation of X_2 and Y can be shown to be composed of two causal paths: C and $A \times B$. Finally, the correlation of X_1 and X_2 is equal to A. This relationship forms three equations:

$$r_{x1x2} = A$$

$$r_{x1y} = B + AC$$

$$r_{x2y} = C + AB$$

We know that A equals .50, so we can substitute this value into the other equations. By now solving these two equations, we get values of $B(b_1) = .33$ and $C(b_2) = .53$. The actual calculations are shown in Figure 11B.1. This approach allows path analysis to solve for any causal relationship based only on the correlations among the constructs and the specified causal model. As you can see from our simple example, if we change the path model in some way, the causal relationships will change as well. Such a change provides the basis for modifying the model to achieve better fit, if theoretically justified.

We have presented only a simple illustration of the process, and interested readers are encouraged to examine other treatments of this topic [25, 27, 49, 62].

Overall Goodness-of-fit Measures for Structural Equation Modeling

Assessing the overall goodness-of-fit for structural equation models is not as straightforward as with other multivariate dependence techniques, such as multiple regression, discriminant analysis, multivariate analysis of variance, or even conjoint analysis. Structural equation modeling has no single statistical test that best describes the "strength" of the model's predictions. Instead, researchers have developed a number of goodness-of-fit measures that when used in combination assess the results from three perspectives: overall fit, comparative fit to a base model, and model parsimony. The discussions that follow present alternative measures for each of these perspectives, along with the methods of calculation for those measures that are not contained in the results and that must be computed separately.

One common question arises in the discussion of each measure: What is an acceptable level of fit? None of the measures (except the Chi-square statistic) have an associated statistical test. Although in many instances guidelines have been suggested, no absolute test is available, and the researcher must ultimately decide whether the fit is acceptable. Bollen [18, p. 275] addresses this issue directly: "Overall, selecting a rigid cutoff for the incremental fit indices is like selecting a minimum R^2 for a regression equation. Any value will be controversial. Awareness of the factors affecting the values and good judgment are the best guides to evaluating their size." This advice applies equally well to the other goodness-of-fit measures.

Before examining the various goodness-of-fit measures, it may be useful to review the derivation of degrees of freedom in structural models. The number of unique data values in the input matrix is s (where $s = \frac{1}{2}(k)(k + 1)$ and k is the total number of indicators for both endogenous and exogenous constructs). The degrees of freedom for any estimated model (df) are then calculated as $df = s - t$, where t is the number of estimated coefficients. If the researcher knows the df for an estimated model and the total number of indicators, then t can be calculated directly as $t = s - df$.

The examination and derivation of goodness-of-fit measures for structural equation modeling has gained widespread interest among academic researchers in recent years, resulting in the continual development of new goodness-of-fit measures. This is reflected in the statistical programs as they are continually modified to provide the most relevant information regarding the estimated model. In our discussions we have focused our attention on the LISREL program because of its widespread application in the social sciences. It has undergone these changes as well. The newest version of LISREL (version 8) substantially expands the number and type of fit indices available directly in the output. For this reason, the following discussion and example data detail the calculations of those measures not provided in earlier versions of the program.

Measures of Absolute Fit

Absolute fit measures determine the degree to which the overall model (structural and measurement models) predicts the observed covariance or correlation matrix. No distinction is made as to whether the model fit is better or worse in the structural or measurement models. Among the absolute fit measures commonly used to evaluate structural equation modeling are the Chi-square statistic, the noncentrality parameter, the goodness-of-fit statistic, the root mean square error, the root mean square error of approximation, and the expected cross-validation index.

Likelihood-Ratio Chi-Square Statistic

The most fundamental measure of overall fit is the likelihood-ratio Chi-square statistic (χ^2), the only statistically based measure of goodness-of-fit available in structural equation modeling [48]. A large value of Chi-square relative to the degrees of freedom signifies that the observed and estimated matrices differ considerably. Statistical significance levels indicate the probability that these differences are due solely to sampling variations. Thus low χ^2 values, which result in significance levels greater than .05 or .01, indicate that the actual and predicted input matrices are *not* statistically different. In this instance, *the analyst is looking for nonsignificant differences because the test is between actual and predicted matrices.* The analyst must remember that this method differs from the customary desire to find statistical significance. However, even statistical nonsignificance does not guarantee that you have identified the "correct" model, but only that this proposed model fits the observed covariances/correlations well. It does not assure the researcher that another model would not fit as well or better. It is recommended that the .05 significance level be the minimum accepted, and it has been suggested that levels of .1 or .2 be exceeded before nonsignificance is confirmed [30].

An important criticism of the χ^2 measure is that it is too sensitive to sample size differences, especially in cases where the sample size exceeds 200 respondents. As sample size increases, this measure has a greater tendency to indicate significant differences for equivalent models. If the sample size becomes large enough, significant differences will be found for any specified model. Moreover, as the sample size nears 100 or goes even lower, the Chi-square test will show acceptable fit (nonsignificant differences in the predicted and observed input matrices) even when *none* of the model relationships are shown to be statistically significant. Thus the Chi-square statistic is quite sensitive in different ways to both small and large sample sizes, and the analyst is encouraged to complement this measure with other measures of fit in all instances. Chi square's use is appropriate for sample sizes between 100 and 200, with the significance test becoming less reliable with sample sizes outside this range.

Noncentrality and Scaled Noncentrality Parameter

The noncentrality parameter (NCP) is the result of the statisticians' search for an alternative measure to the likelihood-ratio Chi-square statistic that is less affected by or independent of the sample size. Statistical theory suggests that a noncentrality χ^2 measure will be less affected by sample size in its representation of the differences between the actual and estimated data matrices [54]. In a LISREL problem, the noncentrality parameter (NCP) can be calculated as

$$\text{Noncentrality Parameter (NCP)} = \chi^2 - \text{Degrees of Freedom}$$

While this measure adjusts the χ^2 by the degrees of freedom of the estimated model, it is still in terms of the original sample size. To "standardize" the NCP, it can be divided by the sample size to obtain the scaled noncentrality parameter (SNCP) [54]. This can be calculated as

$$\text{Scaled Noncentrality Parameter (SNCP)} = \frac{\chi^2 - \text{Degrees of Freedom}}{\text{Sample Size}}$$

This scaled measure is analogous to the average squared Euclidean distance measure between the estimated model and the unrestricted model [54]. For both the unscaled and the scaled parameters, the objective is to minimize the parameter value. Because there is no statistical test for this measure, it is best used in making comparisons between alternative models.

Goodness-of-Fit Index

The goodness-of-fit index (GFI) [45, 48] is another measure provided by LISREL. It is a nonstatistical measure ranging in value from 0 (poor fit) to 1.0 (perfect fit). It represents the overall degree of fit (the squared residuals from prediction compared with the actual data) but is not adjusted for the degrees of freedom. Higher values indicate better fit, but no absolute threshold levels for acceptability have been established.

Root Mean Square Residual

The root mean square residual (RMSR) is the square root of the mean of the squared residuals—an average of the residuals between observed and estimated input matrices. If covariances are used, it is the average residual covariance. If a correlation matrix is used, then the RMSR is in terms of an average residual correlation. The RMSR is more useful for correlations, which are all on the same scale, than for covariances, which may differ from variable to variable, depending on unit of measure. Again, no threshold level can be established, but the researcher can assess the practical significance of the magnitude of the RMSR in light of the research objectives and the observed/actual covariances or correlations [6].

Root Mean Square Error of Approximation

Another measure that attempts to correct for the tendency of the Chi-square statistic to reject any specified model with a sufficiently large sample is the root mean square error of approximation (RMSEA). Similar to the RMSR described above, the value is the discrepancy per degree of freedom. It differs from the RMSR, however, in that the discrepancy is measured in terms of the population, not just the sample used for estimation [59]. The value is representative of the goodness-of-fit that could be expected if the model were estimated in the population, not just the sample drawn for the estimation. Values ranging from .05 to .08 are deemed acceptable.

Expected Cross-Validation Index

The expected cross-validation index (ECVI) is an approximation of the goodness-of-fit the estimated model would achieve in another sample of the same size. While based on the sample covariance matrix, it takes into account the actual sample size and the difference that could be expected in another sample. The ECVI also takes into account the number of estimated parameters for both the structural and measurement models. The ECVI is calculated as

Expected Cross-Validation Index (ECVI) =
$$\frac{\chi^2}{\text{Sample Size} - 1} + 2 \frac{\text{Number of Estimated Parameters}}{\text{Sample Size} - 1}$$

The EVCI has no specified range of acceptable values but is used in comparing between alternative models.

Incremental Fit Measures

The second class of measures compares the proposed model to some baseline model, most often referred to as the **null model**. The null model should be some realistic model that all other models should be expected to exceed. In most cases, the null model is a single-construct model with all indicators perfectly measuring the construct (i.e., this represents the χ^2 value associated with the total variance in the set of correlations or covariances—no explanation at all). There is, however, some disagreement over exactly how to specify the null model in many situations [57].

Adjusted Goodness-of-Fit Index

The first measure of this type is provided by the LISREL program. The adjusted goodness-of-fit (AGFI) is an extension of the GFI, adjusted by the ratio of degrees of freedom for the proposed model to the degrees of freedom for the null model. It is quite similar to the PNFI (see below), and a recommended acceptance level is a value greater than or equal to .90.

Tucker-Lewis Index

The first incremental fix measure is the Tucker-Lewis index (TLI) [61], also known as the Nonnormed Fit Index (NNFI). First proposed as a means of evaluating factor analysis, it has been extended to structural equation modeling. It combines a measure of parsimony (see below) into a comparative index between the proposed and null models, resulting in values ranging from 0 to 1.0. It is expressed as

$$\text{Tucker-Lewis index (TLI)} = \frac{(\chi^2_{\text{NULL}}/\text{df}_{\text{NULL}}) - (\chi^2_{\text{PROPOSED}}/\text{df}_{\text{PROPOSED}})}{(\chi^2_{\text{NULL}}/\text{df}_{\text{NULL}}) - 1}$$

A recommended value is .90 or greater. This measure can also be used for comparing between alternative models by substituting the alternative model for the null model.

Normed Fit Index

One of the more popular measures is the normed fit index (NFI) [12], which is a measure ranging from zero (no fit at all) to 1.0 (perfect fit). Again, it is a relative comparison of the proposed model to the null model. The NFI is calculated as

$$\text{Normed Fit Index (NFI)} = \frac{(\chi^2_{\text{NULL}} - \chi^2_{\text{PROPOSED}})}{\chi^2_{\text{NULL}}}$$

As with the Tucker-Lewis index, there is no absolute value indicating an acceptable level of fit, but a commonly recommended value is .90 or greater.

Other Incremental Fit Measures

A number of other incremental fit measures have been proposed and the newer versions of LISREL includes three in its output. The relative fit index (RFI), the incremental fit index (IFI), and the comparative fit index (CFI) all represent comparisons between the estimated model and a null or independence model. The values lie between zero and 1.0 and larger values indicate higher levels of goodness-of-fit. The interested reader can find the specific details of each measure in selected readings [11, 17, 18].

Parsimonious Fit Measures

These measures relate the goodness-of-fit of the model to the number of estimated coefficients required to achieve this level of fit. Their basic objective is to diagnose whether model fit has been achieved by "over fitting" the data with too many coefficients. This procedure is similar to the "adjustment" of the R^2 in multiple regression. However, because no statistical test is available for these measures, their use in an absolute sense is limited in most instances to comparisons between models.

Parsimonious Normed Fit Index

The first measure in this class is the parsimonious normed fit index (PNFI) [42], a modification of the NFI. The PNFI takes into account the number of degrees of freedom used to achieve a level of fit. Parsimony is defined as achieving higher degrees of fit per degree of freedom used (one df per estimated coefficient). Thus more parsimony is desirable. The PNFI is defined as

$$\text{Parsimonious Normed Fit Index (PNFI)} = \frac{\text{Degrees of Freedom}_{\text{PROPOSED}}}{\text{Degrees of Freedom}_{\text{NULL}}} * \text{NFI}$$

Higher values of the PNFI are better, and its principal use is for the comparison of models with differing degrees of freedom. It is used to compare alternative models, and there are no recommended levels of acceptable fit. However, when comparing between models, differences of .06 to .09 are proposed to be indicative of substantial model differences [64].

Parsimonious Goodness-of-Fit Index

The parsimonious goodness-of-fit index (PGFI) modifies the GFI differently from the AGFI. Where the AGFI's adjustment of the GFI was based on the degrees of freedom in the estimated and null models, the PGFI is based on the parsimony of the estimated model. It adjusts the GFI in the following manner:

$$\text{Parsimonious Goodness-of-Fit Index (PGFI)} = \frac{\text{Degrees of Freedom}_{\text{PROPOSED}}}{1/2 * \text{No. of Manifest Variables} * (\text{No. of Manifest Variables} + 1)} * \text{GFI}$$

The value varies between zero and 1.0, with higher values indicating greater model parsimony.

Normed Chi-Square

Jöreskog [43] proposed that the χ^2 be "adjusted" by the degrees of freedom to assess model fit for various models. This measure can be termed the normed Chi-square and is the ratio of the Chi-square divided by the degrees of freedom. This measure provides two ways to assess inappropriate models: (1) a model that may be "overfitted" thereby capitalizing on chance, which is typified by values less than 1.0, and (2) models that are not yet truly representative of the observed data and thus need improvement, having values greater than an upper threshold, either 2.0 or 3.0 [22] or the more liberal limit of 5.0 [63]. However, because the χ^2 value is the major component of this measure, this measure is subject to the sample size effects discussed earlier with regard to the Chi-square statistic.

Akaike Information Criterion

Another measure based on statistical information theory is the Akaike information criterion (AIC) [1]. Similar to the PNFI, the AIC is a comparative measure between models with differing numbers of constructs. The AIC is calculated as

Akaike Information Criterion (AIC) = χ^2 + 2 * Number of Estimated Parameters

AIC values closer to zero indicate better fit and greater parsimony. A small AIC generally occurs when small χ^2 values are achieved with fewer estimated coefficients. This shows not only a good fit of observed versus predicted covariances/correlations but a model not prone to "overfitting" as well.

Summary

The types and number of goodness-of-fit measures are increasing as researchers continually explore the possibilities of structural equation modeling. The user, however, is faced with the task of not only selecting the appropriate measures but also assessing by admittedly subjective standards whether the model is acceptable. The final effect is an uncertainty as to what is acceptable versus unacceptable, leaving the burden of proof on the researcher rather than a statistically based test.

Prevailing thought holds that the strongest test of any proposed model is through the comparison of the model to any number of proposed models. The incremental fit measures were an attempt to provide a "standard" alternative model termed the null or independence model. But the inability of LISREL or any other structural equation modeling program to ensure that no other model will have a better fit to the data than the proposed model makes a formalized process of comparison between alternative or competing models the strictest test of theory. Researchers are strongly encouraged to examine alternative models not in a model development approach but in a test of competing models to find the "best" representation of the proposed theoretical model.

Table 11C.1 presents a summary of the goodness-of-fit measures for the structural model estimated in Chapter 11. The table demonstrates the derivation and interpretation of each measure as well. As this example illustrates, the researcher looks to the various measures to evaluate differing aspects of the model, and one hopes that all measures would indicate agreement on the level of model acceptability.

In evaluating the set of measures, some general criteria are applicable and indicate models with acceptable fit:

- Nonsignificant χ^2 (at least p > .05, perhaps .10 or .20)
- Incremental fit indices (NFI, TLI) greater than .90
- Low RMSR and RMSEA values based on the use of correlations or covariances
- Parsimony indices that portray the proposed model as more parsimonious than alternative models

As indicated earlier, the researcher should evaluate the proposed models on a series of measures from each type. A consensus should be reached on the acceptability of the model only after examination of the results from the entire set of goodness-of-fit measures.

TABLE 11C.1 Comparison of Goodness-of-Fit Measures for the Structural Model

Structural Model Data:
  Fifteen (15) indicators for five constructs (three exogenous, two endogenous)
  Total degrees of freedom: $\frac{1}{2} * (15 * 16) = 120$
  Number of estimated parameters (structural and measurement models): 35
Proposed Model: $\chi^2 = 178.714$ d.f. = 85   prob. = .000
Null or Independence Model: $\chi^2 = 1040.194$ d.f. = 105   prob. = .000

Evaluation of Structural Model with Goodness-of-Fit Measures

Goodness-of-Fit Measure	Levels of Acceptable Fit	Calculation of Measure	Acceptability[a]
ABSOLUTE FIT MEASURES			
Likelihood ratio Chi-square statistic (χ^2)	Statistical test of significance provided	$\chi^2 = 178.714$[b] significance level: 000	Marginal
Noncentrality parameter (NCP)	Stated in terms of respecified χ^2, judged in comparison to alternative models	NCP = 93.714	Not applicable
Scaled noncentrality parameter (SNCP)	NCP stated in terms of average difference per observation for comparison between models	$SNCP = \dfrac{93.714}{136} = .689$	Not applicable
Goodness-of-fit index (GFI)	Higher values indicate better fit, no established thresholds	GFI = .865[b]	Marginal
Root mean square residual (RMSR)	Stated in terms of input matrix (covariance/correlation), with acceptable levels set by analyst	RMSR = .076[b] WANT SMALL	Marginal
Root mean square error of approximation (RMSEA)	Average difference per degree of freedom expected to occur in the population, not the sample. Acceptable values under .08	RMSEA = .090	Marginal
Expected cross-validation index (ECVI)	The goodness-of-fit expected in another sample of the same size. No established range of acceptable values, used in comparing between models	$ECVI = \dfrac{178.714}{136-1} + \dfrac{2*35}{136-1}$ $= 1.842$	Not applicable
INCREMENTAL FIT MEASURES			
Tucker-Lewis index (TLI) or NNFI	Recommended level: 90	$TLI = \dfrac{(1040.194/105) - (178.714/35)}{(1040.194/105) - 1}$ $= .876$	Marginal
Normed fit index (NFI)	Recommended level: 90	$NFI = \dfrac{1040.194 - 178.714}{1040.194}$ $= .828$	Marginal
Adjusted goodness-of-fit index (AGFI)	Recommended level: 90	AGFI = .810[b]	Marginal

[a] An acceptability level of "Not Applicable" is applied to measures used only in comparison between alternative models.
[b] Measures provided in all versions of LISREL.

TABLE 11C.1 *Continued*

Goodness-of-Fit Measure	Levels of Acceptable Fit	Calculation of Measure	Acceptability
PARSIMONIOUS FIT MEASURES			
Parsimonious Goodness-of-fit index (PGFI)	A respecification of the GFI with higher values reflecting greater model parsimony. Used in comparing between models.	$PGFI = \dfrac{85}{\frac{1}{2}(15*16)} * .865$ $= .613$	Not applicable
Normed Chi-square	Recommended level: Lower limit: 1.0 Upper limit: 2.0/3.0 or 5.0	$Normed\ \chi^2 = \dfrac{178.714}{85} = 2.10$	Marginal
Parsimonious normed fit index (PNFI)	Higher values indicate better fit, used only in comparing between alternative models	$PNFI = \underline{85} * .828 = .670$	Not applicable
Akaike information criterion (AIC)	Smaller positive values indicate parsimony, used in comparing alternative models	$AIC = 178.714 + 2(35)$ $= 248.7$	Not applicable

References

1. Akaike, H. "Factor Analysis and AIC." *Psychometrika* 52 (1987):317–32.
2. Anderson J. C., and D. W. Gerbing. "Some Methods for Respecifying Measurement Models to Obtain Unidimensional Construct Measures." *Journal of Marketing Research* 19 (November 1982):453–60.
3. ———. "Structural Equation Modeling in Practice: A Review and Recommended Two-Step Approach." *Psychological Bulletin* 103, no. 3 (1988):411–23.
4. ———. "Assumptions and Comparative Strengths of the Two-Step Approach: Comment on Fornell and Yi." *Sociological Methods and Research* 20 (1990):321–333.
5. Bagozzi, R. P. *Causal Models in Marketing.* New York: Wiley, 1980.
6. Bagozzi, R. P., and Y. Yi. "On the Use of Structural Equation Models in Experimental Designs." *Journal of Marketing Research* 26 (August 1988):271–84.
7. Bentler, P. M. "Multivariate Analysis with Latent Variables: Causal Modeling." *Annual Review of Psychology* 31 (1980):419–56.
8. ———. "Structural Modeling and Psychometrika: A Historical Perspective on Growth and Achievements." *Psychometrika* 51 (1986):35–51.
9. ———. *Theory and Implementation of EQS, A Structural Equations Program.* Los Angeles: BMDP Statistical Software, 1988.
10. ———. *EQS: Structural Equations Program Manual.* Los Angeles: BMDP Statistical Software, 1992.
11. ———. "Comparative Fit Indexes in Structural Models." *Psychological Bulletin* 107 (1990):238–246.
12. Bentler, P. M., and D. G. Bonett. "Significance Tests and Goodness of Fit in the Analysis of Covariance Structures." *Psychological Bulletin* 88 (1980):588–606.

13. Bentler, P. M., and C. Chou. "Practical Issues in Structural Modeling." *Sociological Methods and Research* 16 (August 1987):78–117.

14. Bentler, P. M., and E. J. C. Wu. *EQS/Windows User's Guide.* Los Angeles: BMDP Statistical Software, 1993.

15. Blalock, H. M. *Conceptualization and Measurement in the Social Sciences.* Beverly Hills, Calif.: Sage, 1982.

16. ———. *Causal Modeling in the Social Sciences.* New York: Academic Press, 1985.

17. Bollen, K. A. "Sample Size and Bentler and Bonnett's Nonnormed Fit Index." *Psychometrika* 51 (1986):375–377.

18. ———. *Structural Equations with Latent Variables.* New York: Wiley, 1989.

19. Bollen, K. A., and K. G. Jöreskog. "Uniqueness Does Not Imply Identification." *Sociological Methods and Research* 14 (1985):155–63.

20. Breckler, S. J. "Applications of Covariance Structure Modeling in Psychology: Cause for Concern?" *Psychological Bulletin* 107, no. 2 (1990):260–73.

21. Burt, R. S. "Interpretational Confounding of Unidimensional Variables in Structural Equation Modeling." *Sociological Methods and Research* 5 (1976):3–51.

22. Carmines, E., and J. McIver. "Analyzing Models with Unobserved Variables: Analysis of Covariance Structures," in *Social Measurement: Current Issues,* ed. G. Bohrnstedt and E. Borgatta. Beverly Hills, Calif.: Sage, 1981.

23. Costner, H. L., and R. Schoenberg. "Diagnosing Indicator Ills in Multiple Indicator Models," in *Structural Equation Models in the Social Sciences,* ed. A. Goldberger and O. Duncan. New York: Seminar Press, 1979.

24. Cronbach, L. J. "Coefficient Alpha and the Internal Structure of Tests." *Psychometrica* 16 (1951):297–334.

25. Darden, W. R. "Review of Behavioral Modeling in Marketing." *Review of Marketing.* Chicago: American Marketing Association, 1981.

26. Dillon, W., A. Kumar, and N. Mulani. "Offending Estimates in Covariance Structure Analysis—Comments on the Causes and Solutions to Heywood Cases." *Psychological Bulletin* 101 (1987):126–35.

27. Duncan, O. D. "Path Analysis: Sociological Examples." *American Journal of Sociology* 72 (1966):1–16.

28. ———. *Introduction to Structural Equation Models.* New York: Academic Press, 1975.

29. Fassinger, R. E. "Use of Structural Equation Modeling in Counseling Psychology Research." *Journal of Counseling Psychology* 34 (1987):425–436.

30. Fornell, C. "Issues in the Application of Covariance Structure Analysis: A Comment." *Journal of Consumer Research* 9 (1983):443–48.

31. Fornell, C., and D. F. Larker. "Evaluating Structural Equations Models with Unobservable Variables and Measurement Error." *Journal of Marketing Research* 18 (February 1981):39–50.

32. Fornell, C., and Y. Yi. "Assumptions of the Two-Step Approach to Latent Variable Modeling." *Sociological Methods and Research* 20 (1992):291–320.

33. ———. "Assumptions of the Two-Step Approach: Reply to Anderson and Gerbing." *Sociological Methods and Research* 20 (1992):334–39.

34. Fraser, C. *COSAN User's Guide.* Toronto: Ontario Institute for Studies in Education, 1980.

35. Gerbing, D. W., and J. C. Anderson. "On the Meaning of Within-Factor Correlated Measurement Errors." *Journal of Consumer Research* 11 (1984):572–80.

36. Glymour, C. *Discovering Causal Structure.* Orlando, Fla.: Academic Press, 1988.

37. Goldberger, A. S., and O. D. Duncan. *Structural Equation Models in the Social Sciences*. New York: Seminar Press, 1973.

38. Hayduk, L. A. *Structural Equation Modeling with LISREL: Essentials and Advances*. Baltimore: John Hopkins University Press, 1987.

39. Heise, D. R. *Causal Analysis*. New York: Wiley, 1975.

40. Hoelter, J. W. "The Analysis of Covariance Structures: Goodness-of-Fit Indices." *Sociological Methods and Research* 11 (1983):325–44.

41. Hunt, S. D. *Marketing Theory: The Philosophy of Marketing Science*. Homewood, Ill.: Irwin, 1990.

42. James, L. R., S. A. Muliak, and J. M. Brett. *Causal Analysis: Assumptions, Models and Data*. Beverly Hills, Calif.: Sage, 1982.

43. Jöreskog, K. G. "A General Approach to Confirmatory Maximum Likelihood Factor Analysis." *Psychometrika* 34 (1969):183–202.

44. ———. "A General Method for Analysis of Covariance Structures." *Biometrika* 57 (1970):239–51.

45. ———. *LISREL VII: Analysis of Linear Structure Relationships by the Method of Maximum Likelihood*. Mooresvill, Ill.: Scientific Software, 1988.

46. ———. *LISREL 7*. Chicago: SPSS, Inc., 1988.

47. ———. *PRELIS: A Program for Multivariate Data Screening and Data Summarization*. Mooresville, Ill.: Scientific Software, 1988.

48. ———. *LISREL 8: Structural Equation Modeling with the SIMPLIS Command Language*. Mooresville, Ill.: Scientific Software, 1993.

49. Kenny, D. A. *Correlation and Causation*. New York: Wiley, 1979.

50. Loehlin, J. C. *Latent Variable Models: An Introduction to Factor, Path and Structural Analysis*. Hillsdale, N. J.: Lawrence Erlbaum, 1987.

51. Long, J. S. *Covariance Structure Models: An Introduction to LISREL*. Beverly Hills, Calif.: Sage, 1983.

52. MacCullum, R. "Specification Searches in Covariance Structure Modeling." *Psychological Bulletin* 100 (1986):107–20.

53. Marsh, H. W., J. R. Balla, and R. P. McDonald. "Goodness-of-Fit Indices in Confirmatory Factor Analysis: The Effect of Sample Size." *Psychological Bulletin* 103 (1988):391–410.

54. McDonald, R. P., and H. W. Marsh. "Choosing a Multivariate Model: Noncentrality and Goodness of Fit." *Psychological Bulletin* 107 (1990):247–55.

55. Mulaik, S. A., L. R. James, J. Van Alstine, N. Bennett, S. Lind, and D. C. Stillwell. "An Evaluation of Goodness of Fit Indices for Structural Equation Models." *Psychological Bulletin* 103 (1989):430–55.

56. Neale, M. C., A. C. Heath, J. K. Hewitt, L. J. Eaves, and D. W. Walker. "Fitting Genetic Models with LISREL: Hypothesis Testing." *Behavior Genetics* 19 (1989):37–49.

57. Sobel, M. E., and G. W. Bohrnstedt. "The Use of Null Models in Evaluating the Fit of Covariance Structure Models," in *Sociological Methodology, 1985*, ed. Nancy Brandon Tuma. San Francisco: Jossey-Bass, 1985.

58. Steenkamp, J. E. M., and H. C. M. van Trijp. "The Use of LISREL in Validating Marketing Constructs." *International Journal of Research in Marketing* 8 (1991): 283–99.

59. Steiger, J. H. "Structural Model Evaluation and Modification: An Interval Estimation Approach." *Multivariate Behavioral Research* 25 (1990):173–80.

60. Tanaka, J. "'How Big is Enough?' Sample Size and Goodness-of-Fit in Structural Equation Models with Latent Variables." *Child Development* 58 (1987):134–46.

61. Tucker, L. R., and C. Lewis. "The Reliability Coefficient for Maximum Likelihood Factor Analysis." *Psychometrika* 38 (1973):1–10.

62. Werts, C. E., and R. L. Linn. "Path Analysis: Psychological Examples." *Psychological Bulletin* 74 (1970):193–212.

63. Wheaton, B., D. Muthen, D. Alwin, and G. Summers. "Assessing Reliability and Stability in Panel Models," in *Sociological Methodology, 1977*, ed. D. Heise. San Francisco: Jossey-Bass, 1977.

64. Williams, L. J., and J. T. Hazer. "Antecedents and Consequences of Organizational Turnover: A Reanalysis Using a Structural Equations Model." *Journal of Applied Psychology* 71 (May 1986):219–31.

65. Wold, Herman, ed. *The Fixed Point Approach to Interdependent Systems*. Amsterdam: North Holland, 1981.

Linking Theory Construction and Theory Testing: Models with Multiple Indicators of Latent Variables

Marie Adele Hughes
R. Leon Price
Daniel W. Marrs

The philosophers of science Hempel (1952) and Feigl (1970) view scientific theory as a complex spatial network in which hypotheses link theoretical constructs one to another, correspondence rules link theoretical constructs to derived and empirical concepts, and derived and empirical concepts are given meaning through operational definitions. An important implication of this view of the structure of theory is that the integration of theory construction and theory testing is of major importance.

A major criterion proposed by philosophers of science to evaluate the adequacy of theoretical constructions is empirical testability (Popper, 1959), verifiability (Dodd, 1968), or confirmability (Clark, 1969). Meeting this standard requires a clear and explicit specification of theoretical construct definitions and operationalizations. However, historically in many social sciences, methodological paradigms that allow the theorist and empirical researcher to forge strong links have had many deficiencies.

One major shortcoming of past procedures arises from the fact that the bulk of research in management (as in other behavioral and social sciences) has addressed relationships between and among theoretical constructs that are not directly observable (e.g., labor productivity, employee satisfaction, management style, and motivation factors). As a consequence, if a test of theory is desired, variables that can be observed

From "Linking Theory Construction and Theory Testing: Models with Multiple Indicators of Latent Variables," Marie Adele Hughes, R. Leon Price, and Daniel W. Marrs, *Academy of Management Review*, vol. 11, no. 1 (1986), pp. 128–144. Reprinted by permission.

must be found to use as proxies for the unobservable constructs.

The inability of the researcher to measure variables of interest directly presents especially challenging problems. Two strategies have commonly been used to address these problems. The first is the careful selection, for each theoretical construct, of a single measurable variable that the researcher believes captures the important facts of the construct. Since in many cases it is unrealistic to expect that a single indicator can provide a valid and reliable measure of a complex construct, a second strategy is the construction of an index formed from some combinations of two or more observable indicator variables.

A major problem inherent in both approaches is that the measured variables (even when in the form of indices) usually contain at least moderate amounts of error. As Goldberger (1971) has pointed out, when such measures are used in linear models (e.g., analysis of variance and covariance, regression, and path analysis models), the coefficients obtained will be biased, most often in unknown degree and direction.

Several statistical models recognize that theoretical constructs of interest often are not directly measurable but must instead be estimated from multiple indicator measures. Among them are: latent structure analysis (Goodman, 1974); partial least squares (Wold, 1980); and the most widely used algorithm, LISREL (Jöreskog, 1973; Jöreskog & Sörbom, 1981). There are two major strengths in these latent variable methods of analysis—one technical, the other conceptual. Technically these models provide researchers with a method both for estimating structural relationships among unobservable constructs and for assessing the adequacy with which those constructs have been measured. Conceptually, the use of these models entails a mode of thinking about theory construction, measurement problems, and data analysis that is helpful in stating theory more exactly, testing theory more precisely, and yielding a more thorough understanding of the data.

Management literature/research has been criticized for failure to integrate theory construction and theory testing. Miner, for example, states:

> . . . theories have suffered badly from measurement failures that remain to this day. (1982, p. 444)
>
> Vague constructs tend to give birth to poor measures, as in the case of contingency theory, or to no measures at all, as in the case of classical management theory. (1982, p. 444)
>
> Although problems of measurement and construct validity are often so closely intertwined that they cannot be separated, there is still sufficient independent evidence that all of the theories suffer somewhat from problems of construct definition. (1980, p. 394)

In fields such as sociology, marketing, and political science, the use of latent variable models to address similar problems is becoming relatively common. However, such techniques have not been widely reported in the management literature. (Exceptions are Bagozzi & Phillips, 1982; Gladstein, 1984; Van de Ven & Walker, 1984; Walker & Weber, 1984).

This paper presents a general introduction to latent variable models in the context of a study of the relationship between evaluation and knowledge of job enrichment. The authors believe that using latent variables to represent theoretical constructs in models, illustrated here by the LISREL methodology, will be used increasingly by management researchers in investigating measurement issues and in testing and refining theories. This discussion may introduce that approach to a larger audience of management scholars.

Overview

An introduction to models with multiple indicators of latent variables will be presented within the context of an example application which permits an explanation of the use of such models in both exploratory and confirmatory research phases.

In the exploratory stages of research, relevant theoretical models are either missing or at a primitive stage of development. Here the researcher's interest is directed at clearly defining the theoretical constructs of interest, providing operational definitions for them, and linking the observed variables to the unobservable theoretical constructs. In this stage, a latent variable methodology can be used to examine various measurement models as an aid in selecting indicator variables, assessing construct validity, and investigating differences among models that specify alternate hypotheses about the relationships among the theoretical constructs of interest. The LISREL algorithm provides several statistics that can be used in an exploratory fashion, for example, to suggest ways in which the theoretical model might be modified, to assess the dimensionality of the constructs, and to estimate the reliabilities of the observed indicator variables in measuring model constructs.

In confirmation studies, the researcher formulates a model representing hypothesized relationships (a) among theoretical constructs and (b) between the theoretical constructs and the measures used to operationally define them. Direct tests involving latent variables (which represent the theoretical constructs) are used to make inferences and to draw conclusions about the theoretical constructs. LISREL provides both overall measures of the fit of the hypothesized model to the data (such as goodness-of-fit statistics) and measures that can be used to test specific elements of the model such as structural parameters.

The LISREL Model

The general LISREL model can be viewed as a synthesis of techniques of econometrics and psychometrics (Goldberger, 1971) that subsumes many models as special cases: for example, regression analysis, path analysis, factor analysis, and structural equation models for directly observed variables.

In the most general form of the LISREL model, the researcher posits a causal structure among a set of unobservable constructs. These constructs are represented by latent variables, which are empirical measures of the constructs. Each latent variable is measured by a set of observable indicator variables. Observable variables may be assumed to be measured with error.

The formal LISREL model describes the structural and measurement assumptions of the researcher and consists of two parts, the latent variable (or structural equation) model and the measurement model. The latent variable model specifies the hypothesized causal structure among the unobserved theoretical constructs. The measurement model specifies how the latent variables are measured in terms of the observed variables and represents the correspondence rules by which the unobservable constructs are related to the observed variables.

In a linear model containing variables that are measured with error, the coefficients represent empirical associations. In contrast, multiple indicator latent variable models permit the explicit modeling and estimation of errors in measurement. Thus, the coefficients in the structural equation model in LISREL represent theoretical cause and effect relationships among the unobservable constructs (as represented by the latent variables) and, as such, are the parameters of interest to the researcher.

Exploring Measurement Issues

In order to emphasize the advantages of LISREL (and similar methodologies) in exploring measurement issues, data were chosen from an existing study of job enrichment by Price (1978) in which there had been no consideration of measurement issues prior to data collection. In the management literature, the use of job enrichment as a motivating technique has been the subject of some controversy. Studies such as Herzberg (1968), Paul, Robertson, and Herzberg (1969), Sirota and Greenwood (1971), Miles (1975), and O'Connor and Barrett (1980) support the concept. Others—for example, Winpisinger (1972), Fein (1974), Blood and Hulin (1967), and Umstot, Bell, and Mitchell (1976)—contend that the relationship between job enrichment and motivation is at best unproven and possibly misleading. The use of latent variable methods, which enhance the integration of theory construction and theory testing, offers management scholars the opportu-

nity to pursue the resolution of this and other similar controversies with the rigor advocated by Miner (1980, 1982).

Price surveyed all presidents or presiding officers of unions or national associations that were on record with the United States Department of Labor, Bureau of Labor Statistics (1976). One hundred fourteen complete questionnaires were obtained from the total population of 229 unions or national associations. Below, six measures of knowledge of job enrichment (K1–K6) and eight measures of attitudes toward job enrichment (A1–A8) that were obtained in the original survey (Table 1) are utilized to illustrate the use of LISREL in exploring measurement issues.

Assessing Measurement Properties

Tests of substantive theory (i.e., hypothesized relationships among theoretical constructs) necessarily involve an "auxiliary measurement theory" (Blalock, 1982, p. 25) concerning relationships among theoretical constructs and their indicators. When the auxiliary measurement theory is strong, empirical analysis can lead to a greater understanding of the phenomenon under investigation. However, weak associations between theoretical constructs and observed variables may lead to incorrect inferences and misleading conclusions about relationships among the underlying theoretical constructs of interest.

Factor analysis has frequently been used for studying measurement properties. Traditionally it has been employed primarily in the exploratory mode—as a technique to help identify constructs that underlie a set of observed variables. However, except for specifying the number of common factors and stating whether all factors are correlated or all factors are uncorrelated, researchers using the exploratory factor model are extremely limited with respect to the specification and testing of hypotheses concerning model structure.

The type of factor model that allows the researcher to specify hypotheses and that provides information to determine whether the observed data confirm the hypothesized model structure is called a confirmatory factor analysis model. LISREL can be used to specify and analyze such models.

TABLE 1 Knowledge and Attitude Questions

Knowledge

I believe that job enrichment, if used properly, could:

K1. Be directly responsible for providing promotion opportunities for our membership.
K2. Assist management in recognizing the individual achievement of our membership.
K3. Make the work more challenging for the individual member.
K4. Provide more responsibility to the individual member.
K5. Create better communication channels between the employee and his supervisors.
K6. Decrease the close supervision by management of our membership.

Attitude

A1. I feel that job enrichment concepts if used properly would benefit our membership.
A2. I feel that most firms with which I bargain do not fully understand the theory and proper application of job enrichment.
A3. I think that job enrichment should be a bargaining issue.
A4. I believe that the benefits claimed for job enrichment in publications have been overstated.
A5. I feel that use of job enrichment benefits management rather than the worker.
A6. I believe that the majority of our membership would favor the application of job enrichment to their jobs.
A7. I believe the real motive of management for implementing job enrichment programs is to get more output for the same pay.
A8. I believe the use of job enrichment instills a sense of pride in the individual's work and product.

Note. Questions were coded using a 5-point Likert-type scale with 1 = strongly agree and 5 = strongly disagree.

Consider, for example, the confirmatory factor analysis model shown in Figure 1. It depicts two latent variables (ξ_1 and ξ_2) and posits that ξ_1 (knowledge of job enrichment) and ξ_2 (attitude toward job enrichment) are unidimensional constructs defined by the observed variables K1–K6 and A1–A8, respectively. The parameter ϕ_{12} allows for correlated constructs (i.e., oblique factors); the λ_i parameters are analogous to factor loadings; the random variables δ_1–δ_{14} represent unique factors or errors in measurement, and the variances and covariances of these measurement errors are parameters of the model.

Flexibility in imposing constraints motivated by substantive theory differentiates confirmatory factor analysis models from traditional exploratory models. The model shown in Figure 1 incorporates the structural constraints that factor loadings of variables across common factors are zero (e.g., the loading of the variable K1 on the latent variable ξ_2) and that correlations among unique factors also are zero.

Assessing the strength of the auxiliary measurement theory involves a consideration of some facets of construct validity that are related to measurement and that can be explored within the framework of confirmatory factor analysis. Construct validity, the degree to which a construct achieves theoretical and empirical meaning within the overall structure of one's theory, is a necessary prerequisite for theory development and testing (Bagozzi, 1980). The Campbell and Fiske (1959) multitrait-multimethod (MTMM) matrix has been the most widely used procedure for construct validation. Multiple indicator latent variable models offer a more rigorous methodology that permits statistical testing and provides additional useful information as well. (See Bagozzi, 1978, and Bagozzi & Phillips, 1982, for a more detailed discussion of the limitations of the MTMM methodology.)

Below, using the model of Figure 1 as a base, an illustration is given of how LISREL can be used to explore two aspects of construct validity—internal consistency (dimensionality and reliability) and discriminant validity—and to assist the researcher in refining the measurement model.

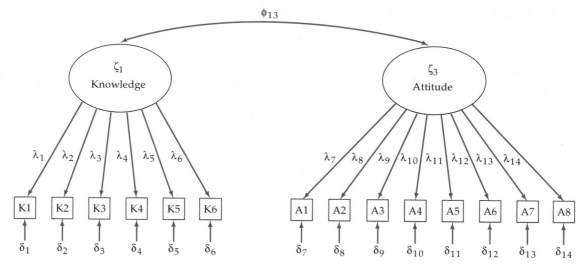

FIGURE 1 Two-construct confirmatory factor analysis model.

Dimensionality

If one is to obtain an unambiguous, meaningful test of a theory, the multiple indicators of each latent variable in the model should converge to measure a single construct; that is, the set of empirical measures of each single construct should, ideally, be unidimensional. The χ^2 statistic provided by the LISREL algorithm can be used to assess dimensionality. The p-value associated with this χ^2 is the probability of obtaining a χ^2 value larger than the value actually obtained under the hypothesis that the model specified is a true reflection of reality. Small p-values indicate that the hypothesized structure is not confirmed by the sample data. For the model in Figure 1, the χ^2 of 255.72 with 76 df ($p = 0.000$) indicates that the observed data does *not* adequately fit the hypothesized model. One reason for this bad fit could be that the sets of measures of the latent variables are not unidimensional. An examination of the reliabilities of the measures can provide additional insights.

Reliability

Empirical measures used to operationalize unobservable constructs should be reliable, that is, as free from random error as possible. Reliable measures assist the researcher in obtaining a clear picture of true relationships; random errors tend to

obscure relationships. In classical test score theory, the reliability of a variable is a measure of the degree of true-score variation relative to the observed-score variation. It is defined as the squared correlation between the true score (i.e., the latent construct for which that variable is assumed to be an indicator) and the observed variable. Reliability can be interpreted as the proportion of the observed variable that is free from error (Lord & Novick, 1968). It would be difficult to justify a proposed indicator of a latent variable in exploratory research if its reliability were less than .50 for in that case more than 50 percent of its variance would be error variance.

Table 2 specifies the reliability of each indicator for the latent variables in the model of Figure 1.

TABLE 2 Reliabilities for Model in Figure 1

Knowledge Indicator	Reliability	Attitude Indicator	Reliability
K1	.59	A1	.83
K2	.68	A2	.00
K3	.71	A3	.10
K4	.51	A4	.31
K5	.67	A5	.29
K6	.34	A6	.52
		A7	.15
		A8	.72

Using a cutoff of .50, it would appear that the latent variable, knowledge, can be reliably operationalized using the five measures K1–K5. The attitude construct presents a more complex problem. Only A1, A6, and A8 have reliabilities above .50. One reason why this may have occurred is that the eight observed variables (A1–A8) may not converge to measure a single underlying attitude construct.

Reformulation of Model

A further examination of the attitude questions in Price's survey (Table 1) shows that only five of the indicators directly address job enrichment attitudes. The three indicators that exhibit high reliabilities in the model of Figure 1 (A1, A6, and A8) are measures of evaluations of job enrichment and its consequences. As such, these three questions reflect an expectancy-value operationalization of attitude. In contrast, questions A5 and A7 refer to attitudes of respondents toward management in its implementation of job enrichment.

The remaining three questions do not appear to be measures of attitude toward job enrichment. Question A2 concerns respondents' opinions of firms' understanding of job enrichment; question A3 concerns bargaining table priorities; and question A4 concerns opinions regarding claimed benefits.

Based on these assessments, the model was reformulated as shown in Figure 2. This model has a χ^2 of 50.44 with 32 degrees of freedom ($p =$ 0.020), which does not indicate a good fit of the data to this model. However, model evaluation should never be based solely on the χ^2 statistic, because it may indicate a bad fit even though specific relationships in the model may reflect initial hypotheses (or vice versa). An assessment of whether the attitude construct as originally formulated comprises the two constructs, attitude toward job enrichment and attitude toward management, thus is dependent on additional considerations.

If the attitude constructs in Figure 2 are highly correlated, then, empirically at least, it is difficult to support the hypothesis that they represent distinct concepts. This is a question of discriminant validity, that is, the degree to which a concept differs from other concepts (Campbell & Fiske, 1959).

A test of discriminant validity can be performed by comparing the model in Figure 2 with a model in which the correlation between the two attitude constructs is restricted to 1. The restricted model is a special case of (i.e., nested in) the Figure 2 model. Two degrees of freedom are lost with this restriction because it logically dictates the additional restriction that $\phi_{12} = \phi_{13}$.

A difference-of-χ^2 test provides statistical evidence of whether the model of Figure 2 provides a significant improvement over the more restricted model. The restricted model has a χ^2 with 34 df of 100.99. The significant difference of χ^2's (50.55 with 2 df) between the model in Figure 2 and the restricted model provides support for the

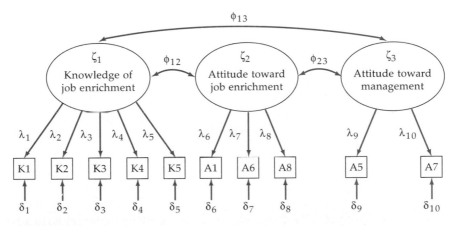

FIGURE 2 Three-construct confirmatory factor analysis model.

reformulation of the attitude construct into two distinct constructs.

Further evidence to support the model in Figure 2 may be obtained by examining the estimates of the λ_i parameters (i.e., the structural relationships between the indicators and their associated latent variables). Conventional tests for the hypothesis that each $\lambda_i = 0$ result in t values ranging from 6.35 to 12.19, supporting the hypothesis that all indicators exhibit a statistically significant positive relationship with the constructs they were hypothesized to measure.

Correlated Errors of Measurement

In any specific research study, some factors may not be explicitly modeled; for example, theories may not be developed to the point of giving a complete model specification, the researcher may be unable to obtain the required measures, or excluded factors may be extraneous to the structural relationships of substantive interest in the investigation. This gives rise to model specification error.

In cases in which two or more indicators of latent variables in a model are systematically influenced by a factor that is not explicitly modeled, correlations may exist between the indicators' measurement errors.

Examination of the knowledge questions remaining in the model suggests that such system-

atic influences may be present in the data analyzed in this paper. Questions K1 and K2 refer to concepts associated with Herzberg's (1969) hygiene factors. References to promotion and recognition for achievement are most likely associated by the respondents with immediate return in monetary form to the union member. Similarly, references to challenge of work, responsibility, and communication in questions K3, K4, and K5 refer to concepts associated with Herzberg's (1969) motivational factors.

Hypothesized correlations between measurement errors are represented by two-headed, curved arrows, which represent relationships that are not explicitly analyzed in the model. The absence of such arrows between δ_i's in Figure 2 indicates that all errors-of-measurement correlations were constrained to a value of zero. The model of Figure 3 (which is identical to Figure 2 with the addition of the measurement error correlations discussed above) has a χ^2 of 22.41 with 28 df ($p = 0.762$), which indicates a good fit of the data to the model.

The addition of correlated measurement errors to a model will nearly always improve goodness of fit; however, it can also mask a true underlying structure (Gerbing & Anderson, 1984). For this reason, measurement error assumptions should be based on substantive considerations and should be made as explicit as possible. In addition, the sensitivity of parameter estimates to the

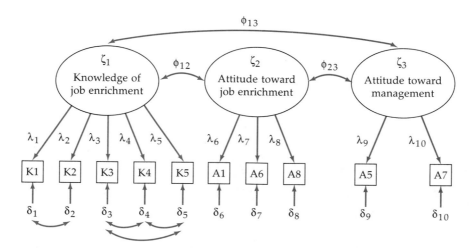

FIGURE 3 Three-construct confirmatory factor analysis model with correlated errors of measurement.

TABLE 3 Standardized Parameter Estimates for Models in Figures 2 and 3

Parameter	Estimates	
	Figure 2	Figure 3
λ_1	.78	.81
λ_2	.77	.80
λ_3	.73	.67
λ_4	.60	.50
λ_5	.81	.77
λ_6	.91	.91
λ_7	.71	.71
λ_8	.84	.84
λ_9	1.21	1.21
λ_{10}	.71	.71
ϕ_{12}	.78	.79
ϕ_{13}	−.31	−.30
ϕ_{23}	−.52	−.52
θ_{12}[a]	0[b]	.01
θ_{34}	0	.19
θ_{35}	0	.06
θ_{45}	0	.14

[a] LISREL notation for correlated errors of measurement.
[b] All correlated errors are constrained to zero in the model of Figure 2.

specification of correlated measurement errors should be investigated. For management theory to advance, it is important that researchers be able to compare findings of their studies with others that differ in settings, time periods, and so on. Instability in underlying structure raises serious questions about the generalizability and comparability of findings. Table 3 contains parameter estimates for both models; they do not differ substantively.

Examination of the *t*-statistics for each pair of measurement errors that were hypothesized to be associated significantly indicates that only the error correlations for question pairs K3–K4 and K4–K5 are statistically significant. In the interest of parsimony, and given the possibility of overfitting the data, the researcher may modify the model to reflect these findings. Alternately, recognizing the possibility of Type II errors inherent in statistical tests, the researcher may retain the model as specified in Figure 3 based on the theoretical and methodological considerations discussed above.

Discussion

In the application of latent variable methods discussed in this section, the process of successively modifying a preliminary specification of latent variables and their operationalizations has been illustrated. One of the major values of latent variable methodology is in providing a framework for theory conceptualization that emphasizes consideration of the issues discussed above during theory construction and in the first stages of planning empirical studies to test theories. Although theoretical considerations guided the development of the final measurement model here, the process was also data-driven; that is, empirical relationships that were manifest in the sample were used to suggest modifications. The process, therefore, was exploratory and should not be viewed as confirming hypotheses concerning the measurement models. Evidence to confirm or disprove hypotheses generated by such exploratory procedures should be obtained by subjecting an independent set of data to a confirmatory analysis.

Parameter Estimation and Testing

For purposes of illustration, it is assumed that the final measurement model developed in the previous section has been deemed acceptable. Although the causal model used at this point for illustration is a simple one, the concepts are generalizable to a variety of models that are useful in management research, such as recursive and nonrecursive structural models for both cross-sectional and longitudinal data, time series models, and models with reciprocal causation structures.

The Causal Model

Figure 4 specifies the causal model that is the subject of the discussion. The latent variable measuring attitudes toward job enrichment is repre-

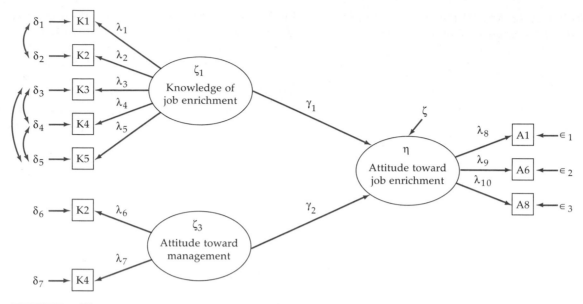

FIGURE 4 Three-construct causal model.

sented by the symbol η; it is endogenous, that is, its value is determined within the model. The latent variables measuring knowledge of job enrichment (represented by the symbol ξ_1) and attitude toward management as it implements job enrichment (represented by the symbol ξ_2) are exogenous, that is, taken as given; their values are determined by factors outside the model. Questions K1–K5 from the survey instrument are used as indicators of the knowledge construct; questions A5 and A7 are indicators of the management attitude construct; and questions A1, A6, and A8 are indicators of the job enrichment attitude construct. As before the λ_i parameters are the structural coefficients linking the latent variables and their indicators.

The parameter γ_1 represents the hypothesized causal relationship between knowledge of job enrichment and attitude toward job enrichment; γ_2 represents the hypothesized causal relationship between attitude toward management and attitude toward job enrichment. The head of the arrow indicates the direction of the hypothesized causal relationship. The parameter ξ represents the error in the structural equation resulting from either random error or systematic influences not explicitly modeled. In equation form, the latent

variable model corresponding to Figure 4 is $\eta = \gamma_1\xi_1 + \gamma_2\xi_2 + \zeta$.

The rationale for the hypothesized structural relationship between knowledge and attitude is based on marketing communications and psychology models of cognition (i.e., knowledge), affect (i.e., attitude/evaluation) and behavior (Cacioppo & Petty, 1979; Kiesler & Sproull, 1982; Lavidge & Stiener, 1961; McGuire 1976; Ray 1973). A positive causal influence of cognition on affect is hypothesized. The job satisfaction literature (Locke, Cartledge, & Knerr, 1970), as well as the job enrichment literature, contains support for that hypothesis. For example, Myers (1971) reports an extreme case in which subsequent to the introduction of job enrichment, workers felt that a union was no longer necessary. Here knowledge preceded and directly (i.e., positively) influenced attitude. Similar support is found in Smith (1973), *Work in America* (1973), and Roche and MacKinnon (1970); there changes in affect occurred after the knowledge base had been established.

The hypothesized relationship between ξ_2 and η is based on the assumption that perceived positive attitudes toward management's motivation for implementing job enrichment programs would cause respondents to have positive attitudes toward job enrichment. Because the indicators of attitude toward management are negatively coded (i.e., positive attitudes are associated with lower scores), the hypothesized sign of the relationship is negative.

Analysis

LISREL provides a hierarchy of goodness-of-fit statistics that can be used to evaluate the quality of the model. Table 4 specifies the overall goodness-of-fit measures and composite reliabilities for the model of Figure 4. The adjusted goodness-of-fit statistic is a measure, adjusted for degrees of freedom, of the relative amount of variance and covariance jointly accounted for by the model. The coefficients of determination are measures of strength of several relationships considered jointly. They are analogous to squared multiple correlations. The structural equation represented by the model in Figure 4 contains only one endogenous variable (η = attitude toward job enrichment). Therefore, the coefficient of determination for the endogenous variable is equivalent to R^2 and measures the proportion of variance in η accounted for by the two exogenous latent variables, ξ_1 (knowledge of job enrichment) and ξ_2 (attitude toward management). A compos-

ite reliability measure is analogous to coefficient alpha and estimates the internal consistency of a latent variable (Bagozzi, 1981); the measure is important because tests of hypothesized causal relationships directly involve latent variables rather than their respective indicators.

The χ^2 statistic for the model in Figure 4 indicates that, overall, the model fits the data very well. Other overall measures also indicate a good fit, and composite reliabilities of the latent variables are high. Table 5 contains estimates of the structural parameters and their associated t-statistics. Both structural parameters are statistically significant with signs in the hypothesized directions.

Additional Considerations

Assessing Goodness-of-Fit

Two cautions should be noted with respect to use of the χ^2 statistic in assessing goodness-of-fit. First, the ability of the χ^2 test to detect departures of the data from the model is dependent on the sample size. For large sample sizes, even when the residuals are quite small, the test very often will indicate that the model should be rejected. The converse occurs with small sample sizes. Various descriptive measures have been proposed that reduce or eliminate this sample size dependency. They include the normal fit index (Bentler & Bonett, 1980) and the ratio of χ^2 to degrees of freedom (Wheaton, Muthén, Alvin, & Summers, 1977).

Second, as mentioned previously, the χ^2 statistic should be considered in an integrated fashion with other test statistics and model parameter estimates in evaluating a model.

TABLE 4 Overall Goodness-of-Fit Measures and Composite Reliabilities of Latent Variables For Model in Figure 4 (Sample Size = 114)

Measure	Value
Chi-square (29 *df*)	
($p = 0.362$)	31.07
Adjusted goodness-of-fit index	0.90
Coefficients of determination for:	
Structural equation	0.68
Endogenous variable	0.90
Exogenous variables	1.00
Composite reliabilities for:	
ξ_1—Knowledge of job enrichment	0.88
ξ_2—Attitude toward management	0.86
η—Attitude toward job enrichment	0.86

TABLE 5 Estimates of Structural Parameters and t-Statistics for Model in Figure 4

Parameter	Unstandardized Estimate	Standardized Estimate	t-value
γ_1	0.78	0.74	7.47
γ_2	−0.26	−0.36	−3.50

Correlation Versus Covariance matrix

Either the covariance or the correlation matrix for the observed variables may be analyzed by LIS-REL. The covariance matrix must be used when comparing structural parameters either across different populations (for the same variables) or across time for the same population. In addition, the χ^2 is a valid test statistic only if the analysis is based on the covariance matrix, and t-values may be inflated when the correlation matrix is analyzed. However, use of the correlation matrix has many advantages, particularly in the early stages of analysis. Management researchers are more accustomed to thinking in terms of correlation coefficients, which can be interpreted in terms of proportion of "explained" variance. In addition, measures of most constructs in management have not been sufficiently standardized or refined to give substantive meaning to the unstandardized coefficients that are obtained when the covariance matrix is analyzed. (See James, Mulaik, & Brett, 1982; Blalock, 1982; and Long, 1983, for additional discussion.)

Table 6 contains the correlation matrix and standard deviations for the variables used in this analysis so that interested readers can reproduce the results. The covariance matrix, which can be calculated from the given information, was used in all analyses in this paper. (A summary of technical details, including model specifications and LISREL code, will be provided by the authors upon request.)

Conclusion

The major conceptual advantage of a methodological paradigm within which both measurement and latent construct linkages can be represented and tested are:

1. Statement of theory is more exact.
2. Testing of theory is more precise.
3. Communication of theory is enhanced.

Use of methods such as LISREL may enhance both theory construction and empirical research in management by focusing attention on the importance of careful consideration of testability in theory formulation and construct development in empirical analysis.

Wider use of latent variable methods, which demand an explicit specification of operationalizations and of hypothesized functional relation-

TABLE 6 Correlations and Standard Deviations

	K1	K2	K3	K4	K5	K6	A1	A2	A3	A4	A5	A6	A7	A8
K1	1.00													
K2	.71	1.00												
K3	.62	.66	1.00											
K4	.44	.50	.71	1.00										
K5	.59	.70	.68	.62	1.00									
K6	.47	.45	.50	.45	.52	1.00								
A1	.61	.58	.60	.48	.54	.32	1.00							
A2	.08	.08	.07	.03	.10	.27	.05	1.00						
A3	.28	.29	.20	.21	.10	.26	.30	.30	1.00					
A4	−.37	−.31	−.31	−.37	−.24	−.05	−.49	.30	.00	1.00				
A5	−.27	−.21	−.28	−.26	−.21	.01	−.47	.28	.11	.70	1.00			
A6	.53	.50	.40	.37	.39	.26	.67	.08	.42	−.37	−.36	1.00		
A7	−.16	−.15	−.21	−.20	−.14	.06	−.31	.35	.16	.59	.65	−.17	1.00	
A8	.58	.58	.54	.50	.56	.34	.78	.09	.31	−.40	−.44	.60	−.33	1.00
Std Dev	1.00	.92	.87	.85	.99	1.08	.99	.85	1.16	1.10	1.23	.98	1.08	.98

ships among constructs, may be valuable to management scholars as they attempt to improve their understanding of complex processes in the field of management.

References

Bagozzi, R. P. (1978) The construct validity of the affective, behavioral, and cognitive components of attitude by analysis of covariance structures. *Multivariate Behavioral Research, 13*, 9–31.

Bagozzi, R. P. (1980) *Causal models in marketing.* New York: Wiley.

Bagozzi, R. P. (1981) An examination of the validity of two models of attitude. *Multivariate Behavioral Research, 16*, 323–369.

Bagozzi, R. P., & Phillips, L. W. (1982) Representing and testing organizational theories: A holistic construal. *Administrative Science Quarterly, 27*, 459–489.

Bentler, P. M., & Bonett, D. G. (1980) Significance tests and goodness-of-fit in the analysis of covariance structures. *Psychological Bulletin, 88*, 588–606.

Blalock, H. M., Jr. (1982) *Conceptualization and measurement in the social sciences.* Beverly Hills, CA: Sage.

Blood, M. R., & Hulin, C. L. (1967) Alienation, environmental characteristics, and worker responses. *Journal of Applied Psychology, 51*, 248–290.

Cacioppo, J. T., & Petty, R. E. (1979) Attitudes and cognitive response: An electrophysiological approach. *Journal of Personality and Social Psychology, 37*, 2181–2199.

Campbell, D. T., & Fiske, D. W. (1959) Convergent and discriminant validation by the multitrait-multimethod matrix. *Psychological Bulletin, 56*, 81–105.

Clark, J. T. (1969) The philosophy of science and the history of science. In M. Clayett (Ed.), *Critical problems in the history of science* (pp. 103–140). Madison: University of Wisconsin Press.

Dodd, S. C. (1968) Systemmetrics for evaluating symbolic systems. *Systemmatics, 6*(1), 27–49.

Feigl, H. (1970) The "orthodox" view of theories: Remarks in defense as well as critique. In M. Radnor & S. Winokur (Eds.), *Minnesota studies in the philosophy of science* (p. 4). Minneapolis: University of Minnesota Press.

Fein, M. (1974) *Motivation for work.* Norcross, GA: Rand McNally.

Gerbing, D. W., & Anderson, J. C. (1984) On the meaning of within-factor correlated measurement errors. *Journal of Consumer Research, 11*, 572–580.

Gladstein, D. L. (1984) Groups in context: A model of task group effectiveness. *Administrative Science Quarterly, 29*, 499–517.

Goldberger, A. S. (1971) Econometrics and psychometrics: A survey of communalities. *Psychometrika, 36*, 83–107.

Goodman, L. A. (1974) Exploratory latent structure analysis using both identifiable and unidentifiable models. *Biometrika, 61*, 215–231.

Hempel, C. G. (1952) Fundamentals of concept formation in empirical science. In *International encyclopedia of unified science* (p. 11) Chicago: University of Chicago Press.

Herzberg, F. (1968) One more time: How do you motivate employees? *Harvard Business Review, 46*(1), 53–62.

Herzberg, F. (1969) *Work and the nature of man* (3rd ed.). Cleveland, OH: World.

James, L. R., Mulaik, S. A., & Brett, J. M. (1982) *Causal analysis: Assumptions, models, and data.* Beverly Hills, CA: Sage.

Jöreskog, K. G. (1973) A general method for estimating a linear structural equation system. In A. S. Goldberger & O. D. Duncan (Eds.), *Structural equation models in the social sciences* (pp. 85–112). New York: Seminar.

Jöreskog, K. G., & Sorböm, D. (1981) *LISREL—User's guide.* Uppsala, Sweden: University of Uppsala.

Kiesler, S., & Sproull, L. (1982) Managerial response to changing perspectives on problem sensing from social cognition. *Administrative Science Quarterly, 27*, 548–570.

Lavidge, R. J., & Steiner, G. A. (1961) A model for predictive measurements of advertising effectiveness. *Journal of Marketing, 25*(6), 59–62.

Locke, E. A., Cartledge, N., & Knerr, C. S. (1970) Studies of the relationship between satisfaction, goal setting, and performance. *Organizational Behavior and Human Performance, 5*, 135–158.

Long, J. S. (1983) *Confirmatory factor analysis.* Sage University Paper Series on Quantitative Applications in the Social Science, 07 033. Beverly Hills, CA, and London: Sage.

Lord, F. M., & Novick, M. R. (1968) *Statistical theories of mental test scores.* Reading, MA: Addison-Wesley.

McGuire, W. J. (1976) Some internal psychological factors influencing consumer choice. *Journal of Consumer Research, 2*, 302–319.

Miles, R. E. (1975) *Theories of management: Implications for organizational behavior and development.* New York: McGraw-Hill.

Miner, J. B. (1980) *Theories of organizational behavior.* Hinsdale, IL: Dryden.

Miner, J. B. (1982) *Theories of organizational structure and process.* Chicago, IL: Dryden.

Myers, M. S. (1971) Overcoming union opposition to job enrichment. *Harvard Business Review, 49*(3), 37–49.

O'Connor, E. J., & Barrett, G. V. (1970) Informational cues and individual differences as determinants of subjective perceptions of task environment. *Academy of Management Journal, 23*, 697–716.

Paul, W. J., Robertson, K. B., & Herzberg, F. (1969) Job enrichment pays off. *Harvard Business Review, 74*(2), 61–78.

Popper, K. R. (1959) *The logic of scientific discovery.* New York: Harper & Row.

Price, R. L. (1978) *An analysis of the perceptions and attitudes of union leaders toward application of job enrichment theory.* Unpublished doctoral dissertation, University of Oklahoma.

Ray, M. L. (1973) *New models for mass communication research.* Beverly Hills, CA: Sage.

Roche, W. J., & MacKinnon, N. L. (1970) Motivating people with meaningful work. *Harvard Business Review,* 48(3), 97–110.

Sirota, D., & Greenwood, J. M. (1971) Understand your overseas work force. *Harvard Business Review,* 49(1), 53–60.

Smith, H. R. (1973) From Moses to Herzberg: An exploration of job de-enrichment. *Proceedings of the 33rd Annual Meeting of the Academy of Management,* 306–311.

Umstot, D. D., Bell, C. H., Jr., & Mitchell, T. R. (1976) Effects of job enrichment and task goals on satisfaction and productivity: Implications for job design. *Journal of Applied Psychology,* 61, 379–394.

United States Department of Labor, Bureau of Labor Statistics. (1976) *Directory of national unions and employee associations,* 1973. Washington, DC: U.S. Government Printing Office.

Van de Ven, A. H., & Walker, G. (1984) The dynamics of interorganizational coordination. *Administrative Science Quarterly.* 29, 598–621.

Walker, G., & Weber, D. (1984) A transaction cost approach to make-or-buy decisions. *Administrative Science Quarterly,* 29, 373–391.

Wheaton B., Muthën, B., Alvin, D. F., & Summers, G. F. (1977) Assessing reliability and stability in panel models. In D. R. Herse (Ed.). *Sociological Methodology,* (pp. 84–136). San Francisco, CA: Jossey-Bass.

Winpisinger, W. (1972) Job enrichment—Another part of the forest. *Proceedings of the Annual Winter Meeting, Industrial Relations Research Association* (pp. 154–159). Toronto, Canada.

Wold, H. (1980) Model construction and evaluation when theoretical knowledge is scarce: Theory and application of partial least squares. In J. Kmenta & J. B. Ramsey (Eds.), *Evaluation of econometric models* (pp. 47–74). New York: Academic Press.

Work in America. (1973) Report of a Special Task Force to the Secretary of Health, Education, and Welfare. Cambridge, MA: M.I.T. Press.

APPENDIX A

Applications of Multivariate Data Analysis

APPENDIX PREVIEW

As noted many times in the text, the acceptance and widespread use of multivariate data analysis was in large part due to the availability of computer programs and access to computers for both academic researchers and business analysts. As we have seen for each technique, statistical programs are needed to handle the computational demands of all but the most trivial problems. The development of computers, particularly personal computers, and of statistical programs is only expected to continue, making even the most sophisticated techniques available to all users with even minimal access to a computer.

In our attempt to foster a working knowledge of multivariate data techniques, this appendix is devoted to providing the control commands necessary to execute the techniques illustrated in the chapters. Whenever possible, the commands for the various analyses are provided for the three most popular statistical packages: SPSS, SAS and BMDP. These statistical packages have gained widespread usage in all types of computers, ranging from mainframes to personal computers. Fortunately, there are few, if any, differences between the control commands necessary for each type of computer. Our examples will be oriented toward personal computer applications, but only slight modifications are necessary for usage on other systems.

Two exceptions to the use of SPSS, SAS or BMDP for our analyses should be noted. First, multidimensional scaling analyses will be performed with a number of popular packages available for both the mainframe and personal computer. These control commands will be documented in a separate section. The other exception is in the area of structural equation modeling. We will employ the popular LISREL program, and the necessary control commands for its execution will be provided as well in a separate section. In both situations, the necessary software is widely available at quite affordable prices.

We assume that the reader has some working knowledge of the program being used. It is beyond the scope of this text to provide a tutorial on the general features of the statistical packages or the commands necessary for general operation of the packages.

We should note that the use of command syntax is lessened dramatically with the advent of personal computer based programs, especially those of a Windows-like nature. The ability of the user to specify the analysis through a series of menus greatly facilitates the use of the programs across a wide range of analysts. The analyst is no longer required to learn the syntax before a single analysis can be performed. Instead, it may be that only in the most advanced instances will the analyst have to deal with the command syntax. We provide the command syntax for these programs to illustrate the actual commands for experienced users and provide guidance to new users if the need arises to utilize command syntax.

Our presentation of the commands for each statistical package contains an annotated set of control commands, where specific commands for each analytical technique are explained and optional commands detailed.

Software Used and Presented

BMDP Statistical Software, Inc., *BMDP Statistical Software Manual, Release 7, vols. 1 and 2*, Los Angeles, CA., 1992.

BMDP Statistical Software, Inc., *BMDP/PC User's Guide, Release 7, vols. 1 and 2*, Los Angeles, CA., 1993.

Joreskog, K. G., and D. Sorbom, *LISREL VII*. Mooresville, Ind.: Scientific Software Inc., 1990.

SAS Institute, Inc., *SAS User's Guide: Basics*, Version 6. Cary, N.C., 1990.

SAS Institute, Inc., *SAS User's Guide: Statistics*, Version 6. Cary, N.C., 1990.

Smith, Scott M., *PC-MDS: A Multidimensional Statistics Package*. Provo, Utah: Brigham Young University, 1989.

SPSS, Inc., *SPSS User's Guide*, 4th ed. Chicago, 1990.

SPSS, Inc., *SPSS Advanced Statistics Guide*, 4th ed. Chicago, 1990.

SPSS, Inc., *SPSS/PC+, Version 5.0*. Chicago, 1992.

SPSS, Inc., *SPSS for Windows, Version 6.0*, Chicago, 1993.

ANNOTATED SPSS CONTROL COMMANDS:

CHAPTER 2: CREATING THE SPSS SYSTEM FILE

```
DATA LIST   /ID 4-6 X1 10-12 X2 16-18
   X3 21-24 X4 28-30 X5 34-36 X6 40-42
   X7 45-48 X8 51 X9 54-57 X10 61-63        ...Identifies variables and column location.
   X11 66 X12 69 X13 72 X14 75.


FORMATS ID X1 X2 X4 X5 X6 X10 (F3.1).
FORMATS X3 X7 X9 (F4.1).                    ....Specifies the format of each variable.
FORMATS X8 X11 X12 X13 X14 (F1.0).


VARIABLE LABELS   ID 'ID'
 /X1 'Delivery Speed'
 /X2 'Price Level'
 /X3 'Price Flexibility'
 /X4 'Manufacturer Image'
 /X5 'Service'
 /X6 'Salesforce Image'
 /X7 'Product Quality'                      ....Specifies a label for each variable.
 /X8 'Firm Size'
 /X9 'Usage Level'
 /X10 'Satisfaction Level'
 /X11 'Specification Buying'
 /X12 'Structure of Procurement' ˙
 /X13 'Type of Industry (SIC)'
 /X14 'Type of Buying Situation' .


VALUE LABELS   X8  0 'SMALL' 1 'LARGE'
 /X11 0 'SPECIFICATION BUYING'
      1 'TOTAL VALUE ANALYSIS'              ....Specifies a label for variable values.
 /X12 0 'DECENTRALIZED' 1 'CENTRALIZED'
 /X13 0 'FIRM TYPE ONE' 1 'FIRM TYPE TWO'
 /X14 1 'NEW TASK' 2 'MODIFIED REBUY'
      3 'STRAIGHT REBUY' .


BEGIN DATA
    1   4.1   .6   6.9   4.7   2.4   2.3   5.2   0   32.0   4.2   1   0   1   1
    2   1.8   3.0   6.3   6.6   2.5   4.0   8.4   1   43.0   4.3   0   1   0   1
    3   3.4   5.2   5.7   6.0   4.3   2.7   8.2   1   48.0   5.2   0   1   1   2
    .    .     .     .     .     .     .     .    .     .     .    .   .   .   .
    .    .     .     .     .     .     .     .    .     .     .    .   .   .   .

    A complete listing of the dataset is provided at the end of this Appendix
    .    .     .     .     .     .     .     .    .     .     .    .   .   .   .

   98   2.0   2.8   5.2   5.0   2.4   2.7   8.4   1   38.0   3.7   0   1   0   1
   99   3.1   2.2   6.7   6.8   2.6   2.9   8.4   1   42.0   4.3   0   1   0   1
  100   2.5   1.8   9.0   5.0   2.2   3.0   6.0   0   33.0   4.4   1   0   0   1
END DATA .

SAVE OUTFILE='HATCO.SYS'.                   ....Saves data as a system file.
```

CHAPTER 2: EXAMINING YOUR DATA

```
EXAMINE X1 TO X7 X9 /STAT=ALL        ....Performs the tests necessary for
       /PLOT=ALL/ID=ID.                  testing the normality of the variables.
```

CHAPTER 3: MULTIPLE REGRESSION ANALYSIS

Multiple Regression

```
REGRESSION DESCRIPTIVES=ALL          ....Initiates the regression procedure and
                                        requests all descriptive statistics.
```

```
/VARIABLES=X1 TO X7 X9               ....Specifies the variables for analysis as
                                        X₁ to X₇ and X₉.
```
/VARIABLES=X1 TO X7 X9 ....Specifies the variables for analysis as X_1 to X_7 and X_9.

```
/CRITERIA=PIN(.05) POUT(.1) TOL(.01) ....Specifies the statistical criteria used
                                        in building the regression equation:
                                        (a) PIN = probability of F-to-enter,
                                        (b) POUT = probability of F-to-remove,
                                        (c) TOL = specified value that applies
                                            to both tolerance tests.
```

```
/STATISTICS=ALL                      ....Prints all summary statistics.
```

```
/DEPENDENT=X9                        ....Specifies the dependent variable as X₉.
```
/DEPENDENT=X9 ....Specifies the dependent variable as X_9.

```
/METHOD=STEPWISE                     ....Specifies the variable selection method
                                        as stepwise.
```

```
/RESIDUALS=DEFAULTS ID(ID)           ....Specifies output of information on
                                        outliers, statistics, histograms, and
                                        normal probability plots.  ID(ID)
                                        specifies id number as the label for
                                        casewise or outlier plots.
```

```
/CASEWISE=ALL SRE MAH SDR COOK LEVER ....Specifies the inclusion of all cases in
                                        the casewise plot and names the
                                        diagnostic variables to be used. Other
                                        diagnostic variables are available.
```

```
/SCATTERPLOT (*RES,*PRED)            ....Specifies variables for scatterplots.
```

```
/PARTIALPLOT=ALL.                    ....Specifies partial regression plots using
                                        all independent variables.
```

Logistic Regression Analysis

```
SET SEED=123456.                    ....Specifies a seed for the random number
                                        generator to generate a holdout sample.

COMPUTE RANDZ=UNIFORM(1)>.60.       ....Computes the variable RANDZ with a
                                        uniform distribution between 0 and 1.

LOGISTIC REGRESSION X11 WITH
    X1, X2, X3, X4, X5, X6, X7      ....Initiates logistic regression with X11 as
                                        the dependent variable with X1 through
                                        X7 as independent variables.

/METHOD=FSTEP                       ....Specifies a stepwise variable selection.

/SELECT=RANDZ EQ 0                  ....Selects cases with RANDZ equal to 0 for
                                        use in model estimation.

/PRINT=ALL                          ....Prints all available output.

/CRITERIA=ITERATE(50)               ....Specifies maximum iterations as 50.

/CASEWISE=PRED PGROUP RESID SRESID
        ZRESID LEVER COOK DFBETA    ....Specifies the diagnostic variables to be
                                        used in the casewise listing.
```

CHAPTER 4: MULTIPLE DISCRIMINANT ANALYSIS

2-Group Discriminant Analysis

```
SET SEED=54321.
COMPUTE RANDZ=UNIFORM(1)>.65.       ....Sets the seed and defines variable RANDZ
                                        for generating a holdout sample.

DSCRIMINANT GROUPS=X11(0,1)         ....Initiates discriminant analysis and
                                        specifies the grouping variable as X11
                                        with a range of values of 0 and 1.

/VARIABLES=X1 TO X7                 ....Specifies the predictor variables used.

/SELECT=RANDZ(0)                    ....Same as in logistic regression, but
                                        different syntax needed.

/METHOD=MAHAL                       ....Specifies the method for selecting
                                        variables for inclusion.

/PRIORS=SIZE                        ....Specifies the prior probabilities of
                                        group membership to be equal to the
                                        sample proportion of cases actually
                                        falling into each group.

/OPTIONS=6 7                        ....Specifies a rotated pattern matrix (6)
                                        and a rotated structure matrix (7).

/STATISTICS=ALL.                    ....Prints all available statistics.
```

3-Group Discriminant Analysis

	The only difference between a two group and three group discriminant analysis is the specification of the grouping variable and its range. It would appear as follows.
DSCRIMINANT GROUPS=X14(1,3)	....Now X_{14} is the grouping variable with values of 1 to 3.

CHAPTER 5:   MULTIVARIATE ANALYSIS OF VARIANCE

Multivariate Analysis of Variance (2 Group)

MANOVA X9 X10 BY X11(0,1)	....Specifies the MANOVA procedure with X_9 and X_{10} as dependent variables and X_{11} as the independent variable with a range of 0 to 1.
/METHOD=SSTYPE(UNIQUE)	....Requests the method of partitioning sums of squares [sstype] corresponding to an unweighted combinations of means [(unique)].
/PLOT=NORMAL,STEMLEAF,BOXPLOTS	....Plot a normal and detrended normal plot, a steam-and-leaf display, and a boxplot for the dependent variables.
/PRINT=CELLINFO(MEANS,CORR,COV),	....Specifies the printed output: (1) cell information of correlation matrices and variance-covariance matrices,
HOMOGENEITY(BARTLETT,BOXM),	(2) homogeneity tests, Bartlett-Box F and Box's M,
SIGNIF(MULTIV,STEPDOWN),	(3) significance tests of mulitvariate F tests for group differences and step-down tests,
PARAMETERS(ESTIM)	(4) estimated parameters, including standard errors, t-tests, and confidence intervals,
ERROR(COR)	(5) significance tests for equality of covariances.
/DESIGN.	....Specifies the structure of the model and must be the last subcommand of the model. Default (as shown) is the full factorial model.

Multivariate Analysis of Variance (3 groups)

The only difference between a two group MANOVA and a three group MANOVA is the specification of the independent variable and its range of values, which is now X_{14} with a range of 1 to 3.

MANOVA X9 X10 BY X14(1,3)

....Specifies the MANOVA procedure with X_9 and X_{10} as dependent variables and X_{14} as the independent variable with a range from 1 to 3.

Multivariate Analysis of Variance (2 Factor)

The only difference between a two group MANOVA and a two factor MANOVA is the specification of the independent variables and their range of values.

MANOVA X9 X10 BY X13 (0,1) X14(1,3)

....Specifies the MANOVA procedure with X_9 and X_{10} as dependent variables and X13 and X_{14} as the independent variables.

CHAPTER 6:   CANONICAL CORRELATION ANALYSIS

```
MANOVA X9 X10 WITH X1 TO X7
/PRINT = ERROR (SSCP COV COR)
        SIGNIF (HYPOTH STEPDOWN)
/DISCRIM =RAW, STAN ESTIM COR ALPHA(1.0)
/RESIDUALS=CASEWISE PLOT
/DESIGN .
```

....SPSS does not have a separate procedure for canonical analysis, but it can be performed through the MANOVA (multivariate analysis of variance) procedure with these commands.

CHAPTER 7:   FACTOR ANALYTIC METHODS

Components Analysis

FACTOR VAR=X1 TO X4 X6 X7

....Specifies the factor analysis procedure of X_1 to X_4, X_6 and X_7. X_5 is omitted after initial analysis.

/CRITERIA=ITERATE(50)

....Specifies the maximum number ofterations for the factor solution, default is 25.

/FORMAT=BLANK(0)

....Controls the displayed format of the factor matrices. BLANK(.30) would not show any variable loading below.30. In this example, all values are shown.

/PRINT=ALL

....Prints all available statistics.

/PLOT=EIGEN ROTAT(1,2)	....Plots eigenvalues in descending order in scree plot. Generate a factor loading plot with factors 1 and 2 as the axes.
/EXTRACT=PC	....Specifies the method of extraction to be principal components.
/ROTAT=VARIMAX.	....Requests the rotation method available. The default is VARIMAX. Other rotation methods (EQUAMAX, QUARTIMAX, OBLIMIN) available with additional /ROTAT commands.

Common Factor Analysis

The primary difference between principal components factor analysis and common factor analysis is the specification of the extraction method, which is now PAF. For this data set, the variable X_5 was dropped in the common factor analysis.

FACTOR VAR=X1 TO X4 X6 X7	....Variables to be analyzed without X_5.
/CRITERIA=ITERATE(150)	....Must increase the number of possible iterations in common factor analysis to ensure convergence.
/EXTRACT=PAF	....Specifies the common factor extraction method (PAF).

CHAPTER 8: CLUSTER ANALYSIS

Hierarchical

CLUSTER X1 TO X7	....Use the cluster procedure with X_1 to X_7.
/MEASURE=SEUCLID	....Specifies the distance measure used, squared Euclidean distance (default).
/METHOD=WARD(WCLUS)	....Clustering method to be used is Ward's method, which requires squared euclidean distances.WCLUS specifies a root name for saving clusters (see next command).
/SAVE=CLUSTER(2,5)	....Saves each case's cluster memberships for the two-, three-, four-, and five-cluster solution. The new variables are WCLUS5, WCLUS4, WCLUS3, and WCLUS2.
/PRINT=CLUSTER(2,5)	....Print cluster membership for each case, from 2 to 5 clusters. Print the distance matrix and agglomeration schedule.
/PLOT=DENDROGRAM.	....Plot the dendrogram procedure.

Non-hierarchical(Prespecified Cluster Seed Points)

```
QUICK CLUSTER X1 TO X7              ....Non-hierarchical clustering of X₁ to X₇.

/INITIAL = (4.40 2.43 1.39 3.22    ....Supplies the initial seed points,
            8.70 6.74 5.09 5.69        reading the cluster centroids for
            2.94 2.87 2.65 2.87        group 1, then group 2, on variables
            5.91 8.10)                 X₁, X₂, ..., X₇.  In this example, the
                                       centroid for group 1 on X₁ is 4.40 while
                                       group 2 has a mean value on X₁ of 2.43.

/CRITERIA = CLUSTERS (2)           ....Specifies two clusters will be formed.

/PRINT = CLUSTER ANOVA             ....Prints ANOVA test for differences of
                                       each variable across clusters.

/SAVE = CLUSTER (NHSCLUS) .        ....Saves the cluster membership in NHSCLUS.
```

Non-hierarchical(Random Selection of Cluster Seed Points)

```
SET SEED= 345678 .                 ....Specifies a seed number for random
                                       number generator to ensure replication.

QUICK CLUSTER X1 TO X7             ....Non-hierarchical clustering of X₁ to X₇.

/INITIAL = SELECT                  ....Randomly select initial cluster centers.

/CRITERIA = CLUSTERS (2)           ....Specifies two clusters will be formed.

/PRINT = CLUSTER ANOVA             ....Same as above.

/SAVE = CLUSTER (NHRCLUS) .        ....Same as above, variable now NHRCLUS.
```

CHAPTER 10: CONJOINT ANALYSIS

Designing The Stimuli: Generating anOrthogonal Fractional Factorial Design

```
                            Control cards used for the automatic generation of
                            an orthogonal set of stimuli in conjoint analysis.
                            Must set the SEED for exact replication.

ORTHOPLAN FACTORS =                ....Uses ORTHOPLAN program to generate an
  MIXTURE  'Product Form'              orthogonal fractional factorial design
           ('Premixed' 'Concentrate'  for five factors, three 3-level factors
            'Powder')                  and 2 two-level factors, labeling each
  NUMAPP   'Num of Applic'             labeling each level for each factor.
           ('50' '100' '200')
  GERMFREE 'Disinfectant'
           ('Yes' 'No')
  BIOPROT  'Biodegradable'
           ('No' 'Yes')
  PRICE    'Price/Applic'
           ('35 cents' '49 cents' '79 cents')
/HOLDOUT = 4 .                     ....4 additional stimuli for holdout sample.
SAVE OUTFILE='CPLAN1.SYS' .        ....Saves the generated plan for later use.
```

Designing The Stimuli: Specifying theOrthogonal Fractional Factorial Design

The control cards necessary to replicate the plan used in the HATCO example. Also example of method to input specified design of stimuli rather than generation as in above method.

```
DATA LIST FREE / MIXTURE  NUMAPP        ....Defines the factors and specifications
                 GERMFREE  BIOPROT              STATUS_: 0 -- use for estimation
                 PRICE                                  1 -- holdout sample
                 STATUS_  CARD_ .                        2 -- choice simulator data
                                               CARD_: card number


BEGIN DATA.
    2.00     3.00     1.00     1.00     1.00     0      1
    3.00     3.00     1.00     1.00     1.00     0      2
    1.00     2.00     1.00     2.00     2.00     0      3
    3.00     3.00     1.00     2.00     2.00     0      4
    3.00     1.00     1.00     1.00     3.00     0      5
    2.00     3.00     2.00     2.00     3.00     0      6
    1.00     2.00     1.00     1.00     3.00     0      7
    1.00     3.00     1.00     1.00     2.00     0      8      ...The levels of
    3.00     2.00     2.00     1.00     2.00     0      9         each factor that
    2.00     1.00     1.00     1.00     2.00     0     10         define each
    3.00     2.00     2.00     1.00     1.00     0     11         stimuli for use
    2.00     2.00     1.00     1.00     3.00     0     12         in estimation,
    1.00     3.00     2.00     1.00     3.00     0     13         validation and
    1.00     1.00     1.00     1.00     1.00     0     14         the choice
    2.00     2.00     1.00     2.00     1.00     0     15         simulator.
    1.00     1.00     2.00     2.00     1.00     0     16
    2.00     1.00     2.00     1.00     2.00     0     17
    3.00     1.00     1.00     2.00     3.00     0     18
    2.00     2.00     1.00     1.00     2.00     1     19
    3.00     2.00     2.00     2.00     1.00     1     20
    3.00     3.00     1.00     2.00     3.00     1     21
    2.00     1.00     2.00     2.00     1.00     1     22
    1.00     1.00     1.00     2.00     3.00     2     23
    2.00     3.00     1.00     2.00     2.00     2     24
    3.00     3.00     1.00     2.00     1.00     2     25
END DATA .

SAVE OUTFILE='CPLAN1.SYS' .                      ....Saves generated
                                                     plan for later
                                                     use.
```

Printing Plancards (Full-Profile Descriptions)

```
GET FILE='CPLAN1.SYS' .                        ....Recalls orthogonal plan.

PLANCARDS FACTORS=MIXTURE NUMAPP               ....Specifies factors to use.
                GERMFREE BIOPROT PRICE

  /FORMAT=BOTH                                 ....Generates both cards and
                                                   listing.

  /TITLE=                                      ....Title appearing on each
    'HYPOTHETICAL INDUSTRIAL CLEANSER )CARD' .     stimuli card.  The
                                                   )CARDplaces the card number
                                                   on each stimuli.
```

Estimating the ConjointAnalysis Model

The control cards necessary to (1) read in the preference data provided by respondents when evaluating the stimuli and (2) estimate the conjoint model.

```
DATA LIST /
  QN 2-5 PROD1     7-8 PROD2   10-11 PROD3  13-14 PROD4   16-17 PROD5 19-20
         PROD6   22-23 PROD7   25-26 PROD8  28-29 PROD9   31-32 PROD10 34-35
         PROD11 37-38 PROD12   40-41 PROD13 43-44 PROD14  46-47 PROD15 49-50
         PROD16 52-53 PROD17   55-56 PROD18 58-59 PROD19  61-62 PROD20 64-65
         PROD21 67-68 PROD22   70-71 .
BEGIN DATA.
  104   4 6 5 4 4 6 4 4 4 4 4 4 4 5 5 4 4 4 4 4 6 6
  107   6 3 5 2 3 1 1 6 6 6 7 4 1 6 6 6 6 1 7 7 1 7
   .   .  .  .  .  .  .  .  .  .  .  .  .  .  .  .  .  .  .  .  .  .  .
        The complete data set is listed at the end of the appendix
   .   .  .  .  .  .  .  .  .  .  .  .  .  .  .  .  .  .  .  .  .  .  .
  417   5 5 2 5 1 1 2 2 2 2 1 1 1 4 6 3 2 1 4 4 2 3
  418   6 7 2 7 1 4 1 3 6 4 7 1 5 2 1 2 4 2 3 7 5 7
END DATA.

CONJOINT PLAN='CPLAN1.SYS'            ....Retrieves the orthogonal plan.

 /FACTORS=                            ....Selects factors in conjoint estimation.
  MIXTURE 'MIXTURE'
          ( 'PREMIXED' 'CONCENTRATE'
            'POWDER' )
  NUMAPP  'NUMBER APP'
          ( '50' '100' '200' )
  GERMFREE 'DISINFECTANT'
          ( 'YES' 'NO' )
  BIOPROT 'BIODEGRADEABLE'
          ( 'NO' 'YES' )
  PRICE 'PRICE PER APP'
          ( '$.35' '$.49' '$.79' )
 /SUBJECT=QN                          ....Selects variable QN as subject ID.
 /SCORE=PROD1 PROD2 PROD3 PROD4 PROD5 ....Specifies the preference response
        PROD6 PROD7 PROD8 PROD9 PROD10    variables. They must be listed in the
        PROD11 PROD12 PROD13 PROD14       order of profiles in the orthogonal
        PROD15 PROD16 PROD17 PROD18       design.
        PROD19 PROD20 PROD21 PROD22
 /UTILITY='UTIL.SYS' .                ....Saves the part-worth estimates.
```

ANNOTATED SAS CONTROL COMMANDS

CHAPTER 1: CREATING THE SAS DATA FILE

```
DATA HATCO;                              ....Specifies a temporary data file name.

INPUT ID 4-6 X1 10-12 X2 16-18 X3 21-24
  X4 28-30 X5 34-36 X6 40-42 X7 45-48
  X8 51 X9 54-57 X10 61-63               ....Identifies variables and column location
  X11 66 X12 69 X13 72 X14 75;

LABEL ID 'ID'
X1 'Delivery Speed'
X2 'Price Level'
X3 'Price Flexibility'
X4 'Manufacturer Image'
X5 'Service'
X6 'Salesforce Image'
X7 'Product Quality'                     ....Specifies a label for each variable.
X8 'Firm Size'
X9 'Usage Level'
X10 'Satisfaction Level'
X11 'Specification Buying'
X12 'Structure of Procurement'
X13 'Type of Industry'
X14 'Type of Buying Situation';

CARDS;
   1   4.1    .6   6.9   4.7   2.4   2.3   5.2   0   32.0   4.2   1   0   1   1
   2   1.8   3.0   6.3   6.6   2.5   4.0   8.4   1   43.0   4.3   0   1   0   1
   .    .     .     .     .     .     .     .    .     .     .    .   .   .   .
   .    .     .     .     .     .     .     .    .     .     .    .   .   .   .
```

A complete listing of the dataset is provided at the end of this Appendix

```
   .    .     .     .     .     .     .     .    .     .     .    .   .   .   .
   .    .     .     .     .     .     .     .    .     .     .    .   .   .   .
  99   3.1   2.2   6.7   6.8   2.6   2.9   8.4   1   42.0   4.3   0   1   0   1
 100   2.5   1.8   9.0   5.0   2.2   3.0   6.0   0   33.0   4.4   1   0   0   1
```

CHAPTER 3: MULTIPLE REGRESSION ANALYSIS

Multiple Regression

```
PROC REG;                                ....Initiates the regression procedure.

MODEL X9=X1-X7
   /ALL SELECTION=STEPWISE PARTIAL;      ....Identifies the regression model to be
                                            used: $X_9$ as the dependent variable, $X_1$
                                            to $X_7$ as independent variables. All
                                            statistics given with the stepwise
                                            entry procedure. Requests partial
                                            regression leverage  plots for each
                                            independent variable.
PLOT R.*P.;                              ....Specifies variables for plotting,
                                            residuals and predicted.
```

Logistic Regression Analysis

PROC CATMOD;   ....Initiates the categorical data modeling.

DIRECT X1-X7;   ....Specifies the independent variables containing design matrix values.

MODEL X11=X1-X7/ML CORRB FREQ ONEWAY
     PRED=PROB NOGLS XPX;   ....Specifies dependent variable as X_{11}, and independent variables, X_1 to X_7, plus optional additional output. For use as logistic regression, one must request ML and NOGLS. This uses maximum-likelihood estimates and suppresses computation of generalized (weighted) least-squares.

CHAPTER 4: MULTIPLE DISCRIMINANT ANALYSIS

2 Group Discriminant Analysis

PROC DISCRIM CAN ALL;   ....Initiates the discriminant analysis procedure with all statistics.

CLASS X11;   ....Specifies X_{11} as dependent variable.

VAR X1-X7;   ....Specifies predictor variables X_1 to X_7.

3 Group Discriminant Analysis

The only modification needed for a three group discriminant analysis is the identification of a new classification variable, X_{14}, a three group variable.

CLASS X14;   ....Selects X_{14} as classification variable.

CHAPTER 5: MULTIVARIATE ANALYSIS OF VARIANCE

Multivariate Analysis of Variance (2 groups)

PROC GLM;   ....Initiates the general linear model (GLM)

CLASS X11;   ....Selects X_{11} as classification variable.

MODEL X9 X10 = X11;   ....Identifies the MANOVA model with X_{11} as the independent variable and X_9, X_{10} as the dependent variables.

MEANS X11 / BON SNK TUKEY;   ....Requests means for each level of X_{11} with Bonferroni t-tests, Student-Newman-Keuls multiple range tests, and Tukey's studentized range test on main effects

MANOVA H=X11 / SUMMARY;   ....Specifies the effect employed as the hypothesis matrices and ANOVA tables for each dependent variable.

Multivariate Analysis of Variance (3 groups)

The only modification needed for a three group MANOVA analysis is the classification variable, X_{14}, a three group variable.

CLASS X14;   ....Selects X_{14} as classification variable.

Multivariate Analysis of Variance (2 Factor)

The only modification needed for a two factor MANOVA analysis is the classification variables, X13 and X_{14}.

CLASS X13 X14;   ....Identifies the classification variables as X13 and X_{14}.

CHAPTER 6: CANONICAL CORRELATION ANALYSIS

PROC CANCORR ALL;   ....Initiates the canonical correlation procedure with all additional output.

VAR X9 X10;   ....Selects X_9 and X_{10} as dependent variables

WITH X1-X7;   ....Selects X_9 to X_{10} as predictor variables

CHAPTER 7: FACTOR ANALYTIC METHODS

Components Analysis

PROC FACTOR ROTATE=P RE;   ....Initiates the factor procedure with a PROMAX rotation of the factor structure. Reorders the printed factor matices by absolute magnitude of each variable's loading on each factor.

VAR X1-X4 X6 X7;   ....Identifies variables for factor analysis as X_1 to X_7, omittng X_5 after initial analysis.

PROC FACTOR R=E RE;   ....Initiates the factor procedure with EQUAMAX rotation of factor structure

VAR X1-X4 X6 X7;   ....Identifies variables for factor analysis as X_1 to X_7, omittng X_5.

PROC FACTOR R=Q RE;   ....Initiates the factor procedure with QUARTIMAX rotation of the factor structure and reordering.

VAR X1-X4 X6 X7;   ....Identifies variables for factor analysis as X_1 to X_7, omittng X_5.

Common Factor Analysis

```
PROC FACTOR METHOD=PRINIT ROTATE=P RE;
VAR X1-X4 X6 X7;
PROC FACTOR METHOD=PRINIT R=E RE;          ....Same as principal components analysis
VAR X1-X4 X6 X7;                               except the method is specified to be
PRINIT,                                        common factor analysis.
PROC FACTOR METHOD=PRINIT R=Q RE;
VAR X1-X4 X6 X7;
```

CHAPTER 8: CLUSTER ANALYSIS

Hierarchical

```
PROC CLUSTER M=WARD;                       ....Initiates the cluster procedure with
                                               Ward's method.

VAR X1-X7;                                 ....Identifies variables for cluster
                                               analysis as $X_1$ to $X_7$.

PROC TREE N=5;                             ....Initiates the tree procedure with the
                                               maximum number of clusters to be
                                               diagramed being 5.  Same as a dendogram.
```

Non-Hierarchical(Prespecified Cluster Seed Points)

```
PROC FASTCLUS MAXC=2 SEED=SEED2
  OUT=NEW DISTANCE;                        ....Initiates the fast cluster procedure,
                                               specifies the maximum number of clusters
                                               to be 2, specifies file (SEED2) with
                                               cluster centers, identifies the new
                                               temporary file to be saved as 'new
                                               distance'.

VAR X1-X7;                                 ....Identifies the variables to be $X_1$ to $X_7$.

PROC ANOVA; CLASS CLUSTER1 CLUSTER2;       ....Implements the ANOVA procedure with the
                                               two clusters as the classification
                                               variable.

MODEL X1-X7 = CLUSTER1 CLUSTER2;           ....Specifies the ANOVA model with
                                               independent variables as clusters and
                                               independent variables of $X_1$ to $X_7$.  This
                                               tests for significant differences
                                               between the clusters on the variables
                                               used.
```

Non-Hierarchical(Random Selection of Cluster Seed Points)

```
PROC FASTCLUS MAXC=2 OUT=NEW DISTANCE;
  VAR X1-X7;                               ....Same as above, but with random selection
                                               of initial cluster centers.

PROC ANOVA; CLASS CLUSTER1 CLUSTER2;
  MODEL X1-X7 = CLUSTER1 CLUSTER2;
RUN;
```

ANNOTATED BMDP CONTROL COMMANDS:

CHAPTER 1: READING THE BMDP DATA FILE

/INPUT TITLE IS 'TITLE'.	....Specifies a title for the output.
FILE='C:\HATCO.DAT'.	....Indicates the data file name.
VARIABLES=15.	....Specifies the number of variables in the data set.
FORMAT=FREE.	....Identifies the use of a freefield format.
MCHAR='.'.	....Specifies the character used to represent missing data.
/VARIABLE NAMES=ID, X1, X2, X3, X4, X5, X6, X7, X8, X9, X10, X11, X12, X13, X14.	....Specifies a name for each variable.

CHAPTER 2: EXAMINING YOUR DATA

Assessing Missing Data (BMDPAM)

/PRINT	....Specifies the following printed output:
COUNT.	Frequencies and percentages
STEM.	Stem and leaf histogram
MISS.	Cases with missing data
MEAN.	Mean and standard deviations
EXTR.	Minimum and maximum values
EZSC.	z-scores of minimum, maximum values
SK.	Skewness and kurtosis per variable.
/END	....Indicates end of command procedure.

Assessing Missing Data (BMDPAM)

/ ESTIMATE	METHOD= REGR. TYPE = ML. CLIMIT=90 .	....Uses regression method to estimate missing data. Maximum likelihood (ML) is used in estimating the correlation matrices, and cases are omitted if more than 90% of variables have missing data.
/ PRINT	MATRICES =	.... Specifies the following printed output:
	PAT,	(a) Pattern of missing data
	FREQ,	(b) Sample size, missing % by variable
	CORRDICH,	(c) Dichotomized correlations
	CORR,	(d) Estimated correlation matrix
	EIGEN,	(e) Eigenvalues of correlation matrix
	EST,	(f) Method of estimating missing data
	DIS,	(g) Mahalanobis D^2 (if METHOD=REGR)
	TTEST .	(h) t-tests of each variable
/ END		

CHAPTER 3: MULTIPLE REGRESSION ANALYSIS

Multiple Regression (BMDP2R)

USE=X1, X2, X3, X4, X5, X6, X7, X9		....Specifies the variables for use.
/REGRESS	DEPEND=X9.	....Identifies the dependent variable as X_9.
	INDEPEND=X1, X2, X3, X4, X5, X6, X7.	....Specifies independent variables X_1 - X_7.
	ENTER=.05, 5.0.	....Designates the F-to-enter limits for forward and backward stepping.
	REMOVE=.04, 4.9.	....Designates the F-to-remove limits for forward and backward stepping.
	TOL=.01.	....Specifies the tolerance limit.

/PRINT   ....Specifies the following printed output:

ANOVA.	(1) Analysis of variance for each step
STEP.	(2) Results at every step
COEF.	(3) Summary of regression coefficients
SUM.	(4) Summary with R, R2, F-to-enter, and F-to-remove at each step
COVA.	(5) Covariance matrix of variables
CORR.	(6) Correlation matrix of variables
DATA.	(7) Input data, residuals, and predicted values
PART.	(8) Stepwise partial correlations
FRAT.	(9) Summary of F-to-enter and F-to-remove values
RREG.	(10) Correlations of coefficients
DIAG=ALL.	....Requests printing of all regression diagnostics for each case

/PLOT   ....Requests the following residual plots:

RESID.	(1) Residuals vs. predicted values
NORM.	(2) Normal probability plot
DNORM.	(3) Detrended normal probability plot
VARIABLE=X_9.	....Specifies the variable to be used in 2 plots: 1) the observed values and predicted values of the dependent variable against the specified variable, and 2) the residuals against the specified variable.
PREP=X1,X2,X3,X4,X5, X6, X7.	....Lists the independent variables to be plotted in partial residual plots

/END   ....Indicates the end of the procedure.

CHAPTER 4:   MULTIPLE DISCRIMINANT ANALYSIS

2-Group Discriminant Analysis (BMDP7M)

USE=X_1, X_2, X_3, X_4, X_5, X_6, X_7, X_{11}.	....Specifies the variables in the analysis.
/TRANSFORM	....Initiates data transformation process.
IF (KASE EQ 1) THEN (N=100. S=60.).	....Designates the random selection of a
USE=-1.	sub-sample of size 60 from the sample of
IF (RNDU (1123)*N LT S)	100.
THEN (USE=1. S=S-1.).	
N=N-1.	
/GROUP VARIABLE=X11.	....Specifies dependent variable as X_{11}.
CODES(X11)=0 TO 1.	....Identifies the values used to assign group membership.
NAMES(X_{11})=SPECBUY, TVA.	....Assigns a name to each group.
/DISC	....Initiates discriminant analysis.
ENTER=1.0, 5.0.	....Designates the F-to-enter limits for forward and backward stepping.
REMOVE=.95, 4.9.	....Designates the F-to-remove limits for forward and backward stepping.
METH=1.	....Selects method of discriminant analysis. Method 1 removes variables from the classification function using F-to-remove limits; method 2 removes variables to reduce Wilks' lambda.
STEP=14.	....Specifies the maximum number of steps.
TOL=.001.	....Specifies the tolerance limit.
/PRINT CORR.	....Prints within-groups correlation matrix.
/END.	....Indicates the end of the procedure.

3-Group Discriminant Analysis (BMDP7M)

	The only difference between a two group and a three group discriminant analysis is the specification of the grouping variable, including its range and labels.
/GROUP VARIABLE=X_{14}.	....Specifies the dependent variable as X_{14}.
CODES(X_{14})=1 TO 3.	....Identifies the values used to assign group membership.
NAMES(X_{14})=NEW, MODIFIED, STRAIGHT.	....Assigns a name to each group.

CHAPTER 5:   MULTIVARIATE ANALYSIS OF VARIANCE

Multivariate Analysis of Variance (2 Group) (BMDP4V)

LABEL=ID.		....Assigns a means of identifying subjects.
USE=X9, X10, X11.		....Specifies the use of X_9, X_{10}, X_{11}.
/PRINT	CELLS.	....Prints statistics (mean, standard deviation, weighted mean, maximum, minimum) for each cell of the design.
	MARG=ALL.	....Prints marginal statistics for all cells
/BETWEEN	FACTORS=X_{11}.	....Specifies X_{11} as the independent variable
	CODES(X_{11})=0 TO 1.	....Identifies values for group membership.
	NAMES(X_{11})=SPEC, TVA.	....Assigns labels to each group.
/WEIGHT	BETW=EQUAL.	....Specifies the desired weighting of groups. (When proportional weighting is desired the command is SIZES.)
/END		....Indicates the end of the procedure.
ANALYSIS	PROCEDURE=FACTORIAL	....Requests an analysis of variance design which includes all main effects and interactions of all factors
	UNISUM.	....Prints a univariate summary table for each dependent variable.
	EST. /	....Prints parameter estimates
	PRIN/	....Prints D-matrices and matrices of mean-contrast coefficients.
END /		....Indicates the end of the procedure.

Multivariate Analysis of Variance (3 Group) (BMDP4V)

		The only difference between a two group and a three group MANOVA is the specification of the independent variable' values and labels.
USE=X9, X10, X14.		....Specifies the variables to be used.
/BETWEEN	FACTORS=X14.	....Specifies the independent variable - X_{14}.
	CODES(X_{14})=1 TO 3.	....Identifies the values used to assign group membership.
	NAMES(X_{14})=NEW, MODIFIED, STRAIGHT.	....Assigns labels to each group.

Multivariate Analysis of Variance (2 Factor) (BMDP4V)

		The only difference between a one factor MANOVA and a two factor MANOVA is the specification of the independent variables' values and labels.

USE=X9, X10, X13, X14. ....Specifies the variables in the analysis.

/BETWEEN FACTORS=X13, X14. ....Specifies X13, X14 as independent variables.

      CODES(X13)=0 TO 1. ....Identifies the values of group membership.

      NAMES(X13)=TYPE1, ...Assigns names to each group of X13.
        TYPE2.

      CODES(X14)=1 TO 3. ....Identifies the values of group membership.

      NAMES(X14)=NEW, ...Assigns names to each group of X14.
      MODIFIED, STRAIGHT.

CHAPTER 6: CANONICAL CORRELATION ANALYSIS (BMDP6M)

USE=X1, X2, X3, X4, X5, X6, X7, ...Specifies the variables in the analysis.
  X9, X10.

/CANON   ....Initiates canonical correlation procedure.

      FIRST=X9, X10. ....Specifies the dependent variables.

      SECOND=X1,X2,X3,X4, ...Specifies the independent variables.
        X5,X6,X7.

/PRINT MATR= ....Prints the following matrices:
      CORR, (a) correlations,
      COVA, (b) covariances,
      LOAD, (c) canonical variable loadings,
      COEF, (d) coefficients of the canonical variables,
      CANV. (e) canonical variable scores.

/END   ....Indicates the end of procedure.

CHAPTER 7:   FACTOR ANALYTIC METHODS

Components Analysis (BMDP4M)

```
USE=X1, X2, X3, X4, X6, X7.        ....Specifies the variables to be used.

/FACTOR        METH=PCA.           ....Specifies method of extraction to be
                                       principal components analysis.

               ITER=100.           ....Specifies the maximum number of iterations
                                       for the factor solution.

/ROTATE        METH=VMAX.          ....Specifies the VARIMAX rotation method.

/PRINT                             ....Requests the following printed output:
               STAND.                 (a) Standard scores for each case.
               COVA.                  (b) Covariance matrix.
               CORR.                  (c) Correlation matrix.
               FSCF.                  (d) Factor score coefficients.
               PART.                  (e) Partial correlations of each pair of
                                          variables.
               RESI.                  (f) Residual correlations.
               CRON.                  (g) Cronbach's alpha.
               ITER.                  (h) Log likelihood at each iteration.

/END                               ....Indicates end of the procedure.
```

Common Factor Analysis (BMDP4M)

The only difference between principal components factor analysis and common factor analysis is the specification of the extraction method.

```
/FACTOR        METH=PFA.           ....Specifies the method of extraction to be
                                       principal factor analysis.
```

CHAPTER 8:   CLUSTER ANALYSIS

Hierarchical (BMDP2M)

```
USE=X1,X2,X3,X4,X5,X6,X7.          ....Specifies the variables to be used.

/PROCEDURE     DIST=SUMOFSQ.        ....Specifies the distance measure to be used,
                                       squared Euclidean distance.

               LINK=SINGLE.         ....Selects clustering method as single linkage.

/PRINT                              ....Prints the following output:
               DIST.                   (a)Initial distances between cases.
               HIST.                   (b) Histogram of case distances.

/END                                ....Indicates the end of the procedure.
```

Non-hierarchical (Prespecified Cluster Seed Points) (BMDPKM)

USE=X1,X2,X3,X4,X5,X6,X7.	....Specifies the variables to be used.
/CLUS	....Initiates clustering procedure.

SEED=4.40, 1.39, 8.70,5.09
  2.94,2.65,5.91 .
SEED=2.43, 3.22, 6.74, 5.69,
  2.87, 2.87, 8.10.
	....Specifies initial cluster seeds,
	  repeated for each cluster.

/PRINT
     DIST.
     MEMB.
	....Prints the following output:
	  (a) Distances between cluster centers.
	  (b) Within cluster statistics and list
	    of cluster members.

/PLOT CLUST.	....Plots a bivariate scatterplot indicating final cluster membership.
/END	....Indicates the end of the procedure.

Non-hierarchical(Random Selection of Cluster Seed Points) (BMDPKM)

The non-hierarchical cluster analysis with random selection of seed points is the same as the non-hierarchical cluster analysis with prespecified cluster seed points, except that when the SEED command is omitted, the procedure will default to random selection.

ANNOTATED PC-MDS CONTROL COMMANDS

CHAPTER 9: MULTIDIMENSIONAL SCALING

Analysis of Similarities data: INDSCAL

```
 3   4   1   1   50   1   0   1   0   0   '12345678' 0 0 1 0 .001   ....Program control cards
18  10  10
(9F3.0)
 4
 6   9
 6   6   1
 6   6   4   8
 3   2   2   3   3
 3   5   2   2   2   4                                    ....Similarity ratings for
 2   2   4   4   4   2   4                                     first of 18 respondents.
 4   2   9   2   8   8   2   2
 8   9   4   2   2   5   9   3   2
 8
 7   6
 1   1   3
 1   1   2   8
 3   1   1   3   1
 1   2   3   1   1   3                                    ....Respondent 2.
 1   1   2   1   1   2   1
 1   4   1   1   7   1   1   1
 1   6   2   1   1   2   1   1   1
 9
 8   7
 5   5   7
 6   6   7   8
 7   5   5   7   6
 6   7   7   6   6   7
 6   6   8   6   6   7   5
 5   8   9   7   8   8   7   6
 8   8   7   7   7   8   9   7   7
 7
 6   4
 3   3   5
 2   2   3   8
 3   3   3   4   3
 3   5   4   3   3   4
 3   3   5   3   4   5   3
 3   4   8   4   8   4   3   3
 7   6   3   3   3   4   8   2   3
 5
 2   7
 2   2   1
 3   2   8   5
 3   3   2   2   2
 2   2   2   2   2   4
 2   2   2   2   2   1   2
 2   7   7   1   6   5   2   1
 5   5   2   1   1   1   6   2   2
```

```
8
7  4
1  1  2
1  1  2  8
2  1  1  1  1
1  2  2  1  1  1
1  1  2  1  1  1  1
1  6  8  1  8  6  1  1
7  9  1  1  1  1  6  1  1
6
4  3
1  1  3
1  1  3  6
4  1  1  4  1
2  4  4  1  1  4
1  1  4  2  1  3  1
2  3  7  2  7  4  1  2
8  6  1  2  2  2  6  2  2
8
7  2
1  1  2
1  1  2  7
2  4  1  1  1
1  1  1  1  1  2
1  1  2  1  1  2  1
1  6  6  1  2  2  1  1
2  2  1  1  1  4  6  1  1
3
3  4
2  2  2
2  2  2  6
2  2  2  4  2
2  3  3  3  3  3
3  3  3  3  3  2  2
2  5  7  2  8  7  4  2
7  8  5  2  2  4  8  2  2
9
9  3
1  7  2
2  2  7  9
2  2  2  7  2
2  2  2  2  2  7
2  2  2  2  2  7  3
3  2  8  2  8  7  2  2
8  8  2  2  2  2  8  2  2
3
3  6
1  1  1
1  1  2  4
3  1  1  2  1
1  1  4  2  1  2
1  1  1  1  1  2  1
1  1  7  2  7  6  3  2
5  8  2  2  2  2  7  2  1
```

```
8
8  6
6  5  8
6  6  6  8
6  6  6  8  6
6  5  6  5  5  8
6  6  6  7  6  6  6
6  5  8  6  8  7  4  5
7  7  5  6  6  6  8  5  5
9
8  6
1  1  5
3  3  6  8
6  2  2  5  3
3  6  6  1  1  5
3  3  6  4  4  7  3
3  8  8  2  6  5  5  6
6  6  3  4  4  2  8  2  2
7
4  4
2  1  1
1  1  3  5
6  1  1  1  1
1  2  5  1  1  1
1  1  2  1  1  1  1
1  3  3  1  7  4  1  1
4  7  1  1  1  1  5  1  1
8
7  7
5  6  7
6  7  5  7
6  5  5  7  6
7  5  6  5  5  6
5  6  5  5  5  6  5
5  7  6  5  8  7  6  6
7  8  6  5  5  6  8  6  6
7
7  5
1  1  2
1  1  2  5
5  1  1  2  1
1  2  5  1  1  2
1  1  2  2  2  1  1
1  7  2  1  7  7  1  1
7  8  1  1  1  2  7  1  1
4
4  5
2  2  2
2  2  4  8
4  2  2  4  2
2  3  4  2  2  2
2  2  4  2  2  4  2
2  4  6  4  5  4  3  3
5  5  3  3  3  5  6  3  3
```

```
6
7  9
1  1  8
1  1  6  7
6  1  1  8  1
1  6  6  1  1  8
1  1  7  1  1  7  1
1  7  9  1  9  9  2  1
9  9  1  1  1  6  9  1  1
```

Interpretation of the MDS Solution: PROFIT

```
1  10   2   8   0  0  3  0.0                    ....Program control cards.
(2F9.0)
    .03698   -.41766
    .12935   -.44803
    .39184   -.16141
    .05099    .44502                    ....Stimulus coordinates obtained
    .36999    .38176                         from INDSCAL analysis.
   -.32953    .26374
   -.49305   -.09691
   -.15564    .20935
    .39511    .15367
   -.39604   -.32954
(10F6.0)
Property 1
  6.94   7.17   7.67   3.22   4.78   5.11   6.56   1.61   8.78   3.17   ....Average evaluation
Property 2                                                                    of each object
  4.00   1.83   6.33   7.67   6.00   5.78   5.50   6.11   7.50   4.17         on 8 attributes
Property 3
  6.94   5.67   3.39   3.67   3.67   6.94   6.44   7.22   4.94   6.11
Property 4
  4.00   3.39   7.33   6.11   7.50   4.22   7.17   4.33   8.22   5.56
Property 5
  5.16   3.47   6.41   5.88   6.06   4.94   5.29   4.82   8.35   4.65
Property 6
  5.11   1.22   5.78   7.89   6.56   3.83   4.28   6.94   8.67   4.72
Property 7
  5.33   3.72   6.33   5.56   6.39   4.72   5.28   5.22   7.33   5.11
Property 8
  4.17   1.56   6.06   8.22   7.72   4.28   3.89   6.33   7.72   5.06
HATCO
Firm B
Firm C
Firm D
Firm E
Firm F                                                           ....Labels.
Firm G
Firm H
Firm I
Firm J
```

Incorporating Preferences into the MDS Solution: PREFMAP

```
  10   2   6   0   1   0   2   4   0   1   1   25   0   0   1              ....Program control cards.
(2F9.0)
      .03698   -.41766
      .12935   -.44803
      .39184   -.16141
      .05099    .44502
      .36999    .38176                                        ....Stimulus coordinates
     -.32953    .26374                                           from INDSCAL analysis.
     -.49305   -.09691
     -.15564    .20935
      .39511    .15367
     -.39604   -.32954
(2X,10F6.0)
   1     2     3     5     6     7     4    10     8     1     9
   2     5     2     7     6     9     3     4     1    10     8
   3     4     1     8     7     6     9     3     5    10     2        ....Preference
   4     4     3    10     2     7     8     6     1     9     5            ratings of 10
   5     4     2     7     9     6     1     5     8    10     3            objects by 6
   6     4     1     8     7     9     3     5     2    10     6            subjects.
HATCO
Firm B
Firm C
Firm D
Firm E
Firm F                                                          ....Labels.
Firm G
Firm H
Firm I
Firm J
SUBJ 1
SUBJ 2
SUBJ 3
SUBJ 4
SUBJ 5
SUBJ 6
```

Developing Perceptual Maps With Crosstabulated Data by Correspondence Analysis:
CORRESP

```
 8  10
(10F3.0)
  4   3   1  13   9   6   3  18   2  10
 15  16  15  11  11  14  16  12  14  14
 15  14   6   4   4  15  14  13   7  13
 16  13   8  13   9  17  15  16   6  12
 14  14  10  11  11  14  12  13  10  14
  7  18  13   4   9  16  14   5   4  16
  6   6  14  10  11   8   7   4  14   4
 15  18   9   2   3  15  16   7   8   8
Product Quality
Strategic Orientation
Service
Delivery Speed
Price Level
Sales Force Image
Price Flexibility
Mfgr. Image
HATCO
Firm A
Firm B
Firm C
Firm D
Firm E
Firm F
Firm G
Firm H
Firm I
```

....Program control cards.

....Crosstabulated data of attributes by objects (firms).

....Labels.

ANNOTATED LISREL VII CONTROL COMMANDS

CHAPTER 11:   STRUCTURAL EQUATION MODELING

Confirmatory Factor Analysis: Initial Model Specification

```
CONFIRMATORY FACTOR ANALYSIS          ....Title Card.
DA NI=7 NO=100 MA=KM                  ....Specifies data file for number of
                                          variables (7), sample size (100) and
                                          data type (KM = correlation).

KM  FU FI=C:\HATCO.COR     FO=5        ....Reads data file from disk.
(7F9.0)

SELECT                                ....Selects correlations for analysis from
  1 2 3 4 6 7 /                           entire matrix.  Note that variable 5 is
                                          omitted.

MO NX=6 NK=2 PH=ST TD=SY,FI           ....Model card defines number of exogenous
                                          indicators (6), number of exogenous
                                          constructs (2), and characteristics of
                                          associated matrices.

LA                                    ....Labels for variables in input matrix.
  'DelvSpd' 'PriceLvl' 'PriceFlx'
  'MfgImage' 'Service' 'SalesImg' 'Quality'

LK                                    ....Labels for exogenous constructs.
'Strategy' 'Image'

PA LX                                 ....Pattern matrix specifying loadings of
1 (1 0)                                   indicvators on exogenous constructs.
1 (1 0)                                   This format is suggested as it
1 (1 0)                                   corresponds directly to the familiar
1 (0 1)                                   format of factor analysis and text
1 (0 1)                                   discussions.
1 (1 0)

FR TD(1,1) TD (2,2) TD(3,3) TD(4,4) TD(5,5) TD(6,6)   ...."Frees" the indicator error
                                          terms for estimation.

OU  SS TV RS MI                       ....Output card: requests standardized
                                          solution, t-values,residuals, and
                                          modification indices.
```

Confirmatory Factor Analysis: Model Respecification

```
CONFIRMATORY FACTOR ANALYSIS
DA NI=7 NO=100 MA=KM
KM  FU FI=C:\HATCO.COR      FO=5
(7F9.0)
SELECT
  1 2 3 4 6 7 /
MO NX=6 NK=2 PH=ST TD=SY,FI
LA
  'DelvSpd' 'PriceLvl' 'PriceFlx' 'MfgImage'   ....Cards same as in the earlier model.
  'Service' 'SalesImg' 'Quality'
LK
'Strategy' 'Image'
PA LX
1 (1 0)
1 (1 0)
1 (1 0)
1 (0 1)
1 (0 1)
1 (1 0)
FR TD(1,1) TD (2,2) TD(3,3) TD(5,5) TD(6,6)


VA .005 TD(4,4)                              ....Specifies error variance of variable
                                                 4 to be .005 as remedy for Heywood
                                                 case.


OU  SS TV RS MI                              ....Same as in earlier model.
```

Confirmatory Factor Analysis: Estimation of Null Model

```
CONFIRMATORY FACTOR ANALYSIS -- NULL MODEL
DA NI=7 NO=100 MA=KM
KM  FU FI=C:\HATCO.COR      FO=5
(7F9.0)
SELECT                                       ....Cards same as earlier models.
  1 2 3 4 6 7 /

MO NX=6 NK=1 PH=ST TD=SY,FI                  ....Specifies single exogenous construct
                                                 for null model.

LA
  'DelvSpd' 'PriceLvl' 'PriceFlx' 'MfgImage'   ....Same as earlier.
  'Service' 'SalesImg' 'Quality'

LK                                           ....Labels single factor as Null Model.
'Null Mod'

PA LX                                        ....Specifies no loadings for indicators
6 (0)                                            (see below).

VA 1.0 LX(1,1) LX(2,1) LX(3,1) LX(4,1) LX(5,1) LX(6,1)
FR TD(1,1) TD (2,2) TD(3,3) TD(4,4) TD(5,5) TD(6,6)    ....Specifies that construct
                                                           loadings for all indicators
                                                           equals 1.0 (no measurement
                                                           error) and frees error
                                                           terms for estimation.
```

Structural Equation Model (Path Model): Model Estimation

```
CAUSAL MODEL WITH MULTIPLE INDICATORS
DA  NI=15 NO=136 MA=KM              ....Specifies a correlation file with 15
KM FU FILE=C:\STRUC1.COR FO=5          variables and sample size of 136 to
(8F6.4/7F6.4)                          to be read from disk.

MO NX=13 NK=3 NY=2 NE=2 GA=FU,FI PS=SY,FI C   ....Model consists of 13 indicators for
   BE=FU,FI TE=SY,FI PH=SY,FR             three exogenous constructs and two
                                          endogenous constructs with one
                                          indicator each (total=15).
                                          Associated matrices also defined.

LA
'USAGE' 'SATISFAC' 'PRODQUAL' 'INVACCUR'
'TECHSUPT' 'NEWPROD' 'DELIVERY'
'MKTLEADR' 'PRDVALUE' 'LOWPRICE' 'NEGOTIAT'   ....Labels for variables in input
'MUTUALTY' 'INTEGRTY' 'FLEXBLTY' 'PROBRES'        matrix.

LK  'FIRMPROD' 'PRICEFAC' 'RELATFAC'    ....Exogenous construct labels.
LE  'USAGE' 'SATISFAC'                  ....Endogenous construct labels.

PA LX
1 (0 0 0)
1 (1 0 0)
1 (1 0 0)
1 (1 0 0)                          ....Specification of measurement model for
1 (1 0 0)                              that V1, exogenous indicators.  Note
1 (1 0 0)                              that V1, V7, and V10 have no loading
1 (0 0 0)                              since each will be set to 1.0 to control
1 (0 1 0)                              for cale invariance (see below).
1 (0 1 0)
1 (0 0 0)
1 (0 0 1)
1 (0 0 1)
1 (0 0 1)
PA GA
1 (1 1 1)                          ....Specify exogenous coefficients for
1 (0 0 0)                              structural equations.
PA BE
1 (0 0)                            ....Specify endogenous coefficients for
1 (1 0)                                structural equations.
PA PHI
1                                  ....Correlations among exogenous constructs.
1 1
1 1 1
PA PS
1                                  ....No correlations among endogenous
0 1                                    constructs.

VA 1 LX(1,1) LX(7,2) LX(10,3) LY(1,1) LY(2,2)   ....Set indicator loadings to 1.0
                                                    to control scale invariance.

VA 0.00 TE(2,2) TE(1,1)            ....Set measurement error to zero for single
                                       item indicators for endogenous
                                       constructs.

OU SE TV RS SS MI AD=OFF           ....Specify output.
```

Structural Equation Model (Path Model): Null Model Estimation

```
CAUSAL MODEL WITH MULTIPLE INDICATORS--NULL MODEL
DA  NI=15 NO=136 MA=KM
KM FU FILE=C:\STRUC1.COR FO=5
(8F6.4/7F6.4)
SELECT
   1 2 3 4 5 6 7 8 9 10 11 12 13 14 15/          .... Same as earlier except with
MO NX=15 NK=1 TD=DI,FR PH=SY,FR                       new variables.
LA  'USAGE' 'SATISFAC' 'PRODQUAL' 'INVACCUR'
'TECHSUPT' 'NEWPROD' 'DELIVERY' 'MKTLEADR' 'PRDVALUE'
'LOWPRICE' 'NEGOTIAT' 'MUTUALTY' 'INTEGRTY' 'FLEXBLTY' 'PROBRES'
PA LX
15 (1)
OU SE TV RS SS MI
```

Structural Equation Model (Path Model): Competing Model (COMPMOD1)

```
CAUSAL MODEL WITH MULTIPLE INDICATORS -- COMPMOD1
DA  NI=15 NO=136 MA=KM
KM FU FILE=C:\STRUC1.COR FO=5
(8F6.4/7F6.4)
MO NX=13 NK=3 NY=2 NE=2 GA=FU,FI C
   PS=SY,FI BE=FU,FI TE=SY,FI PH=SY,FR
LA  'USAGE' 'SATISFAC' 'PRODQUAL' 'INVACCUR' 'TECHSUPT' 'NEWPROD' 'DELIVERY'
'MKTLEADR' 'PRDVALUE' 'LOWPRICE' 'NEGOTIAT'
'MUTUALTY' 'INTEGRTY' 'FLEXBLTY' 'PROBRES'
LK  'FIRMPROD' 'PRICEFAC' RELATFAC'
LE  'USAGE' 'SATISFAC'
PA LX
1 (0 0 0)
1 (1 0 0)
1 (1 0 0)                                    ....Same as earlier structural model.
1 (1 0 0)
1 (1 0 0)
1 (1 0 0)
1 (0 0 0)
1 (0 1 0)
1 (0 1 0)
1 (0 0 0)
1 (0 0 1)
1 (0 0 1)
1 (0 0 1)
PA GA                                        ....Specifies that exogenous constructs are
1 (1 1 1)                                        constructs now related to all endogenous
1 (1 1 1)                                        constructs in the structural equations.
PA BE
1 (0 0)
1 (1 0)
PA PHI                                       ....Same as earlier model.
1
1 1
1 1 1
PA PS
1
0 1
VA 1 LX(1,1) LX(7,2) LX(10,3) LY(1,1) LY(2,2)
VA 0.00 TE(2,2) TE(1,1)
OU SE TV RS SS MI AD=OFF
```

Correlation Matrix: Confirmatory Factor Analysis (HATCO.COR)

```
 1.000000
 -.349225   1.00000
  .509295  -.487213  1.00000
  .050414   .272187  -.116104  1.000000
  .611901   .512981   .066617   .298677  1.000000
  .077115   .186243  -.034316   .788225   .240808  1.000000
 -.482631   .469746  -.448112   .199981  -.055161   .177294  1.000000
```

Correlation Matrix: Strucutral Model Estimation (STRUC1.COR)

```
1.0000 .4112 .2878 .3594 .2683 .2116 .2498 .3046
 .3284 .2679 .1423 .3289 .5187 .5100 .3411
 .41121.0000 .1676 .1585 .1413 .0810 .0603 .1270
 .1326 .0459 .1043 .0625 .0768 .0896 .1732
 .2878 .16761.0000 .7845 .6763 .5813 .6323 .6903
 .2931 .1844 .2890 .3266 .4130 .3297 .1540
 .3594 .1585 .78451.0000 .6370 .6222 .6436 .6667
 .2626 .1245 .2492 .2924 .3851 .2718 .1728
 .2683 .1413 .6763 .63701.0000 .6266 .5378 .5507
 .3365 .2296 .2993 .1119 .2647 .1919 .1818
 .2116 .0810 .5813 .6222 .62661.0000 .6986 .6251
 .2898 .2604 .2194 .2923 .3456 .1514 .1526
 .2498 .0603 .6323 .6436 .5378 .69861.0000 .6923
 .2065 .1096 .1550 .3258 .2593 .2098 .1070
 .3046 .1270 .6903 .6667 .5507 .6251 .69231.0000
 .1740 .2536 .2986 .2526 .2622 .2254 .1018
 .3284 .1326 .2931 .2626 .3365 .2898 .2065 .1740
1.0000 .3009 .3071 .1790 .3472 .3479 .2718
 .2679 .0459 .1844 .1245 .2296 .2604 .1096 .2536
 .30091.0000 .6760 .1137 .2251 .2590 .0522
 .1423 .1043 .2890 .2492 .2993 .2194 .1550 .2986
 .3071 .67601.0000 .1727 .2048 .1616 .0933
 .3289 .0625 .3266 .2924 .1119 .2923 .3258 .2526
 .1790 .1137 .17271.0000 .4105 .5550 .2018
 .5187 .0768 .4130 .3851 .2647 .3456 .2593 .2622
 .3472 .2251 .2048 .41051.0000 .5318 .4112
 .5100 .0896 .3297 .2718 .1919 .1514 .2098 .2254
 .3479 .2590 .1616 .5550 .53181.0000 .4247
 .3411 .1732 .1540 .1728 .1818 .1526 .1070 .1018
 .2718 .0522 .0933 .2018 .4112 .42471.0000
```

HATCO DATASETS

HATCO Database (X_1 to X_{14})

1	4.1	.6	6.9	4.7	2.4	2.3	5.2	0	32.0	4.2	1	0	1	1
2	1.8	3.0	6.3	6.6	2.5	4.0	8.4	1	43.0	4.3	0	1	0	1
3	3.4	5.2	5.7	6.0	4.3	2.7	8.2	1	48.0	5.2	0	1	1	2
4	2.7	1.0	7.1	5.9	1.8	2.3	7.8	1	32.0	3.9	0	1	1	1
5	6.0	.9	9.6	7.8	3.4	4.6	4.5	0	58.0	6.8	1	0	1	3
6	1.9	3.3	7.9	4.8	2.6	1.9	9.7	1	45.0	4.4	0	1	1	2
7	4.6	2.4	9.5	6.6	3.5	4.5	7.6	1	46.0	5.8	1	0	1	1
8	1.3	4.2	6.2	5.1	2.8	2.2	6.9	1	44.0	4.3	0	1	0	2
9	5.5	1.6	9.4	4.7	3.5	3.0	7.6	0	63.0	5.4	1	0	1	3
10	4.0	3.5	6.5	6.0	3.7	3.2	8.7	1	54.0	5.4	0	1	0	2
11	2.4	1.6	8.8	4.8	2.0	2.8	5.8	0	32.0	4.3	1	0	0	1
12	3.9	2.2	9.1	4.6	3.0	2.5	8.3	0	47.0	5.0	1	0	1	2
13	2.8	1.4	8.1	3.8	2.1	1.4	6.6	1	39.0	4.4	0	1	0	1
14	3.7	1.5	8.6	5.7	2.7	3.7	6.7	0	38.0	5.0	1	0	1	1
15	4.7	1.3	9.9	6.7	3.0	2.6	6.8	0	54.0	5.9	1	0	0	3
16	3.4	2.0	9.7	4.7	2.7	1.7	4.8	0	49.0	4.7	1	0	0	3
17	3.2	4.1	5.7	5.1	3.6	2.9	6.2	0	38.0	4.4	1	1	1	2
18	4.9	1.8	7.7	4.3	3.4	1.5	5.9	0	40.0	5.6	1	0	0	2
19	5.3	1.4	9.7	6.1	3.3	3.9	6.8	0	54.0	5.9	1	0	1	3
20	4.7	1.3	9.9	6.7	3.0	2.6	6.8	0	55.0	6.0	1	0	0	3
21	3.3	.9	8.6	4.0	2.1	1.8	6.3	0	41.0	4.5	1	0	0	2
22	3.4	.4	8.3	2.5	1.2	1.7	5.2	0	35.0	3.3	1	0	0	1
23	3.0	4.0	9.1	7.1	3.5	3.4	8.4	0	55.0	5.2	1	1	0	3
24	2.4	1.5	6.7	4.8	1.9	2.5	7.2	1	36.0	3.7	0	1	0	1
25	5.1	1.4	8.7	4.8	3.3	2.6	3.8	0	49.0	4.9	1	0	0	2
26	4.6	2.1	7.9	5.8	3.4	2.8	4.7	0	49.0	5.9	1	0	1	3
27	2.4	1.5	6.6	4.8	1.9	2.5	7.2	1	36.0	3.7	0	1	0	1
28	5.2	1.3	9.7	6.1	3.2	3.9	6.7	0	54.0	5.8	1	0	1	3
29	3.5	2.8	9.9	3.5	3.1	1.7	5.4	0	49.0	5.4	1	0	1	3
30	4.1	3.7	5.9	5.5	3.9	3.0	8.4	1	46.0	5.1	0	1	0	2
31	3.0	3.2	6.0	5.3	3.1	3.0	8.0	1	43.0	3.3	0	1	0	1
32	2.8	3.8	8.9	6.9	3.3	3.2	8.2	0	53.0	5.0	1	1	0	3
33	5.2	2.0	9.3	5.9	3.7	2.4	4.6	0	60.0	6.1	1	0	0	3
34	3.4	3.7	6.4	5.7	3.5	3.4	8.4	1	47.0	3.8	0	1	0	1
35	2.4	1.0	7.7	3.4	1.7	1.1	6.2	1	35.0	4.1	0	1	0	1
36	1.8	3.3	7.5	4.5	2.5	2.4	7.6	1	39.0	3.6	0	1	1	1
37	3.6	4.0	5.8	5.8	3.7	2.5	9.3	1	44.0	4.8	0	1	1	2
38	4.0	.9	9.1	5.4	2.4	2.6	7.3	0	46.0	5.1	1	0	1	3
39	.0	2.1	6.9	5.4	1.1	2.6	8.9	1	29.0	3.9	0	1	1	1
40	2.4	2.0	6.4	4.5	2.1	2.2	8.8	1	28.0	3.3	0	1	1	1
41	1.9	3.4	7.6	4.6	2.6	2.5	7.7	1	40.0	3.7	0	1	1	1
42	5.9	.9	9.6	7.8	3.4	4.6	4.5	0	58.0	6.7	1	0	1	3
43	4.9	2.3	9.3	4.5	3.6	1.3	6.2	0	53.0	5.9	1	0	0	3
44	5.0	1.3	8.6	4.7	3.1	2.5	3.7	0	48.0	4.8	1	0	0	2
45	2.0	2.6	6.5	3.7	2.4	1.7	8.5	1	38.0	3.2	0	1	1	1
46	5.0	2.5	9.4	4.6	3.7	1.4	6.3	0	54.0	6.0	1	0	0	3
47	3.1	1.9	10.0	4.5	2.6	3.2	3.8	0	55.0	4.9	1	0	1	3
48	3.4	3.9	5.6	5.6	3.6	2.3	9.1	1	43.0	4.7	0	1	1	2
49	5.8	.2	8.8	4.5	3.0	2.4	6.7	0	57.0	4.9	1	0	1	3
50	5.4	2.1	8.0	3.0	3.8	1.4	5.2	0	53.0	3.8	1	0	1	3
51	3.7	.7	8.2	6.0	2.1	2.5	5.2	0	41.0	5.0	1	0	0	2
52	2.6	4.8	8.2	5.0	3.6	2.5	9.0	1	53.0	5.2	0	1	1	2
53	4.5	4.1	6.3	5.9	4.3	3.4	8.8	1	50.0	5.5	0	1	0	2
54	2.8	2.4	6.7	4.9	2.5	2.6	9.2	1	32.0	3.7	0	1	1	1
55	3.8	.8	8.7	2.9	1.6	2.1	5.6	0	39.0	3.7	1	0	0	1
56	2.9	2.6	7.7	7.0	2.8	3.6	7.7	0	47.0	4.2	1	1	1	2

57	4.9	4.4	7.4	6.9	4.6	4.0	9.6	1	62.0	6.2	0	1	0	2
58	5.4	2.5	9.6	5.5	4.0	3.0	7.7	0	65.0	6.0	1	0	0	3
59	4.3	1.8	7.6	5.4	3.1	2.5	4.4	0	46.0	5.6	1	0	1	3
60	2.3	4.5	8.0	4.7	3.3	2.2	8.7	1	50.0	5.0	0	1	1	2
61	3.1	1.9	9.9	4.5	2.6	3.1	3.8	0	54.0	4.8	1	0	1	3
62	5.1	1.9	9.2	5.8	3.6	2.3	4.5	0	60.0	6.1	1	0	0	3
63	4.1	1.1	9.3	5.5	2.5	2.7	7.4	0	47.0	5.3	1	0	1	3
64	3.0	3.8	5.5	4.9	3.4	2.6	6.0	0	36.0	4.2	1	1	1	2
65	1.1	2.0	7.2	4.7	1.6	3.2	10.0	1	40.0	3.4	0	1	1	1
66	3.7	1.4	9.0	4.5	2.6	2.3	6.8	0	45.0	4.9	1	0	0	2
67	4.2	2.5	9.2	6.2	3.3	3.9	7.3	0	59.0	6.0	1	0	0	3
68	1.6	4.5	6.4	5.3	3.0	2.5	7.1	1	46.0	4.5	0	1	0	2
69	5.3	1.7	8.5	3.7	3.5	1.9	4.8	0	58.0	4.3	1	0	0	3
70	2.3	3.7	8.3	5.2	3.0	2.3	9.1	1	49.0	4.8	0	1	1	2
71	3.6	5.4	5.9	6.2	4.5	2.9	8.4	1	50.0	5.4	0	1	1	2
72	5.6	2.2	8.2	3.1	4.0	1.6	5.3	0	55.0	3.9	1	0	1	3
73	3.6	2.2	9.9	4.8	2.9	1.9	4.9	0	51.0	4.9	1	0	0	3
74	5.2	1.3	9.1	4.5	3.3	2.7	7.3	0	60.0	5.1	1	0	1	3
75	3.0	2.0	6.6	6.6	2.4	2.7	8.2	1	41.0	4.1	0	1	0	1
76	4.2	2.4	9.4	4.9	3.2	2.7	8.5	0	49.0	5.2	1	0	1	2
77	3.8	.8	8.3	6.1	2.2	2.6	5.3	0	42.0	5.1	1	0	0	2
78	3.3	2.6	9.7	3.3	2.9	1.5	5.2	0	47.0	5.1	1	0	1	3
79	1.0	1.9	7.1	4.5	1.5	3.1	9.9	1	39.0	3.3	0	1	1	1
80	4.5	1.6	8.7	4.6	3.1	2.1	6.8	0	56.0	5.1	1	0	0	3
81	5.5	1.8	8.7	3.8	3.6	2.1	4.9	0	59.0	4.5	1	0	0	3
82	3.4	4.6	5.5	8.2	4.0	4.4	6.3	0	47.0	5.6	1	1	1	2
83	1.6	2.8	6.1	6.4	2.3	3.8	8.2	1	41.0	4.1	0	1	0	1
84	2.3	3.7	7.6	5.0	3.0	2.5	7.4	0	37.0	4.4	1	1	0	1
85	2.6	3.0	8.5	6.0	2.8	2.8	6.8	1	53.0	5.6	0	1	0	2
86	2.5	3.1	7.0	4.2	2.8	2.2	9.0	1	43.0	3.7	0	1	1	1
87	2.4	2.9	8.4	5.9	2.7	2.7	6.7	1	51.0	5.5	0	1	0	2
88	2.1	3.5	7.4	4.8	2.8	2.3	7.2	0	36.0	4.3	1	1	0	1
89	2.9	1.2	7.3	6.1	2.0	2.5	8.0	1	34.0	4.0	0	1	1	1
90	4.3	2.5	9.3	6.3	3.4	4.0	7.4	0	60.0	6.1	1	0	0	3
91	3.0	2.8	7.8	7.1	3.0	3.8	7.9	0	49.0	4.4	1	1	1	2
92	4.8	1.7	7.6	4.2	3.3	1.4	5.8	0	39.0	5.5	1	0	0	2
93	3.1	4.2	5.1	7.8	3.6	4.0	5.9	0	43.0	5.2	1	1	1	2
94	1.9	2.7	5.0	4.9	2.2	2.5	8.2	1	36.0	3.6	0	1	0	1
95	4.0	.5	6.7	4.5	2.2	2.1	5.0	0	31.0	4.0	1	0	1	1
96	.6	1.6	6.4	5.0	.7	2.1	8.4	1	25.0	3.4	0	1	1	1
97	6.1	.5	9.2	4.8	3.3	2.8	7.1	0	60.0	5.2	1	0	1	3
98	2.0	2.8	5.2	5.0	2.4	2.7	8.4	1	38.0	3.7	0	1	0	1
99	3.1	2.2	6.7	6.8	2.6	2.9	8.4	1	42.0	4.3	0	1	0	1
100	2.5	1.8	9.0	5.0	2.2	3.0	6.0	0	33.0	4.4	1	0	0	1

Pretest of HATCO Database (X_1 to X_{14}) with Missing Data (Denoted with a .)
 (Read with same format as HATCO database)

201	3.3	0.9	8.6	4.0	2.1	1.8	6.3	0	41.0	4.5	1	0	0	2
202	.	0.4	.	2.5	1.2	1.7	5.2	0	35.0	3.3	1	0	0	1
203	3.0	.	9.1	7.1	3.5	3.4	.	0	55.0	5.2	1	1	0	3
204	.	1.5	.	4.8	1.9	2.5	7.2	1	36.0	.	0	1	0	1
205	5.1	1.4	.	4.8	3.3	2.6	3.8	0	49.0	4.9	1	0	0	2
206	4.6	2.1	7.9	5.8	3.4	2.8	4.7	0	49.0	5.9	1	0	1	3
207	.	1.5	.	4.8	1.9	2.5	7.2	1	36.0	.	0	1	0	1
208	5.2	1.3	9.7	6.1	3.2	3.9	6.7	0	54.0	5.8	1	0	1	3
209	3.5	2.8	9.9	3.5	3.1	1.7	5.4	0	49.0	5.4	1	0	1	3
210	4.1	3.7	5.9	.	.	.	.	.	.	.	0	1	0	2

211	3.0	2.8	7.8	7.1	3.0	3.8	7.9	0	49.0	4.4	1	1	1	2
212	4.8	1.7	7.6	4.2	3.3	1.4	5.8	0	39.0	5.5	1	0	0	2
213	3.1	.	.	7.8	3.6	4.0	5.9	0	43.0	5.2	1	1	1	2
214	.	2.7	5.0	.	2.2	.	.	1	.	3.6	.	1	.	1
215	4.0	0.5	6.7	4.5	2.2	2.1	5.0	0	31.0	4.0	1	0	1	1
216	.	1.6	6.4	5.0	.	2.1	8.4	1	25.0	3.4	0	1	1	1
217	6.1	0.5	9.2	4.8	3.3	2.8	7.1	0	60.0	5.2	1	0	1	3
218	.	2.8	5.2	5.0	.	2.7	8.4	1	38.0	3.7	0	1	0	1
219	3.1	2.2	6.7	6.8	2.6	2.9	.	1	.	4.3	0	1	0	1
220	6.5	.	9.0	7.0	3.2	3.7	8.0	0	33.0	5.4	1	0	0	1
221	.	1.6	.	4.8	2.0	2.8	.	0	32.0	4.3	1	0	0	1
222	3.9	2.2	.	4.6	.	2.5	8.3	0	47.0	5.0	1	0	1	2
223	2.8	1.4	8.1	3.8	2.1	1.4	6.6	1	39.0	4.4	0	1	0	1
224	.	.	8.6	5.7	2.7	3.7	6.7	0	.	5.0	1	0	1	1
225	4.7	1.3	.	.	3.0	2.6	6.8	0	54.0	5.9	1	0	0	3
226	3.4	2.0	9.7	4.7	2.7	1.7	4.8	0	49.0	4.7	1	0	0	3
227	3.2	.	5.7	5.1	3.6	2.9	6.2	0	.	4.4	1	1	1	2
228	.	1.8	7.7	.	3.4	1.5	5.9	0	40.0	5.6	1	0	0	2
229	5.3	1.4	9.7	6.1	.	3.9	6.8	0	54.0	5.9	1	0	1	3
230	4.7	1.3	9.9	6.7	3.0	2.6	6.8	0	55.0	6.0	1	0	0	3
231	3.7	0.7	8.2	6.0	2.1	2.5	.	0	41.0	5.0	1	0	0	2
232	.	.	8.2	5.0	3.6	2.5	9.0	1	53.0	5.2	0	1	1	2
233	4.5	.	.	5.9	.	.	8.8	1	50.0	.	0	.	0	.
234	2.8	2.4	6.7	4.9	2.5	2.6	9.2	1	32.0	3.7	0	1	1	1
235	3.8	0.8	8.7	2.9	1.6	.	5.6	0	39.0	.	1	0	0	1
236	2.9	2.6	7.7	7.0	2.8	3.6	7.7	0	47.0	4.2	1	1	1	2
237	4.9	.	7.4	6.9	4.6	4.0	9.6	1	62.0	6.2	0	1	0	2
238	.	2.5	9.6	5.5	4.0	3.0	7.7	0	65.0	6.0	1	0	0	3
239	4.3	1.8	7.6	5.4	3.1	2.5	4.4	0	46.0	5.6	1	0	1	3
240	.	1.5	9.9	2.7	1.3	1.2	1.7	1	50.0	5.0	0	1	1	2
241	3.1	1.9	.	4.5	.	3.1	3.8	0	54.0	4.8	1	0	1	3
242	5.1	1.9	9.2	5.8	3.6	2.3	4.5	0	60.0	6.1	1	0	0	3
243	4.1	1.1	9.3	5.5	2.5	2.7	7.4	0	47.0	5.3	1	0	1	3
244	3.0	3.8	5.5	4.9	3.4	2.6	6.0	0	.	4.2	1	1	1	2
245	.	2.0	.	4.7	.	3.2	.	1	.	3.4	0	.	1	.
246	3.7	1.4	9.0	.	2.6	2.3	6.8	0	45.0	4.9	1	0	0	2
247	4.2	2.5	9.2	6.2	3.3	3.9	7.3	0	59.0	6.0	1	0	0	3
248	.	.	6.4	5.3	3.0	2.5	7.1	1	46.0	4.5	0	1	0	2
249	5.3	.	8.5	3.7	3.5	1.9	4.8	0	58.0	4.3	1	0	0	3
250	.	3.7	.	5.2	3.0	2.3	9.1	1	49.0	4.8	0	1	1	2
251	3.0	3.2	6.0	5.3	3.1	3.0	8.0	1	43.0	3.3	0	1	0	1
252	2.8	3.8	8.9	6.9	3.3	3.2	8.2	0	53.0	5.0	1	1	0	3
253	.	2.0	9.3	5.9	3.7	2.4	4.6	0	60.0	6.1	1	0	0	3
254	3.4	3.7	6.4	5.7	3.5	3.4	8.4	1	47.0	3.8	0	1	0	1
255	.	1.0	.	3.4	1.7	1.1	6.2	1	35.0	4.1	0	1	0	1
256	.	3.3	7.5	4.5	2.5	2.4	7.6	1	39.0	3.6	0	1	1	1
257	3.6	.	.	5.8	3.7	2.5	9.3	1	44.0	4.8	0	1	1	2
258	4.0	0.9	9.1	5.4	2.4	2.6	7.3	0	46.0	5.1	1	0	1	3
259	.	2.1	6.9	5.4	1.1	2.6	8.9	1	29.0	3.9	0	1	1	1
260	.	2.0	6.4	4.5	2.1	2.2	8.8	1	28.0	3.3	0	1	1	1
261	3.6	.	.	6.2	4.5	.	.	1	.	.	.	1	1	2
262	5.6	2.20	8.2	3.1	4.0	1.6	5.3	0	55.0	3.9	1	0	1	3
263	3.6	.	9.9	.	.	.	.	.	.	4.9	1	0	0	3
264	5.2	1.3	9.1	4.5	3.3	2.7	7.3	0	60.0	5.1	1	0	1	3
265	3.0	2.0	6.6	6.6	2.4	2.7	8.2	1	41.0	4.1	0	1	0	1
266	4.2	2.4	9.4	4.9	3.2	2.7	8.5	0	49.0	5.2	1	0	1	2
267	3.8	0.8	.	.	2.2	2.6	5.3	0	42.0	5.1	1	0	0	2
268	3.3	2.6	9.70	3.30	2.9	1.5	5.2	0	47.0	.	1	0	1	3
269	.	1.9	.	4.50	1.5	3.1	9.9	1	39.0	3.3	0	1	1	1
270	4.5	1.6	8.7	4.6	3.1	2.1	6.8	0	56.0	5.1	1	0	0	3

Conjoint Data Set: Evaluations of 22 Stimuli by 100 respondents

```
104  4 6 5 4 4 6 4 4 4 4 4 4 5 5 4 4 4 4 4 6 6
107  6 3 5 2 3 1 1 6 6 6 7 4 1 6 6 6 6 1 7 7 1 7
109  7 7 6 7 3 3 3 6 3 5 2 3 2 2 6 3 3 3 6 3 3 3
110  5 5 7 7 1 5 1 1 4 5 5 5 1 5 5 2 6 1 3 5 1 5
120  7 7 1 6 1 1 1 1 1 5 5 1 1 1 7 1 1 1 5 6 1 1
123  7 7 5 5 1 1 1 2 2 1 7 1 1 2 7 7 7 1 7 7 1 7
129  7 7 7 7 7 7 7 7 6 7 6 7 6 7 7 6 6 7 7 6 7 6
133  5 7 2 7 2 2 2 3 3 2 5 2 1 6 5 2 2 2 2 6 2 7
135  7 7 2 5 1 1 1 2 2 2 7 1 1 7 4 5 2 1 2 5 2 5
144  6 7 3 6 6 2 3 3 5 5 7 7 2 2 7 3 6 5 5 5 5 3
150  7 7 5 5 3 6 3 4 7 7 5 2 5 7 6 7 4 3 6 7 3 6
155  7 7 3 6 2 1 1 5 1 5 1 3 2 2 7 1 1 1 5 1 1 1
156  7 6 5 3 5 5 2 6 5 2 7 2 5 6 7 5 5 6 6 7 2 6
161  6 6 5 5 1 1 1 3 2 5 2 1 1 5 7 3 1 1 5 2 1 2
162  7 6 3 6 3 5 3 6 3 2 6 6 2 5 7 4 2 2 4 7 6 5
167  4 5 3 3 2 3 3 2 6 5 6 3 3 3 4 2 3 5 3 6 5 5
168  7 7 3 2 2 2 1 1 2 3 7 1 1 7 7 7 3 2 3 7 2 7
170  6 6 4 2 1 3 2 1 2 5 7 3 3 1 7 2 6 2 4 7 2 5
173  6 6 4 4 3 3 3 5 2 4 6 3 3 5 5 5 4 3 4 6 5 6
174  6 5 6 6 6 1 3 6 1 6 1 5 1 6 7 1 1 5 6 1 3 1
178  7 7 2 5 1 1 1 1 1 5 1 1 1 1 6 1 1 1 5 3 2 1
180  7 7 2 6 6 1 2 2 2 5 2 7 1 7 7 2 1 5 7 2 7 2
181  6 6 3 5 4 2 2 3 3 5 6 4 2 3 6 4 6 2 6 5 4 6
187  6 7 6 7 2 2 2 5 5 6 6 3 1 4 7 5 2 3 6 7 3 6
190  6 6 5 5 1 1 1 3 5 5 7 1 1 3 6 3 5 1 5 5 1 6
192  7 7 5 6 2 1 2 2 2 2 2 2 1 4 5 1 1 2 5 3 3 3
193  7 7 6 6 2 1 6 6 1 5 4 5 2 3 7 1 2 2 7 2 7 2
194  2 6 6 2 6 2 2 4 2 4 2 4 3 4 6 1 4 6 4 2 6 2
195  5 5 5 5 5 4 5 5 4 5 4 5 4 5 5 4 4 4 4 5 4
197  6 7 5 6 3 3 4 6 4 4 6 3 7 4 4 4 5 3 3 7 6 6
198  7 7 5 7 4 6 5 7 5 4 2 7 2 4 7 1 1 2 5 2 6 2
200  5 4 2 2 3 5 5 4 5 3 4 2 2 2 3 1 2 3 5 4 3 1
209  6 6 1 2 2 2 1 2 5 2 5 1 1 2 6 1 1 1 2 7 1 4
211  5 7 5 7 2 2 1 5 5 2 5 1 2 2 7 5 2 2 1 7 7 7
222  7 7 3 3 2 2 5 5 3 6 5 2 2 6 7 3 4 2 5 5 2 5
225  7 7 4 6 2 2 2 6 2 4 3 3 1 4 5 2 2 3 5 3 4 2
229  7 7 7 7 1 4 5 3 6 2 6 3 1 2 2 7 1 4 1 1 5 1
231  5 5 2 2 1 1 1 5 1 2 6 2 1 3 5 2 2 1 6 4 1 3
233  6 5 2 1 7 5 6 2 1 7 5 3 3 3 7 6 2 1 4 4 1 4
234  5 5 4 5 4 4 5 5 4 4 4 4 4 5 4 4 5 6 4 5 4
235  5 7 6 6 2 3 6 3 3 6 7 6 3 7 7 6 6 3 5 5 3 5
236  4 6 2 4 1 1 1 2 2 2 4 1 1 5 7 2 1 1 2 2 1 2
240  2 2 5 7 2 6 2 2 4 2 2 2 2 2 7 6 1 5 4 6 6 4
246  7 3 6 4 2 6 5 6 5 5 5 4 3 3 2 2 3 3 1 1 4 1
249  4 4 6 7 2 1 2 5 5 4 3 2 1 4 6 4 2 2 6 4 2 3
250  7 4 3 3 4 5 3 3 3 4 7 3 5 5 4 5 4 4 3 4 5 7
251  4 4 1 4 1 1 1 2 2 2 2 1 1 4 3 2 1 1 1 3 5 2
254  5 7 2 1 1 1 1 3 1 3 3 1 1 2 4 1 1 1 3 1 1 1
258  7 1 1 3 1 2 1 1 3 1 7 1 3 1 7 1 1 3 4 4 1 4
260  5 5 5 5 5 2 4 5 2 5 2 6 2 5 5 1 2 2 5 2 2 2
261  6 6 4 6 2 2 2 4 6 6 6 2 2 4 6 4 4 2 4 4 2 4
266  6 7 7 5 5 6 4 2 3 3 3 2 1 4 3 4 2 1 2 1 1
271  7 5 4 3 3 2 3 4 5 6 6 3 4 4 2 2 2 1 2 1 1 1
272  7 7 6 5 6 6 3 3 4 5 5 6 2 2 3 4 1 4 1 3 3 1
277  6 6 5 5 4 3 4 3 3 3 5 4 1 2 3 1 2 2 1 1 1
285  5 4 2 3 5 5 4 3 4 2 4 4 3 3 1 1 5 2 2 1 2 1
287  3 5 2 2 1 1 1 1 1 1 3 1 1 2 2 3 1 1 1 5 1 2
```

288	7	4	4	3	4	4	3	5	6	6	2	3	3	3	2	2	1	2	3	3	1	1
289	5	4	2	4	4	5	5	2	4	2	4	5	3	3	3	2	3	1	3	3	2	2
300	6	5	4	5	2	2	2	4	4	5	5	2	2	5	6	3	5	2	5	5	2	5
302	6	6	6	7	5	3	4	5	3	5	3	5	3	5	6	2	2	4	7	3	7	3
303	6	6	5	7	4	2	2	3	3	3	3	6	2	3	7	2	3	5	6	3	7	2
306	5	5	6	7	5	2	4	4	2	5	1	4	2	6	7	3	3	5	7	3	7	2
308	7	7	3	6	3	3	3	3	3	3	6	3	6	6	6	7	3	3	3	5	7	5
309	7	6	5	7	3	1	1	3	3	3	1	2	2	6	6	3	2	2	4	1	4	3
310	6	7	3	7	3	3	3	4	4	4	4	5	2	5	7	2	2	7	5	2	6	3
317	7	7	6	6	3	3	3	5	3	5	3	4	3	5	5	3	3	5	5	3	2	3
318	3	4	5	6	5	2	5	6	1	6	2	5	3	6	7	4	2	5	6	4	6	4
319	1	5	6	5	2	1	4	3	2	6	2	5	5	4	6	4	6	4	6	4	6	4
323	4	4	1	6	1	6	1	4	1	4	4	1	4	6	6	6	4	4	1	4	1	4
324	6	7	7	6	6	2	5	6	2	5	5	7	3	6	7	1	2	7	7	2	7	5
325	7	7	5	5	5	7	6	5	4	3	3	3	4	6	7	7	5	7	6	5	7	7
330	4	5	5	6	3	6	3	4	5	4	6	5	4	4	6	5	4	7	5	5	6	4
336	3	1	4	1	4	4	5	5	5	5	3	2	5	7	4	6	5	4	2	3	1	4
339	6	6	2	4	1	1	1	2	2	6	2	1	1	6	7	1	2	1	3	4	1	3
348	3	3	1	1	1	1	1	1	1	1	3	1	1	3	5	6	1	1	1	6	1	6
350	6	6	3	7	5	2	3	5	2	5	2	5	2	5	5	2	2	5	6	3	6	3
352	6	7	4	4	1	1	1	4	4	4	6	1	1	6	7	1	1	1	4	1	1	1
353	7	7	5	5	3	3	3	5	1	3	1	1	1	7	7	3	3	3	5	1	1	7
354	5	5	4	6	1	1	2	4	3	5	2	2	6	6	6	6	5	2	6	6	5	6
356	4	5	2	4	1	1	1	4	3	2	5	2	1	3	5	7	2	1	3	6	2	6
363	7	6	4	6	1	1	1	5	3	6	7	2	6	5	7	6	5	2	6	5	5	6
366	5	5	6	3	3	3	3	4	4	3	4	2	2	5	6	4	2	1	6	4	3	3
368	7	7	6	6	3	1	2	4	2	5	3	2	1	3	6	3	2	1	3	2	2	2
370	6	3	5	5	3	2	4	6	5	5	2	4	4	6	7	5	4	4	6	3	2	3
372	7	6	7	5	5	3	5	5	2	6	2	5	3	6	7	2	3	4	7	3	2	3
381	1	3	3	2	2	1	7	2	2	4	4	3	3	2	2	1	2	2	3	3	4	2
382	3	3	3	2	2	1	2	4	7	5	2	2	1	2	3	3	3	4	4	6	1	1
385	4	7	7	5	7	5	7	5	6	4	7	1	4	3	4	4	4	3	1	2	1	1
394	6	6	4	5	1	2	3	7	6	7	1	1	1	6	2	3	6	4	5	4	5	5
396	7	7	5	5	1	1	1	5	5	5	7	1	1	7	7	5	1	5	7	1	7
399	4	5	2	1	1	1	1	1	2	2	4	1	1	4	5	2	1	1	1	4	1	2
401	6	6	3	3	1	1	1	3	1	3	5	1	1	5	7	6	1	1	4	5	1	5
403	5	5	6	5	1	1	1	3	2	5	2	2	1	0	5	4	2	1	5	4	5	2
412	7	7	6	6	1	1	1	5	1	6	1	1	1	7	6	1	1	1	5	1	1	1
413	1	1	3	1	1	1	1	1	1	1	1	1	1	1	1	1	1	1	3	1	1	1
414	6	6	5	7	3	4	3	5	2	5	2	2	3	4	6	2	1	2	5	4	3	2
416	5	7	2	6	2	2	1	4	3	1	6	3	2	1	7	2	1	1	4	5	2	2
417	5	5	2	5	1	1	2	2	2	2	1	1	1	4	6	3	2	1	4	4	2	3
418	6	7	2	7	1	4	1	3	6	4	7	1	5	2	1	2	4	2	3	7	5	7

Index

A

A priori test   257, 281
Absolute fit measures   683–85
Adaptive conjoint models (ACA)   5, 582
Additive model   557, 570
Adjusted coefficient of determination   79–80, 119
Adjusted goodness-of-fit index (AGFI)   652, 686
Agglomeration coefficient   442–43, 448–49
Agglomerative method   421, 437
Aggregate analysis
   in conjoint analysis   503, 563, 579, 587–89
   in multidimensional scaling   485, 495
Akaike information criterion (AIC)   688
Algorithm   421
All-available approach (missing data)   33, 48
All possible subset regression   80, 117
Alpha (significance level)   1, 10–12, 257
ALSCAL   5, 497
ANACOR   515
Analysis of variance (ANOVA)   2, 14, 258, 262–63
   calculation of effects   262–63
   computer example, three group   294–97
   computer example, two group   285–88
Analysis sample
   in conjoint analysis   578, 580
   in multiple discriminant analysis   179, 195
Anti-image correlation matrix   365, 374
Assignment matrix: See Classification matrix
Assumptions of multivariate analysis   62–75
   absence of correlated errors   69, 113–14
   computer example   71–75
   homoscedasticity   66–68, 113
   independence of observations   275
   individual variables versus the variate   64, 110
   linearity   68–69, 113–14
   normality   64–66, 114
Attribute-based methods: See Decompositional methods
Attribute-free methods: See Compositional methods
Attribute vectors: See Vector representations
Average linkage   421, 440

B

Backward elimination   80, 116
Balanced design: See Factorial design

Bartlett's test of sphericity   365, 374
Bartlett's V: See Bartlett's test of sphericity
Bayesian regression   128
Beta   1, 10–12, 258
Beta coefficient   80, 125
Beta matrix (SEM)   673
Between-group variance   261
Binary variable
   as dependent variable in logistic regression   130–31
   as dummy variable   81, 109
Blocking factor   258, 270
Biserial correlation   618, 637
Bivariate partial correlation   1, 14
Bonferroni inequality   258, 281
Box's M test
   in multiple discriminant analysis   196–97
   in multivariate analysis of variance   258, 275, 289
Box plot
   definition   33, 40
   use   285, 287, 293, 299, 300, 303, 307

C

CA   515
Canonical correlation   327
Canonical correlation analysis (CCA)   14, 326–46
   applications   328
   assumptions   332
   computer example   339–44
   cross-loadings   337, 343
   decision process   330–31
   derivation of canonical functions   333–36
   hypothetical example   328–30
   interpretation of canonical functions   336–38
   loadings   337, 343
   multivariate test statistics   334, 340
   objectives   330
   orthogonality of functions   333
   redundancy   334–36, 340–41
   statistical significance   340
   validation   338–39
   weights   336, 342
Canonical cross-loadings   337, 343
Canonical function   327, 330, 340–41
Canonical loadings   327, 337, 343

Canonical roots   327, 333
Canonical structure matrix   341
Canonical variates   327, 330, 340–43
Canonical weights   327, 336–37
Categorical variable   179
    measurement properties   7
    use in multiple regression   109–10
    use in multivariate analysis of variance
        257, 269
Causal relationship   618, 626
Causation   618, 629
Censored data   33
Centroid
    in cluster analysis   421
    in discriminant analysis   179
Centroid method (CA)   421, 440
Chance criterion in discriminant analysis
    maximum chance   180, 203, 220, 227
    Press's Q statistic   205, 221, 227
    proportional chance   204, 220, 227
Choice-based conjoint approach   558, 583–85
    advantages and disadvantages   583–84
    hypothetical example   584–85
Choice set   558, 583
Choice-simulator   558, 580–81, 591–92
    computer example   591–92
    estimation   581, 592
    uses   581
City block approach   421, 432
Classification accuracy: See Classification matrix
Classification analysis: See Cluster analysis
Classification matrix
    in logistic regression   132
    in multiple discriminant analysis   179, 202–03
Cluster analysis   16, 365, 371, 420–56
    agglomeration coefficient   442–43, 448–49
    algorithms   437
    applications   423–24
    assumptions   435–36
    choosing between hierarchical and
        nonhierarchical   441–42
    computer example, hierarchical procedure
        445–55
    computer example, nonhierarchical procedure
        453–55
    data standardization   434–35
    decision process   425–27
    deriving clusters   436–43
    design of research   428–35
    graphical portrayal of results   449–52
    hierarchical procedures   437–40
    hypothetical example   424–25
    interpretation   443–44
    linkage approaches   437–39
    nonhierarchical procedures   440–41

Cluster analysis (*cont.*)
    number of clusters   442–43
    objectives   426–28
    outlier detection   429
    respecification   443
    similarity measures   429–34
    validation   444
    variable selection   427–28
Cluster centroid   421
Cluster seeds   421, 440
Coefficient of determination (R^2)
    adjusted coefficient of determination   119
    calculation   92, 96
    definition   80, 92
    in linear probability models   132
    statistical significance   119
    use   96, 137–40
Collinearity: Also see Multicollinearity   80, 93,
    151, 152
Combinatorial methods (Multiple regression)   118–19
Common factor analysis   365, 375
Common variance   365, 375
Communality   365, 376
Comparison tests: See Multiple comparison tests
Competing models
    as a modeling strategy   618, 626
    use of   643–44, 666–68
Competing models strategy   618, 626
Complete case approach   33, 48
Complete linkage   422, 439
Component analysis   365, 375
Composite reliability   641
Composition rule   558, 570
Compositional method
    in conjoint analysis   558, 562
    in multidimensional scaling   485, 497
Condition index   151, 153
Confidence intervals   124–25
Confirmatory analysis   618
Confirmatory factor analysis: See Structural equation
    modeling
Confirmatory modeling strategy   618, 625
Confusion data   485
Confusion matrix: See Classification matrix
Conjoint analysis   15, 556–93
    aggregate versus disaggregate analysis   563,
        587–89
    applications   562
    assumptions   577
    attribute importance   579
    calculation of part-worths   595–98
    choice-based approach   582–85
    choice simulator   580–81, 591–92
    comparison to other multivariate methods
        562–64

Conjoint analysis (*cont.*)
    composition rules 570–71
    computer example 585–92
    data collection 572
    dealing with a large number of factors 581–82
    decision process 564–66
    designing stimuli 564, 568–70
    estimation techniques 578
    factorial versus fractional stimuli design 574–76
    hypothetical example 561–62
    interpretation 579
    objectives 564–68
    part-worth utility 561, 571–72, 595–99
    presentation methods 572–74
    profitability analysis with 580
    segmentation with 580
    selecting preference (dependent) measures 570–71, 576–77
    types of relationships 564, 571–72
    validation 579–80
Conjoint simulator: See Choice simulator
Constant variance of the error term: See Homoscedasticity
Construct (SEM) 618, 629
    endogenous 619, 631
    exogenous 619, 631
Contingency table 485, 513
Contrast 258, 283
Cook's distance 151, 157
CORRAN 515
Correlation coefficient 80, 88
Correlation matrix
    in examining the data 38–40
    in factor analysis 365
    in multiple regression 127, 136
    in structural equation modeling 636–37
CORRESP 515
Correspondence analysis 16–17, 485, 513–16
    assumptions 514
    computer example 523–27
    estimation 514
    input data matrix 513
    interpretation 515
    multiple correspondence analysis 486
    number of dimensions 515
    objectives 513
    research design 514
    validation 515
COSAM (SEM) 637
Covariance structure analysis: See Structural equation modeling
Covariate analysis: See Multivariate analysis of variance
Covariates: See Multivariate analysis of variance

COVRATIO 151, 157
Criterion variable: See Dependent variable
Criterion validity 422, 444
Critical Z value: See Cutting score, optimum
Cronbach alpha 618, 641
Cross-loading; See Canonical cross-loadings
Crosstabulation table 485
Cutting score 180, 200–02
    calculation 200, 218
    optimum 200
    use in classification 183, 199–202

D
Databases for computer examples 27–29
Data transformations 33–34, 69–71
    achieving homoscedasticity 70
    achieving linearity 70
    achieving normality 70
    creating dummy variables 109–10
    guidelines 70–71
    interaction and moderator effects 107–09
    polynomials 106–07
Decision process for multivariate model building
    definition 24–26
    in canonical correlation analysis 330–31
    in cluster analysis 425–27
    in conjoint analysis 564–66
    in factor analysis 368–69
    in multidimensional scaling 492–94
    in multiple discriminant analysis 191–93
    in multiple regression 97–99
    in multivariate analysis of variance 266–67
    in structural equation modeling 626–33
Decompositional method
    in conjoint analysis 563
    in multidimensional scaling 485, 496
Decompositional model 558, 562
Degenerate solution 486, 504
Degrees of freedom
    in multiple regression 80, 119
    in multivariate analysis of variance 263
    in structural equation modeling 618, 638, 640, 682–83
Dendrogram 422, 438
Dependence technique 1, 17
Dependent variable 1, 17
    in canonical correlation analysis 330, 332
    in conjoint analysis 576–77
    in multiple discriminant analysis 182, 194
    in multiple regression 81, 85
    in multivariate analysis of variance 257, 263, 269
Derived measures 486, 501
DFBETA 151, 156

DFFIT   151, 157
Dichotomous data: See Binary variable
Dimensions   486, 488
    in correspondence analysis   515
    in factor analysis   377–80
    in multidimensional scaling   485, 504–06,
        510–12
    in multiple discriminant analysis   183, 187–89
    in multivariate analysis of variance   257
Disaggregate analysis
    in conjoint analysis   563, 587–89
    in multidimensional scaling   486, 495
Discriminant analysis: See Multiple discriminant
    analysis
Discriminant coefficient   180
Discriminant function
    in multiple discriminant analysis   180, 182
    in multivariate analysis of variance   258, 265
Discriminant loadings   180, 206
Discriminant score   180
Discriminant weight   180, 206
Disordinal interaction   258, 271
Disparities   486, 505
Dissimilarity data: See Similarity data
Distance measures (Cluster analysis)
    calculation of   431–33
    city block measure   432
    Euclidean distance measure   432
    impact of variable scales   433
    Mahalanobis distance   434
    use as similarity measure   430–31
Divisive method   422, 438
Dummy variable   1, 81, 109–10
    use   109
    effects coding   110
    indicator coding   110
Duncan's multiple range test   282

E
Effect size   1, 10–12, 258, 278
Effects coding   81, 110
Eigenvalue
    in canonical correlation   327, 333
    in factor analysis   365–66, 377
    in multiple regression   151, 153
Endogenous construct   619, 631
Entropy Group   422, 425
EQS (SEM)   5, 637
Equality of covariance matrices among groups
    in multiple discriminant analysis   196–97, 213
    in multivariate analysis of variance   275–76,
        289, 299
EQUIMAX rotation   384

Error
    constant variance in error term: See
        Homoscedasticity
    control of experiment-wide error rate   266
    independence of error terms   142
    measurement error in structural equation
        modeling   619, 623–24
    normality of error term distribution   114
    residual in multiple regression   84, 111
    specification error   2, 23, 162, 629
    standard error of the coefficient   137
    standard error of the estimate   90–91,
        137, 261
Error variance in factor analysis   365
Euclidean distance   422, 432
Exogenous construct   619, 631
Experimental design   258
Exploratory analysis   619
Exploratory factor analysis: See Confirmatory factor
    analysis
External analysis (MDS)   508, 509
Extreme value: See Outlier

F
F-ratio
    calculation   119
    in multivariate analysis of variance   262
F-test: See F-ratio
Factor
    in ANOVA/MANOVA   259
    in conjoint analysis   558, 560
    in factor analysis   365, 367
Factor analysis (FA)   16, 364–406
    applications   367–68, 392–93
    assumptions   374–75
    common factor versus components analysis
        375–77
    comparison to cluster analysis   372–73
    computer example of common factor analysis
        402–04
    computer example of components analysis
        391–02
    decision process   368–69
    deriving factors   375
    design of analysis   372–74
    factor loadings   384–88
    factor scores   390–91, 400
    hypothetical example   368
    interpretation of factors   379–88
    measures of sampling adequacy   374
    naming of factors   380, 386–88, 392–93
    number of factors   377–79, 394–96, 397–98
    objectives   368–72

Factor analysis (FA) (*cont.*)
    Q-type versus R-type   372–73
    rotation of factors   380–84, 396–97, 398
    sample size requirements   373–74
    selecting surrogate variables   389–90
    statistical significance of loadings   384–85
    validation   388–89
    variable selection   373
Factor indeterminancy   365, 376
Factor loadings   366, 380
Factor matrix   366, 372
Factor rotation   366, 380
Factor scores   366, 390
Factorial design
    in conjoint analysis   558, 574, 586
    in multivariate analysis of variance   259, 270,
        303–06
Fixed predictor variables (Multiple regression)   105
FMATCH   512
Follow-up tests: See Multiple comparison tests
Fractional factorial design   558, 574, 586–87
Full-profile method   559, 572
Furthest neighbor: See Complete linkage

G

Gamma matrix (SEM)   673
GFI index (SEM)   684
Global test: See Multiple comparison tests
Glyphs   41
Goodness-of-fit (SEM)   619, 637, 640, 682–88
    adjusted goodness-of-fit index (AGFI)   686
    Akaike information criterion (AIC)   688
    computer example, confirmatory factor
        analysis 651–54
    computer example, structural modeling
        661–64
    empirical example   689–90
    expected cross-validation index (ECVI)   685
    goodness-of-fit index (GFI)   684
    likelihood-ratio chi-square statistic   683–84
    measurement model fit   634–35, 641–43,
        652–54
    noncentrality parameter   684
    normed chi square   687–88
    normed fit index (NFI)   686
    overall model fit   639–40
    parsimonious goodness-of-fit index (PGFI)
        687
    parsimonious normed fit index (PNFI)   687
    root mean square error of approximation
        (RMSEA)   685
    root mean square residual (RMSR)   684–85
    scaled noncentrality parameter   684

Goodness-of-fit (SEM) (*cont.*)
    structural model fit   642
    Tucker-Lewis index (TLI)   686
Greatest characteristic root (gcr)
    in canonical correlation analysis   334
    in multivariate analysis of variance   259,
        265, 277

H

Hat matrix   151, 155
Heteroscedasticity: See Homoscedasticity
Heywood cases   619, 640
Hierarchical clustering procedures   422, 437–40
Histogram   34, 37–38
Hit ratio   180, 203
Hold-out sample: Also see Validation sample
    in conjoint analysis   578, 580
    in multiple discriminant analysis   180, 195
HOMALS   515
Homoscedasticity   34, 66–68
    computer example   74
    graphical tests   68
    in canonical correlation analysis   332
    in factor analysis   374
    in multiple regression   81, 113
    remedies for heteroscedasticity   68
    statistical tests   68
Hotelling's T^2 259, 264, 288–89
Hotelling's trace
    in canonical correlation analysis   334
    in multiple discriminant analysis   198
    in multivariate analysis of variance   277
Hybrid conjoint models   582

I

Iconic representations in graphical displays   41
Ideal point   486, 506
    internal versus external estimation   508
    point versus vector representation   509, 511,
        521–22
Identification   619, 638
Ignorable missing data   34, 44
Importance performance grid   486, 498
Imputation methods for missing data   34, 47–50,
    55–57
    all-available approach   48
    case substitution   48–49
    cold deck imputation   49
    complete case approach   48
    computer example   55–57
    mean substitution   49
    model based procedures   50

Imputation methods for missing data (*cont.*)
  multiple methods 49
  rationale 47
  regression methods 49, 56–57
Incremental fit measures 685–86
Independence model 619
Independence of error terms: See Error
Independent variable 1, 17
  in canonical correlation analysis 327, 330, 332
  in conjoint analysis 557, 560, 563, 567–70
  in multiple discriminant analysis 179, 194–95
  in multiple regression 81, 85
  in multivariate analysis of variance 257,
    263–64, 270–73
Index of fit 486, 506
Indicator 1, 9, 15, 619, 632
Indicator coding 81, 110
INDSCAL 5, 496, 497, 518, 519
Influential observations 81, 143–45, 151–65
  computer example 143–45, 157–65
  Cook's distance 157
  COVRATIO 157, 160
  DFBETA 156, 162
  DFFIT 157, 162
  hat matrix 155
  identification through residuals 154–55
  leverage points 155–56
  Mahalanobis distance 155–56
  remedies 145, 157
  studentized delated residuals 156
Initial configuration (MDS) 486, 503
Interacting variables: See Interaction effect
Interaction effect
  in multiple regression 107–09
  in conjoint analysis 559, 570, 598
  in multivariate analysis of variance 259,
    271–73, 308–10
Interattribute correlation 559, 569
Intercept 81–82, 89
Interdependence technique 1, 17
Internal analysis (MDS) 508
Inter-object similarity 422
Interpretational confounding (SEM) 635
Interval scales 1, 8

J
Jackknife analysis 210

K
K-means clustering 440
Kurtosis 34, 65
KYST 5, 497

L
Lambda-X matrix (SEM) 673–74, 677
Lambda-Y matrix (SEM) 673–74, 677
Latent construct/variable 619
Latent roots: See Eigenvalue
Latent root criterion 377
Latent variable analysis: See Structural equation
    modeling
Latent variables 623
Level of independent variable in conjoint analysis
    559, 560
Levene test 68, 285
Leverage points 82, 152, 155
Likelihood-ratio chi-square statistic (SEM) 683–84
Likelihood value 82, 132
Linear combinations: See Variate
Linear composites: See Variate
Linear probability model 14, 82, 130–33
  goodness-of-fit 132
  interpreting coefficients 131–32
  similarity to multiple discriminant analysis
    196–97
  similarity to multiple regression 196
  statistical significance of coefficients 133
  using a binary dependent variable 130–31
Linearity 34, 68–69, 74–75
  computer example 74–75
  graphical analysis of residuals 111–13
  identification 69
  in factor analysis 374
  in multiple regression 82, 106–07, 111–13,
    142
  in multivariate analysis of variance 276
  in structural equation modeling 639
  polynomials 106–07
  remedies for nonlinearity 69
Linkage methods (cluster analysis) 438–40
  average linkage 440
  centroid method 440
  complete linkage 439
  single linkage 438
  Ward's method 440
LINMAP 578
LISREL
  notation of 672–79
  use in structural equation modeling 631, 637,
    672
Log likelihood 132
Logistic regression: See Linear probability model
Logit analysis: See Linear probability model

M
Mahalanobis D^2 distance
  in cluster analysis 422, 434

Mahalanobis D^2 distance (*cont.*)
    in multiple discriminant analysis 198
    in multiple regression 152, 155
    in outlier detection 59
Main effects
    in conjoint analysis 559, 595
    in multivariate analysis of variance 259, 270
MANCOVA: See Multivariate analysis of variance
Manifest variable 619, 623
MANOVA: See Multivariate analysis of variance
MAPWISE 515
Maximum chance criterion: See Chance criterion,
    maximum
Maximum likelihood criterion: See Wilks' lambda
MDPREF 497, 508
MDSCAL 497, 508
Measure of sampling adequacy 366, 374
Measurement error 1, 8–9
    in multiple regression 82, 102
    in structural equation modeling 619, 623–24
Measurement model (SEM) 620, 624
    computer example 649–54
    goodness-of-fit 634–35, 641–43, 652–54
    LISREL notation 647–48, 674
    similarity to factor analysis 632–33
Measurement scales 6–8
Metric analysis (MDS) 499–500
Metric data 1
Metric scale 7–8
MINISSA 497
Missing at random (MAR) 34, 45
Missing completely at random (MCAR) 34, 45
Missing data 34, 43–57
    computer example 50–57
    diagnosis 45–46
    ignorable vs. nonignorable missing data 44
    imputation methods 47–50
    MAR (missing at random) 45
    MCAR (missing completely at random) 45
    missing data processes 34, 43–45
    reasons for 43–44
    remedies of 46–50
Model 620, 624
Modeling strategy in SEM 620, 625–26
    competing models 626
    confirmatory modeling 625
    model development 626
Moderator effect 82, 107
Modification indices 620, 644
MONONOVA 578
Multicollinearity 2, 23, 152–54
    assessment with a variance-decomposition
        matrix 153
    calculation of shared and unique variance
        94–95

Multicollinearity (*cont.*)
    computer example 153–54
    impact on shared and unique variance 93
    in cluster analysis 436
    in multiple discriminant analysis 197, 206
    in multiple regression analysis 82, 93, 126–
        28, 146–47, 152
    in multivariate analysis of variance 276
    tolerance 146, 152, 153
    variance inflation factor (VIF) 146, 152, 153
Multidimensional scaling (MDS) 16, 484–527
    aggregate versus disaggregate analysis
        495–96
    applications 488–89
    assumptions 502–03
    comparison to other multivariate methods
        491–92
    compositional methods 497–98
    computer example 516–27
    correspondence analysis 513–16, 523–27
    data collection 500–02
    decision process 492–94
    decompositional methods 496–97
    degenerate solution 486, 504
    determining dimensionality 504–06, 518–19
    hypothetical example 489–92
    ideal points 506–08
    incorporating preference data 506–10, 517,
        522–23
    index of fit 486, 506
    internal versus external analysis 508
    interpretation of dimensions 510–12, 520–22
    measuring similarity 500–01
    metric versus nonmetric methods 499–500
    objectives 492–96
    perceived versus objective dimensions 488
    point representation of ideal points 509–10
    research design 496–502
    selecting objects for analysis 494, 499
    similarity versus preference data 495
    stress measure 505
    validation 512–13, 526–27
    vector representation of attributes 511,
        521–22
    vector representation of ideal points 509–10
Multiple comparison tests
    a priori tests 281, 282–83
    Bonferroni inequality 281
    planned comparisons: See A priori tests
    post hoc tests 259, 281–82
Multiple correspondence analysis: See
    Correspondence analysis
Multiple discriminant analysis (MDA) 13–14, 178–237
    analogy with regression and ANOVA 183–84
    applications 181–82

Multiple discriminant analysis (MDA) (*cont.*)
    assumptions 196–97, 212, 213
    classification accuracy 203–05, 219–20, 227–29
    classification matrices 202–03, 217–20, 228
    classification procedures 200–02
    computational methods 197–98
    computer example, three group 221–37
    computer example, two group 211–21
    decision process 191–93
    derivation of discriminant functions 187–89, 212–16, 223–24
    discriminant loadings 180–206
    discriminant weights 180, 206
    geometric representation 187–88
    graphical display of results 208–09, 234–36
    hypothetical example 184–86, 188–91
    interpretation of results 205–09, 220–21
    multiple discriminant functions 207–09
    objectives 191–94
    potency index 207–08, 234–35
    Press's Q 205, 220, 229
    probabilities of classification 200–01
    rotation of functions 207
    sample division 195–96, 211
    sample size requirements 195, 211
    statistical significance 198, 215–17, 226–28
    validation of results 209–10
    variable selection 194–95, 211
Multiple R 135
Multiple regression (MR) 13, 78–149
    applications 79
    assumptions 110–14, 118, 134–35, 142–43
    beta coefficients 125–26, 146
    computer example 133–48
    decision process 97–99
    degrees of freedom 119
    dummy variables for nonmetric data 109–10
    hypothetical example 133–48
    influential observations 143–45, 151–65
    interaction or moderator effects 107–09
    interpretation 124–28
    model validation 128–29
    multicollinearity 93, 126–28, 146–47
    nonlinear effects 106–07
    objectives 97–103
    polynomial transformations 106–07
    power 103–06, 135
    prediction versus explanation 98–101
    PRESS statistic 128–29
    regression coefficients 125, 137, 145–46
    research design 103–10
    sample size requirements 103–05
    selecting dependent and independent variables 101–03

Multiple regression (MR) (*cont.*)
    statistical significance 118–20
    stepwise estimation 116–17
    unique and shared variance 94–95
    using a single independent variable 88–92
    using several independent variables 92–96
    validation 128–29, 147–48
    variable selection 115–18
Multiple t-tests: See Multiple comparisons
MULTISCALE 497
Multivariate analysis 2, 3–26
Multivariate analysis of variance (MANOVA) 14, 256–313
    assumptions 274–76, 285
    blocking factor 270
    computer example, three groups 293–302
    computer example, two groups 284–93
    computer example, factorial design 302–11
    control of experimentwide error rate 266
    covariates (MANCOVA) 273–74, 280
    decision process 266–67
    dependent variables 269
    factorial design of treatments 259, 270
    interaction effects 271–74, 309–10
    interpretation 280–83
    multivariate tests 277–78, 300–01, 306–08
    objectives 268–69
    planned comparisons 282
    post hoc tests 281–82
    power 278–80
    repeated measures 274
    replication 283
    research design 269–74
    sample size requirements 269–70
    significance testing 277–78
    statistical tests of the dependent variables 281–83
    stepdown analysis 281, 292–93, 308
    univariate tests (ANOVA) 262–63
    validation 283–84
Multivariate graphical display 34, 40–42
Multivariate measurement 2, 9
Multivariate model building approach 24–26
Multivariate normal distribution
    in multivariate analysis of variance 259, 276
    in structural equation modeling 639
Multivariate profiles
    graphical displays 41–42
    use in examining data 40
Multivariate statistical techniques
    definition 1, 21
    distinction between dependence and interdependence techniques 17
    selection 17–21

N

Nearest neighbor: See Single linkage method
Nested models 620, 644
Newman-Kuels test 282
Nominal scale 7
Non-hierarchical clustering procedures 422,
    440–41
      optimization techniques 441
      parallel threshold method 441
      sequential threshold method 440
      use of cluster seed points 440
Nonmetric analysis: See Multidimensional scaling
    analysis
Nonmetric data 2
Nonmetric scale 7
Normal distribution 35
Normal probability plot 35, 64–65, 82, 114
Normality of error term distribution: See Error
Normality of variables 35, 64–66
      computer example 72–73
      graphical analysis 64–66
      in canonical correlation analysis 332
      in factor analysis 374
      in multiple regression 143
      in multivariate analysis of variance 276
      in structural equation modeling 639
      kurtosis 65
      remedies for nonnormality 66
      skewness 66
      statistical tests 66
Normalized distance function 422, 435
Normed chi-square (SEM) 687
Normed fit index (NFI) (SEM) 686
Null hypothesis 259, 261
Null model 620
Null plot 82, 111
Numerical taxonomy: See Cluster analysis

O

Object 486, 488
Objective dimensions 487, 488
Oblique factor rotation 366, 382
Odds ratio 131
Offending estimates (SEM) 620, 639
Optimizing procedure 422, 441
Optimum cutting score 180
Ordinal interaction 259, 271
Ordinal scales 7
Orthogonal factor rotation 366, 382
Orthogonality
      in canonical correlation analysis 328, 333
      in conjoint analysis 559, 569, 757
      in factor analysis 366, 382
      in multivariate analysis of variance 259, 265

Outlier 57–62
      bivariate detection 59, 60–61
      computer example 60–62
      description and profiling 59–60
      in cluster analysis 429
      in multiple regression 83
      in multivariate analysis of variance 276–77
      multivariate detection 59, 62
      remedy 60, 62
      univariate detection 58
Overall tests: See Multiple comparison tests

P

Pairwise comparison method 560, 572
Parallel threshold method 422, 441
Parameter 83
Parsimonious fit index (PFI) (SEM) 686
Parsimonious fit measures (SEM) 686–88
Parsimony
      in multivariate techniques 23
      in structural equation modeling 620, 640
Part correlation coefficient 83, 94
Partial correlation coefficient 83, 94, 116, 138
Partial F (or t) values
      in multiple discriminant analysis 214, 226
      in multiple regression analysis 83, 116
Partial regression plot 83, 113, 159
Part worth 560–61, 587–89, 595–98
      calculation 595–98
      evaluation 587–89
      interaction between 598–99
      relationship between 564, 571–72
Path analysis 621
      calculation of coefficients 680–81
      use with path diagram 680
Path diagram 621, 629–31
      assumptions 680
      computation of coefficients 681
      computer example 657–58
      conversion to structural equations 631–34,
        658
      intercorrelation among constructs 630
      representing causal relationships 629–31
Perceived dimensions 487, 488
Percentage of variance criterion 378
Perceptual map 487, 488
Phi matrix (SEM) 673, 674
Pilliai's criterion
      in canonical correlation analysis 334
      in multiple discriminant analysis 198
      in multivariate analysis of variance 277
Planned comparison 259, 282
PLS 637
Point representations 509–10

Polar extremes   180, 194
Polychoric correlation   621, 637
POLYCON   497
Polynomials   83, 106–07
Polyserial correlation   621, 637
Post hoc tests: See Multiple comparison tests
Potency index   180, 207, 233–34
Power   2, 10–12
    components   10–12
    impact of sample size   11–12, 279
    in multiple regression analysis   84, 104–06
    in multivariate analysis of variance   260,
        278–79
    relationship to Type I and Type II error
        10, 278
Practical significance   2, 22
Prediction error   84
Predictive validity   444
Predictor variable: See Independent variable
Preference data (MDS)   487, 495, 501–02, 508–10
    data collection   501–02
    direct ranking method   502
    external versus internal analysis   508–09
    paired comparison method   502
    point versus vector representation   509–10
Preference structure   560, 561
PREFMAP   5, 497, 508, 522
PRELIS   639
Press statistic
    in multiple discriminant analysis   181
    in multiple regression   83, 128
Principal component analysis: see Factor analysis;
    Component analysis
Prior probabilities   200–02
Profile diagram   422
PROFIT   511, 520
Projections   487
Proportional chance criterion   181
Psi matrix (SEM)   673–74

Q
Q analysis: See Cluster analysis
"Q" factor analysis   366, 371
Qualitative variable: See Nonmetric data
Quantitative data: See Metric data

R
"R" factor analysis   366, 371
R² statistic: See Coefficient of determination
Random predictor variable (multiple regression)
    105
Ratio scale   1, 8
Redundancy index (CCA)   328, 335–36

Regression analysis: See Multiple regression
Regression coefficient   84, 89
Regression coefficient variance-decomposition matrix
    152, 153
Regression variate   84, 86
Reliability   2, 9
    calculation of   642
    in structural equation modeling   621, 623, 641
Repeated measures   260, 274
Replication   260, 283
Residual   35, 38
    in multiple regression analysis   84, 111
    in structural equation modeling   644
Response style effect   422, 435
Ridge regression   128
Root mean square error of approximation (RMSEA)
    685
Root mean square residual (RMSR) (SEM)   684–85
Rotation of results
    EQUIMAX   384
    in factor analysis   382–84, 396–98, 404
    in multiple discriminant analysis   207, 229–30
    oblique rotation   382, 398
    orthogonal rotation   382
    QUARTIMAX   383
    VARIMAX   384, 396, 397, 404
Row centering standardization   423, 435
Roy's greatest characteristic root: See Greatest
    characteristic root

S
Sample division: See Multiple discriminant analysis
Sample size   11, 22
    in factor analysis   373–74
    in multiple discriminant analysis   195
    in multiple regression   103–05
    in multivariate analysis of variance
        269–70
    in structural equation modeling   637
Scatterplot   35, 38–40, 59
Scheffe's test of contrasts   282
Scree test
    in factor analysis   378–79
    in multidimensional scaling   505
Scree test criterion   378, 395, 403
Self-explicated conjoint models   582
Semipartial correlation   84, 94
Sequential search approaches   115–17
Sequential threshold method   423, 440
Significance level   260, 261
Similarity data (MDS)   487, 495, 500–01
    comparison of paired objects   500–01
    confusion data   501
    data collection   500–01

Similarity data (MDS) (*cont.*)
    derived measures 501
    difference from preferences 502
Similarity measures (cluster analysis) 430–34
    association measures 434
    city-block distance 432
    correlation measures 430
    distance measures 430–34
    Euclidean distance 432
    Mahalanobis distance 434
Similarity scale 487, 490
Simple regression 84, 86, 88–92
Simultaneous estimation
    in multiple discriminant analysis 181, 197
    in multiple regression 115
Single linkage method 423, 438
Skewness 35, 66
Spatial maps: See Perceptual map 487
Specific variance 366, 375
Specification error 2, 23
    in multiple regression analysis 84, 102
    in structural equation modeling 621, 629
Split sample validation
    in multiple discriminant analysis 181, 209–10
    in multiple regression 147–48
Standard error: See Error 260, 261
Standard error of coefficient: See Error
Standard error of the estimate: See Error
Standardization 84, 434–35
Statistical error 10–13
Statistical relationship 84, 101
Statistical significance 10, 22
Stem and leaf diagram 35, 38
Step-down analysis 260, 281, 292–93
Stepwise estimation
    in multiple discriminant analysis 181, 197,
        214–17, 224–26
    in multiple regression 84, 116–17
Stimulus 560
Stress measure 487, 505
Structure correlations: See Discriminant loadings
Structural equation modeling (SEM) 15, 621,
    616–71
    absolute goodness-of-fit measures 683–85
    assumptions 639
    causal relationships in 626–30
    competing or nested models in 643–44,
        666–68
    computer example, confirmatory factor
        analysis 645–55
    computer example, structural model 655–70
    computing path coefficients 680–81
    decision process 626–33
    estimation methods 637
    identification of model 638–39

Structural equation modeling (SEM) (*cont.*)
    incremental goodness-of-fit measures 685–86
    input matrix 636–37
    interpretation 644–45
    LISREL notation 672–79
    mathematical notation 672–79
    measurement model 632–34
    measurement model goodness-of-fit 641–42,
        664
    modeling strategy 625–26
    overall goodness-of-fit measures 639–40,
        661–64
    parsimonious goodness-of-fit measures
        686–88
    path diagrams in 629–31
    reliability estimates of constructs 634–35,
        641–43
    role of theory 624–25
    sample size 637
    specification of construct reliabilities 634–35
    structural model 632
    structural model goodness-of-fit 642,
        665–66
Structural model (SEM) 621
    computer example 655–70
    conversion from path diagram 632–34,
        674 75
    goodness-of-fit 661–64
    LISREL notation 675
Structure loadings matrix: See Discriminant loadings
Studentized residual 85, 111, 152, 155
Subjective clustering: See Confusion data
Subjective evaluation 487, 505
Summated scales 2, 9
    in factor analysis 390–91, 400–02
    in structural equation modeling 634–35
Sum of squared errors 85, 87, 119
Sum of squares regression 85, 91, 119

T
t-statistic 260, 261
t-test 260, 261, 285–88
Territorial map 208, 233
Tetrachoric correlation 621, 637
Theory 621, 624–25
Theta-delta matrix (SEM) 673–74, 678
Theta-epsilon matrix (SEM) 673–74, 678
Tolerance
    in multiple discriminant analysis 181
    in multiple regression 85, 127, 152, 153
Total sum of squares 85, 91
Trace 366
Trade-off method 560, 572

Transformation: See Data transformation
Treatment   2, 14
        in conjoint analysis   560
        in multivariate analysis of variance   260,
            261
Treatment levels: See Level or treatment
Tree-graph: See Dendrogram
Tucker-Lewis index (TLI) (SEM)   686
Tukey's HSD (honestly significant difference)   282
Tukey's LSD (least significant difference)   282
Type I error   2, 10, 260, 261
Type II error   2, 10, 260, 278
Typology: See Cluster analysis

U
U method   210
U statistic: See Wilks' lambda   260
Unfolding   487, 508
Unidimensionality   622, 641
Unique variance
        impact of multicollinearity   93–95
        in factor analysis   375–76
Univariate analysis   5
Univariate analysis of variance: See Analysis of
        variance
Unobserved construct/variable: See Latent construct/
        variable

V
Validation sample
        in canonical correlation analysis   338
        in linear probability models   132
        in multiple discriminant analysis   195, 207–10
        in multiple regression analysis   128
Validity   3, 9, 622, 641
Variance criterion, percentage of: See Percentage of
        variance criterion

Variance-covariance matrix
        equality between groups in discriminant
            analysis   196–97, 213
        equality between groups in MANOVA
            275–76, 289, 299
        use as input matrix in structural equation
            modeling   636–37
Variance extracted measure   622, 641
        calculation   642
        in structural equation modeling   641, 642
Variance inflation factor (VIF)   85, 127, 152, 153
Variate: Also see Linear combination   3, 5–6, 35, 64
        in canonical correlation analysis   327, 330
        in cluster analysis   421, 423, 427–28
        in conjoint analysis   558, 563
        in factor analysis   367
        in multiple discriminant analysis   181, 182
        in multiple regression   86, 110
        in multivariate analysis of variance   260, 263,
            264
VARIMAX rotation   366
Vector
        in multidimensional scaling   487, 509
        in multiple discriminant analysis   181, 208
        in multivariate analysis of variance   260
Vector representation (MDS)   509–11, 521–22
Vertical icicle diagram   423

W
Wald statistic   85, 133
Ward's method   423, 440
Wilks' lambda
        in multiple discriminant analysis   198
        in multivariate analysis of variance   261, 277
Within-case standardization   423, 435
Within-group variance   262